Glencoe

Algebra

Concepts and Applications

 **Glencoe McGraw-Hill**

New York, New York   Columbus, Ohio   Woodland Hills, California   Peoria, Illinois

GLENCOE Online

Visit the Glencoe Mathematics Internet Site for
Algebra: Concepts and Applications at

www.algconcepts.glencoe.com

You'll find:

- Research Helps
- Data Updates
- Career Data
- Investigations
- Review Activities
- Test Practice

 links to Web sites relevant to Problem-Solving Workshops, Investigations, Math In the Workplace features, exercises, and much more!

Glencoe/McGraw-Hill

A Division of The **McGraw·Hill** Companies

Send all inquiries to:
Glencoe/McGraw-Hill
8787 Orion Place
Columbus, OH 43240

ISBN: 0-07-821347-9

3 4 5 6 7 8 9 10   027/055   09 08 07 06 05 04 03 02 01

Dear Students, Teachers, and Parents,

Algebra: Concepts and Applications is designed to help you learn algebra and apply it to the real world. Throughout the text, you will be given opportunities to make connections from concrete models to abstract concepts. The real-world photographs and realistic data will help you see algebra in your world. You will also have plenty of opportunities to review and use arithmetic and geometry concepts as you study algebra. And for those of you who love a good debate, you will find plenty of opportunities to communicate your understanding of algebra.

We know that most of you haven't yet decided which careers you would like to pursue, so we've also included a little career guidance. This text offers real examples of how mathematics is used in many types of careers.

You may have to take an end-of-course exam for algebra, a proficiency test for graduation, the SAT, and/or the ACT. When you enter the workforce, you may also have to take job placement tests that include a section on mathematics. All of these tests include algebra problems. Because all Algebra 1 concepts are covered in this text, this program will prepare you for all of those tests.

Each day, as you use **Algebra: Concepts and Applications,** you will see the practical value of algebra. You will grow to appreciate how often algebra is used in ways that relate directly to your life. You will have meaningful experiences that will prepare you for the future. If you don't already see the importance of algebra in your life, you soon will!

Sincerely,
The Authors

Jerry Cummins

Carol E. Malloy

Kay McClain

Yvonne M. Mojica

Jack Price

Contents in Brief

Jerry Cummins

Staff Development Specialist
Bureau of Education and
    Research
State of Illinois
President, National Council of
    Supervisors of Mathematics
Western Springs, IL

Carol Malloy

Assistant Professor of
    Mathematics Education
University of North Carolina
    at Chapel Hill
Chapel Hill, NC

Kay McClain

Lecturer
George Peabody College
Vanderbilt University
Nashville, TN

Yvonne Mojica

Mathematics Teacher and
    Mathematics Department
    Chairperson
Verdugo Hills High School
Tujunga, CA

Jack Price

Professor, Mathematics
    Education
California State Polytechnic
    University
Pomona, CA

Academic Consultants and Teacher Reviewers

Each of the Academic Consultants read all 15 chapters, while each Teacher Reviewer read two chapters. The Consultants and Reviewers gave suggestions for improving the Student Editions and the Teacher's Wraparound Editions.

Academic Consultants

Richie Berman, Ph.D.
Mathematics Lecturer &
  Supervisor
University of California at
  Santa Barbara
Santa Barbara, California

Judith Cubillo
Mathematics Teacher &
  Department Chairperson
Northgate High School
Walnut Creek, California

Mary C. Enderson
Faculty
Middle Tennessee State
  University
Murfreesboro, Tennessee

Alan G. Foster
Former Mathematics Teacher
  & Department Chairperson
Addison Trail High School
Addison, Illinois

Deborah A. Haver, Ed.D.
Assistant Principal
Great Bridge Middle School
Chesapeake, Virginia

Nicki Hudson
Mathematics Teacher
West Linn High School
West Linn, Oregon

Daniel Marks, Ph.D.
Associate Professor of
  Mathematics
Auburn University at
  Montgomery
Montgomery, Alabama

Donald McGurrin
Senior Administrator for
  Secondary Mathematics
Wake County Public Schools
Raleigh, North Carolina

C. Vincent Pané, Ed.D.
Associate Professor of
  Education
Molloy College
Rockville Centre, New York

Marianne Weber
National Mathematics
  Consultant
St. Louis, Missouri

Teacher Reviewers

Breta J. Brown
Mathematics Teacher
Hillsboro High School
Hillsboro, North Dakota

Kimberly A. Brown
Mathematics Teacher
McGehee High School
McGehee, Arkansas

Helen Carpini
Mathematics Department
  Chairperson
Middletown High School
Middletown, Connecticut

Terry Cepaitis
Mathematics Resource
  Teacher
Anne Arundel County
  Public Schools
Annapolis, Maryland

Kindra R. Cerfoglio
Mathematics Teacher
Reed High School
Sparks, Nevada

Tom Cook
Mathematics Department
  Chairperson
Carlisle High School
Carlisle, Pennsylvania

Donna L. Cooper
Mathematics Department
  Chairperson
Walter E. Stebbins High
  School
Riverside, Ohio

James D. Crawford
Instructional Coordinator—
  Mathematics
Manchester Memorial
  High School
Manchester, New Hampshire

David A. Crine
Mathematics Department
  Chairperson
Basic High School
Henderson, Nevada

Carol Damiano
Mathematics Teacher
Morton High School
Hammond, Indiana

Douglas D. Dolezal, Ph.D.
Mathematics Educator/
  Methods Instructor
Crete High School/Doane
  College
Crete, Nebraska

Richard F. Dube
Mathematics Supervisor
Taunton High School
Taunton, Massachusetts

Dianne Foerster
Mathematics Department
  Chairperson
Riverview High School
Riverview, Florida

Victoria G. Fortenberry
Mathematics Teacher
Lincoln High School
Tallahassee, Florida

Candace Frewin
Mathematics Teacher
East Lake High School
Tarpon Springs, Florida

Linda Glover
Mathematics Department
  Chairperson
Conway High School East
Conway, Arkansas

Karin Sorensen Grandone
Mathematics Department
  Supervisor
Bremen District 228
Midlothian, Illinois

R. Emilie Greenwald
Mathematics Teacher
Worthington Kilbourne
  High School
Worthington, Ohio

John Scott Griffith
Mathematics Department
  Chairperson
New Castle Middle School
New Castle, Indiana

Rebecca M. Gummerson
Mathematics Department
  Chairperson
Burley High School
Burley, Idaho

T. L. Watanabe Hall
Mathematics Teacher
Northwood Junior
  High School
Kent, Washington

T. B. Harris
Mathematics Teacher
Luray High School
Luray, Virginia

Jerome D. Hayden
Mathematics Department
  Chairperson
McLean County Unit District 5
Normal, Illinois

Karlene M. Hubbard
Mathematics Teacher
Rich East High School
Park Forest, Illinois

Joseph Kavanaugh
Academic Head of
  Mathematics
Scotia-Glenville Central
  School District
Scotia, New York

Ruth C. Keefe
Mathematics Department
  Chairperson
Litchfield High School
Litchfield, Connecticut

Roger M. Marchegiano
Supervisor of Mathematics
Bloomfield School District
Bloomfield, New Jersey

Marilyn Martau
Mathematics Teacher (Retired)
Lakewood High School
Lakewood, Ohio

Jane E. Morey
Mathematics Department
  Chairperson
Washington High School
Sioux Falls, South Dakota

Grace Clover Mullen
Talent Search Tutor
Dabney Lancaster
  Community College
Lexington, Virginia

Laurie D. Newton
Mathematics Teacher
Crossler Middle School
Salem, Oregon

Rinda Olson
Mathematics Department
  Chairperson
Skyline High School
Idaho Falls, Idaho

Catherine E. Oppio
Mathematics Teacher
Truckee Meadows
  Community College
  High School
Reno, Nevada

Peter Pace
Mathematics Teacher
Life Center Academy
Burlington, New Jersey

LaVonne Peterson
Instructional Leader
Minico High School
Rupert, Idaho

Table Of Contents

Lesson 1–1, page 7

Lesson 2–5, page 78

Lesson 3–4, page 115

Chapter ③ Addition and Subtraction Equations 92

Lesson 4–2, page 150

Chapter 4 Multiplication and Division Equations.............................138

Lesson 5–5, page 217

Lesson 6–3, page 254

Chapter 7 Linear Equations

Lesson 7–4, page 303

In the Workplace

Standardized Test Practice

Hands-On Algebra......362

*inter*NET
CONNECTION

Graphing Calculator Exploration......338

Lesson 8-6, page 365

Lesson 9–5, page 405

Lesson 10–3, page 434

Chapter 11 Quadratic and Exponential Functions

Lesson 11–2, page 467

Math In the Workplace

Standardized Test Practice

Hands-On Algebra 489

*inter*NET CONNECTION

Graphing Calculator Exploration 471, 491

Photo Graphic

Lesson 12–3, page 517

Chapter 12 Inequalities

Math In the Workplace

Standardized Test Practice

Hands-On Algebra

*inter*NET CONNECTION

Lesson 13–4, page 566

Chapter 14 Radical Expressions 598

Lesson 14–2, page 608

Lesson 15–3, page 654

Chapter 15 Rational Expressions and Equations

Preparing for Standardized Test Success

The **Preparing for Standardized Tests** pages at the end of each chapter have been created in partnership with **The Princeton Review**, the nation's leader in test preparation materials, to help you get ready for the mathematics portions of your standardized tests. On these pages, you will find strategies for solving problems and test-taking advice to help you maximize your score.

It is important to remember that there are many different standardized tests given by schools and states across the country. Find out as much as you can about your test. Start by asking your teacher and counselor for any information, including practice materials, that may be available to help you prepare.

To help you get ready for these tests, do the **Standardized Test Practice** question in each lesson. Also review the concepts and techniques contained in the **Preparing for Standardized Tests** pages at the end of each chapter listed below. This will help you become familiar with the types of math questions that are asked on various standardized tests.

The **Preparing for Standardized Tests** pages are part of a complete test preparation course offered in this text. The test items on these pages were written in the same style as those in state proficiency tests and standardized tests like ACT and SAT. The 15 topics are closely aligned with those tests, the algebra curriculum, and this text. These topics cover all of the types of problems you will see on these tests.

Chapter	Mathematics Topic	Pages
1	Number Concept Problems	48–49
2	Data Analysis Problems	90–91
3	Number Concept Problems	136–137
4	Statistics Problems	184–185
5	Expression and Equation Problems	234–235
6	Probability and Counting Problems	280–281
7	Algebra Word Problems	332–333
8	Pythagorean Theorem Problems	378–379
9	Percent Problems	416–417
10	Function and Graph Problems	454–455
11	Polynomial and Factoring Problems	500–501
12	Angle, Triangle, and Quadrilateral Problems	546–547
13	Perimeter, Area, and Volume Problems	596–597
14	Systems of Equations Problems	634–635
15	Right Triangle Problems	680–681

With some practice, studying, and review, you will be ready for standardized test success. Good luck from Glencoe/McGraw-Hill and The Princeton Review! The Princeton Review is not affiliated with Princeton University nor Educational Testing Service.

Graphing Calculator Quick Reference Guide

Throughout this text, **Graphing Calculator Explorations** have been included so you can use technology to solve problems. These activities use the TI–83 Plus graphing calculator. Graphing calculators have a wide variety of applications and features. If you are just beginning to use a graphing calculator, you will not need to use all of its features. This page is designed to be a quick reference for the features you will need to use as you study from this text.

To darken or lighten the screen:	[2nd] [▲] or [2nd] [▼]
To clear an entry:	[CLEAR]
To get to the home screen:	[2nd] [QUIT]
To recall an entry:	[2nd] [ENTRY]
To recall an answer:	[2nd] [ANS]
To turn the calculator off:	[2nd] [OFF]

Task	Keystrokes
Using tables	[2nd] [TABLE]
Use lists	[STAT] [ENTER]
Find the mean and median of listed data	[STAT] [▶] [ENTER] [ENTER]
Solve equations	[MATH] 0
Set the viewing window	[WINDOW]
Plot points	[2nd] [DRAW] [▶] [ENTER]
Enter an equation	[Y=]
Enter an inequality symbol	[2nd] [TEST]
Graph an equation	[GRAPH]
Zoom in	[ZOOM] 2
Zoom out	[ZOOM] 3
Graph in the viewing window x: $[-10, 10]$ and y: $[-10, 10]$	[ZOOM] 6
Graph an equation with integer coordinates	[ZOOM] 8
Place a statistical graph in a good viewing window	[ZOOM] 9
Trace a graph	[TRACE]
Find the intersection of two graphs	[2nd] [CALC] 5
Shade an inequality	[2nd] [DRAW] 7
Enter a program	[PRGM] [▶] [▶] [ENTER]

The Language of Algebra

▶ What You'll Learn in Chapter 1:

- to translate words into algebraic expressions and equations *(Lesson 1–1)*,
- to use the order of operations to evaluate expressions *(Lesson 1–2)*,
- to use properties of real numbers to simplify expressions *(Lessons 1–3 and 1–4)*,
- to use the four-step plan to solve problems *(Lesson 1–5)*,
- to collect and organize data using sampling and frequency tables *(Lesson 1–6)*, and
- to construct and interpret line graphs, histograms, and stem-and-leaf plots *(Lesson 1–7)*.

Problem-Solving Workshop

Project

As a new video club member, you can choose seven movies. Each movie costs 1¢ plus $1.69 for shipping and handling. Within three years, you must order at least five more movies, each at the regular club price, plus the same shipping fee.

Type of Movie	Regular Club Price
Children's	$12.99
New Release	$24.99
All-Time Favorite	$16.99
Classic	$8.99

Suppose you buy a total of 12 movies. What is the lowest possible average cost per movie? the highest possible average cost per movie?

Working on the Project

Work with a partner and choose a strategy to help analyze and solve the problem. Here are some questions to help you get started.

- How much do you pay for the first shipment of seven 1¢ movies?
- What are the least and greatest amounts you can spend on the five required regular-priced movies?

Technology Tools

- Use a **spreadsheet** to calculate the average cost of the movies.
- Use a **word processor** to write your newspaper article.

 Research For more information about CD clubs, visit: www.algconcepts.glencoe.com

Presenting the Project

Write an article for the school newspaper discussing the advantages and disadvantages of joining a video, book, or CD club.

- Research prices at retail stores or an Internet site. What would you pay for the same number of videos, books, or CDs required by the club?
- Show how the average cost per video, book, or CD changes as you buy more at the regular club price.

▶ Strategies

Look for a pattern.

Draw a diagram.

Make a table.

Work backward.

Use an equation.

Make a graph.

Guess and check.

Writing Expressions and Equations

What You'll Learn

You'll learn to translate words into algebraic expressions and equations.

Why It's Important

Communication You can use expressions to represent the cost of your long-distance phone calls.
See Exercise 43.

Suppose a candy bar costs 45 cents. Then 45×2 is the cost of 2 candy bars, 45×3 is the cost of 3 candy bars, and so on. Generally, the cost of any number of candy bars is *45 cents times the number of bars*. We can represent this situation with an **algebraic expression**.

$$\underbrace{45\ cents}_{45}\ \ \underbrace{times}_{\times}\ \ \underbrace{the\ number\ of\ bars}_{n}$$

In the expression $45 \times n$, the letter n is called a **variable** because it stands for an unknown number. Its value *varies*. An algebraic expression contains at least one variable and at least one mathematical operation, as shown in the examples below.

$$h \div 3 \qquad 5n + 1 \qquad \frac{r}{t} - 1 \qquad xy \qquad 4 \times a$$

A **numerical expression** contains only numbers and mathematical operations. For example, $6 + 2 \div 1$ is a numerical expression.

In an expression involving multiplication, the quantities being multiplied are called **factors**, and the result is the **product**.

$$4 \times 5 \times 8 = 160$$

factors product

Reading Algebra

When one factor in a product is a variable, the multiplication sign is usually omitted. Read *4a* as *four a*.

To write a multiplication expression such as $4 \times a$, a raised dot or parentheses can be used. A fraction bar can be used to represent division.

$$\left.\begin{array}{l} 4 \cdot a \\ 4(a) \\ (4)(a) \\ 4a \end{array}\right\} \quad \boxed{means \Rightarrow} \quad 4 \times a \quad \bigg| \quad \frac{t}{2} \quad \boxed{means \Rightarrow} \quad t \div 2$$

To solve verbal problems in mathematics, you may have to translate words into algebraic expressions. The chart below shows some of the words and phrases used to indicate mathematical operations.

Addition	Subtraction	Multiplication	Division
plus	minus	times	divided by
the sum of	the difference of	the product of	the quotient of
increased by	decreased by	multiplied by	the ratio of
more than	less than	at	per
added to	fewer than	of	
the total of	subtracted from		

Write an algebraic expression for each verbal expression.

① the sum of m and 18

$m + 18$

② g divided by y

$g \div y$ or $\dfrac{g}{y}$

Your Turn

a. 26 decreased by w

b. 4 more than 8 times k

Life Science Link

Real World

A certain kangaroo can travel 30 feet in a single leap.

③ Write a numerical expression to represent the distance it can travel if it leaps 4 times.

$30 \cdot 4$ or $30(4)$

④ Write an algebraic expression to represent the distance it can travel if it leaps x times.

$30 \cdot x$ or $30x$

You can also translate algebraic expressions into verbal expressions.

Examples

Write a verbal expression for each algebraic expression.

⑤ $32 - b$

32 less b
b less than 32
the difference of 32 and b
b subtracted from 32
32 decreased by b

⑥ $(y \div 4) + 9$

y divided by 4, plus 9
the quotient of y and 4, increased by 9
9 added to the ratio of y and 4

Your Turn

c. $15v$

d. $r - \dfrac{t}{d}$

An **equation** is a sentence that contains an equals sign (=). Some words used to indicate the equals sign are in the chart at the right. An equation may contain numbers, variables, or algebraic expressions.

Equality	
equals	is equal to
is	is the same as
is equivalent to	is as much as
	is identical to

Examples

Write an equation for each sentence.

⑦ Three times g equals 21.

$3g = 21$

⑧ Five more than twice n is 15.

$2n + 5 = 15$

Your Turn

e. A number k divided by 4 is equal to 18.

Write a sentence for each equation.

9 $x - 2 = 14$
Two less than x
is equivalent to 14.

10 $7y + 6 = 34$
Seven times y increased
by 6 is 34.

Your Turn

f. $4b - 5 = 3$

Check for Understanding

Communicating Mathematics

Math Journal

Study the lesson. Then complete the following.

1. **Write** three examples of numerical expressions and three examples of algebraic expressions.
2. **Write** three examples of equations.
3. **Write** about a real-life application that can be expressed using an algebraic expression or an equation.

Vocabulary

algebraic expression
variable
numerical expression
factors
product
equation

Guided Practice

Write an algebraic expression for each verbal expression.
(Examples 1 & 2)

4. t more than s
5. the product of 7 and m
6. 11 decreased by the quotient of x and 2

Write a verbal expression for each algebraic expression.
(Examples 5 & 6)

7. $\dfrac{7}{q}$

8. $3\ell - 9$

Write an equation for each sentence. *(Examples 7 & 8)*

9. A number m added to 6 equals 17.
10. Ten is the same as four times r minus 6.

Write a sentence for each equation. *(Examples 9 & 10)*

11. $5 + r = 15$

12. $\dfrac{4p}{3} = 12$

13. **Biology** The family of great white sharks has w different species. The blue shark family has nine times w plus three different species. Write an algebraic expression to represent the number of species in the blue shark family. *(Example 4)*

Practice

Write an algebraic expression for each verbal expression.

14. twelve less than y **15.** the product of r and s

16. the quotient of t and 5   **17.** three more than five times a

18. p plus the quotient of 9 and 5   **19.** the difference of 1 and n

20. f divided by y **21.** ten plus the product of h and 1

22. seven less than the quotient of j and p

Write a verbal expression for each algebraic expression.

23. $9x$ **24.** $11 + b$ **25.** $6 - y$

26. $2m + 1$ **27.** $\dfrac{3}{r} - 8$ **28.** $16 - rt$

Write an equation for each sentence.

29. Three plus w equals 15.   **30.** Five times r equals 7.

31. Two is equal to seven divided by x.

32. Five less than the product of two and g equals nine.

33. Three minus the product of five and y is the same as two times z.

34. The quotient of 19 and j is equal to the total of a and b and c.

Write a sentence for each equation.

35. $3r = 18$ **36.** $g + 7 = 3$ **37.** $h = 10 - i$

38. $6v - 2 = 8$ **39.** $\dfrac{t}{4} = 16$ **40.** $10z + 7 = \dfrac{6}{r}$

41. Choose a variable and write an equation for *Four times a number minus seven equals the sum of 15 and c and two times the number.*

Applications and Problem Solving

Real World

42. Biology   A smile requires 26 fewer muscles than a frown. Let f represent the number of muscles it takes for a frown and let s represent the number of muscles for a smile. Then write an equation to represent the number of muscles a person uses to smile.

43. Communication   A long-distance telephone call costs 20¢ for the first minute plus 10¢ for each additional minute.

   a. Write an expression for the total cost of a call that lasts 15 minutes.

   b. Write an expression for the total cost of a call that lasts m minutes.

44. Critical Thinking   The ancient Hindus enjoyed number puzzles like the one below.   **Source:** *Mathematical History*

> If 4 is added to a certain number, the result divided by 2, that result multiplied by 5, and then 6 subtracted from that result, the answer is 29. Can you find the number?

   a. Choose a variable and write an algebraic equation to represent the puzzle.

   b. Is your answer in part a the only correct way to write the equation? Explain.

Extra Practice   See p. 692.

Math
In the Workplace

What You'll Learn
You'll learn to use the order of operations to evaluate expressions.

Why It's Important
Business
Businesspeople use the order of operations to determine the cost of renting a car. *See Exercise 18.*

Trisha scored 9 points in each of her first five basketball games. In the sixth game, she scored 4 points. What is the total number T of points?

number of points per game		*number of games*		*points in the 6th game*		*total number of points*
9	·	5	+	4	=	T

Method 1 | **Method 2**

$9 \cdot 5 + 4 = 45 + 4$ *Multiply 9 and 5.* | $9 \cdot 5 + 4 = 9 \cdot 9$ *Add 5 and 4.*
$\qquad = 49$ *Add 45 and 4.* | $\qquad = 81$ *Multiply 9 and 9.*

Did Trisha score 49 points or 81 points? The values are different because in Method 1 we multiplied and then added. In Method 2 we added and then multiplied. To find the correct value of an expression, follow the **order of operations**.

Order of Operations	1. Find the values of expressions inside grouping symbols, such as parentheses ( ), brackets [ ], and as indicated by fraction bars.
	2. Do all multiplications and/or divisions from left to right.
	3. Do all additions and/or subtractions from left to right.

According to the order of operations, do multiplication and then addition. So, the value of the expression in Method 1 above is correct. Trisha scored 49 points.

Examples

Find the value of each expression.

1 $38 - 5 \cdot 6$

$38 - 5 \cdot 6 = 38 - 30$ *Multiply 5 and 6.*
$\qquad = 8$ *Subtract 30 from 38.*

2 $\dfrac{4 \times 9}{26 - 8}$

$\dfrac{4 \times 9}{26 - 8} = \dfrac{36}{18}$ *Evaluate the numerator and the denominator separately.*
$\qquad = 2$ *Divide 36 by 18.*

Your Turn

a. $7 \cdot 4 + 7 \cdot 3$ **b.** $12 \div 3 \cdot 5 - 4$ **c.** $\dfrac{6 + 12}{5(3) - 13}$

The order of operations is useful in solving problems in everyday life.

Example

Finance Link

Real World

③ As a 16-year old, Trent Eisenberg ran his own consulting company called *F1 Computer*. Suppose he charged a flat fee of $50, plus $25 per hour. One day he worked 2 hours for one customer and the next day he worked 3 hours for the same customer. Find the value of the expression $50 + 25(2 + 3)$ to find the total amount of money he earned.

Source: *Scholastic Math*

$$50 + 25(2 + 3) = 50 + 25(5) \quad \textit{Do the operation in parentheses first.}$$
$$= 50 + 125 \quad \textit{Multiply 25 and 5.}$$
$$= 175 \quad \textit{Add 50 and 125.}$$

Trent earned $175.

In algebra, certain statements or *properties* are true for any number. Four properties of equality are listed in the table below.

Property of Equality	Symbols	Numbers
Substitution	If $a = b$, then a may be replaced by b.	If $9 + 2 = 11$, then $9 + 2$ may be replaced by 11.
Reflexive	$a = a$	$21 = 21$
Symmetric	If $a = b$, then $b = a$.	If $10 = 4 + 6$, then $4 + 6 = 10$.
Transitive	If $a = b$ and $b = c$, then $a = c$.	If $3 + 5 = 8$ and $8 = 2(4)$, then $3 + 5 = 2(4)$.

Examples

Name the property of equality shown by each statement.

④ If $9 + 3 = 12$, then $12 = 9 + 3$.

Symmetric Property of Equality

⑤ If $z = 8$, then $z \div 4 = 8 \div 4$.

Substitution Property of Equality   *z is replaced by 8.*

Your Turn

d. $7 - c = 7 - c$
e. If $10 - 3 = 4 + 3$ and $4 + 3 = 7$, then $10 - 3 = 7$.

These properties of numbers may help to find the value of expressions.

Property	Words	Symbols	Numbers
Additive Identity	When 0 is added to any number a, the sum is a.	For any number a, $a + 0 = 0 + a = a$.	$45 + 0 = 45$ $0 + 6 = 6$ *0 is the identity.*
Multiplicative Identity	When a number a is multiplied by 1, the product is a.	For any number a, $a \cdot 1 = 1 \cdot a = a$.	$12 \cdot 1 = 12$ $1 \cdot 5 = 5$ *1 is the identity.*
Multiplicative Property of Zero	If 0 is a factor, the product is 0.	For any number a, $a \cdot 0 = 0 \cdot a = 0$.	$7 \cdot 0 = 0$ $0 \cdot 23 = 0$

When two or more sets of grouping symbols are used, simplify within the innermost grouping symbols first.

Example

6 **Find the value of $5[3 - (6 \div 2)] + 14$. Identify the properties used.**

$5[3 - (6 \div 2)] + 14$
$= 5[3 - 3] + 14$ *Substitution Property of Equality*
$= 5(0) + 14$ *Substitution Property of Equality*
$= 0 + 14$ *Multiplicative Property of Zero*
$= 14$ *Additive Identity*

Your Turn   **f.** $(22 - 15) \div 7 \cdot 9$ **g.** $8 \div 4 \cdot 6(5 - 4)$

You can also apply the properties of numbers to find the value of an algebraic expression. This is called **evaluating** an expression. Replace the variables with known values and then use the order of operations.

Examples

Evaluate each expression if $a = 9$ and $b = 1$.

7 $7 + \left(\dfrac{a}{b} - 9\right)$

$7 + \left(\dfrac{a}{b} - 9\right) = 7 + \left(\dfrac{9}{1} - 9\right)$ *Replace a with 9 and b with 1.*
$= 7 + (9 - 9)$ *Substitution Property of Equality*
$= 7 + 0$ *Substitution Property of Equality*
$= 7$ *Additive Identity*

8 $(a + 4) - 3 \cdot b$

$(a + 4) - 3 \cdot b = (9 + 4) - 3 \cdot 1$ *Replace a with 9 and b with 1.*
$= (13) - 3 \cdot 1$ *Substitution Property of Equality*
$= 13 - 3$ *Multiplicative Identity*
$= 10$ *Substitution Property of Equality*

Your Turn   **Evaluate each expression if $m = 8$ and $p = 2$.**

h. $6 \cdot p - m \div p$ **i.** $[m + 2(3 + p)] \div 2$

Communicating Mathematics

Study the lesson. Then complete the following.

1. **Name** two of the three types of grouping symbols discussed in this lesson.

2. **Translate** the verbal expression *six plus twelve divided by three* and *the sum of six and twelve divided by three* into numerical expressions by using grouping symbols. Evaluate the expressions and explain why they are different.

Math Journal

3. Label a section of your math journal "Toolbox." Record all properties given in this course, beginning with this lesson.

Guided Practice

⏱ **Getting Ready** State which operation to perform first.

Sample: $3 + 2 \cdot 4$ **Solution:** Multiply 2 and 4.

4. $8 \div 4 \cdot 2$ 5. $12 - 6 \cdot 2$
6. $5(7 + 7)$ 7. $(10 - 4) \div 3$

Find the value of each expression. *(Examples 1–3)*

8. $7 \cdot 4 + 3$ 9. $4(1 + 5) \div 8$ 10. $18 \div [3(11 - 8)]$

Name the property of equality shown by each statement. *(Examples 4 & 5)*

11. If $5 + 2n = 5 + 3$ and $5 + 3 = 2 \cdot 4$, then $5 + 2n = 2 \cdot 4$.

12. If $\frac{y}{2} = 19$, then $19 = \frac{y}{2}$.

Find the value of each expression. Identify the property used in each step. *(Example 6)*

13. $8(4 - 8 \div 2)$ 14. $5(2) \cdot (15 \div 15)$

Evaluate each algebraic expression if $q = 4$ and $r = 1$. *(Examples 7 & 8)*

15. $4(q - 2r)$ 16. $\frac{7q}{r + 3}$ 17. $r + \frac{q}{2} \cdot 6$

18. **Car Rental** The cost to rent a car is given by the expression $25d + 0.10m$, where d is the number of days and m is the number of miles. If Teresa rents the car for five days and drives 300 miles, what is the cost? *(Examples 7 & 8)*

Exercises

Practice

Find the value of each expression.

19. $36 \div 4 + 5$ 20. $16 - 4 \cdot 4$ 21. $4 + 7 \cdot 2 + 8$

22. $42 - (24 \div 2) + 10$ 23. $42 - 24 \div (2 + 10)$ 24. $24 \div 12 \div 2 \cdot 5$

25. $\frac{7(3 + 6)}{3}$ 26. $\frac{4(8 - 2)}{2 \times 2}$ 27. $38 - [3(9 + 1)]$

Name the property of equality shown by each statement.

28. If $x + 3 = 5$ and $x = 2$, then $2 + 3 = 5$.

29. $8t - 1 = 8t - 1$

30. If $6 = 3 + 3$, then $3 + 3 = 6$.

31. $\frac{20 - 2}{9} = \frac{18}{9}$

32. If $4 \cdot (7 - 7) = 4 \cdot 0$ and $4 \cdot 0 = 0$, then $4 \cdot (7 - 7) = 0$.

33. $a + 1 = 15x$ and $15x = 30$, so $a + 1 = 30$.

Find the value of each expression. Identify the property used in each step.

34. $7(10 - 1 \cdot 3)$

35. $8(9 - 3 \cdot 2)$

36. $19 - 15 \div 5 \cdot 2$

37. $10(6 - 5) - (20 \div 2)$

38. $\frac{9 \cdot 9 - 1}{3(1 + 2) - 1}$

39. $6(12 - 48 \div 4) + 7 \cdot 1$

Evaluate each algebraic expression if $j = 5$ and $s = 2$.

40. $7j - 3s$

41. $j(3s + 4)$

42. $j + 5s - 7$

43. $\frac{9 \cdot 4 + 5 \cdot s}{7 - j}$

44. $\frac{14 + s}{2(j - 1)}$

45. $\frac{4js}{s - 1}$

46. $50 \div js + 6$

47. $(3s - j)(5s - j)$

48. $[3j - s(4 + s)] \div 3$

49. $2[16 - (j - s)]$

50. a. Write an algebraic expression for *nine added to the quantity three times the difference of a and b.*

  b. Let $a = 4$ and $b = 1$. Evaluate the expression in part a.

Applications and Problem Solving

51. Real Estate The Phams own a $150,000 home in Rochester, New York, and plan to move to San Diego, California. How much will a similar home in San Diego cost? Evaluate the expression $150,000 \div a \times b$ for $a = 79$ and $b = 164$ to find the answer to the nearest dollar.
Source: *USA Today, 1999*

52. Sports A person's handicap in bowling is usually found by subtracting the person's average a from 200, multiplying by 2, and dividing by 3.

  a. Write an algebraic expression for a handicap in bowling.

  b. Find a person's handicap whose average is 170.

53. Gardening Mr. Martin is building a fence around a rectangular garden, as shown at the left. Evaluate the expression $2\ell + 2w$, where ℓ represents the length and w represents the width, to find how much fencing he needs.

10 ft

12 ft

54. Critical Thinking The symbol $<$ means "is less than." Are the following properties of equality true for statements containing this symbol? Give examples to explain.

   a. Reflexive   **b.** Symmetric   **c.** Transitive

Mixed Review

Write a sentence for each equation. *(Lesson 1–1)*

55. $x + 8 = 12$ **56.** $2y = 16$ **57.** $25 \div n = 5$

Write an equation for each sentence. *(Lesson 1–1)*

58. Six more than g is 22.

59. Three times c equals 27.

60. Two is the same as the quotient of 8 and x.

61. b increased by 10 and then decreased by 1 is equivalent to 18.

62. Time How many seconds are there in a day? *(Lesson 1–1)*

   a. Write an expression to answer this question.

   b. Evaluate the expression.

63. History Lincoln's Gettysburg Address began "Four score and seven years ago, . . ." *(Lesson 1–1)*

   a. A score is 20. Write a numerical expression for the phrase.

   b. Evaluate the expression to find the number of years.

64. Standardized Test Practice At the movie theater, the price for an adult ticket a is $1.50 less than two times the price of a student ticket s. Choose the algebraic expression that represents the price of an adult ticket in terms of the price of a student ticket.
(Lesson 1–1)

  A $1.50 - 2s$ **B** $2(s - 1.50)$

  C $2s + 2(1.50)$ **D** $2s - 1.50$

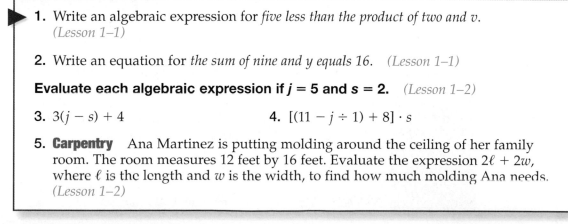

Quiz 1   Lessons 1–1 and 1–2

1. Write an algebraic expression for *five less than the product of two and v.*
(Lesson 1–1)

2. Write an equation for *the sum of nine and y equals 16.* *(Lesson 1–1)*

Evaluate each algebraic expression if $j = 5$ and $s = 2$. *(Lesson 1–2)*

3. $3(j - s) + 4$ **4.** $[(11 - j \div 1) + 8] \cdot s$

5. Carpentry Ana Martinez is putting molding around the ceiling of her family room. The room measures 12 feet by 16 feet. Evaluate the expression $2\ell + 2w$, where ℓ is the length and w is the width, to find how much molding Ana needs. *(Lesson 1–2)*

Math In the Workplace

What You'll Learn

You'll learn to use the commutative and associative properties to simplify expressions.

Why It's Important

Construction

Lumber yards use the commutative and associative properties to determine the amount of wood to order. *See Exercise 25.*

In Bloomington, Illinois, a group of grain silos was converted into a large climbing gym. Suppose Jerome climbs a 35-foot route and then a 50-foot route. Suppose Danielle climbs the 50-foot route first and then the 35-foot route.

Total Distance

Jerome	$35 + 50 = 85$ feet
Danielle	$50 + 35 = 85$ feet

Although they climbed the routes in a different order, Jerome and Danielle both climbed the same total distance. This illustrates the **Commutative Property of Addition**.

Commutative Property of Addition	**Words:**	The order in which two numbers are added does not change their sum.
	Symbols:	For any numbers a and b, $a + b = b + a$.
	Numbers:	$5 + 7 = 7 + 5$

Likewise, the order in which you multiply numbers does not matter.

Commutative Property of Multiplication	**Words:**	The order in which two numbers are multiplied does not change their product.
	Symbols:	For any numbers a and b, $a \cdot b = b \cdot a$.
	Numbers:	$3 \cdot 10 = 10 \cdot 3$

Some expressions are easier to evaluate if you group or *associate* certain numbers. Look at the expression below.

$16 + 7 + 3 = 16 + (7 + 3)$ *Group 7 and 3.*

$\qquad\qquad = 16 + 10$ *Add 7 and 3.*

$\qquad\qquad = 26$ *Add 16 and 10.*

This is an application of the **Associative Property of Addition**.

Associative Property of Addition	**Words:**	The way in which three numbers are grouped when they are added does not change their sum.
	Symbols:	For any numbers a, b, and c, $(a + b) + c = a + (b + c)$.
	Numbers:	$(24 + 8) + 2 = 24 + (8 + 2)$

The Associative Property also holds true for multiplication.

Associative Property of Multiplication	**Words:**	The way in which three numbers are grouped when they are multiplied does not change their product.
	Symbols:	For any numbers a, b, and c, $(a \cdot b) \cdot c = a \cdot (b \cdot c)$.
	Numbers:	$(9 \cdot 4) \cdot 25 = 9 \cdot (4 \cdot 25)$

Examples

Name the property shown by each statement.

1 $4 \cdot 11 \cdot 2 = 11 \cdot 4 \cdot 2$ Commutative Property of Multiplication

2 $(n + 12) + 5 = n + (12 + 5)$ Associative Property of Addition

Your Turn

    a. $(5 \cdot 4) \cdot 3 = 5 \cdot (4 \cdot 3)$ **b.** $16 + t + 1 = 16 + 1 + t$

You can use the Commutative and Associative Properties to simplify and evaluate algebraic expressions. To **simplify** an expression, eliminate all parentheses first and then add, subtract, multiply, or divide.

Example

3 **Simplify the expression $15 + (3x + 8)$. Identify the properties used in each step.**

$$15 + (3x + 8) = 15 + (8 + 3x) \quad \textit{Commutative Property of Addition}$$
$$= (15 + 8) + 3x \quad \textit{Associative Property of Addition}$$
$$= 23 + 3x \quad \textit{Substitution Property}$$

Your Turn

Simplify each expression. Identify the properties used in each step.

    c. $7 + 2a + 6 + 9$ **d.** $(x \cdot 5) \cdot 20$

Example

Geometry Link

Real World

4 **The volume of a box can be found using the expression $\ell \times w \times h$, where ℓ is the length, w is the width, and h is the height. Find the volume of a box whose length is 30 inches, width is 6 inches, and height is 5 inches.**

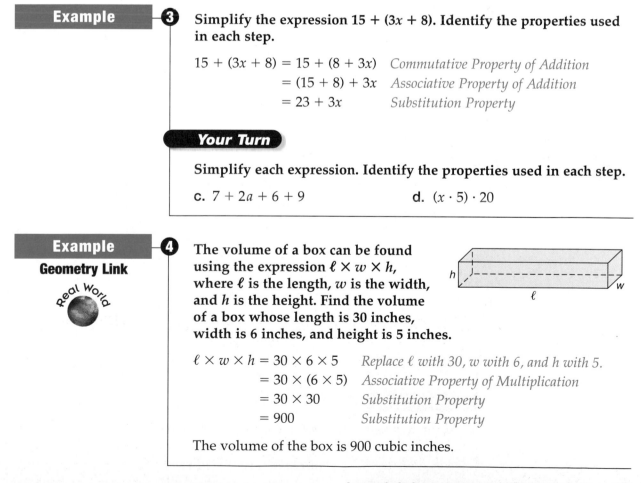

$$\ell \times w \times h = 30 \times 6 \times 5 \quad \textit{Replace } \ell \textit{ with 30, w with 6, and h with 5.}$$
$$= 30 \times (6 \times 5) \quad \textit{Associative Property of Multiplication}$$
$$= 30 \times 30 \quad \textit{Substitution Property}$$
$$= 900 \quad \textit{Substitution Property}$$

The volume of the box is 900 cubic inches.

Whole numbers are the numbers 0, 1, 2, 3, 4, and so on. When you add whole numbers, the sum is always a whole number. Likewise, when you multiply whole numbers, the product is a whole number. This is an example of the **Closure Property**. We say that the whole numbers are *closed* under addition and multiplication.

| **Closure Property of Whole Numbers** | **Words:** | Because the sum or product of two whole numbers is also a whole number, the set of whole numbers is closed under addition and multiplication. |
| | **Numbers:** | $2 + 5 = 7$, and 7 is a whole number. $2 \cdot 5 = 10$, and 10 is a whole number. |

Are the whole numbers closed under division? Study these examples.

$$2 \div 1 = 2 \quad \textit{whole number}$$
$$28 \div 4 = 7 \quad \textit{whole number}$$
$$5 \div 3 = \frac{5}{3} \quad \textit{fraction}$$

It is impossible to list every possible division expression to prove that the Closure Property holds true. However, we can easily show that the statement is false by finding one **counterexample**. A counterexample is an example that shows the statement is not true. Consider $5 \div 3$ or $\frac{5}{3}$. While 5 and 3 are whole numbers, $\frac{5}{3}$ is not. So, the statement *The whole numbers are closed under division* is false.

| **Example** | **5** | State whether the statement *Division of whole numbers is commutative* is *true* or *false*. If false, provide a counterexample. |

Write two division expressions using the Commutative Property and check to see whether they are equal.

$6 \div 3 \stackrel{?}{=} 3 \div 6$ *Evaluate each expression separately.*

$\quad 2 \neq \frac{1}{2} \qquad 6 \div 3 = 2 \text{ and } 3 \div 6 = \frac{1}{2}$

We found a counterexample, so the statement is false. Division of whole numbers is *not* commutative.

Your Turn

e. State whether the statement *Subtraction of whole numbers is associative* is *true* or *false*. If false, provide a counterexample.

Check for Understanding

Communicating Mathematics

Study the lesson. Then complete the following.

1. **Describe** what is meant by the statement *The whole numbers are closed under multiplication.*

2. **Write** an equation that illustrates the Commutative Property of Addition.

3. 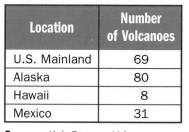 Abeque says that the expression $(7 \cdot 2) + 5$ equals $7 \cdot (2 + 5)$ because of the Associative Properties of Addition and Multiplication. Jessie disagrees with her. Who is correct? Explain.

Guided Practice

Name the property shown by each statement. *(Examples 1 & 2)*

4. $27 + 59 = 59 + 27$

5. $(8 + 7) + 3 = 8 + (7 + 3)$

Simplify each expression. Identify the properties used in each step. *(Example 3)*

6. $(n \cdot 2) \cdot 10$

7. $(3 + p + 47)(7 - 6)$

8. State whether the statement *Whole numbers are closed under subtraction* is *true* or *false*. If false, provide a counterexample. *(Example 5)*

Active volcano in Hawaii

9. **Geology** The table shows the number of volcanoes in the United States and Mexico. *(Example 4)*

   a. Find the total number of volcanoes in these two countries mentally.

   b. Describe the properties you used to add the numbers.

Location	Number of Volcanoes
U.S. Mainland	69
Alaska	80
Hawaii	8
Mexico	31

Source: *Kids Discover Volcanoes,* 1996

Exercises

Practice

Name the property shown by each statement.

10. $(9 \cdot 5) \cdot 20 = 9 \cdot (5 \cdot 20)$

11. $a + 14 = 14 + a$

12. $r \cdot s = s \cdot r$

13. $4 \times 15 \times 25 = 4 \times 25 \times 15$

14. $(7 + 5) + 5 = 7 + (5 + 5)$

15. $c \cdot (d \cdot 10) = (c \cdot d) \cdot 10$

Simplify each expression. Identify the properties used in each step.

16. $h + 1 + 9$

17. $(r \cdot 30) \cdot 5$

18. $17 + k + 23$

19. $6 + (3 + y)$

20. $2 \cdot (19p)$

21. $2 \cdot j \cdot 7$

State whether each statement is *true* or *false*. If false, provide a counterexample.

22. Subtraction of whole numbers is commutative.

23. Division of whole numbers is associative.

Applications and Problem Solving

Real World

24. Sports The table shows the point values for different plays in football. The following expression represents the total possible points for a team in a game.

$$6t + 1x + 3f + 2c + 2s$$

If a team scores 3 touchdowns, 2 extra points, 2 field goals, and 2 safeties, how many total points are scored?

Type of Score	Number of Points
touchdown, *t*	6
extra point, *x*	1
two-point conversion, *c*	2
field goal, *f*	3
safety, *s*	2

25. Construction Lumber mills sell wood to lumberyards in *board feet*. The expression shown below represents the number of board feet in a stack of wood.

$$\frac{\text{inches thick} \times \text{inches wide} \times \text{feet long}}{12}$$

Find the number of board feet if the stack of wood is 10 inches thick, 12 inches wide, and 10 feet long.

26. Geography The Chattahoochee and Savannah rivers form natural boundaries for the state of Georgia.

   a. Write an expression to approximate the total length of Georgia's borders using the map at the right.

   b. Evaluate the expression that you wrote in part a, listing any properties that you used.

27. Critical Thinking Use a counterexample to show that subtraction of whole numbers is not associative.

Mixed Review

Find the value of each expression. *(Lesson 1–2)*

28. $16 \div 2 \cdot 5 \times 3$ **29.** $48 \div [2(3 + 1)]$ **30.** $25 - \frac{1}{3}(18 - 9)$

Evaluate each expression if *a* = 4 and *b* = 11. *(Lesson 1–2)*

31. $196 \div [a(b - a)]$ **32.** $\frac{ab}{a - 2}$

33. Standardized Test Practice Which of the following is the value of $3t - 5q(r + 1)$, if $q = 2$, $r = 0$, and $t = 11$? *(Lesson 1–2)*

   A 23   **B** 52   **C** 53   **D** 33

Extra Practice   See p. 692.

What You'll Learn
You'll learn to use the Distributive Property to evaluate expressions.

Why It's Important
Shopping Cashiers use the Distributive Property when they total customers' groceries.
See Exercise 40.

Members of the Cousteau Society work to raise funds for ocean exploration, research, and conservation. Professional scuba divers can buy their own tanks for $128 each plus $12 for shipping each tank. How much would two tanks cost?

Method 1
Total Cost = 2 × (cost of tank plus shipping)
= 2 × (128 + 12)
= 2(140) or 280

Method 2
Total Cost = (2 × cost of tank) + (2 × cost of shipping)
= (2 × 128) + (2 × 12)
= 256 + 24 or 280

Using both methods, the cost is $280. This example shows how the **Distributive Property** can be applied to simplifying expressions.

Distributive Property	**Symbols:**	For any numbers a, b, and c, $a(b + c) = ab + ac$ and $a(b - c) = ab - ac$.
	Numbers:	$2(5 + 3) = (2 \cdot 5) + (2 \cdot 3)$ $2(5 - 3) = (2 \cdot 5) - (2 \cdot 3)$

It does not matter whether a is placed to the left or to the right of the expression in parentheses. So, $(b + c)a = ba + ca$ and $(b - c)a = ba - ca$.

Examples

Simplify each expression.

1 $3(x + 7)$

$3(x + 7) = (3 \cdot x) + (3 \cdot 7)$ *Distributive Property*
$= 3x + 21$ *Substitution Property*

2 $5(2n + 8)$

$5(2n + 8) = (5 \cdot 2n) + (5 \cdot 8)$ *Distributive Property*
$= 10n + 40$ *Substitution Property*

Your Turn

a. $6(a + b)$ **b.** $(1 + 3t)9$

A **term** is a number, variable, or product or quotient of numbers and variables.

Examples of Terms		Not Terms	
7	7 is a number.	$7 + x$	$7 + x$ is the sum of two terms.
t	t is a variable.	$8rs + 7y + 6$	$8rs + 7y + 6$ is the sum of three terms.
$5x$	$5x$ is a product.	$x - y$	$x - y$ is the difference of two terms.

The numerical part of a term that contains a variable is called the **coefficient**. For example, the coefficient of $2a$ is 2. **Like terms** are terms that contain the same variables, such as $2a$ and $5a$ or $7xy$ and $3xy$.

Consider the expression $5b + 3b + x + 12x$.
- There are four terms.
- The like terms are $5b$ and $3b$, x and $12x$.
- The coefficients are shown in the table.

Term	Coefficient
$5b$	5
$3b$	3
x	1
$12x$	12

The Distributive Property allows us to combine like terms. If $a(b + c) = ab + ac$, then $ab + ac = a(b + c)$ by the Symmetric Property of Equality.

$$2n + 7n = (2 + 7)n \quad \textit{Distributive Property}$$
$$= 9n \quad \textit{Substitution Property}$$

The expressions $2n + 7n$ and $9n$ are called **equivalent expressions** because their values are the same for any value of n. You can write an algebraic expression in **simplest form** by writing an equivalent expression having no like terms and no parentheses.

Examples

Simplify each expression.

3 $4x + 9x$

$$4x + 9x = (4 + 9)x \quad \textit{Distributive Property}$$
$$= 13x \quad \textit{Substitution Property}$$

4 $a + 7b + 3a - 2b$

$$a + 7b + 3a - 2b = a + 3a + 7b - 2b \quad \textit{Commutative Property } (+)$$
$$= (a + 3a) + (7b - 2b) \quad \textit{Associative Property } (+)$$
$$= (1 + 3)a + (7 - 2)b \quad \textit{Distributive Property}$$
$$= 4a + 5b \quad \textit{Substitution Property}$$

Your Turn

  c. $5st + 2st$ **d.** $6 + y + 3z + 4y$

You can use the Distributive Property to solve problems in different, and possibly simpler, ways.

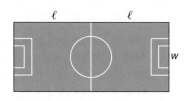

Example **5**

Sports Link

Real World

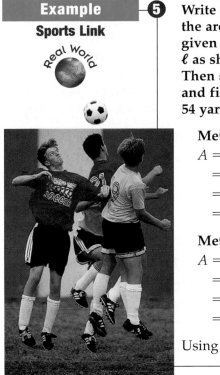

Write an equation representing the area A of a soccer field given its width w and length ℓ as shown in the diagram. Then simplify the expression and find the area if w is 54 yards and ℓ is 60 yards.

Method 1

$\begin{aligned} A &= w(\ell + \ell) & \text{\textit{Multiply the total length by the width.}} \\ &= 54(60 + 60) & \text{\textit{Replace }} w \text{\textit{ with 54 and }} \ell \text{\textit{ with 60.}} \\ &= 54(120) & \text{\textit{Substitution Property}} \\ &= 6480 & \text{\textit{Substitution Property}} \end{aligned}$

Method 2

$\begin{aligned} A &= w\ell + w\ell & \text{\textit{Add the areas of the smaller rectangles.}} \\ &= 54(60) + 54(60) & \text{\textit{Replace }} w \text{\textit{ with 54 and }} \ell \text{\textit{ with 60.}} \\ &= 3240 + 3240 & \text{\textit{Substitution Property}} \\ &= 6480 & \text{\textit{Substitution Property}} \end{aligned}$

Using either method, the area of the soccer field is 6480 square yards.

Check for Understanding

Communicating Mathematics

Study the lesson. Then complete the following.

1. **Write** an algebraic expression having five terms. One term should have a coefficient of three. Also, include two pairs of like terms.

2. **Explain** why $3xy$ is a term but $3x + y$ is not a term.

3. **Determine** which two expressions are equivalent. Explain how you determined your answer.
   a. $20n + 3p$ b. $16n + p - 4n + 2p$
   c. $16n + 4p + 4n$ d. $12p + 3n$
   e. $12n + 3p$ f. $20n - p - 16n + 2p$

> **Vocabulary**
> term
> coefficient
> like terms
> equivalent expressions
> simplest form

Guided Practice

⊙ Getting Ready **Name the like terms in each list of terms.**

Sample: $3c, a, ab, 5, 2c$ **Solution:** $3c, 2c$

4. $5m, 2n, 7n$ 5. $8, 8p, 9p, 9q$
6. $4h, 10gh, 8, 2h$ 7. $6b, 6bc, bc$

Simplify each expression. *(Examples 1–4)*

8. $5x + 9x$

9. $4y + 2 - 3y$

10. $2(5g + 3g)$

11. $3(4 - 6m)$

12. $8(2s + 7)$

13. $(3a + 5t) + (4a + 2t)$

14. School Every student at Miller High School must wear a uniform. Suppose shirts or blouses cost $18 and skirts or pants cost $25. *(Example 5)*

    a. If 250 students buy a uniform consisting of a shirt or blouse and a skirt or pants, write an expression representing the total cost.

    b. Find the total cost.

Exercises •

Practice

Simplify each expression.

15. $16f + 5f$

16. $9a + 6a$

17. $4r - r$

18. $3 + 7 - 2st$

19. $4g - 2g + 6$

20. $5a + 7a + 8b + 5b$

21. $14x + 6y - y - 8x$

22. $3(2n + 10)$

23. $3(5am - 4)$

24. $6(5q + 3w - 2w)$

25. $4a + 7b + (3a - 2b)$

26. $13x - 1 - 8x + 6$

27. $2y + y + y$

28. $bp + 25bp + p$

29. $2(15xy + 8xy)$

30. $5(n + 2r) + 3n$

31. $(r + 2s)3 - 2s$

32. $3(2v + 5m) + 2(3v - 2m)$

33. Write $5(2n + 3r) + 4n + 3(r + 2)$ in simplest form. Indicate the property that justifies each step.

34. Is the statement $2 + (s \cdot t) = (2 + s) \cdot (2 + t)$ *true* or *false*? Find values for s and t to show that the statement may be true. Otherwise, find a counterexample to show that the statement is false.

35. What is the value of $6y$ decreased by the quantity $2y$ plus 1 if y is equivalent to 3?

36. What is the sum of $14xy$, xy, and $5xy$ if x equals 1 and y equals 4?

Applications and Problem Solving

37. Sports Rich bought two baseballs for $4 each and two basketballs for $22 each. What is the total cost? Use the Distributive Property to solve the problem in two different ways.

38. Retail Marie and Mark work at a local department store. Each earns $6.25 per hour. Maria works 24 hours per week, and Mark works 32 hours per week. How much do the two of them earn together each week?

39. Health If an adult male's height is h inches over five feet, his approximate normal weight is given by the expression $6.2(20 + h)$.

    a. What should the normal weight of a 5'9" male be?

    b. How many more pounds should the normal weight of a man that is 6'2" tall be than a man that is 5'9" tall?

40. Shopping Luanda went to the grocery store and bought the items in the table below.

Item	Cost per Item	Quantity
can of soup	$0.99	4
can of corn	$0.49	4
bag of apples	$2.29	4
box of crackers	$3.29	2
jar of jelly	$2.69	2

a. Use the Distributive Property to write an expression representing the cost of the items.

b. Find the change Luanda will receive if she gives the clerk $30.

41. Theater A school's drama club is creating a stage backdrop with a city theme for a performance. The students sketched a model of buildings as shown at the right.

a. How many square feet of cardboard will they need to make the buildings? Use the expression ℓw, where ℓ is the length and w is the width of each rectangle, to find the area of each rectangle. Then add to find the total area.

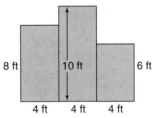

8 ft   10 ft   6 ft

4 ft   4 ft   4 ft

b. Show how to use the Distributive Property as another method in finding the total area of the buildings.

42. Critical Thinking Use the Distributive Property to write an expression that is equivalent to $3ax + 6ay$.

Mixed Review

Name the property shown by each statement. *(Lesson 1–3)*

43. $8(2 \cdot 6) = (2 \cdot 6)8$

44. $(7 + 4) + 3 = 7 + (4 + 3)$

45. If $19 - 3 = 16$, then $16 = 19 - 3$. *(Lesson 1–2)*

Find the value of each expression. *(Lesson 1–2)*

46. $8 + 6 \div 2 + 2$

47. $3(6 - 32 \div 8)$

48. Write an equation for the sentence *Eighteen decreased by d is equal to f.* *(Lesson 1–1)*

49. Sound Some toys are 30 decibels louder than jets during takeoff. Suppose jets produce d decibels of noise during takeoff. *(Lesson 1–1)*

a. Write an expression to represent the number of decibels produced by the loud toys.

b. Evalute the expression if $d = 140$.

50. Standardized Test Practice Which of the following is an algebraic expression for *six times a number decreased by 17*? *(Lesson 1–1)*

A $6n + 17$ **B** $6n - 17$ **C** $17 + 6n$ **D** $17 - 6n$

Extra Practice See p. 693.

Math In the Workplace

What You'll Learn

You'll learn to use a four-step plan to solve problems.

Why It's Important

Savings The four-step plan is useful for finding the amount of interest earned in a bank account.
See Exercise 18.

A survey found that students spend most of their money on snacks. What if you saved 50¢ every school day rather than buying a snack? The chart shows the money you would save by high school graduation.

Saving Just 50 Cents a School Day . . .

If you start saving in this grade . . .	. . . you'll have this amount at high school graduation*
6th grade	$630
7th grade	$540
8th grade	$450
9th grade	$360

*Based on 180 school days in each year × 50 cents a day = $90 a school year.
Source: *Zillions,* Sept/Oct 1999

You would save even more if you deposited the money in a bank account that pays *interest*. You find simple interest by using the **formula** $I = prt$. A formula is an equation that states a rule for the relationship between quantities.

I = interest
p = principal, or amount deposited
r = interest rate, written as a decimal
t = time in years

Using a formula is just one way to solve problems. Any problem can be solved using a problem-solving plan.

1. **Explore**   Read the problem carefully. Identify the information that is given and determine what you need to find.

2. **Plan**   Select a strategy for solving the problem. Some strategies are shown at the right. If possible, estimate what you think the answer should be before solving the problem.

3. **Solve**   Use your strategy to solve the problem. You may have to choose a variable for the unknown, and then write an expression. Be sure to answer the question.

4. **Examine**   Check your answer. Does it make sense? Is it reasonably close to your estimate?

Problem-Solving Strategies

Look for a pattern.

Draw a diagram.

Make a table.

Work backward.

Use an equation or formula.

Make a graph.

Guess and check.

Example

Savings Link

Real World

1 **Suppose you deposit $220 into an account that pays 3% interest. How much money would you have in the account after five years?**

Explore   What do you know?
- The amount of money deposited is $220.
- The interest rate is 3% or 0.03.
- The time is 5 years.

Prerequisite Skills Review
Decimals and Percents, p. 689

What do you need to find?
• the amount of money, including interest, at the end of five years

Plan   **What is the best strategy to use?**
Use the formula $I = prt$ and substitute the known values. Add this amount to the original deposit.

Estimate: 1% of $220 is $2.20. So, 3% of $220 is about $3 \times \$2$ or $6 per year. This will be $30 in five years. You should have approximately $220 + 30$ or $250 in five years.

Solve $I = prt$ *Interest Formula*
$I = 220 \cdot 0.03 \cdot 5$ or 33 $p = 220, r = 0.03, and\ t = 5$

You will earn $33 in interest, so the total amount after five years is $220 + \$33$ or $253.

Examine   Is your answer close to your estimate?
Yes, $253 is close to $250, so the answer is reasonable.

Your Turn

Science   Use $F = 1.8C + 32$ to change degrees Celsius C to degrees Fahrenheit F. Find the temperature in degrees Fahrenheit if it is 29°C.

Another important problem-solving strategy is *using a model*. In the activity below, you will use a model to find a formula for the surface area of a rectangular box.

Hands-On Algebra

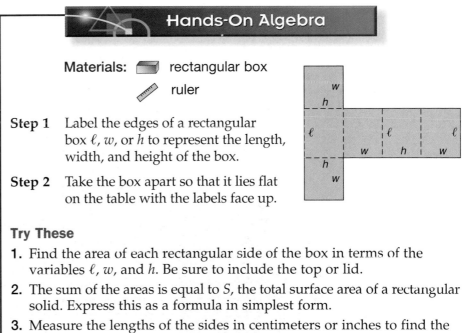

Materials: ▱ rectangular box
📏 ruler

Step 1   Label the edges of a rectangular box ℓ, w, or h to represent the length, width, and height of the box.

Step 2   Take the box apart so that it lies flat on the table with the labels face up.

Try These

1. Find the area of each rectangular side of the box in terms of the variables ℓ, w, and h. Be sure to include the top or lid.

2. The sum of the areas is equal to S, the total surface area of a rectangular solid. Express this as a formula in simplest form.

3. Measure the lengths of the sides in centimeters or inches to find the values of ℓ, w, and h.

4. Use the formula in Exercise 2 to find the total surface area of your box.

Graphing Calculator Tutorial
See pp. 724–727.

Example

Money Link

Real World

2 **How many ways can you make 25¢ using dimes, nickels, and pennies?**

Explore A quarter is worth 25¢. How many ways can you make 25¢?

Plan Make a chart listing every possible combination.

Solve

Coin	Number											
Dimes	2	2	1	1	1	1	0	0	0	0	0	0
Nickels	1	0	3	2	1	0	5	4	3	2	1	0
Pennies	0	5	0	5	10	15	0	5	10	15	20	25

There are 12 ways to make 25¢.

Examine Check that all of the combinations total 25¢ and that there are no other possible combinations. *The solution checks.*

You can use a graphing calculator to solve problems involving formulas.

Graphing Calculator Exploration

The area of a trapezoid is $A = \frac{1}{2}h(a + b)$, where h is the height and a and b are the lengths of the bases. Use a graphing calculator to find the area of trapezoid *JKLM*.

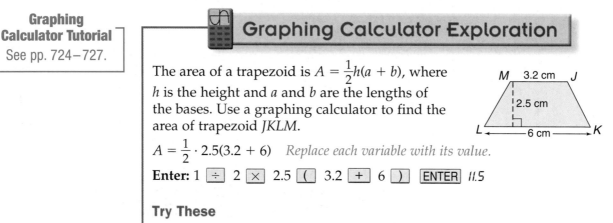

$A = \frac{1}{2} \cdot 2.5(3.2 + 6)$ *Replace each variable with its value.*

Enter: 1 $\div$ 2 $\times$ 2.5 (3.2 + 6) ENTER *11.5*

Try These

1. Find the area of trapezoid *JKLM* if the height is 15 centimeters and the bases remain the same.
2. Find the area of a trapezoid if base *a* is 14 inches long, base *b* is 10 inches long, and the height is 7 inches.

The chart below summarizes the properties that you have learned in this chapter. The properties are useful when you are solving problems.

The following properties are true for any numbers *a*, *b*, and *c*.		
Property	**Addition**	**Multiplication**
Commutative	$a + b = b + a$	$ab = ba$
Associative	$(a + b) + c = a + (b + c)$	$(ab)c = a(bc)$
Identity	$a + 0 = 0 + a = a$ 0 is the identity.	$a \cdot 1 = 1 \cdot a = a$ 1 is the identity.
Zero		$a \cdot 0 = 0 \cdot a = 0$
Distributive	$a(b + c) = ab + ac$ and $a(b - c) = ab - ac$	
Substitution	If $a = b$, then a may be substituted for b.	

Check for Understanding

Study the lesson. Then complete the following.

Vocabulary

formula

1. **List** three reasons for "looking back" when examining the answer to a problem.
2. **Write** a problem in which you need to find the surface area of a rectangular solid. Then solve the problem.

Guided Practice

⏱ Getting Ready   For each situation, answer the related questions.

Phoebe and Hai picked fourteen pints of raspberries in three hours. Hai picked five more pints than Phoebe.

Samples:	Solutions:
How many pints were picked in all?	14
How long did Phoebe and Hai work?	3 hours
Who picked more pints?	Hai
If Phoebe picked x pints, how many did Hai pick?	$x + 5$

Carlos went to a music store and bought 2 more rock CDs than jazz CDs and 3 fewer country CDs than rock CDs. Including 1 classical CD, he bought eight CDs.

3. Did Carlos buy more country than rock?
4. Which type of CD did he buy the most of?
5. If he bought n jazz CDs, how many rock CDs did he buy?

6. **Geometry**   The perimeter P of a rectangle is the sum of two times the length ℓ and two times the width w. *(Example 1)*

   6 cm

   14 cm

   a. Write a formula for the perimeter of a rectangle.
   b. What is the perimeter of the rectangle shown above?

7. **Money**   Nate has $267 in bills. None of the bills is greater than $10. He has eleven $10 bills. He has seven fewer $5 bills than $1 bills.
   a. How many $5 and $1 bills does he have?
   b. Describe the problem-solving strategy that you used to solve this problem.   *(Example 2)*

8. **Shopping**   Two cans of vegetables together cost $1.08. One of them costs 10¢ more than the other.   *(Example 2)*
   a. Would 2 cans of the less expensive vegetable cost more or less than $1.08?
   b. How much would 3 cans of each cost?

Exterior

Exercises •

Practice

Solve each problem. Use any strategy.

9. Craig is 24 years younger than his mother. Together their ages total 56 years. How old is each person? Explain how you found your answer.

10. Moira has $500 in the bank at an annual interest rate of 4%. How much money will she have in her account after two years?

11. Joanne has 20 books on crafts and cooking. She has 6 more cookbooks than craft books. How many of each does she have?

12. How many ways are there to make 20¢ using dimes, nickels, and pennies?

13. Six Explorer Scouts from different packs met for the first time. They all shook hands with each other when they met.

    a. Make a chart or draw a diagram to represent the problem.

    b. How many handshakes were there in all?

    c. The number of handshakes h can also be found by using the formula $h = \dfrac{p(p-1)}{2}$, where p represents the number of people. How many handshakes would there be among 12 people?

    d. Which method would you prefer to solve the problem: using a chart, a diagram, or a formula?

Applications and Problem Solving

Real World

14. **Savings** The table shows the cost of leasing a car for 36 months. Which option is a better deal? Explain.

Type of Fee	Option A Cost ($)	Option B Cost ($)
monthly payment	99	168
monthly tax	6	7
bank fee	495	0
down payment	1956	0
license plates	75	0

15. **Weather** Meteorologists can predict when a storm will hit their area by examining the travel time of the storm system. To do this, they use the following formula.

distance from storm (miles) ÷ speed of storm (miles per hour) = travel time of storm (hours)

It is 4:00 P.M., and a storm is heading towards the coast at a speed of 30 miles per hour. The storm is about 150 miles from the coast. What time will the storm hit?

16. Geometry The area of a triangle is one-half times the product of the base b and the height h.

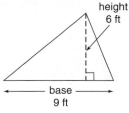

height
6 ft

base
9 ft

   a. Write a formula for the area of a triangle.

   b. Find the area of the triangle.

17. History Distance traveled d equals the product of rate r and time t.

   a. Write the formula for distance.

   b. In 1936, the *Douglas DC-3* became the first commercial airliner to transport passengers when it flew nonstop from New York to Chicago. If it flew at an average of 190 miles per hour and the flight lasted approximately 3.7 hours, find the distance it flew.

18. Savings Refer to the application at the beginning of the lesson.

   a. Suppose you saved 50¢ each school day (180 days) while you were in eighth grade. How much would you have to deposit in a bank account at the end of the school year?

   b. Suppose you deposited your "school year savings" in an account with a 4% annual interest rate. How much money would you have in your account after a year?

   c. Suppose you saved the 50¢ each school day (180 days) while you were a freshman in high school. Find the sum of this amount and the money already in your account from the eighth grade.

   d. How much money would you have in your account after the second year?

   e. Repeat steps c and d for your junior and senior years. How much money would you have in your account by the time you graduated from high school? Compare this amount to that listed in the chart in the application at the beginning of the lesson.

19. Critical Thinking Refer to Exercise 13. In a meeting, there were exactly 190 handshakes. How many people were at the meeting?

Douglas DC-3

Mixed Review

Simplify each expression. *(Lesson 1–4)*

20. $21x - 10x$

21. $5b + 3b$

22. $3(x + 2y)$

23. $9a + 15(a + 3)$

24. State the property shown by $4(ab) = (4a)b$. *(Lesson 1–3)*

25. Standardized Test Practice The top of a volleyball net is 7 feet 11 inches from the floor. The bottom of the net is 4 feet 8 inches from the floor. How wide is the volleyball net? *(Lesson 1–3)*

   A 3 feet 3 inches   **B** 2.31 feet

   C 7 feet 1 inch   **D** 6 feet

Extra Practice See p. 693.

Investigation

Materials

- 5 sheets of different-colored paper
- paper punch
- 4 three-inch square slips of paper
- 4 two-inch squares of paper

Logical Reasoning

Mathematicians use logical reasoning to discover new ideas and solve problems. *Inductive* and *deductive* reasoning are two forms of reasoning. Let's investigate them to find out how they differ.

Investigate

1. Use the colored paper and the paper punch to make at least 25 dots of each color. You will use these dots to explore patterns.

   a. *Triangular numbers* are represented by the number of dots needed to form different-sized triangles. Use your colored dots to form the first four triangular numbers shown below.

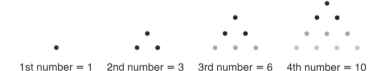

   1st number = 1   2nd number = 3   3rd number = 6   4th number = 10

   b. Draw the fifth triangular number. Do you see a pattern? Use the pattern to write the next five triangular numbers.

   c. In Step 1b, you used **inductive reasoning**, where a conclusion is made based on a pattern or past events.

2. **Deductive reasoning** is the process of using facts, rules, definitions, or properties in a logical order. You use deductive reasoning to reach valid conclusions.

   If-then statements, called **conditionals**, are commonly used in deductive reasoning. Consider the following conditional.

   If <u>I visit the island of Kauai</u>, then <u>I am in Hawaii</u>.

   The portion of the sentence following *if* is called the **hypothesis**, and the part following *then* is called the **conclusion**. This conditional is true since Kauai is a Hawaiian island.

   a. Use the following information to reach a valid conclusion.
      Conditional: If I visit the island of Kauai, then I am in Hawaii.
      Given: I visit the island of Kauai.

b. On a two-inch square, write the hypothesis "I visit the island of Kauai." On a three-inch square, write the conclusion "I am in Hawaii." Place the two-inch square inside the three-inch square.

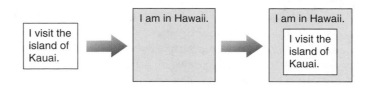

c. Place your pencil on the given statement, "I visit the island of Kauai." Since the pencil is also contained within the square with the conclusion, "I am in Hawaii," the conclusion is valid.

d. Repeat Step 2c, only this time, suppose the given statement is, "I am in Hawaii." You can place your pencil in any region marked by the ×'s in the diagram. You may be in Hawaii, visiting Maui, not the island of Kauai. You cannot reach the conclusion "I visit the island of Kauai" using the conditional and the given information.

Extending the Investigation

In this extension, you will continue to investigate inductive and deductive reasoning.

1. The first four *square numbers* are 1, 4, 9, and 16. Use the colored dots to make the first five square numbers. Use inductive reasoning to list the first ten square numbers.

2. The first four *pentagonal numbers* are 1, 5, 12, and 22. Use the colored dots to make the first five pentagonal numbers. Use inductive reasoning to list the first ten pentagonal numbers.

3. For each problem, identify the hypothesis and conclusion. Then use squares as shown above to determine whether a valid conclusion can be made from the conditional and given information.
  a. Conditional: If the living organism is a grizzly bear, then it is a mammal.
       Given: The living organism is a grizzly bear.
  b. Conditional: If Aislyn is in the Sears Tower, then she is in Chicago.
       Given: Aislyn is in Chicago.

4. Write a paragraph explaining the difference between inductive and deductive reasoning. Include an example of each type of reasoning.

Presenting Your Investigation

Here are some ideas to help you present your conclusions to the class.

- Make a poster showing the triangular, square, and pentagonal numbers.

- Include a description of the patterns you observed.

*inter*NET
CONNECTION   **Investigation** For more information on logical reasoning, visit: www.algconcepts.glencoe.com

Math In the Workplace

What You'll Learn
You'll learn to collect and organize data using sampling and frequency tables.

Why It's Important
Marketing
Businesses use surveys to collect data in order to test new ideas.
See Example 3.

In a survey, a research group randomly selected a group of people who go to radio station Web sites. The people were asked what they were looking for at the sites. The top five reasons are shown in the table.

Why Go to Radio Web Sites?
1. community events information
2. concert information
3. titles and artists of songs
4. listen to station
5. see advertisers' products

Source: *American Demographics,* August 1999

The research group used **sampling** because it is a convenient way to gather **data**, or information. It would have been impractical to try to question all radio Web site visitors. A **sample** is a small group that is used to represent a much larger **population**. Three important characteristics of a good sample are listed below.

Sampling Criteria	A good sample is: • representative of the larger population, • selected at random, and • large enough to provide accurate data.

A survey can be biased and give false results if these criteria are not followed. Note that there is no given number to make the sample large enough. You must consider each situation individually.

Example
Health Link

Real World

① **One hundred people in Lafayette, Colorado, were asked to eat a bowl of oatmeal every day for a month to see whether eating a healthy breakfast daily could help reduce cholesterol. After 30 days, 98 of those in the sample had lower cholesterol. Is this a good sample? Explain.** **Source:** Quaker Oats, 1999

If the people were randomly chosen, then this is a good sample. Also, the sample appears to be large enough to be representative of the population. For example, the results of two or three people would not have been enough to make any conclusions.

Your Turn

Determine whether each is a good sample. Explain.

a. Two hundred students at a school basketball game are surveyed to find the students' favorite sport.

b. Every other person leaving a supermarket is asked to name their favorite soap.

One way to organize data is by using a **frequency table**. In a frequency table, you use **tally marks** to record and display the frequency of events.

2 **In an experiment, students "charged" balloons by rubbing them with wool. Then they placed the balloons on a wall and counted the number of seconds they remained. The class results are shown in the chart at the right. Make a frequency table to organize the data.**

Static Electricity Time (s)				
15	52	26	22	25
26	29	33	36	20
43	21	30	39	34
35	27	29	42	35
16	18	21	21	40

Step 1 Make a table with three columns: Time (s), Tally, and Frequency. Add a title.

Step 2 It is sometimes helpful to use *intervals* so there are fewer categories. In this case, we are using intervals of size 10.

Step 3 Use tally marks to record the times in each interval.

Step 4 Count the tally marks in each row and record this number in the Frequency column.

Static Electricity		
Time (s)	Tally	Frequency
15–24	JHT III	8
25–34	JHT IIII	9
35–44	JHT II	7
45–54	I	1

Your Turn

c. Make a frequency table to organize the data in the chart at the right.

Noon Temperature (°C)					
32	30	18	29	20	14
21	32	36	15	19	10
16	22	25	30	26	21

In Example 2, suppose the science teacher wanted to know how many balloons stayed on the wall *no more than 44 seconds*. To answer this question, use a **cumulative frequency table** in which the frequencies are accumulated for each item.

Static Electricity		
Time (s)	Frequency	Cumulative Frequency
15–24	8	8
25–34	9	17
35–44	7	**24**
45–54	1	25

8 + 9 = 17
17 + 7 = 24
24 + 1 = 25

From the cumulative frequency table, we see that 24 balloons stayed on the wall for 44 seconds or less. Or, 24 balloons stayed on the wall no more than 44 seconds.

Once you have summarized data in a frequency table or in a cumulative frequency table, you can analyze the information and make conclusions.

Example **3**

Marketing Link

Real World

Owners of a fast-food restaurant are looking for a new location. They counted the number of people who passed by the proposed location one day during lunchtime. The frequency table at the right shows the results of their sampling.

Age of People	Tally	Frequency
under 13	JHT II	7
teens	JHT JHT	10
20s	JHT JHT JHT III	18
30s	JHT JHT JHT JHT JHT JHT JHT JHT II	42
40s	JHT JHT JHT JHT JHT JHT JHT I	36
50s	JHT JHT JHT IIII	19
60s	JHT JHT I	11

A. Which two groups of people passed by the location most frequently?

adults in their 30s and 40s

B. If the restaurant is an ice cream shop aimed at teens during their lunchtimes, is this a good location for the restaurant? Explain.

Since very few teens pass by the location compared to adults, the owners should probably look for another location.

Check for Understanding

Communicating Mathematics

Study the lesson. Then complete the following.

1. **Explain** the difference between a frequency table and a cumulative frequency table.

2. **List** some examples of how a survey might be biased.

Vocabulary

sampling
data
sample
population
frequency table
tally marks
cumulative frequency table

Guided Practice

Determine whether each is a good sample. Explain.   *(Example 1)*

3. Four people out of 500 are randomly chosen at a senior assembly to find the percent of seniors who drive to school.

4. Six hundred randomly chosen pea seeds are used to determine whether wrinkled seeds or round seeds are more common.

Refer to the chart at the right.

5. Make a frequency table to organize the data. *(Example 2)*

6. What number of goals was scored most frequently? *(Example 3)*

7. How many times did the team score 8 goals? *(Example 3)*

8. How many more times did the soccer team score six goals than three goals? *(Example 3)*

Number of Soccer Goals Scored This Season		
1	2	5
1	6	2
6	8	4
2	4	5
5	1	3
4	7	2
2	6	4

9. **Technology** When lines of cars get too long at some traffic lights, computers override the signals to turn the lights green and allow the cars to move. A cycle is the number of seconds it takes a light to change from red back to red. The frequency table below shows different traffic light cycles during one afternoon. *(Examples 2 & 3)*

Cycle (s)	Tally	Frequency
80	JHT JHT JHT JHT JHT JHT III	33
90	JHT JHT JHT JHT JHT JHT JHT JHT II	42
100	JHT JHT JHT JHT JHT JHT JHT JHT JHT JHT JHT JHT	60
110	JHT JHT JHT JHT JHT	25

a. Which cycle occurred the most?

b. Make a cumulative frequency table of the data.

c. If the standard cycle for a traffic light is 100 seconds, how many times during this period was the cycle less than the standard?

Exercises

Practice

Determine whether each is a good sample. Describe what caused the bias in each poor sample. Explain.

10. Thirty people standing in a movie line are asked to name their favorite actor.

11. Police stop every fifth car at a sobriety checkpoint.

12. Every other household in a neighborhood of 240 homes is surveyed to determine how many people in the area recycle.

13. Every other household in a neighborhood of 20 homes is surveyed to determine the country's favorite presidential candidate.

14. Every third student on a class roster is surveyed to determine the average number of hours students in the class spend on a computer.

15. All people leaving a sporting goods store are asked to name their favorite golfer.

16. Refer to the chart at the right.

    a. Make a frequency table to organize the data.

    b. How many fewer sausage pizzas were ordered than cheese pizzas?

    c. Suppose x mushroom pizzas were also ordered. Write an expression representing the total number of mushroom, vegetable, and pepperoni pizzas ordered. Write the expression in simplest form.

Pizzas Ordered				
C	C	C	P	C
C	P	P	V	S
S	P	C	P	C
S	S	C	P	P
C	V	P	C	C
V	S	S	C	C

C = cheese, P = pepperoni,
S = sausage, V = vegetable

17. Refer to the chart at the right.

    a. Make a frequency table to organize the data.

    b. What was the most common score?

    c. Suppose each S represents the score and each F represents the frequency for that score. Explain why the formula below determines the class *average* A for this quiz.

$$A = \frac{(S_1 \cdot F_1) + (S_2 \cdot F_2) + \ldots + (S_8 \cdot F_8)}{30}$$

    d. Find the class average for the quiz.

Quiz Scores (out of 10 points)				
9	8	8	9	8
8	10	6	7	8
10	8	8	9	9
7	9	7	8	10
9	6	7	8	9
9	10	8	9	8

Reading Algebra

Read S_1 as S *sub 1*. The 1 is called a *subscript*.

18. Refer to the chart at the right.

    a. Make a cumulative frequency table to organize the data.

    b. In how many games were there at least three home runs?

    c. In how many games were there no more than four home runs?

Number of Home Runs in a Game per Month				
2	2	3	5	1
0	1	2	6	3
3	1	1	4	5
2	3	4	3	0
1	0	1	1	2

19. Why do you suppose a coffee and bagel shop would want to locate where a lot of people walk past the store between 7:00 A.M. and 10:30 A.M.?

20. **Health** When you have a blood test taken for your health, why does the technician only take a few vials of your blood? Use the terms you learned in this lesson to explain your answer.

21. **Marketing** A new cola drink is out on the market. Name three places where the cola company could set up taste tests to determine interest in the drink.

22. Entertainment The frequency table at the right shows students' favorite types of movies in one class.

Favorite Type of Movie		
Movie	**Tally**	**Frequency**
Adventure	JHT II	7
Comedy	JHT IIII	9
Horror	IIII	4
Drama	III	3

a. Suppose you invite students in this class to a party. What type of movie would you show? Explain.

b. In another class, three times more students favored drama, and two fewer favored comedy. Write an expression to find the total number of people in that class who favored drama and comedy. Then find the number.

23. Critical Thinking Suppose someone takes a phone survey from a large random sample of people. Do you think that the wording of a question or the surveyor's tone of voice can affect the responses and cause biased results? Explain.

Mixed Review

24. An adult bus ticket and a child's bus ticket together cost $2.40. The adult fare is twice the child's fare. What is the adult's fare? Use any strategy to solve the problem. *(Lesson 1–5)*

25. Travel What distance can a car travel in 5 hours at a constant rate of 55 miles per hour? Use a diagram or the formula $d = rt$ to solve the problem. *(Lesson 1–5)*

26. Simplify the expression $16a + 21a + 30b - 7b$. *(Lesson 1–4)*

27. Write a verbal expression for $x + 9$. *(Lesson 1–1)*

28. Open-Ended Test Practice Write an algebraic expression for *4 times n less 3*. *(Lesson 1–1)*

Quiz 2   Lessons 1–3 through 1–6

▶ **Simplify each expression. Identify the properties used in each step.** *(Lesson 1–3)*

1. $11 + 2a + 6$ **2.** $4 \cdot (8t)$

3. Health Your optimum exercise heart rate per minute is given by the expression $0.7(220 - a)$, where a is your age. Choose a value for a and find your optimum exercise heart rate. *(Lesson 1–4)*

4. Fitness Lorena runs for 30 minutes each day. Find the distance she runs if she averages 660 feet per minute. Use the formula $d = rt$. *(Lesson 1–5)*

5. Biology Make a frequency table to organize the data in the chart at the right. Which eye color occurs the least? *(Lesson 1–6)*

Eye Color				
H	B	B	G	U
H	G	G	B	B
B	G	U	G	H
H	U	B	B	G
H	B	U	B	B

B = brown, U = blue,
G = green, H = hazel

Math In the Workplace

What You'll Learn
You'll learn to construct and interpret line graphs, histograms, and stem-and-leaf plots.

Why It's Important
Research
Researchers collect data and use graphs to help them make predictions.
See Exercises 4–6.

The graph shows the yearly attendance at movie theaters. For example, in 1994, approximately 1.3 billion people went to the movies.

Graphs are a good way to display and analyze data. The graph at the right is a **line graph**. It shows trends or changes over time. There are no holes in the graph and every point on the graph has meaning. To construct a line graph, include the following items.

1. a title
2. a label on each axis describing the variable that it represents
3. equal intervals on each axis

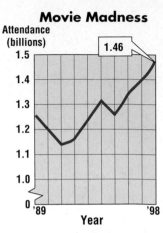

Movie Madness

Attendance (billions)

1.46

Source: *USA Today,* 1999

Example ①

Travel Link

The number of annual visitors to the Grand Canyon is given in the table at the right. Construct a line graph of the data. Then use the graph to predict the number of annual visitors to the Grand Canyon in the year 2010.

Step 1 Draw a horizontal axis and a vertical axis and label them as shown below. Include a title.

Step 2 Plot the points.

Step 3 Draw a line by connecting the points.

You can see from the graph that the general trend is that the number of visitors to the Grand Canyon increases steadily every ten years. A good prediction for the year 2010 might be about 6 or 6.5 million people.

Year	Grand Canyon Visitors (millions)
1960	1.0
1970	2.2
1980	2.5
1990	3.5
2000*	5.0

Source: Grand Canyon National Park, 1999
*estimated

Grand Canyon Visitors

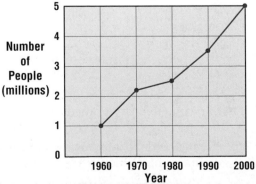

Number of People (millions)

a. The table at the right shows the approximate U.S. consumption of bottled water per person. Construct a line graph of the data. Then use it to predict the amount of bottled water each person will drink in the year 2005.

Year	Bottled Water (gallons)
1975	1
1980	2.5
1985	5
1990	7
1995	12
2000*	16

Source: *Nutrition Action*, April, 1999
*estimated

Another type of graph that is used to display data is a **histogram**. A histogram uses data from a frequency table and displays it over equal intervals. To construct a histogram, include the same three items as the line graph: title, axes labels, and equal intervals. In a histogram, all of the bars should be the same width with no space between them.

Example **2**

Physical Science Link

Real World

The frequency table is from Example 2 in Lesson 1–6. It shows the various time intervals that "charged" balloons remained stuck to the wall. Construct a histogram of the data.

Static Electricity		
Time (s)	Tally	Frequency
15–24	JHT III	8
25–34	JHT IIII	9
35–44	JHT II	7
45–54	I	1

Step 1 Draw a horizontal axis and a vertical axis and label them as shown below. Include the title.

Step 2 Show equal intervals given in the frequency table on the horizontal axis. Show equal intervals of 1 on the vertical axis.

Step 3 For each time interval, draw a bar whose height is given by the frequency.

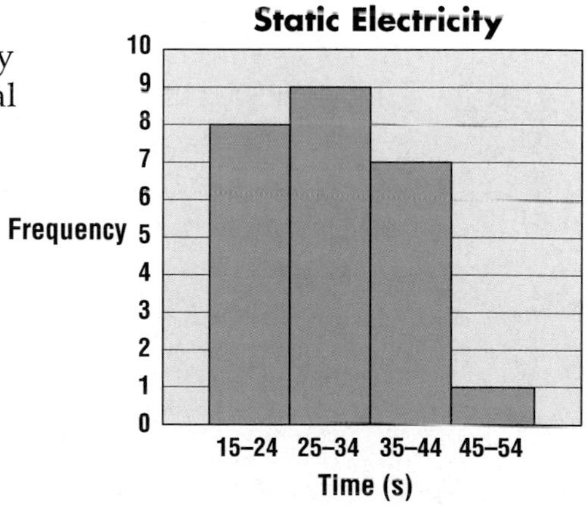

The histogram gives a better visual display of the data than the frequency table. In Lesson 1–6, we used cumulative frequency tables to organize data. Likewise, we can construct **cumulative frequency histograms**.

The ages of people who participated in a recent survey are shown in the table at the right. Construct a cumulative frequency histogram to display the data.

Survey		
Age	Tally	Frequency
1–10	卌 III	8
11–20	卌 IIII	9
21–30	卌	5
31–40	卌 卌 II	12
41–50	卌 III	8

First, make a cumulative frequency table. Then construct a histogram using the cumulative frequencies for the bar heights. Remember to label the axes and include the title.

Survey		
Age	Frequency	Cumulative Frequency
1–10	8	8
11–20	9	17
21–30	5	22
31–40	12	34
41–50	8	42

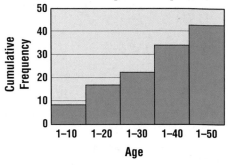

Survey Participants

Your Turn

b. Construct a cumulative frequency histogram of the data in Example 2.

Another way to display data is a **stem-and-leaf plot**.

The greatest common place value for each data item is used to form the *stem*.

Stem	Leaf
1	1 6
2	1 3 9
3	
4	5 5

$2\,|\,3 = 23$

The *leaves* are formed by the next greatest place value.

In this case, the tens digits are the stems. The ones digits are the leaves. Write the leaves in order from least to greatest.

A *key* is always included. This shows how the digits are related.

Info Graphic

In the stem-and-leaf plot at the right, the data are represented by three-digit numbers. In this case, use the digits in the first two place values to form the stems. For example, the values for 102, 114, 115, 108, 139, 127, 125, and 131 are shown in the stem-and-leaf plot at the right.

Stem	Leaf
10	2 8
11	4 5
12	5 7
13	1 9

$11\,|\,5 = 115$

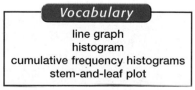
④ The table shows the class results on a 50-question test. Make a stem-and-leaf plot of the grades.

Class Scores					
29	37	48	40	17	34
28	43	37	35	49	29
13	29	42	45	37	46

The tens digits are the stems, so the stems are 1, 2, 3, and 4. The ones digits are the leaves.

Stem	Leaf
1	7 3
2	9 8 9 9
3	7 4 7 5 7
4	8 0 3 9 2 5 6

$3\,|\,7 = 37$

Now arrange the leaves in numerical order to make the results easier to observe and analyze.

Stem	Leaf
1	3 7
2	8 9 9 9
3	4 5 7 7 7
4	0 2 3 5 6 8 9

$3\,|\,7 = 37$

What were the highest and lowest scores?
49 and 13

Which score occurred most frequently?
29 and 37, three times each

How many students received a score of 35 or better?
11 students

Your Turn

c. Make a stem-and-leaf plot of the quiz grades below.
54, 55, 60, 42, 41, 75, 50, 68, 62, 54, 70, 50

Check for Understanding

Communicating Mathematics

Study the lesson. Then complete the following.

1. **Explain** the differences between the use of line graphs and histograms.

2. **Identify** each essential part of a correctly drawn line graph or histogram.

Vocabulary
line graph
histogram
cumulative frequency histograms
stem-and-leaf plot

3.  Marcia says that a histogram works as well as a line graph to show trends over time. Manuel says that a histogram shows intervals, not trends. Who is correct? Explain.

Guided Practice

The table at the right shows the percent of homes in California with on-line access. *(Example 1)*

4. Make a line graph of the data.

5. Between which two years was the growth of on-line access the greatest?

6. Predict the percent of homes with on-line access in the year 2001.

Year	Percent of Homes On-Line
1997	19
1998	25
1999	28
2000*	31

Source: Pacific Telesis, 1998
*estimated

Refer to the histogram at the right.

7. Determine the length of most maple leaves. *(Example 2)*

8. How many leaves were sampled? *(Example 2)*

9. Construct a cumulative frequency histogram of the data. *(Example 3)*

10. How many of the leaves were no more than 15 centimeters long? *(Example 3)*

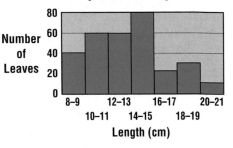

Leaf Lengths in a Maple Tree Population

11. Weather The stem-and-leaf plot at the right shows the daily high temperatures for March, 1999, in McComb, Mississippi. *(Example 4)* **Source:** *The Weather Underground*

a. What was the highest temperature?

b. On how many days was the high temperature in the 70s?

c. What temperature occurred most frequently?

Daily High Temperatures (°F)

Stem	Leaf
4	8
5	4 9 9
6	3 3 3 3 4 4 8 8 8
7	0 0 2 2 2 3 3 5 5 9 9 9
8	1 1

$4 \mid 8 = 48°F$

Exercises

● ● ● ● ● ● ● ● ● ● ●

Practice

The percent of unemployment among workers ages 16 to 19 is shown at the right.

12. Make a line graph of the data.

13. When was unemployment at its highest?

14. Describe the general trend in unemployment among teens ages 16 to 19.

Year	Percent of Working Teens Unemployed
1992	17
1993	19
1994	18
1995	17.5
1996	17
1997	15
1998	13

Source: U.S. Labor Dept., 1998

In a survey, men and women were asked how long they were willing to stay on hold when calling a customer service representative about a product they purchased. The results are shown in the table at the right.

How Long on Hold?		
Time (min)	Percent of Men	Percent of Women
0–1	28	18
2–3	32	36
4–5	23	27
6–7	7	10
8+	10	9

Source: Bruskin/Goldring for Inference

15. Make a histogram showing the men's responses.

16. Make a histogram showing the women's responses.

17. How do your histograms compare?

18. Who do you think would hang up the phone sooner, men or women?

The stem-and-leaf plot at the right gives the number of catches of the NFL's leading pass receiver for each season through 1998.

Stem	Leaf
6	0 1 2 6 7
7	1 1 1 2 3 3 3 5 7 8
8	2 5 8 8 9
9	0 0 1 2 2 2 3 5
10	0 0 0 1 4 6 8 8
11	2
12	2 3 $12|3 = 123$

19. What was the greatest number of catches during a season?

20. How many seasons are represented?

21. What number of catches occurred most frequently?

22. How many leading pass receivers have at least 90 catches?

23. Critical Thinking *Back-to-back stem-and-leaf plots* are used to compare two sets of data. The back-to-back stem-and-leaf plot below compares the performance of two algebra classes on their first test. Which class do you believe did better on the test? Why do you think so?

First Period	Stem	Second Period
9 8	5	
9 9 8 7	6	7 8
7 7 6 5 3	7	2 4 4 4 5 5 8
5 4 1	8	1 3 4 5 7
8 8 6 5 2	9	0 1 1 3 8
		$8|9 = 89$

Mixed Review

Determine whether each is a good sample. *(Lesson 1–6)*

24. A survey is taken in Alaska to determine how much money an average family in the United States spends on heating their home.

25. In a survey, every third name in the phone book is called and the person answering is interviewed.

26. Write a formula for the perimeter P of a square with side s in simplest form. *(Lesson 1–5)*

27. Write $5x + 3(x - y)$ in simplest form. *(Lesson 1–4)*

28. Standardized Test Practice Evaluate $12 \cdot 6 + 3 \cdot 2 - 8$. *(Lesson 1–2)*

   A 142   **B** −144   **C** 70   **D** 120

Extra Practice See p. 694.

Understanding and Using the Vocabulary

After completing this chapter, you should be able to define each term, property, or phrase and give an example of each.

*inter*NET
CONNECTION **Review Activities**
For more review activities, visit:
www.algconcepts.glencoe.com

Algebra

algebraic expression *(p. 4)*
coefficient *(p. 20)*
equation *(p. 5)*
equivalent expressions *(p. 20)*
evaluating *(p. 10)*
factors *(p. 4)*
formula *(p. 24)*
like terms *(p. 20)*
numerical expression *(p. 4)*
order of operations *(p. 8)*
product *(p. 4)*
simplest form *(p. 20)*
simplify *(p. 15)*

term *(p. 20)*
variable *(p. 4)*
whole numbers *(p. 16)*

Statistics

cumulative frequency
   histogram *(p. 39)*
cumulative frequency
   table *(p. 33)*
data *(p. 32)*
frequency table *(p. 33)*
histogram *(p. 39)*
line graph *(p. 38)*
population *(p. 32)*

sample *(p. 32)*
sampling *(p. 32)*
stem-and-leaf plot *(p. 40)*
tally marks *(p. 33)*

Logic

conclusion *(p. 30)*
conditional *(p. 30)*
counterexample *(p. 16)*
deductive reasoning *(p. 30)*
hypothesis *(p. 30)*
if-then statement *(p. 30)*
inductive reasoning *(p. 30)*

Choose the correct term to complete each sentence.

1. A (coefficient, term) is a number, a variable, or a product or quotient of numbers and variables.
2. The result of two numbers multiplied together is the (factor, product).
3. A(n) (numerical expression, algebraic expression) contains variables.
4. According to the (order of operations, like terms), you do multiplication before addition.
5. A (counterexample, hypothesis) shows that a statement is not always true.
6. Some examples of (like terms, whole numbers) are $2x$, $10x$, and $-6x$.
7. A (sample, variable) is a group used to represent a much larger population.
8. Any sentence that contains an equals sign is a(n) (equation, formula).
9. Using (sampling, frequency tables) is a way to organize data.
10. A (histogram, stem-and-leaf plot) makes it easier to identify specific data items.

Skills and Concepts

Objectives and Examples	Review Exercises

• Lesson 1–1 Translate words into algebraic expressions and equations.

Write an algebraic expression for the verbal expression *7 decreased by the quantity x divided by 2.*

$7 - (x \div 2)$ or $7 - \dfrac{x}{2}$

Write an algebraic expression.

11. the product of 5 and n
12. the sum of 2 and three times x

Write an equation for each sentence.

13. Six less than two times y equals 14.
14. The quotient of 20 and x is 4.

Objectives and Examples

Review Exercises

• **Lesson 1–2** Use the order of operations to evaluate expressions.

$2 \cdot 7 + 2 \cdot 3 = 14 + 6$
$= 20$

Find the value of each expression.

15. $3 + 8 \div 2$

16. $12 \div 4 + 15 \cdot 3$

17. $29 - 3(9 - 4)$

18. $4(11 + 7) - 9 \cdot 8$

19. Find the value of $3ac - b$ if $a = 6$, $b = 9$, and $c = 1$.

• **Lesson 1–3** Use the commutative and associative properties to simplify expressions.

Name the property shown by
$3 + x + 2 = 3 + 2 + x$. Then simplify.

$3 + x + 2 = 3 + 2 + x$ *Commutative (+)*
$= 5 + x$ *Substitution*

Name the property shown by each statement. Then simplify.

20. $6 + (7 + b) = (6 + 7) + b$

21. $2 \cdot c \cdot 10 = 2 \cdot 10 \cdot c$

22. $9 \cdot (5 \cdot f) = (9 \cdot 5) \cdot f$

23. $x(5 + 4) = (5 + 4)x$

24. $3 + a + 8 = 3 + 8 + a$

25. $(g + 1) + 2 = g + (1 + 2)$

• **Lesson 1–4** Use the Distributive Property to evaluate expressions.

$5b + 3(b + 2) = 5b + 3 \cdot b + 3 \cdot 2$
$= 5b + 3b + 6$
$= (5 + 3)b + 6$
$= 8b + 6$

Simplify each expression.

26. $4(8 + y)$

27. $7(v - 1)$

28. $10x + x$

29. $h(2 + a)$

30. $5z + 2z - 6$

31. $10 + 3(4 - d)$

• **Lesson 1–5** Use the four-step plan to solve problems.

Explore   What do you know? What are you trying to find?

Plan   How will you go about solving this? What problem-solving strategy could you use?

Solve   Carry out your plan. Does it work? Do you need another plan? If necessary, choose a variable for an unknown and write an expression.

Examine   Check your answer. Does it make sense? Is it reasonably close to your estimate?

Use the four-step plan to solve each problem.

32. Finance   Mr. Rockwell deposited $1000 in an account that pays 2% interest. How much money would he have in the account after ten years?

33. School   Jamal is typing a three-page report for school. He thought he could finish typing the report with approximately 400 words per page in 2 hours. After $1\frac{1}{2}$ hours, he had finished 2 pages.

a. How many words are in his paper?

b. About how many words had Jamal typed in $1\frac{1}{2}$ hours?

Chapter 1 Study Guide and Assessment

Objectives and Examples

- **Lesson 1–6** Collect and organize data using sampling and frequency tables.

Make a frequency table for the data
{1, 4, 3, 4, 0, 2, 3, 1, 0, 2, 0, 2, 0, 4, 0, 0, 4, 1, 2}.

Number	Tally	Frequency
0	JHT I	6
1	III	3
2	IIII	4
3	II	2
4	IIII	4

Review Exercises

Use the frequency table at the left to answer each question.

34. How many numbers are in the sample?
35. Which number occurs most frequently?
36. How many times does the number 2 occur?
37. Make a cumulative frequency table from the data.
38. How many times does a number less than 2 occur?
39. How many times does a number greater than or equal to 2 occur?

- **Lesson 1–7** Construct and interpret line graphs, histograms, and stem-and-leaf plots.

Construct a histogram for the data {10, 10, 10, 10, 11, 11, 12, 13, 14, 14, 14, 15, 15}.

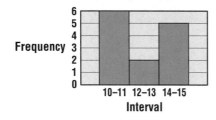

Use the histogram at the left to answer each question.

40. How large is each interval?
41. Which interval has the most data?
42. How many numbers have a value greater than 11?
43. Make a cumulative histogram from the data.
44. How many numbers are in the sample?
45. How many numbers have a value less than 14?

Applications and Problem Solving

46. **Geometry** Write an equation to represent the perimeter *P* of the figure below. Then solve for *P* if $x = 9$ and $y = 5$. *(Lesson 1–5)*

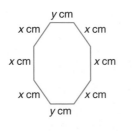

47. **Testing** The stem-and-leaf plot below shows the scores from a driver's test. *(Lesson 1–7)*

   a. What were the highest and lowest scores?

   b. Which score occurred most frequently?

   c. How many people received a score of 76 or better?

Stem	Leaf
6	2 8
7	4 5 5 6
8	0 4 8 8 8
9	2 4 7
10	0 $7|4 = 74$

1. **Explain** why we use the order of operations in mathematics.

2. **List** three like terms with the variable k.

Write an algebraic expression for each verbal expression.

3. x increased by 12   4. the quotient of 5 and y 5. 1 less than 8 times p

Use the order of operations to find the value of each expression.

6. $13 + 4 \cdot 5$ 7. $12 + 6 \div 3 - 4$ 8. $3(8 + 2) - 7$

Evaluate each expression if $h = 8$, $j = 3$, and $k = 2$.

9. $k(4 + j) + 6$ 10. $\dfrac{h + k}{h - j}$

Name the property shown by each statement.

11. If $11 - 7 + x$, then $7 + x = 11$.   12. $28 \cdot 1 = 28$
13. $(r \cdot 9) \cdot 3 = r \cdot (9 \cdot 3)$ 14. $10 + b = b + 10$
15. $6(m + 2) = 6 \cdot m + 6 \cdot 2$

Simplify each expression.

16. $n + 5n$ 17. $6x - 4x + 9y - 4y$ 18. $4(2s + 8t - 1)$

19. **Sports** Danny stayed late after every basketball practice and shot 5 free throws. The chart shows how many free throws he made out of 5 for each night of practice.

Free Throws (out of 5)			
1	4	0	2
1	3	3	2
4	3	3	2
5	3	2	2
1	0	4	3

   a. Make a frequency table to organize the data.
   b. If Danny has basketball practice 5 days a week, how many weeks did he stay late, shooting free throws?
   c. What number of free throws did he make most often?
   d. How many times did he not make any free throws?
   e. How many times did he make all 5 free throws?

20. **Communication** The line graph shows the growth in sales of prepaid calling cards.
   a. Between which two years was growth in sales the greatest?
   b. Predict the number of sales for the year 2002.

Sales of Prepaid Calling Cards

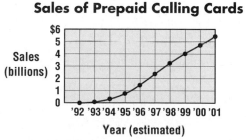

Sales (billions)

Year (estimated)

Source: Atlantic ACM, 1998

Number Concept Problems

Standardized tests include many questions written with realistic settings. Read each question carefully. Be sure you understand the situation and what the question asks.

A calculator can help, but you can often find the answer faster with a pencil and your own math skills. Since standardized tests are timed, you will want to find the correct answers as quickly as possible.

THE PRINCETON REVIEW

Translate words into arithmetic symbols.

is	$=$
of	$\times$
per	$\div$

Proficiency Test Example

Mrs. Lopez estimates that $\frac{2}{3}$ of the families in her neighborhood will participate in the annual garage sale. If there are 225 families in her neighborhood, how many families does she expect to participate?

A 75   **B** 150
C 175   **D** 220

Hint Estimate the answer before making any calculations.

Solution First estimate. Since $\frac{2}{3}$ is more than one half, more than one half of the 225 families will participate. One half of 225 is about 112. So, choice A is not possible.

The word *of* ($\frac{2}{3}$ *of* the families) tells you to use multiplication.

$\frac{2}{3} \cdot 225 = \frac{2(225)}{3}$ *Multiply.*

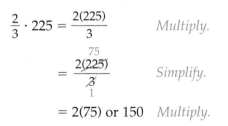

$= \frac{2(\overset{75}{\cancel{225}})}{\underset{1}{\cancel{3}}}$ *Simplify.*

$= 2(75)$ or 150   *Multiply.*

If you use your calculator, multiply 2 by 225, and then divide the answer by 3.

Two-thirds of the 225 families is 150 families. So, the answer is B.

SAT Example

Jan drove 144 miles between 10:00 A.M. and 12:40 P.M. What was her average speed in miles per hour?

Hint Pay attention to the units of measure.

Solution From 10:00 A.M. to 12:40 P.M. is 2 hours and 40 minutes. You need time in *hours*. Convert minutes to hours.

2 hours 40 minutes $= 2\frac{40}{60}$ hours or $2\frac{2}{3}$ hours

$\dfrac{144}{2\frac{2}{3}} = \dfrac{144}{\frac{8}{3}}$ *Divide the miles by time.*
             Rename $2\frac{2}{3}$ as $\frac{8}{3}$.

$= 144 \cdot \dfrac{3}{8}$ *Multiply by the reciprocal of $\frac{8}{3}$.*

$= \overset{18}{\cancel{144}} \cdot \dfrac{3}{\underset{1}{\cancel{8}}}$ *Divide by the GCF, 8.*

$= 18(3)$ or 54   *Multiply.*

The answer is 54 miles per hour. Record it on the grid.

- Start with the *left* column.

- Write the answer in the boxes at the top. Write one digit in each column.

- Mark the corresponding oval in each column.

- *Never* grid a mixed number; change it to a fraction or a decimal.

After you work each problem, record your answer on the answer sheet provided or on a sheet of paper.

1. Grant purchased a shirt for $29.95 and 2 pairs of socks for $2.95 a pair. The sales tax on these purchases was $2.42. What was the total amount Grant spent?

   A $35.32 **B** $35.85

   C $38.27 **D** $39.37

2. Ariel adds $\frac{1}{2}$ cup of flour to a bowl that already has $3\frac{2}{3}$ cups of flour. How many total cups of flour will be in the bowl?

   A $7\frac{1}{3}$ **B** $4\frac{1}{6}$ **C** 4 **D** $3\frac{1}{6}$

3. The number 1134 is divisible by all of the following except—

   A 3. **B** 6. **C** 9. **D** 12. **E** 14.

4. Dr. Hewson has 758 milliliters of a solution to use for a class lab experiment. She divides the solution evenly among 32 students. If 22 milliliters are left after the experiment, how much of the solution did she give each student?

   A 23.0 mL **B** 24.2 mL

   C 33.0 mL **D** 35.9 mL

5. For shipping and handling, a company charges $2.75 in addition to $1.25 for each $10 ordered. Which equation represents the cost c for shipping an order worth $50?

   A $\frac{c}{50} = 2.75 + 1.25$

   B $c + 2.75 = 1.25 + \frac{50}{10}$

   C $c = 2.75 + 1.25\left(\frac{50}{10}\right)$

   D $c = \frac{50}{10}(2.75) + 1.25$

6. Baseballs are packed one dozen per box. There are 208 baseballs to be packed. How many more baseballs will be needed to fill the last, partially filled box?

   A 0 **B** 4 **C** 8 **D** 12 **E** 18

7. Franco is making a casserole. The recipe uses 8 cups of macaroni and serves 12 people. How many cups of macaroni does the recipe use per person?

   A $\frac{2}{3}$ **B** 1 **C** $\frac{1}{2}$ **D** $\frac{1}{3}$

8. Use the commutative and associative properties to compute the product.

   $$2 \cdot 4 \cdot 2.5 \cdot 15 \cdot 5 \cdot 10$$

   A 1500 **B** 12,000

   C 15,000 **D** 120,000

Open-Ended Questions

9. **Grid-In** The daily newspaper always follows a particular format. Each even-numbered page contains 6 articles, and each odd-numbered page contains 7 articles. If today's paper has 36 pages, how many articles does it contain?

10. The average annual snowfall in Denver, Colorado, is 59.8 inches. How many feet of snow can Denver residents expect in the next 4 years?

    Part A List the operations you use to solve this problem. Calculate the answer to the nearest hundredth of a foot. Show your work.

    Part B Round your answer to the nearest foot.

*inter*NET
CONNECTION **Test Practice** For additional test practice questions, visit:
www.algconcepts.glencoe.com

Integers

▶ What You'll Learn in Chapter 2:

- to graph integers on a number line and points on a coordinate plane *(Lessons 2–1 and 2–2)*,
- to compare and order integers *(Lesson 2–1)*, **and**
- to add, subtract, multiply, and divide integers *(Lessons 2–3, 2–4, 2–5, and 2–6)*.

Oak

41

Lake Michigan

Michigan

Chicago

Wells

La Salle

Clark

Dearborn

State

Huron

Fairbanks

Ontario

Ohio

Grand

Illinois

Hubbard

Lake Shore

Chicago River

Problem-Solving Workshop

Project

Sears Tower, the Magnificent Mile, Lake Shore Drive, Wrigley Field—
you're in Chicago to see the sights! Suppose you are at the intersection
of Illinois Street and Wells Street. You need to meet your group at the
intersection of Ohio Street and Dearborn Street. How many ways are
there to walk there if you can only cross streets at intersections and
can't backtrack?

Working on the Project

Work with a partner and choose a strategy. Develop
a plan. Here are some suggestions to help you get
started.

- As you leave Illinois and Wells, which
  intersections could you walk to next if you can't
  backtrack?
- Use grid paper to label the intersections with
  letters or numbers.
- Draw the different routes you can take on the
  grid paper.

> **Strategies**
>
> **Look for a pattern.**
>
> **Draw a diagram.**
>
> **Make a table.**
>
> **Work backward.**
>
> **Use an equation.**
>
> **Make a graph.**
>
> **Guess and check.**

Technology Tools

- Use a **word processor** to write an explanation of your solution.
- Use **drawing software** to draw your routes.

*inter*NET
CONNECTION   **Research**   For more information about Chicago, visit:
www.algconcepts.glencoe.com

Presenting the Project

Make a poster showing the various routes. Include an explanation of your
strategy for solving this problem. Make sure your explanation includes the
following:

- a discussion of the number of routes there would be if you had to meet your
  group one block past Ohio and Dearborn at Ontario and Dearborn, and
- a conjecture about how many routes there are between any two
  intersections on the map.

What You'll Learn

You'll learn to graph integers on a number line and to compare and order integers.

Why It's Important

Meteorology
Weather forecasters use integers to determine wind chill.
See Exercise 49.

The summer of 1999 brought drought conditions to parts of Georgia and South Carolina. One way to measure a drought is with the Palmer Drought Severity Index. It compares the actual amount of rain, sleet, and snow received with the amount expected.

Palmer Drought Severity Index

4	extreme moist spell
3	very moist spell
2	unusual moist spell
1	moist spell
0	near normal
−1	mild drought
−2	moderate drought
−3	severe drought
−4	extreme drought

Source: South Carolina Department of Natural Resources, 1999

interNET CONNECTION

Data Update For the latest information on the Palmer Drought Severity Index in your area, visit:
www.algconcepts.
glencoe.com

The numbers of the Palmer Drought Severity Index can also be shown on a **number line**.

Info Graphic

The set of whole numbers is often represented on a number line, as shown below.

0 1 2 3 4 5 6 7

A number line is drawn by choosing a starting position on a line, usually 0, and marking off equal distances from that point.

The arrowhead indicates that the line and the set of numbers continue indefinitely.

A **negative number** is a number less than zero. To include negative numbers on a number line, extend the line to the left of zero and mark off equal distances. Negative whole numbers are members of the set of **integers**. So, integers can also be represented on a number line.

Integers	**Words:**	Integers are the negative numbers −1, −2, −3, −4, ... and whole numbers 0, 1, 2, 3, 4, ...
	Symbols:	{..., −4, −3, −2, −1, 0, 1, 2, 3, 4, ...}
	Model:	

$$\xleftarrow{\qquad \text{negative} \qquad | \qquad \text{positive} \qquad}$$

−4 −3 −2 −1   0 +1 +2 +3 +4

Zero is neither negative nor positive.

Sets of numbers can also be represented by **Venn diagrams**.

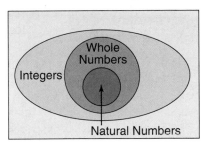

natural numbers	1, 2, 3, 4, ...
whole numbers	0, 1, 2, 3, ...
integers	... −3, −2, −1, 0, 1, 2, 3, ...

Notice that every natural number is also a whole number. Natural numbers are a *subset* of whole numbers. Similarly, whole numbers are a subset of integers.

To **graph** a set of integers, locate the points named by those numbers on a number line and place a dot on the number line. The number that corresponds to a point is called the **coordinate** of that point.

Examples

1 Name the coordinates of *A*, *B*, and *C*.

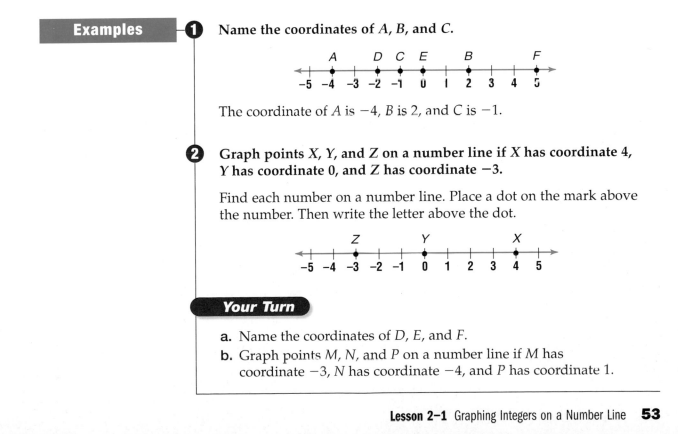

The coordinate of *A* is −4, *B* is 2, and *C* is −1.

2 Graph points *X*, *Y*, and *Z* on a number line if *X* has coordinate 4, *Y* has coordinate 0, and *Z* has coordinate −3.

Find each number on a number line. Place a dot on the mark above the number. Then write the letter above the dot.

Your Turn

a. Name the coordinates of *D*, *E*, and *F*.

b. Graph points *M*, *N*, and *P* on a number line if *M* has coordinate −3, *N* has coordinate −4, and *P* has coordinate 1.

The numbers on a number line increase as you move to the right and decrease as you move to the left. When two integers are graphed on a number line, the number to the right is always greater. The number to the left is always less than the other.

Reading Algebra

The symbol always points to the lesser number.

Words: 3 is greater than -2.
Symbols: $3 > -2$

Words: -2 is less than 3.
Symbols: $-2 < 3$

Examples

Replace each ● with $<$ or $>$ to make a true sentence.

3 $4 ● -1$

4 is to the right of -1 on the number line. So, $4 > -1$.

4 $-5 ● -3$

-5 is to the left of -3 on the number line. So, $-5 < -3$.

Your Turn

c. $-1 ● -2$ **d.** $2 ● -2$ **e.** $0 ● 1$

Integers are used to compare numbers in many everyday applications.

Example
Meteorology Link

5 The table shows the average high temperatures for January in selected cities. Order the temperatures from least to greatest.

Graph each integer on a number line. Use the first letter of each city name to label the points.

City	Temperature (°C)
Boston, MA	2
Chicago, IL	-2
Detroit, MI	-1
Juneau, AK	3
New York, NY	4
St. Louis, MO	-3

Write the integers as they appear on the number line from left to right.

$-3°, -2°, -1°, 2°, 3°, 4°$ are in order from least to greatest.

Looking at the graphs of 4 and −4 on a number line, you can see that they are the same number of units from 0. We say that they have the same **absolute value**.

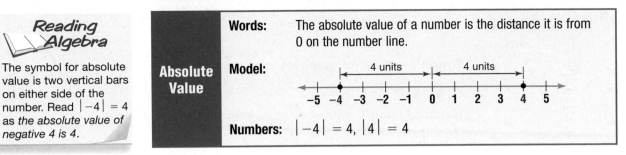

	Words:	The absolute value of a number is the distance it is from 0 on the number line.
Absolute Value	**Model:**	*(number line showing 4 units from −4 to 0 and 4 units from 0 to 4)*
	Numbers:	$|-4| = 4, \ |4| = 4$

Examples

Evaluate each expression.

6 $|-3|$

$|-3| = 3$ *The graph of −3 is 3 units away from 0.*

7 $|-5| - |2|$

$|-5| - |2| = 5 - 2$ *The absolute value of −5 is 5.*

$\qquad\qquad\ = 3$ *The absolute value of 2 is 2.*

Your Turn

f. $|9|$ **g.** $|-2| + |-6|$ **h.** $|15| - |-4|$

Check for Understanding

Communicating Mathematics

Study the lesson. Then complete the following.

1. **Describe** a situation in the real world where negative integers are used.

2. **Draw** a number line from −6 to 6. Graph two points whose coordinates have the same absolute value.

3. Tiffany says that 0 is a negative number. Ramon says that 0 is a positive number. Who is correct? Explain.

Vocabulary

number line
negative number
integers
Venn diagrams
natural numbers
graph
coordinate
absolute value

Guided Practice

Getting Ready **Write an integer for each situation.**

Sample: 10 feet below sea level **Solution:** −10

4. 4 degrees above zero
5. a loss of 6 pounds
6. 3 inches less rain than normal
7. a salary increase of $150

Name the coordinates of each point. *(Example 1)*

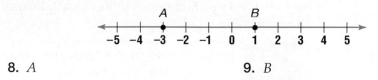

8. *A*   **9.** *B*

Graph each set of numbers on a number line. *(Example 2)*

10. {−3, 1, 4}   **11.** {5, 0, −4}

Replace each ● with < or > to make a true sentence.
(Examples 3 & 4)

12. −8 ● −5   **13.** −4 ● 2   **14.** 9 ● −7

Evaluate each expression. *(Examples 6 & 7)*

15. $|-8| + |-2|$ **16.** $|-7| - |4|$

17. Meteorology The table gives the record low temperatures for each month at the Grand Canyon Airport in Arizona. Order the temperatures from least to greatest. *(Example 5)*

Month	J	F	M	A	M	J	J	A	S	O	N	D
Temperature (°F)	−22	−17	−7	9	10	26	35	35	22	13	−1	−14

Source: *The Weather Almanac*

Exercises

Practice

Name the coordinate of each point.

18. *C*   **19.** *D*   **20.** *E*
21. *F*   **22.** *G*   **23.** *H*

Graph each set of numbers on a number line.

24. {2, 3, 5}   **25.** {−1, −3, 4}   **26.** {−2, 4, 0}
27. {−3, −2, 1}   **28.** {−2, −1, 0, 1}   **29.** {−4, −3, −2, −1}

Replace each ● with < or > to make a true sentence.

30. 4 ● −4   **31.** 0 ● −2   **32.** −2 ● −1
33. 2 ● −3   **34.** −10 ● 1   **35.** −15 ● −10
36. −5 ● $|-5|$ **37.** $|4|$ ● −4   **38.** $|-6|$ ● $|-3|$

Evaluate each expression.

39. $|-6|$ **40.** $|10|$ **41.** $|-5| - |3|$
42. $|-7| + |-2|$ **43.** $|14| - |-5|$ **44.** $|-13| + |-17|$

45. Which is greater, −5 or −3?

46. Order $-3, -4, 0, 1, -5,$ and 3 from least to greatest.

47. Order $-25, 78, -36, 14,$ and -14 from greatest to least.

Applications and Problem Solving

48. Population In 1990, the population of North Carolina was 2 million greater than the average of all 50 state populations. The population of Nevada was 4 million less than the average state population. Write an integer for each situation.

49. Meteorology *Windchill factor* is an estimate of the cooling effect the wind has on a person in cold weather.

Windchill Factor (°F)

Wind Speed (mph)	Actual Temperature (F°)						
	30	20	10	0	−10	−20	−30
0	30	20	10	0	−10	−20	−30
5	27	16	6	−5	−15	−26	−36
10	16	4	−9	−21	−33	−46	−58
15	9	−5	−18	−36	−45	−58	−72

Source: BMFA, 1999

a. Find the windchill factor when the actual temperature is 0° with a wind speed of 15 mph.

b. What is the windchill factor when the actual temperature is −20° with a wind speed of 5 mph?

c. Which is less: the windchill factor in part a or part b?

50. Critical Thinking Determine whether each statement is *true* or *false*. If *false*, give a counterexample.

a. Every integer is a whole number.

b. Every whole number is an integer.

Mixed Review

History Refer to the table for Exercises 51–53.

Heights of United States Presidents					
Height (in.)	63–65	66–68	69–71	72–74	75–77
Number	1	9	13	18	1

51. Make a histogram of the data. *(Lesson 1–7)*

52. Make a cumulative frequency table for the data. *(Lesson 1–6)*

53. How many presidents were at least six feet tall? *(Lesson 1–6)*

Simplify each expression. *(Lesson 1–4)*

54. $5x + 6x$

55. $9a - 3a$

56. $9x - x + 7x$

57. $3m + 2n + 4m$

58. $3r + 2s + 2r + s$

59. $5x + 12y - y + 2x$

60. Standardized Test Practice You have two more sisters than brothers. If you have s sisters, which equation could be used to find b, the number of brothers you have? *(Lesson 1–1)*

A $s = b - 2$ **B** $s - 2 = b$ **C** $b = s + 2$ **D** $b = 2s$

Extra Practice   See p. 694.

2-2 The Coordinate Plane

Math In the Workplace

What You'll Learn

You'll learn to graph points on a coordinate plane.

Why It's Important

Meteorology
Weather forecasters at the National Hurricane Center use a coordinate system to track hurricanes. *See Exercise 36.*

The streets in downtown San Francisco are laid out using a grid. The grid makes it easy to give directions from one point to another. Suppose your friend asked you how to get from the Japan Center to Lafayette Park.

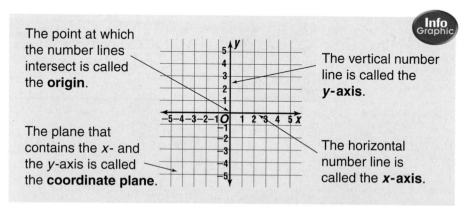

The directions *east* and *north* tell how to move from the starting point. So, you might tell your friend to go 1 block east and then 5 blocks north. *Would your friend have gotten to the same point on the map if you had said to go 5 blocks east and then 1 block north?*

In mathematics, you locate a point on a **coordinate system** that is similar to the grid above. The coordinate system is formed by the intersection of two number lines that meet at right angles at their zero points.

Info Graphic

The point at which the number lines intersect is called the **origin**.

The plane that contains the *x*- and the *y*-axis is called the **coordinate plane**.

The vertical number line is called the **y-axis**.

The horizontal number line is called the **x-axis**.

Japan Center, San Francisco

The directions *east* and *north* tell you how to locate a point on a map. In mathematics, an **ordered pair** of numbers is used to locate any point on a coordinate plane. For example, the directions to Lafayette Park could be written as (1, 5), where the first number represents the distance east, and the second number represents the distance north.
Notice that (1, 5) and (5, 1) will not get you to the same point on the map.

The first number in an ordered pair is called the **x-coordinate**. It corresponds to a number on the x-axis. The second number is called the **y-coordinate**. It corresponds to a number on the y-axis.

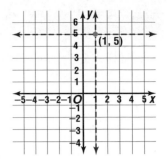

$$(1, 5)$$

x-coordinate ⌐ ⌐ y-coordinate

Examples

Write the ordered pair that names each point.

1 *A*

- Start at the origin. Move left on the x-axis to find the x-coordinate of point *A*. The x-coordinate is −2.
- Move up along the grid lines to find the y-coordinate. The y-coordinate is 4.

The ordered pair for point *A* is (−2, 4).

2 *B*

The x-coordinate is 4, and the y-coordinate is −1. The ordered pair for point *B* is (4, −1).

3 *C*

Point *C* is the origin. The ordered pair for the origin is (0, 0).

Your Turn

a. *D* **b.** *E* **c.** *F* **d.** *G*

A point can be named by both a letter and its ordered pair. For example, *P*(2, 3) means point *P* has an x-coordinate of 2 and a y-coordinate of 3. To graph an ordered pair on a coordinate plane, draw a dot at the point that corresponds to the ordered pair. This is called *plotting* the point.

Example

4 **Graph *P*(2, 3) on a coordinate plane.**

You can assume that each unit on the axes represents 1 unit.

- Start at the origin, *O*.
- The x-coordinate is 2. So, move 2 units to the right.
- The y-coordinate is 3. Move 3 units up and draw a dot.
- Label the dot with the letter *P*.

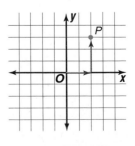

Example 5

Graph $Q(-3, 0)$ on a coordinate plane.

- Start at the origin, O.
- The x-coordinate is -3. So, move 3 units to the left.
- The y-coordinate is 0. So the dot is placed on the axis.

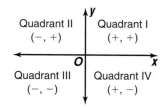

Your Turn

Graph each point on a coordinate plane.

e. $R(2, -4)$ **f.** $S(-1, 4)$ **g.** $T(0, -3)$

The x-axis and the y-axis separate the coordinate plane into four regions, called **quadrants**. The quadrants are numbered as shown at the right. Note that the axes are not located in any of the quadrants.

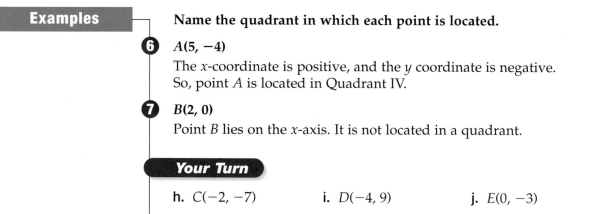

Name the quadrant in which each point is located.

6 $A(5, -4)$

The x-coordinate is positive, and the y coordinate is negative. So, point A is located in Quadrant IV.

7 $B(2, 0)$

Point B lies on the x-axis. It is not located in a quadrant.

Your Turn

h. $C(-2, -7)$ **i.** $D(-4, 9)$ **j.** $E(0, -3)$

You can use ordered pairs to show how data are related.

Example 8
Biology Link

Dolphins can swim at 30 mph over long distances. Let x represent the number of hours. Then, $30x$ represents the total distance traveled. Evaluate the expression to find the distances traveled in 1, 2, and 3 hours. Then graph the ordered pairs (time, distance).

The data are graphed in the first quadrant because both values are positive.

Time (hours)	Distance (miles)
x	$30x$
1	30
2	60
3	90

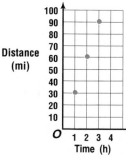

Graphing Calculator Exploration

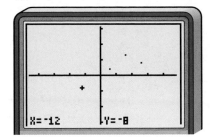

You can plot points on a graphing calculator.

Step 1 First press ZOOM 8 ENTER
to display a coordinate grid.

Step 2 Press 2nd [DRAW] ▶
ENTER. Use the arrow keys
to move the cursor to each
desired location. Press
ENTER to plot the point.

Try These

1. Choose four ordered pairs such that the sum of their x- and y-coordinates is 5. Graph them.

2. What do you notice about the graphs of the points?

Check for Understanding

Communicating Mathematics

Study the lesson. Then complete the following.

1. **Explain** how to graph $(-5, 1)$ on a coordinate plane.

2. **Name** an ordered pair whose graph satisfies each condition.

   a. located in Quadrant IV

   b. not located in any quadrant

Math Journal

3. **Draw** a coordinate system and label the origin, x-axis, y-axis, and quadrants.

Guided Practice

Write the ordered pair that names each point. *(Examples 1–3)*

   4. F **5.** G

Graph each point on a coordinate plane.
(Examples 4 & 5)

   6. $R(-5, 2)$ **7.** $S(0, -2)$

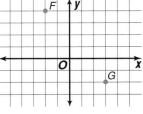

Name the quadrant in which each point is located. *(Examples 6 & 7)*

   8. $D(-9, 1)$ **9.** $E(0, -6)$

10. Biology   A tortoise is one of the slowest animals on land. It travels at an average speed of only 20 feet per minute. *(Example 8)*

   a. Find the distance traveled in 2, 4, and 6 minutes.

   b. Graph the ordered pairs (time, distance).

Practice

Write the ordered pair that names each point.

11. *A*   12. *B*   13. *C*
14. *D*   15. *E*   16. *F*

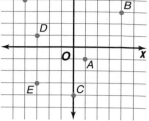

Graph each point on a coordinate plane.

17. $J(-3, 3)$ 18. $K(4, 0)$ 19. $L(4, -3)$
20. $M(3, 1)$ 21. $N(-1, -2)$ 22. $P(0, 5)$

Name the quadrant in which each point is located.

23. $(-3, -4)$ 24. $(6, -2)$ 25. $(0, 4)$
26. $(11, 15)$ 27. $(-15, 25)$ 28. $(-18, 0)$

29. What point lies on both the *x*-axis and the *y*-axis?

30. Graph three ordered pairs in which the *x*- and *y*-coordinates are equal. Describe the graph.

If the graph of A(x, y) satisfies the given conditions, name the quadrant in which point A is located.

31. $x > 0, y > 0$ 32. $x < 0, y < 0$ 33. $x > 0, y < 0$

Applications and Problem Solving

34. **Geometry**   Graph the points $A(-1, 1)$, $B(4, 1)$, $C(4, 0)$, and $D(-1, 0)$ on the same coordinate plane. Connect the points in alphabetical order and then connect *A* and *D*. Describe the figure.

35. **Entertainment**   It costs $3 to rent a video for a day.
    a. Find the total cost of renting 1, 3, and 5 videos for a day.
    b. Graph the ordered pairs (number of videos, cost).
    c. Make a prediction about the location of the graph of (4, 12). Check your prediction by graphing (4, 12).

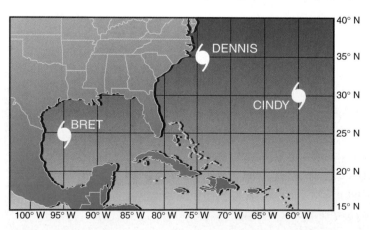

36. **Meteorology**   Weather forecasters use a coordinate system composed of latitude (horizontal) and longitude (vertical) lines to locate hurricanes. For example, the position of Hurricane Dennis is 35°N latitude and 74°W longitude, or (35°N, 74°W). Write the position of each hurricane as an ordered pair.
    a. Bret   b. Cindy

37. Geometry   A *vertex* of a triangle is a point where two sides of the triangle meet.

   a. Identify the coordinates of the vertices in the triangle at the right.

   b. Multiply each *x*- and *y*-coordinate of the vertices by 2 and graph the new ordered pairs. Connect the points.

   c. Compare the two figures. Write a sentence that tells how the figures are the same and how they are different.

38. Critical Thinking   Where are all of the possible locations for the graph of (x, y) if $x = y$?

Mixed Review

39. Geography   The Caribbean Sea has an average depth of 8685 feet below sea level. Use an integer to express this depth.   *(Lesson 2–1)*

40. Compare and contrast a histogram and a cumulative frequency histogram.   *(Lesson 1–7)*

Name the property shown by each statement.   *(Lesson 1–3)*

41. $15 + 4 = 4 + 15$ **42.** $a(bc) = (ab)c$

43. $3 + 4 = 7, 7$ is a whole number   **44.** $4 + (5 + 6) = 4 + (6 + 5)$

45. Standardized Test Practice   Evaluate $8x - 3y$ if $x = 2$ and $y = 3$.
(Lesson 1–2)

   A 7   **B** 25   **C** 39   **D** 57

Quiz 1   Lessons 2–1 and 2–2

▶ **Replace each ● with < or > to make a true sentence.**   *(Lesson 2–1)*

   1. 3 ● −2   **2.** 0 ● −5   **3.** −6 ● −2

   4. Order −6, −10, 10, 5, −7, and 0 from least to greatest.   *(Lesson 2–1)*
   5. Evaluate $\left|-3\right| + \left|-8\right|$.   *(Lesson 2–1)*

Name the quadrant in which each point is located.   *(Lesson 2–2)*

   6. $A(4, -2)$ **7.** $B(-5, -5)$
   8. $C(0, -4)$ **9.** $D(-8, 6)$

10. Entertainment   It costs $4 to buy a student ticket to the movies.
   a. Find the cost of 2, 4, and 5 tickets.
   b. Graph the ordered pairs (number of tickets, cost).   *(Lesson 2–2)*

Extra Practice   See p. 694.

What You'll Learn
You'll learn to add integers.

Why It's Important
Banking Banks use integers in checking accounts.
See Example 5.

Have you ever walked across a carpeted floor and gotten shocked as you reached out to touch a doorknob? This is an example of static electricity, which is caused by a buildup of electrons on your body.

Electrons have a negative charge, and protons have a positive charge. They can be used as a model for the addition of integers. One way to add integers is to use the 1-tiles from a set of algebra tiles. You can think of the negative tiles as electrons and the positive tiles as protons.

Find 3 + 2.
Combine 3 positive tiles with 2 positive tiles on a mat.

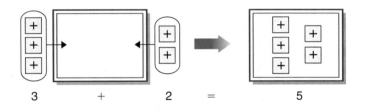

There are 5 positive tiles on the mat. Therefore, $3 + 2 = 5$.

Find $-3 + (-2)$.
Combine 3 negative tiles with 2 negative tiles.

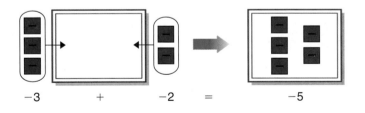

There are 5 negative tiles on the mat. Therefore, $-3 + (-2) = -5$.

You can also add integers on a number line. Start at 0. Positive integers are represented by arrows pointing *right*. Negative integers are represented by arrows pointing *left*.

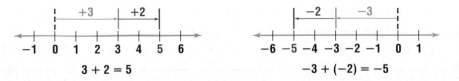

These and other similar examples suggest the following rule for adding integers with the same sign.

| Adding Integers with the Same Sign | **Words:** | To add integers with the same sign, add their absolute values. Give the result the same sign as the integers. |
| | **Numbers:** | $3 + 2 = 5$, $-3 + (-2) = -5$ |

Examples

Find each sum.

1 $4 + 5$

$4 + 5 = 9$ *Both numbers are positive, so the sum is positive.*

2 $-6 + (-2)$

$-6 + (-2) = -8$ *Both numbers are negative, so the sum is negative.*

Your Turn

a. $8 + 9$ **b.** $-2 + (-4)$ **c.** $-5 + (-10)$ **d.** $11 + 6$

What is the result when you add two numbers like 3 and -3?

Start at zero. Move 3 units to the right. From there, move 3 units to the left.

$3 + (-3) = 0$

You can also use tiles. When one positive tile is paired with one negative tile, the result is a **zero pair**. You can remove zero pairs from the mat because removing zero does not change the value.

In electricity, when you remove a proton-electron pair from a collection of protons and electrons, the electrical charge is not changed.

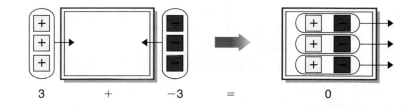

The models above show $3 + (-3) = 0$. If the sum of two numbers is 0, the numbers are called **opposites** or **additive inverses**.

-3 is the additive inverse, or opposite, of 3. $3 + (-3) = 0$
7 is the additive inverse, or opposite, of -7. $-7 + 7 = 0$

Additive Inverse Property	**Words:**	The sum of any number and its additive inverse is 0.
	Symbols:	$a + (-a) = 0$
	Numbers:	$3 + (-3) = 0, -7 + 7 = 0$

In the following activity, you'll use tiles to find a rule for adding two integers with different signs.

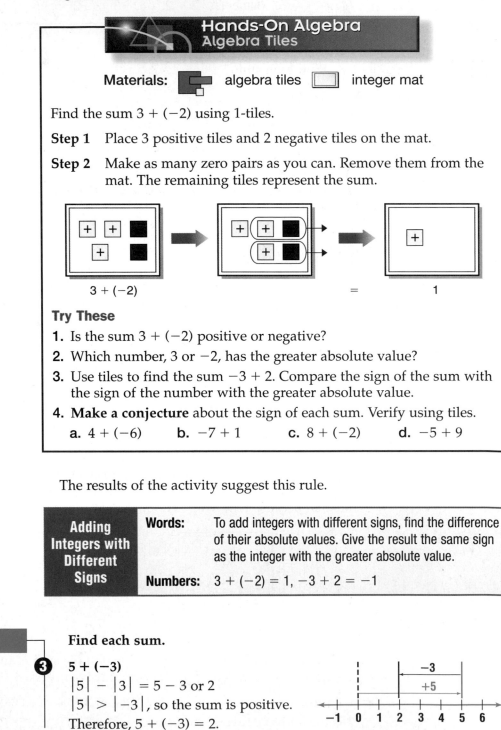

Hands-On Algebra
Algebra Tiles

Materials: ▨ algebra tiles ▢ integer mat

Find the sum $3 + (-2)$ using 1-tiles.

Step 1 Place 3 positive tiles and 2 negative tiles on the mat.

Step 2 Make as many zero pairs as you can. Remove them from the mat. The remaining tiles represent the sum.

$$3 + (-2) \qquad = \qquad 1$$

Try These

1. Is the sum $3 + (-2)$ positive or negative?
2. Which number, 3 or -2, has the greater absolute value?
3. Use tiles to find the sum $-3 + 2$. Compare the sign of the sum with the sign of the number with the greater absolute value.
4. **Make a conjecture** about the sign of each sum. Verify using tiles.
      a. $4 + (-6)$ **b.** $-7 + 1$ **c.** $8 + (-2)$ **d.** $-5 + 9$

The results of the activity suggest this rule.

| **Adding Integers with Different Signs** | **Words:** | To add integers with different signs, find the difference of their absolute values. Give the result the same sign as the integer with the greater absolute value. |
| | **Numbers:** | $3 + (-2) = 1, -3 + 2 = -1$ |

Examples

Find each sum.

3 $5 + (-3)$
$|5| - |3| = 5 - 3$ or 2
$|5| > |-3|$, so the sum is positive.
Therefore, $5 + (-3) = 2$.

4 $4 + (-6)$

$|-6| - |4| = 6 - 4$ or 2

$|-6| > |4|$, so the sum is negative.

Therefore, $4 + (-6) = -2$.

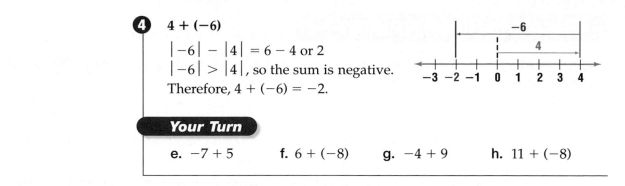

Your Turn

e. $-7 + 5$ **f.** $6 + (-8)$ **g.** $-4 + 9$ **h.** $11 + (-8)$

Example

Banking Link

Real World

5 **Talisa opened a checking account with a deposit of \$25. During the next two weeks, she wrote checks for \$20 and \$15 and made a deposit of \$30. Find the balance in her account.**

Explore   You know that Talisa made deposits of \$25 and \$30. She wrote checks for \$20 and \$15. You want to find the balance in her account.

Plan   Deposits are represented by positive integers ($+25$ and $+30$). Checks are represented by negative integers (-20 and -15). Write an addition sentence and solve.

Solve   Let x represent the balance in her account.

$x = 25 + (-20) + (-15) + 30$

$x = 5 + (-15) + 30$ *$25 + (-20) = 5$*

$x = -10 + 30$ *$5 + (-15) = -10$*

$x = 10$ *$-10 + 30 = 20$*

The balance in Talisa's account is \$20.

Examine   Addition of integers is commutative. So, you can check the solution by adding the integers in a different order. One way is to group all of the positive numbers and all of the negative numbers.

$x = 25 + 30 + (-20) + (-15)$

$x = 55 + (-35)$ *$25 + 30 = 55; -20 + (-15) = -35$*

$x = 20$ ✓

Look Back

Commutative
Property:
Lesson 1–3

You can use the rules for adding integers to simplify expressions.

Example

6 **Simplify $5x + (-3x)$.**

$5x + (-3x) = [5 + (-3)]x$ *Use the Distributive Property.*

           $= 2x$ *$5 + (-3) - 2$*

Your Turn

Simplify each expression.

i. $-8y + 3y$ **j.** $6m + 4m + (-2m)$ **k.** $-5x + 4x$

Check for Understanding

Communicating Mathematics

Study the lesson. Then complete the following.

1. **Show** how to find the sum of -5 and -3 on a number line.

2. **Explain** why -10 and 10 are additive inverses.

3. **Draw** a diagram that shows how to find the sum of -4 and 6 using tiles.

Math Journal

4. **Write** a paragraph that describes how to add two integers. Be sure to include examples with your description.

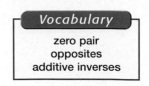

Vocabulary

zero pair
opposites
additive inverses

Guided Practice

Getting Ready **Tell whether each sum is** *positive* **or** *negative*.

Sample 1: $-4 + (-3)$
Solution: Both integers are negative, so the sum is negative.

Sample 2: $-9 + 11$
Solution: $|11| > |-9|$, so the sum is positive.

5. $5 + 12$
6. $12 + (-15)$
7. $-3 + (-7)$
8. $-3 + 9$
9. $-5 + (-2)$
10. $-8 + 12$

Find each sum. *(Examples 1–4)*

11. $7 + 9$
12. $-2 + (-8)$
13. $8 + (-9)$
14. $-12 + 15$
15. $-10 + 5$
16. $11 + (-2)$

Simplify each expression. *(Example 6)*

17. $4x + (-2x)$
18. $-9y + (-2y)$
19. $3a + (-4a) + 3a$

20. **Games** On a famous TV game show, contestants earn money for each correct answer and lose money for each incorrect answer. Suppose a contestant answered questions worth \$100, \$200, and \$400 correctly, but answered questions worth \$300, \$300, and \$400 incorrectly. What was the contestant's score? *(Example 5)*

Exercises

Practice

Find each sum.

21. $3 + 9$
22. $8 + 6$
23. $5 + 16$
24. $-3 + (-10)$
25. $-5 + (-6)$
26. $-11 + (-7)$
27. $-13 + 5$
28. $12 + (-7)$
29. $-6 + 15$
30. $6 + (-6)$
31. $5 + (-18)$
32. $-9 + (-9)$
33. $-15 + 7$
34. $16 + (-11)$
35. $-10 + (-11)$
36. $30 + (-15)$
37. $-20 + (-35)$
38. $-40 + 26$
39. $8 + (-5) + 10$
40. $3 + 15 + (-6)$
41. $-10 + (-4) + (-8)$
42. $15 + 7 + (-7) + (-13)$
43. $-6 + 12 + (-11) + 1$
44. $17 + (-21) + 10 + (-17)$

45. Find the value of y if $y = -3 + 2$.

46. What is the value of w if $-7 + (-2) = w$?

47. Find the value of b if $b = 3 + (-6)$.

Simplify each expression.

48. $-9a + 3a$

49. $-5x + (-10x)$

50. $-16y + 15y$

51. $-11m + 14m$

52. $4z + (-3z)$

53. $8c + (-8c)$

54. $-8b + 4b + (-2b)$

55. $3y + 8y + (-3y)$

56. $-2n + (-4n) + 3n$

Evaluate each expression if $x = -4$, $y = -5$, and $z = 4$.

57. $x + 4 + (-9)$

58. $-7 + y + z$

59. $|x| + y$

60. Golf In golf, a score of 0 is called *even par*. One over par is represented by $+1$, and one under par is represented by -1. In the 1999 U.S. Open, Tiger Woods had scores represented by -2, $+1$, $+2$, and 0. What was his final score?

61. Geometry The graphs of $A(2, 3)$, $B(3, -3)$, and $C(-3, -2)$ are connected with line segments to form a triangle.

   a. Add 2 to each y-coordinate and draw another triangle.

   b. How did the position of the triangle change?

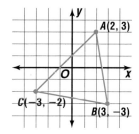

62. Critical Thinking Refer to Exercise 61. What change would you make to the ordered pairs so that the triangle would move to the right?

Mixed Review

Name the quadrant in which each point is located. *(Lesson 2–2)*

63. $A(6, -5)$

64. $B(-2, -2)$

65. $C(-5, 3)$

66. $(0, 4)$

Write an integer for each situation. *(Lesson 2–1)*

67. a debt of \$5

68. 2 inches more rain than normal

69. a loss of 10 yards

70. a deposit of \$17

71. maintaining your present weight

72. Marketing One hundred people were surveyed outside a movie theater to determine the favorite leisure-time activity for a large population. Is this a representative sample? Explain your reasoning. *(Lesson 1–6)*

73. Standardized Test Practice Use the pattern in the perimeter P of each rectangle to determine the perimeter of a rectangle made up of ten unit squares. *(Lesson 1–5)*

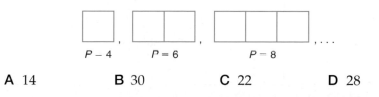

$P - 4$ $P = 6$ $P - 8$

   A 14   **B** 30   **C** 22   **D** 28

Extra Practice See p. 695.

What You'll Learn

You'll learn to subtract integers.

Why It's Important

Budgeting Families often need to find the difference between the amount of money in a budget and the actual amount spent. *See Exercise 48.*

The element lithium has 3 protons in its nucleus and has 3 electrons in the region surrounding the nucleus. Since there is an equal number of positive and negative charges, lithium is electrically neutral. When lithium loses one of its electrons, it becomes positively charged. This can be used as a model for subtraction of integers.

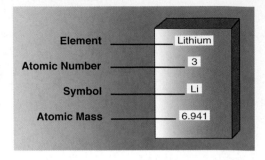

You can use algebra tiles to see how addition and subtraction of integers are related.

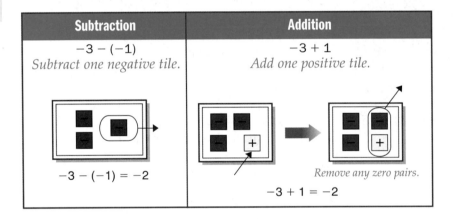

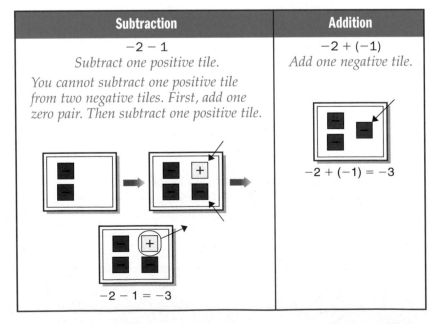

The examples on the previous page suggest that subtracting an integer is the same as adding the additive inverse or opposite of the integer.

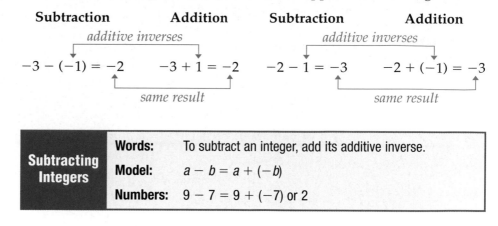

| Subtraction | Addition | Subtraction | Addition |

$-3 - (-1) = -2$ $-3 + 1 = -2$ $-2 - 1 = -3$ $-2 + (-1) = -3$

same result   *same result*

Subtracting Integers	**Words:**	To subtract an integer, add its additive inverse.
	Model:	$a - b = a + (-b)$
	Numbers:	$9 - 7 = 9 + (-7)$ or 2

Examples

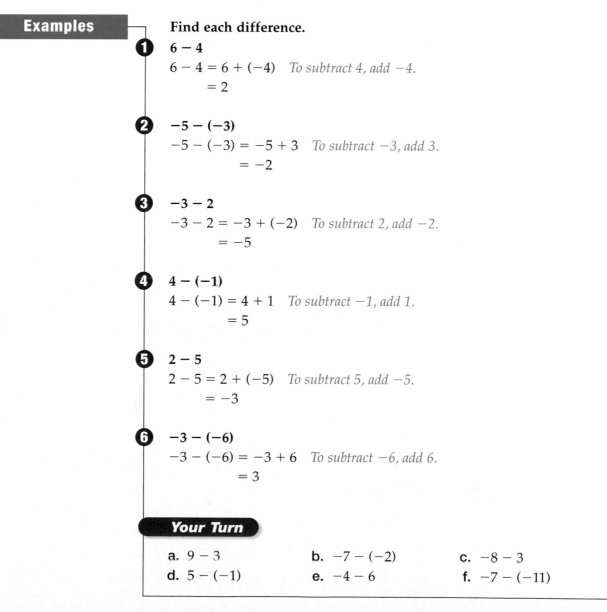

Find each difference.

1 $6 - 4$

$6 - 4 = 6 + (-4)$ *To subtract 4, add −4.*
$\quad\ = 2$

2 $-5 - (-3)$

$-5 - (-3) = -5 + 3$ *To subtract −3, add 3.*
$\qquad\quad = -2$

3 $-3 - 2$

$-3 - 2 = -3 + (-2)$ *To subtract 2, add −2.*
$\qquad = -5$

4 $4 - (-1)$

$4 - (-1) = 4 + 1$ *To subtract −1, add 1.*
$\qquad\ = 5$

5 $2 - 5$

$2 - 5 = 2 + (-5)$ *To subtract 5, add −5.*
$\qquad = -3$

6 $-3 - (-6)$

$-3 - (-6) = -3 + 6$ *To subtract −6, add 6.*
$\qquad\quad = 3$

Your Turn

a. $9 - 3$ **b.** $-7 - (-2)$ **c.** $-8 - 3$

d. $5 - (-1)$ **e.** $-4 - 6$ **f.** $-7 - (-11)$

When you evaluate expressions, it is helpful to write any subtraction expressions as addition expressions first.

Examples

7 Evaluate $x - y$ if $x = -2$ and $y = 1$.

$\begin{aligned} x - y &= -2 - 1 && \textit{Replace x with } -2 \textit{ and y with 1.}\\ &= -2 + (-1) && \textit{Write } -2 - 1 \textit{ as } -2 + (-1).\\ &= -3 && -2 + (-1) = -3 \end{aligned}$

8 Evaluate $a - b + c$ if $a = 6$, $b = -2$, and $c = -6$.

$\begin{aligned} a - b + c &= 6 - (-2) + (-6) && \textit{Replace a with 6, b with } -2, \textit{ and c with } -6.\\ &= 6 + 2 + (-6) && \textit{Write } 6 - (-2) \textit{ as } 6 + 2.\\ &= 8 + (-6) && 6 + 2 = 8\\ &= 2 && 8 + (-6) = 2 \end{aligned}$

Your Turn

g. Evaluate $m - n$ if $m = 5$ and $n = -3$.

h. Evaluate $w - x + y - z$ if $w = -5$, $x = -7$, $y = 10$, and $z = -5$.

Integers are often used to show how data has changed for a given time.

Example

Population Link

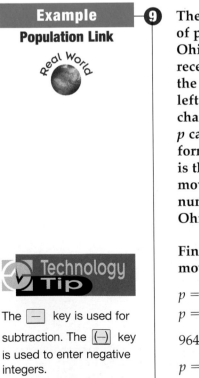

9 The map shows the number of people who moved to Ohio from Indiana in a recent year. It also shows the number of people who left Ohio for Indiana. The change in Ohio's population p can be found by using the formula $p = m - l$, where m is the number of people moving to Ohio and l is the number of people leaving Ohio.

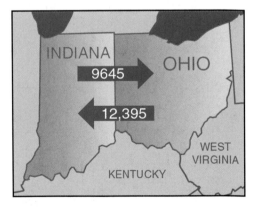

Find the net change in Ohio's population resulting from people moving to and from Indiana.

$p = m - l$

$p = 9645 - 12{,}395$ *Replace m with 9645 and l with 12,395.*

9645 ⊟ 12395 ENTER −2750

$p = -2750$ Ohio's population decreased by 2750 people.

Check for Understanding

Communicating Mathematics

Study the lesson. Then complete the following.

1. **Explain** how additive inverses are used in subtraction.
2. **Draw** a diagram using algebra tiles that shows how $2 - 5$ and $2 + (-5)$ have the same result.

Guided Practice

🕐 **Getting Ready**   **Write each expression as an addition expression.**

Sample: $4 - (-3)$ **Solution:** $4 + 3$

3. $10 - 3$ 4. $-2 - (-5)$ 5. $-4 - 8$

Find each difference.   *(Examples 1–6)*

6. $8 - 2$ 7. $-6 - (-4)$ 8. $-5 - 4$

9. $7 - (-4)$ 10. $8 - 11$ 11. $-4 - (-9)$

Evaluate each expression if $a = -2$, $b = 6$, $c = -3$, and $d = -1$.
(Examples 7 & 8)

12. $a - b$ 13. $b - c + d$

14. **Population**   In a recent year, 5899 people moved to Ohio from West Virginia, and 5394 people left Ohio for West Virginia. Find the net change in Ohio's population.   *(Example 9)*

Exercises

Practice

Find each difference.

15. $15 - 2$ 16. $11 - 6$ 17. $14 - 7$

18. $-9 - (-3)$ 19. $-10 - (-2)$ 20. $-15 - (-4)$

21. $-10 - 3$ 22. $-8 - 4$ 23. $-9 - 2$

24. $5 - (-2)$ 25. $5 - (-11)$ 26. $9 - (-8)$

27. $4 - 10$ 28. $9 - 16$ 29. $0 - 9$

30. $-4 - (-10)$ 31. $0 - (-12)$ 32. $-8 - (-14)$

33. Find the value of x if $3 - (-4) = x$.
34. What is the value of y if $y = -3 - (-12)$?
35. Find the value of v if $v = 2 - 19$.

Evaluate each expression if $x = 10$, $y = -7$, $z = -10$, and $w = 12$.

36. $x - y$ 37. $y - z$ 38. $15 - w$

39. $7 - x + y$ 40. $x - z - w$ 41. $x + z - w$

Simplify each expression.

42. $5y - 2y$ 43. $20n - (-5n)$ 44. $4a - 9a + 3a$

45. What is the difference of 25 and -25?
46. Write $a - (-b)$ as an addition expression.

47. **Meteorology** The record high temperature in Minneapolis-St. Paul, Minnesota, is 108°F. The record low temperature is 142°F lower. What is the record low temperature?

Expenditures for July

Expenses	Amount Budgeted (dollars)	Amount Spent (dollars)
Food	160	175
Electric	45	44
Telephone	35	41
Heating Fuel	50	15
Water	25	32
Cable TV	25	25

48. **Budgeting** The table shows the Thomas family's budget and expense summary for food and household utilities for July.

   a. For each item, find the difference between the budgeted amount and the amount spent.

   b. What does a negative difference indicate?

   c. Was the total amount spent for these items more or less than the amount budgeted? by how much?

49. **Critical Thinking** Determine whether each statement is *true* or *false*. If *false*, give a counterexample.

   a. Subtraction of integers is commutative.

   b. Subtraction of integers is associative.

   c. The set of integers is closed under the operation of subtraction.

Mixed Review

Find each sum. *(Lesson 2–3)*

50. $16 + (-5)$

51. $-12 + (-8)$

52. $9 + (-15)$

53. $-24 + (-3)$

54. $18 + 6$

55. $-12 + 4$

56. **Communications** A new long-distance plan charges a flat rate of 5¢ per minute. *(Lesson 2–2)*

   a. Find the amount spent for calls of 5, 8, and 10 minutes.

   b. Graph the ordered pairs (time, cost).

Replace each ● with < or > to make a true sentence. *(Lesson 2–1)*

57. $2 ● -3$

58. $-4 ● -8$

59. $-15 ● -14$

60. **Open-Ended Test Practice** The table shows the record high temperatures for each state in the United States. Make a histogram of the data. Use 100–104, 105–109, 110–114, 115–119, 120–124, 125–129, and 130–134 as categories for the histogram. *(Lesson 1–7)*

Record High Temperatures (°F)									
112	100	128	120	134	118	106	110	109	112
100	118	117	116	118	121	114	114	105	109
107	112	114	115	118	117	118	125	106	110
122	108	110	121	113	120	119	111	104	111
120	113	120	117	105	110	118	112	114	114

Source: *World Almanac,* 1999

Extra Practice See p. 695.

What You'll Learn

You'll learn to multiply integers.

Why It's Important

Health Health workers use negative integers to describe declining death rates. *See Exercise 48.*

Health professionals recommend that adults on weight-loss programs try to lose an average of 2 pounds per week. At this rate, how much weight would a person expect to lose by the end of 3 weeks?

A weight loss of 2 pounds per week can be represented by the integer -2. Since the weight loss occurs in each of 3 weeks, this situation can be represented by the multiplication problem $3(-2)$. You can find this product by using a number line. *Remember that multiplication is repeated addition.*

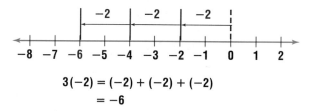

$$3(-2) = (-2) + (-2) + (-2)$$
$$= -6$$

Therefore, $3(-2) = -6$.

What happens if the order of the factors is changed to $(-2)3$? The Commutative Property of Multiplication guarantees that $3(-2) = (-2)3$. Therefore, $-2(3) = -6$.

In $3(-2) = -6$ and $-2(3) = -6$, one factor is positive, one factor is negative, and the product is negative. These examples suggest the following rule for multiplying two integers with different signs.

Multiplying Two Integers with Different Signs	**Words:**	The product of two integers with different signs is negative.
	Numbers:	$3(-2) = -6$, $-2(3) = -6$

Examples

Find each product.

1 $6(-8)$

$6(-8) = -48$ *The factors have different signs. The product is negative.*

2 $-5(9)$

$-5(9) = -45$

Your Turn

a. $10(-3)$ **b.** $-7(7)$ **c.** $15(-3)$

You already know that the product of two positive numbers is positive. What is the sign of the product of two negative numbers? Consider the product $-2(-3)$.

Look Back

Multiplicative Property of Zero: Lesson 1–4

$0 = -2(0)$	*Multiplicative Property of Zero*
$0 = -2[3 + (-3)]$	*Replace 0 with 3 + (−3) or any zero pair.*
$0 = -2(3) + (-2)(-3)$	*Distributive Property*
$0 = \;\; -6 \;\; + \;\;\;\; ?$	$-2(3) = -6$

By the Additive Inverse Property, $-6 + 6 = 0$. Therefore, $-2(-3)$ must be equal to 6. This example suggests the following rule for multiplying two integers with the same sign.

Multiplying Two Integers with the Same Sign	**Words:**	The product of two integers with the same sign is positive.
	Numbers:	$2(3) = 6$, $-2(-3) = 6$

Examples

Find each product.

3 **15(2)**

$15(2) = 30$ *The factors have the same sign. The product is positive.*

4 **−5(−6)**

$-5(-6) = 30$

Your Turn

d. $11(9)$ **e.** $-6(-7)$ **f.** $-10(-8)$

To find the product of three or more numbers, multiply the first two numbers. Then multiply the result by the next number, until you come to the end.

Examples

Find each product.

5 **8(−10)(−4)**

$8(-10)(-4) = -80(-4)$ *$8(-10) = -80$*

$\qquad\qquad\quad = 320$ *$-80(-4) = 320$*

6 **5(−3)(−2)(−2)**

$5(-3)(-2)(-2) = -15(-2)(-2)$ *$5(-3) = -15$*

$\qquad\qquad\qquad = 30(-2)$ *$-15(-2) = 30$*

$\qquad\qquad\qquad = -60$ *$30(-2) = -60$*

Your Turn

g. $-2(-3)(4)$ **h.** $6(-2)(3)$ **i.** $(-1)(-5)(-2)(-3)$

You can use the rules for multiplying integers to evaluate algebraic expressions and to simplify expressions.

Examples — **7** Evaluate $2xy$ if $x = -4$ and $y = -2$.

$$\begin{aligned} 2xy &= 2(-4)(-2) && \textit{Replace x with } -4 \textit{ and y with } -2. \\ &= -8(-2) && 2(-4) = -8 \\ &= 16 && -8(-2) = 16 \end{aligned}$$

8 Simplify $(2a)(-5b)$.

$$\begin{aligned} (2a)(-5b) &= (2)(a)(-5)(b) && 2a = (2)(a); -5b = (-5)(b) \\ &= (2)(-5)(a)(b) && \textit{Commutative Property} \\ &= -10ab && (2)(-5) = -10; (a)(b) = ab \end{aligned}$$

Your Turn

j. Evaluate $-5n$ if $n = -7$. **k.** Simplify $12(-3z)$.

Example — **9**
Geometry Link

The graphs of $A(3, 5)$, $B(1, 2)$, and $C(5, -1)$ are connected with line segments to form a triangle. Multiply each x-coordinate by -1 and redraw the triangle. Describe how the position of the triangle changed.

Reading Algebra

Read A' as A prime. A' corresponds to point A on the original triangle.

$A(3, 5) \rightarrow (3 \times -1, 5) \rightarrow A'(-3, 5)$
$B(1, 2) \rightarrow (1 \times -1, 2) \rightarrow B'(-1, 2)$
$C(5, -1) \rightarrow (5 \times -1, -1) \rightarrow C'(-5, -1)$

The graph of $A'B'C'$ is shown in green. It is the same size and shape as triangle ABC, but it was reflected, or flipped, over the y-axis.

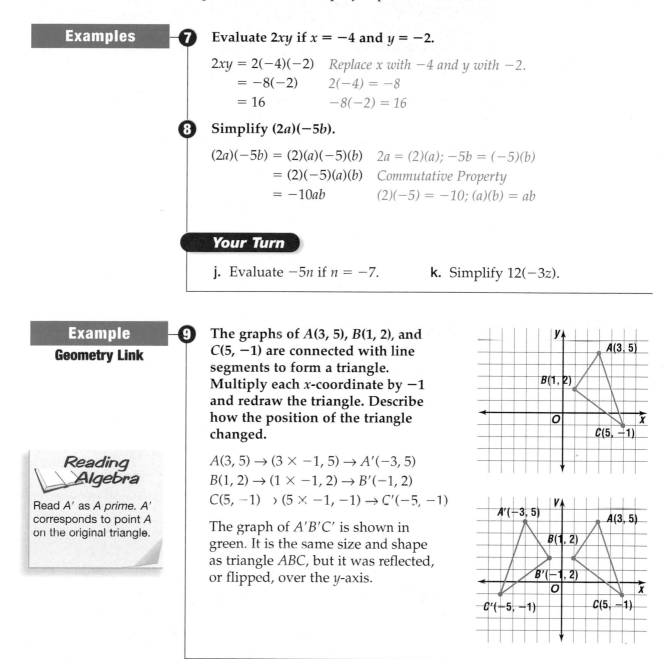

Check for Understanding

Communicating Mathematics

Study the lesson. Then complete the following.

1. **Write** the multiplication sentence represented by the model.
2. **Name** the property that allows you to write $-5(6)$ as $6(-5)$.

Guided Practice

Find each product. *(Examples 1–6)*

3. $2(-6)$

4. $-4(9)$

5. $10(8)$

6. $-7(-11)$

7. $2(-6)(-3)$

8. $4(-1)(-5)(-2)$

Evaluate each expression if $a = -4$ and $b = -6$. *(Example 7)*

9. $-7a$

10. $-3ab$

Simplify each expression. *(Example 8)*

11. $9(-2x)$

12. $(-3m)(-2n)$

13. Geometry The graphs of $A(4, 2)$, $B(-3, 4)$, and $C(-1, 1)$ are connected with line segments to form a triangle. *(Example 9)*
  a. Multiply each y-coordinate by -1 and redraw the triangle.
  b. Describe how the position of the triangle changed.

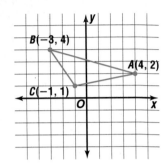

Exercises

Practice

Find each product.

14. $5(8)$

15. $12(-4)$

16. $-1(-1)$

17. $9(-1)$

18. $-6(5)$

19. $3(15)$

20. $5(-15)$

21. $13(0)$

22. $-8(-9)$

23. $-3(8)$

24. $-12(-5)$

25. $-13(3)$

26. $3(-2)(4)$

27. $-1(-3)(9)$

28. $-2(-2)(-2)$

29. $3(4)(-7)$

30. $-2(4)(-5)(2)$

31. $-1(-1)(1)(-1)$

32. Find the value of a if $a = -3(14)$.

33. What is the value of n if $n = (-11)(-9)$?

34. Find the value of p if $12(-10) = p$.

Evaluate each expression if $x = 2$, $y = -3$, and $z = -5$.

35. $-4x$

36. $7xy$

37. xyz

38. $2y + z$

39. $5x - y$

40. $3y + 4z$

Simplify each expression.

41. $4(-2a)$

42. $-8(5m)$

43. $(-4m)(-8n)$

44. What is the product of -3, -4, and -5?

45. Evaluate $8a - 2b$ if $a = -2$ and $b = 3$.

Applications and Problem Solving

46. Patterns Find the next term in the pattern $-1, 2, -4, 8, \ldots$

47. Oceanography A research submarine descends to the ocean floor at a rate of 100 feet per minute. Write a multiplication equation that tells how far the submarine moves in 5 minutes.

48. Health From 1995 through 1998, deaths from AIDS decreased by an average of about 11,000 per year. If 49,351 people died in 1995, about how many died in 1998?

49. Geometry $A(-5, 0)$, $B(-3, -5)$, and $C(-1, -2)$ are connected with line segments to form a triangle.

    a. Multiply each x- and y-coordinate by -1 and draw another triangle.

    b. Describe how the position of the triangle changed.

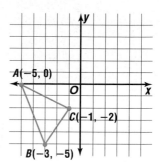

50. Critical Thinking If the product of three integers is negative, what can you conclude about the signs of the integers? Write a rule for determining the sign of the product of three nonzero integers.

Mixed Review

Evaluate each expression if $a = -3$, $b = 7$, $c = -8$, and $d = -15$. *(Lessons 2–3 & 2–4)*

51. $a + b$	**52.** $b - (-1)$	**53.** $c - (-3)$
54. $c + d$	**55.** $d + b$	**56.** $d - b$
57. $5 - b$	**58.** $u - b$	**59.** $c + 8$

60. Chemistry The melting point of several common elements are shown. Which element has the lowest melting point? *(Lesson 2–1)*

Element	Melting Point (°F)
Helium	−458
Hydrogen	−435
Mercury	−38
Oxygen	−361

Exercise 60

61. Standardized Test Practice Which verbal expression represents the algebraic expression $5x - 3$? *(Lesson 1–1)*

    A three minus five times a number x

    B a number x decreased by three

    C three less than five times a number x

    D five more than a number x minus three

Quiz 2   Lessons 2–3 through 2–5

Find each sum, difference, or product. *(Lessons 2–3, 2–4, & 2–5)*

1. $-5 + (-2)$	**2.** $3 - 8$	**3.** $-4(-8)$
4. $6(-9)$	**5.** $9 - (-4)$	**6.** $-10 + 5$
7. $-15(3)$	**8.** $18 + (-2)$	**9.** $-11 - (-2)$

10. Meteorology The temperature between the ground and 11 kilometers above the ground drops about 7°C for each kilometer of increase in altitude. Suppose the ground temperature is 0°C. Find the temperature 2 kilometers above the ground. *(Lesson 2–5)*

Bits, Bytes, and BUGS!

Materials

 calculator

Matrices

The computer bug has bitten the United States and the world! Here are some facts about the number of personal computers shipped in the United States in 1996 and 1997.

Shipments of Personal Computers

Manufacturer	1996	1997
C	3,417,360	5,035,118
P	3,030,398	2,776,144
I	2,196,318	2,738,588
D	1,790,755	2,930,235
A	1,687,161	1,276,249
G	1,666,706	2,219,395

Source: *The Wall Street Journal Almanac, 1999*

Let's use these facts to investigate a mathematical tool called a matrix.

Investigate

> **Reading Algebra**
>
> The plural of *matrix* is *matrices*.

1. A **matrix** is a rectangular arrangement of numbers in rows and columns. Each number in a matrix is called an **element**. The data about the computers shipped in 1996 could be organized in a matrix as shown. Write the data for 1997 as a matrix.

$$\begin{bmatrix} 3,417,360 \\ 3,030,398 \\ 2,196,318 \\ 1,790,755 \\ 1,687,161 \\ 1,666,706 \end{bmatrix}$$

2. Two matrices can be added as shown below.

$$\begin{bmatrix} 2 & -3 \\ -1 & 8 \\ 0 & 5 \end{bmatrix} + \begin{bmatrix} -5 & 3 \\ 7 & -7 \\ -10 & 3 \end{bmatrix} = \begin{bmatrix} 2+(-5) & -3+3 \\ -1+7 & 8+(-7) \\ 0+(-10) & 5+3 \end{bmatrix} = \begin{bmatrix} -3 & 0 \\ 6 & 1 \\ -10 & 8 \end{bmatrix}$$

   a. Write your own rule for adding two matrices.

   b. Use matrix addition to find the total number of personal computers shipped by each manufacturer in 1996 and 1997.

3. Two matrices can be subtracted as shown below.

$$\begin{bmatrix} -1 & 0 \\ 4 & -2 \end{bmatrix} - \begin{bmatrix} 1 & 5 \\ -3 & 6 \end{bmatrix} = \begin{bmatrix} -1-1 & 0-5 \\ 4-(-3) & -2-6 \end{bmatrix} = \begin{bmatrix} -2 & -5 \\ 7 & -8 \end{bmatrix}$$

   a. Write your own rule for subtracting two matrices.

   b. Use matrix subtraction to find how many more personal computers were shipped by each manufacturer in 1997 than in 1996.

   c. What do negative elements indicate?

4. You can multiply any matrix by a number. A number outside a matrix is called a **scalar**. When **scalar multiplication** is performed, each element is multiplied by the scalar, and a new matrix is formed.

$$6 \begin{bmatrix} 8 & -2 & 10 \\ -5 & 4 & 6 \end{bmatrix} = \begin{bmatrix} 6(8) & 6(-2) & 6(10) \\ 6(-5) & 6(4) & 6(6) \end{bmatrix} = \begin{bmatrix} 48 & -12 & 60 \\ -30 & 24 & 36 \end{bmatrix}$$

Suppose the computer industry predicted a 20% increase in shipments compared to the number of shipments in 1997. Use scalar multiplication to find the predicted number of computer shipments. (*Hint:* Multiply the matrix by 1.2 to show an increase of 20%.)

Extending the Investigation

In this extension, you will investigate how matrices are used in the real world. Here are some suggestions.

- The matrices below show the sales and expenses for two different companies for 2000 and 2001. Use the information in the matrices to find a matrix that shows each company's profits in 2000 and 2001. (*Hint:* Profits = Sales – Expenses)

	Sales (million dollars) 2000	2001		Expenses (million dollars) 2000	2001
Company A →	4761	6471		4362	5917
Company B →	5061	3483		4904	4838

- Find some data that can be organized using matrices. Then write a problem using the data that can be solved by adding, subtracting, or using scalar multiplication.

Presenting Your Conclusions

Here are some ideas to help you present your conclusions to the class.

- Prepare a poster presenting matrices in a creative manner. Show how you solved problems using matrices.
- Make a booklet of your problems, matrices, and solutions.

*inter*NET **Investigation** For more information on matrix
CONNECTION addition, visit: www.algconcepts.glencoe.com

What You'll Learn
You'll learn to divide integers.

Why It's Important
Farming The average change in the number of farms can be found by dividing integers.
See Exercise 43.

Global warming is causing the glaciers in Glacier National Park to melt at a faster rate than projected. Some scientists believe that all of these glaciers will be gone in the next 50 to 70 years.

Suppose a glacier has receded 6 meters in the last 2 weeks. On average, how much has it receded each week? This situation can be represented by the division expression $-6 \div 2$. *The expression $-6 \div 2$ means to separate six negative tiles into 2 groups.*

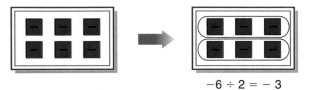

$$-6 \div 2 = -3$$

Therefore, $-6 \div 2 = -3$. Is a negative integer divided by a positive integer always negative? Recall that division is related to multiplication.

$$-6 \div 2 = -3 \qquad 2 \times (-3) = -6$$

Study the following pairs of related sentences. Look for a pattern in the signs.

| Related Sentences ||
Multiplication	Division
$2 \times (-3) = -6$	$-6 \div 2 = -3$
$-2 \times (-3) = 6$	$6 \div (-2) = -3$
$-2 \times 3 = -6$	$-6 \div (-2) = 3$
$2 \times 3 = 6$	$6 \div 2 = 3$

Signs are different.
  The quotients are negative.
Signs are the same.
  The quotients are positive.

The pattern suggests the following rule for dividing integers.

Dividing Integers	Words:	The quotient of two integers with the same sign is positive.
	Numbers:	$6 \div 2 = 3, -6 \div (-2) = 3$
	Words:	The quotient of two integers with different signs is negative.
	Numbers:	$-6 \div 2 = -3, 6 \div (-2) = -3$

Leconte Glacier

Find each quotient.

1 $-10 \div 2$

The signs are different.
The quotient is negative.
$-10 \div 2 = -5$

2 $-32 \div (-8)$

The signs are the same.
The quotient is positive.
$-32 \div (-8) = 4$

Your Turn

a. $-9 \div 3$ **b.** $-20 \div (-4)$ **c.** $16 \div (-2)$

Recall that fractions are another way of showing division.

Example

3 Evaluate $\dfrac{6x}{y}$ if $x = -4$ and $y = 8$.

$\dfrac{6x}{y} = \dfrac{6(-4)}{8}$ *Replace x with −4 and y with 8.*

$\quad = \dfrac{-24}{8}$ *$6(-4) = -24$*

$\quad = -3$ *$\dfrac{-24}{8}$ means $-24 \div 8$.*

Your Turn

Evaluate each expression if $x = -3$ and $y = 6$.

d. $-12 \div x$ **e.** $\dfrac{xy}{-2}$ **f.** $\dfrac{36}{3x}$

Example

Media Link

Real World

4 The table shows the number of CDs and cassettes that were shipped in 1990 and 1997. What was the average change in the number of cassettes that were shipped for each of those seven years?

Recording Media Shipped (millions)

Medium	1990	1997
CDs	289	751
Cassettes	440	174

Source: *Statistical Abstract of the United States,* 1998

First, find the change in the number of cassettes that were shipped.
$174 - 440 = -266$ *There were 266 fewer cassettes shipped in 1997 than in 1990.*

To find the average change, divide -266 by 7.
$-266 \div 7 = -38$

The average change in the number of cassettes that were shipped was -38 per year. This means that each year there were about 38,000,000 fewer cassettes shipped than the year before.

Communicating Mathematics

Study the lesson. Then complete the following.

1. **Write** two division sentences related to the multiplication sentence $-5 \times 2 = -10$.

2. **You Decide?** Joel claims that a positive number divided by a negative number is a positive number. Abbey claims that a negative number divided by a negative number is a negative number. Who is correct? Explain.

Guided Practice

Find each quotient. *(Examples 1 & 2)*

3. $-55 \div 11$ 4. $-14 \div (-2)$ 5. $15 \div (-3)$

6. $16 \div 4$ 7. $-20 \div (-5)$ 8. $\dfrac{-8}{2}$

Evaluate each expression if $a = 3$, $b = -12$, and $c = -6$. *(Example 3)*

9. $-24 \div a$ 10. $\dfrac{ab}{9}$ 11. $\dfrac{6b}{c}$

12. **Business**   In July, 1998, there were about 6,200,000 people who were unemployed in the United States. Twelve months later, this figure had dropped to 5,900,000. What was the average change in unemployment for each of the last twelve months?   *(Example 4)*

Exercises

Practice

Find each quotient.

13. $-12 \div (-12)$ 14. $-18 \div 3$ 15. $36 \div 6$
16. $-10 \div (-2)$ 17. $30 \div (-5)$ 18. $15 \div 5$
19. $-25 \div (-5)$ 20. $-21 \div 7$ 21. $45 \div (-5)$
22. $24 \div (-24)$ 23. $-20 \div (-2)$ 24. $-72 \div 9$
25. $64 \div (-8)$ 26. $-48 \div (-4)$ 27. $-40 \div 8$
28. $\dfrac{-49}{-7}$ 29. $\dfrac{60}{-5}$ 30. $\dfrac{-26}{2}$

31. Find the value of a if $-42 \div 7 = a$.
32. What is the value of m if $m = -81 \div (-9)$?
33. Find the value of w if $w = 85 \div (-17)$.

Evaluate each expression if $x = 5$, $y = -6$, $z = 2$, and $w = -3$.

34. $18 \div y$ 35. $y \div z$ 36. $\dfrac{x}{5}$

37. $\dfrac{-4w}{2}$ 38. $\dfrac{x - z}{w}$ 39. $\dfrac{y - 8}{z}$

40. What is the quotient of -42 and -7?
41. Divide 100 by -50.

42. Animals Experts estimate that there were about 100,000 tigers living 100 years ago. Today, there are only about 6000. What was the average change in tiger population for each of the last 100 years?

43. Farming The table shows the number of farms in California according to their size.

a. Find the average yearly change in the number of farms that are between 50 and 179 acres in size.

b. For which size farm is the average change a positive number?

California Farms (number)

Acres	1992	1997
1–9	21,485	20,662
10–49	26,089	24,250
50–179	13,883	13,288
180–499	7512	7270
500–999	3702	3572
1000 or more	4998	5084

Source: Census of Agriculture, 1997

44. Media Refer to the table on page 83.

a. What was the average change in the number of CDs that were shipped for each of the seven years from 1990 to 1997?

b. If this trend continues, estimate the number of CDs that will be shipped in 2005.

45. Energy A measure called *degree days* is used to estimate the energy needed for heating on cold days. The formula $d = 65 - \frac{h+l}{2}$ can be used to find degree days. In the formula, d represents degree days, h represents the high temperature of a given day, and l represents the low temperature of that day. Find the degree days for a day in which the high temperature was $-2°F$ and the low temperature was $-16°F$.

46. Critical Thinking Explain why division by zero is not possible.

Mixed Review

Find each sum, difference, or product. *(Lessons 2–3, 2–4, & 2–5)*

47. $9(-6)$ **48.** $-11 + (-4)$ **49.** $9 - (-7)$

50. $15 + (-25)$ **51.** $-10(-8)$ **52.** $8 - 10$

53. $-7 - (-5)$ **54.** $-8(9)$ **55.** $-16 + 20$

56. Swimming A competition swimming pool is 75 feet long and 72 feet wide. It is filled to a depth of 6 feet. Use the formula $V = \ell wh$, where ℓ is the length, w is the width, and h is the depth, to find the volume V in cubic feet of water in the pool. *(Lesson 1–5)*

57. Standardized Test Practice Which property of real numbers allows you to conclude that if $2t + 4t = 36$, then $4t + 2t = 36$? *(Lesson 1–3)*

A Distributive Property   **B** Commutative Property

C Associative Property   **D** Additive Inverse Property

Study Guide and Assessment

Understanding and Using the Vocabulary

After completing this chapter, you should be able to define each term, property, or phrase and give an example or two of each.

*inter***NET**
CONNECTION **Review Activities**
For more review activities, visit:
www.algconcepts.glencoe.com

absolute value *(p. 55)*	matrix *(p. 80)*	scalar multiplication *(p. 81)*
additive inverse *(p. 65)*	natural numbers *(p. 53)*	Venn diagrams *(p. 53)*
coordinate *(p. 53)*	negative numbers *(p. 52)*	*x*-axis *(p. 58)*
coordinate plane *(p. 58)*	number line *(p. 52)*	*x*-coordinate *(p. 59)*
coordinate system *(p. 58)*	opposites *(p. 65)*	*y*-axis *(p. 58)*
element *(p. 80)*	ordered pair *(p. 58)*	*y*-coordinate *(p. 59)*
graph *(p. 53)*	origin *(p. 58)*	zero pair *(p. 65)*
integers *(p. 52)*	quadrants *(p. 60)*	

Complete each sentence using a term from the vocabulary list.

1. On a number line, the numbers to the left of zero are ___?___.

2. The ___?___ is the plane that contains the *x*-axis and the *y*-axis.

3. If the sum of two numbers is 0, the numbers are called ___?___.

4. The ___?___ is the first number in an ordered pair.

5. The distance a number is from 0 on the number line is the ___?___.

6. The numbers 1, 2, 3, 4, . . . are ___?___.

7. Whole numbers are a subset of ___?___.

8. The ___?___ of a point is the number corresponding to that point on a number line.

9. A(n) ___?___ is used to locate any point on a coordinate plane.

10. The *x*-axis and *y*-axis separate the coordinate plane into four regions called ___?___.

Skills and Concepts

Objectives and Examples	Review Exercises
• **Lesson 2–1** Graph integers on a number line and compare and order integers. Replace the ● with < or > to make a true sentence. $$7 \bullet -3$$ 7 is to the right of −3 on the number line, so 7 > −3.	Replace each ● with < or > to make a true sentence. **11.** 0 ● −5 **12.** −3 ● 3 **13.** −9 ● −7 **14.** $\lvert -12 \rvert$ ● −12 **15.** Order −4, 7, 4, −2, −3, and 0 from least to greatest. **16.** Order −15, −23, −18, and −20 from greatest to least.

Objectives and Examples

Review Exercises

- **Lesson 2-2** Graph points on a coordinate plane.

Write the ordered pair that names each point and name the quadrant in which each point is located.

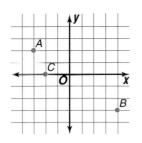

$A(-3, 2)$, II
$B(4, -3)$, IV
$C(-2, 0)$, none

Write the ordered pair that names each point.

17. P
18. Q
19. N
20. M

Name the quadrant in which each point is located.

21. $(6, 10)$ **22.** $(-4, 8)$
23. $(0, -12)$ **24.** $(13, -7)$

- **Lesson 2-3** Add integers.

Find $-2 + (-3)$.
Both numbers are negative, so the sum is negative.
$-2 + (-3) = -5$

Find $4 + (-12)$.
$|-12| > |4|$, so the sum is negative.
$4 + (-12) = -8$

Find each sum.

25. $8 + (-14)$ **26.** $-7 + 5$
27. $-8 + (-2)$ **28.** $-8 + 8$
29. $23 + (-18)$ **30.** $-14 + (-12)$
31. $-10 + 3 + (-6) + 8$
32. $7 + (-5) + (-7) + 15$

Simplify each expression.

33. $7x + (-5x)$ **34.** $-4y + (-y)$
35. $14m + (-10m)$ **36.** $-31x + 27x$

- **Lesson 2-4** Subtract integers.

Find $7 - (-3)$.
$7 - (-3) = 7 + 3$ *To subtract -3, add 3.*
$\qquad = 10$

Find $-4 - 8$.
$-4 - 8 = -4 + (-8)$ *To subtract 8, add -8.*
$\qquad = -12$

Find each difference.

37. $6 - 14$ **38.** $-11 - (-5)$
39. $4 - (-5)$ **40.** $-3 - 5$
41. $-6 - (-2)$ **42.** $10 - (-10)$

Evaluate each expression if $x = 3$, $y = -5$, and $z = -1$.

43. $2 - x$ **44.** $y - z$
45. $x + y - z$ **46.** $x - y + z$

Chapter 2 Study Guide and Assessment

Objectives and Examples

- **Lesson 2–5** Multiply integers.

  $-6(-4) = 24$ *The integers have the same sign, so the product is positive.*

  $3(-5) = -15$ *The integers have different signs, so the product is negative.*

Review Exercises

Find each product.

47. $-7(-5)$ **48.** $8(-4)$

49. $-3(-2)(6)$ **50.** $-1(-4)(-5)$

Evaluate each expression if $a = -3$ and $b = -6$.

51. $-9a$ **52.** $-7ab$

Simplify each expression.

53. $-7(6m)$ **54.** $(-3x)(-15y)$

- **Lesson 2–6** Divide integers.

  $-9 \div (-3) = 3$ *The integers have the same sign, so the quotient is positive.*

  $\dfrac{8}{-2} = -4$ *The integers have different signs, so the quotient is negative.*

Find each quotient.

55. $42 \div (-6)$ **56.** $-63 \div -7$

57. $-24 \div (4)$ **58.** $-40 \div (-5)$

Evaluate each expression if $a = -4$, $b = -2$, and $c = 3$.

59. $\dfrac{6a}{c}$ **60.** $\dfrac{a + b}{2c}$

Applications and Problem Solving

61. Banking Mikaela opened a checking account by depositing $250. She later wrote a check for $25 for the phone bill and $32 for a magazine subscription. Then Mikaela received $20 for her birthday and deposited it into her account. What was her balance after her birthday deposit? *(Lesson 2–3)*

63. Scuba Diving A scuba diver descends at a rate of 40 feet per minute. Write a multiplication equation that tells how far the scuba diver moves in 2 minutes. *(Lesson 2–5)*

62. Geometry The graphs of $M(5, 4)$, $N(-4, 3)$, and $P(0, 1)$ are connected with line segments to form a triangle. *(Lesson 2–5)*

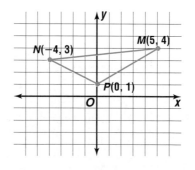

   a. Multiply each y-coordinate by -1 and draw another triangle.

   b. Describe how the position of the triangle changed.

CHAPTER 2 **Test**

1. **Write** two division sentences related to the multiplication sentence $6 \times (-7) = -42$.
2. **Graph** $\{4, -2, 1\}$ on a number line.

Replace each ● with < or > to make a true sentence.

3. $-5 ● -8$

4. $9 ● -2$

Graph each point on a coordinate plane and name the quadrant in which each point is located.

5. $X(4, 3)$

6. $M(-3, 2)$

7. $A(0, -4)$

Find each sum, difference, product, or quotient.

8. $-16 + 9$

9. $5 - (-2)$

10. $-3 - 5$

11. $4(-6)$

12. $-14 \div (-7)$

13. $-8(-7)$

14. $\dfrac{-32}{8}$

15. $\dfrac{-25}{-5}$

16. $-8 - 2$

17. $-7 + 5 + (-12)$

18. $8 + (-14) + (-6)$

19. $5(-2)(3)$

Evaluate each expression if $m = 5$, $n = -8$, and $p = -3$.

20. $n - p$

21. $m + n$

22. $2(n)$

23. $p - m + n$

24. $\dfrac{n}{-2}$

25. $\dfrac{m + n}{p}$

Simplify each expression.

26. $3x - 8x$

27. $10y - (-3y)$

28. $9(-4x)$

29. $-2(-5y)$

30. $(-7m)(3n)$

31. $3x + 4y + x - 2y$

32. **Weather** The table shows record high and low temperatures for some cities for the month of November.
   a. Find the differences in temperatures for each city.
   b. Which city had the greatest difference in its record temperatures?

33. **Sports** The Tigers football team had a gain of 7 yards on their first run. They lost 3 yards on their second run and gained 12 yards on their third run. What was the total gain or loss of yardage in the three runs?

Record Temperatures in November (°F)		
City	High	Low
Atlanta, GA	84	3
Boston, MA	78	15
Columbus, OH	80	5
Duluth, MN	70	−23
Juneau, AK	56	−5
Kahului, HI	93	55

Exercise 32

Data Analysis Problems

You will need to create and interpret frequency tables as well as data graphs. This includes bar graphs, histograms, line graphs, and stem-and-leaf plots.

THE PRINCETON REVIEW

Read the graph or table *before* you read the question.

Proficiency Test Example

Use the information on movie-making costs in the table. Make two line graphs on one grid, one for average production costs and the other for average marketing costs. Title the graph, label the axes, use appropriate scales, and accurately graph the data.

Movie-Making Costs ($ millions)

Year	Average Production Costs	Average Marketing Costs
1980	9.4	4.3
1985	16.8	6.5
1990	26.8	12.0
1995	36.3	17.7

Hint In open-ended questions, you may need to construct a graph, draw a diagram, or explain your answer.

Solution The *x*-axis shows the years. Decide on a scale for the *y*-axis, which represents costs. Since the lowest cost is 4.3 and the highest is 36.6, use a scale of 0 − 40 with intervals of 5. Mark each point (year, cost). Connect the points with line segments.

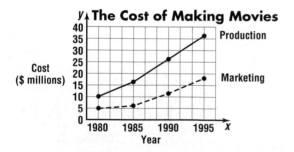

SAT Example

The graph below represents the amount of money each person earns per day. How many days must Andy work to earn as much as Jill would earn in four days?

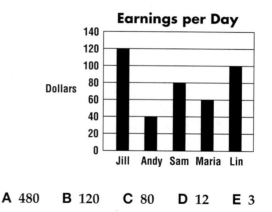

Earnings per Day

A 480   **B** 120   **C** 80   **D** 12   **E** 3

Hint Look carefully at the graph. Read the title and labels. The range of the *y*-axis scale is 140.

Solution Calculate the amount that Jill earns in four days. The graph shows that she earns $120 per day. In four days, she will earn 4 × $120 or $480.

Now calculate how many days it will take Andy to earn the amount of $480. The graph shows that Andy earns $40 per day.

Divide 480 by 40.

$$\frac{\$480}{\$40 \text{ per day}} = 12 \text{ days}$$

Andy needs to work 12 days to earn $480. So, the answer is D.

After you work each problem, record your answer on the answer sheet provided or on a sheet of paper.

1. One winter night the temperature dropped 3° every hour. If the temperature was 0° at midnight, what was the temperature at 4:00 A.M.?

   A −12° **B** −15° **C** 12° **D** 32°

2. Which of the following numbers, when subtracted from −8, gives a result greater than −8?

   A −2 **B** 0 **C** 2 **D** 3

3. How many even integers are there between −4 and 4?

   A 2 **B** 3 **C** 4 **D** 6 **E** 8

4. If hot dogs are sold in packs of 10 and buns are sold in packs of 12, what is the smallest number of each you can buy to have no extra hot dogs or buns?

   A 30 **B** 60 **C** 90 **D** 120

5. Use the stem-and-leaf plot below to find the greatest number of fat grams of any hamburger shown in the data.

Hamburgers	Stem	Chicken
	0	9
9 8 0	1	0 1 2 5 5 7 9
7 2	2	
5 2 0	3	*1|2 = 12 g*

   A 53 **B** 91 **C** 35 **D** 21

6. Use the line graph to find the year when the income was closest to $20,000.

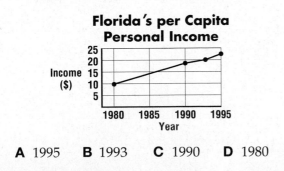

   Florida's per Capita Personal Income

   Income ($)

   A 1995 **B** 1993 **C** 1990 **D** 1980

7. The frequency table shows the number of books that each student read over the summer. Which statement is correct?

Number of Books	Tally	Frequency
0	III	3
1	ЖШ II	7
2	ЖШ IIII	9
3	III	3
4	I	1

   A The students read 4 different books.

   B The most students read only 2 books.

   C Two students read 9 books.

   D There are a total of 21 students.

8. Which expression has the greatest value?

   A −38 × (−10) **B** −38 × 10

   C 38 × (−10) **D** 1 × 38

Open-Ended Questions

9. **Grid-In** What is the value of

   $$\frac{-2 \times 4}{36 \div 2 - 5 \times 2}?$$

10. The table below shows the winners of the first 16 World Cup soccer competitions.

    Winners of the First 16 World Cup Soccer Competitions

Year	Champion	Year	Champion
1930	Uruguay	1970	Brazil
1934	Italy	1974	West Germany
1938	Italy	1978	Argentina
1950	Uruguay	1982	Italy
1954	West Germany	1986	Argentina
1958	Brazil	1990	West Germany
1962	Brazil	1994	Brazil
1966	England	1998	France

    Part A Construct a frequency table of the World Cup soccer champions.

    Part B Use the frequency table to make a bar graph.

*inter***NET** CONNECTION **Test Practice** For more test practice, visit: www.algconcepts.glencoe.com

CHAPTER 3

Addition and Subtraction Equations

▶ ## What You'll Learn in Chapter 3:

- to compare, order, add, and subtract rational numbers *(Lessons 3–1 and 3–2)*,
- to find the mean, median, mode, and range of a set of data *(Lesson 3–3)*,
- to determine whether a given number is a solution of an equation *(Lesson 3–4)*,
- to solve addition and subtraction equations by using models and by using the properties of equality *(Lessons 3–5 and 3–6)*, and
- to solve equations involving absolute value *(Lesson 3–7)*.

Problem-Solving Workshop

Project

At a certain amusement park, you can buy a 50-ticket pass to use at any time. The table shows the number of tickets needed for some of the rides.

You can use all 50 tickets by riding one of the 5-ticket rides 10 times. In what other ways can you use exactly 50 tickets?

Ride	Tickets Needed
The Mighty Axe	7
Log Chute	6
Kite-Eating Tree	5
Mystery Mine Ride	7
Ripsaw Roller Coaster	7
Screaming Yellow Eagle	6
Skyscraper Ferris Wheel	5
Tumbler	6

Working on the Project

Work with a partner to solve the problem.

- If you ride each ride once, can you use exactly 50 tickets?
- Suppose you ride six 5-ticket rides. Can you use all of the remaining tickets with none left over if you do not ride any more 5-ticket rides?

Technology Tools

- Use a **spreadsheet** to find the number of tickets a person can use for different combinations of rides.
- Use a **word processor** to design an ad campaign.

> ### Strategies
> **Look for a pattern.**
> **Draw a diagram.**
> **Make a table.**
> **Work backward.**
> **Use an equation.**
> **Make a graph.**
> **Guess and check.**

*inter*NET
CONNECTION **Research** For more information about amusement parks, visit: www.algconcepts.glencoe.com

Presenting the Project

Suppose the 50-ticket pass costs $25 and an all-day pass for use by one person costs $18.95. Design an ad campaign that answers these questions.

- Suppose the amusement park is open from 10:00 A.M. to 9 P.M. About how many tickets could a person use during that time period?
- What are the advantages and disadvantages of buying a 50-ticket pass? an all-day pass?

What You'll Learn
You'll learn to compare and order rational numbers.

Why It's Important
Shopping Knowing how to compare rational numbers will help you determine better buys.
See Example 6.

The shutter speed of a camera is the amount of time in seconds that the shutter on the lens is open to allow light to hit the film. Various shutter speeds of a camera are listed below.

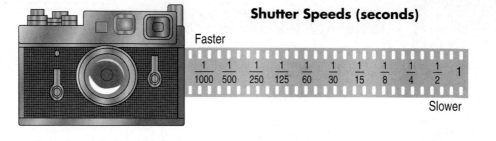

Shutter Speeds (seconds)

Faster

$\frac{1}{1000}$ $\frac{1}{500}$ $\frac{1}{250}$ $\frac{1}{125}$ $\frac{1}{60}$ $\frac{1}{30}$ $\frac{1}{15}$ $\frac{1}{8}$ $\frac{1}{4}$ $\frac{1}{2}$ 1

Slower

The shutter speeds are examples of **rational numbers**.

Rational Number	**Words:**	A rational number is any number that can be expressed as a fraction where the numerator and denominator are integers and the denominator is not zero.
	Symbols:	$\frac{a}{b}$, where a and b are integers and $b \neq 0$

Some examples of rational numbers and their form as a fraction are listed in the table below.

Rational Number	-2	$3\frac{1}{4}$	0.625	0	$-0.66\overline{6}$
Form $\frac{a}{b}$	$-\frac{2}{1}$	$\frac{13}{4}$	$\frac{5}{8}$	$\frac{0}{1}$	$-\frac{2}{3}$

Rational numbers can be graphed on a number line in the same manner as integers.

-1 -0.75 $-\frac{1}{2}$ -0.25 0 $\frac{1}{4}$ 0.5 $\frac{3}{4}$ 1

Graphing rational numbers on a number line helps you to compare them. Just as with integers, the numbers on a number line increase in value as you move to the right and decrease in value as you move to the left.

The following statements can be made about the graph shown above.

Words		Symbols
The graph of -1 is to the left of the graph of $-\frac{1}{2}$. -1 is less than $-\frac{1}{2}$.		$-1 < -\frac{1}{2}$
The graph of 1 is to the right of -0.25. 1 is greater than -0.25.		$1 > -0.25$
The graph of $\frac{1}{4}$ is to the left of 0.5. $\frac{1}{4}$ is less than 0.5.		$\frac{1}{4} < 0.5$

A mathematical sentence that uses $<$ and $>$ to compare two expressions is called an **inequality**. When you compare two numbers, the following property applies.

Comparison Property	**Words:**	For any two numbers a and b, exactly one of the following sentences is true.
		$a < b$ $\qquad$ $a > b$ $\qquad$ $a = b$
	Numbers:	$2 < 3$ $\qquad$ $2 > 3$ $\qquad$ $2 = 3$

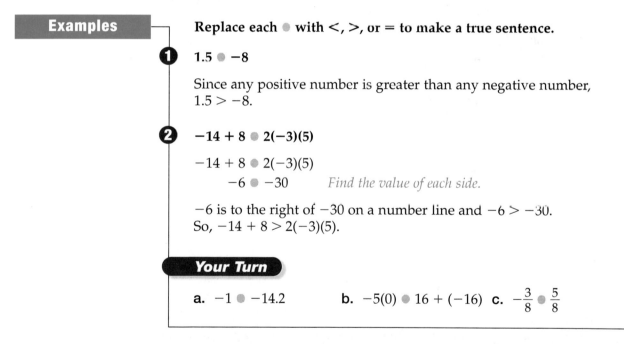

Examples

Replace each ● with $<$, $>$, or $=$ to make a true sentence.

1 **1.5 ● -8**

Since any positive number is greater than any negative number, $1.5 > -8$.

2 **$-14 + 8$ ● $2(-3)(5)$**

$-14 + 8$ ● $2(-3)(5)$
$\qquad -6$ ● -30 $\qquad$ *Find the value of each side.*

-6 is to the right of -30 on a number line and $-6 > -30$.
So, $-14 + 8 > 2(-3)(5)$.

Your Turn

a. -1 ● -14.2 $\qquad$ **b.** $-5(0)$ ● $16 + (-16)$ $\qquad$ **c.** $-\frac{3}{8}$ ● $\frac{5}{8}$

To compare two fractions with different denominators, you can use **cross products**. When two fractions are compared, the cross products are the products of the diagonal terms.

$$\frac{2}{7} \, ● \, \frac{3}{8}$$

$8(2)$ ● $7(3)$ $\qquad$ *Find the cross products.*

$16 < 21$

If $16 < 21$, then $\frac{2}{7} < \frac{3}{8}$.

This is illustrated in the following property.

<table>
<tr><td rowspan="2">**Comparison Property for Rational Numbers**</td><td>**Symbols:**</td><td>For any rational numbers $\frac{a}{b}$ and $\frac{c}{d}$, with $b > 0$ and $d > 0$:
1. If $\frac{a}{b} < \frac{c}{d}$, then $ad < bc$, and
2. If $ad < bc$, then $\frac{a}{b} < \frac{c}{d}$.</td></tr>
<tr><td>**Numbers:**</td><td>1. If $\frac{2}{3} < \frac{3}{4}$, then $8 < 9$.
2. If $8 < 9$, then $\frac{2}{3} < \frac{3}{4}$.</td></tr>
</table>

This property also holds true if < is replaced by > or =.

Replace each ● with <, >, or = to make a true sentence.

3
$$\frac{3}{8} \bullet \frac{4}{10}$$

$\frac{3}{8} \, \frac{4}{10}$

$10(3) \bullet 8(4)$ *Find the cross products.*

$30 < 32$

So, $\frac{3}{8} < \frac{4}{10}$.

4
$$-\frac{5}{6} \bullet -\frac{3}{4}$$

$-\frac{5}{6} \, -\frac{3}{4}$

$4(-5) \bullet 6(-3)$

$-20 < -18$

So, $-\frac{5}{6} < -\frac{3}{4}$.

Technology Tip

You can use a calculator to express rational numbers as decimals so they can be easily compared.

Your Turn

d. $\frac{4}{5} \bullet \frac{7}{8}$

e. $-\frac{5}{10} \bullet -\frac{3}{5}$

Another way to compare rational numbers is to express them as terminating or repeating decimals.

Example

5 **Write $\frac{3}{8}, \frac{1}{3}$, and $\frac{2}{5}$ in order from least to greatest.**

$\frac{3}{8} = 0.375$ *This is a terminating decimal.*

$\frac{1}{3} = 0.3333\ldots$ or $0.\overline{3}$ *This is a repeating decimal.*

$\frac{2}{5} = 0.4$ *This is a terminating decimal.*

In order from least to greatest, the decimals are $0.\overline{3}$, 0.375, 0.4. So, the fractions in order from least to greatest are $\frac{1}{3}, \frac{3}{8}, \frac{2}{5}$.

Reading Algebra

A bar over a number indicates that it repeats indefinitely. The number $0.\overline{3}$ is read as *zero point three repeating*.

Your Turn **Write the numbers in each set from least to greatest.**

f. $\frac{4}{8}, \frac{9}{10}, \frac{2}{3}$

g. $-\frac{1}{3}, -\frac{5}{8}, -\frac{1}{6}$

You can use the cost per unit, or **unit cost**, to compare the costs of similar items. Many people use unit cost to comparison shop. The least unit cost is the best buy.

$$\text{unit cost} = \text{total cost} \div \text{number of units}$$

Example
Shopping Link
Real World

6 **Rolando needs to buy colored pencils. The cost of a package of 12 pencils is $6.39. A package of 24 pencils costs $12.89. Which is the better buy? Explain.**

Find the unit cost of each package. In each case, the unit cost is expressed in cents per pencil.

unit cost of package of 12:
$6.39 \div 12 = 0.5325$ or about $0.53 per pencil

unit cost of package of 24:
$12.89 \div 24 \approx 0.5371$ or about $0.54 per pencil

Since $0.53 < $0.54, the package of 12 pencils is the better buy.

Check for Understanding

Communicating Mathematics

Study the lesson. Then complete the following.

1. **Write** three examples of rational numbers. Include one in decimal form and one negative number.

2. **Give an example** of two different fractions whose cross products are equal.

Math Journal

3. **Create** a list of at least five commonly used fraction-decimal equivalencies such as $\frac{1}{2} = 0.5$. Explain why memorizing them is useful.

> **Vocabulary**
> rational number
> inequality
> cross products
> unit cost

Guided Practice

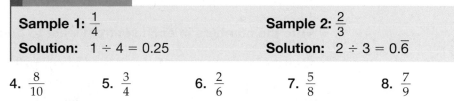

Getting Ready **Express each fraction as a decimal.**

Sample 1: $\frac{1}{4}$	**Sample 2:** $\frac{2}{3}$
Solution: $1 \div 4 = 0.25$	**Solution:** $2 \div 3 = 0.\overline{6}$

4. $\frac{8}{10}$ 5. $\frac{3}{4}$ 6. $\frac{2}{6}$ 7. $\frac{5}{8}$ 8. $\frac{7}{9}$

9. Graph the set of numbers $\left\{-3, 1.5, \frac{1}{4}, 0, -2.\overline{6}\right\}$ on a number line. Use it to explain why $-3 < -2.\overline{6}$.

Replace each ● with <, >, or = to make a true sentence.
(Examples 1–4)

10. $0 ● -4.3$

11. $15 - 24 ● -3(2)(-6)$

12. $\dfrac{1}{2} ● \dfrac{1}{10}$

13. $\dfrac{2}{8} ● \dfrac{1}{4}$

Write the numbers in each set from least to greatest. *(Example 5)*

14. $-\dfrac{2}{3}, -\dfrac{7}{8}, -\dfrac{3}{5}$

15. $\dfrac{4}{5}, \dfrac{7}{10}, \dfrac{6}{8}$

16. Shopping Thompson's Market sells a 25-pound bag of Happy Chow dog food for $15.49 and a 30-pound bag of Super Chow dog food for $17.39. Which brand is the better buy? Explain. *(Example 6)*

Exercises ● ● ● ● ● ● ● ● ● ● ● ● ● ● ● ● ●

Practice

Replace each ● with <, >, or = to make a true sentence.

17. $-12 ● -2$

18. $8.2 ● -7$

19. $-0.88 ● -0.86$

20. $-3 ● 3(0)$

21. $-5 - 3 ● 9$

22. $-8 + 3 ● 7(2)(-3)$

23. $\dfrac{1}{8} ● 0.124$

24. $\dfrac{4}{5} ● 0.8$

25. $-\dfrac{3}{8} ● \dfrac{3}{4}$

26. $-\dfrac{1}{5} ● -\dfrac{2}{10}$

27. $\dfrac{4}{6} ● \dfrac{2}{5}$

28. $\dfrac{5}{8} ● \dfrac{5}{6}$

29. Compare the numbers -2.002 and -2.02 using an inequality.

30. Write an inequality that compares $\dfrac{5}{8}$ and $\dfrac{2}{3}$.

31. Using a number line, explain why $-\dfrac{8}{10} > -\dfrac{5}{6}$.

Write the numbers in each set from least to greatest.

32. $0.2, -\dfrac{5}{6}, \dfrac{4}{5}$

33. $\dfrac{1}{2}, \dfrac{5}{8}, 0.6$

34. $-\dfrac{3}{5}, -\dfrac{8}{10}, \dfrac{4}{6}$

35. $-\dfrac{1}{4}, -\dfrac{2}{6}, -\dfrac{1}{8}$

36. $\dfrac{3}{10}, -\dfrac{2}{8}, -\dfrac{1}{3}$

37. $\dfrac{3}{5}, 0.\overline{6}, \dfrac{3}{8}$

Which is the better buy? Explain.

38. a 16-ounce bottle of juice for $0.89 or a 20-ounce bottle for $1.09

39. a dozen eggs for $1.59 or a package of 18 eggs for $2.49

40. a 25-pack of computer diskettes for $9.95 or a 30-pack for $12.97

41. an $8\frac{1}{3}$-yd by 12-in. roll of Ruppert's Wrap aluminum foil for $1.29, a $16\frac{2}{3}$-yd by 12-in. roll for $2.89, or a 25-yd by 12-in. roll for $2.99

Applications and Problem Solving

42. Tools Julie is using a socket wrench to tighten a bolt on her car. She needs to use the next size smaller than the $\frac{7}{16}$-inch socket. Which socket should she use, $\frac{1}{2}$-inch, $\frac{3}{8}$-inch, or $\frac{9}{16}$-inch?

43. Life Science Scientists have identified over 360,000 species of beetles. The table shown lists the average lengths of various beetles.

Species	Firefly	Confused Flour Beetle	Elm Leaf Beetle	Mealworm	Lady Beetle	Japanese Beetle
Length (in.)	$\frac{3}{4}$	$\frac{1}{8}$	$\frac{1}{4}$	$\frac{5}{8}$	$\frac{1}{3}$	$\frac{1}{2}$

a. Which beetles are larger than the Lady Beetle?

b. Name the beetle whose length is half that of the Japanese Beetle.

44. Critical Thinking Does $\frac{8}{6.4}$ name a rational number? Explain.

Japanese Beetle

Mixed Review

Find each quotient. *(Lesson 2–6)*

45. $-42 \div 6$ **46.** $-16 \div (-4)$ **47.** $-\frac{12}{3}$

Evaluate each expression if $a = 4$, $b = -5$, and $c = 2$. *(Lesson 2–5)*

48. $-3b$ **49.** $6ac$ **50.** $a + bc$

51. Geography The deepest lake in the world is Lake Baikal, in Siberia. Its deepest point is 5315 feet deep. If its surface is 1493 feet above sea level, how many feet below sea level is the deepest point? *(Lesson 2–3)*

52. Standardized Test Practice Use the graph to name the coordinates of point *F*. *(Lesson 2–2)*

A $(-2, 3)$ **B** $(3, 2)$

C $(-2, -3)$ **D** $(2, 3)$

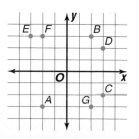

Math In the Workplace

What You'll Learn

You'll learn to add and subtract rational numbers.

Why It's Important

Agriculture Knowing how to add and subtract rational numbers can help you solve agriculture problems.
See Example 4.

Jason Wilson has kept track of the level of water in his farm's pond over the past five years. He compared the water level to the overall average. What was the net change in the water level for this time period?
This problem will be solved in Example 4.

To add rational numbers, you can use the same rules you used to add integers.

Year	Level Above or Below Average (feet)
1997	2
1998	$1\frac{1}{3}$
1999	$-\frac{3}{4}$
2000	$-2\frac{1}{2}$
2001	$1\frac{1}{4}$

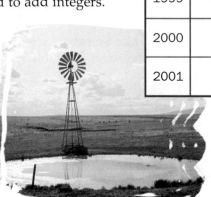

Examples

Prerequisite Skills Review

Operations with Decimals, p. 684

Find each sum.

1 $-4.8 + (-8.7)$

$$-4.8 + (-8.7) = -(|-4.8| + |-8.7|) \quad \text{\textit{The numbers have the same sign.}}$$
$$= -(4.8 + 8.7) \quad\quad\quad\quad \text{\textit{Add their absolute values.}}$$
$$= -13.5$$

2 $3\frac{2}{3} + \left(-4\frac{1}{6}\right)$

$$3\frac{2}{3} + \left(-4\frac{1}{6}\right) = 3\frac{4}{6} + \left(-4\frac{1}{6}\right) \quad \text{\textit{The LCD is 6. Replace } } 3\frac{2}{3} \text{ \textit{with} } 3\frac{4}{6}.$$

$$= -\left(\left|-4\frac{1}{6}\right| - \left|3\frac{4}{6}\right|\right) \quad \text{\textit{The signs differ. Subtract the lesser absolute value from the greater.}}$$

$$= -\left(4\frac{1}{6} - 3\frac{4}{6}\right) \quad \text{\textit{The sign is negative. Why?}}$$

$$= -\left(3\frac{7}{6} - 3\frac{4}{6}\right) \quad \text{\textit{Replace } } 4\frac{1}{6} \text{ \textit{with} } 3\frac{7}{6}. \text{ \textit{Then subtract.}}$$

$$= -\frac{3}{6} \text{ or } -\frac{1}{2}$$

Your Turn

a. $-0.76 + (-1.34)$

b. $-1\frac{3}{4} + \left(-4\frac{1}{8}\right)$

When adding three or more rational numbers, you can use the Commutative and Associative Properties to rearrange the addends.

Example — ③ Find 14.8 + (−7.2) + 30.7.

$$14.8 + (-7.2) + 30.7$$
$$= (14.8 + 30.7) + (-7.2) \quad \textit{Comm. \& Assoc. Properties } (+)$$
$$= 45.5 + (-7.2) \qquad\qquad \textit{Add.}$$
$$= 38.3$$

Your Turn Find each sum.

c. $28.3 + (-56.1) + 32.4 + (-75.1)$ d. $-\dfrac{4}{3} + \dfrac{5}{8} + \left(-\dfrac{7}{3}\right)$

Example — ④

Agriculture Link

Real World

Prerequisite Skills Review
Adding and Subtracting Fractions, p. 687

Refer to the application at the beginning of the lesson. Find the net change in the water level of the pond.

To find the net change in the water level, add.

$$2 + 1\frac{1}{3} + \left(-\frac{3}{4}\right) + \left(-2\frac{1}{2}\right) + 1\frac{1}{4}$$

Group positive numbers together and negative numbers together.

$$2 + 1\frac{1}{3} + \left(-\frac{3}{4}\right) + \left(-2\frac{1}{2}\right) + 1\frac{1}{4}$$
$$= 2 + 1\frac{1}{3} + 1\frac{1}{4} + \left(-\frac{3}{4}\right) + \left(-2\frac{1}{2}\right)$$
$$= 2 + 1\frac{4}{12} + 1\frac{3}{12} + \left(-\frac{9}{12}\right) + \left(-2\frac{6}{12}\right) \quad \textit{The LCD is 12.}$$
$$= \frac{24}{12} + \frac{16}{12} + \frac{15}{12} + \left(-\frac{9}{12}\right) + \left(-\frac{30}{12}\right)$$
$$= \frac{55}{12} + \left(-\frac{39}{12}\right)$$
$$= \frac{16}{12}$$
$$= \frac{16 \div 4}{12 \div 4} \quad \textit{Divide 16 and 12 by their GCF, 4. Then simplify.}$$
$$= \frac{4}{3} \text{ or } 1\frac{1}{3}$$

You subtract rational numbers the same way as you subtract integers.

Example — ⑤ Find −4.5 − 6.8.

$$-4.5 - 6.8 = -4.5 + (-6.8) \quad \textit{To subtract 6.8, add } -6.8.$$
$$= -11.3 \qquad\qquad \textit{The numbers have the same sign.}$$
$$\qquad\qquad\qquad\quad \textit{Add their absolute values.}$$

Your Turn Find each difference.

e. $-72.5 - 81.3$ f. $7\frac{3}{8} - \left(-4\frac{1}{3}\right)$

6 Evaluate $a - b$ if $a = -\dfrac{7}{8}$ and $b = -\dfrac{3}{5}$.

$$a - b = -\dfrac{7}{8} - \left(-\dfrac{3}{5}\right) \quad \textit{Replace a with } -\dfrac{7}{8} \textit{ and b with } -\dfrac{3}{5}.$$

$$= -\dfrac{7}{8} + \dfrac{3}{5} \quad \textit{To subtract } -\dfrac{3}{5}, \textit{add } \dfrac{3}{5}.$$

$$= -\dfrac{35}{40} + \dfrac{24}{40} \quad \textit{The LCD is 40.}$$

$$= -\dfrac{11}{40}$$

Your Turn

g. Evaluate $x - y$ if $x = 25.8$ and $y = -13.9$.

h. If $m = -\dfrac{3}{4}$ and $n = -\dfrac{2}{5}$, what is the value of $m - n$?

Check for Understanding

Communicating Mathematics

Study the lesson. Then complete the following.

1. **Explain** how the sum $-4.78 + 10.23 + (-7.04) + 0.92$ can be solved in more than one way.

2. **Describe** an example of adding rational numbers you can find in a newspaper.

Guided Practice

Find each sum or difference. *(Examples 1–5)*

3. $4.5 + (-3.6)$ **4.** $-1.6 - 3.8$ **5.** $-5.6 - 9.45$

6. $-5\dfrac{7}{8} - 2\dfrac{3}{4}$ **7.** $4\dfrac{1}{2} + \left(-7\dfrac{3}{4}\right) + 3\dfrac{1}{4}$

Evaluate each expression if $a = -\dfrac{1}{2}$, $b = -0.85$, $c = 1.36$, and $d = -\dfrac{5}{6}$.

8. $b - c$ **9.** $a - d$

10. Stock Market On Monday, stock in JAB Corporation rose $3\dfrac{1}{8}$ points. The next day, the stock dropped $1\dfrac{3}{4}$ points. What was the net change in the price of the stock? *(Example 6)*

Exercises

Practice

Find each sum or difference.

11. $-4.7 + (-8.6)$ **12.** $89.3 + (-14.4)$

13. $-2.68 - 3.14$ **14.** $-18.7 + 12.0 + (-9.2)$

15. $-0.23 - 0.13 + (-0.9)$ **16.** $-3.12 + 4.33 + (-1.89)$

17. $-18.9 + (-3.15) - 7.43$ **18.** $-0.8 + 3.5 + (-7.6) + 2.8$

19. $-\dfrac{1}{2} + \dfrac{1}{8}$ **20.** $-\dfrac{3}{5} - \left(-\dfrac{5}{8}\right)$ **21.** $3\dfrac{4}{5} - 9\dfrac{1}{10}$

22. $-\dfrac{5}{6} + \dfrac{5}{8} - \left(-\dfrac{1}{2}\right)$ **23.** $-\dfrac{2}{7} + \dfrac{2}{5} + \dfrac{3}{7}$

24. $\dfrac{1}{4} + 2 + \left(-\dfrac{3}{4}\right)$ **25.** $-2\dfrac{3}{4} + 5\dfrac{1}{2} - \left(-\dfrac{3}{8}\right)$

Evaluate each expression if $a = -\dfrac{2}{5}$, $b = 1\dfrac{3}{4}$, $c = 14.6$, and $d = -5.9$.

26. $a + b$ **27.** $a - b$ **28.** $c - d$ **29.** $1 - d$

30. Find the value of $y - 0.5$ if $y = -0.8$.

31. Evaluate $x - 3\dfrac{1}{3}$ if $x = -2\dfrac{5}{6}$.

32. Find the value of a if $a = \dfrac{3}{4} + \left(-\dfrac{4}{5}\right) - \dfrac{2}{5}$.

Applications and Problem Solving

33. Personal Finance   On Wednesday, Lakeesha Burton wrote checks for $289.53, $312.41, and $76.89. On Thursday, she deposited $210.08 and $315.17 into her checking account. What is the net increase or decrease in Lakeesha's account?

34. Sports   The game of *quoits* is similar to horseshoes, but it is played with metal rings. The object of the game is to throw the quoit onto or as close as possible to a vertical metal pole. The dimensions of the quoit are shown at the right. What is the diameter of the hole in the center of the quoit?

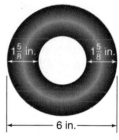

35. Critical Thinking   The net snow pack level in the Sierra Nevadas over the past five years has been $1\dfrac{1}{4}$ feet above normal. The following snow pack levels were recorded for the first four years: $1\dfrac{1}{2}$ feet above normal, $1\dfrac{7}{8}$ feet below normal, $\dfrac{3}{4}$ foot below normal, and $2\dfrac{1}{8}$ feet above normal. What was the snow pack level for the fifth year?

Mixed Review

36. Shopping   Which is a better buy: a 184-gram can of peanuts for $1.09 or a 340-gram can for $1.99? Explain. *(Lesson 3–1)*

Evaluate each expression if $g = -4$, $h = 5$, and $j = -6$. *(Lesson 2–6)*

37. $-20 \div (h + g)$ **38.** $\dfrac{j - 9}{h}$

39. Simplify $(4b)(-3c)$. *(Lesson 2–5)*

40. Music   Monica has 16 CDs that are either country or jazz. She has 8 more jazz CDs than country CDs. How many of each does Monica have? *(Lesson 1–5)*

41. Simplify $5y + 2(7 + 3y)$. *(Lesson 1–4)*

42. Standardized Test Practice   Find the value of $16 \div [2(13 - 11)]$. *(Lesson 1–2)*

Extra Practice   See p. 696.   **Lesson 3-2** Adding and Subtracting Rational Numbers   **103**

3-3 Mean, Median, Mode, and Range

Math In the Workplace

What You'll Learn
You will learn to find the mean, median, mode, and range of a set of data.

Why It's Important
Meteorology A meteorologist can compare climates by comparing the mean, median, mode, and range of a set of temperature data.
See Example 5.

Do you like to snack on potato chips, pretzels, and popcorn? Well, if so, you're not alone. Most Americans enjoy eating snack food. The table below shows how much snack food Americans consume.

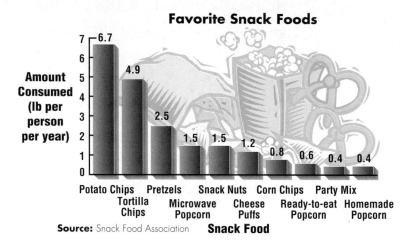

Favorite Snack Foods

Amount Consumed (lb per person per year)

Potato Chips 6.7
Tortilla Chips 4.9
Pretzels 2.5
Microwave Popcorn 1.5
Snack Nuts 1.5
Cheese Puffs 1.2
Corn Chips 0.8
Ready-to-eat Popcorn 0.6
Party Mix 0.4
Homemade Popcorn 0.4

Source: Snack Food Association **Snack Food**

Sometimes, it is helpful to have one number to describe a set of data. This number is called a **measure of central tendency** because it represents the center or middle of the data. The most commonly-used measures of central tendency are the **mean**, **median**, and **mode**.

Mean	The mean, or *average*, of a set of data is the sum of the data divided by the number of pieces of data.

Example 1

Food Link

Real World

Prerequisite Skills Review
Operations with Decimals, p. 684

Find the mean of the snack food data.

First, find the sum of the amounts consumed. Then divide by the number of items of data. In this case there are 10 items of data.

$$\text{mean} = \frac{6.7 + 4.9 + 2.5 + 1.5 + 1.5 + 1.2 + 0.8 + 0.6 + 0.4 + 0.4}{10}$$

$$= \frac{20.5}{10} \text{ or } 2.05 \quad \text{The mean of the data is 2.05 pounds.}$$

Your Turn Find the mean of each set of data.

a. 19, 21, 18, 17, 18, 22, 46

b.
Stem	Leaf
7	3 5 6
8	2 2 4
9	0 4 7 9
10	5 8
11	4 6 *9│4 = 94*

Notice that the amount of potato chips consumed is far greater than the amounts of other snacks consumed. Because the mean is an average of several numbers, a single number that is so much greater or less than the others can affect the mean a great deal. In such cases, the mean becomes less representative of the values in a set of data.

The median is another measure of central tendency.

Median	The median of a set of data is the middle number when the data in the set are arranged in numerical order.

Example

Food Link

Real World

2 **Find the median of the snack food data.**

Arrange the numbers in order from least to greatest.

0.4   0.4   0.6   0.8 |1.2 ↑ 1.5| 1.5   2.5   4.9   6.7
                            median

Since there is an even number of data items, the median is the mean of the two middle values, 1.5 and 1.2. *If there were an odd number of data items, the middle one would be the median.*

$$\text{median} = \frac{1.5 + 1.2}{2} = \frac{2.7}{2} \text{ or } 1.35$$

The median is 1.35 pounds. *The number of values that are greater than the median is the same as the number of values that are less than the median.*

Your Turn   **Find the median of each set of data.**

   c. 4, 6, 12, 5, 8   **d.** 10, 3, 17, 1, 8, 6, 12, 15

Graphing Calculator Tutorial
See pp. 724–727.

Graphing Calculator Exploration

You can use a graphing calculator to find the mean and median of the snack food data. First press ⬚STAT⬚ ⬚ENTER⬚. Under L1, enter in the numerical data. Once all of the data are entered, press ⬚STAT⬚ ⬚▶⬚ ⬚ENTER⬚ ⬚ENTER⬚. Statistics will appear on the screen. The first statistic, $\overline{\times}$, is the mean. Scroll down, and you will find the median given by the label "Med."

Try These

1. Use a graphing calculator to find the mean and median of the snack food data. Did you get the same answers?

2. An *outlier* is an item that is much greater or much less than the other data. In the snack food data, potato chips, at 6.7 pounds per person per year, is an outlier. Find the mean and median of the data without potato chips. Is there a significant difference in your answers? Explain.

A third measure of central tendency is the mode.

Mode	The mode of a set of data is the number that occurs most often in the set.

Sometimes a set of data has only one mode. In other cases, some data sets have no number that occurs more often than the other numbers in the set. When this happens, the set has no mode.

Example

Food Link

❸ **Find the mode of the snack food data.**

Look for the number that occurs most often.

0.4   0.4   0.6   0.8   1.2   1.5   1.5   2.5   4.9   6.7

In this set, 0.4 and 1.5 each appear twice. So, the set of data has two modes, 0.4 pound and 1.5 pounds.

Your Turn **Find the mode of each set of data.**

e. 7, 19, 9, 4, 7, 2   **f.** 300, 34, 40, 50, 60

The snack food data has a mean of 2.05 pounds, a median of 1.35 pounds, and two modes of 0.4 pound and 1.5 pounds. As you can see, the mean, median, and mode are rarely the same value.

Measures of central tendency may not give an accurate description of a set of data. Often, **measures of variation** are used to describe the *distribution* of the data. A common measure of variation is the **range**.

Range	The range of a set of data is the difference between the greatest and the least values of the set.

To find the range of the snack food data, subtract the least value of the data set from the greatest value.

The greatest value is 6.7.
The least value is 0.4.
So, the range of the data is 6.7 − 0.4 or 6.3.

Example

❹ **Find the range of the data set {19, 21, 18, 17, 18, 22, 46}.**

The greatest value is 46. The least value is 17.
So, the range is 46 − 17 or 29.

Your Turn **Find the range of each set of data.**

g. 4, 6, 12, 5, 8   **h.**

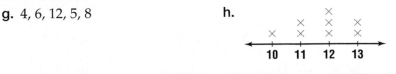

Measures of central tendency and measures of variation can be used to compare two sets of data.

Example

Meteorology Link

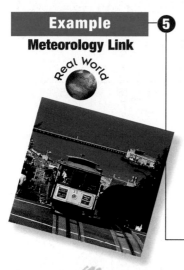

5 The table below shows yearly temperature data for the cities of St. Louis, Missouri, and San Francisco, California. How does the climate of St. Louis compare to the climate of San Francisco?

City	Mean	Median	Mode	Range
St. Louis	55.3°	57°	34°, 75°	50°
San Francisco	56.8°	56.5°	49°, 55°, 61°	15°

The mean and median temperatures show that the temperatures for the two cities are similar. Yet, the modes and range suggest that the temperatures in St. Louis vary more widely than in San Francisco. So, the climate of San Francisco is less extreme than that of St. Louis.

Check for Understanding

Communicating Mathematics

Study the lesson. Then complete the following.

1. **Illustrate** with an example of how the mean of a set of data is affected by an extremely high or low value.

2. **Describe** the steps you would take to find the mean, median, mode, and range of the data below.

3. **YOU Decide** Eric says that his mean test score is always a good measure of how he is doing in class. Sonia disagrees. Who is correct? Explain.

Vocabulary

measure of central tendency
mean
median
mode
measure of variation
range

Guided Practice

Getting Ready Find each quotient without using a calculator.

Sample 1: 156 ÷ 8

Solution:
```
       19.5
   8)156
     −8
     76
    −72
     40
    −40
      0
```

Sample 2: 98.7 ÷ 3

Solution:
```
       32.9
   3)98.7
    −9
     8
    −6
     2 7
    −2 7
       0
```

4. 87 ÷ 6

5. 126 ÷ 3

6. 2.04 ÷ 8

7. 0.453 ÷ 3

8. 35.6 ÷ 2

9. 189 ÷ 4

Find the mean, median, mode, and range of each set of data.
(Examples 1–4)

10. 26, 30, 45, 61, 68

11. 12.5, 11.2, 12.4, 12.1, 12.5, 32.3

12. 3.8, 6, 4.5, 9, 6.2, 4.9, 7, 7.6, 3.8, 4, 7

13.

Stem	Leaf
5	5 6 7 8
6	0 0 1
7	2 2 2 7|2 = 72

14.

15. Sales The district manager of a cellular phone company plans on promoting one of two candidates to sales manager. The quarterly sales figures for the two employees for the last two years are listed in the table below.   *(Example 5)*

Name	Quarterly Sales (thousands of dollars)							
Ms. Diaz	200.8	190.9	200.0	210.0	200.1	200.5	200.7	210.0
Mr. Cruz	210.0	180.1	191.2	239.2	210.8	190.8	180.9	210.0

    a. Find the mean, median, mode, and range of the quarterly sales data for each employee.

    b. Compare the data. Based on sales, who should be promoted to sales manager? Explain your reasoning.

Exercises   • • • • • • • • • • • • • • • • •

Practice

Find the mean, median, mode, and range of each set of data.

16. 7, 9, 7, 9, 8, 9, 7

17. 10, 3, 14, 1, 8, 7, 20

18. 1.5, 1.2, 1.1, 1.3, 1.6, 1.1

19. 0.3, 0.3, 0.3, 0.1

20. 45, 32, 17, 65, 80, 55

21. 200, 24, 20, 40, 20

22. 95, 76, 88, 82, 73, 65, 76, 76, 84, 90

23. 10.2, 8.8, 10.0, 9.4, 12.6, 10.2, 9.8, 12.0

24.

25.

26.

Stem	Leaf
5	6 6 7
6	0 1 1 8
7	0 2 4 4 4
8	0 9 7|2 = 72

27.

Stem	Leaf
18	3 3 4
19	0 0 2 5
20	0 8
21	0 0 20|8 = 20.8

28. Find the median of the set of data $12.99, $5.89, $6.75, $2.45, $9.25, and $5.88.

29. Give an example of a set of data that has no mode.

Write a set of data with six numbers that satisfies each set of conditions.

30. The mean is greater than all but one of the numbers.

31. The mean is smaller than all but one of the numbers.

32. Sports The batting averages for 10 players on a baseball team are 0.234, 0.253, 0.312, 0.333, 0.281, 0.240, 0.183, 0.222, 0.297, and 0.275. Find the mean batting average for these players.

33. School Jack's last four test scores in French were 88, 77, 81, and 83. What must he score on the next test so that his average is exactly 85?

34. Zoos The table shows the number of acres for the five largest zoos in the United States.

Zoo	San Diego Wild Animal Park	Minnesota Zoo	Miami Metrozoo	Bronx Zoo	Albuquerque Biological Park
Acres	2200	500	300	265	240

Source: *World Almanac, 1998*

a. What was the mean size of the zoos?

b. Does the mean size accurately describe the values in the set of data? Explain.

35. Business The salaries of the 12 employees at The Hanson Company are $28,500, $32,000, $29,500, $31,200, $28,600, $38,500, $20,100, $85,000, $36,000, $25,350, $26,500, and $19,850.

a. Find the mean, median, and mode of the data.

b. Suppose the president is interviewing an applicant. Should the president quote the mean, median, or mode as the "average" salary? Explain.

c. If the employees are asking for a pay raise, should they quote the mean, median, or mode as their "average" salary? Explain.

36. Critical Thinking List six numbers whose mean is 50, median is 40, and mode is 20.

Mixed Review

Find each sum or difference. *(Lesson 3–2)*

37. $-2.5 - 7.4$ **38.** $6\frac{1}{3} + \left(-3\frac{3}{4}\right)$ **39.** $\frac{3}{4} + \left(-\frac{5}{4}\right) - \frac{1}{2}$

40. Write an inequality that compares $-\frac{8}{17}$ and $\frac{1}{9}$. *(Lesson 3–1)*

41. Graph the set of points $\{-3, 4, 2, -2, 0, -1\}$ on a number line. *(Lesson 2–1)*

42. Open-Ended Test Practice You have 40 pieces of fencing that are each 1 meter long. What is the largest rectangular area you can enclose with these pieces of fencing? (*Hint*: The formula for perimeter is $P = 2(\ell + w)$, and the formula for area is $A = \ell w$, where ℓ is the length and w is the width.) *(Lesson 1–5)*

Sugar and Pizza and Everything Nice!

Materials

sugar cubes or wooden cubes

Arithmetic Sequences

A **sequence** is a set of numbers in a specific order.

Investigate

1. Use cubes to form the four figures shown. Examine how Figure 2 is different from Figure 1, Figure 3 from Figure 2, and so on.

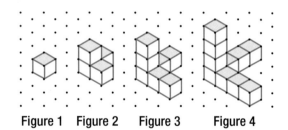

Figure 1   Figure 2   Figure 3   Figure 4

a. Make a table like the one below. Write the numbers 1–10 in columns 1 and 2. Record the number of cubes in each of the four figures. Build a fifth figure and record the number of cubes used. Look for a pattern and complete column 3.

Figure Number	Term Number	Number of Cubes	New Number of Cubes— Previous Number of Cubes
1	1	1	—
2	2	4	4 − 1 = ?

b. Find the difference between the new number of cubes and the number of cubes from the previous row to complete column 4.

c. List the numbers in column 3 from least to greatest, with commas separating the numbers. This list is a sequence. Each number in the sequence is called a **term**.

d. The sequence in part c is called an **arithmetic sequence** because the difference between any two consecutive terms is always the same. What is this difference?

2. An arithmetic sequence can be written as a formula in several ways.

a. Copy and complete a table like the one below for terms 1–10.

Term Number	Term Value	Form 1	Form 2	Form 3
1	1	1	1	$1 + (0 \cdot 3)$
2	4	$1 + 3$	$1 + 3$	$1 + (1 \cdot 3)$
3	7	$4 + 3$	$1 + 3 + 3$	$1 + (2 \cdot 3)$

b. A **recursive formula** for a sequence describes the new term as it relates to the previous term. Look at the column labeled "Form 1." For this arithmetic sequence, you can describe a recursive formula as *term value = previous term value + 3*. Describe Forms 2 and 3.

3. A pizza chain offers the deal *Buy one pizza at regular price and get additional pizzas for only $4 each*. The first pizza costs $7.99.

a. Copy and complete a table like the one below for terms 1–10.

Pizza Number	Term Number	Cost of Pizzas	New Cost of Pizza— Previous Cost of Pizza
1	1	$7.99	—
2	2	$11.99	$11.99 − $7.99 = ?

b. Write a sequence representing the cost of the pizzas. Then write a recursive formula representing the sequence.

Extending the Investigation

In this extension, you will examine other sequences and their formulas.

1. Use cubes to make the figures shown.

2. Make a table similar to the one in Exercise 1.

3. Write a recursive formula for the sequence.

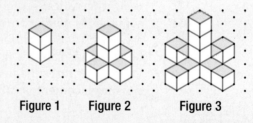

Figure 1   Figure 2   Figure 3

Presenting Your Investigation

Here are some ideas to help you present your conclusions to the class.

• Make a brochure showing the sequences you wrote in this investigation. Include your tables, formulas, and any diagrams.

• Describe a situation that results in a sequence similar to those in this investigation. Make a poster showing tables, formulas, and diagrams for this sequence.

*inter*NET
CONNECTION   **Investigation** For more information on sequences, visit: www.algconcepts.glencoe.com

What You'll Learn

You'll learn to determine whether a given number is a solution of an equation.

Why It's Important

Health Knowing how to find the solution of an equation can help you determine your normal blood pressure. *See Example 3.*

In a recent year, Americans ate 26 million hot dogs in major league ballparks. The graph shows American's favorite hot dog toppings. Note that each of the sentences below is either true or false.

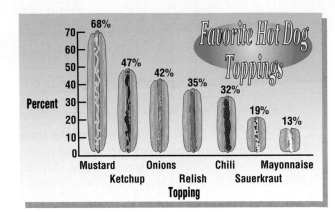

Source: ICR Survey Research Group

Statement	True or False?
The least favorite topping in the survey is mayonnaise.	true
Ketchup is the favorite topping in the survey.	false
More people in the survey prefer onions than chili on their hot dog.	true

A **statement** is any sentence that is either true or false, but not both. Look at the sentences below. Are they true or false?

Words	Symbols
A number *m* minus 5 is equal to 12.	$m - 5 = 12$
20 equals a number *y* plus 4.	$20 = y + 4$
Eight increased by a number *d* is 6.	$8 + d = 6$

Mathematical sentences like these are called **open sentences**. An open sentence is neither true nor false until the variable has been replaced by a value. A set of numbers from which replacements for a variable may be chosen is called a **replacement set**. Finding the replacements for the variable that results in a true sentence is called **solving** the open sentence. This replacement is called a **solution** of the open sentence.

Examples

1 Find the solution of $m - 5 = 12$ if the replacement set is {15, 16, 17, 18}.

Value for *m*	$m - 5 = 12$	True or False?
15	$15 - 5 \stackrel{?}{=} 12$	false
16	$16 - 5 \stackrel{?}{=} 12$	false
17	$17 - 5 \stackrel{?}{=} 12$	true ✓
18	$18 - 5 \stackrel{?}{=} 12$	false

Since 17 makes the sentence $m - 5 = 12$ true, the solution is 17.

② Find the solution of $2n + 3 = 9$ if the replacement set is {2, 3, 4, 5}.

Value for n	$2n + 3 = 9$	True or False?
2	$2(2) + 3 \stackrel{?}{=} 9$	false
3	$2(3) + 3 \stackrel{?}{=} 9$	true ✓
4	$2(4) + 3 \stackrel{?}{=} 9$	false
5	$2(5) + 3 \stackrel{?}{=} 9$	false

Therefore, the solution of $2n + 3 = 9$ is 3.

Your Turn

a. Find the solution of $3n - 2 = 13$ if the replacement set is {4, 5, 6, 7}.
b. Find the solution of $-2x + 1 = 7$ if the replacement set is {-3, -2, 2, 3}.

Open sentences and replacement sets are used frequently in real-life situations.

Example

Health Link

Real World

③ The normal systolic blood pressure p of a woman who is A years old is given by the formula $p = \dfrac{A(A + 5)}{100} + 107$. Ms. Adams recently had her annual physical. Her blood pressure was normal. If Ms. Adams' blood pressure was 121, how old is she: 34, 35, 36, or 37?

You need to find the replacement for the variable that results in a true sentence. First, replace p with 121.

$$121 = \frac{A(A + 5)}{100} + 107$$

Then replace A in the formula with each of the replacements.

Value for A	$121 = \dfrac{A(A + 5)}{100} + 107$	True or False?
34	$121 \stackrel{?}{=} \dfrac{34(34 + 5)}{100} + 107$	false
35	$121 \stackrel{?}{=} \dfrac{35(35 + 5)}{100} + 107$	true ✓
36	$121 \stackrel{?}{=} \dfrac{36(36 + 5)}{100} + 107$	false
37	$121 \stackrel{?}{=} \dfrac{37(37 + 5)}{100} + 107$	false

The solution is 35. So, Ms. Adams is 35 years old.

Blood pressure cuff

Look Back

Equations:
Lesson 1–1

Recall that an open sentence containing an equals sign, =, is an *equation*. Sometimes, you can solve an equation by simply applying the order of operations.

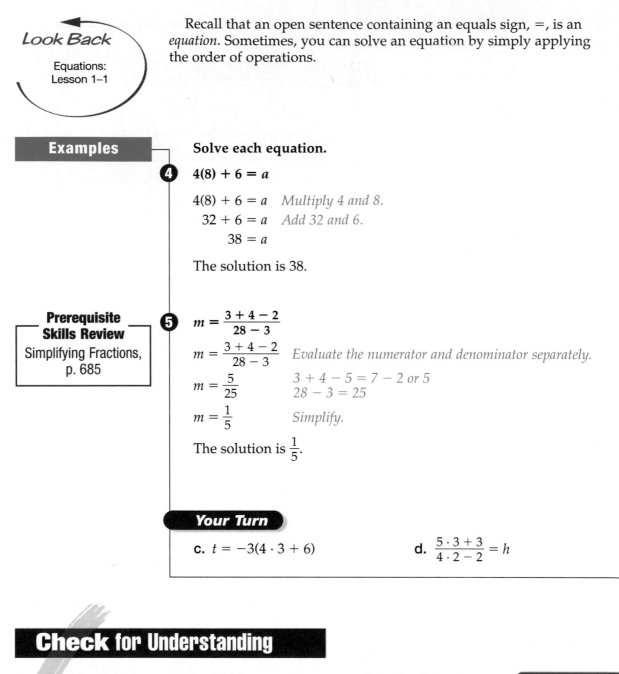

Examples

Solve each equation.

4 $4(8) + 6 = a$

$4(8) + 6 = a$ *Multiply 4 and 8.*
$32 + 6 = a$ *Add 32 and 6.*
$38 = a$

The solution is 38.

Prerequisite Skills Review
Simplifying Fractions, p. 685

5 $m = \dfrac{3 + 4 - 2}{28 - 3}$

$m = \dfrac{3 + 4 - 2}{28 - 3}$ *Evaluate the numerator and denominator separately.*

$m = \dfrac{5}{25}$ $\quad\begin{aligned}3 + 4 - 5 &= 7 - 2 \text{ or } 5\\ 28 - 3 &= 25\end{aligned}$

$m = \dfrac{1}{5}$ *Simplify.*

The solution is $\dfrac{1}{5}$.

Your Turn

c. $t = -3(4 \cdot 3 + 6)$

d. $\dfrac{5 \cdot 3 + 3}{4 \cdot 2 - 2} = h$

Check for Understanding

Communicating Mathematics

Study the lesson. Then complete the following.

1. **Describe** the difference between a statement and an open sentence.

2. **Define** the term *solution*.

3. **Explain** how to find the solution of $2n - 3 = 25$ if the replacement set is {12, 13, 14, 15}.

Vocabulary

statement
open sentences
replacement set
solving
solution

Guided Practice

Find the solution of each equation if the replacement sets are $a = \{5, 6, 7\}$, $b = \{-1, 0, 1\}$, and $c = \{-2, -1, 0, 1\}$. *(Examples 1 & 2)*

4. $6 = c + 8$

5. $2a + 5 = 17$

6. $\dfrac{3 + 15}{6} = 3b$

Solve each equation. *(Examples 4 & 5)*

7. $12 - 0.3 = y$ **8.** $p = 5(10) + 2 - 4$ **9.** $n = \dfrac{24 \div 8 + 1}{15 \div 3}$

10. Health   The normal systolic blood pressure p of a man who is A years old is given by the formula $p = \dfrac{A(3A - 10)}{500} + 120$. How old is Mr. Nichols if his normal blood pressure is 122: 18, 19, 20, or 21? *(Example 3)*

Exercises   • • • •   • • • • • • • • • • • • • • •

Practice

Find the solution of each equation if the replacement sets are
$x = \{2, 3, 4, 5\}$, $m = \{-5, -4, -3\}$, **and** $d = \{-1, 0, 1, 2\}$.

11. $x + 5 = 9$ **12.** $2d = -2$ **13.** $8 - m = 11$

14. $-11 = -7 + m$ **15.** $4x + (-2) = 10$ **16.** $9d - 20 = -20$

17. $6x + 7 = 31$ **18.** $-23 = 3d - 4(5)$ **19.** $3 - m = 8$

20. $\dfrac{3 + 15}{m} - 6 = -12$ **21.** $\dfrac{6d}{4} + 3 = 3d$ **22.** $\dfrac{2(x - 2)}{3} = \dfrac{4}{7 - 5}$

Solve each equation.

23. $6 \cdot 7 = y$ **24.** $g = 14.8 - 3.75$ **25.** $h = 5 + 2 \cdot 3$

26. $-7.7 - 5.2 = k$ **27.** $3 + 18 \div 6 = w$ **28.** $m = -2(5) + 4 \cdot 5$

29. $n = (18 \div 6) - 4$ **30.** $20 - 3 \cdot 3 - 14 = m$ **31.** $7 \cdot 3 - 15 \div 5 = b$

32. $\dfrac{7 \cdot 5 + 5}{(9 \cdot 3) - 7} = p$ **33.** $a = \dfrac{14 + 7}{2 \cdot 2}$ **34.** $\dfrac{4 \cdot 7 - 4}{2 \div 2 + 5} = d$

35. Find the solution of $2(7) - 12 = v$.

36. What is the value of g if $g = \dfrac{24 \div 2}{6}$?

Applications and Problem Solving

37. Recreation   At Sinclair's Balloon Rental, the cost of renting a hot air balloon is given by the formula $C = 85 + 36h$, where C is the cost in dollars and h is the number of hours. Michael and Keisha paid a total of $301 for a balloon ride. Did they rent the balloon for 5, 6, 7, or 8 hours?

38. Geometry   To find the area of a trapezoid, you can use the formula $A = \frac{1}{2}h(a + b)$, where h is the height of the trapezoid and a and b are the lengths of the bases. If a trapezoid has an area of 35 square feet and bases of 4 feet and 10 feet, is the height 3, 4, 5, or 6 feet?

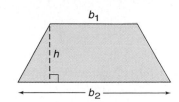

39. **Life Science** At birth, a baby blue whale weighs 4000 pounds and gains 200 pounds each day while nursing. The formula $W = 4000 + 200d$ represents this situation, where W is the weight of the baby and d is the number of days it has nursed. How many days has a baby blue whale been nursing if it weighs 17,000 pounds: 63, 64, or 65?

40. **Critical Thinking** Write four different open sentences that each have 3 as a solution.

Mixed Review

Find the mean, median, mode, and range of each set of data. *(Lesson 3–3)*

41. 38, 42, 46, 40, 40

42. 3.1, 3.2, 3.3, 3.4, 3.5, 3.6, 3.7, 3.8

Evaluate each expression if $x = -\frac{1}{3}$, $y = \frac{1}{2}$, and $z = 3\frac{1}{4}$. *(Lesson 3–2)*

43. $2 - x$

44. $y + z$

45. $x - y - z$

46. **Crafts** Thomas buys $\frac{2}{3}$ yard of red ribbon, $\frac{3}{8}$ yard of plaid fabric, and $\frac{1}{2}$ yard of striped fabric. Write the lengths in order from least to greatest. *(Lesson 3–1)*

47. **Standardized Test Practice** Which of the following is equal to the expression $[(8 - 7) - (6 - 5)] - [(8 - 7 - 6) - 5]$? *(Lesson 2–4)*

A -31 B -11 C 0 D 10

Quiz 1   Lessons 3–1 through 3–4

▶ Replace each ● with <, >, or = to make a true sentence. *(Lesson 3–1)*

1. -16.3 ● -16.03

2. $3(-4) + 7$ ● $-12 + 6$

3. $\frac{2}{3}$ ● $\frac{4}{6}$

4. Which is the better buy, a 12-pack of soda for $2.79 or a 24-pack of soda for $5.69? *(Lesson 3–1)*

5. Find $3.54 - (-4.78) + 1.02 + (-7.65)$. *(Lesson 3–2)*

Find the mean, median, mode, and range for each set of data. *(Lesson 3–3)*

6. 45, 12, 67, 22, 34

7.
Stem	Leaf
1	1 3 7
2	1 6 7
3	0 0 4 6
4	4 7 $1|1 = 1.1$

8.

```
                          ×
                       ×  ×
           ×  ×  ×  ×  ×
         ┼──┼──┼──┼──┼──┼──→
           6  7  8  9  10
```

9. Find the solution of $3d - 7 = -13$ if the replacement set is $\{-4, -3, -2, -1\}$. *(Lesson 3–4)*

10. Solve $m = \frac{2 + 4 \cdot 6}{3 + 10} - 8$. *(Lesson 3–4)*

Extra Practice   See p. 697.

3-5 Solving Equations by Using Models

Math In the Workplace

What You'll Learn

You'll learn to solve addition and subtraction equations by using models.

Why It's Important

Zoology Knowing how to solve equations can help in finding unknown values.
See Example 3.

Did you know that the largest bear is the Alaskan brown bear? When fully grown, it has a height of 9 feet. The smallest bear is the sun bear. The Alaskan brown bear is 6 feet longer than the sun bear. What is the height of the sun bear? *This problem will be solved in Example 3.*

The height of the sun bear can be found by solving an equation. You can use algebra tiles to solve equations.

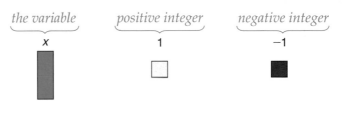

the variable   *positive integer*   *negative integer*
x 1 -1

Examples

Use algebra tiles to solve each equation.

1 $x + (-2) = -7$

Step 1 Model $x + (-2) = -7$ by placing 1 green tile and 2 red square tiles on one side of the mat to represent $x + (-2)$. Place 7 red square tiles on the other side to represent -7.

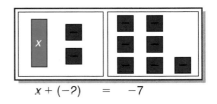

$x + (-2) = -7$

Step 2 To get the green tile by itself, remove 2 red square tiles from each side.

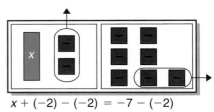

$x + (-2) - (-2) = -7 - (-2)$

Step 3 The green tile on the left side of the mat is matched with 5 red square tiles. Therefore, $x = -5$.

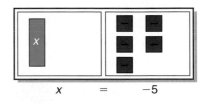

$x = -5$

(continued on the next page)

2 $x - 3 = 5$

Step 1 Write the equation in the form $x + (-3) = 5$. Place 1 green tile and 3 red square tiles on one side of the mat to represent $x + (-3)$. Place 5 yellow square tiles on the other side of the mat to represent $+5$.

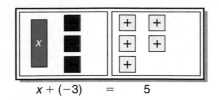

$x + (-3) \qquad = \qquad 5$

Step 2 To get the green tile by itself, you need to remove 3 red square tiles from each side. Since there are no red square tiles on the right side of the mat, you will need to add 3 yellow square tiles to each side to make 3 zero pairs on the left side of the mat.

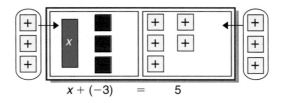

$x + (-3) \qquad = \qquad 5$

Step 3 Remove the zero pairs.

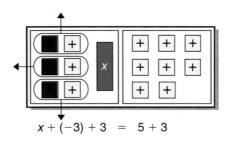

$x + (-3) + 3 \quad = \quad 5 + 3$

Step 4 The green tile is matched with 8 yellow square tiles. Therefore, $x = 8$.

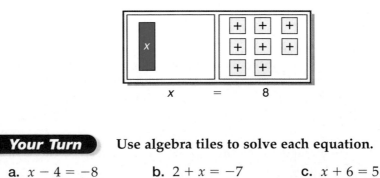

$x \qquad = \qquad 8$

Your Turn   Use algebra tiles to solve each equation.

a. $x - 4 = -8$ **b.** $2 + x = -7$ **c.** $x + 6 = 5$

Alaskan Brown Bear

In Examples 1 and 2, you solved each equation for the variable. This means you isolated the variable on one side of the equation.

Example

Zoology Link

3 **Refer to the application at the beginning of the lesson. Find the length of the sun bear.**

Explore   You know that the Alaskan brown bear is 6 feet longer than the sun bear. You need to find the length of the sun bear.

Plan   Let x represent the length of the sun bear. Translate the problem into an equation using the variable x. Model the equation and then solve.

Solve

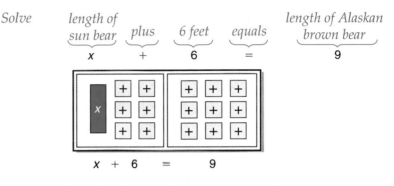

length of sun bear	plus	6 feet	equals	length of Alaskan brown bear
x	$+$	6	$=$	9

$$x + 6 = 9$$

To solve for x, subtract 6 from each side.

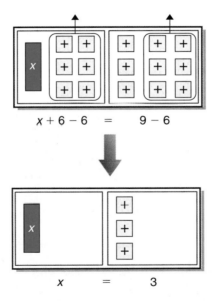

$$x + 6 - 6 = 9 - 6$$

$$x = 3$$

The solution is 3. So, the length of the sun bear is 3 feet.

Examine   Check the solution.

The length of the Alaskan brown bear is equal to the length of the sun bear plus 6 feet. The length of the sun bear is 3 feet, and 3 feet plus 6 feet is 9 feet. ✓

The answer is correct.

Sun Bear

Check for Understanding

Communicating Mathematics

Study the lesson. Then complete the following.

1. **Show** how to model $6 + x = -2$ using algebra tiles.

2. **Explain** why the model below represents $x - (-8) = -6$.

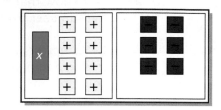

Guided Practice

⏱ Getting Ready **Write an equation for each model.**

Sample 1:

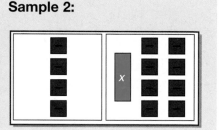

Solution: $x + 7 = -5$

Sample 2:

Solution: $-4 = x + (-8)$

3.

4.

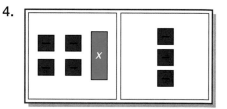

5.

6.

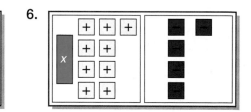

Solve each equation. Use algebra tiles if necessary. *(Examples 1 & 2)*

7. $m + 2 = -2$ **8.** $3 = n - 6$ **9.** $t - 4 = -8$

10. $y + (-3) = -1$ **11.** $4 = x - 2$ **12.** $-5 + c = -5$

13. Find the value of h if $-4 = h - 7$.

14. Sales Jeff Simons sold 27 cars last month. This is 36 fewer cars than he sold during the same time period one year ago. What were his sales one year ago? *(Example 3)*

   a. Write an equation that can be used to find Jeff Simons' sales one year ago.

   b. What were Jeff Simons' sales one year ago?

Exercises • • • •

Practice

Solve each equation. Use algebra tiles if necessary.

15. $-2 + y = -8$ **16.** $w - 1 = 3$ **17.** $n + 5 = -2$
18. $7 = q - 4$ **19.** $p - 6 = 2$ **20.** $3 = z + 10$
21. $-3 + k = -3$ **22.** $4 = g + 6$ **23.** $2 + f = -1$
24. $10 = 8 + b$ **25.** $-2 = d + (-3)$ **26.** $x + 6 = -23$
27. $-7 = a + (-9)$ **28.** $k - 7 = -8$ **29.** $p - 9 = 7$
30. $-12 + r = -2$ **31.** $v + 9 = 2$ **32.** $-13 = c + (-4)$

33. What is the value of m if $8 + m = -13$?

34. If $a + (-14) = -16$, what is the value of a?

35. When 6 is subtracted from d, the result is 5. Find the value of d.

Applications and Problem Solving

Bobby Labonte

36. Sports   The 1998 NASCAR season consisted of 33 races. Out of the 33 races, driver Jeff Gorden finished in one of the top five positions 26 times. This number is 15 more than the number of times driver Bobby Labonte finished in the top five positions.   **Source:** NASCAR

  a. Write an equation that represents this situation.

  b. How many times did Bobby Labonte finish in one of the top five positions?

37. Weather   Jaclyn needs to record the outside temperature for her science project. At 6:00 P.M., the temperature was 8°F. The difference between the temperature she recorded in the morning and the evening temperature was −2°F.

  a. Write an equation that Jaclyn could use to determine the morning temperature.

  b. What was the morning temperature?

38. Critical Thinking   Explain how algebra tiles can be used to solve the equation $2n + 1 = 7$.

Mixed Review

Find the solution of each equation if the replacement set is $x = \{-3, -2, -1, 0, 1, 2, 3\}$.   *(Lesson 3–4)*

39. $18 + x = 20$ **40.** $x \div 9 = 0$ **41.** $5x - 7 = -2$

42. Sports   The stem-and-leaf plot shows Tanya's bowling scores for the past eight games. Find Tanya's mean bowling score.   *(Lesson 3–3)*

Stem	Leaf
9	0 1
10	4 5 8
11	5 6 9 $10\,|\,5 = 105$

43. Find the product of −9 and −6.   *(Lesson 2–5)*

44. Simplify $(4m - 5n) + (3m + 3n)$.   *(Lesson 1–4)*

45. Identify the property shown by $24 + 7 = 7 + 24$.   *(Lesson 1–3)*

46. Standardized Test Practice   Choose the expression that represents *5 subtracted from the sum of a number and 2.*   *(Lesson 1–1)*

  A $5 - x + 2$ **B** $x + 2 - 5$

  C $2x - 5$ **D** $5 - 2x$

Extra Practice   See p. 697.

What You'll Learn

You'll learn to solve addition and subtraction equations by using the properties of equality.

Why It's Important

History Knowing how to solve equations can help you find important dates in history. *See Exercise 17.*

The teams that make it to the semi-finals of the NCAA basketball championships are called the Final Four. The graph shows the schools with the most Final Four appearances. Notice that North Carolina and UCLA have each reached the Final Four 14 times.

The Men's Final Four

North Carolina	14
UCLA	14
Kentucky	13
Duke	13
Kansas	10
Ohio State	10

Source: NCAA

If North Carolina and UCLA both reach the Final Four in each of the next 3 years, they would still have an equal number of Final Four appearances.

$$
\begin{array}{cc}
NC & UCLA \\
14 = 14 & \\
14 + 3 = 14 + 3 & \text{\textit{Add 3 to each side.}} \\
17 = 17 &
\end{array}
$$

This example illustrates the **Addition Property of Equality**.

Addition Property of Equality	**Words:**	If you add the same number to each side of an equation, the two sides remain equal.
	Symbols:	For any numbers a, b, and c, if $a = b$, then $a + c = b + c$.
	Numbers:	If $x = 2$, then $x + 3 = 2 + 3$.

Note that c can be positive, negative, or 0. Look at the equations below. In the equation on the left, $c = 3$. In the equation on the right, $c = -3$.

$$a + c = b + c \qquad\qquad\qquad a + c = b + c$$
$$15 + 3 = 15 + 3 \qquad\qquad 15 + (-3) = 15 + (-3)$$

When the same number is added to each side of an equation, the result is an **equivalent equation**. Equivalent equations have the same solution.

$$x + 2 = 6 \qquad \text{\textit{The solution of the equation is 4.}}$$

$$x + 2 + 3 = 6 + 3 \qquad \text{\textit{Using the Addition Property of Equality, add 3 to each side.}}$$

$$x + 5 = 9 \qquad \text{\textit{The solution of this equation is also 4.}}$$

You can use the Addition Property of Equality to solve equations. Model the equation $x - 3 = 5$ using algebra tiles. Each side of the mat represents a side of the equation. When you add 3 tiles to each side, the result is an equivalent equation, $x = 8$.

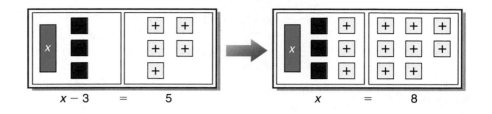

$$x - 3 \quad = \quad 5 \qquad\qquad x \quad = \quad 8$$

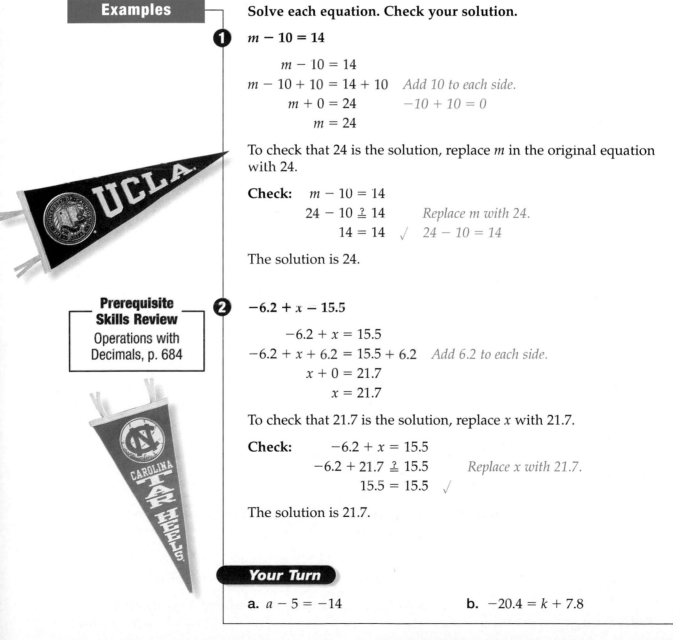

Prerequisite Skills Review
Operations with Decimals, p. 684

Examples

Solve each equation. Check your solution.

1 $m - 10 = 14$

$$m - 10 = 14$$
$$m - 10 + 10 = 14 + 10 \quad \textit{Add 10 to each side.}$$
$$m + 0 = 24 \qquad \textit{-10 + 10 = 0}$$
$$m = 24$$

To check that 24 is the solution, replace m in the original equation with 24.

Check: $m - 10 = 14$
$$24 - 10 \overset{?}{=} 14 \quad \textit{Replace m with 24.}$$
$$14 = 14 \quad \checkmark \quad \textit{24 - 10 = 14}$$

The solution is 24.

2 $-6.2 + x - 15.5$

$$-6.2 + x = 15.5$$
$$-6.2 + x + 6.2 = 15.5 + 6.2 \quad \textit{Add 6.2 to each side.}$$
$$x + 0 = 21.7$$
$$x = 21.7$$

To check that 21.7 is the solution, replace x with 21.7.

Check: $-6.2 + x = 15.5$
$$-6.2 + 21.7 \overset{?}{=} 15.5 \quad \textit{Replace x with 21.7.}$$
$$15.5 = 15.5 \quad \checkmark$$

The solution is 21.7.

Your Turn

a. $a - 5 = -14$

b. $-20.4 = k + 7.8$

In addition to the Addition Property of Equality, there is also a
Subtraction Property of Equality.

Subtraction Property of Equality	**Words:**	If you subtract the same number from each side of an equation, the two sides remain equal.
	Symbols:	For any numbers a, b, and c, if $a = b$, then $a - c = b - c$.
	Numbers:	If $x = 5$, then $x - 3 = 5 - 3$.

Example **3** Solve $9 + a = -3$. Check your solution.

$$9 + a = -3$$
$$9 + a - 9 = -3 - 9 \quad \textit{Subtract 9 from each side.}$$
$$a = -12$$

Check: $9 + a = -3$
$9 + (-12) \stackrel{?}{=} -3$ *Replace a with −12.*
$-3 = -3 \quad \checkmark$ *Add −12 to 9.*

The solution is -12.

Sometimes the Subtraction Property of Equality can be used instead of
the Addition Property of Equality. Recall that subtracting a number is the
same as adding its inverse.

Example **4**

Art Link

Real World

**Romanian Constantin Brancusi and American David Smith are
successful twentieth-century sculptors. Brancusi's *Bird in Space* and
Smith's *Cubi XVIII* were created in 1918 and 1964, respectively. Use
the equation $1918 + y = 1964$ to find the number of years between
these two sculptures.**

Method 1: Use the Subtraction Property of Equality.

$$1918 + y = 1964$$
$$1918 + y - 1918 = 1964 - 1918 \quad \textit{Subtract 1918 from each side.}$$
$$y = 46$$

Check: $1918 + y = 1964$
$1918 + 46 \stackrel{?}{=} 1964$ *Replace y with 46.*
$1964 = 1964 \quad \checkmark$

Method 2: Use the Addition Property of Equality.

$$1918 + y = 1964$$
$$1918 + y + (-1918) = 1964 + (-1918) \quad \textit{Add −1918 to each side.}$$
$$y = 46 \quad \textit{The answers are the same.}$$

The solution is 46. So, the sculptures were created 46 years apart.

Bird in Space

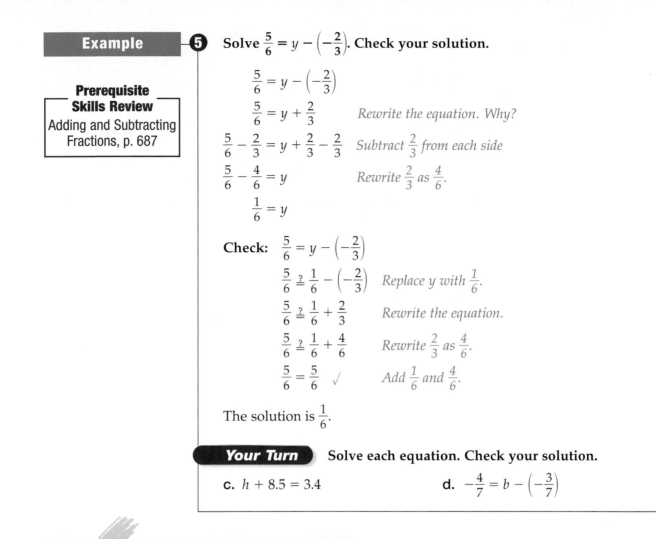

Example

5 Solve $\frac{5}{6} = y - \left(-\frac{2}{3}\right)$. Check your solution.

$$\frac{5}{6} = y - \left(-\frac{2}{3}\right)$$

$$\frac{5}{6} = y + \frac{2}{3} \qquad \textit{Rewrite the equation. Why?}$$

$$\frac{5}{6} - \frac{2}{3} = y + \frac{2}{3} - \frac{2}{3} \qquad \textit{Subtract } \frac{2}{3} \textit{ from each side}$$

$$\frac{5}{6} - \frac{4}{6} = y \qquad \textit{Rewrite } \frac{2}{3} \textit{ as } \frac{4}{6}.$$

$$\frac{1}{6} = y$$

Check: $\frac{5}{6} = y - \left(-\frac{2}{3}\right)$

$$\frac{5}{6} \stackrel{?}{=} \frac{1}{6} - \left(-\frac{2}{3}\right) \qquad \textit{Replace } y \textit{ with } \frac{1}{6}.$$

$$\frac{5}{6} \stackrel{?}{=} \frac{1}{6} + \frac{2}{3} \qquad \textit{Rewrite the equation.}$$

$$\frac{5}{6} \stackrel{?}{=} \frac{1}{6} + \frac{4}{6} \qquad \textit{Rewrite } \frac{2}{3} \textit{ as } \frac{4}{6}.$$

$$\frac{5}{6} = \frac{5}{6} \quad \checkmark \qquad \textit{Add } \frac{1}{6} \textit{ and } \frac{4}{6}.$$

The solution is $\frac{1}{6}$.

Prerequisite Skills Review
Adding and Subtracting Fractions, p. 687

Your Turn Solve each equation. Check your solution.

c. $h + 8.5 = 3.4$

d. $-\frac{4}{7} = b - \left(-\frac{3}{7}\right)$

Check for Understanding

Communicating Mathematics

Study the lesson. Then complete the following.

1. **Explain** how you can show that -4 is a solution of $12 + m = 8$.

2. **Write** two equivalent equations.

3. **Complete** the following statement. Justify your answer.
   If $x - 5 = 12$, then $x - 2 = $ ___?___ .

4.  Malcom thinks that $y - (-24) = -36$ is equivalent to $y + (-24) = -36$. Adita disagrees. Who is correct, and why?

Vocabulary
equivalent equations

Guided Practice

⏱ **Getting Ready** State the number you would add to or subtract from each side of the equation to solve it.

Sample 1: $c + 16 = 14$ **Solution:** Subtract 16 from each side.
Sample 2: $-3 + h = -13$ **Solution:** Add 3 to each side.

5. $y + 21 = -7$ 6. $-10 + p = 25$ 7. $34 = d + (-9)$
8. $14 = 16 + s$ 9. $u - (-9.3) = -53.1$ 10. $8 = a + (-5)$

Solve each equation. Check your solution. *(Examples 1–2 & 4–5)*

11. $h + 8 = -7$ **12.** $4 = -5 + d$ **13.** $2.75 + x = 3.85$

14. $x + 12 = -32$ **15.** $h - \frac{1}{2} = -\frac{2}{3}$ **16.** $8 = x - (-12)$

17. History The state of Kentucky entered the union in 1792. Oregon entered the union 67 years later. *(Example 3)*

  a. Write an equation that could be used to find the year Oregon entered the union.

  b. When did Oregon enter the union?

Exercises •

Practice

Solve each equation. Check your solution.

18. $k + (-8) = 29$ **19.** $y - (-7) = 3$ **20.** $-9 = 8 + q$

21. $b - 14 = -11$ **22.** $-21 = 5 + c$ **23.** $-3 + m = 10$

24. $27 + y = -49$ **25.** $7 + w = -6$ **26.** $11 = r + (-9)$

27. $-15 + y = -63$ **28.** $-39 = n - 24$ **29.** $52 = 18 + j$

30. $k - 14.6 = 9.2$ **31.** $3.4 = m + 7.2$ **32.** $-3.41 + n = 0.23$

33. $\frac{5}{6} = g - \left(-\frac{1}{6}\right)$ **34.** $\frac{3}{4} + y = \frac{7}{8}$ **35.** $g + \left(-\frac{4}{9}\right) = \frac{2}{3}$

36. Solve for s if $7 + s = -10$.

37. What is the solution of the equation $g + (-2) = 11$?

Write an equation and solve.

38. A number increased by -22 is 45. Find the number.

39. What number decreased by -15 is 20?

40. The sum of a number and 8 is -47. What is the number?

Applications and Problem Solving

41. Hardware Today, the cost of a 50-pound box of nails is about $32.50. This is $27.16 more than the cost of the same size box of nails sold in 1800.

Source: *Family Handyman*

  a. Write an equation that could be used to find the cost of a 50-pound box of nails sold in 1800.

  b. In 1800, how much did a 50-pound box of nails cost?

42. Lighthouses There are an estimated 850 lighthouses in the United States. The tallest is located in Cape Hatteras, North Carolina. It is 27 feet taller than the one located in Cape Lookout, North Carolina. The height of the Cape Lookout lighthouse is 169 feet.

   a. Write an equation that could be used to find the height of the lighthouse located in Cape Hatteras, North Carolina.

   b. What is the height of the lighthouse?

43. Critical Thinking What value of x makes $x + x = x$ true?

Mixed Review

44. Sports On the first day of an LPGA golf tournament, professional golfer Pearl Sinn's score was $+4$ or 4 over par. Her score on the second day was added to the first day's score, and the total was $+1$ or 1 over par. What was Sinn's score on the second day? *(Lesson 3–5)*

   a. Let s represent Sinn's score on the second day. Then translate the problem into an equation using the variable s.

   b. Solve for s to find Sinn's score on the second day.

Solve each equation. *(Lesson 3–4)*

45. $3.75 - 0.5 = m$

46. $x = -4(8) + 6$

47. $(-15 \div -3) + 1 = y$

48. $t = [3 - (-5)] \cdot 7$

Pearl Sinn

49. Statistics Find the range of the set of data $5.25, $4.39, $1.36, $8.12, $2.90. *(Lesson 3–3)*

50. Find the quotient of 36 and -4. *(Lesson 2–6)*

51. Standardized Test Practice Choose the set of numbers that is correctly ordered from greatest to least. *(Lesson 2–1)*

   A $\{-3, 2, 1, 0\}$ **B** $\{5, -4, 0, -1\}$

   C $\{-4, -3, -2, 0\}$ **D** $\{7, 4, -1, -3\}$

Quiz 2   Lessons 3–5 and 3–6

Solve each equation. Check your solution. *(Lessons 3–5 & 3–6)*

1. $y - 3 = -7$

2. $6 = x + 9$

3. $17.8 = s + (-5.2)$

4. $-\dfrac{5}{6} = p - \dfrac{1}{2}$

5. Dining In Sacramento, California, the average cost of a fast-food meal was $3.89 in a recent year. This was 62 cents less than the average cost of a fast-food meal in Chicago, Illinois. What was the average price of a fast-food meal in Chicago? **Source:** Runzheimer International, 1999

   a. Choose a variable and write an equation that can be used to find the average cost of a fast-food meal in Chicago.

   b. Find the average cost of a fast-food meal in Chicago.

3-7 Solving Equations Involving Absolute Value

Hydrogen can exist as a solid, liquid, or gas. For hydrogen to be a liquid, its temperature must be within 2° of −257°C. What are the least and greatest temperatures for hydrogen to remain a liquid? *This problem will be solved in Example 4.*

What You'll Learn

You'll learn to solve equations involving absolute value.

Why It's Important

Science Knowing how to solve equations involving absolute value can help you determine greatest and least values. *See Example 4.*

Some equations involve absolute value. Consider the graph of the equation $|a| = 4$.

$$|a| = 4$$

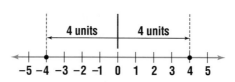

The distance from 0 to a is 4 units. Therefore, $a = -4$ or $a = 4$. When an equation has more than one solution, the solutions are often written as a set $\{a, b\}$. The solution set is $\{-4, 4\}$.

Equations that involve absolute value can be solved by using a number line or by writing them as compound sentences and solving them.

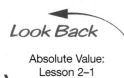

Look Back

Absolute Value: Lesson 2–1

Example

1 **Solve $|x - 3| = 5$. Check your solution.**

Method 1: Use a graph.

$|x - 3| = 5$ means the distance between x and 3 is 5 units. So, to find x on the number line, start at 3 and move 5 units in either direction.

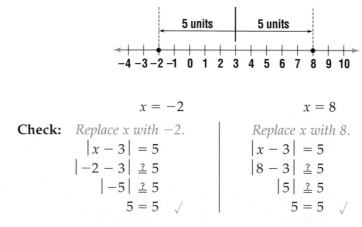

$$x = -2 \qquad\qquad x = 8$$

Check: *Replace x with −2.*   *Replace x with 8.*

$$|x - 3| = 5 \qquad\qquad |x - 3| = 5$$
$$|-2 - 3| \stackrel{?}{=} 5 \qquad\qquad |8 - 3| \stackrel{?}{=} 5$$
$$|-5| \stackrel{?}{=} 5 \qquad\qquad |5| \stackrel{?}{=} 5$$
$$5 = 5 \;\checkmark \qquad\qquad 5 = 5 \;\checkmark$$

The solution set is $\{-2, 8\}$.

Method 2: Write and solve a compound sentence.

$|x - 3| = 5$ also means $x - 3 = 5$ or $x - 3 = -5$.

$$x - 3 = 5 \qquad\qquad \text{or} \qquad\qquad x - 3 = -5$$
$$x - 3 + 3 = 5 + 3 \quad \textit{Add 3 to each side.} \quad x - 3 + 3 = -5 + 3$$
$$x = 8 \qquad\qquad\qquad\qquad\qquad x = -2$$

The answers for both methods are the same. This verifies the solution.

Your Turn

a. $|c - 4| = 2$

b. $6 = |5 + h|$

When solving an equation involving an absolute value symbol, you can think of the absolute value bars as grouping symbols.

Example

2 **Solve $|a + 6| + 5 = 12$. Check your solution.**

To solve the equation, first simplify the expression.

$$|a + 6| + 5 = 12$$
$$|a + 6| + 5 - 5 = 12 - 5 \quad \textit{Subtract 5 from each side.}$$
$$|a + 6| = 7$$

Next, write a compound sentence and solve it.

$$a + 6 = 7 \qquad\qquad \text{or} \qquad\qquad a + 6 = -7$$
$$a + 6 - 6 = 7 - 6 \quad \textit{Subtract 6 from each side.} \quad a + 6 - 6 = -7 - 6$$
$$a = 1 \qquad\qquad\qquad\qquad\qquad a = -13$$

Check: *Replace a with 1.* | *Replace a with −13.*

$$|a + 6| + 5 = 12 \qquad\qquad\qquad |a + 6| + 5 = 12$$
$$|1 + 6| + 5 \stackrel{?}{=} 12 \qquad\qquad |-13 + 6| + 5 \stackrel{?}{=} 12$$
$$|7| + 5 \stackrel{?}{=} 12 \qquad\qquad\qquad |-7| + 5 \stackrel{?}{=} 12$$
$$7 + 5 \stackrel{?}{=} 12 \qquad\qquad\qquad\quad 7 + 5 \stackrel{?}{=} 12$$
$$12 = 12 \; \checkmark \qquad\qquad\qquad\quad 12 = 12 \; \checkmark$$

The solution set is $\{1, -13\}$.

Your Turn **Solve each equation. Check your solution.**

c. $|m + 5| - 4 = 18$

d. $13 = |-8 + d| + 2$

Sometimes an equation has no solution. For example, $|x| = -6$ is never true. The absolute value of a number is always positive or zero. So, there is no replacement for x that will make the sentence true. The solution set has no members. It is called the **empty set** and is symbolized by $\{ \}$ or $\varnothing$.

Example ─ **3**   Solve $|d| + 7 = 2$. Check your solution.

First, simplify the expression.

$$|d| + 7 = 2$$
$$|d| + 7 - 7 = 2 - 7 \quad \textit{Subtract 7 from each side.}$$
$$|d| = -5$$

This sentence can never be true. The solution is the empty set or $\varnothing$.

Your Turn

e. $3 = |s - 4| + 9$

f. $|w| - 18 = -6$

You can use equations that involve absolute value to solve real-world problems.

Example   **4**   **Refer to the application at the beginning of the lesson. Write and then solve an equation that could be used to find the least and greatest temperatures at which hydrogen is a liquid.**

Science Link

Let t represent temperature. Then t differs from -257 by exactly 2 degrees. Write an equation that represents the least and greatest temperatures.

The difference between t and -257   *is*   *2 degrees.*

$$|t - (-257)| \qquad\qquad = \qquad 2$$

$$t + 257 = 2 \qquad \textit{Rewrite the equation.} \qquad t + 257 = -2$$
$$t + 257 - 257 = 2 - 257 \qquad \textit{Subtract 257.} \qquad t + 257 - 257 = -2 - 257$$
$$t = -255 \qquad\qquad\qquad\qquad t = -259$$

The solutions are -255 and -259. So, the least temperature for hydrogen to remain a liquid is $-259°C$, and the greatest temperature is $-255°C$.

Check for Understanding

Communicating Mathematics

Study the lesson. Then complete the following.

1. Write an absolute value equation that has no solution.

2. Draw a graph that can be used to show the solution of $|d - 2| = 1$.

Math Journal

3. Write a few sentences explaining why an absolute value equation can have two solutions.

┌─ *Vocabulary* ─┐
│   empty set   │
└───────────────┘

Guided Practice

Solve each equation. Check your solution.   *(Examples 1–3)*

4. $|a| = 6$

5. $|w| - 2 = 1$

6. $|x + 1| = 4$

7. $-5 = |w - 9|$

8. $|3 + g| = 7$

9. $|h - 3| + 6 = 16$

10. The absolute value of the quantity 6 plus a number is 2. What is the number? *(Example 3)*

11. **Sports** Ricardo's bowling score was within 6 points of his average score of 145. Write and then solve an equation that could be used to find the least and greatest bowling score Ricardo could have received in a game. *(Example 4)*

Exercises •

Practice

Solve each equation. Check your solution.

12. $|\ell| = -8$
13. $|b| = 2$
14. $4 + |x| = 7$
15. $|y| - 10 = -4$
16. $|x - 6| = 8$
17. $|10 + n| = -2$
18. $1 = |2 + p|$
19. $|-1 + y| = 0$
20. $8 = |-9 + q|$
21. $|s| + 12 = 6$
22. $-9 = |2 + h|$
23. $11 = |10 + d|$
24. $17 = |y + (-3)|$
25. $|z - (-6)| = 13$
26. $9 = 4 + |k + 1|$
27. $|b - 2| + 5 = 16$
28. $6 + |-5 + f| = 18$
29. $|2 + r| + 4 = 1$

30. How many solutions exist for $|a + 3| = 6$?
31. How many solutions exist for $-2 = |3 + c|$?
32. How many solutions exist for $|4 + b| = 0$?

Applications and Problem Solving

For Exercises 33 and 34, write and solve an equation.

33. **Real Estate** Tamika Samuel is going to place an offer on a home. If she comes within $10,000 of the asking price of $165,000, she has a good chance of getting the home. Write and solve an equation to find out what her greatest and least offers should be.

34. **Shipping** The Jones Packaging Company needs to ship a crate of books to a book store. For shipping, the crate must weigh within 6 kilograms of 40 kilograms. If the crate alone weighs 8 kilograms, what are the greatest and least amounts the books can weigh?

35. **Critical Thinking** Which graph represents the solution set of the equation $1 = -2 + |c + 1|$?

A ⟵+─┼─●─┼─┼─┼─┼─┼─●─┼─┼─⟶
   −5−4−3−2−1 0 1 2 3 4 5

B ⟵┼─┼─┼─┼─┼─┼─┼─┼─●─┼─┼─⟶
   −5−4−3−2−1 0 1 2 3 4 5

C ⟵┼─┼─●─┼─●─┼─┼─┼─┼─┼─⟶
   −5−4−3−2−1 0 1 2 3 4 5

D ⟵┼─┼─●─┼─┼─┼─┼─┼─●─┼─⟶
   −5−4−3−2−1 0 1 2 3 4 5

Mixed Review

Solve each equation. Check your solution. *(Lesson 3–6)*

36. $13 + w = -6$
37. $t - 3 = 45$
38. $11.3 = h - 5.7$

39. What is the value of x if $9 + x = -5$? *(Lesson 3–5)*

40. If $y - (-8) = 12$, what is the value of y? *(Lesson 3–5)*

41. **Standardized Test Practice** A store stocks five different brands of cookies priced at $2.25, $2.50, $2.00, $2.25, and $1.85. What is the difference between the mode price and the mean price? *(Lesson 3–3)*
   A $0.00   B $0.08   C $2.25   D $4.42

Understanding and Using the Vocabulary

After completing this chapter, you should be able to define each term, property, or phrase and give an example or two of each.

*inter*NET
CONNECTION **Review Activities**
For more review activities, visit:
www.algconcepts.glencoe.com

empty set *(p. 129)*
equivalent equations *(p. 122)*
cross products *(p. 95)*
inequality *(p. 95)*
mean *(p. 104)*
measure of central tendency *(p. 104)*

measure of variation *(p. 106)*
median *(p. 104)*
mode *(p. 104)*
open sentences *(p. 112)*
range *(p. 106)*
rational numbers *(p. 94)*
replacement set *(p. 112)*

solution *(p. 112)*
solving *(p. 112)*
statement *(p. 112)*
unit cost *(p. 97)*

Choose the correct term to complete each sentence.

1. The (median, mean) is the middle number when data are arranged in numerical order.

2. The (range, unit cost) is used to compare the price of similar items.

3. The difference between the greatest and least values of a set is the (mode, range).

4. When the same number is subtracted from each side of an equation, the result is a(n) (replacement set, equivalent equation).

5. A (rational number, mean) can be expressed in the form $\frac{a}{b}$, where $b \neq 0$.

6. A solution set with no members is the (replacement set, empty set).

7. An (inequality, equivalent equation) uses the symbols $<$ and $>$ to compare two expressions.

8. (Measures of central tendency, measures of variation) describe the distribution of data.

9. A (solution, rational number) can always be expressed as a fraction.

10. You can use (cross products, open sentences) to compare two fractions with different denominators.

Skills and Concepts

Objectives and Examples	**Review Exercises**

• **Lesson 3–1** Compare and order rational numbers.

Replace the ● with $<$, $>$, or $=$ to make a true sentence.

$$5 \cdot 2 \; \bullet \; 3 \cdot 4$$

If $10 < 12$, then $\frac{2}{3} < \frac{4}{5}$.

Replace each ● with $<$, $>$, or $=$ to make a true sentence.

11. $-9 \; \bullet \; -11$ 12. $-13 \; \bullet \; 13$

13. $0.35 \; \bullet \; -3.5$ 14. $2.2 \; \bullet \; 2.20$

15. $\frac{3}{8} \; \bullet \; \frac{3}{4}$ 16. $-\frac{7}{4} \; \bullet \; -1.75$

17. Write $-1.1, \frac{1}{8}, 0.25, 0$ in order from least to greatest.

Chapter 3 Study Guide and Assessment

Objectives and Examples

- **Lesson 3–2** Add and subtract rational numbers.

  Find the difference.
  $$-0.3 - (-1.8) = -0.3 + 1.8$$
  $$= 1.5$$

Review Exercises

Find each sum or difference.

18. $-4.5 + (-8.1)$ 19. $3.7 - (-6.8)$

20. $\frac{8}{9} - \frac{1}{3}$ 21. $-\frac{13}{7} + \frac{6}{7}$

22. $-\frac{4}{3} + \frac{5}{6} + \left(-\frac{7}{3}\right)$

- **Lesson 3–3** Find the mean, median, mode, and range of a set of data.

  Find the mean, median, mode, and range of $2, 4, 5, 6, 6, 13, 14, 14, 14, 16,$ and 16.

  mean: $\frac{110}{11} = 10$

  median: The 6th or middle value is 13.
  mode: The most frequent number is 14.
  range: $16 - 2$ or 14

Find the mean, median, mode, and range of each set of data.

23. $6, 8, 6, 5, 5, 7, 9, 2, 7, 5, 3$

24. $8.6, 7.5, 9.9, 5.1, 7.1$

25.

Stem	Leaf
0	3 6 7 7 7 7 9
1	0 0 0 0 0 0 3 4 4 6 6 7 7 9 9
2	0 1
3	1 $3\,|\,1 = 31$

- **Lesson 3–4** Determine whether a given number is a solution of an equation.

  Find the solution of $x + (-3) = 2$ if the replacement set is $x = \{4, 5, 6\}$.

  The solution is 5, since $5 + (-3) = 2$.

Find the solution of each equation if the replacement sets are $x = \{-3, -2, -1\}$, $y = \{0, 2, 4\}$, and $z = \{1, 3, 5\}$.

26. $3x = -6$ 27. $5y - 4 = -4$

28. $\frac{3}{5}z = 3$ 29. $\frac{2+8}{x} + 9 = -1$

Solve each equation.

30. $a = 21 - 5$ 31. $5(6) - 3(-5) = w$

- **Lesson 3–5** Solve addition and subtraction equations using algebra tiles.

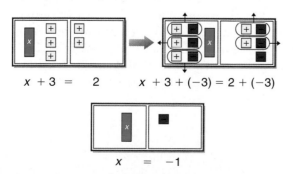

$x + 3 = 2$ $x + 3 + (-3) = 2 + (-3)$

$x = -1$

Solve each equation. Use algebra tiles if necessary.

32. $x + (-3) = 5$ 33. $4 = 6 + y$

34. $-7 + z = 3$ 35. $p + 2 = -2$

36. $s - (-6) = 1$ 37. $-1 = g + (-1)$

38. What is the value of h if $h + 4 = 2$?

39. A number increased by 4 is 11. Write an equation. Then solve for the number.

Objectives and Examples

- **Lesson 3–6** Solve addition and subtraction equations by using the properties of equality.

Solve $m + 16 = 8$. Check your solution.

$$m + 16 = 8$$
$$m + 16 - 16 = 8 - 16 \quad \textit{Subtract 16 from}$$
$$m = -8 \qquad \textit{each side.}$$

Check:
$$m + 16 = 8$$
$$-8 + 16 \stackrel{?}{=} 8 \quad \textit{Replace m with } -8.$$
$$8 = 8 \quad \checkmark$$

Review Exercises

Solve each equation. Check your solution.

40. $z + 15 = -9$ **41.** $19 = y + 7$

42. $p + (-7) = 31$ **43.** $m - (-4) = 21$

44. $r - (-1.2) = -7.3$ **45.** $j + (-2.4) = 6.6$

46. $\frac{1}{2} + t = \frac{1}{2}$ **47.** $-\frac{1}{6} = v - \frac{2}{3}$

Write an equation and solve.

48. Some number added to -16 is equal to 39. What is the number?

49. Twelve more than a number is -108. What is the number?

- **Lesson 3–7** Solve equations involving absolute value.

Solve $|x + 6| = 14$. Check your solution.

$$x + 6 = 14 \qquad \text{or} \qquad x + 6 = -14$$
$$x + 6 - 6 = 14 - 6 \qquad x + 6 - 6 = -14 - 6$$
$$x = 8 \qquad\qquad x = -20$$

Check:
$$|8 + 6| = 14 \qquad |-20 + 6| = 14$$
$$|14| \stackrel{?}{=} 14 \qquad |-14| \stackrel{?}{=} 14$$
$$14 = 14 \quad \checkmark \qquad 14 = 14 \quad \checkmark$$

Solve each equation. Check your solution.

50. $|x| = 22$

51. $|c| - 13 = -5$

52. $7 = 10 + |y|$

53. $6 = |-2 + f|$

54. $|h - 1| = 0$

55. $|a - 3| = -8$

56. Write an equation to represent the situation *The distance between a number x and -4 is 10 units.*

Applications and Problem Solving

57. Shopping The approximate prices for 16 different types of athletic shoes are shown on the line plot below. *(Lesson 3–3)*

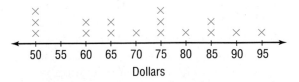

Find the mean, median, mode, and range for the prices.

58. Manufacturing A certain bolt used in lawn mowers will work properly only if its diameter differs from 2 centimeters by no more than 0.04 centimeter. Write and then solve an equation that could be used to find the least and greatest diameter of the bolt. *(Lesson 3–7)*

1. **List** at least two methods you could use to compare fractions with unlike denominators.

2. **Describe** the process of finding the mean, median, mode, and range of a set of data.

Replace each ● with <, >, or = to make a true sentence.

3. $-4 ● -3$

4. $-1.01 ● -1.1$

5. $\dfrac{3}{5} ● \dfrac{8}{15}$

6. Write $-\dfrac{1}{2}, -\dfrac{2}{9},$ and $-\dfrac{3}{8}$ in order from least to greatest.

Find each sum or difference.

7. $-4.0 - 2.8$

8. $0.32 - (-0.45)$

9. $6.1 + (-3.2)$

10. $\dfrac{7}{10} + \left(-\dfrac{3}{10}\right)$

11. $\dfrac{1}{5} - \left(-\dfrac{3}{4}\right)$

12. $\dfrac{3}{8} + 2\dfrac{1}{2}$

13. **Finance** Sally recorded the amount of money she spent on lunch for five school days. She recorded the amounts in the table at the right.
    a. Find Sally's mean, median, mode, and range of lunch money.
    b. Explain which measure of central tendency is the least useful in describing Sally's lunch costs.

Day	Cost
Monday	$0.99
Tuesday	$1.50
Wednesday	$3.99
Thursday	$0.99
Friday	$2.28

Solve each equation.

14. $10 \div 2 + 6 = x$

15. $8(0.3) - 0.5 = y$

16. $a = \dfrac{2 \cdot 6 + 4}{8 \div 2}$

Solve each equation. Use algebra tiles if necessary.

17. $g + 7 = 2$

18. $-6 + h = -1$

Solve each equation. Check your solution.

19. $u - (-76) = 44$

20. $5.2 + f = 16.4$

21. $\dfrac{1}{2} = b - \dfrac{7}{8}$

22. $3 + |c| = 6$

23. $|m - 4| = 1$

24. $-12 = |t - (-7)|$

25. Taryn visited the "Guess Your Age" booth at the state fair. If the person in the booth could not guess Taryn's age within three years, Taryn would win a stuffed animal. Taryn is 16 years old. Write and solve an equation that could be used to find the person's greatest and least guesses, without Taryn's winning a prize.

Number Concept Problems

Standardized tests include many questions that use fractions and decimals. You'll need to convert between fractions and decimals. You'll also need to add, subtract, multiply, and divide fractions and decimals. Here are some terms to remember.

reciprocal   repeating decimal

THE PRINCETON REVIEW

Be sure you practice using your calculator before the test. Look at each number you enter to avoid mistakes.

Proficiency Test Example

Which of the following numbers, when divided by $\frac{1}{2}$, gives a result less than $\frac{1}{2}$?

A $\frac{2}{8}$ **B** $\frac{7}{12}$ **C** $\frac{2}{3}$ **D** $\frac{5}{24}$

Hint In some multiple-choice questions, you can check each answer choice to see if it is correct.

Solution Recall that dividing by a fraction is equivalent to multiplying by its reciprocal. Dividing by $\frac{1}{2}$ is the same as multiplying by 2. Remember that 2 is also written as $\frac{2}{1}$.

A $\frac{2}{8} \times \frac{2}{1} = \frac{4}{8}$ or $\frac{1}{2}$ **B** $\frac{7}{12} \times \frac{2}{1} = \frac{7}{6}$ or $1\frac{1}{6}$

C $\frac{2}{3} \times \frac{2}{1} = \frac{4}{3}$ or $1\frac{1}{3}$ **D** $\frac{5}{24} \times \frac{2}{1} = \frac{10}{24}$ or $\frac{5}{12}$

Decide which of the four resulting numbers is less than $\frac{1}{2}$. Answer choice A is equal to $\frac{1}{2}$, choices B and C are greater than $\frac{1}{2}$, but choice D is less than $\frac{1}{2}$.

$$\frac{5}{12} < \frac{6}{12} = \frac{1}{2}$$

The answer is D.

SAT Example

If $\frac{2}{11}$ is written as a decimal carried out to 100 places, what is the sum of the first 50 digits to the right of the decimal point?

A 100   **B** 225   **C** 350   **D** 450   **E** 900

Hint Your calculator can be useful for some types of questions.

Solution First write the fraction as a decimal. Use either long division or your calculator.

$$\frac{2}{11} = 0.1818181818 \ldots \text{ or } 0.\overline{18}$$

Since this is a repeating decimal, the first 50 digits are 25 pairs of the digits 1 and 8.

The question asks for the *sum* of the first 50 digits. The sum of each pair of 1 and 8 is 9. The sum of 25 pairs is 25×9 or 225.

The answer is B.

After you work each problem, record your answer on the answer sheet provided or on a sheet of paper.

1. In Store X, which color button costs the most per individual unit?

Price of Buttons in Store X	
Color	**Cost**
Black	$2 per 5 buttons
Blue	$2 per 6 buttons
Brown	$3 per 8 buttons
Orange	$4 per 12 buttons
Red	$4 per 7 buttons

A black   B blue   C brown

D orange   E red

2. You burn 15 additional Calories each hour by standing instead of sitting. Suppose you burn x Calories per hour while sitting. How many Calories would you burn in one hour of standing?

A $15x$ B $x + 15$

C $x - 15$ D $15 - x$

3. The total number of students in a class is represented by $5x - 6$. If $2x + 6$ represents the number of boys, which expression represents the number of girls?

A $3x$ B $12 - 3x$

C $3x - 12$ D $3x + 12$

4. Order $-\frac{3}{4}, \frac{3}{4},$ and $-\frac{4}{3}$ from least to greatest.

A $\frac{3}{4}, -\frac{4}{3}, -\frac{3}{4}$ B $-\frac{4}{3}, \frac{3}{4}, -\frac{3}{4}$

C $-\frac{3}{4}, -\frac{4}{3}, \frac{3}{4}$ D $-\frac{4}{3}, -\frac{3}{4}, \frac{3}{4}$

5. What is the GCF of the numbers in the sequence 6, 12, 18, 24, 30, . . . ?

A 3   B 6   C 12   D 18

6. Two years ago, Luna was ℓ years of age. How old will Luna be 2 years from now?

A $\ell + 4$ B $\ell + 2$ C 2ℓ D ℓ

7. If you evaluate the following expression according to the order of operations, in what order should the operations be performed?

$$12.5 \times 3 + 5 + 20 - 12$$

A $\times, +, +, -$ B $+, -, \times, +$

C $+, \times, -, +$ D $-, +, +, \times$

8. Which expression finds multiples of 5?

A $x + 5$ B $x - 5$ C $5x$ D $5 - x$

Open-Ended Questions

9. **Grid-In**   Yvette's goal is to run 30 miles a week. So far this week, she has run 6.5, 5.2, 7.8, 3, and 6.9 miles. How many more miles does she need to run to reach her goal?

10. Band members Scott and Ben are taking the bus to a football game. The bus leaves at 5:45 P.M. They have to arrive 15 minutes before the bus leaves to load instruments. It takes Scott 10 minutes to drive to Ben's house and 25 minutes to drive from there to the school. What is the latest time that Scott can leave his house?

Part A   Explain the steps you use in solving this problem.

Part B   Calculate the latest time Scott can leave home.

Test Practice   For additional test practice questions, visit: www.algconcepts.glencoe.com

CHAPTER 4

Multiplication and Division Equations

▶ **What You'll Learn in Chapter 4:**

- to multiply and divide rational numbers *(Lessons 4–1 and 4–3)*,
- to use tree diagrams or the Fundamental Counting Principle to count outcomes *(Lesson 4–2)*,
- to solve multiplication and division equations by using the properties of equality *(Lesson 4–4)*,
- to solve equations involving more than one operation *(Lesson 4–5)*,
- to solve equations with variables on both sides *(Lesson 4–6)*, and
- to solve equations with grouping symbols *(Lesson 4–7)*.

Problem-Solving Workshop

Project

The United States government began issuing nine-digit Social Security numbers in 1935. An example is 123-45-6789. Each person can have only one number and the numbers are never reused. When do you think the government will run out of new numbers?

Year	Population	Births
1930	123,202,624	2,618,000
1940	132,164,569	2,559,000
1950	151,325,798	3,632,000
1960	179,323,175	4,258,000
1970	203,302,031	3,731,000
1980	226,542,203	3,612,000
1990	248,709,873	4,158,000

Working on the Project

Work with a partner and choose a strategy to solve the problem.

- How many Social Security numbers are possible if each of the nine digits can be 0 through 9?
- Estimate the maximum number of Social Security numbers that could have been issued in 1935.
- Estimate the total number of people who had Social Security numbers by the year 2000 if every U.S. citizen had a Social Security number.

▶ **Strategies**

Look for a pattern.

Draw a diagram.

Make a table.

Work backward.

Use an equation.

Make a graph.

Guess and check.

Technology Tools

- Use a **calculator** to find the total number of unique Social Security numbers.
- Use **spreadsheet software** to estimate the total number of Social Security numbers that had been issued by the year 2000.

*inter*NET **Research** For more information about Social Security numbers, visit:
CONNECTION www.algconcepts.glencoe.com

Presenting the Project

Write a one-page paper describing your results. Include the following:

- the total number of Social Security numbers possible,
- estimates for the total number of Social Security numbers issued by the year 2000 and the number issued each year after the year 2000, and
- what year you think there will be no Social Security numbers left to issue.

What You'll Learn

You'll learn to multiply rational numbers.

Why It's Important

Health One way to determine safe backpack weights is to multiply rational numbers.
See Exercise 16.

When the moon, Earth, and sun are lined up, high tides are higher than normal, and low tides are lower than normal. Suppose low tide decreases 0.2 meter per day for four days. What is the total change?

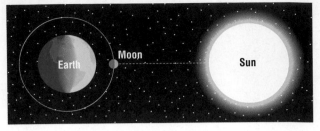

We can represent the decrease of 0.2 meter per day as −0.2. The total change after four days can be found by multiplying.

Number of days: 4
Change per day: −0.2

$$\underbrace{4}_{\substack{number \\ of\ days}} \quad \underbrace{\times}_{times} \quad \underbrace{-0.2}_{\substack{change \\ per\ day}}$$

$4 \times (-0.2) = -0.2 + (-0.2) + (-0.2) + (-0.2)$ *Multiplication is repeated*
$\qquad\qquad = -0.8$ *addition.*

In four days, the total change in low tide is −0.8 meter. This means that low tide has decreased 0.8 meter.

Look Back

Multiplying Integers: Lesson 2–5

This example illustrates the rule for multiplying two rational numbers.

Multiplying Two Rational Numbers	Words	Numbers
Different Signs	The product of two rational numbers having different signs is negative.	1.2(−4.5) = −5.4 −1.2(4.5) = −5.4
Same Sign	The product of two rational numbers having the same sign is positive.	3(2.8) = 8.4 −3(−2.8) = 8.4

Examples

Prerequisite Skills Review
Operations with Decimals, p. 684

Find each product.

1 **2.3(−4)**

2.3(−4) = −9.2   *The factors have different signs. The product is negative.*

2 **−3.5(−0.8)**

−3.5(−0.8) = 2.8   *The factors have the same sign. The product is positive.*

Your Turn

a. −5(−1.3)   **b.** −2.4(7.5)   **c.** 8 · (−3.2)

Physical Science Link

3 A skydiver jumps from 12,000 feet. Solve the equation $h = 12,000 + (0.5)(-32.1)(576)$ to find his height after he free-falls for 24 seconds.

$h = 12,000 + (0.5)(-32.1)(576)$

$h = 12,000 + (-16.05)(576)$ *Multiply 0.5 and -32.1.*

$h = 12,000 + (-9244.8)$ *Multiply -16.05 and 576.*

$h = 2755.2$ *Add 12,000 and -9244.8.*

After 24 seconds, the skydiver's height is 2755.2 feet.

You can use grid paper to model the multiplication of fractions.

High Tide

Low Tide

Hands-On Algebra
Models

Materials: grid paper   colored pencils   straightedge

Multiply $\frac{3}{7}$ and $\frac{1}{5}$.

Step 1   Use a straightedge to draw a rectangle that has seven rows and five columns.

Step 2   Shade three of the seven rows yellow to represent $\frac{3}{7}$.

Step 3   Shade one column blue to represent $\frac{1}{5}$.

Try These

1. How many of the squares are shaded green?

2. What fraction of the total number of squares in the rectangle is shaded green?

3. The portion of squares that is shaded green represents the product of the fractions. What is the product of $\frac{3}{7}$ and $\frac{1}{5}$?

4. Use grid paper to model $\frac{3}{4}$ and $\frac{5}{8}$. Then use your model to find the product of $\frac{3}{4}$ and $\frac{5}{8}$.

5. What do you notice about the numerators and the denominators in $\frac{3}{4}$ and $\frac{5}{8}$ compared to the product of these fractions?

The Hands-On Algebra activity suggests the following rule for multiplying fractions.

Multiplying Fractions	Words:	To multiply fractions, multiply the numerators and multiply the denominators.
	Symbols:	$\dfrac{a}{b} \cdot \dfrac{c}{d} = \dfrac{ac}{bd}$, where $b, d \neq 0$
	Numbers:	$\dfrac{1}{5} \cdot \dfrac{3}{4} = \dfrac{1 \cdot 3}{5 \cdot 4}$ or $\dfrac{3}{20}$

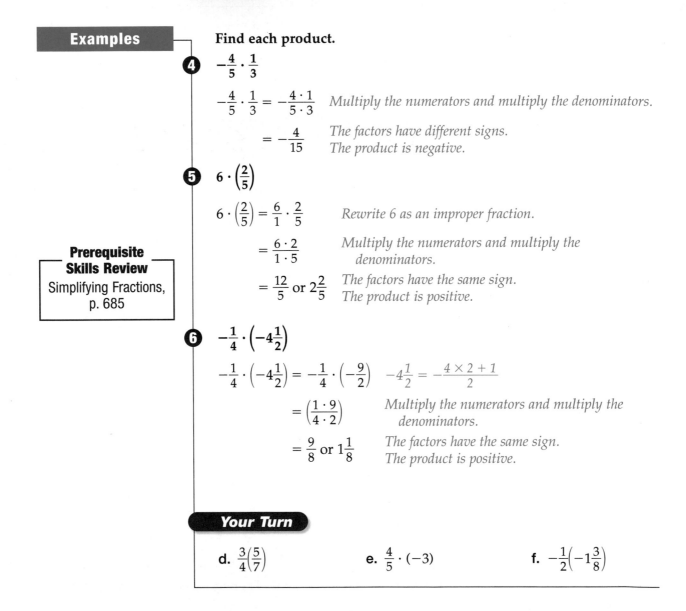

Examples

Find each product.

4 $-\dfrac{4}{5} \cdot \dfrac{1}{3}$

$-\dfrac{4}{5} \cdot \dfrac{1}{3} = -\dfrac{4 \cdot 1}{5 \cdot 3}$ *Multiply the numerators and multiply the denominators.*

$\phantom{-\dfrac{4}{5} \cdot \dfrac{1}{3}} = -\dfrac{4}{15}$ *The factors have different signs. The product is negative.*

5 $6 \cdot \left(\dfrac{2}{5}\right)$

$6 \cdot \left(\dfrac{2}{5}\right) = \dfrac{6}{1} \cdot \dfrac{2}{5}$ *Rewrite 6 as an improper fraction.*

$\phantom{6 \cdot \left(\dfrac{2}{5}\right)} = \dfrac{6 \cdot 2}{1 \cdot 5}$ *Multiply the numerators and multiply the denominators.*

$\phantom{6 \cdot \left(\dfrac{2}{5}\right)} = \dfrac{12}{5}$ or $2\dfrac{2}{5}$ *The factors have the same sign. The product is positive.*

Prerequisite Skills Review
Simplifying Fractions, p. 685

6 $-\dfrac{1}{4} \cdot \left(-4\dfrac{1}{2}\right)$

$-\dfrac{1}{4} \cdot \left(-4\dfrac{1}{2}\right) = -\dfrac{1}{4} \cdot \left(-\dfrac{9}{2}\right)$ $-4\dfrac{1}{2} = -\dfrac{4 \times 2 + 1}{2}$

$\phantom{-\dfrac{1}{4} \cdot \left(-4\dfrac{1}{2}\right)} = \left(\dfrac{1 \cdot 9}{4 \cdot 2}\right)$ *Multiply the numerators and multiply the denominators.*

$\phantom{-\dfrac{1}{4} \cdot \left(-4\dfrac{1}{2}\right)} = \dfrac{9}{8}$ or $1\dfrac{1}{8}$ *The factors have the same sign. The product is positive.*

Your Turn

d. $\dfrac{3}{4}\left(\dfrac{5}{7}\right)$ **e.** $\dfrac{4}{5} \cdot (-3)$ **f.** $-\dfrac{1}{2}\left(-1\dfrac{3}{8}\right)$

Study these products.

$-1(5) = -5$ *−1 times 5 equals −5.*

$(-1)(-2.4) = 2.4$ *−1 times −2.4 equals 2.4.*

Notice that multiplying a number by -1 results in the opposite of the number. This suggests the **Multiplicative Property of -1**.

Multiplicative Property of -1	**Words:** The product of -1 and any number is the number's additive inverse.	
	Symbols: $-1 \cdot a = -a$	**Numbers:** $-1 \cdot \frac{2}{7} = -\frac{2}{7}$

Knowing how to multiply rational numbers is essential in simplifying algebraic expressions and solving equations.

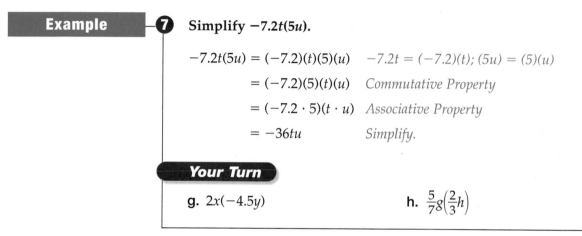

Example

7 Simplify $-7.2t(5u)$.

$$-7.2t(5u) = (-7.2)(t)(5)(u) \quad -7.2t = (-7.2)(t); (5u) = (5)(u)$$
$$= (-7.2)(5)(t)(u) \quad \textit{Commutative Property}$$
$$= (-7.2 \cdot 5)(t \cdot u) \quad \textit{Associative Property}$$
$$= -36tu \quad \textit{Simplify.}$$

Your Turn

g. $2x(-4.5y)$

h. $\frac{5}{7}g\left(\frac{2}{3}h\right)$

If you add any two rational numbers, is the sum always a rational number? What happens if you subtract or multiply rational numbers? In all three cases, the result is a rational number.

Closure Property of Rational Numbers	Because the sum, difference, or product of two rational numbers is also a rational number, the set of rational numbers is closed under addition, subtraction, and multiplication.

For example, you can add, subtract, or multiply $\frac{5}{7}$ and $\frac{1}{7}$, and the result is a rational number. *Check this.*

Check for Understanding

Communicating Mathematics

Study the lesson. Then complete the following.

1. **State** whether $1.2bc$ is *sometimes*, *always*, or *never* a rational number if b and c are rational numbers. Explain.
2. **Determine** whether the product of three negative rational numbers can ever be positive. Explain.
3. **Write** a multiplication equation that can be shown using the model.

Exercise 3

Guided Practice

⏱ **Getting Ready** **Find each product.**

Sample: $(-4)(-6)$ **Solution:** $(-4)(-6) = 24$

4. $-1 \cdot 8$ **5.** $3(-3)$ **6.** $-7 \cdot (-4)$

Find each product. *(Examples 1–6)*

7. $2.1(-7)$ **8.** $(-1.8)(-3.5)$ **9.** $0.5(-4.6)$

10. $-\frac{1}{3}\left(-\frac{4}{7}\right)$ **11.** $-3 \cdot \frac{3}{4}$ **12.** $2\frac{3}{5}\left(\frac{1}{4}\right)$

Simplify each expression. *(Example 7)*

13. $3\left(\frac{1}{5}y\right)$ **14.** $-\frac{1}{4} \cdot \frac{7}{10}k$ **15.** $-2.5c(-2.4d)$

16. Health Students whose backpacks are too heavy could develop spinal problems. The maximum recommended backpack weight y is given by the equation $y = \frac{3}{20}x$, where x is a student's weight. Would a 12-pound backpack be too heavy for a student who weighs 110 pounds? Explain. **Source:** *Zillions, 1999* *(Example 5)*

Exercises   • • • • •   • • • • • • • • • • • •

Practice

Find each product.

17. $4 \cdot 6.1$ **18.** $3.8(8)$ **19.** $7.1(-1)$

20. $(-0.5)(-5.6)$ **21.** $-6.0(8.5)$ **22.** $-0.4(1.5)$

23. $-2.9 \cdot 10$ **24.** $-7.3 \cdot 0$ **25.** $(-12.8)(-1)(4.5)$

26. $\frac{1}{5}\left(\frac{2}{3}\right)$ **27.** $-1 \cdot \left(-\frac{5}{8}\right)$ **28.** $-\frac{4}{7} \cdot \frac{3}{5}$

29. $(0)\left(-\frac{6}{7}\right)$ **30.** $-\frac{3}{5}\left(-\frac{9}{10}\right)$ **31.** $\frac{7}{3}\left(-\frac{3}{7}\right)$

32. $\left(-\frac{3}{5}\right)\left(\frac{-2}{7}\right)(-1)$ **33.** $3 \cdot \frac{7}{8}$ **34.** $\frac{6}{7}\left(-2\frac{1}{3}\right)$

Simplify each expression.

35. $2(-1.5x)$ **36.** $0.6(-15y)$

37. $(4r)(2.2s)$ **38.** $\frac{5}{6}r(-18s)$

39. $\left(\frac{3}{5}a\right)\left(\frac{2}{5}b\right)$ **40.** $\frac{1}{3}y\left(-\frac{1}{4}z\right)$

41. $-\frac{1}{2}s\left(-\frac{1}{2}\right) + \frac{2}{3}\left(\frac{3}{2}s\right)$ **42.** $\frac{5}{6}\left(\frac{3}{4}a\right) - \left(\frac{1}{8}a\right)\left(-\frac{1}{3}\right)$

43. Solve $x = 2(-3.2) - 0.5(-4.8)$ to find the value of x.

44. Evaluate $7a(-4.5b) + (-2ab)$ if $a = 2$ and $b = -1$.

Applications and Problem Solving

45. **Purchases**   Teri wants to buy a bicycle with the money she earns from her after-school job. If she can save $22.50 a week, how much money will she have after six weeks?

46. **Space**   A day on Jupiter is much shorter than a day on Earth because Jupiter rotates on its axis about $2\frac{2}{5}$ times as fast as Earth rotates on its axis. During our month of June, how many days will Jupiter have?

47. **History**   Galileo (1564–1642) was a mathematician and scientist who is said to have performed experiments with falling objects from the Leaning Tower of Pisa. Suppose Galileo dropped a stone from the tower and it took $3\frac{1}{2}$ seconds to reach the ground. Solve $d = -\frac{1}{2}\left(9\frac{4}{5}\right)\left(3\frac{1}{2}\right)\left(3\frac{1}{2}\right)$ to find the distance d in meters that the stone fell. *A negative answer means that the final position is lower than the initial position.*

48. **Critical Thinking**   Describe the values of a and b for each of the following to be true.

   a. ab is positive.   **b.** ab is negative.   **c.** $ab = 0$

Mixed Review

Solve each equation. Check your solution.   *(Lesson 3–7)*

49. $|t| = -5$ 50. $2 + |x| = 6$ 51. $3 = |-9 - a|$

52. **Aviation**   A traffic helicopter descended 160 meters to observe road conditions. It leveled off at 225 meters.   *(Lesson 3–6)*

   a. Let a represent the original altitude. Write an equation to represent this problem.

   b. What was its original altitude?

Find each sum or difference.   *(Lesson 3–2)*

53. $3.5 + (-2.1)$ 54. $-1.7 - 4.0$ 55. $-\frac{3}{4} + \frac{1}{8}$

56. Find $24 \div (-6)$.   *(Lesson 2–6)*

57. **Height**   The stem-and-leaf plot shows the heights of 30 students.   *(Lesson 1–7)*

   a. What is the height of the tallest student?

   b. Which height occurs most frequently?

Class Heights

Stem	Leaf
5	7 8 8 9 9 9
6	0 0 1 2 2 2 3 3 4 4 4 4 4 5 5 7 8 9 9
7	0 0 2 3 3 $7|2 = 72\ in.$

58. **Standardized Test Practice**   The expression $(220 - y) \times 0.8 \div 4$ gives the target 15-second heart rate for an athlete y years old during a workout. Find the target 15-second heart rate for a 17-year-old athlete during a workout.   *(Lesson 1–2)*

   A 27   **B** 40.6   **C** 216.6   **D** 219.7

Extra Practice   See p. 698.

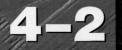

What You'll Learn

You'll learn to use tree diagrams or the Fundamental Counting Principle to count outcomes.

Why It's Important

Manufacturing Car manufacturers count outcomes to determine the number of different key combinations that are possible.
See Exercise 23.

Baroness Martine de Beausoleil was a French scientist in the 17th century who spent 30 years studying geology and mathematics. She determined which rocks were valuable by the minerals they contained. Some other ways to classify rocks are by their texture and by their color.

How many different rocks are possible having the characteristics shown in the table? We can represent this situation by using a **tree diagram**.

Rock Characteristics		
Mineral	**Texture**	**Color**
gold	glassy	brown
quartz	dull	gray
iron		
silver		

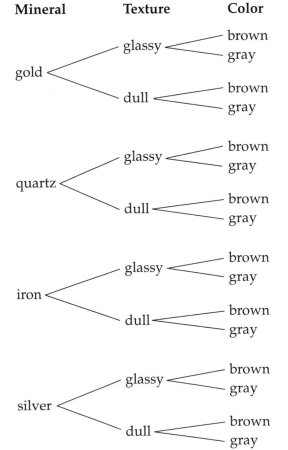

Mineral	Texture	Color	Outcome
gold	glassy	brown	gold, glassy, brown
		gray	gold, glassy, gray
	dull	brown	gold, dull, brown
		gray	gold, dull, gray
quartz	glassy	brown	quartz, glassy, brown
		gray	quartz, glassy, gray
	dull	brown	quartz, dull, brown
		gray	quartz, dull, gray
iron	glassy	brown	iron, glassy, brown
		gray	iron, glassy, gray
	dull	brown	iron, dull, brown
		gray	iron, dull, gray
silver	glassy	brown	silver, glassy, brown
		gray	silver, glassy, gray
	dull	brown	silver, dull, brown
		gray	silver, dull, gray

Gold

By using the tree diagram, we find that there are 16 different rocks possible. These results are called **outcomes**. For example, the first outcome is a rock that contains gold, has a glassy texture, and is brown. The list of all the possible outcomes is called the **sample space**.

Example

1 The Greenwood Dance Company has three lead female dancers, Teresa, Kendra, and Angelica, and two lead male dancers, Keith and Jamal. How many different ways can the director choose one female dancer and one male dancer for the next production?

We can make a tree diagram to find the number of combinations.

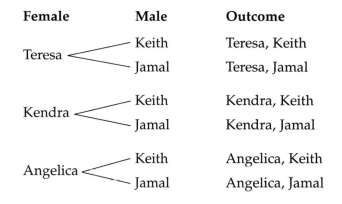

There are six outcomes in the sample space. So, there are six different ways that the director can choose the leads for the next production.

Your Turn

a. The Ice Cream Parlor offers the choices below for making sundaes. Draw a tree diagram to find the number of different sundaes that can be made.

Ice Cream	Topping	Whipped Cream
chocolate vanilla rocky road	chocolate butterscotch	yes no

An **event** is a subset of the possible outcomes. In Example 1, the choice of female dancers is one event, and the choice of male dancers is another event. Notice that the product of the number in each event is 3 · 2 or 6. Finding the number of possible outcomes by multiplying involves the **Fundamental Counting Principle**.

Fundamental Counting Principle	If event *M* can occur in *m* ways and is followed by event *N* that can occur In *n* ways, then the event *M* followed by event *N* can occur in $m \times n$ ways.

Example

Clothing Link

2 How many different kinds of school sweatshirts are possible?

There are 3 styles, 2 colors, and 5 sizes, so the number of different sweatshirts is $3 \times 2 \times 5$ or 30.

Style	Color	Size
school name school logo team graphic	red tan	small medium large 1X 2X

Your Turn

b. Catina wants to buy a red sweatshirt in size large or 1X in any style. How many choices does she have?

Check for Understanding

Communicating Mathematics

Study the lesson. Then complete the following.

1. **Explain** in your own words the Fundamental Counting Principle.

2. **Represent** a real-life situation by using a tree diagram. Include a description of each event and a list of all the outcomes in the sample space.

3. **You Decide** Ling says that if three coins are tossed, then the number of outcomes with 2 heads and 1 tail is the same as the number with 1 head and 2 tails. Lorena thinks he is wrong. Who is correct and why?

Vocabulary

tree diagram
outcomes
sample space
event

Guided Practice

Getting Ready Suppose you spin the spinner twice. Determine whether each is an *outcome* or a *sample space*.

Sample 1: (red, blue)

Solution: Spinning a red and then a blue (red, blue) is an outcome because it is only one possible result.

Sample 2: (red, red), (red, blue), (blue, red), (blue, blue)

Solution: This is a sample space because it is a list of all possible outcomes.

Determine whether each is an *outcome* or a *sample space* for the given experiment.

4. (5, 2, 2); rolling a number cube three times

5. (H, T); tossing a coin once

6. (H, H), (H, T), (T, H), (T, T); tossing a coin twice

7. (4, 10, J, 3, 7); choosing five cards from a standard deck

8. Suppose you can order a burrito or a taco with beef, chicken, or bean filling. Find the number of possible outcomes by drawing a tree diagram. *(Example 1)*

9. Suppose you roll a die twice. Find the number of possible outcomes by using the Fundamental Counting Principle. *(Example 2)*

10. **Bicycles** Enrique wants to buy a bicycle, but he is having trouble deciding what kind to buy. *(Example 1)*

Type	Speed	Handlebars
road	21	touring
mountain	16	racing
	14	
	10	

    a. Draw a tree diagram to represent his choices.

    b. How many different kinds of bicycles are possible?

    c. How many different kinds of bicycles have racing handlebars and 16 or greater gear speeds?

Exercises • • • • • • • • • • • • • • • • • •

Practice

Find the number of possible outcomes by drawing a tree diagram.

11. three tosses of a coin

12. choosing one marble from each box shown in the table at the right

Box A	Box B	Box C
blue	green	green
green	red	white

13. Anita purchased five T-shirts and three pairs of jeans for school, as shown in the table at the right. How many different T-shirt and jeans outfits are possible?

T-Shirt	Jeans
red	black
orange	white
green	blue
white	
striped	

Find the number of possible outcomes by using the Fundamental Counting Principle.

14. different cars with options shown at the right

15. possible sequences of answers on a 5-question true-false quiz

Color	Engine	Transmission
red	4-cylinder	manual
blue	6-cylinder	automatic
white		
green		

16. the number of different 1-topping pizzas with the choices shown at the right

Crust	Size	Topping	
thin	individual	pepperoni	onion
regular	small	mushroom	olive
thick	medium	sausage	pepper
	large		

17. Refer to Example 1. How many different ways can the director choose one female dancer and one male dancer if there are five female dancers and six male dancers?

18. Suppose the spinner at the right is spun twice.

   a. Let R represent red, B represent blue, and Y represent yellow. Write all of the outcomes in the sample space as ordered pairs.
   b. How many outcomes have at least one blue?

19. Write a situation that fits the tree diagram below.

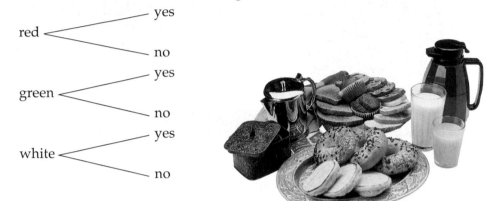

red — yes
red — no
green — yes
green — no
white — yes
white — no

Applications and Problem Solving

Real World

20. Dining A free continental breakfast offers one type of bread and one beverage.
   a. How many different breakfasts are possible?
   b. How many of the breakfasts include a muffin?

Bread	Beverage
bagel	tea
bran muffin	coffee
English muffin	juice
white toast	milk
wheat toast	
raisin toast	

21. History In 1869, Fanny Jackson Coppin became the first African-American school principal. One of her favorite teaching poems began:

> A noun is the name of anything,
> As _school_, or _garden_, _hoop_, or _swing_.
> Adjectives tell the kind of noun,
> As _great_, _small_, _pretty_, _white_, or _brown_.

How many two-word phrases can you make using one of the underlined nouns and one of the underlined adjectives from the poem?

22. Family Valerie and Jessie just got married. They hope to eventually have two girls and a boy, in any order.
   a. How many combinations of three children are possible?
   b. How many outcomes will give them two girls and a boy?
   c. In how many of the outcomes from part b will the girls be born in consecutive order?

23. **Keys** A car manufacturer makes keys with six sections.

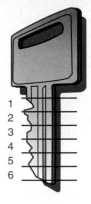

   a. Until the 1960s, there were only two different patterns for each section. How many different keys were possible?

   b. After the 1960s, the car manufacturer made keys having three different patterns for each section. How many different keys were possible after the 1960s?

24. **Critical Thinking** The president, vice president, secretary, and treasurer of the Drama Club are to be seated in four chairs for a yearbook picture. How many different seating arrangements are possible?

Mixed Review

Find each product. *(Lesson 4–1)*

25. $-3 \cdot 5.4$

26. $-7.2(-1.5)$

27. $\dfrac{3}{5} \cdot \dfrac{2}{7}$

28. **Manufacturing** A certain bolt used in lawn mowers will work properly only if its diameter differs from 2 centimeters by exactly 0.04 centimeter. *(Lesson 3–7)*

   a. Let d represent the diameter. Write an equation to represent this problem.

   b. What are the least and greatest diameters for this bolt?

29. Solve $x - \dfrac{1}{3} = -\dfrac{3}{4}$. Check your solution. *(Lesson 3–6)*

30. **Weather** In the graph at the right, how many degrees did the temperature rise from 1:00 to 5:00? *(Lesson 2–4)*

31. **Standardized Test Practice** Ayani has $600 in the bank at an annual interest rate of 3%. How much money will he have in his account after four years? (*Hint:* Use the formula for simple interest $I = prt$.) *(Lesson 1–5)*

   A $72

   B $672

   C $607

   D $720

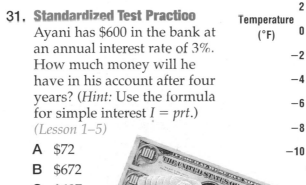

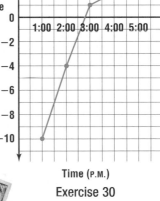

Time (P.M.)

Exercise 30

Extra Practice See p. 699.

Permutations and Combinations

Materials

- index cards
- green, blue, pink, and yellow self-adhesive notes
- graphing calculator

Mandy makes and sells stuffed rabbits at craft shows. There are three sets of pieces in the rabbit pattern, and each set is cut from a different color fabric. One set of pieces is for the body, one is for the arms, and the other is for the legs. How many different rabbits can she make?

Investigate

1. Label the index cards Set 1, Set 2, and Set 3.

   a. Suppose Mandy has three different fabrics. Let a green, a blue, and a pink self-adhesive note represent the three fabrics.

   b. Lay the cards on a table and place a different color note on each.

This represents one arrangement of colors.

Set 1 (Body)	Set 2 (Arms)	Set 3 (Legs)

   c. Find and record all of the different arrangements that are possible using each color only once. Two different arrangements are shown.

Arrangement	Set 1 (body)	Set 2 (arms)	Set 3 (legs)
A	blue	green	pink
B	blue	pink	green

   d. An arrangement of objects in which order is important is called a **permutation**. How many permutations of the three colors are there?

   e. Suppose the order of the three colors does not matter. For example, a rabbit with a green body, blue arms, and pink legs is considered the same as a rabbit with a blue body, pink arms, and green legs. An arrangement of objects in which order is not important is called a **combination**. How many combinations are possible using green, blue, and pink fabrics?

2. Suppose Mandy has four different fabrics. Use a yellow self-adhesive note to represent the fourth fabric.

a. Use the index cards to find the number of different rabbits that are possible when there are four colors from which to choose. Record your results in a new table. Remember that you can use only three colors at a time. Two possible arrangements are shown below.

Arrangement	Set 1 (body)	Set 2 (arms)	Set 3 (legs)
A	green	blue	pink
B	yellow	blue	pink

How many permutations are possible?

b. How many different rabbits are possible if the order of the colors is not important?

Extending the Investigation

In this extension, you will use formulas to find permutations and combinations. The mathematical notation 4! means $4 \cdot 3 \cdot 2 \cdot 1$. The symbol 4! is read *four* **factorial**. n! means the product of all counting numbers beginning with n and counting backward to 1.

- The number of permutations of n objects taken r at a time is defined as $_nP_r = \dfrac{n!}{(n-r)!}$.

  The number of permutations of 4 colors taken 3 at a time can be written $_4P_3$. Find $_4P_3$ and compare your answer to the answer in Exercise 2a.

- The number of combinations of n objects taken r at a time is defined $_nC_r = \dfrac{n!}{(n-r)!r!}$.

  The number of combinations of 4 colors taken 3 at a time can be written $_4C_3$. Find $_4C_3$ and compare your answer to the answer in Exercise 2b.

- Use the formulas to find the number of permutations and combinations of 5 colors taken 3 at a time and 6 colors taken 3 at a time.

- We define 0! as 1. **Make a conjecture** comparing $_nP_n$ and $_nC_n$.

Presenting Your Conclusions

Here are some ideas to help you present your conclusions to the class.

- Make a pamphlet showing the permutations and combinations for the rabbit pieces and fabrics.

- Include diagrams and tables as needed. Show how the formulas can be used to make the calculations.

*inter*NET **Investigation** For more information on permutations
CONNECTION and combinations, visit: www.algconcepts.glencoe.com

Math In the Workplace

What You'll Learn
You'll learn to divide rational numbers.

Why It's Important
Cooking To adjust recipes for different serving sizes, cooks must know how to divide rational numbers.
See Example 5.

The recipe makes two dozen cookies. If you want to make only one dozen, you would divide the recipe by 2. How much sugar would you need to make one dozen cookies?
This problem will be solved in Example 5.

Raspberry Almond Cookies

1 cup butter	2 cups flour
$\frac{2}{3}$ cup sugar	$\frac{1}{2}$ cup raspberry jam
$\frac{1}{2}$ tsp. almond extract	

When dividing rational numbers, you can use the rules below to determine the sign of the quotient. They are the same as the rules for dividing integers.

Dividing Rational Numbers	Words	Numbers
Different Signs	The quotient of two numbers having different signs is negative.	$-2.1 \div 7 = -0.3$ $2.1 \div (-7) = -0.3$
Same Sign	The quotient of two numbers having the same sign is positive.	$4.5 \div 0.9 = 5$ $-4.5 \div (-0.9) = 5$

Examples

Prerequisite Skills Review
Operations with Decimals, p. 684

Find each quotient.

1 $-6 \div 0.5$
$-6 \div 0.5 = -12$ *Numbers have different signs. The quotient is negative.*

2 $-7.4 \div (-2)$
$-7.4 \div (-2) = 3.7$ *Numbers have the same sign. The quotient is positive.*

Your Turn

a. $16 \div (-2.5)$ **b.** $-3.9 \div 3$ **c.** $-8.4 \div (-1.2)$

Two numbers whose product is 1 are called **multiplicative inverses** or **reciprocals**.

$\frac{7}{8}$ and $\frac{8}{7}$ are reciprocals because $\frac{7}{8} \cdot \frac{8}{7} = 1$.

-3 and $-\frac{1}{3}$ are reciprocals because $-3 \cdot \left(-\frac{1}{3}\right) = 1$.

a and $\frac{1}{a}$, where $a \neq 0$, are reciprocals because $a \cdot \frac{1}{a} = 1$.

These examples demonstrate the **Multiplicative Inverse Property**.

	Words:	The product of a number and its multiplicative inverse is 1.
Multiplicative Inverse Property	Symbols:	For every number $\frac{a}{b}$, where $a, b \neq 0$, there is exactly one number $\frac{b}{a}$ such that $\frac{a}{b} \cdot \frac{b}{a} = 1$.
	Numbers:	$-\frac{4}{9}$ and $-\frac{9}{4}$ are multiplicative inverses.

In Lesson 2–4, you learned that subtracting a number is the same as adding its additive inverse. In the same way, dividing by a number is the same as multiplying by its multiplicative inverse or reciprocal.

	Words:	To divide a fraction by any nonzero number, multiply by its reciprocal.
Dividing Fractions	Symbols:	$\frac{a}{b} \div \frac{c}{d} = \frac{a}{b} \cdot \frac{d}{c}$, where $b, c, d \neq 0$
	Numbers:	$\frac{3}{4} \div \frac{2}{5} = \frac{3}{4} \cdot \frac{5}{2}$ or $\frac{15}{8}$

Use the rules for dividing rational numbers to determine the sign of the quotient.

Examples

Prerequisite Skills Review
Simplifying Fractions, p. 685

Find each quotient.

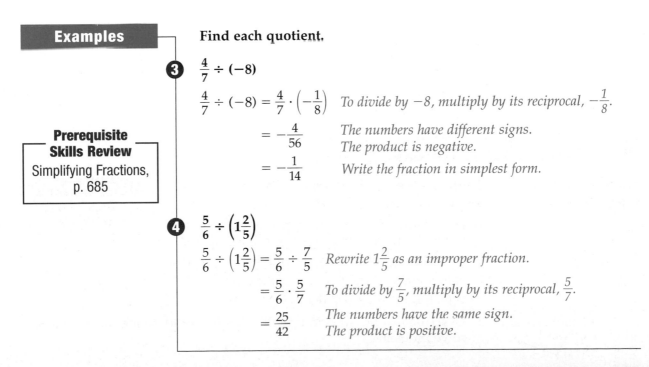

3 $\frac{4}{7} \div (-8)$

$\frac{4}{7} \div (-8) = \frac{4}{7} \cdot \left(-\frac{1}{8}\right)$ *To divide by -8, multiply by its reciprocal, $-\frac{1}{8}$.*

$= -\frac{4}{56}$ *The numbers have different signs. The product is negative.*

$= -\frac{1}{14}$ *Write the fraction in simplest form.*

4 $\frac{5}{6} \div \left(1\frac{2}{5}\right)$

$\frac{5}{6} \div \left(1\frac{2}{5}\right) = \frac{5}{6} \div \frac{7}{5}$ *Rewrite $1\frac{2}{5}$ as an improper fraction.*

$= \frac{5}{6} \cdot \frac{5}{7}$ *To divide by $\frac{7}{5}$, multiply by its reciprocal, $\frac{5}{7}$.*

$= \frac{25}{42}$ *The numbers have the same sign. The product is positive.*

Example ⑤

Cooking Link

Real World

Refer to the application at the beginning of the lesson. Find the amount of sugar needed if you divide the recipe by 2.

$\left(\begin{array}{c}\text{amount} \\ \text{of sugar}\end{array}\right) \div 2 = \dfrac{2}{3} \div 2$ *The recipe calls for $\dfrac{2}{3}$ cup of sugar.*

$\qquad = \dfrac{2}{3} \cdot \dfrac{1}{2}$ *To divide by 2, multiply by its reciprocal, $\dfrac{1}{2}$.*

$\qquad = \dfrac{2}{6} \text{ or } \dfrac{1}{3}$ *Write the fraction in simplest form.*

To make one dozen cookies, you need $\dfrac{1}{3}$ cup of sugar.

Your Turn

d. $14 \div -\dfrac{2}{3}$ **e.** $-\dfrac{2}{7} \div \left(-\dfrac{5}{9}\right)$ **f.** $\dfrac{1}{6} \div \left(2\dfrac{4}{5}\right)$

You can use what you know about dividing rational numbers to evaluate algebraic expressions.

Example ⑥

Evaluate $\dfrac{x}{4}$ if $x = \dfrac{3}{5}$.

$\dfrac{x}{4} = \dfrac{\frac{3}{5}}{4}$ *Replace x with $\dfrac{3}{5}$.*

$\qquad = \dfrac{3}{5} \div 4$ *Rewrite the fraction as a division sentence.*

$\qquad = \dfrac{3}{5} \cdot \dfrac{1}{4}$ *To divide by 4, multiply by its reciprocal, $\dfrac{1}{4}$.*

$\qquad = \dfrac{3}{20}$ *Multiply the numerators and multiply the denominators.*

Your Turn   Evaluate if $x = \dfrac{3}{5}$.

g. $\dfrac{6}{x}$ **h.** $-\dfrac{x}{2}$ **i.** $\dfrac{4}{7x}$

Check for Understanding

Communicating Mathematics

Study the lesson. Then complete the following.

1. **Determine** whether the reciprocal of a negative rational number can ever be positive. Explain.

2. **Compare** a positive rational number n with the value of $n \div \dfrac{1}{2}$ and with the value of $n \cdot \dfrac{1}{2}$. Which is the greatest? the least?

Math Journal

3. **Describe** real-life situations in which you would divide fractions to solve problems.

Vocabulary

multiplicative inverses
reciprocals

⏱ Getting Ready **Name the reciprocal of each number.**

Sample: $-\dfrac{1}{4}$ **Solution:** $-\dfrac{1}{4}\left(-\dfrac{4}{1}\right) = 1$, so the reciprocal is $-\dfrac{4}{1}$ or -4.

4. 3   **5.** $-\dfrac{2}{3}$ **6.** $2\dfrac{3}{4}$ **7.** $-\dfrac{1}{x}, x \neq 0$

Find each quotient. *(Examples 1–5)*

8. $-6 \div 1.5$ **9.** $3.8 \div 19$ **10.** $-4.7 \div (-0.5)$

11. $-\dfrac{1}{2} \div \left(\dfrac{7}{3}\right)$ **12.** $4 \div \left(\dfrac{7}{8}\right)$ **13.** $2\dfrac{2}{5} \div \dfrac{1}{3}$

Evaluate each expression if $a = \dfrac{1}{4}$ and $b = -\dfrac{2}{3}$. *(Example 6)*

14. $\dfrac{a}{3}$ **15.** $\dfrac{5}{b}$ **16.** $\dfrac{a}{b}$

17. Lunch Students working on decorations for homecoming ordered pizza for lunch. They ordered five 8-slice pizzas, and each person ate $2\dfrac{1}{2}$ slices. There was no pizza left over. How many students were there? *(Example 4)*

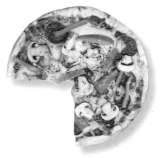

Exercises • • • • • • • • • • • • • • • • • • •

Find each quotient.

18. $8 \div (-1.6)$ **19.** $-7.5 \div 3$ **20.** $9.6 \div 3.2$

21. $2.7 \div -27$ **22.** $0.4 \div -0.4$ **23.** $15.6 \div 1.3$

24. $\dfrac{1}{8} \div \dfrac{1}{7}$ **25.** $0 \div \dfrac{3}{8}$ **26.** $\dfrac{3}{4} \div \left(-\dfrac{4}{5}\right)$

27. $-\dfrac{5}{6} \div \dfrac{5}{6}$ **28.** $\dfrac{1}{-7} \div \dfrac{4}{9}$ **29.** $-\dfrac{2}{3} \div \left(-\dfrac{5}{7}\right)$

30. $-\dfrac{7}{9} \div \left(-\dfrac{9}{7}\right)$ **31.** $5 \div \dfrac{1}{9}$ **32.** $-\dfrac{1}{5} \div 10$

33. $\dfrac{-3}{4} \div 8$ **34.** $-14 \div \left(-2\dfrac{4}{5}\right)$ **35.** $-3\dfrac{1}{2} \div \dfrac{1}{6}$

Evaluate each expression if $j = -\dfrac{1}{2}$, $k = \dfrac{3}{4}$, and $n = \dfrac{1}{5}$.

36. $\dfrac{3}{n}$ **37.** $\dfrac{9}{k}$ **38.** $\dfrac{k}{2}$

39. $\dfrac{-j}{5}$ **40.** $\dfrac{1}{2n}$ **41.** $\dfrac{n}{j}$

42. $\dfrac{k}{j}$ **43.** $\dfrac{2n}{j}$ **44.** $\dfrac{jk}{n}$

45. Find the quotient of -12 and $-\frac{2}{3}$.

46. Solve the equation $t = -\frac{7}{18} \div 3\frac{1}{2}$ to find the value of t.

Applications and Problem Solving

47. **Track**   Mr. Vance is training for a 10-kilometer race in a senior citizen track meet. Suppose $3\frac{1}{4}$ laps equals 1 kilometer and he wants to finish the race in 65 minutes. In how many minutes will he have to run each lap?

48. **Nutrition**   A 20-ounce bottle of soda contains $16\frac{2}{3}$ teaspoons of sugar. A 12-ounce can of soda contains $16\frac{2}{3} \div \frac{5}{3}$ teaspoons of sugar. How much sugar is in a 12-ounce can of soda?

49. **Arts**   The performing arts series at Central State University offers a package ticket to students for $62.50. If there are five performances, how much do students pay for each performance?

50. **Critical Thinking**

   a. Is the division of two rational numbers always a rational number? If not, give a counterexample.

   b. Is the set of rational numbers closed under division? Explain.

   c. What number has no reciprocal? Explain.

Mixed Review

51. **Uniforms**   The school dress code consists of a uniform, a blouse, and socks. A sweater may be added, if necessary. Find the number of different outfits that students can wear while still following the dress code.   *(Lesson 4–2)*

Uniform	Blouse	Socks	Sweater
skirt	white	white	yes
jumper	yellow	blue	no
	blue	gray	
	pink		

52. **Sales**   Ms. Ortiz wants to buy a washer and dryer from Bargain Appliances. If she buys on the "90 days same as cash" plan, she will have to make 5 equal payments of $146.12 each. If she writes a check for each payment, what is the net effect on her checking account? *(Lesson 4–1)*

Find the mean, median, mode, and range for each set of data. *(Lesson 3–3)*

53. 6, 8, 8, 8, 6, 4, 3, 7, 4

54. 9, 4, 13, 11, 7, 6, 20

55. 20, 11, 20, 23, 20, 29

Replace each ● with <, >, or = to make a true sentence.
(Lesson 3–1)

56. -3.6 ● 0 **57.** $\frac{3}{10}$ ● 0.3 **58.** $-7(-4)$ ● $7 - 15$

Which is the better buy? Explain. *(Lesson 3–1)*

59. a $\frac{1}{2}$-pound bag of cashews for \$3.15 or $\frac{3}{4}$-pound bag for \$5.19

60. three liters of soda for \$2.25 or two liters for \$1.98

61. a 48-ounce bottle of dishwashing liquid for \$2.69 or a 22-ounce bottle for \$1.09

62. Evaluate $16 + k$ if $k = -11$. *(Lesson 2–3)*

63. Standardized Test Practice Write the ordered pair that names point Q. *(Lesson 2–2)*

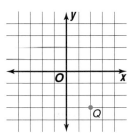

  A $(2, -3)$ **B** $(-2, -3)$

  C $(-3, 2)$ **D** $(-3, -2)$

Quiz 1   Lessons 4–1 through 4–3

Find each product. *(Lesson 4–1)*

1. $9.4 \cdot (-2)$ **2.** $-\frac{3}{8}\left(-\frac{1}{7}\right)$ **3.** $4 \cdot \frac{5}{8}$

Find the number of possible outcomes. *(Lesson 4–2)*

4. choosing a sandwich, a side, and a beverage from the table

5. different three-digit security codes using the numbers 1, 2, and 3 (You can use the numbers more than once.)

6. rolling a number cube three times

Sandwich	Side	Beverage
ham	French fries	milk
turkey	cole slaw	tea
egg salad	fruit	soda
		juice

Exercise 4

Find each quotient. *(Lesson 4–3)*

7. $-8.1 \div (-3)$ **8.** $\frac{1}{2} \div \left(-\frac{3}{4}\right)$ **9.** $2\frac{1}{3} \div \frac{3}{4}$

10. World Records The longest loaf of bread ever baked was 2132 feet $2\frac{1}{2}$ inches. If this loaf were cut into $\frac{1}{2}$-inch wide slices, how many slices of bread would there have been? *(Lesson 4–3)*

Extra Practice See p. 699. **Lesson 4–3** Dividing Rational Numbers **159**

Math In the Workplace

What You'll Learn

You'll learn to solve multiplication and division equations by using the properties of equality.

Why It's Important

Plumbing Plumbers solve equations to determine correct pipe weights. *See Exercise 36.*

In 1999, Nunavut (NU na voot) became Canada's newest territory in fifty years. In Nunavut, 15,000 people are under age 25. This is about $\frac{3}{5}$ of the total population. What is the population of Nunavut? *This problem will be solved in Example 8.*

One way to solve algebraic equations is to use the **Division Property of Equality**.

Beaufort Sea

North Magnetic Pole

Nunavut

Hudson Bay

CANADA

Division Property of Equality	**Words:**	If you divide each side of an equation by the same nonzero number, the two sides remain equal.
	Symbols:	For any numbers a, b, and c, with $c \neq 0$, if $a = b$, then $\frac{a}{c} = \frac{b}{c}$.
	Numbers:	If $3x = 12$, then $\frac{3x}{3} = \frac{12}{3}$.

Examples

Solve each equation. Check your solution.

1 $3y = 45$

$$3y = 45$$
$$\frac{3y}{3} = \frac{45}{3} \quad \textit{Divide each side by 3.}$$
$$y = 15$$

Check: $3y = 45$
$$3(15) \stackrel{?}{=} 45 \quad y = 15$$
$$45 = 45 \quad \checkmark$$

2 $16 = -2h$

$$16 = -2h$$
$$\frac{16}{-2} = \frac{-2h}{-2} \quad \textit{Divide each side by } -2.$$
$$-8 = h$$

Check: $16 = -2h$
$$16 \stackrel{?}{=} -2(-8) \quad h = -8$$
$$16 = 16 \quad \checkmark$$

Prerequisite Skills Review

Operations with Decimals, p. 684

3 $7.4a = -37$

$$7.4a = -37$$
$$\frac{7.4a}{7.4} = \frac{-37}{7.4} \quad \textit{Divide each side by 7.4.}$$
$$a = -5 \quad \textit{Check by substituting into the original equation.}$$

Sometimes, you will have to write and solve an equation to solve a problem.

Example ④

Consumer Link

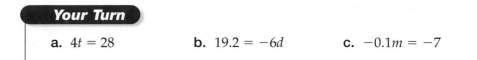

Real World

Xue Wu is in charge of buying sandwiches for the hiking trip. She has $24, and sandwiches cost $3.80 each. How many sandwiches can she buy?

Explore   You know the sandwiches cost $3.80 and Xue Wu has $24.

Plan   Let *s* represent the number of sandwiches and write an equation to represent the problem.

price per sandwich	times	number of sandwiches	equals	total cost
3.80	×	s	=	24

Solve   Solve the equation for *s*.

$$3.80s = 24$$

$$\frac{3.80s}{3.80} = \frac{24}{3.80} \qquad \textit{Divide each side by 3.80.}$$

$$s \approx 6.3$$

Since she can't buy part of a sandwich, Xue Wu has enough money to buy 6 sandwiches.

Examine   Does the answer make sense? Round 3.80 to 4 and multiply by 6.

$$4 \times 6 = 24$$

So 6 sandwiches would cost about $24.

Another useful property in solving equations is the **Multiplication Property of Equality**.

Multiplication Property of Equality	**Words:**	If you multiply each side of an equation by the same number, the two sides remain equal.
	Symbols:	For any numbers *a*, *b*, and *c*, if $a = b$, then $ac = bc$.
	Numbers:	If $\frac{1}{5}x = 8$, then $5\left(\frac{1}{5}x\right) = 5(8)$.

Solve each equation. Check your solution.

5 $\dfrac{g}{4} = 9$

$\dfrac{g}{4} = 9$ **Check:** $\dfrac{g}{4} = 9$

$4\left(\dfrac{g}{4}\right) = 4(9)$ *Multiply each side by 4.* $\dfrac{36}{4} \overset{?}{=} 9$ $g = 36$

$g = 36$ $9 = 9$ ✓

6 $-5 = -\dfrac{1}{6}t$

$-5 = -\dfrac{1}{6}t$ **Check:** $-5 = -\dfrac{1}{6}t$

$-6(-5) = -6\left(-\dfrac{1}{6}t\right)$ *Multiply each side by −6.* $-5 \overset{?}{=} -\dfrac{1}{6}(30)$

$30 = t$ $-5 = -5$ ✓

7 $-\dfrac{2}{3}x = 12$

$-\dfrac{2}{3}x = 12$

$-\dfrac{3}{2}\left(-\dfrac{2}{3}x\right) = -\dfrac{3}{2}(12)$ *Multiply each side by $-\dfrac{3}{2}$.*

$x = -18$ *Check by substituting into the original equation.*

Your Turn

d. $\dfrac{t}{4} = 8$ **e.** $\dfrac{5}{6}a = 25$ **f.** $36 = -\dfrac{3}{4}x$

---

8 Refer to the application at the beginning of the lesson. There are 15,000 people in Nunavut under age 25. This is about $\dfrac{3}{5}$ of the total population. What is the population of Nunavut?

Let n represent the total population. Then $\dfrac{3}{5}n = 15,000$ represents the problem.

$\dfrac{3}{5}n = 15,000$

$\dfrac{5}{3}\left(\dfrac{3}{5}n\right) = \dfrac{5}{3}(15,000)$ *Multiply each side by $\dfrac{5}{3}$. Why?*

$n = 25,000$

The total population of Nunavut is 25,000.

Communicating Mathematics

Study the lesson. Then complete the following.

1. **Describe** how to solve $4x = 36$.
2. **List** the steps you take when you check the solution of $4x = 36$.
3. **Write** a multiplication equation whose solution is -7.

Guided Practice

Solve each equation. Check your solution. *(Examples 1–8)*

4. $8x = -24$

5. $-5v = -40$

6. $6 = 0.3y$

7. $\frac{r}{15} = 3$

8. $-7 = \frac{f}{8}$

9. $-9 = -\frac{9}{4}n$

10. **Geometry** The area of the rectangle is 51.2 square centimeters. Find the length. *(Example 3)*

6.4 cm

x cm

Exercises

Practice

Solve each equation. Check your solution.

11. $4c = 24$

12. $9p = 63$

13. $30 = 2x$

14. $-6a = 60$

15. $-8t = 56$

16. $-3 = -5n$

17. $14 = 0.5x$

18. $0 = 3.9c$

19. $2.2r = 11$

20. $\frac{9}{2}t = 1$

21. $\frac{1}{2}y = 32$

22. $\frac{b}{3} = -5$

23. $10 = \frac{a}{-7}$

24. $-\frac{2}{7}x = -6$

25. $-18 = -\frac{2}{3}p$

26. $-\frac{8}{5}p = 40$

27. $-\frac{24}{5}k = 12$

28. $-\frac{63}{10} = -\frac{21}{10}b$

29. What is the solution of $-18.4 = -9.2n$?

30. Solve $\frac{k}{6} = -6$. Then check your solution.

31. Find the value of r in the equation $\left(-4\frac{1}{2}\right)r = 36$.

32. If $7h - 5 = 4$, then $21h - 15 = $ ___?___ .

Write an equation and solve.

33. Eight times a number x is 112. What is the number?

34. The number -154 equals the product of negative seven times a number p. What is the number?

35. Three fifths of a number y is negative 9. What is the number?

36. Plumbing Two meters of copper tubing weigh 0.25 kilogram. How much do 50 meters of the same tubing weigh?

37. Entertainment James is in charge of purchasing student tickets for an Atlanta Braves baseball game. Each student paid $6 per ticket, and he collected $288.

  a. Write an equation representing the number of students n who bought tickets.

  b. How many students are going to the game?

38. Geometry Find the length of the base b of the triangle if its area A is 182 square centimeters. Use the formula $A = \frac{1}{2}bh$.

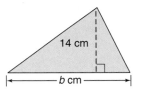

14 cm

b cm

39. Technology Engineering students at The Ohio State University built an experimental electric race car that went around a 1.1-mile lap in 36 seconds. How fast did it go? Find the rate r in miles per hour by using the formula $d = rt$, where d is the distance and t is the time. (*Hint:* Convert seconds to hours.)

40. Critical Thinking Explain why the Multiplicative and Division Properties of Equality for fractions can be thought of as one property.

Mixed Review

Find each quotient. *(Lesson 4–3)*

41. $4 \div 0.2$ **42.** $-8.1 \div 9$ **43.** $\frac{1}{4} \div \frac{3}{7}$

Solve each equation. *(Lesson 3–4)*

44. $n = 5 + 16 \div 4$ **45.** $2.6(5) - 13 = x$ **46.** $\frac{5 + 3}{3 \cdot 3} = t$

47. Animals Giant deep-sea squids can be up to 19 yards long. How many feet long can they be? *(Lesson 2–5)*

Squid

48. Sports A field hockey playing field is shown in the diagram. *(Lesson 1–4)*

  a. Write an equation representing the area A of the field.

  b. Simplify the expression and find the area if a is 60 yards and b is 25 yards.

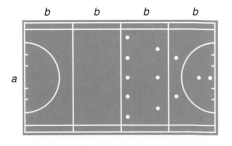

b b b b

a

49. Standardized Test Practice Write an equation for the following sentence. Three less than six times y equals 14. *(Lesson 1–1)*

  A $3 + 6y = 14$ **B** $3 - 6y = 14$

  C $6y = 14 - 3$ **D** $6y - 3 = 14$

Extra Practice   See p. 699.

4-5 Solving Multi-Step Equations

Math In the Workplace

What You'll Learn

You'll learn to solve equations involving more than one operation.

Why It's Important

Shopping Multi-step equations are used to calculate purchase orders.
See Exercise 40.

Trains that float on magnets above rails are being developed in Japan. They are designed to run at a maximum speed of 341 miles per hour. This is 71 miles per hour faster than twice the speed of Japan's bullet trains. How fast are the bullet trains in Japan? *This problem will be solved in Example 4.*

Some equations require more than one step to solve. To solve a problem that has more than one operation, the best strategy is to undo the operations in reverse order. In other words, work backward.

Examples

Solve each equation. Check your solution.

1 $\frac{x}{3} + 5 = 14$

$$\frac{x}{3} + 5 = 14$$

$$\frac{x}{3} + 5 - 5 = 14 - 5 \quad \textit{Subtract 5 from each side.}$$

$$\frac{x}{3} = 9$$

$$3\left(\frac{x}{3}\right) = 3(9) \quad \textit{Multiply each side by 3.}$$

$$x = 27$$

Check: $\quad \frac{x}{3} + 5 = 14 \quad \textit{Original equation}$

$$\frac{27}{3} + 5 \stackrel{?}{=} 14 \quad \textit{Replace x with 27.}$$

$$9 + 5 \stackrel{?}{=} 14 \quad \textit{Simplify } \frac{27}{3}.$$

$$14 = 14 \quad \checkmark \quad \textit{Add 9 and 5.}$$

2 $2k - 7 = 23$

$$2k - 7 = 23$$

$$2k - 7 + 7 = 23 + 7 \quad \textit{Add 7 to each side.}$$

$$2k = 30$$

$$\frac{2k}{2} = \frac{30}{2} \quad \textit{Divide each side by 2.}$$

$$k = 15 \quad \textit{Check the solution.}$$

3 Solve $\dfrac{h-7}{2.5} = -10$. Check your solution.

$$\frac{h-7}{2.5} = -10$$

$$2.5\left(\frac{h-7}{2.5}\right) = 2.5(-10) \quad \textit{Multiply each side by 2.5.}$$

$$h - 7 = -25$$

$$h - 7 + 7 = -25 + 7 \quad \textit{Add 7 to each side.}$$

$$h = -18$$

Check: $\qquad \dfrac{h-7}{2.5} = -10$

$$\frac{-18-7}{2.5} \stackrel{?}{=} -10 \quad \textit{Replace h with -18.}$$

$$\frac{-25}{2.5} \stackrel{?}{=} -10$$

$$-10 = -10 \quad \checkmark$$

Your Turn   Solve each equation. Check your solution.

a. $11 + 9v = 119$ **b.** $\dfrac{a}{8} - 5.2 = 3$ **c.** $\dfrac{b+4}{7} = 6$

Technology Link

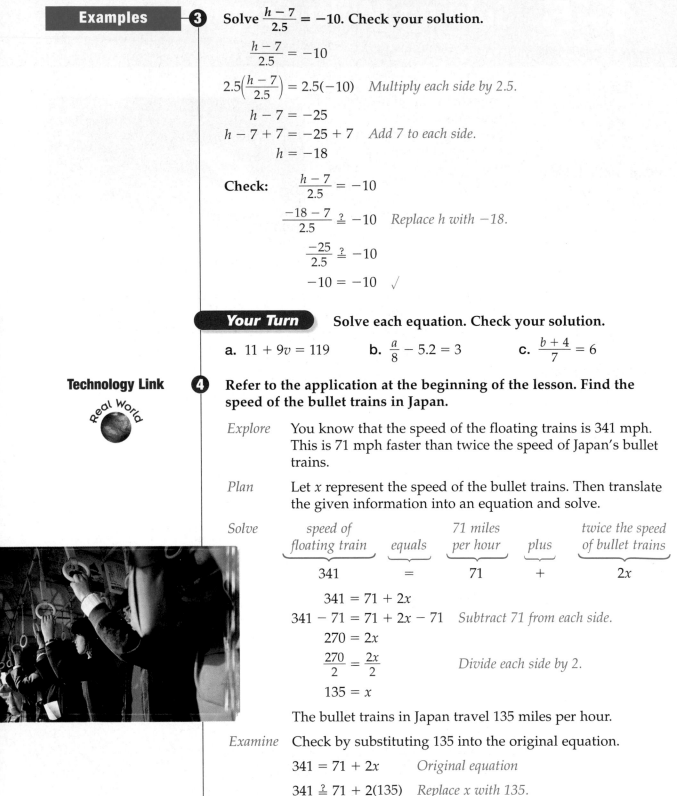

Real World

4 **Refer to the application at the beginning of the lesson. Find the speed of the bullet trains in Japan.**

Explore   You know that the speed of the floating trains is 341 mph. This is 71 mph faster than twice the speed of Japan's bullet trains.

Plan   Let x represent the speed of the bullet trains. Then translate the given information into an equation and solve.

Solve

$\underbrace{\text{speed of floating train}}$	$\underbrace{\text{equals}}$	$\underbrace{\text{71 miles per hour}}$	$\underbrace{\text{plus}}$	$\underbrace{\text{twice the speed of bullet trains}}$
341	=	71	+	$2x$

$$341 = 71 + 2x$$

$$341 - 71 = 71 + 2x - 71 \quad \textit{Subtract 71 from each side.}$$

$$270 = 2x$$

$$\frac{270}{2} = \frac{2x}{2} \qquad\qquad \textit{Divide each side by 2.}$$

$$135 = x$$

The bullet trains in Japan travel 135 miles per hour.

Examine   Check by substituting 135 into the original equation.

$$341 = 71 + 2x \qquad \textit{Original equation}$$

$$341 \stackrel{?}{=} 71 + 2(135) \quad \textit{Replace x with 135.}$$

$$341 \stackrel{?}{=} 71 + 270 \quad \textit{Multiply 2 and 135.}$$

$$341 = 341 \quad \checkmark \qquad \textit{Add 71 and 270.}$$

You can use a graphing calculator to solve multi-step equations.

Graphing Calculator Tutorial
See pp. 724–727.

Technology Tip

To edit or replace an equation, press ▲ until the equation editor is displayed. Then edit or replace the equation.

Graphing Calculator Exploration

Solve $2 = \dfrac{x-5}{4}$ using a graphing calculator.

Step 1 Rewrite the equation so that one side is equal to 0.

$$2 = \frac{x-5}{4}$$

$$2 - 2 = \frac{x-5}{4} - 2 \quad \textit{Subtract 2 from each side.}$$

$$0 = \frac{x-5}{4} - 2$$

Step 2 Press MATH and select 0:Solver.

Step 3 Enter the equation after the 0 =. Press (X,T,θ,n − 5) ÷ 4 − 2 ENTER.

```
(X-5)/4-2=0
■X=13
 bound={-1ε99,1...
■left-rt=0
```

Step 4 Press ALPHA [SOLVE].

The solution is 13.

Try These

1. What would you enter into the calculator to solve $3x - 4 = 6$?
2. Work through the examples in this lesson with a calculator.

Consecutive integers are integers in counting order, such as 3, 4, and 5. Beginning with an even integer and counting by two will result in *consecutive even integers*. For example, $-6, -4, -2, 0,$ and 2 are consecutive even integers. Beginning with an odd integer and counting by two will result in *consecutive odd integers*. For example, $-1, 1, 3,$ and 5 are consecutive odd integers.

Consecutive Integers		
Integers	**Even Integers**	**Odd Integers**
$n = 1$	$n = 2$	$n = 3$
$n + 1 = 2$	$n + 2 = 4$	$n + 2 = 5$
$n + 2 = 3$	$n + 4 = 6$	$n + 4 = 7$
⋮	⋮	⋮

Example

Number Theory Link

Real World

5 **Find three consecutive even integers whose sum is −18.**

Explore You know that the sum of three consecutive even integers is −18. You need to find the three numbers.

Plan Let n represent the first even integer. Then $n + 2$ represents the second even integer, and $n + 4$ represents the third. *Why?*

(continued on the next page)

Solve $n + (n + 2) + (n + 4) = -18$ *Write an equation.*

$3n + 6 = -18$ *Add like terms.*

$3n + 6 - 6 = -18 - 6$ *Subtract 6 from each side.*

$3n = -24$

$\dfrac{3n}{3} = \dfrac{-24}{3}$ *Divide each side by 3.*

$n = -8$

$n = -8$ $n + 2 = -8 + 2$ $n + 4 = -8 + 4$

$= -6$ $= -4$

The numbers are -8, -6, and -4.

Examine   Check by adding the three numbers.

$-8 + (-6) + (-4) \stackrel{?}{=} -18$

$-18 = -18$ ✓

Your Turn

d. Find three consecutive integers whose sum is 27.

Check for Understanding

Communicating Mathematics

Study the lesson. Then complete the following.

Vocabulary

consecutive integers

1. **Describe** how you work backward to solve multi-step equations.

2. **Write** an equation that requires more than one operation to solve.

3. 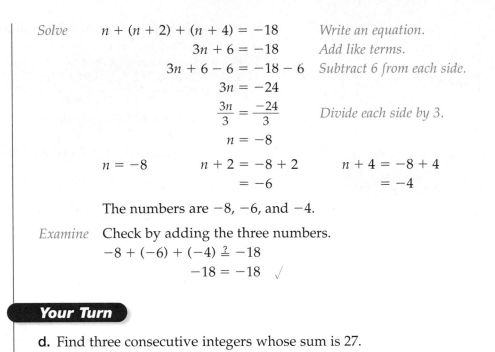 Soto solved the equation $\dfrac{t + 16}{4} = 1$ using the steps at the right. Jean does not agree with his answer. Who is correct? Explain.

$\dfrac{t + 16}{4} = 1$

$t + 16 = 1$

$t + 16 - 16 = 1 - 16$

$t = -15$

Guided Practice

⏱ **Getting Ready**   **State the first step in solving each equation.**

Sample: $3x - 4 = -10$ **Solution:** Add 4 to each side.

4. $7 + 6v = 43$ **5.** $11 = \dfrac{g}{8} - 5$ **6.** $\dfrac{-y - 2}{3} = 15$

Solve each equation. Check your solution.   *(Examples 1–4)*

7. $6t - 1 = 11$ **8.** $7 = 4 - \dfrac{n}{3}$ **9.** $-5r + 2 = 27$

10. $-5.6 + 4d = 2$ **11.** $2 = \dfrac{f + 8}{-6}$ **12.** $\dfrac{7n - 1}{8} = 6$

13. **Animals** A teacher in Wisconsin came up with a formula that more accurately gives a cat's or dog's age in people years. *(Example 1)*

> Take your cat's/dog's age, subtract 1, multiply by 6, add 21, and that will give you the equivalent in people years.

Source: *The Mathematics Teacher*

a. Write an equation to represent the age of a dog or cat in people years y if the dog or cat is x years old.

b. How old is a dog if he is 17 in people years?

Exercises •

Practice

Solve each equation. Check your solution.

14. $6 = 4n + 2$

15. $8 + 3k = 5$

16. $3b - 7 = 2$

17. $15 = 1 - 2t$

18. $12 = -5h + 2$

19. $-13 - 9y = -13$

20. $8 + 1.6a = 0$

21. $0.3x + 3 = 4.8$

22. $6.5 = 2x + 4.1$

23. $27 = 3 + 2.5t$

24. $-4 - 0.7m = 2.3$

25. $-6z + 8 = -40.6$

26. $7 = \frac{x}{2} + 5$

27. $\frac{y}{3} - 6 = 4$

28. $8 + \frac{c}{-4} = 12$

29. $28 = 7 - \frac{3}{2}t$

30. $\frac{3 + n}{7} = -5$

31. $-10 = \frac{s - 8}{-6}$

32. What is the solution of $6 = \frac{s - 8}{-7}$?

33. Find the value of y in the equation $-2.5y - 4 = 14$.

For Exercises 34–36, write an equation and solve each problem.

34. Five more than three-fourths of a number c is 26. Find the number.

35. The steps below are applied to a number n to get $\frac{1}{3}$. What is the number?
    - subtract 2
    - divide by -3

36. Start with the number x. If you multiply the number by 2, divide by 5 and add 20, you get 10. What is the value of x?

Applications and Problem Solving

37. **Geometry** Find the value of x if the perimeter of the square is 20 inches.

38. **Number Theory** Four consecutive odd integers have a sum of 56.
    a. Define a variable and write an equation to represent the problem.
    b. Find the numbers.

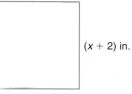

$(x + 2)$ in.

Exercise 37

39. Consumer Spending Dwayne was checking his expenses and found that in eight visits to the barbershop, he had spent $143 for haircuts. Of that amount, $15 was tips.

    a. Write an equation to represent the cost c of the haircuts.

    b. How much does Dwayne pay for each haircut before the tip?

40. Shopping Tanisa and her sister ordered a total of 5 pairs of shoes from a catalog. All of the shoes were the same price. They had to pay $10 for shipping and handling. If their total purchase was $170, how much did each pair of shoes cost?

41. Gardening Mr. Green purchased 39 plants for his garden. His purchase included an equal number of ferns, irises, and gladiolas, as well as 12 impatiens.

    a. Write an equation representing the number of items he purchased.

    b. How many ferns did he buy?

Impatiens

42. Critical Thinking According to legend, Islamic mathematician al-Khowarizmi (810 A.D. – ?) stated in his will that if he had a son, the son should receive twice as much as al-Khowarizmi's wife when he died. If he had a daughter, the daughter should receive half as much as al-Khowarizmi's wife. However, al-Khowarizmi had twins, a boy and a girl. What fraction of his estate should each one get? Explain how you found your answer.

Mixed Review

Geometry **Find each missing measure.** *(Lesson 4–4)*

43. $A = 15$ in^2 **44.** $A = 49$ cm^2 **45.** $A = 27$ m^2

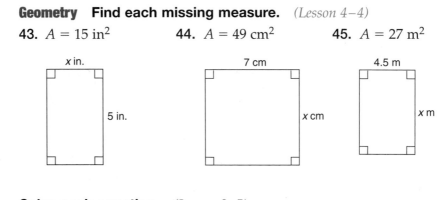

Solve each equation. *(Lesson 3–5)*

46. $-1 + x = -5$ **47.** $n - 3 = 1$ **48.** $t + 4 = -2$

49. Sequences Find the next term in the pattern $-1, 3, -9, 27, \ldots$. *(Lesson 2–5)*

50. Open-Ended Test Practice Graph two positive integers and two negative integers on a number line. *(Lesson 2–1)*

Extra Practice See p. 700.

4-6 Variables on Both Sides

What You'll Learn

You'll learn to solve equations with variables on both sides.

Why It's Important

Sports Equations comparing athletes' performances have variables on both sides.
See Example 3.

On the Fahrenheit scale, freezing occurs at 32°. On the Celsius scale, freezing occurs at 0°. The formula $F = \frac{9}{5}C + 32$ gives the temperature in degrees Fahrenheit when the Celsius temperature is known. Is the temperature ever the same on both scales? *You will solve this problem in Exercise 36.*

Many equations contain variables on both sides. To solve these equations, first use the Addition or Subtraction Property of Equality to write an equivalent equation that has all of the variables on one side. Then solve.

Examples

Solve each equation. Check your solution.

1 $5x = x - 12$

$$5x = x - 12$$
$$5x - x = x - 12 - x \quad \textit{Subtract x from each side.}$$
$$4x = -12$$
$$\frac{4x}{4} = \frac{-12}{4} \quad\quad \textit{Divide each side by 4.}$$
$$x = -3$$

Check:

$$5x = x - 12 \quad \textit{Original equation}$$
$$5(-3) \stackrel{?}{=} -3 - 12 \quad \textit{Replace x with -3.}$$
$$-15 = -15 \quad \checkmark$$

2 $16 + \frac{1}{4}n = \frac{3}{4}n$

$$16 + \frac{1}{4}n = \frac{3}{4}n$$
$$16 + \frac{1}{4}n - \frac{1}{4}n = \frac{3}{4}n - \frac{1}{4}n \quad \textit{Subtract } \frac{1}{4}n \textit{ from each side.}$$
$$16 = \frac{1}{2}n \quad\quad \frac{3}{4}n - \frac{1}{4}n = \frac{2}{4}n \textit{ or } \frac{1}{2}n$$
$$2(16) = 2\left(\frac{1}{2}n\right) \quad \textit{Multiply each side by 2.}$$
$$32 = n \quad\quad \textit{Check the solution.}$$

(continued on the next page)

a. $a + 9 = 4a$ **b.** $\frac{2}{3}n = \frac{1}{3}n - 2$ **c.** $5k - 2.4 = 3k + 1$

You can solve an equation with a variable on both sides to determine when a quantity that is increasing and one that is decreasing will be the same.

Example

Swimming Link

Real World

③ **In the 1996 Olympics, the winning times for the 100-meter butterfly were 52.3 seconds for men and 59.1 seconds for women. Suppose the men's times decrease 0.1 second per year and the women's times decrease 0.3 second per year. Solve $52.3 - 0.1x = 59.1 - 0.3x$ to find when men and women would have the same winning times.**

$$52.3 - 0.1x = 59.1 - 0.3x$$
$$52.3 - 0.1x + 0.3x = 59.1 - 0.3x + 0.3x \quad \textit{Add 0.3x to each side.}$$
$$52.3 + 0.2x = 59.1$$
$$52.3 + 0.2x - 52.3 = 59.1 - 52.3 \qquad \textit{Subtract 52.3 from each side.}$$
$$0.2x = 6.8$$
$$\frac{0.2x}{0.2} = \frac{6.8}{0.2} \qquad\qquad \textit{Divide each side by 0.2.}$$
$$x = 34$$

If the rates continue, men and women would have the same winning times 34 years after 1996, or in 2030.

Some equations have no solution. This means that there is no value for the variable that will make the equation true. Other equations may have every number as the solution. An equation that is true for every value of the variable is called an **identity**.

Examples

Solve each equation.

④ **$5x + 3 = 8 + 5x$**

$$5x + 3 = 8 + 5x$$
$$5x + 3 - 5x = 8 + 5x - 5x \quad \textit{Subtract 5x from each side.}$$
$$3 = 8$$

The equation has no solution.   *3 = 8 is never true.*

⑤ **$3 + 3t - 5 = 3t - 2$**

$$3 + 3t - 5 = 3t - 2$$
$$3t - 2 = 3t - 2 \quad \textit{Simplify.}$$

The equation is an identity.   *3t − 2 = 3t − 2 is true for all values of t.*

Your Turn

d. $2t + 4 - t = 4 + t$ **e.** $16h + 7 = 16 + 16h$

Communicating Mathematics

Study the lesson. Then complete the following.

1. **List** the steps you would take to solve the equation $3x + 5 = -2x - 16$.

2. **Explain** why it is important to check your solutions.

3. **Explain** the difference between an equation that is an identity and an equation with no solution.

Guided Practice

Getting Ready **State the first step in solving each equation.**

Sample: $9r = -10 + 7r$ **Solution:** Subtract $7r$ from each side.

4. $2y = 4 - 3y$

5. $0.5p + 4 = 1.5p$

6. $\dfrac{-y - 6}{3} = 7y$

Solve each equation. Check your solution. *(Examples 1–5)*

7. $10x = 3x + 14$

8. $2r - 21 = 3 - 4r$

9. $8y + 1 = 1 + 8y$

10. $7 - 6x = 41 - 6x$

11. $3.4a + 3 = 2a - 4$

12. $\dfrac{1}{3}m - 10 = \dfrac{2}{3}m + 4$

13. **Number Theory** Three times a number n is 24 less than five times the number. *(Example 1)*

   a. Write an equation to represent the problem.

   b. Find the value of n.

Exercises

Practice

Solve each equation. Check your solution.

14. $10h = 8h + 6$

15. $7u - 19 = 6u$

16. $6x + 2 = 3x - 1$

17. $5t - 9 = -3t + 7$

18. $2n + 15 = 2n + 20 - 5$

19. $3x + 6 = 7x - 4$

20. $8y - 5 = -y + 2$

21. $18 + 3k = 22 + 3k$

22. $12 + 2.5a = 5a$

23. $a + 3.1 = 6.6 - 4a$

24. $6.2y + 7 = 3y - 1$

25. $3.2x - 1.8 = 6 + 11x$

26. $4.5 - x = 2x - 8.1$

27. $1.6 - 4.3y = -5.2 - 2.3y$

28. $\dfrac{1}{2}d = 2 - \dfrac{1}{2}d$

29. $\dfrac{1}{5}p - 4 = \dfrac{2}{5}p$

30. $\dfrac{1}{3}z - 4 = -\dfrac{1}{3}z + 18$

31. $\dfrac{1}{9}p - 9 = \dfrac{4}{9}p - 29$

32. Solve $\frac{3}{4}n + 16 = 2 - \frac{1}{8}n$.

33. Find the solution of $18 - 3.8x = 7.5 - 1.3x$.

34. Eight times a number n is 51 more than five times the number.

    a. Write an equation to represent the problem.

    b. Find the number.

35. Geometry The length of the rectangle is twice the width.

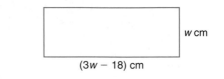

w cm

$(3w - 18)$ cm

    a. Write an equation to represent the problem.

    b. What are the dimensions of the rectangle?

36. Temperature Refer to the application at the beginning of the lesson. Is the temperature ever the same on both the Fahrenheit and Celsius scales? To find a temperature when both are the same, let $F = C$ in the equation $F = \frac{9}{5}C + 32$ and solve $C = \frac{9}{5}C + 32$.

37. Critical Thinking Write an equation with variables on both sides that has a solution of 6.

Mixed Review

38. Solve $3 + 6k = 45$. Check your solution. *(Lesson 4–5)*

Solve each equation. Check your solution. *(Lesson 4–4)*

39. $-7t = 63$ **40.** $0 = 4.5c$ **41.** $-\frac{2}{3}x = -8$

42. School Supplies How many different kinds of transparent tape are possible? *(Lesson 4–2)*

Transparent Tape		
Type	**Size**	**Color**
permanent	$\frac{1}{2}$ in.	clear
removable		lime
double-sided	$\frac{3}{4}$ in.	purple
	1 in.	pink

43. Entertainment In order to raise money for a trip to see a Broadway musical, the music club has set up a lottery using three numbers. The first number is between 1 and 4, inclusive. The second number is between 3 and 8, inclusive. The third number is between 5 and 15, inclusive. How many different lottery numbers are possible? *(Lesson 4–2)*

Solve each equation. Check your solution. *(Lesson 3–6)*

    44. $8 + g = 17$ **45.** $b + (-5) = 26$ **46.** $k - 6 = -11$

47. Health Most people drink more than the recommended eight glasses of beverage each day. However, some beverages actually make a person thirsty. The graph shows the daily average glasses of beverages per person. Find the difference between the glasses of water and the glasses of caffeine soda that a person drinks on average per day. *(Lesson 3–2)*

Some Drinks Leave You Thirsty

Quenches Your Thirst

Water 4.6
Juice 1.4
Milk 1.3
No-caffeine soda 0.6

Leaves You Thirsty

Coffee/tea 2.8
Caffeine soda 1.3
Alcohol 0.8

Source: Yankelovich for New York Hospital Nutrition Information Center, 1998

48. Standardized Test Practice The frequency table shows the results of a survey that asked students how many times a week they buy lunch at school. Determine which statement is correct. *(Lesson 1–6)*

A Most students buy lunch four times a week.

B The greatest number of students buys lunch three times a week.

C The least number of students buys lunch every day.

D Five students buy lunch twice a week.

Buying Lunch		
Number of Days	**Tally**	**Frequency**
0	III	3
1	ЖГ	5
2	ЖГ I	6
3	ЖГ ЖГ	10
4	ЖГ III	8
5	ЖГ	5

Quiz 2   Lessons 4–4 through 4–6

▶ **Solve each equation. Check your solution.**

1. $-3t = 18$

2. $\dfrac{c}{-10} = -5$

3. $\dfrac{5}{7}d = 45$ *(Lesson 4–4)*

4. $2g + 11 = -7$

5. $\dfrac{b}{4} - 11 = 7$

6. $-9 = \dfrac{d + 5}{3}$ *(Lesson 4–5)*

7. $5n + 4 = 7n - 20$

8. $8t + 15 = -9t - 2$

9. $6a - 4 = 15 + 6a$ *(Lesson 4–6)*

10. Basketball According to NBA standards, a basketball should bounce back $\dfrac{2}{3}$ the distance from which it is dropped. How high should a basketball bounce when it is dropped from a height of $3\dfrac{3}{4}$ feet? *(Lesson 4–4)*

Math In the Workplace

What You'll Learn
You'll learn to solve equations with grouping symbols.

Why It's Important
Painting Painters can solve equations with grouping symbols to determine how much paint to buy.

A gallon of paint covers about 350 square feet. Figure the room's square footage by multiplying the combined wall lengths by wall height and subtracting 15 square feet for each window and door.

Source: *The Family Handyman Magazine Presents Handy Hints for Home, Yard, and Workshop, 1998*

Suppose a room measures 16 feet long by 12 feet wide by 7 feet high and has two windows and a door. How many gallons of paint are needed to paint the room? To solve this problem, first find the square footage of the room. *The sum of the wall lengths is the perimeter of the room.*

$$\underbrace{x}_{\substack{square \\ footage}} = \underbrace{2(16 + 12)}_{\substack{perimeter \\ of\ room}} \cdot \underbrace{7}_{\substack{wall \\ height}} - \underbrace{(15 + 15 + 15)}_{window\ +\ window\ +\ door}$$

Now solve the equation to find x.

$x = 2(16 + 12) \cdot 7 - (15 + 15 + 15)$	
$x = 2(28) \cdot 7 - 45$	*Simplify inside the parentheses.*
$x = 56 \cdot 7 - 45$	*Multiply 2 and 28.*
$x = 392 - 45$	*Multiply 56 and 7.*
$x = 347$	*Subtract 45 from 392.*

The room has an area of 347 square feet. Therefore, only one gallon of paint is needed to paint the room.

In solving the equation above, the first step was to remove the parentheses. This is often the first step in solving any equation that contains grouping symbols.

Examples

Solve each equation. Check your solution.

❶ $8 = 4(3x + 5)$

$8 = 4(3x + 5)$	
$8 = 12x + 20$	*Distributive Property*
$8 - 20 = 12x + 20 - 20$	*Subtract 20 from each side.*
$-12 = 12x$	
$\dfrac{-12}{12} = \dfrac{12x}{12}$	*Divide each side by 12.*
$-1 = x$	

Check: $8 = 4(3x + 5)$

$8 \stackrel{?}{=} 4[3(-1) + 5]$ *Replace x with −1.*

$8 \stackrel{?}{=} 4(-3 + 5)$

$8 \stackrel{?}{=} 4(2)$

$8 = 8$ ✓

2 $2(3n - 5) - 3(1 - n) = 5$

$2(3n - 5) - 3(1 - n) = 5$

$6n - 10 - 3 + 3n = 5$ *Distributive Property*

$9n - 13 = 5$ *Add like terms.*

$9n - 13 + 13 = 5 + 13$ *Add 13 to each side.*

$9n = 18$

$\dfrac{9n}{9} = \dfrac{18}{9}$ *Divide each side by 9.*

$n = 2$

Check: $2(3n - 5) - 3(1 - n) = 5$

$2[3(2) - 5] - 3(1 - 2)] \stackrel{?}{=} 5$ *Replace n with 2.*

$2(6 - 5) - 3(-1) \stackrel{?}{=} 5$

$2(1) + 3 \stackrel{?}{=} 5$

$5 = 5$ ✓

Your Turn

a. $7 = 3(x + 1) - 2$ **b.** $4(t + 5) + 6(2t - 3) = 12$

Building Link

Real World

3 The Rock and Roll Hall of Fame in Cleveland, Ohio, has a trapezoid-shaped cinema with an area of about 4675 square feet. Find x, the approximate width of the movie screen in the cinema.

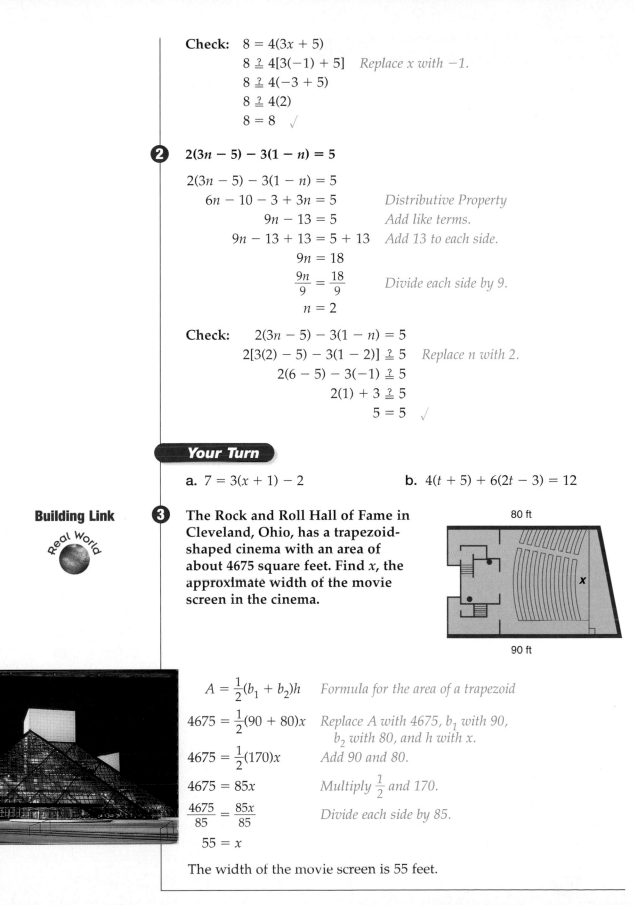

80 ft

x

90 ft

$A = \dfrac{1}{2}(b_1 + b_2)h$ *Formula for the area of a trapezoid*

$4675 = \dfrac{1}{2}(90 + 80)x$ *Replace A with 4675, b_1 with 90, b_2 with 80, and h with x.*

$4675 = \dfrac{1}{2}(170)x$ *Add 90 and 80.*

$4675 = 85x$ *Multiply $\dfrac{1}{2}$ and 170.*

$\dfrac{4675}{85} = \dfrac{85x}{85}$ *Divide each side by 85.*

$55 = x$

The width of the movie screen is 55 feet.

Check for Understanding

Study the lesson. Then complete the following.

1. **State** the first step in solving $w - 8(w + 15) = -9$.

2. **Describe** a way to remove grouping symbols from an equation.

3. **Write** a summary of the techniques you learned in this chapter for solving equations.

Guided Practice

Solve each equation. Check your solution. *(Examples 1–3)*

4. $12 = 2(k + 1)$

5. $3n - 2 = 3(n + 1) - 5$

6. $7 + 6x = 2(10 + 3x)$

7. $3p - 8(1 + p) = 12$

8. $5(2k - 3) + 2(k + 4) = 17$

9. $5 - \frac{1}{2}(b - 6) = 4$

10. **Geometry** Find the value of x in the trapezoid if its area is 39 square centimeters. *(Example 3)*

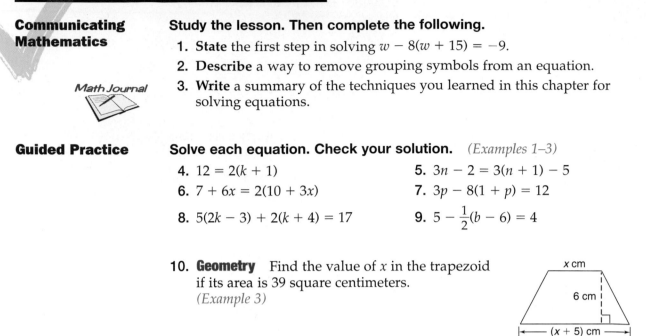

Exercises · · · · · · · · · · · · · · · · · · ·

Practice

Solve each equation. Check your solution.

11. $3(a + 2) = 24$

12. $14 = 2(4n - 1)$

13. $7 = 5(x - 4) + 2$

14. $6(2 + y) - 4 = -10$

15. $-2(6 + p) = 8$

16. $5(z - 1) = 5z$

17. $t + 4(1 - t) = 9t$

18. $3(b - 2) - 4(2b + 5) = 9$

19. $18 = 2(g + 4) + 3(2g - 6)$

20. $7(y - 2) + 8 = 3(y - 4) - 2$

21. $2.4h + 1 = 7(2h - 4)$

22. $2(f + 3) - 2.7 = 3(f + 1.1)$

23. $6(3.5 + r) - 1 = 7r - r$

24. $4(5 - x) - 2.1(6x) = -4.9$

25. $4(2a - 8) = 7a + 10$

26. $3 + \frac{5}{8}x = 2(x - 4)$

27. $3(1 - 4k) = \frac{1}{2}(-24k + 6)$

28. $\frac{5(n - 1)}{3} = n + 1$

29. Find the solution of $2[3 + 4(g + 1)] = -2$. Check your solution.

30. What is the value of y in $6(3 + 5y) + 5(y - 8) = 13$?

**Applications and
Problem Solving**

Real World

31. **Geometry** The area of a trapezoid is 70 square inches. Find the lengths of the bases if base 1 is four more than twice the length of base 2, and the altitude is 5 inches.

32. Gardening Carlos is building a fence around his garden to keep out the rabbits. One side of the garden is against the garage, as shown in the diagram at the right. He has 88 feet of fencing that fit around the three other sides.

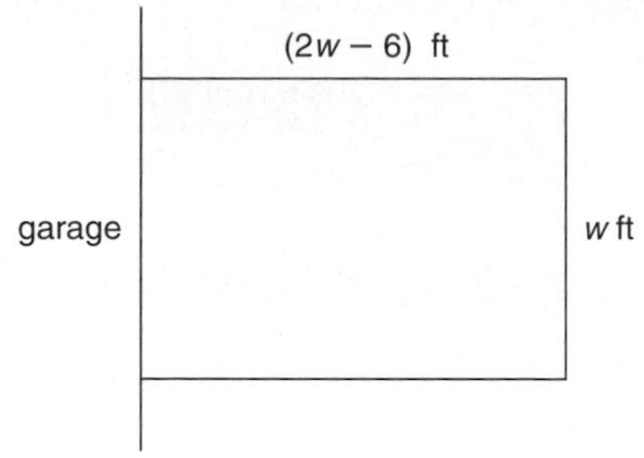

$(2w - 6)$ ft

garage

w ft

a. Write an equation to find the width w of the garden.

b. What are the dimensions of the garden?

33. Number Theory Twice the greater of two consecutive odd integers is 13 less than three times the lesser. Find the integers.

34. Critical Thinking *Complementary angles* are two angles whose measures have a sum of 90°. Each angle is said to be the *complement* of the other angle. The measure of angle A is one-half the measure of its complement. Find the measure of angle A and the measure of its complement.

A

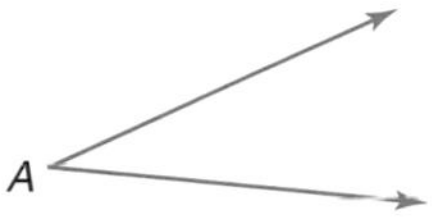

Mixed Review

Solve each equation. Check your solution. *(Lesson 4–6)*

35. $2n + 3 = 6 - n$

36. $5t - 12 = 8t$

37. Number Theory Four less than one-half of a number x is 7. *(Lesson 4–5)*

a. Write an equation to represent the problem.

b. What is the number?

38. Evaluate $\dfrac{ab}{-3}$ if $a = 6$ and $b = -5$. *(Lesson 2–6)*

39. Astronomy The table gives the average temperatures for six planets. Order the temperatures from least to greatest. *(Lesson 2–1)*

Planet	Average Temperature (°F)
Jupiter	−229
Neptune	−364
Pluto	−382
Saturn	−292
Uranus	−369
Venus	867

40. Standardized Test Practice A pitcher's earned-run average is given by the formula $ERA = 9\left(\dfrac{a}{b}\right)$, where a is the number of earned runs the pitcher has allowed and b is the number of innings pitched. Find a pitcher's ERA to the nearest hundredth if he has allowed 23 earned runs in 180 innings. *(Lesson 1–5)*

A 0.01   **B** 0.13   **C** 1.15   **D** 7.83

Understanding and Using the Vocabulary

After completing this chapter, you should be able to define each term, property, or phrase and give an example or two of each.

*inter*NET CONNECTION **Review Activities**
For more review activities, visit:
www.algconcepts.glencoe.com

Algebra
consecutive integers *(p. 167)*
identity *(p. 172)*
multiplicative inverses *(p. 154)*
reciprocals *(p. 154)*

Counting
combination *(p. 152)*
factorial *(p. 153)*
event *(p. 147)*
outcomes *(p. 146)*

permutation *(p. 152)*
sample space *(p. 146)*
tree diagram *(p. 146)*

Complete each sentence using a term from the vocabulary list.

1. A subset of the possible outcomes of a sample space is called a(n) ___?___.
2. Two numbers whose product is 1 are called ___?___.
3. To find the sample space of a particular situation, a(n) ___?___ may be used.
4. Integers in counting order are called ___?___.
5. An equation that is true for every value of the variable is called a(n) ___?___.
6. The ___?___ is the list of all the possible outcomes.
7. The results shown by a tree diagram are called ___?___.
8. The numbers $-\frac{4}{3}$ and $-\frac{3}{4}$ are called ___?___.
9. To find the possible number of ___?___ by multiplying, the Fundamental Counting Principle is used.
10. The equation $\frac{1}{2}x + 1 = \frac{1}{2}x + 1$ is an example of a(n) ___?___.

Skills and Concepts

Objectives and Examples	Review Exercises

• **Lesson 4–1** Multiply rational numbers.

$-3.2(-4) = 12.8$ *The signs are the same. The product is positive.*

$\frac{4}{5}\left(-\frac{2}{9}\right) = -\frac{4 \cdot 2}{5 \cdot 9}$

$= -\frac{8}{45}$ *The signs are different. The product is negative.*

Find each product.
11. $6(3.8)$
12. $-9.2 \cdot 0$
13. $-1\left(\frac{3}{4}\right)$
14. $-\frac{5}{6}\left(-\frac{1}{3}\right)$

Simplify each expression.
15. $-2.7(4y)$
16. $5a(3.4b)$
17. $6s\left(-\frac{1}{7}\right)$
18. $\left(-\frac{3}{8}m\right)\left(-\frac{3}{5}n\right)$

Objectives and Examples

- **Lesson 4–2** Use tree diagrams or the Fundamental Counting Principle to count outcomes.

Shirt	Pants	Socks
red	black	red
green	blue	green
white		white

There are 3 shirts, 2 pants, and 3 socks, so the number of different combinations is $3 \times 2 \times 3$ or 18.

Review Exercises

19. A sandwich can be made with wheat or white bread. It can have ham, turkey, or beef. Find the number of possible outcomes by drawing a tree diagram.

20. Find the number of different team jerseys.

Shirt	Stripe	Numbers
white	green	green
blue	yellow	black
	gold	gold
	red	red

- **Lesson 4–3** Divide rational numbers.

Evaluate $\frac{t}{3}$ if $t = \frac{1}{5}$.

$$\frac{t}{3} = \frac{\frac{1}{5}}{3} \qquad \textit{Replace } t \textit{ with } \frac{1}{5}.$$

$$= \frac{1}{5} \div 3$$

$$= \frac{1}{5} \cdot \frac{1}{3} \text{ or } \frac{1}{15}$$

Find each quotient.

21. $5 \div (-2.5)$

22. $-7.2 \div (-1.6)$

23. $\frac{4}{5} \div \frac{3}{2}$

24. $3\frac{1}{2} \div \left(-\frac{1}{2}\right)$

Evaluate each expression if $x = \frac{2}{3}$ and $y = -\frac{1}{2}$.

25. $\frac{y}{4}$

26. $\frac{5}{x}$

27. $\frac{y}{x}$

- **Lesson 4–4** Solve multiplication and division equations by using the properties of equality.

$$\frac{x}{3} = -12$$

$$3\left(\frac{x}{3}\right) = 3(-12) \quad \textit{Multiply each side by 3.}$$

$$x = -36$$

Solve each equation. Check your solution.

28. $5n = 30$

29. $16 = 2a$

30. $-k = 5$

31. $-8y = 32$

32. $3t = 4.2$

33. $-2x = -24.6$

34. $\frac{m}{4} = -6$

35. $\frac{5}{9}r = 5$

- **Lesson 4–5** Solve equations involving more than one operation.

$$\frac{x}{4} + 6 = 18$$

$$\frac{x}{4} + 6 - 6 = 18 - 6 \quad \textit{Subtract 6 from each side.}$$

$$4\left(\frac{x}{4}\right) = 4(12) \quad \textit{Multiply each side by 4.}$$

$$x = 48$$

Solve each equation. Check your solution.

36. $7t + 3 = 24$

37. $1 + 6y = -11$

38. $8 - 2b = 10$

39. $-4y + 1 = 1$

40. $2.1 - 3r = 4.8$

41. $-1.8 = 5.1w - 12$

42. $14 = 6 + \frac{x}{5}$

43. $\frac{4 + x}{7} = -3$

Chapter 4 Study Guide and Assessment

Objectives and Examples	Review Exercises

• **Lesson 4–6** Solve equations with variables on both sides.

$$7x = 4x - 3$$
$$7x - 4x = 4x - 3 - 4x \quad \textit{Subtract 4x from}$$
$$3x = -3 \qquad\qquad \textit{each side.}$$
$$\frac{3x}{3} = \frac{-3}{3} \quad \textit{Divide each side by 3.}$$
$$x = -1$$

Solve each equation. Check your solution.

44. $3a = a + 4$

45. $2t + 4 = 10 + t$

46. $11 + 4m = 4m + 8$

47. $5x - 5.2 = 9.4 + 3x$

48. $6 + \frac{1}{3}t = -2 + \frac{2}{3}t$

• **Lesson 4–7** Solve equations with grouping symbols.

$$5(x - 7) = 10$$
$$5x - 35 = 10 \qquad\qquad \textit{Distributive Property}$$
$$5x - 35 + 35 = 10 + 35 \quad \textit{Add 35 to each side.}$$
$$5x = 45$$
$$\frac{5x}{5} = \frac{45}{5} \qquad\qquad \textit{Divide each side by 5.}$$
$$x = 9$$

Solve each equation. Check your solution.

49. $6a = 2(4 - a)$

50. $10 = 5(n + 3)$

51. $4(3h - 2) = 7h + 2$

52. $2(1 + 2x) + 3(x - 4) = 11$

53. $\frac{1}{3}(12 - 6t) = 4 - 2t$

Applications and Problem Solving

54. Dining With each shrimp, salmon, or perch dinner at the Seafood Palace, you may have soup or salad. You may also have broccoli, baked potato, or rice. How many different dinners are possible? *(Lesson 4–2)*

55. Space Halley's Comet flashes through the sky every 76.3 years. How many times will it appear in 381.5 years? *(Lesson 4–3)*

56. Number Theory Four more than two-thirds of a number *n* is zero. *(Lesson 4–5)*
   a. Write an equation to represent the problem.
   b. What is the value of *n*?

57. Geometry The width of the rectangle is equal to one-half the length. What is the width of the rectangle? *(Lesson 4–6)*

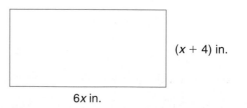

$(x + 4)$ in.

$6x$ in.

1. **Write** an equation using grouping symbols that results in $x = -5$.

2. **Draw** a tree diagram to show the different calculators that are possible from the choices given in the table.

3. **Explain** whether the equation $3(2x - 3) - 1 = 6x - 10$ is an identity or has no solution.

Brand	Type	Power
A	scientific	solar
B	graphing	battery
C		

Exercise 2

Simplify each expression.

4. $7.2(-4x)$

5. $-3.9(-10t)$

6. $-\frac{4}{5}a\left(\frac{2}{3}b\right)$

7. $\frac{2}{3}\left(\frac{1}{2}g\right) + \frac{1}{3}g$

8. Find the number of possible outfits by drawing a tree diagram.

Shirt	Pants
red	blue
yellow	black
green	gray

Find the number of possible outcomes by using the Fundamental Counting Principle.

9. rolling an eight-sided die twice

10. spinning the spinner three times

Evaluate each expression if $f = \frac{1}{2}$, $g = \frac{2}{5}$, and $h = -\frac{1}{3}$.

11. $\frac{f}{5}$

12. $\frac{3}{h}$

13. $\frac{g}{f}$

14. $\frac{h}{g}$

Solve each equation.

15. $5n = 35$

16. $18.4 = -2m$

17. $\frac{3}{4}x = 6$

18. $20 = -\frac{5}{4}a$

19. $3r + 1 = 7$

20. $13 - 4y = 29$

21. $10 = 7 + \frac{5}{8}w$

22. $7b + 8 = 5b - 4$

23. $2(h - 3) = 5 + 13h$

24. **Number Theory** Find three consecutive even integers whose sum is -12.

25. **Geometry** Find the dimensions of the rectangle if the perimeter is 220 feet.

$2w - 40$

w

Statistics Problems

Many standardized test questions provide a table or a list of numerical data to calculate the *mean, median, mode,* and *range.*

$$\text{mean} = \frac{\text{sum of the numbers}}{\text{number of numbers}}$$

THE PRINCETON REVIEW

On standardized tests, *average* is the same as *mean.*

Proficiency Test Example

The average speeds of the winners of the Indianapolis 500 from 1989 to 1998 are listed in the table. What is the median of the speeds?

Indianapolis 500 Winners			
Year	Average Speed (mph)	Year	Average Speed (mph)
1989	167.6	1994	160.9
1990	186.0	1995	153.6
1991	176.5	1996	148.0
1992	134.5	1997	145.8
1993	157.2	1998	145.2

A 153.6   B 155.4
C 157.53   D 157.2

Hint The median of an even number of values is the average (mean) of the middle two values when they are arranged in numeric order.

Solution Don't be confused by the phrase *average speed*; think of it as the winning speed. Write the winning speeds in order from least to greatest.

134.5, 145.2, 145.8, 148.0, <u>153.6</u>, <u>157.2</u>, 160.9, 167.6, 176.5, 186.0

There are ten values, so the median is the mean of the two middle values.

$$\frac{153.6 + 157.2}{2} = \frac{310.8}{2} \text{ or } 155.4$$

The median is 155.4 mph. The answer is B.

SAT Example

A survey of the town of Niceville found a mean of 3.2 persons per household and a mean of 1.2 televisions per household. If 48,000 people live in Niceville, how many televisions are there in Niceville?

A 15,000   B 16,000   C 18,000
D 40,000   E 57,600

Hint Use your calculator for large numbers.

Solution Solve this problem one step at a time.

You need to find the number of televisions. The survey information is based on households, so first calculate the number of households.

number of people	divided by	people per household	equals	number of households
48,000	÷	3.2	=	15,000

The survey information says there is a mean (average) of 1.2 televisions per household. Multiply the number of households by the televisions per household.

number of households	times	televisions per household	equals	number of televisions
15,000	×	1.2	=	18,000

The answer is C.

After you work each problem, record your answer on the answer sheet provided or on a sheet of paper.

1. A company plans to open a new fitness center. It conducts a survey of the number of hours people exercise weekly. The results for 12 people chosen at random are 7, 2, 4, 8, 3, 0, 3, 1, 0, 5, 3, and 4 hours. What is the mode of the data?

   A 3   **B** 3.3   **C** 4

   D There is no mode.

2. For their science fiction book reports, students must choose one book from a list of eight. The books have the following numbers of pages: 272, 188, 164, 380, 442, 216, 360, and 262. What is the median number of pages?

   A 267   **B** 285.5   **C** 303   **D** 442

3. $\dfrac{0.5 + 0.5 + 0.5 + 0.5}{4} =$

   A 0.05   **B** 0.125   **C** 0.5

   D 1   **E** 2.0

4. The graph shows the percent of men and women who held two jobs. In what year was the total percent of men and women holding two jobs greatest?

   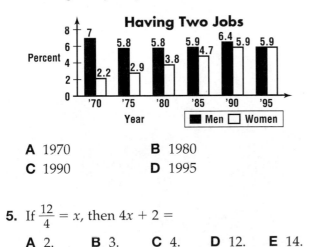

   A 1970   **B** 1980
   C 1990   **D** 1995

5. If $\dfrac{12}{4} = x$, then $4x + 2 =$

   A 2.   **B** 3.   **C** 4.   **D** 12.   **E** 14.

6. Over nine games, a baseball team had an average of 8 runs per game. In the first seven games, they had an average of 6 runs per game. They scored the same number of runs in each of the last two games. How many runs did they score in the last game?

   A 5   **B** 15   **C** 26   **D** 30   **E** 46

7. The stem-and-leaf plot shows the populations of ten cities in Oklahoma. Which statement is correct?

Stem	Leaf
35	6
36	2 4
37	5 6 7
38	1 2 3 6 $35\,|\,6 = 35{,}600$

   A The mode is 38,000.
   B The range of the data is 35,600.
   C Most of these cities have fewer than 36,000 people.
   D The median population is 37,650.

8. If $5 + \dfrac{n}{4} = 9$, then what is the value of n?

   A 3.5   **B** 4   **C** 16   **D** 56

Open-Ended Questions

9. **Grid-In**   If $3x = 12$, then $8 \div x = ?$

10. The number of daily visitors to the local history museum during the month of June is shown below.

    11 19 61 18 43 22 18 14 21 8
    19 41 36 16 16 14 24 31 64 7
    29 24 27 33 31 71 89 61 41 34

    Part A   Construct a frequency table that represents the data.
    Part B   Construct a histogram from the frequency table.

Test Practice   For additional test practice questions, visit: www.algconcepts.glencoe.com

Proportional Reasoning and Probability

▶ What You'll Learn in Chapter 5:

- to solve proportions *(Lesson 5–1)*,
- to solve problems involving scale drawings and models *(Lesson 5–2)*,
- to solve problems by using the percent proportion and the percent equation *(Lessons 5–3, 5–4, and 5–5)*, and
- to find the probability of simple events, mutually exclusive events, and inclusive events *(Lessons 5–6 and 5–7)*.

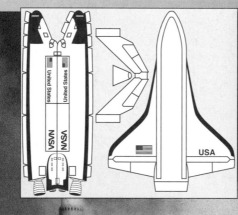

Problem-Solving Workshop

Project

You are volunteering at a community youth club after school. This week, the students are making paper models of the space shuttle. The pattern that you have is 2 inches wide and 2.25 inches long. How can you enlarge the pattern to fit on an 11-inch by 14-inch sheet of paper if the copy machine will only enlarge up to 150% in increments of 1%?

Working on the Project

Work with a partner and choose a strategy. Develop a plan. Here are some suggestions to help you get started.

- If you enlarge each dimension of the pattern 150% on the copier, find the new dimensions of the pattern.
- Find the largest factor by which you can multiply both dimensions and still fit the pattern on an 11-inch by 14-inch sheet of paper.

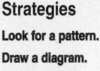

Strategies

Look for a pattern.

Draw a diagram.

Make a table.

Work backward.

Use an equation.

Make a graph.

Guess and check.

Technology Tools

- Use a **spreadsheet** to calculate various enlargements.

 Research For more information about the space shuttle, visit:
www.algconcepts.glencoe.com

Presenting the Project

Make a poster that shows the different enlargements you used. In addition, answer these questions.

- Suppose you need to enlarge a pattern 300% of its original size. Would enlarging the pattern 150% and then enlarging the new pattern 150% work? Explain your reasoning.
- Can you make the 2-inch by 2.25-inch pattern fit exactly on an 11-inch by 14-inch sheet of paper? Why or why not?
- How does the problem change if a copier can enlarge up to 200%?

Math In the Workplace

What You'll Learn

You'll learn to solve proportions.

Why It's Important

Cooking Caterers use proportions to adjust their recipes. *See Exercise 50.*

Two slices of a large pepperoni pizza have 15 grams of fat, while two slices of a large cheese pizza have only 10 grams of fat. How many times more fat is in the pepperoni pizza than the cheese pizza? To find out, divide 15 by 10.

$$15 \div 10 = 1.5$$

So, two slices of pepperoni pizza have 1.5 times as much fat as two slices of cheese pizza.

In mathematics, a **ratio** is a comparison of two numbers by division. The ratio above can be expressed in the following ways.

$$15 \text{ to } 10 \qquad 15{:}10 \qquad \frac{15}{10}$$

You can express $\frac{15}{10}$ in simplest form as $\frac{3}{2}$.
An equation stating that two ratios are equal is called a **proportion**. So, $\frac{15}{10} = \frac{3}{2}$ is a proportion.
Every proportion has two cross products.

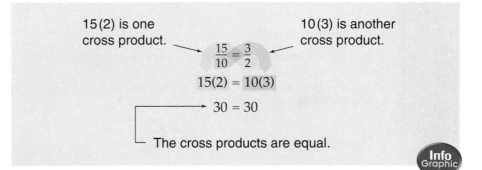

15(2) is one cross product.

10(3) is another cross product.

$$\frac{15}{10} = \frac{3}{2}$$

$$15(2) = 10(3)$$

$$30 = 30$$

The cross products are equal.

Info Graphic

In the proportion, $b \neq 0$ and $d \neq 0$.

Property of Proportions	
Words:	The cross products of a proportion are equal.
Symbols:	If $\frac{a}{b} = \frac{c}{d}$, then $ad = bc$. If $ad = bc$, then $\frac{a}{b} = \frac{c}{d}$.
Numbers:	If $\frac{15}{10} = \frac{3}{2}$, then $15(2) = 10(3)$.
	If $15(2) = 10(3)$, then $\frac{15}{10} = \frac{3}{2}$.

You can use cross products to solve proportions. Sometimes you'll use the Distributive Property.

Examples

Solve each proportion.

1 $\dfrac{5}{3} = \dfrac{75}{n}$

$$\dfrac{5}{3} = \dfrac{75}{n}$$

$5n = 3(75)$ *Find the cross products.*

$5n = 225$

$\dfrac{5n}{5} = \dfrac{225}{5}$ *Divide each side by 5.*

$n = 45$

Check: $\dfrac{5}{3} = \dfrac{75}{n}$

$\dfrac{5}{3} \stackrel{?}{=} \dfrac{75}{45}$ *Replace n with 45.*

$5(45) \stackrel{?}{=} 3(75)$ *Find the cross products.*

$225 = 225$ $\checkmark$

The solution is 45.

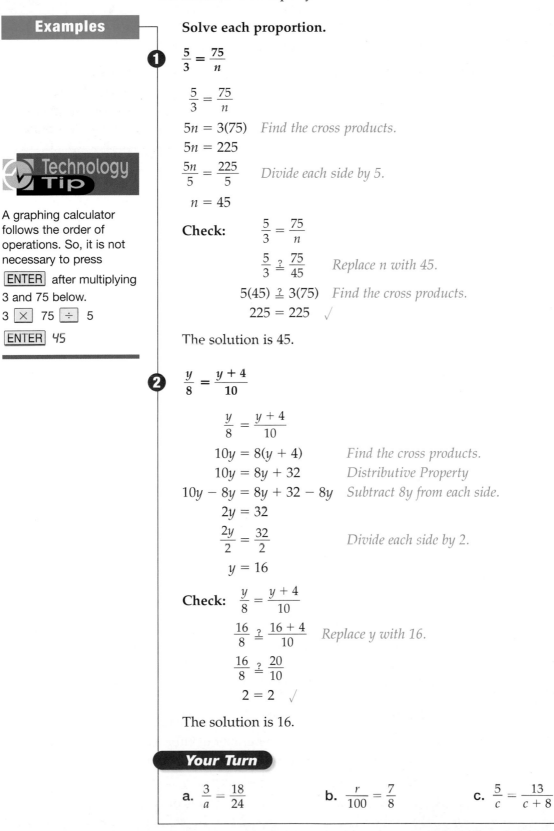

Technology Tip

A graphing calculator follows the order of operations. So, it is not necessary to press ENTER after multiplying 3 and 75 below.

3 × 75 ÷ 5

ENTER 45

2 $\dfrac{y}{8} = \dfrac{y + 4}{10}$

$$\dfrac{y}{8} = \dfrac{y + 4}{10}$$

$10y = 8(y + 4)$ *Find the cross products.*

$10y = 8y + 32$ *Distributive Property*

$10y - 8y = 8y + 32 - 8y$ *Subtract 8y from each side.*

$2y = 32$

$\dfrac{2y}{2} = \dfrac{32}{2}$ *Divide each side by 2.*

$y = 16$

Check: $\dfrac{y}{8} = \dfrac{y + 4}{10}$

$\dfrac{16}{8} \stackrel{?}{=} \dfrac{16 + 4}{10}$ *Replace y with 16.*

$\dfrac{16}{8} \stackrel{?}{=} \dfrac{20}{10}$

$2 = 2$ $\checkmark$

The solution is 16.

Your Turn

a. $\dfrac{3}{a} = \dfrac{18}{24}$

b. $\dfrac{r}{100} = \dfrac{7}{8}$

c. $\dfrac{5}{c} = \dfrac{13}{c + 8}$

You can use proportions to find equivalent measurements within the English or metric systems.

Examples

Convert each measurement as indicated.

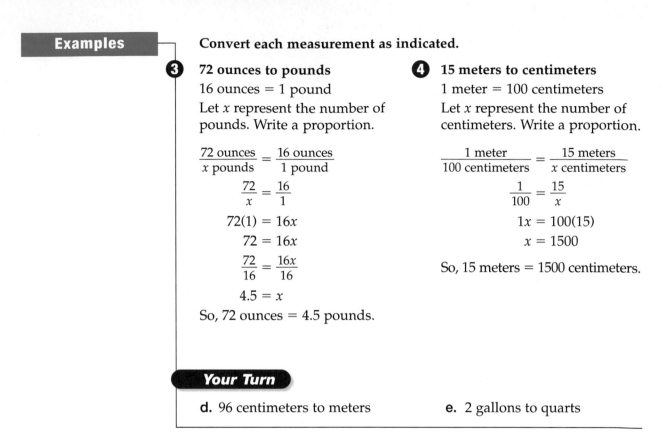

3 **72 ounces to pounds**
16 ounces = 1 pound
Let x represent the number of pounds. Write a proportion.

$$\frac{72 \text{ ounces}}{x \text{ pounds}} = \frac{16 \text{ ounces}}{1 \text{ pound}}$$

$$\frac{72}{x} = \frac{16}{1}$$

$$72(1) = 16x$$

$$72 = 16x$$

$$\frac{72}{16} = \frac{16x}{16}$$

$$4.5 = x$$

So, 72 ounces = 4.5 pounds.

4 **15 meters to centimeters**
1 meter = 100 centimeters
Let x represent the number of centimeters. Write a proportion.

$$\frac{1 \text{ meter}}{100 \text{ centimeters}} = \frac{15 \text{ meters}}{x \text{ centimeters}}$$

$$\frac{1}{100} = \frac{15}{x}$$

$$1x = 100(15)$$

$$x = 1500$$

So, 15 meters = 1500 centimeters.

Your Turn

d. 96 centimeters to meters   **e.** 2 gallons to quarts

A **rate** is a ratio of two measurements having different units of measure. For example, 120 miles in 2 hours is a rate. When a rate is simplified so it has a denominator of 1, it is called a **unit rate**. Since 60 miles per hour is the same as 60 miles per 1 hour, it is a unit rate.

To solve problems involving rates, you can use **dimensional analysis**. This is the process of carrying units throughout a computation.

Example
Science Link
Real World

Note that cubic centimeters can be cancelled.

5 **Density is a measure of the amount of matter that occupies a given volume. The density of wood is 0.71 gram per cubic centimeter. Suppose you have a piece of wood whose volume is 50 cubic centimeters. How many grams of wood does it contain?**

$$0.71 \text{ gram per cubic centimeter} = \frac{0.71 \text{ gram}}{1 \text{ cubic centimeter}}$$

Multiply the unit rate by the number of cubic centimeters of wood.

$$\frac{0.71 \text{ gram}}{1 \text{ cubic centimeter}} \cdot \frac{50 \text{ cubic centimeters}}{1} = \frac{35.5 \text{ grams}}{1} \text{ or } 35.5 \text{ grams}$$

So, the piece of wood contains 35.5 grams of wood.

You could also solve this problem by solving the proportion
$$\frac{400\ miles}{8\ hours} = \frac{700\ miles}{h\ hours}.$$

6 Juanita's family drove 400 miles in 8 hours. The next day, they need to drive another 700 miles. At the same rate, how long will it take them to drive the 700 miles?

Explore   You know that the family drove 400 miles in 8 hours. You need to find how long it will take to drive 700 miles farther.

Plan   Write the rate 400 miles in 8 hours as a unit rate. Then divide 700 miles by the unit rate.

Solve $\dfrac{400\ miles}{8\ hours} = \dfrac{50\ miles}{1\ hour}$ *The unit rate is 50 miles per hour.*

$$700\ miles \div \frac{50\ miles}{1\ hour} = \frac{700\ miles}{1} \cdot \frac{1\ hour}{50\ miles}$$

$$= \frac{\overset{14}{\cancel{700\ miles}}}{1} \cdot \frac{1\ hour}{\underset{1}{\cancel{50\ miles}}} \quad \textit{Note that the units cancel.}$$

$$= 14\ hours$$

Examine   It takes 8 hours to travel 400 miles. A trip of 700 miles would take nearly twice as long. So, 14 hours is reasonable.

Check for Understanding

Communicating Mathematics

Study the lesson. Then complete the following.

1. **Explain** how to use the Property of Proportions to solve $\dfrac{x}{12} = \dfrac{5}{6}$.

2. **Write** two ratios that form a proportion and two other ratios that do *not* form a proportion.

Vocabulary
ratio
proportion
rate
unit rate
dimensional analysis

Guided Practice

Getting Ready   **Find the cross products for each proportion.**

Sample: $\dfrac{2}{8} = \dfrac{4}{16}$ **Solution:** $2(16) = 8(4)$

3. $\dfrac{75}{100} = \dfrac{3}{4}$ 4. $\dfrac{16}{12} = \dfrac{12}{9}$ 5. $\dfrac{8}{12} = \dfrac{10}{15}$

Solve each proportion.   *(Examples 1 & 2)*

6. $\dfrac{6}{2} = \dfrac{9}{a}$ 7. $\dfrac{3}{5} = \dfrac{b}{100}$ 8. $\dfrac{5}{c} = \dfrac{25}{35}$

9. $\dfrac{2}{1.9} = \dfrac{4}{x}$ 10. $\dfrac{y}{4} = \dfrac{y+2}{5}$ 11. $\dfrac{z}{5} = \dfrac{z+2}{7}$

Convert each measurement as indicated.   *(Examples 3 & 4)*

12. 3 pounds to ounces   13. 2000 grams to kilograms

14. Agriculture A farmer gets 300 bushels of corn for each acre of land that he plants. At this rate, how many bushels will he expect to get if he plants 90 acres of corn? *(Example 5)*

15. Travel Katie knows that she can drive her car an average of 40 miles while using one gallon of gasoline. How many gallons of gasoline will she need for a trip of 180 miles? *(Example 6)*

Exercises •

Practice

Solve each proportion.

16. $\dfrac{14}{6} = \dfrac{21}{m}$

17. $\dfrac{5}{4} = \dfrac{y}{12}$

18. $\dfrac{5}{8} = \dfrac{x}{100}$

19. $\dfrac{3}{a} = \dfrac{18}{24}$

20. $\dfrac{18}{27} = \dfrac{t}{3}$

21. $\dfrac{8}{20} = \dfrac{30}{x}$

22. $\dfrac{6}{1} = \dfrac{1200}{r}$

23. $\dfrac{45}{5} = \dfrac{81}{p}$

24. $\dfrac{5}{g} = \dfrac{6}{3}$

25. $\dfrac{2}{3} = \dfrac{7}{y}$

26. $\dfrac{a}{3.5} = \dfrac{14}{7}$

27. $\dfrac{n}{2} = \dfrac{0.7}{0.4}$

28. $\dfrac{10}{y} = \dfrac{2.5}{4}$

29. $\dfrac{0.35}{3} = \dfrac{c}{18}$

30. $\dfrac{y+2}{y} = \dfrac{8}{6}$

31. $\dfrac{8}{x+3} = \dfrac{5}{x}$

32. $\dfrac{x+6}{5} = \dfrac{x}{4}$

33. $\dfrac{x}{x+4} = \dfrac{3}{5}$

34. $\dfrac{5}{x} = \dfrac{9}{x+4}$

35. $\dfrac{x-7}{x+8} = \dfrac{2}{3}$

36. $\dfrac{4}{x-3} = \dfrac{5}{x+5}$

37. Are $\dfrac{15}{2}$ and $\dfrac{7.5}{1}$ equivalent ratios? Explain your reasoning.

38. Find the value of x that makes $\dfrac{18}{9} = \dfrac{28}{x}$ a proportion.

Convert each measurement as indicated.

39. 5 feet to inches

40. 3 quarts to pints

41. 3 liters to milliliters

42. 2.4 grams to milligrams

43. 16 quarts to gallons

44. 4200 pounds to tons

45. 520 meters to kilometers

46. 1600 milliliters to liters

Applications and Problem Solving

47. Health An average adult's body contains about 5 quarts of blood. If a person donates 1 pint of blood, how many pints are left?

48. City Planning A water treatment plant can filter about 350,000 gallons of water in seven days. At this rate, about how many gallons of water can be filtered in 30 days?

49. Measurement Is a 2-quart pitcher large enough to hold 1 batch of fruit punch?

50. Cooking A recipe for 72 chocolate chip cookies needs $4\frac{1}{2}$ cups of flour. How much flour is needed to make 48 cookies?

> **Fruit Punch**
>
> Mix together:
> $2\frac{1}{2}$ cups cherry juice
> 2 cups orange juice
> $1\frac{1}{2}$ cups pineapple juice
> 3 cups ginger ale

Exercise 49

51. Construction A batch of concrete is made by mixing 1 part cement, $2\frac{1}{2}$ parts sand, and $3\frac{1}{2}$ parts gravel. How many pounds of sand and gravel are needed if there are 300 pounds of cement?

52. Education A student-to-teacher ratio is used to compare the number of students at a university with the number of faculty.

a. To the nearest tenth, express the student-to-teacher ratio for each school as a decimal.

b. Rank the universities in the chart in order from greatest student-to-teacher ratio to least.

University	Students	Faculty
Texas	48,857	2517
Ohio State	48,278	4151
UCLA	35,558	3276
Georgia	29,693	3075
Tennessee	25,397	1453

Source: *The World Almanac*, 1999

Data Update For the latest information on university enrollments, visit: www.algconcepts. glencoe.com

53. Critical Thinking If $\frac{6}{a} = \frac{b}{8}$, what happens to the value of b as the value of a increases?

Mixed Review

Solve each equation. *(Lessons 4–6 and 4–7)*

54. $5a - 3 = 8a + 12$

55. $10n + 6 = 8n - 6$

56. $6x + 7 = 8x - 9$

57. $3y - 4 = -2y - 14$

58. $4(m - 2) = 6m$

59. $-4(a - 2) = 10$

60. $6(z + 2) - 4 = -10$

61. $3x - 2(x + 3) = 6$

62. Food The Pasta Palace offers a choice of six types of pasta with five types of sauce. Each dinner comes with or without a meatball. How many different dinners are possible? *(Lesson 4–?)*

63. Find the product of $\frac{2}{3}$ and $-\frac{3}{5}$. *(Lesson 4–1)*

64. Standardized Test Practice For what value(s) of x is the equation $|x - 2| = 5$ true? *(Lesson 3–7)*

  A −3 only   B 7 only   C −3 and 7   D −7 and 3

Extra Practice See p. 701.

5-2 Scale Drawings and Models

What You'll Learn
You'll learn to solve problems involving scale drawings and models.

Why It's Important
Movies Movie directors use models for special effects. *See Example 2.*

Leonardo da Vinci was commissioned to construct a huge bronze horse for the Duke of Milan. Unfortunately, he died before he could complete it.

Five hundred years later, master sculptor Nina Akamu created an 8-foot clay model from da Vinci's sketches. The model was then enlarged by sections to become the 24-foot *Il Cavallo*—the largest bronze horse in the world.

A **scale drawing** or **scale model** is used to represent an object that is too large or too small to be drawn or built at actual size. The **scale** is the ratio of a length on the drawing or model to the corresponding length of the real object.

Il Cavallo

$$\begin{array}{l}\text{length of clay model} \rightarrow \\ \text{length of bronze horse} \rightarrow\end{array} \quad \frac{8 \text{ ft}}{24 \text{ ft}} = \frac{1 \text{ ft}}{3 \text{ ft}}$$

The scale is 1 foot = 3 feet or 1:3.
If the scale uses the same units, it is not necessary to include them.

In the following activity, you will investigate scales with rectangles.

Hands-On Algebra

Materials: grid paper

Step 1 Draw a rectangle that is 4 units wide and 6 units long on your grid paper.

Step 2 Draw a second rectangle in which the ratio of the sides of the first rectangle to the second rectangle is 1:2.

Try These

1. What are the lengths of the sides of the second rectangle?
2. Draw two rectangles so that the scale is 1:3.
3. Draw two rectangles so that the scale is 2:3.
4. Draw a rectangle that is 5 units by 12 units. Then draw another rectangle that is 15 units by 36 units. What is the scale?

One of the most common types of scale drawings is a map.

1 **The scale on a map of Texas is 1 inch = 50 miles. Find the actual distance between Dallas and Houston if the distance between them on the map is 4.5 inches.**

Use the scale and the distance given on the map to write a proportion.

$$\frac{1 \text{ inch}}{50 \text{ miles}} = \frac{4.5 \text{ inches}}{x \text{ miles}} \quad \begin{matrix} \longleftarrow \textit{map distance} \\ \longleftarrow \textit{actual distance} \end{matrix}$$

$$1x = 50(4.5) \qquad \textit{Find the cross products.}$$

$$x = 225$$

The distance between Dallas and Houston is about 225 miles.

Your Turn

Find the actual distance between San Antonio and Houston if the distance between them on the map is $3\frac{3}{4}$ inches.

You can determine the scale if you know the dimensions of the model and the dimensions of the actual object.

2 **The ship _Titanic_, which sank in 1912, was 880 feet long. When a movie was made about it more than 80 years later, a 44-foot-long model of the ship was created for special effects shots. Find the scale that was needed to design other parts of the model.**

Write the ratio of the length of the model to the length of the ship. Then solve a proportion in which the length of the model is 1 foot and the length of the ship is x feet.

$$\frac{44 \text{ feet}}{880 \text{ feet}} = \frac{1 \text{ foot}}{x \text{ feet}} \quad \begin{matrix} \longleftarrow \textit{model length} \\ \longleftarrow \textit{actual length} \end{matrix}$$

$$44x = 880(1) \quad \textit{Find the cross products.}$$

$$44x = 880$$

$$\frac{44x}{44} = \frac{880}{44} \qquad \textit{Divide each side by 44.}$$

$$x = 20$$

The scale is 1 foot = 20 feet or 1:20.

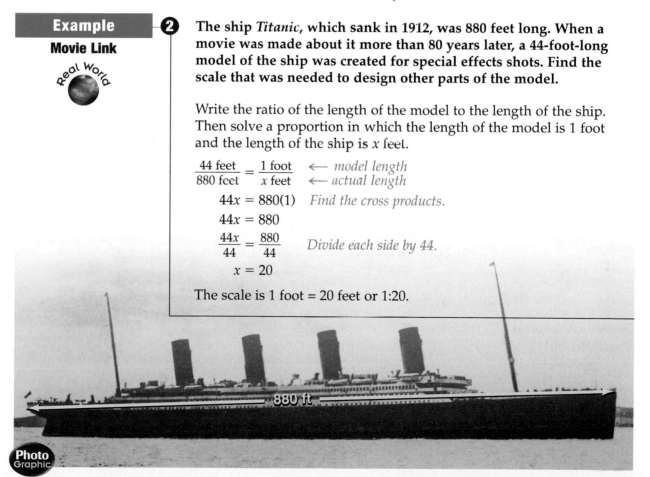

880 ft

Photo
Graphic

Check for Understanding

Communicating Mathematics

Study the lesson. Then complete the following.

1. **Sketch** two rectangles in which the ratio of the sides of the first rectangle to the sides of the second rectangle is 1:2.

2. **Name** two careers in which you would use scale drawings or scale models.

Vocabulary

scale drawing
scale model
scale

Guided Practice

On a map, the scale is 1 inch = 50 miles. Find the actual distance for each map distance. *(Example 1)*

	From	To	Map Distance
3.	Sacramento, CA	Fresno, CA	3 inches
4.	Kansas City, MO	St. Louis, MO	$4\frac{3}{4}$ inches

5. **Transportation** A scale model of a Boeing 747 jumbo jet is 2 meters long. If an actual jet is 70.4 meters long, find the scale of the model. *(Example 2)*

Exercises

Practice

On a map, the scale is 1 inch = 40 miles. Find the actual distance for each map distance.

	From	To	Map Distance
6.	Tampa, FL	Orlando, FL	2 inches
7.	Albany, NY	Syracuse, NY	3 inches
8.	Jacksonville, FL	Pensacola, FL	8 inches
9.	Washington, DC	Richmond, VA	2.5 inches
10.	Cleveland, OH	Pittsburgh, PA	$2\frac{3}{4}$ inches
11.	Hartford, CT	New Haven, CT	$\frac{3}{4}$ inch

Applications and Problem Solving

12. **Bridges** The 1500-foot Natchez Trace Bridge in Natchez, Mississippi, is the longest precast segmental arch bridge in North America. If a model of the bridge is 12 inches long, find the scale of the model.

13. **Model Cars** In a scale model of a racing car, the wheels have a diameter of 2 inches. If the actual wheels have a diameter of 30 inches, find the scale of the model.

14. **Architecture** On a blueprint of a house, one bedroom is 4 inches long. When the house is built, the room will be 16 feet long. Find the scale of the blueprint.

15. **Life Science** In a photograph in a science textbook, the length of a ladybug is 2.5 centimeters. The scale of the photograph is 1 centimeter = 0.2 centimeter. What is the actual length of the ladybug?

16. **Travel** On a map of Colorado, the scale is 1 centimeter = 40 kilometers. The distance on the map between the northern and southern boundaries of Colorado is 11.5 centimeters.

   a. How many kilometers is it from the northern boundary to the southern boundary?

   b. Suppose you travel at an average speed of 80 kilometers per hour. How long will it take you to travel through the state along Route 25? Round your answer to the nearest hour.

17. **Critical Thinking** The state of Pennsylvania is 332 miles long and 179 miles wide. Suppose you want to draw a map of Pennsylvania on an $8\frac{1}{2}$-by-11 inch piece of paper. Find the scale that will allow you to draw the map as large as possible.

Mixed Review

18. **Monuments** A statue of Crazy Horse, the famed Oglala chief, has been under construction for more than 50 years. The sculptors use a model to calculate how to carve the statue. The length of the model is 18.8 feet, the height of the model is 16.6 feet, and the height of the statue is 563 feet. To the nearest foot, what is the length of the statue? *(Lesson 5–1)*

Solve each equation. *(Lesson 4–7)*

19. $3(-2x + 5) = 7$ 20. $-2(3y - 4) = 3 - y$ 21. $2(3x + 1) - 8x = 10$

22. **Standardized Test Practice** What relationship exists between the x- and y-coordinates of each of the points shown on the graph? *(Lesson 2–2)*

   A They are opposites.

   B Their sum is 2.

   C The y-coordinate is 1 more than the x-coordinate.

   D Their sum is 1.

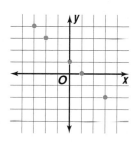

5-3 The Percent Proportion

Math In the Workplace

What You'll Learn
You'll learn to solve problems by using the percent proportion.

Why It's Important
Health Nurses use percents to calculate the correct dosage of medicines.
See Exercise 42.

Nurses, doctors, and pharmacists often deal with quantities that are expressed as percents. **Percent** is a ratio that compares a number to 100. Percent also means *per hundred* or *hundredths*. For example, a 5% glucose solution contains 5 grams of glucose in 100 milliliters of solution.

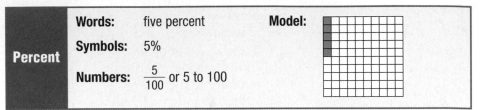

Percent	**Words:**	five percent	**Model:**
	Symbols:	5%	
	Numbers:	$\frac{5}{100}$ or 5 to 100	

You can express a fraction or a ratio as a percent by using a proportion in which one denominator is 100.

Examples

Express each fraction or ratio as a percent.

1 $\frac{2}{5}$ of the circle is shaded.

$$\frac{2}{5} = \frac{r}{100} \quad r \text{ is the percent.}$$
$$2(100) = 5r \quad \text{Find the cross products.}$$
$$200 = 5r$$
$$\frac{200}{5} = \frac{5r}{5} \quad \text{Divide each side by 5.}$$
$$40 = r$$

So, $\frac{2}{5} = 40\%$.

2 Nearly 3 out of 8 people switch channels when a commercial is on television. **Source:** *American Demographics*, September, 1999

$$\frac{3}{8} = \frac{r}{100} \quad \text{Write a proportion.}$$
$$3(100) = 8r \quad \text{Find the cross products.}$$
$$300 = 8r$$
$$\frac{300}{8} = \frac{8r}{8} \quad \text{Divide each side by 8.}$$
$$37.5 = r \quad \text{Nearly 37.5\% of people switch channels.}$$

Your Turn

a. $\frac{73}{50}$

b. 2 out of 3

In Example 1, the number of parts shaded, 2, is called the **percentage** (*P*). It is being compared to the total number of parts, 5, which is called the **base** (*B*).

$$\text{percentage} \rightarrow \frac{2}{5} = \frac{40}{100} \leftarrow \text{percent}$$
$$\text{base} \rightarrow$$

This is an example of a **percent proportion**.

Percent Proportion	If *P* is the percentage, *B* is the base, and *r* is the percent, the percent proportion is $\frac{P}{B} = \frac{r}{100}$.

Examples

Reading Algebra

In percent problems, the base follows the word *of*. It is *not* always the larger number.

3 **Forty is what percent of 160?**

$\dfrac{P}{B} = \dfrac{r}{100}$ *Use the percent proportion. Write "what percent" as $\frac{r}{100}$.*

$\dfrac{40}{160} = \dfrac{r}{100}$ *Replace P with 40 and B with 160.*

$40(100) = 160r$ *Find the cross products.*

$4000 = 160r$

$\dfrac{4000}{160} = \dfrac{160r}{160}$ *Divide each side by 160.*

$25 = r$

So, 40 is 25% of 160.

4 **50% of what number is 42.5?**

$\dfrac{P}{B} = \dfrac{r}{100}$ *Use the percent proportion.*

$\dfrac{42.5}{B} = \dfrac{50}{100}$ *Replace P with 42.5 and r with 50.*

$42.5(100) = 50B$ *Find the cross products.*

$4250 = 50B$

$\dfrac{4250}{50} = \dfrac{50B}{50}$ *Divide each side by 50.*

$85 = B$

So, 50% of 85 is 42.5.

Your Turn

c. 7 is what percent of 20?

d. 60 is 15% of what number?

e. Find 25% of 66.

f. What number is 10% of 88?

Businesses can use a time study to determine what percent of an employee's time is spent on various activities.

5 The manager of The Furniture Store conducted a time study of the employees in its warehouse. The chart at the right shows the average number of hours spent on each activity during the workday. What percent of the time do the employees spend on each activity?

Activity	Time (hours)
Loading furniture	3
Delivering furniture	4
Lunch	1

The employees work $3 + 4 + 1$ or 8 hours each day. This is the base. To find each percent, write and solve the percent proportion for each activity.

Loading: $\dfrac{3}{8} = \dfrac{r}{100}$ 3 ⟨×⟩ 100 ⟨÷⟩ 8 ⟨ENTER⟩ *37.5*

Delivering: $\dfrac{4}{8} = \dfrac{r}{100}$ 4 ⟨×⟩ 100 ⟨÷⟩ 8 ⟨ENTER⟩ *50*

Lunch: $\dfrac{1}{8} = \dfrac{r}{100}$ 1 ⟨×⟩ 100 ⟨÷⟩ 8 ⟨ENTER⟩ *12.5*

The results are summarized in the chart.

Note that the total of the percent column is 100%.

Activity	Time (hours)	Percent
Loading furniture	3	37.5
Delivering furniture	4	50.0
Lunch	1	12.5

Data are often displayed using circle graphs. A **circle graph** is a graph that shows the relationship between parts of the data and the whole.

6 Make a circle graph of the data in Example 5.

Loading furniture accounts for 37.5% of the time. A circle is composed of 360°. To find the number of degrees for this section of the graph, find 37.5% of 360. Repeat this process for the other sections. *Replace B with 360 and r with each percent.*

Loading: $\dfrac{P}{360} = \dfrac{37.5}{100}$ 360 ⟨×⟩ 37.5 ⟨÷⟩ 100 ⟨ENTER⟩ *135*

Delivering: $\dfrac{P}{360} = \dfrac{50}{100}$ 360 ⟨×⟩ 50 ⟨÷⟩ 100 ⟨ENTER⟩ *180*

Lunch: $\dfrac{P}{360} = \dfrac{12.5}{100}$ 360 ⟨×⟩ 12.5 ⟨÷⟩ 100 ⟨ENTER⟩ *45*

Use a compass to draw a circle. Then draw a radius. Use a protractor to draw a 135° angle. *You can start with any of the angles.*

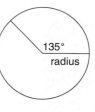

135°
radius

From the new radius, draw a 180° angle. The remaining section should be 45°. Label each section of the graph with the category and percent. Give the graph a title.

The Furniture Store Employee Time Study

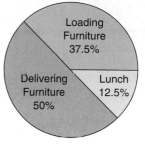

Loading Furniture 37.5%

Delivering Furniture 50%

Lunch 12.5%

Check for Understanding

Communicating Mathematics

Study the lesson. Then complete the following.

1. **Explain** what P, B, and r represent in the percent proportion.

2. **YOU Decide?** Brian and Julia are trying to find what percent 30 is of 20. Brian uses the proportion $\frac{30}{20} = \frac{r}{100}$. Julia uses the proportion $\frac{20}{30} = \frac{r}{100}$. Who is correct? Explain your reasoning.

Guided Practice

⊕ Getting Ready Write a proportion that can be used to find each number.

Sample: 5 is what percent of 40?   **Solution:** $\frac{5}{40} = \frac{r}{100}$

3. What percent of 24 is 12?
4. 80 is 75% of what number?
5. Find 45% of 60.
6. What percent of 50 is 100?

Express each fraction or ratio as a percent. *(Examples 1–2)*

7. 3 out of 5
8. $\frac{1}{3}$
9. 7 to 10

10. In a recent year, 17 out of 20 students used a personal computer as a reference source.

Use the percent proportion to find each number. *(Examples 3 & 4)*

11. 30 is what percent of 120?

12. 16 is 40% of what number?

13. Find 20% of 82.

14. Find 250% of 8.

15. **Government** In the 106th Congress, there were 45 Democratic Senators and 55 Republican Senators. **Source:** *World Almanac, 1999*

 a. What percent of the Senators were Democrats? What percent were Republicans? *(Example 5)*

 b. Make a circle graph of the data. *(Example 6)*

Exercises • • • • • • • • • • • • • • • • • • •

Practice

Express each fraction or ratio as a percent.

16. $\frac{1}{4}$

17. 5 out of 8

18. 9 to 10

19. 4 to 5

20. $\frac{5}{4}$

21. $\frac{9}{12}$

22. 9 out of 20

23. 3 to 2

24. 1 out of 6

25. Three of 5 people have eaten in an Italian restaurant during the past six months.

26. Two-thirds of all households own their homes.

27. About 3 of every 50 people in the United States are between the ages of 14 and 17.

Use the percent proportion to find each number.

28. 60 is what percent of 150?

29. 9 is what percent of 25?

30. 15 is 20% of what number?

31. 50% of what number is 95?

32. What number is 60% of 5?

33. Find 10% of 125.

34. What percent of 50 is 75?

35. What number is 125% of 48?

36. 5% of what number is 6.5?

37. 1 is what percent of 200?

38. Find 0.1% of 450.

39. 18.6 is 200% of what number?

Applications and Problem Solving

Real World

40. **Education** If you answer 35 problems correctly on a 40-problem test, what percent did you answer correctly?

41. **Pharmacy** A peroxide solution contains 5 milliliters of peroxide in 10 milliliters of solution. What is the percent of peroxide in the solution?

42. **Nursing** A 5% glucose solution has been prepared by dissolving 5 grams of glucose in 100 milliliters of solution. How many grams of glucose are in 20 milliliters of the solution?

43. Business About 21 million people give floral gifts for Mother's Day. The chart shows the percent of fresh-cut flowers that are typically purchased. Make a circle graph of the data.

Mother's Day Flowers (percent)

Mixed Bouquets	47
Roses	23
Carnations	16
Single-Stem Flowers	14

Source: California Cut Flower Commission, 1999

44. Critical Thinking Describe a real-life situation in which the percentage is greater than the base.

Mixed Review

45. Maps The distance from St. Louis, Missouri, to Dallas, Texas, is 630 miles. On a map, the distance is 9 inches. What is the scale of the map? *(Lesson 5–2)*

Convert each measurement as indicated. *(Lesson 5–1)*

46. 54 inches to yards

47. 2.5 kilometers to meters

48. Health An ulcer medication has 300 milligrams in 2 tablets. How many milligrams are in 3 tablets? *(Lesson 5–1)*

Add or subtract. *(Lesson 3–2)*

49. $-1.5 - 3.7$

50. $\frac{3}{8} - \left(-\frac{1}{8}\right)$

51. $\frac{3}{5} + \left(-\frac{1}{5}\right)$

52. Standardized Test Practice Which set of numbers is *not* closed under the operation of multiplication? *(Lesson 1–3)*

A real numbers

B negative integers

C positive integers

D whole numbers

Quiz 1 — Lessons 5–1 through 5–3

Solve each proportion. *(Lesson 5–1)*

1. $\frac{9}{25} = \frac{x}{75}$

2. $\frac{z}{9} = \frac{z + 4}{21}$

3. Surveying A civil engineer uses the scale 1 inch = 40 feet to show a parcel of land. How many linear feet of land are represented by a 2.5-inch line on the drawing? *(Lesson 5–2)*

Use the percent proportion to find each number. *(Lesson 5–3)*

4. Six is what percent of 15?

5. 95 is 10% of what number?

What You'll Learn

You'll learn to solve problems by using the percent equation.

Why It's Important

Business Employers use the percent equation to calculate taxes.
See Example 3.

If you have money in a savings account, you earn interest. If you buy new jeans, you pay sales tax. Interest and taxes are situations that involve percent. You can use the percent proportion to solve these kinds of problems. However, it is usually easier to write the percent proportion as an equation.

$$\frac{P}{B} = \frac{r}{100} \quad \text{Start with the percent proportion.}$$

$$\frac{P}{B} = R \quad \text{Replace } \frac{r}{100} \text{ with } R.$$

$$\frac{P}{B}B = RB \quad \text{Multiply each side of the equation by } B.$$

$$P = RB$$

The equation $P = RB$ is called the **percent equation**. In this equation, R is the **rate**. The rate is the decimal form of the percent.

Percent Equation		
	Words:	The percentage is equal to the rate times the base.
	Symbols:	$P = RB$, where P is the percentage, B is the base, and R is the rate.

The percent equation is easier to use when the rate and base are known. However, the percent equation can be used to solve any percent problem.

Examples

1 **Find 4% of $160.**

$P = RB$ *Use the percent equation.*
$P = 0.04(160)$ *Replace R with 0.04 and B with 160.*
$P = 6.4$ *0.04* $\times$ *160* $\boxed{\text{ENTER}}$ *6.4*

So, 4% of $160 is $6.40.

Reading Algebra

To write a percent as a decimal, move the decimal point two places to the left and drop the percent sign.

$4\% = 04\% = 0.04$

$60\% = 60\% = 0.6$

2 **12 is 60% of what number?**

$P = RB$ *Use the percent equation.*
$12 = 0.6B$ *Replace P with 12 and R with 0.6.*
$\dfrac{12}{0.6} = \dfrac{0.6B}{0.6}$ *Divide each side by 0.6.*
$20 = B$ *12* $\boxed{\div}$ *0.6* $\boxed{\text{ENTER}}$ *20*

So, 12 is 60% of 20.

Your Turn

a. Find 62% of 120.
b. 75% of what number is 12?

Many real-world problems can be solved by using the percent equation.

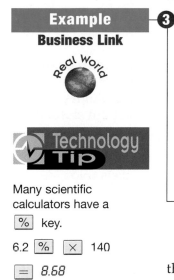
3 **The Federal Insurance Contributions Act (FICA) requires employers to deduct 6.2% of your income for social security taxes. Suppose your weekly pay is $140. What amount would be deducted from your pay for social security taxes?**

To find the amount deducted, find 6.2% of 140.

$P = RB$ *Use the percent equation.*

$P = 0.062(140)$ *Replace R with 0.062 and B with 140.*

$P = 8.68$ *0.062* ☒ *140* ENTER *8.68*

So, $8.68 would be deducted from your pay.

Technology Tip

Many scientific calculators have a ☒ % key.

6.2 % ☒ 140

= *8.68*

Percents are also used in simple interest problems. **Simple interest** is the amount paid or earned for the use of money. If you have a savings account, you earn interest. If you borrow money either through a loan or with a credit card, you pay interest.

The formula $I = prt$ is used to solve problems involving interest.

- I represents the interest,
- p represents the amount of money invested or borrowed, which is called the *principal*,
- r represents the annual interest rate, and
- t represents the time in years.

Example

Banking Link

Real World

4 **Rodney Turner is opening a savings account that earns 4% annual interest. He wants to earn at least $50 in interest after 2 years. How much money should he invest in order to earn $50 in interest?**

$I = prt$

$50 = p(0.04)(2)$ *Replace I with 50, r with 0.04, and t with 2.*

$50 = 0.08p$ *0.04 × 2 = 0.08*

$\dfrac{50}{0.08} = \dfrac{0.08p}{0.08}$ *Divide each side by 0.08.*

$625 = p$ *50* ÷ *0.08* ENTER *625*

Rodney should invest at least $625 to earn $50 in interest.

Your Turn

c. Jessica deposited $3000 in a savings account that pays an interest rate of 6%. How long should she leave the money in the account if she wants to earn $90 in interest?

Mixture problems involve combining two or more parts into a whole. The parts that are combined usually have a different price or a different percent of something.

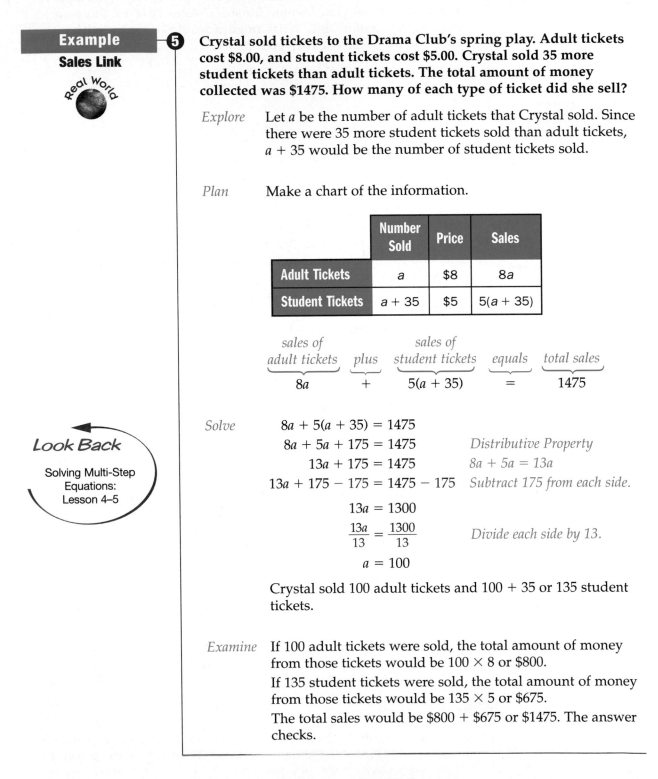

Example

5

Sales Link

Real World

Crystal sold tickets to the Drama Club's spring play. Adult tickets cost $8.00, and student tickets cost $5.00. Crystal sold 35 more student tickets than adult tickets. The total amount of money collected was $1475. How many of each type of ticket did she sell?

Explore Let *a* be the number of adult tickets that Crystal sold. Since there were 35 more student tickets sold than adult tickets, $a + 35$ would be the number of student tickets sold.

Plan Make a chart of the information.

	Number Sold	Price	Sales
Adult Tickets	a	$8	$8a$
Student Tickets	$a + 35$	$5	$5(a + 35)$

sales of adult tickets	*plus*	*sales of student tickets*	*equals*	*total sales*
$8a$	$+$	$5(a + 35)$	$=$	1475

Solve

$$8a + 5(a + 35) = 1475$$

$8a + 5a + 175 = 1475$ *Distributive Property*

$13a + 175 = 1475$ $8a + 5a = 13a$

$13a + 175 - 175 = 1475 - 175$ *Subtract 175 from each side.*

$$13a = 1300$$

$$\frac{13a}{13} = \frac{1300}{13}$$ *Divide each side by 13.*

$$a = 100$$

Crystal sold 100 adult tickets and 100 + 35 or 135 student tickets.

Look Back

Solving Multi-Step Equations: Lesson 4–5

Examine If 100 adult tickets were sold, the total amount of money from those tickets would be 100 × 8 or $800.

If 135 student tickets were sold, the total amount of money from those tickets would be 135 × 5 or $675.

The total sales would be $800 + $675 or $1475. The answer checks.

Mixture problems occur often in chemistry.

Example

Chemistry Link

Real World

6 Kelsey is doing a chemistry experiment that calls for a 30% solution of copper sulfate. She has 40 milliliters of a 25% solution. How many milliliters of a 60% solution should she add to obtain the required 30% solution?

Let x represent the amount of 60% solution to be added. Since she started with 40 milliliters of solution, the final solution will have $40 + x$ milliliters.

	Amount of Solution (mL)	Amount of Copper Sulfate
25% solution	40	0.25(40)
60% solution	x	0.60x
30% solution	40 + x	0.30(40 + x)

$$\underbrace{\begin{array}{c}\textit{amount of}\\\textit{copper sulfate}\\\textit{in 25\% solution}\end{array}}\;\overbrace{\textit{plus}}\;\underbrace{\begin{array}{c}\textit{amount of}\\\textit{copper sulfate}\\\textit{in 60\% solution}\end{array}}\;\overbrace{\textit{equals}}\;\underbrace{\begin{array}{c}\textit{amount of}\\\textit{copper sulfate}\\\textit{in mixture}\end{array}}$$

$$0.25(40) \qquad + \qquad 0.60x \qquad = \qquad 0.30(40 + x)$$

$$\begin{aligned}
0.25(40) + 0.60x &= 0.30(40 + x) \\
10 \quad + 0.6x &= 12 + 0.3x && \textit{Distributive Property} \\
10 + 0.6x - 0.3x &= 12 + 0.3x - 0.3x && \textit{Subtract } 0.3x \textit{ from each side.} \\
10 + 0.3x &= 12 \\
10 + 0.3x - 10 &= 12 - 10 && \textit{Subtract 10 from each side.} \\
0.3x &= 2 \\
\frac{0.3x}{0.3} &= \frac{2}{0.3} && \textit{Divide each side by 0.3.} \\
x &\approx 6.7 && \text{2 } \boxed{\div} \text{ 0.3 } \boxed{\text{ENTER}} \text{ 6.666666667}
\end{aligned}$$

Kelsey should add about 6.7 milliliters of the 60% solution.

Check for Understanding

Communicating Mathematics

Study the lesson. Then complete the following.

1. **Explain** how r and R are different in the percent proportion and the percent equation.

2. **Write** a percent equation that you could use to find the following. 20% of what amount is $500?

3. **You Decide** Akira and Nikki are using the percent equation to find 5% of 17.99. Akira multiplies 17.99 by 5. Nikki multiplies 17.99 by 0.05. Who is correct? Explain.

Vocabulary

percent equation
rate
simple interest
mixture problem

Use the percent equation to find each number. *(Examples 1 & 2)*

4. Find 90% of 200.
5. What number is 38% of 110?
6. 60 is 80% of what number?
7. 220 is 40% of what number?

8. Business Employers are required to deduct 1.45% of your income for Medicare taxes. If your weekly pay is $100, what amount would be deducted from your pay? *(Example 3)*

9. Banking Miguel bought a $500 certificate of deposit that paid an annual interest rate of 5%. If the certificate was issued for three years, how much simple interest will he earn? *(Example 4)*

10. Business Marian works at The Daily Grind coffee shop. She makes a blend of coffee by mixing hazelnut, which sells for $4 a pound, and chocolate almond, which sells for $7 a pound. How many pounds of chocolate almond should she mix with 10 pounds of hazelnut if she wants to sell the mixture for $6 a pound? *(Examples 5 & 6)*

Exercises

Practice

Use the percent equation to find each number.

11. Find 40% of 280.
12. What number is 5% of 120?
13. What number is 65% of 180?
14. Find 200% of 90.
15. 18 is 75% of what number?
16. 110 is 20% of what number?
17. 250 is 40% of what number?
18. 60 is 150% of what number?
19. Find 110% of 80.
20. 90 is 200% of what number?
21. What number is 400% of 16?
22. Find 5% of 3200.
23. 25 is 40% of what number?
24. Find 0.5% of 240.

Applications and Problem Solving

Real World

25. Banking How much interest will Marissa earn if she invests $250 at an annual rate of 4% for 5 years?

26. Banking How long will it take Mr. Albany to earn $75 if he invests $3000 at an annual rate of 5%?

27. Statistics The graph shows ways that teenagers ages 12 to 17 get spending money.

    a. Suppose you survey a group of 150 15-year-olds. Write a percent equation to predict how many students get a regular allowance.

    b. Solve the equation. Round to the nearest whole number.

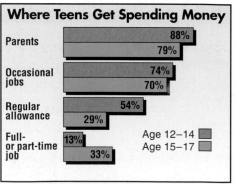

Where Teens Get Spending Money

Parents	88%
	79%
Occasional jobs	74%
	70%
Regular allowance	54%
	29%
Full- or part-time job	13%
	33%

Age 12–14 ▮
Age 15–17 ▯

Source: *ICR TeenEXCEL, 1998*

28. Credit Cards   A credit card company charges an annual rate of 18% on the unpaid balance on credit card accounts. Suppose you have an unpaid balance of $800. How much monthly interest will you be charged? (*Hint:* First find the monthly interest rate.)

29. Business   Great Smoky Mountains Tours conducts hiking trips along the Appalachian Trail. The owner charges $250 for adults and $175 for children. He is able to take groups of 15 people on a single trip. If he wants to earn $3150 for his next trip to pay for some new equipment, how many of the 15 people on the trip should be adults?

30. Chemistry   Roberto needs a solution that is 30% silver nitrate. He has 12 ounces of a solution that is 25% silver nitrate. How many ounces of a 40% silver nitrate solution should he add to make a solution that is 30% silver nitrate?

31. Critical Thinking   If $a\%$ of $b = x$ and $b\%$ of $a = y$, what is the relationship between the value of x and the value of y?

Mixed Review

32. Health   Nutritionists recommend a diet in which less than 30% of your total Calories come from fat. Which of the subs listed in the chart have less than 30% of their Calories from fat?   (*Lesson 5–3*)

The Sub Shop		
Kind of Sub	**Total Calories**	**Calories from Fat**
Italian	522	288
Meatball	459	198
Turkey	322	90
Roast Beef	345	108
Ham & Cheese	322	81

33. On a map, the scale is 1 inch = 60 miles. Find the actual distance for each map distance.   (*Lesson 5–2*)

	From	To	Map Distance
a.	Memphis, TN	Nashville, TN	3 inches
b.	Shreveport, LA	Jackson, MS	$3\frac{1}{2}$ inches
c.	Little Rock, AR	Memphis, TN	$2\frac{1}{4}$ inches

Solve each equation.   (*Lesson 4–6*)

34. $12x + 15 = 35 + 2x$

35. $3y + 10 = 2y - 21$

36. $6 - 8a = 20a + 20$

37. $7n - 13 = 3n + 7$

38. Open-Ended Test Practice   List a set of data that includes at least five numbers for which the mean is greater than the median.   (*Lesson 3–3*)

BUILDING BRIDGES

Materials

calculator

Box-and-Whisker Plots

The table shows the lengths of the longest suspension bridges.

Name of Bridge (inside U.S.)	Length (ft)	Name of Bridge (outside U.S.)	Length (ft)
Verrazano-Narrows	4260	Akashi Kaikyo	6529
Golden Gate	4200	Izmit Bay	5472
Mackinac Straits	3800	Storebaelt	5328
George Washington	3500	Humber	4626
Tacoma Narrows II	2800	Jiangyin Yangtze	4543
San Francisco-Oakland	2310	Tsing Ma	4518
Bronx-Whitestone	2300	Hardanger Fjord	4347
Delaware Memorial	2150	Hoga Kusten	3970
Seaway Skyway	2150	High Coast	3969
Walt Whitman	2000	Minami Bisan-Seto	3668
Ambassador International	1850	Second Bosporus	3576

Source: *Information Please Almanac, 1999*

Let's use a **box-and-whisker plot** to investigate these data.

Investigate

1. To make box-and-whisker plots for these data, you need to find five important values for each set of data: the median, the two extremes, the upper quartile (UQ), and the lower quartile (LQ).

   a. First find the median (M) of each list.

   b. Consider only the upper half of each list. Find the medians. These numbers are the **upper quartiles**.

   c. Consider only the bottom half of each list. Find the medians. These numbers are the **lower quartiles**.

   d. The greatest (GV) and least values (LV) in each list are called the **extremes**. Find the extremes.

Look Back

Median:
Lesson 3–3

2. On a sheet of paper, draw a number line.

   a. To make a box-and-whisker plot for the U.S. bridges, use the five values from Exercise 1a–1d. Mark dots above their coordinates.

   b. Complete the box-and-whisker plot as shown below.

The lines outside the box are called <u>*whiskers*</u>.

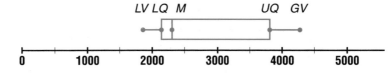

3. Make a box-and-whisker plot for the bridges outside the U.S.

Extending the Investigation

In this extension, you will compare the box-and-whisker plots for the two sets of data. You will also investigate how changes in the data affect a box-and-whisker plot.

1. Use the box-and-whisker plots to answer these questions.

   a. What percent of the data lies below the median?

   b. What percent of the data lies inside the box?

   c. What percent of the data lies in one whisker?

   d. What percent of bridges outside the U.S. are longer than the longest U.S. bridge? Explain your reasoning.

2. When data are arranged in order from least to greatest, you can describe the data using percentiles. A **percentile** is the point below which a given percent of the data lies. For example, 50% of the data falls below the median. So, the median is the 50th percentile for the data.

   a. Which U.S. bridge is at the 50th percentile?

   b. Which bridge outside the U.S. is at the 25th percentile?

   c. Which U.S. bridge is at the 75th percentile?

3. Suppose a 7000-foot bridge is built in the United States.

   a. Redraw the box-and-whisker plot, including this value.

   b. Describe the change in the box-and-whisker plot.

   c. How do these data change your comparisons of the bridges?

Presenting Your Conclusions

Here are some ideas to help you present your conclusions to the class.

- Prepare a poster displaying your box-and-whisker plots. Write a comparison of the data.

- Find two related sets of data that you would like to compare with box-and-whisker plots. Draw the plots. Then prepare an oral presentation using the plots as visual aids.

*inter***NET** **CONNECTION** **Investigation** For more information on box-and-whisker plots, visit: www.algconcepts.glencoe.com

Percent of Change

What You'll Learn

You'll learn to solve problems involving percent of increase or decrease.

Why It's Important

Retail Sales Percent of decrease is used when stores have sales.
See Example 4.

In 1990, the average cost of one dozen large eggs was $1.00. By 1997, the cost had increased to $1.17. The increase in the cost of goods and services is called *inflation*. **Source:** *Statistical Abstract of the United States, 1998*

To find the inflation rate, write a ratio that compares the amount of increase to the original price.

amount of increase: $1.17 − $1.00 = $0.17 or 17¢
original price: $1.00 or 100¢

$$\text{amount of increase} \rightarrow \quad \frac{17}{100} = 17\% \quad \leftarrow \textit{inflation rate}$$
$$\text{original price} \rightarrow$$

So, the cost of one dozen eggs increased 17% from 1990 to 1997.

When an increase or decrease is expressed as a percent, the percent is called the **percent of increase** or the **percent of decrease**. The earlier amount is always used as the base in the percent equation.

Examples

Find the percent of increase or decrease. Round to the nearest percent.

❶ original: 25
new: 29

Find the amount of increase.
29 − 25 = 4

Use the percent proportion.

$$\frac{P}{B} = \frac{r}{100}$$

$$\frac{4}{25} = \frac{r}{100}$$

$4(100) = 25r$ *Cross products*

$400 = 25r$

$$\frac{400}{25} = \frac{25r}{25} \quad \begin{array}{l}\textit{Divide each side}\\ \textit{by 25.}\end{array}$$

$16 = r$

The percent of increase is 16%.

❷ original: 18
new: 12

Find the amount of decrease.
18 − 12 = 6

Use the percent proportion.

$$\frac{P}{B} = \frac{r}{100}$$

$$\frac{6}{18} = \frac{r}{100}$$

$6(100) = 18r$ *Cross products*

$600 = 18r$

$$\frac{600}{18} = \frac{18r}{18} \quad \begin{array}{l}\textit{Divide each side}\\ \textit{by 18.}\end{array}$$

$33 \approx r$

The percent of decrease is about 33%.

Your Turn

a. original: 12
    new: 20

b. original: 50
    new: 49

Two applications of percent of change are sales tax and discounts. **Sales tax** is a tax that is added to the cost of the item. It is an example of a percent of increase. **Discount** is the amount by which the regular price of an item is reduced. It is an example of a percent of decrease.

❸ **Ms. Cruz bought a car for $12,500. A state sales tax of 4% is then added to the price of the car. What was the total price?**

Method 1	**Method 2**
First, use the percent equation to find the sales tax.	A sales tax of 4% means that Ms. Cruz will pay 100% + 4% or 104% of the price of the car.
$P = RB$	
$P = 0.04(12{,}500)$	
$P = 500$	Use the percent equation to find the total price.
	$P = RB$
Then, add the $500 sales tax to $12,500.	$P = 1.04(12{,}500)$
$12{,}500 + 500 = 13{,}000$	$P = 13{,}000$

The total price will be $13,000.

❹ **All shoes at The Runner's Place are on sale at a 25% discount. If a pair of running shoes originally cost $65, what is the sale price?**

Method 1

First, use the percent equation to find the discount.
$P = RB$
$P = 0.25(65)$
$P = 16.25$
Then, subtract the $16.25 discount from $65.
$65 - 16.25 = 48.75$

Method 2

A discount of 25% means that the buyer will pay 100% − 25% or 75% of the selling price. Use the percent equation to find the sale price.
$P = RB$
$P = 0.75(65)$
$P = 48.75$
The sale price of the running shoes is $48.75.

Your Turn

c. What is the total cost of a basketball that sells for $45 if the sales tax rate is 7%?

d. All long-sleeve T-shirts are on sale for 40% off. If the original price was $19.95, what is the discount price?

Sometimes discount stores advertise an additional discount. For example, suppose all merchandise in an outlet store is 50% off retail prices every day. If the store has a sale where you can take another 25% off, there are two ways you might interpret this.

- The discounts can be successive. That is, 50% is taken off the original price and then 25% is taken off the resulting price.
- The discounts can be combined. That is, the discount is 50% + 25% or 75% of the original price.

BLOW-OUT SALE

25% off already low prices

Graphing Calculator Tutorial
See pp. 724–727.

Graphing Calculator Exploration

This graphing calculator program finds the cost of an item for successive discounts and combined discounts. Enter the original price and each discount (as a decimal) when prompted.

```
PROGRAM:DISCOUNT
:Disp "ORIG. PRICE"
:Input P
:Disp "1ST DISCOUNT?"
:Input A
:Disp "2ND DISCOUNT?"
:Input B
:P(1-A)(1-B) → C
:P(1-(A + B)) → D
:Disp "SUCCESSIVE
 DISCOUNT", C
:Disp "COMBINED
DISCOUNT", D
```

Try These

Copy and complete the table.

	Price	First Discount	Second Discount	Sale Price Successive Discount	Sale Price Combined Discount
1.	$ 49.00	20%	10%		
2.	$185.00	25%	10%		
3.	$ 12.50	30%	20%		
4.	$156.95	30%	15%		

5. What is the relationship between the sale price using successive discounts and the sale price using combined discounts?
6. Which type of discount is usually used in stores?

Percent of change is often used to describe how data change from one year to another. Sometimes the percent of increase is greater than 100%.

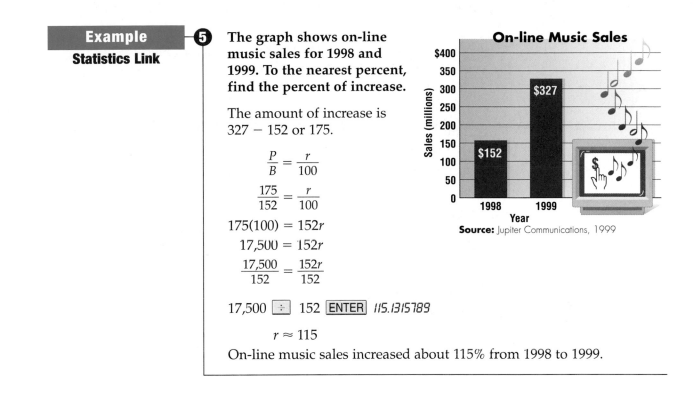

5 The graph shows on-line music sales for 1998 and 1999. To the nearest percent, find the percent of increase.

The amount of increase is 327 − 152 or 175.

$$\frac{P}{B} = \frac{r}{100}$$

$$\frac{175}{152} = \frac{r}{100}$$

$$175(100) = 152r$$

$$17,500 = 152r$$

$$\frac{17,500}{152} = \frac{152r}{152}$$

17,500 ⊡ 152 ENTER *115.1315789*

$$r \approx 115$$

On-line music sales increased about 115% from 1998 to 1999.

On-line Music Sales

Source: Jupiter Communications, 1999

Check for Understanding

Communicating Mathematics

Study the lesson. Then complete the following.

1. **Describe** two different methods for finding the total price of an item once you know the sales tax rate and the cost of the item.

2. **Explain** how a percent of increase can be greater than 100%.

Math Journal

3. **Find** an example of percent of change in a newspaper or magazine. Then write and solve a problem using your example.

Vocabulary

percent of increase
percent of decrease
sales tax
discount

Guided Practice

⏱ **Getting Ready** Write a percent proportion to find the percent of change.

Sample: original: 15, new: 20   **Solution:** $\frac{5}{15} = \frac{r}{100}$

4. original: 25
   new: 23

5. original: 14
   new: 18

6. original: 30
   new: 70

Find the percent of increase or decrease. Round to the nearest percent. *(Examples 1 & 2)*

7. original: 20
   new: 16

8. original: 8
   new: 20

The cost of an item and a sales tax rate are given. Find the total price of each item to the nearest cent. *(Example 3)*

9. in-line skates: $90; 5%

10. make-up: $14.95; 4%

The original cost of an item and a discount rate are given. Find the sale price of each item to the nearest cent. *(Example 4)*

11. computer: $1200; 10%

12. tennis balls: $4.50; 25%

13. **Sales** Between 1998 and 1999, retail sales over the Internet increased from $6 to $11 billion. Find the percent of increase to the nearest percent. **Source:** The Direct Marketing Association, 1999 *(Example 5)*

Exercises • • • • • • • • • • • • • • • • • • •

Practice

Find the percent of increase or decrease. Round to the nearest percent.

14. original: 10
    new: 12

15. original: 50
    new: 56

16. original: 30
    new: 12

17. original: 500
    new: 420

18. original: 15
    new: 36

19. original: 48
    new: 112

The cost of an item and a sales tax rate are given. Find the total price of each item to the nearest cent.

20. television: $500; 6%

21. CD player: $40; 7%

22. dress: $55; 6%

23. jeans: $38; 4%

24. CD: $17.99; 5%

25. book: $25.99; 2.5%

The original cost of an item and a discount rate are given. Find the sale price of each item to the nearest cent.

26. shoes: $40; 25%

27. tires: $280; 5%

28. earrings: $12; 40%

29. watch: $35; 15%

30. video: $24.99; 30%

31. radio: $18.95; 25%

32. What number is 30% less than 40?

33. Find the percent of increase from $18 to $36.

Applications and Problem Solving

34. **Retail Sales** Georgia bought a new belt that was on sale for 15% off. If the original cost was $24, what was the sale price?

35. Travel The table shows the number of new passports issued in the United States. Between which two years was the percent of increase greater: 1996 to 1997, or 1997 to 1998? Explain your reasoning.

New Passports (millions)	
1996	5.5
1997	6.3
1998	6.5

Source: U.S. State Department, 1999

36. Biology Between 1987 and 1998, the number of breeding pairs of bald eagles increased from 2.2 thousand pairs to 5.7 thousand pairs. Find the percent of increase to the nearest percent.
Source: Fish and Wildlife Service, 1999

Bald Eagle

37. Critical Thinking An amount is increased by 10%. The result is decreased by 10%. Is the final result less than, greater than, or equal to the original amount?

Mixed Review

38. Real Estate A *commission* is an amount of money that is paid for selling a product. Tom Augustus sells real estate at a 7% commission. Last week, his sales totaled $90,000. What was his commission? *(Lesson 5–4)*

Express each fraction or ratio as a percent. *(Lesson 5–3)*

39. 5 out of 8   **40.** 2 to 5   **41.** $\frac{7}{20}$

Convert each measurement as indicated. *(Lesson 5–1)*

42. 15 inches to feet   **43.** 1500 grams to kilograms

44. Standardized Test Practice Which point corresponds to a number that *cannot* be obtained by subtracting two integers? *(Lesson 2–4)*

   A *A*   **B** *B*   **C** *C*   **D** *D*

Quiz 2   Lessons 5–4 and 5–5

▶ **Use the percent equation to find each number.** *(Lesson 5–4)*

1. Find 4% of 625.   **2.** What percent of 50 is 6?
3. 15 is 75% of what number?   **4.** Find 105% of 80.

5. Economics The inflation rate for groceries is 5%. How much will a cart of groceries cost next year if it costs $134.10 this year? Round to the nearest cent. *(Lesson 5–5)*

Extra Practice   See p. 702.

Accountant

Accountants prepare and analyze financial reports. Some accountants track the depreciation of their company's property and inventory. *Depreciation* is the decrease in the value of an item because of its age. There are different ways to calculate depreciation.

- *Straight-Line Method:* The depreciation is the same each year.
- *Double-Declining-Balance Method:* The depreciation is determined by a percent of decrease.

Suppose a media production company purchased equipment for $240,000. The company estimates that the equipment will last for 5 years. At that time, the value of the equipment will be $20,000. The table and graph show the value each year for a straight-line (SL) depreciation of $44,000 and a double-declining-balance (DDB) depreciation of 40%.

Value (dollars)		
Year	SL	DDB
0	240,000	240,000
1	196,000	144,000
2	152,000	86,400
3	108,000	51,840
4	64,000	31,104
5	20,000	20,000

Value
(1000 dollars)

Key:
—— SL
—— DDB

Media Equipment

Year

1. Why do you think the first method is called the straight-line method?
2. For which year is the difference between the straight-line value and the double-declining-balance value the greatest? How is this shown on the graph?

FAST FACTS About Accountants

Working Conditions

- usually work in a comfortable environment
- generally work a 40-hour week, but some work 50 hours a week or more

Education

- most have a college degree in accounting
- knowledge of computers a necessity

Earnings

Starting Salaries, 1996

Bachelor's Degree $29,400

Master's Degree $33,000

Source: Bureau of Labor Statistics, 1999

*inter*NET
CONNECTION
Career Data For the latest information on accountants, visit:
www.algconcepts.glencoe.com

What You'll Learn
You'll learn to find the probability and odds of a simple event.

Why It's Important
Manufacturing
Inspectors use experimental probability when they find defective items. *See Examples 2 & 3.*

Backgammon is a game played with two dice. If you roll doubles, you get to roll again. Is there a good chance that you will roll doubles?

The table shows all of the possible outcomes when you roll a pair of dice. The highlighted outcomes are doubles.

	1	2	3	4	5	6
1	(1, 1)	(1, 2)	(1, 3)	(1, 4)	(1, 5)	(1, 6)
2	(2, 1)	(2, 2)	(2, 3)	(2, 4)	(2, 5)	(2, 6)
3	(3, 1)	(3, 2)	(3, 3)	(3, 4)	(3, 5)	(3, 6)
4	(4, 1)	(4, 2)	(4, 3)	(4, 4)	(4, 5)	(4, 6)
5	(5, 1)	(5, 2)	(5, 3)	(5, 4)	(5, 5)	(5, 6)
6	(6, 1)	(6, 2)	(6, 3)	(6, 4)	(6, 5)	(6, 6)

There are 36 possible outcomes. If the dice are fair, each outcome is equally likely to occur. Of those 36 outcomes, 6 are doubles. You can measure the chances of an event happening with **probability**.

Reading Algebra

P(doubles) is read as the *probability of rolling doubles*.

Probability	**Words:**	The probability of an event is a ratio that compares the number of favorable outcomes to the number of possible outcomes.
	Symbols:	$P(\text{event}) = \dfrac{\text{number of favorable outcomes}}{\text{number of possible outcomes}}$
	Numbers:	$P(\text{doubles}) = \dfrac{6}{36} \text{ or } \dfrac{1}{6}$

The probability that an event will happen is between 0 and 1 inclusive.

- A probability of 0 means that the event is impossible.
- A probability of 1 means the event is certain to happen.
- The closer a probability is to 1, the more likely it is to happen.

impossible $\frac{1}{6}$ equally likely   certain

0               $\frac{1}{2}$ or 0.5   1

0%   50%   100%

— probability of rolling doubles

Info Graphic

When all outcomes have an equally likely chance of happening, the outcomes happen at **random**.

Example

Population Link

① A population distribution for California is shown. If a person is chosen at random, what is the probability that the person is age 65 or older?

There are 4 million people who are age 65 or older. The total population is $3 + 6 + 3 + 5 + 5 + 4 + 2 + 4$ or 32 million.

$P(65 \text{ or older})$

$\quad = \dfrac{\text{number of people age 65 or older}}{\text{total population}}$

$\quad = \dfrac{4}{32} \text{ or } \dfrac{1}{8}$

The probability of choosing a person age 65 or older is $\dfrac{1}{8}$ or 12.5%.

Population of California

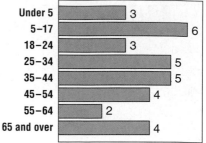

Age

Under 5	3
5–17	6
18–24	3
25–34	5
35–44	5
45–54	4
55–64	2
65 and over	4

Number (millions)

Source: U.S. Census Bureau, 1999

The probability that was found in Example 1 is called the **theoretical probability**. Theoretical probability is what *should* occur. What *actually* occurs when we conduct an experiment is called the **experimental probability**.

Reading Algebra

The **empirical probability** is the most accurate probability based upon repeated trials in an experiment.

Hands-On Algebra

Materials: paper bag ▯ 20 two-inch pieces of paper

Work with a partner.

Step 1 Mark 4 slips of paper with an X, 7 slips of paper with a Y, and 9 slips of paper with a Z.

Step 2 Put the slips of paper in the bag and mix well.

Step 3 Draw one slip of paper from the bag and record its letter.

Step 4 Return the slip of paper to the bag and mix well. Repeat Steps 3 and 4 until you have completed 20 trials.

Try These

1. Calculate the experimental probability of choosing each letter. Express each probability as a percent.
2. Calculate the theoretical probability of choosing each letter.
3. Compare the experimental probability with the theoretical probability. How similar are they?

Experimental probability is often used by quality-control inspectors.

2 **A quality-control inspector for Office Suppliers checked a sample of 250 marking pens and found that 14 of them were defective. Find the experimental probability of choosing a defective pen.**

$$P(\text{defective pen}) = \frac{\text{number of defective pens}}{\text{total number of pens}}$$

$$= \frac{14}{250} \text{ or } \frac{7}{125}$$

3 **If the percent of defective pens in Example 2 is greater than 6%, production will be stopped. Should the quality-control inspector stop production?**

First, change $\frac{7}{125}$ to a percent. Then compare it to 6%.

$\dfrac{P}{B} = \dfrac{r}{100}$ *Use the percent proportion.*

$\dfrac{7}{125} = \dfrac{r}{100}$ *Replace P with 7 and B with 125.*

$7(100) = 125r$ *Find the cross products.*

$700 = 125r$

$\dfrac{700}{125} = \dfrac{125r}{125}$ *Divide each side by 125.*

$5.6 = r$ 700 ⌷÷⌷ 125 ⌷ENTER⌷ 5.6

The sample contains 5.6% defective pens. Since this is less than 6%, the quality-control inspector should not stop production.

Another way to measure the chance of an event occurring is with **odds**.

Odds	**Words:**	The odds of an event occurring is a ratio that compares the number of favorable outcomes to the number of unfavorable outcomes.
	Symbols:	$\text{Odds} = \dfrac{\text{number of favorable outcomes}}{\text{number of unfavorable outcomes}}$

4 **A bag contains 6 red marbles, 3 blue marbles, and 1 yellow marble. Find the odds of choosing a red marble.**

There are 6 red marbles. So, there are 6 favorable outcomes.

There are 3 + 1 or 4 marbles that are *not* red. So, there are 4 unfavorable outcomes.

odds of choosing a red marble = 6:4 or 3:2

Your Turn

Find the odds of choosing a blue marble.

Check for Understanding

Study the lesson. Then complete the following.

1. **Compare and contrast** theoretical probability and experimental probability.

2. **Describe** how to find the odds of an event happening.

3. Can the odds of an event happening be greater than 1? **Explain** why or why not.

Vocabulary

probability
random
theoretical probability
experimental probability
empirical probability
odds

Guided Practice

Refer to the application at the beginning of the lesson. Find the probability of each outcome if a pair of dice is rolled. *(Example 1)*

4. an even number on the first die

5. a sum of 2

Find the odds of each outcome if a die is rolled. *(Example 4)*

6. a number greater than 4

7. an even number

8. **Basketball** Michael made 7 free throws out of 8 attempts during the last basketball game. *(Examples 2 & 3)*

   a. What was the experimental probability of making a free throw?

   b. Express the probability as a percent.

Exercises • • • • • • • • • • • • • • • • • •

Practice

Refer to the application at the beginning of the lesson. Find the probability of each outcome if a pair of dice are rolled.

9. an odd number on the first die

10. a sum of 12

11. a sum of 5

12. an even sum

13. a sum greater than 12

14. a sum less than 13

Find the odds of each outcome if the spinner at the right is spun.

15. an even number

16. greater than 2

17. *not* a 5

18. red

19. red or blue

20. *not* red

21. What is the probability that a month picked at random starts with the letter *J*?

22. If the probability that an event happens is $\frac{1}{2}$, find the odds that it happens.

23. **Manufacturing** A quality-control inspector checked three samples of sweaters during the day. Each sample contained 40 pieces. The chart lists the number of sweaters that were defective.

Quality Control Results	
Sample (time)	Number Defective
9:00	2
1:00	1
4:00	5

   a. For each sample, find the experimental probability of choosing a defective sweater.

   b. Production is stopped if the percent of defective sweaters is greater than 5%. For which sample(s) should production be stopped?

24. **Marketing** The local video store advertises that 1 out of 4 customers will receive a free box of popcorn when they rent a video.

   a. What are the odds of receiving free popcorn?

   b. What are the odds *against* receiving free popcorn?

   c. At the end of the first day, 15 customers out of 75 had received free popcorn. Find the experimental probability.

   d. Did the store give away as much free popcorn as it advertised?

25. **Allowance** The graph shows the weekly allowance for students in grades 6 through 12.

   a. If a student is chosen at random, what is the probability the student receives $5 or more as a weekly allowance?

   b. What are the odds that a student chosen at random receives no allowance?

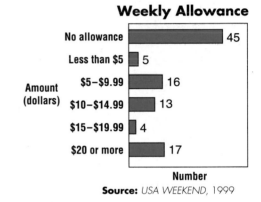

Weekly Allowance

No allowance — 45
Less than $5 — 5
$5–$9.99 — 16
$10–$14.99 — 13
$15–$19.99 — 4
$20 or more — 17

Amount (dollars) / Number

Source: *USA WEEKEND*, 1999

26. **Critical Thinking** The **complement** of an event's occurring is the event's *not* occurring. If the probability that an event will occur is $\frac{3}{5}$, what is the probability of its complement?

Mixed Review

The original cost of an item and a discount rate are given. Find the sale price of each item to the nearest cent. (*Lesson 5–5*)

27. basketball: $45, 20%

28. television: $399, 10%

29. **Banking** How much interest will Ben earn if he invests $1000 at a rate of 6% for 3 years? (*Lesson 5–4*)

30. What percent of 25 is 30? (*Lesson 5–3*)

31. **Standardized Test Practice** Which expression is equivalent to $-2(x + 5) + 6(x + 5)$? (*Lesson 2–5*)

   A $4(x + 5)$ **B** $4x + 40$ **C** $8x + 40$ **D** $4x - 20$

What You'll Learn

You'll learn to find the probability of mutually exclusive and inclusive events.

Why It's Important

Marketing
Restaurant owners can use probability to make predictions about their business. *See Exercise 21.*

The Born Loser®

THE BORN LOSER reprinted by permission of Newspaper Enterprise Association, Inc.

In the comic above, the weather forecaster predicted a 100% chance of rain for the weekend. Do you think this forecast is correct? *This problem will be solved in Example 3.*

The comic is an example of a **compound event**. Compound events consist of two or more simple events that are connected by the words *and* or *or*. Let's investigate a case where simple events are connected by the word *and*.

Hands-On Algebra

Materials: 2 paper bags   red and yellow counters

Work with a partner.

Step 1 Place a red counter and a yellow counter in each bag.

Step 2 Without looking, remove one counter from each bag. Record the color combination in the order that you drew the counters. Return the counters to their respective bags.

Step 3 Repeat 99 times. Count and record the number of red/red, red/yellow, yellow/red, and yellow/yellow combinations.

Try These

Estimate the probability of each outcome.

1. P(red *and* red)   2. P(red *and* yellow)
3. P(yellow *and* red)   4. P(yellow *and* yellow)

Reading Algebra

P(red *and* yellow) means the probability of choosing red from the first bag and yellow from the second bag.

Choosing a counter from bag 1 did not affect choosing a counter from bag 2. These events are called **independent events** because the outcome of one event does not affect the outcome of the other event.

Look Back

Tree diagrams:
Lesson 4–2

You can analyze the experiment on the previous page with a tree diagram.

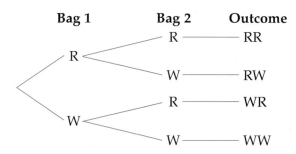

Bag 1	Bag 2	Outcome
R	R	RR
R	W	RW
W	R	WR
W	W	WW

There are four equally-likely outcomes. So, the probability of choosing white on the first draw *and* white on the second draw is $\frac{1}{4}$. You can also multiply to find the probability of two independent events.

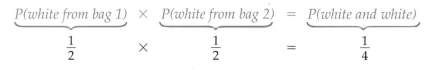

$$\underbrace{P(white\ from\ bag\ 1)}_{\frac{1}{2}} \times \underbrace{P(white\ from\ bag\ 2)}_{\frac{1}{2}} = \underbrace{P(white\ and\ white)}_{\frac{1}{4}}$$

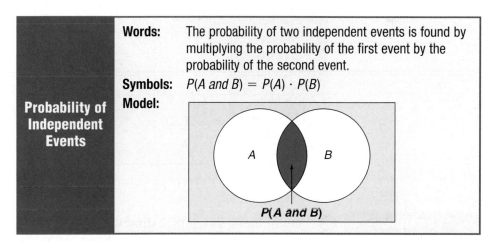

Probability of Independent Events

Words: The probability of two independent events is found by multiplying the probability of the first event by the probability of the second event.

Symbols: $P(A\ and\ B) = P(A) \cdot P(B)$

Model:

P(A and B)

Example ❶ **Two dice are rolled. Find the probability that an odd number is rolled on the first die and the number 4 is rolled on the second.**

$P(\text{odd number}) = \frac{3}{6}$ or $\frac{1}{2}$

$P(4) = \frac{1}{6}$

$P(\text{odd number } and \text{ 4}) = \frac{1}{2} \cdot \frac{1}{6}$ or $\frac{1}{12}$

Your Turn

a. Two dice are rolled. Find the probability that an even number is rolled on the first die and a number greater than 4 on the second die.

Two events can also be connected by the word *or*. For example, consider the probability of drawing a jack or a queen from a standard deck of 52 cards. Since a card cannot be both a jack and a queen, the events are **mutually exclusive**. That is, both events cannot occur at the same time.

The probability of two mutually exclusive events is found by adding.
$P(\text{jack or queen}) = P(\text{jack}) + P(\text{queen})$

$$= \frac{4}{52} + \frac{4}{52}$$

$$= \frac{8}{52} \text{ or } \frac{2}{13}$$

The probability of drawing a jack or a queen is $\frac{2}{13}$.

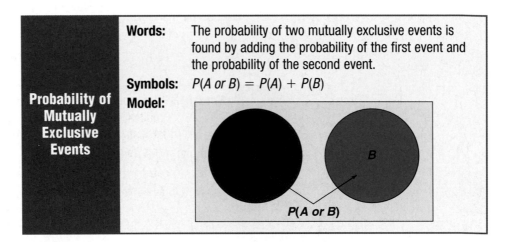

Probability of Mutually Exclusive Events	**Words:**	The probability of two mutually exclusive events is found by adding the probability of the first event and the probability of the second event.
	Symbols:	$P(A \text{ or } B) = P(A) + P(B)$
	Model:	*P(A or B)*

Example ➋ Jamal has 4 quarters, 2 dimes, and 4 nickels in his pocket. He takes one coin from his pocket at random. What is the probability that the coin is either a quarter or a dime?

A coin cannot be both a quarter and a dime, so these are mutually exclusive events. Find the sum of the individual probabilities.

$P(\text{quarter } or \text{ dime}) = P(\text{quarter}) + P(\text{dime})$

$$= \frac{4}{10} + \frac{2}{10}$$

$$= \frac{6}{10} \text{ or } \frac{3}{5}$$

The probability of choosing a quarter or a dime is $\frac{3}{5}$.

Your Turn

b. Find the probability of Jamal's choosing a quarter or a nickel.

Sometimes events are connected by the word *or*, but they are not mutually exclusive. For example, refer to the comic on page 224. There is a 40% chance of rain on Saturday and a 60% chance of rain on Sunday. These are not mutually exclusive because it might rain on *both* Saturday and Sunday. They are called **inclusive** events.

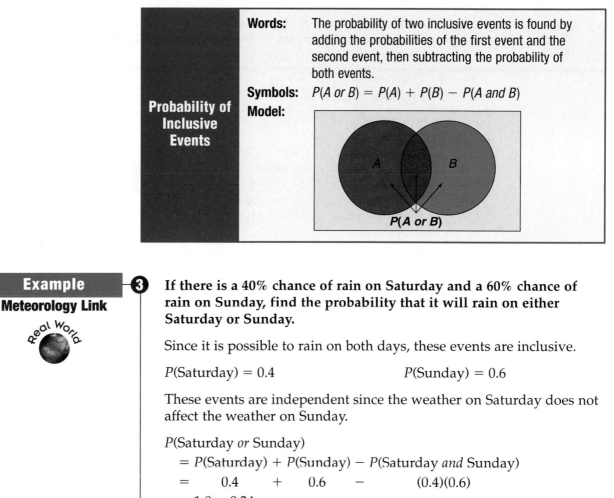

Probability of Inclusive Events

Words: The probability of two inclusive events is found by adding the probabilities of the first event and the second event, then subtracting the probability of both events.

Symbols: $P(A \text{ or } B) = P(A) + P(B) - P(A \text{ and } B)$

Model:

A   B

$P(A \text{ or } B)$

Example ❸

Meteorology Link

Real World

If there is a 40% chance of rain on Saturday and a 60% chance of rain on Sunday, find the probability that it will rain on either Saturday or Sunday.

Since it is possible to rain on both days, these events are inclusive.

$P(\text{Saturday}) = 0.4$ $P(\text{Sunday}) = 0.6$

These events are independent since the weather on Saturday does not affect the weather on Sunday.

$P(\text{Saturday or Sunday})$
$= P(\text{Saturday}) + P(\text{Sunday}) - P(\text{Saturday and Sunday})$
$= \quad 0.4 \quad + \quad 0.6 \quad - \quad\quad (0.4)(0.6)$
$= 1.0 - 0.24$
$= 0.76 \text{ or } 76\%$

The probability that it will rain on the weekend is 76%.

Check for Understanding

Communicating Mathematics

Math Journal

Study the lesson. Then complete the following.

1. **Explain** the difference between $P(A \text{ and } B)$ and $P(A \text{ or } B)$.

2. **Compare and contrast** independent events, mutually exclusive events, and inclusive events. Give an example of each.

Vocabulary

compound events
independent events
mutually exclusive
inclusive

Guided Practice

A die is rolled and the spinner is spun. Find the probability of each event. *(Example 1)*

3. $P(3 \text{ and blue})$

4. $P(\text{even and red})$

A card is drawn from a standard deck of cards. Determine whether each event is *mutually exclusive* or *inclusive*. Then find each probability. *(Examples 2 & 3)*

5. $P(\text{jack or king})$

6. $P(\text{ace or red card})$

7. $P(\text{jack or diamond})$

8. $P(\text{diamond or club})$

9. **Savings** The chart shows the probability that students in grades 6 through 12 are saving for certain items. What is the probability that a student picked at random is saving for a CD player or a computer? *(Example 3)*

What Students Save For	
Item	**Probability**
Car	31%
College	27%
Clothing	19%
CDs or CD player	15%
Trip or vacation	6%
Computer or software	6%

Source: *USA WEEKEND, 1999*

Exercises • • • • • • • • • • • • • • • • • •

Practice

A card is drawn from a deck of ten cards numbered 1 through 10. The card is replaced in the deck and another card is drawn. Find the probability of each outcome.

10. $P(5 \text{ and then a } 3)$

11. $P(\text{two even numbers})$

12. $P(\text{two numbers greater than } 4)$

13. $P(6 \text{ and then an odd number})$

14. $P(\text{an odd number and then an even number})$

15. $P(\text{a number greater than } 7 \text{ and then a number less than } 6)$

16. What is the probability of tossing a coin three times and getting heads each time?

Determine whether each event is *mutually exclusive* or *inclusive*. Then find each probability.

17. There are 3 books about the American Revolution, 2 books about the Civil War, and 4 books about World War II on a shelf. If a book is selected at random, what is the probability of choosing a book about the Civil War or World War II?

18. A card is drawn from a deck of cards. What is the probability that it is a black card or a face card?

19. **Sports**   The probability that
Mark McGwire hit a home run
during an official time at bat
during the 1999 season was
0.125. Find the probability
that he hit a home run on
two consecutive times at bat.
Express your answer as a
decimal rounded to the
nearest thousandth.

20. **Games**   A radio station sponsored The Birthday Game, in which it
would award $1 million to any caller whose birthday matched a
certain month, day, and year. If there were 62 possible years, find
the probability that a caller wins $1 million. Assume 30 days in
each month.

21. **Marketing**   The graph shows the percent
of adults who say they've eaten at certain
types of restaurants during the last six
months.

   a. What is the probability that a person
   chosen at random has eaten in both a
   Chinese and Mexican restaurant?

   b. Suppose no one has eaten at both a
   French and a Japanese restaurant. What
   was the probability that a person ate at
   either a French or a Japanese restaurant?

   c. What is the probability that a person
   chosen at random has eaten at either an
   American or Italian restaurant?

Restaurant Choices	
Kind	**Percent**
American	70
Chinese	60
French	10
Italian	61
Japanese	18
Mexican	57
Thai	11

Source: *American
Demographics,* 1999

22. **Critical Thinking**   A standard domino set has 28 tiles. Seven of
these tiles have the same number of dots on each side and are called
doubles. As the game begins, each player draws one tile. What is the
probability that the first and second players each draw doubles?

Mixed Review

Determine the probability of each outcome.   *(Lesson 5–6)*

23. A month is chosen at random that ends in *-ber.*

24. A die is rolled and shows a number greater than 1.

25. A brown-haired student is chosen at random from a class of
25 students. There are 15 brown-haired students in the class.

26. **Population**   In 1998, the population of North America was about
300 million. It is expected to increase to about 390 million by 2050. By
what percent is the population expected to increase?   *(Lesson 5–5)*

27. **Standardized Test Practice**   Evaluate the expression $|3| + |-5|$.
   (Lesson 2–1)

   A 2   **B** -2 **C** -8 **D** 8

Understanding and Using the Vocabulary

After completing this chapter, you should be able to define each term, property, or phrase and give an example or two of each.

*inter*NET
CONNECTION   **Review Activities**
For more review activities, visit:
www.algconcepts.glencoe.com

Algebra

base *(p. 199)*
dimensional analysis *(p. 190)*
discount *(p. 213)*
mixture problems *(p. 206)*
percent *(p. 198)*
percent equation *(p. 204)*
percent of decrease *(p. 212)*
percent of increase *(p. 212)*
percent proportion *(p. 199)*
percentage *(p. 199)*
proportion *(p. 188)*
rate *(pp. 190, 204)*
ratio *(p. 188)*
sales tax *(p. 213)*
scale *(p. 194)*
scale drawing *(p. 194)*
scale model *(p. 194)*
simple interest *(p. 205)*
unit rate *(p. 190)*

Statistics

box-and-whisker plot *(p. 210)*
circle graph *(p. 200)*
extremes *(p. 210)*
lower quartile *(p. 210)*
percentile *(p. 211)*
upper quartile *(p. 210)*

Probability

complement *(p. 223)*
compound event *(p. 224)*
empirical probability *(p. 220)*
experimental probability
  (p. 220)
inclusive *(p. 227)*
independent events *(p. 224)*
mutually exclusive *(p. 226)*
odds *(p. 221)*
probability *(p. 219)*
random *(p. 220)*
theoretical probability *(p. 220)*

Choose the correct term to complete each sentence.

1. The equation $\frac{2}{3} = \frac{10}{15}$ is a (proportion, ratio).

2. A ratio is a comparison of two numbers by (multiplication, division).

3. A probability of 0 means that an event is (certain to happen, impossible).

4. The probability that occurs when you conduct an experiment is called the (empirical probability, theoretical probability).

5. The (odds, probability) that an event will occur is the ratio of the number of favorable outcomes to the number of unfavorable outcomes.

6. In the percent proportion $\frac{3}{8} = \frac{r}{100}$, 8 is the (base, percentage).

7. A (unit rate, ratio) always has a denominator of 1.

8. In a percent equation, the base is (always, not always) the largest number.

9. (Discount, Sales tax) is an example of a percent of decrease.

10. If two events cannot occur at the same time, they are (mutually exclusive, inclusive).

Chapter 5 Study Guide and Assessment

Skills and Concepts

Objectives and Examples	Review Exercises

• Lesson 5–1   Solve proportions.

$\dfrac{7}{4} = \dfrac{n}{2}$

$14 = 4n$ *Find the cross products.*
$3.5 = n$ *Divide each side by 4.*

Solve each proportion.

11. $\dfrac{5}{8} = \dfrac{n}{72}$ **12.** $\dfrac{r}{11} = \dfrac{35}{55}$

13. $\dfrac{x}{x-1} = \dfrac{4}{3}$ **14.** $\dfrac{z-7}{6} = \dfrac{z+3}{7}$

• Lesson 5–2   Solve problems involving scale drawings and models.

In a drawing, 3 cm = 45 m. Find the length of a line representing 75 meters.

$\dfrac{3 \text{ centimeters}}{45 \text{ meters}} = \dfrac{x \text{ centimeters}}{75 \text{ meters}}$

$225 = 45x$ *Find the cross products.*
$5 = x$ *Divide each side by 45.*

On a map, 2 inches = 5 miles. Find the actual distance for each map distance.

15. 4 inches   **16.** 5 inches
17. 12 inches   **18.** 9 inches

19. A scale model of the Statue of Liberty is 10 inches high. If the Statue of Liberty is 305 feet tall, find the scale of the model.

• Lesson 5–3   Solve problems by using the percent proportion.

75 is what percent of 250?

$\dfrac{75}{250} = \dfrac{r}{100}$ *Use the percent proportion.*
$7500 = 250r$ *Find the cross products.*
$30 = r$ *Divide each side by 250.*

So, 75 is 30% of 250.

Use the percent proportion to find each number.

20. What number is 60% of 80?
21. 21 is 35% of what number?
22. 7 is what percent of 56?
23. 60 is what percent of 40?
24. Find 12% of 5200.
25. 15 is 30% of what number?

• Lesson 5–4   Solve problems by using the percent equation.

Find 15% of 82.

$P = RB$ *Use the percent equation.*
$P = 0.15(82)$ *Replace R with 0.15 and*
$P = 12.3$ *B with 82.*

So, 15% of 82 is 12.3.

Use the percent equation to find each number.

26. 54 is 150% of what number?
27. Find 45% of 18.
28. What is 8% of 80?
29. 21 is 14% of what number?
30. What percent of 34 is 17?

Objectives and Examples

Review Exercises

• **Lesson 5–5** Solve problems involving percent of increase or decrease.

original: $120   new: $108
amount of decrease: $120 − $108 = $12

$$\frac{12}{120} = \frac{r}{100}$$

$1200 = 120r$ *Find the cross products.*
$10 = r$ *Divide each side by 120.*

The percent of decrease is 10%.

Find the percent of increase or decrease. Round to the nearest percent.

31. original: 8
new: 10

32. original: 10
new: 15

33. original: 18
new: 12

34. original: 800
new: 300

35. Find the percent of increase from $25 to $50.

36. What number is 20% less than 80?

• **Lesson 5–6** Find the probability of a simple event.

Find the probability of randomly choosing the letter I in the word PITTSBURGH.

$$\frac{\text{number of favorable outcomes}}{\text{number of possible outcomes}} = \frac{1}{10}$$

One letter from the word MISSISSIPPI is chosen at random. Find each probability.

37. $P(M)$

38. $P(I)$

39. $P(\text{consonant})$

40. $P(\text{vowel})$

41. Find the probability that a letter chosen at random from the word OHIO is *not* an O.

• **Lesson 5–7** Find the probability of mutually exclusive and inclusive events.

mutually exclusive:

$P(A \text{ or } B) = P(A) + P(B)$

inclusive:

$P(A \text{ or } B) = P(A) + P(B) − P(A \text{ and } B)$

A six-sided die is rolled. Determine whether each event is *mutually exclusive* or *inclusive*. Then find each probability.

42. $P(\text{even or less than 5})$

43. $P(6 \text{ or odd})$

Applications and Problem Solving

44. Food Service Of the students surveyed, 40% chose pizza as their favorite lunch. If there are 1250 students in the school, how many would you expect to order pizza? *(Lesson 5–4)*

45. Consumerism Inez is buying a sound system that costs $399. Since she is an employee of the store, she receives a 15% discount. How much will Inez pay for the sound system? *(Lesson 5–5)*

1. **Write** the percent proportion and the percent equation. Explain the meaning of P, B, r, and R.

2. **Compare and contrast** experimental probability and theoretical probability.

Solve each proportion.

3. $\dfrac{3}{x} = \dfrac{12}{16}$

4. $\dfrac{9}{y} = \dfrac{15}{10}$

5. $\dfrac{x}{5} = \dfrac{x + 2}{6}$

6. $\dfrac{3}{3a + 2} = \dfrac{3}{8}$

Find each number.

7. 20% of 40 is what number?

8. 12 is what percent of 60?

9. 20 is what percent of 16?

10. 23 is 25% of what number?

11. Find 120% of 32.

12. What is 35% of 60?

13. **Measurement**   Convert 42 inches to feet.

14. **Recycling**   When 2000 pounds of paper are recycled or reused, 17 trees are saved. How many trees would be saved if 8000 pounds of paper are recycled?

15. **Hobbies**   Model railroads are scaled-down models of real trains. The scale on an HO model is 1 inch = 87 inches. An HO model of a modern diesel locomotive is 8 inches long. How long is the real locomotive?

16. **Banking**   How long will it take Mr. Roberts to earn $1500 if he invests $5000 at a rate of 6%?

17. **Shopping**   The Just Skates sporting goods store advertises that all in-line skates are on sale for 20% off the regular price. Find the sale price of a pair of skates that cost $160.

18. **Taxes**   What is the cost of a pair of jeans that sells for $49 if the sales tax rate is 6%?

19. **Football**   A quarterback threw 18 completed passes out of 30 attempts. Find the experimental probability of making a completed pass. Express as a percent.

20. **Probability**   A die is rolled. What is the probability of rolling a 5 or a number greater than 3?

Preparing for Standardized Tests

Expression and Equation Problems

All standardized tests include questions that ask you to evaluate expressions and solve equations. You'll need to calculate with positive and negative integers, as well as with fractions and decimals. Be sure that you know and can apply the properties of equality.

THE PRINCETON REVIEW

You can use a strategy called "backsolving" on some multiple-choice questions. To use this strategy, substitute each answer choice into the expression or equation to determine which is correct.

Proficiency Test Example

Solve $y = -14x - 5$ if $x = -1$.

A -19 **B** -9 **C** 9 **D** 19

Hint Work carefully with negative numbers. Apply the rules for adding and subtracting integers.

Solution   Substitute -1 for x and evaluate the right side of the equation.

$y = -14x - 5$
$y = -14(-1) - 5$ *Substitution*
$y = 14 - 5$ $-14(-1) = 14$
$y = 9$ $14 - 5 = 9$

The answer is C.

You can check your answer by replacing y with 9 and x with -1 in the original equation. If the statement is true, then your answer is correct.

$9 = -14(-1) - 5$
$9 = 14 + 5$
$9 = 9$ ✓

SAT Example

If $2 + a = 2 - a$, then $a =$

A -1 **B** 0 **C** 1 **D** 2 **E** 4

Hint Use the properties of equality to solve an equation.

Solution

$2 + a = 2 - a$
$2 + a + (-2) = 2 - a + (-2)$ *Add -2 to each side.*
$\qquad a = -a \qquad 2 + (-2) = 0$
$\qquad a + a = -a + a$ *Add a to each side.*
$\qquad 2a = 0 \qquad -a + a = 0$
$\qquad a = 0 \qquad$ *Divide each side by 2.*

The answer is B.

Alternate Solution   Use the strategy known as "backsolving." Substitute each of the answer choices for a and see which value makes the equation true.

A $2 + (-1) = 2 - (-1)$ $1 = 3$ *not true*
B $2 + 0 = 2 - 0$ $2 = 2$ *true*
C $2 + 1 = 2 - 1$ $3 = 1$ *not true*
D $2 + 2 = 2 - 2$ $4 = 0$ *not true*
E $2 + 4 = 2 - 4$ $6 = -2$ *not true*

After you work each problem, record your answer on the answer sheet provided or on a sheet of paper.

1. The time that a traffic light remains yellow is given by the formula $t = \frac{1}{8}s + 1$, where t is the time in seconds and s is the speed limit (mph). If the speed limit is 40 mph, how long will a light stay yellow?

   A 5 seconds

   B 6 seconds

   C 7 seconds

   D 8 seconds

2. The heights of Monica's sunflowers are listed below. What is the median height, in inches, of the sunflowers?

Height of Sunflowers	
2 ft 4 in.	3 ft 11 in.
2 ft 9 in.	4 ft 5 in.
4 ft 3 in.	3 ft 5 in.

   A 41   **B** 42.2

   C 44   **D** 49

3. If $9b = 81$, then $3 \times 3b =$

   A 9.   **B** 27.   **C** 81.

   D 243.   **E** 729.

4. The Tigers baseball team scored four more runs than the Bears. The number of runs the Bears scored is represented by n. Which expression represents the number of runs the Tigers scored?

   A $n - 4$ **B** $n + 4$

   C $4 - n$ **D** $4n$

interNET CONNECTION **Test Practice** For additional test practice questions, visit: www.algconcepts.glencoe.com

5. The formula $F = \frac{9}{5}C + 32$ is used to convert between degrees Celsius and degrees Fahrenheit. If the temperature is 82° F, what is the temperature in degrees Celsius?

   A 27.8   **B** 45.6

   C 63.3   **D** 90

6. Which expression is equivalent to $\frac{4 + 8x}{12x}$?

   A $\frac{1 + 2x}{3x}$ **B** $\frac{1 + 8x}{3x}$ **C** 1

   D $\frac{8}{3}$ **E** $\frac{1 + 2x}{3}$

7. The Huang family had weekly grocery bills of $105, $115, $120, and $98 last month. What was their mean (average) weekly grocery bill last month?

   A $101.50   **B** $102.00

   C $109.50   **D** $117.50

8. The equation $y = \frac{1}{3}x - \frac{3}{8}$ relates x and y. When y is $-\frac{1}{2}$, what is the value of x?

   A $-\frac{1}{24}$ **B** $-\frac{13}{24}$

   C $-2\frac{5}{8}$ **D** $-\frac{3}{8}$

Open-Ended Questions

9. **Grid-In** If $\frac{x + 2x + 3x}{2} = 6$, find the value of x.

10. Consider $3(x - 2) + 4(2x + 1) - 2(1 - x)$.

    Part A Simplify the expression completely. Show your work.

    Part B List the properties you used.

▶ What You'll Learn in Chapter 6:

- to show relations as sets of ordered pairs, as tables, and as graphs *(Lesson 6–1)*,
- to solve linear equations for a given domain *(Lesson 6–2)*,
- to graph linear relations *(Lesson 6–3)*,
- to determine whether a given relation is a function *(Lesson 6–4)*, and
- to solve problems involving direct and inverse variations *(Lessons 6–5 and 6–6)*.

Problem-Solving Workshop

Project

What features would you want in a new car? The table compares some of the features of certain cars. In this project, you will graph these features and determine whether the graphs represent functions or relations. You will also determine if any relationships exist among the features.

Feature	Car A	Car B	Car C	Car D	Car E
Price	$19,616	$21,710	$24,870	$24,045	$21,165
Engine Size (liters)	2.4	2.4	3.8	2.5	3.8
Horsepower	150	141	200	168	190
Fuel Economy (mpg)	26	25	23	22	23
Curb Weight (lb)	3054	2943	3525	3415	3264
Wheelbase (in.)	104.1	98.9	101.1	106	101.3

Working on the Project

Work with a partner and choose a strategy to solve the problem. Here are some suggestions.

- Write and graph a set of ordered pairs (curb weight, engine size). Is the relation a function?
- Describe any relationships that exist between curb weight and engine size.

Strategies

Look for a pattern.

Draw a diagram.

Make a table.

Work backward.

Use an equation.

Make a graph.

Guess and check.

Technology Tools

- Use a **graphing calculator** or **software** to create graphs.
- Use a **word processor** to write about any relationships between features.

interNET **Research** For more information about cars, visit:
CONNECTION www.algconcepts.glencoe.com

Presenting the Project

Prepare a portfolio of the graphs. Be sure to address the following:

- whether the graph is a function or relation, and
- whether there appears to be a relationship between the two features and why.

What You'll Learn
You'll learn to show relations as sets of ordered pairs, as tables, and as graphs.

Why It's Important
Demographics
Population growth data can be expressed as a relation.
See Example 3.

When marine mammals wash ashore, they are called *stranded* animals. Whales, dolphins, and sea turtles are among those found stranded. The table shows the estimated number of sea turtles found stranded on the Texas Gulf Coast over several months in 1999.

These data can also be represented by a set of ordered pairs, as shown below. Each first coordinate represents the month, and the second coordinate is the number of stranded sea turtles.

Sea Turtles Stranded	
Month	**Number**
January	22
February	29
March	43
April	76
May	41
June	23

Source: HEART

$$\{(1, 22), (2, 29), (3, 43), (4, 76), (5, 41), (6, 23)\}$$

The first coordinate in an ordered pair is usually called the **x-coordinate**. The second coordinate is the **y-coordinate**.

Each ordered pair can be graphed.

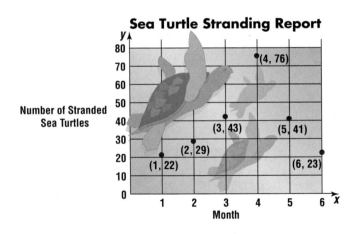

A set of ordered pairs, such as the one for the sea turtle data, is a **relation**. The set of all first coordinates of the ordered pairs, or x-coordinates, is called the **domain** of the relation. The set of all second coordinates, or y-coordinates, is called the **range**.

Domain and Range of a Relation	The domain of a relation is the set of all first coordinates from the ordered pairs of the relation.
	The range of the relation is the set of all second coordinates from the ordered pairs of the relation.

For the relation of the sea turtle data, the domain is {1, 2, 3, 4, 5, 6}, and the range is {22, 29, 43, 76, 41, 23}.

A relation, such as {(−1, −1), (0, 2), (1, 3)}, can be shown in several ways.

Ordered Pairs

(−1, −1)

(0, 2)

(1, 3)

Table

x	y
−1	−1
0	2
1	3

Graph

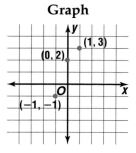

Examples

❶ **Express the relation {(−2, 4), (2, 5), (3, 3), (4, 2), (5, −1)} as a table and as a graph. Then determine the domain and range.**

x	y
−2	4
2	5
3	3
4	2
5	−1

The domain is {−2, 2, 3, 4, 5}, and the range is {4, 5, 3, 2, −1}.

❷ **Express the relation shown on the graph as a set of ordered pairs and in a table. Then find the domain and range.**

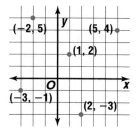

The set of ordered pairs for the relation is {(−3, −1), (−2, 5), (1, 2), (2, −3), (5, 4)}.

x	y
−3	−1
−2	5
1	2
2	−3
5	4

The domain is {−3, −2, 1, 2, 5}, and the range is {−1, 5, 2, −3, 4}.

(continued on the next page)

a. Express the relation {(−4, −1), (−3, 1), (0, 3), (2, −6), (5, 5)} as a table and as a graph. Then determine the domain and range.

b. Express the relation shown on the graph as a set of ordered pairs and in a table. Then find the domain and range.

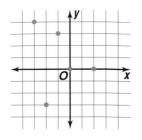

In real-life situations, you may need to select a range of values for the *x*- or *y*-axis that does not begin with 0.

 Example ③

Demographics Link

*inter***NET**
CONNECTION

Data Update For the latest information on population, visit:
www.algconcepts. glencoe.com

Among the 50 states, Kentucky ranks 24th in population. The table shows the population of Kentucky since 1930.

A. Determine the domain and range of the relation.

B. Graph the relation.

C. During which decade was there the greatest increase in population?

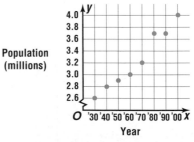

Kentucky State Capitol Building

Year	1930	1940	1950	1960	1970	1980	1990	2000 (est.)
Population (millions)	2.6	2.8	2.9	3.0	3.2	3.7	3.7	4.0

Source: U.S. Census

A. The domain is {1930, 1940, 1950, 1960, 1970, 1980, 1990, 2000}. The range is {2.6, 2.8, 2.9, 3.0, 3.2, 3.7, 4.0}.

B. The *x*-coordinates need to go from 1930 to 2000. There is no need to begin the scale at 0. The *y*-coordinates need to include values from 2.5 to 4.0. You can include 0 and use units of 0.2.

C. By looking at the graph, you can see that the greatest increase in population was during the 1970s.

Communicating Mathematics

Study the lesson. Then complete the following.

1. **Describe** three ways to express a relation.

2. **Describe** the following relation.

{(1, Washington), (2, Adams), (3, Jefferson), . . .}

Math Journal

3. **Write** a sentence explaining the difference between the domain and the range of a relation.

Vocabulary
relation
domain
range
x-coordinate
y-coordinate

Guided Practice

Express each relation as a table and as a graph. Then determine the domain and the range. *(Example 1)*

4. {(−4, 2), (−2, 0), (0, 2), (2, 4)}

5. {(−3, −3.5), (−1, 3.9), (0, 2), (5, 2.5)}

Express each relation as a set of ordered pairs and in a table. Then determine the domain and the range. *(Example 2)*

6.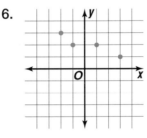

7.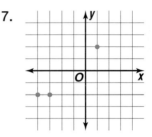

8. **Economics** For his economics class, Alonso is researching the growth of the national debt since 1970. He organized his results in a table. *(Example 3)*

a. Determine the domain and range of the relation.

b. Graph the relation.

c. Predict the national debt in 2000. Check your answer on the Internet.

National Debt	
Year	Amount (billions)
1970	$389
1975	$577
1980	$930
1985	$1946
1990	$3233
1995	$4974

Source: Bureau of the Public Debt

Exercises •

Practice

Express each relation as a table and as a graph. Then determine the domain and the range.

9. {(4, 3), (−2, 3), (−2, 4), (4, 4)}   10. {(2, 3.5), (2.9, 1), (4.5, 7.5)}

11. {(−2, 0), (3, −7), (2, −5), (−6, 3), (1, 5)}

12. {(3.1, 0), (4, 1.5), (−1, −1), (−5.5, 1.9)}

13. $\left\{\left(-\frac{1}{2}, \frac{1}{2}\right), (1, 0), \left(-\frac{1}{2}, -5\right)\right\}$ 14. $\left\{\left(\frac{1}{2}, 3\right), \left(\frac{1}{2}, 1\right), (0, -1), \left(\frac{3}{4}, -2\right)\right\}$

Express each relation as a set of ordered pairs and in a table. Then determine the domain and the range.

15.

16.

17.

18.

19.

20.

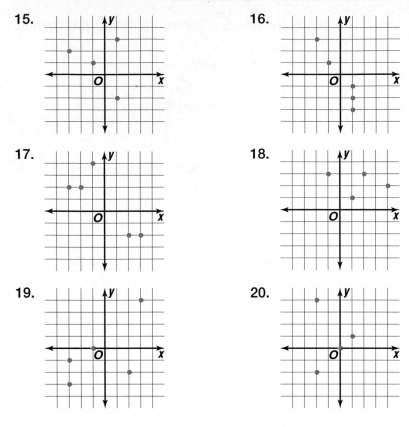

Express each relation as a set of ordered pairs.

21.

x	y
−1	1
0	2
1	3
2	4

22.

x	y
6	−12
5	−10
4	−8
3	−6

23.

x	y
0	0
−1	−0.5
1	0.5
−2	−1

Applications and Problem Solving

24. **Counting** A die is rolled, and the spinner shown is spun.

   a. Write a relation that shows the possible outcomes. Describe the data as the set of ordered pairs (number, color).

   b. Determine the domain and range of the relation.

   c. Use this relation to find the number of ways to roll an even number and land on blue.

25. **Food** Kelly buys a dozen bagels to share with her co-workers. She knows that her co-workers prefer blueberry bagels to plain bagels. So, she buys at least twice as many blueberry bagels as plain bagels.

   a. Write a relation to show the different possibilities. (*Hint:* Let the domain represent the number of blueberry bagels.)

   b. Express the relation in a table.

26. Entertainment The size of a movie on the screen and the distance the projector is from the screen can be expressed as a relation. The diagram below shows that if the distance between the screen and the projector is 3 units, the size of the picture is 9 squares.

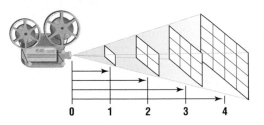

a. Express the relation as a set of ordered pairs.

b. Describe the pattern that exists between the distance the projector is from the screen and the size of the picture.

c. Find the size of the picture if the projector is 12 units from the screen.

d. Graph the relation. Describe the graph.

27. Critical Thinking There are 8 counters numbered 1 through 8 in a bag. Two counters are chosen at random without replacement.

a. What possible sums could you get?

b. Describe the data as the set of ordered pairs (sum, number of ways to occur).

Mixed Review

28. Probability A card is drawn from a standard deck of cards. Find the probability of choosing an ace or a jack. *(Lesson 5–7)*

29. Odds A bag contains 4 yellow marbles, 3 green marbles, and 3 red marbles. Find the odds of choosing a yellow marble. *(Lesson 5–6)*

30. Food The Dessert Factory offers the choices shown for ordering a piece of cheesecake. Draw a tree diagram to find the number of different cheesecakes that can be ordered. *(Lesson 4–2)*

Cheesecake	Topping
original	blueberry
chocolate chip	strawberry
fudge swirl	raspberry
	cherry

Evaluate each expression. *(Lesson 2–1)*

31. $-|-24|$

32. $|8| - |-12|$

33. Standardized Test Practice The stem-and-leaf plot shows the scores for Neshawn's bowling team. What are the highest and lowest scores? *(Lesson 1–7)*

Stem	Leaf
15	3 5 9
16	5 5 8
17	2 3 5 6

$17|3 = 173$

A 15, 17   **B** 153, 176

C 153, 172   **D** 150, 170

Extra Practice See p. 703.

Math In the Workplace

What You'll Learn

You'll learn to solve linear equations for a given domain.

Why It's Important

Sales Knowing how to solve linear equations can help salespeople determine their monthly income. *See Exercise 38.*

Have you ever wondered why plastic foam cups keep hot chocolate hot and iced tea cold? R-values are used to measure a material's ability to insulate. The higher the R-value, the better the material's ability to insulate.

The R-value of foam cups can be found by the equation $y = 4.2x$, where x is the thickness, in inches, of the material. The table below shows the R-value for certain thicknesses of cups.

x (thickness, in.)	4.2x	y (R-value)	(x, y)
0.1	4.2(0.1)	0.42	(0.1, 0.42)
0.25	4.2(0.25)	1.05	(0.25, 1.05)
0.5	4.2(0.5)	2.1	(0.5, 2.1)
0.75	4.2(0.75)	3.15	(0.75, 3.15)
1.0	4.2(1.0)	4.2	(1.0, 4.2)

Each ordered pair (x, y) is a *solution* of the equation $y = 4.2x$. The equation $y = 4.2x$ is an example of an **equation in two variables**.

Solution of an Equation in Two Variables	If a true statement results when the numbers in an ordered pair are substituted into an equation in two variables, then the ordered pair is a solution of the equation.

Example

1 **Which of the ordered pairs (0, 1), (2, 3), (−1, 1), or (3, 5) are solutions of $y = 2x − 1$?**

Make a table. Substitute the x and y values of each ordered pair into the equation.

x	y	y = 2x − 1	True or False?
0	1	1 = 2(0) − 1 1 = −1	false
2	3	3 = 2(2) − 1 3 = 3	true √
−1	1	1 = 2(−1) − 1 1 = −3	false
3	5	5 = 2(3) − 1 5 = 5	true √

A true statement results when the ordered pairs (2, 3) and (3, 5) are substituted into the equation.

Ordered pairs (2, 3) and (3, 5) are solutions of the equation $y = 2x − 1$. The set of solutions to the problem is called the **solution set**.

Your Turn

a. Which of the ordered pairs (1, 1), (0, 2), (−2, 8), or (−1, −5) are solutions of $y = -3x + 2$?

Since the solutions of an equation in two variables are ordered pairs, such an equation describes a relation. The set of values of x is the domain of the relation. The set of corresponding values of y is the range of the relation.

Example 2

Solve $y = 3x$ if the domain is {−2, −1, 0, 1, 3}. Graph the solution set.

Make a table. Substitute each value of x into the equation to determine the corresponding values of y.

x	3x	y	(x, y)
−2	3(−2)	−6	(−2, −6)
−1	3(−1)	−3	(−1, −3)
0	3(0)	0	(0, 0)
1	3(1)	3	(1, 3)
3	3(3)	9	(3, 9)

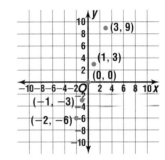

The solution set is {(−2, −6), (−1, −3), (0, 0), (1, 3), (3, 9)}.

Your Turn

b. Solve $y = -4x$ if the domain is {−2, −1, 0, 1, 2}. Graph the solution set.

Sometimes it is helpful to solve an equation for y before substituting each domain value into the equation. This makes creating a table of values easier.

Example 3

Solve $4x + 2y = 8$ if the domain is {−2, −1, 0, 1, 2}. Graph the solution set.

First, solve the equation for y in terms of x.

$4x + 2y = 8$

$\quad 2y = 8 - 4x$ *Subtract 4x from each side.*

$\quad \dfrac{2y}{2} = \dfrac{8 - 4x}{2}$ *Divide each side by 2.*

$\quad\quad y = \dfrac{8}{2} - \dfrac{4x}{2}$ *Write $\dfrac{8 - 4x}{2}$ as the difference of fractions.*

$\quad\quad y = 4 - 2x$

(continued on the next page)

Now, substitute each value of x from the domain to determine the corresponding values of y.

x	$4 - 2x$	y	(x, y)
-2	$4 - 2(-2)$	8	$(-2, 8)$
-1	$4 - 2(-1)$	6	$(-1, 6)$
0	$4 - 2(0)$	4	$(0, 4)$
1	$4 - 2(1)$	2	$(1, 2)$
2	$4 - 2(2)$	0	$(2, 0)$

The solution set is $\{(-2, 8), (-1, 6), (0, 4), (1, 2), (2, 0)\}$.

Your Turn

c. Solve $3x + 6y = 12$ if the domain is $\{-3, -2, -1, 5, 7\}$.

You can also solve an equation for given range values.

Example

4 **Find the domain of $y = 2x - 5$ if the range is $\{-3, -1, 1, 3\}$.**

Make a table. Substitute each value of y into the equation. Then solve each equation to determine the corresponding values of x.

y	$y = 2x - 5$	x	(x, y)
-3	$-3 = 2x - 5$	1	$(1, -3)$
-1	$-1 = 2x - 5$	2	$(2, -1)$
1	$1 = 2x - 5$	3	$(3, 1)$
3	$3 = 2x - 5$	4	$(4, 3)$

The domain is $\{1, 2, 3, 4\}$.

Your Turn

d. Find the domain of $y = -3x$ if the range is $\{-6, -3, 0, 3, 6\}$.

Sometimes variables other than x and y are used in an equation. In this text, the values of the variable that comes first alphabetically are from the domain.

$$m = 2n - 5$$
domain   range

$$3d = c$$
range   domain

$$2a + 3b = 6$$
domain   range

Example

Geometry Link

5 The equation $2w + 2\ell = P$ can be used to determine the perimeter of a rectangle. Suppose a rectangle has a perimeter of 48 inches. Find the possible dimensions of the rectangle given the domain values {3, 4, 8, 12}.

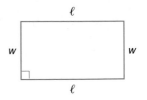

Assume that the values of ℓ come from the domain. Therefore, the equation should be solved for w in terms of ℓ.

$2w + 2\ell = P$

$2w + 2\ell = 48$ *Replace P with 48.*

$\qquad 2w = 48 - 2\ell$ *Subtract 2ℓ from each side.*

$\qquad \dfrac{2w}{2} = \dfrac{48 - 2\ell}{2}$ *Divide each side by 2.*

$\qquad\qquad w = \dfrac{48}{2} - \dfrac{2\ell}{2}$ *Write $\dfrac{48 - 2\ell}{2}$ as the difference of fractions.*

$\qquad\qquad w = 24 - \ell$

Now, substitute the values of ℓ from the domain to determine the corresponding values of w.

ℓ	$24 - \ell$	w	(ℓ, w)
3	24 − 3	21	(3, 21)
4	24 − 4	20	(4, 20)
8	24 − 8	16	(8, 16)
12	24 − 12	12	(12, 12)

The dimensions of the rectangle could be (3, 21), (4, 20), (8, 16), or (12, 12).

Check for Understanding

Communicating Mathematics

Study the lesson. Then complete the following.

1. **Explain** why (2, 5) is a member of the solution set of $y = 2x + 1$.

2. **Tell** what set of numbers make up the range of $y = x + 2$ if the domain is {0, 1, 2, 4}.

3. **You Decide** Lorena and Dan are solving the equation $3n + 2m = 11$ for the domain {−1, 3, 6, 7}. Lorena says to replace n with each of these values. Dan disagrees. He says to replace m with each value. Who is correct? Explain your reasoning.

Vocabulary

equation in two variables
solution set

Guided Practice

Which ordered pairs are solutions of each equation? *(Example 1)*

4. $3a + b = 8$ **a.** (4, −4)   **b.** (2, 2)   **c.** (8, 0)   **d.** (3, 1)

5. $2c + 3d = 11$ **a.** (3, 1)   **b.** (4, −1)   **c.** (1, 3)   **d.** (−2, 5)

Solve each equation if the domain is {−2, −1, 0, 1, 2}. Graph the solution set. *(Examples 2 & 3)*

6. $4x = y$

7. $y = -5x - 3$

8. $2x + 2 = y$

9. $5 + 2b = 3a$

10. Find the domain of $y = x + 3$ if the range is {−4, −1, 0, 1, 2}. *(Example 4)*

11. **Geometry**   Refer to Example 5. Suppose a rectangle has a perimeter of 56 inches. Find the possible dimensions of the rectangle given the domain values {3, 4, 8, 12}. *(Example 5)*

Exercises •

Practice

Which ordered pairs are solutions of each equation?

12. $2p - 5q = 1$ **a.** $(7, 3)$ **b.** $(2, 1)$ **c.** $(-7, -3)$ **d.** $(-2, -1)$

13. $3g = h + 7$ **a.** $(2, 4)$ **b.** $(2, 3)$ **c.** $(-1, 2)$ **d.** $(2, -1)$

14. $3x + 3y = 0$ **a.** $(2, -2)$ **b.** $(-2, 2)$ **c.** $(1, -1)$ **d.** $(-1, 1)$

15. $8 - 2n = 4m$ **a.** $(0, 2)$ **b.** $(2, 0)$ **c.** $(1, -2)$ **d.** $(0.5, -3)$

16. $2c + 4d = 8$ **a.** $(0, 2)$ **b.** $(-2, 1)$ **c.** $(2, 0)$ **d.** $(-3, 0.5)$

17. $8u - 4 = 3v$ **a.** $(4, 2)$ **b.** $\left(0, \frac{1}{2}\right)$ **c.** $(2, 4)$ **d.** $\left(\frac{2}{3}, \frac{3}{4}\right)$

Solve each equation if the domain is {−1, 0, 1, 2, 3}. Graph the solution set.

18. $y = 2x$

19. $y = -3x$

20. $y = 7 - x$

21. $x - y = 6$

22. $y = 2x + 1$

23. $y = -5x + 5$

24. $3y = 9x + 3$

25. $2m - 2n = 4$

26. $3 - 5r = 2s$

27. $12 = 3b + 6a$

28. $6w - 3u = 36$

29. $4 = 3a - b$

30. Name an ordered pair that is a solution of $y = 6x - 2$.

Find the domain of each equation if the range is {−2, 0, 2, 4}.

31. $y = 2x - 2$

32. $y = 6 - 2x$

33. $2y = 4x$

34. Find the value of d if a solution of $y - dx = 1$ is $(2, 5)$.

35. A solution of $4 = ay + 2x$ is $(-1, 2)$. What is the value of a?

Applications and Problem Solving

36. **Shopping**   Cedric has 90¢ to spend on school supplies. Erasers cost 10¢ each, and pencils cost 15¢ each. Suppose r represents the number of erasers and p represents the number of pencils. The equation $10r + 15p = 90$ represents this situation.

   a. Determine the ordered pairs (p, r) that satisfy the equation if the domain is {0, 2, 4, 6}.

   b. Express the relation in a table.

   c. Graph the relation.

37. Life Science The equation $y = 0.025w$ gives the weight of the brain of an infant at birth, where w is the birth weight in pounds. The table shows the brain weights of five newborns.

a. Find the birth weight of each infant.

b. Express the relation as a set of ordered pairs.

c. Graph the relation.

w	y
?	0.15
?	0.17
?	0.18
?	0.2
?	0.22

38. Sales Mr. Richardson is a sales representative for an electronics store. Each month, he receives a salary of $1800 plus a 6% commission on sales over his target amount. Suppose s represents his sales over the target amount. The equation $t = 1800 + 0.06s$ can be used to determine his total monthly income. In July, his sales will either be $800, $1300, or $2000 over the target amount. Find the possible total incomes for July.

39. Critical Thinking The sum of the measures of the two acute angles in a right triangle is 90°. In a given triangle, one acute angle is twice as large as the other acute angle.

a. Write two equations that represent these two relations.

b. Determine the ordered pair that is a solution of both equations.

Mixed Review

40. Express the relation shown in the table as a set of ordered pairs and as a graph. Then find the domain and range. *(Lesson 6–1)*

x	y
−4	−4
−3	0
2	5
5	−3

Exercise 40

41. Consumerism At Crafter's Corner, all paint supplies are on sale at a 30% discount. A set of paint brushes normally sells for $15.70. What is the sale price? *(Lesson 5–5)*

42. Banking McKenna invests $580 at a rate of 6% for 5 years. How much interest will she earn? *(Lesson 5–4)*

Solve each equation.

43. $5(4 - 2c) + 3(c + 9) = 12$ *(Lesson 4–7)*

44. $4n - 9 = 5 + 3n$ *(Lesson 4–6)*

45. $3x - 7 = -10$ *(Lesson 4–5)*

46. Simplify $5(7ab - 4ab)$. *(Lesson 1–4)*

47. Standardized Test Practice Name the property of equality shown by the statement *If 6 + 2 = 8, then 8 = 6 + 2.* *(Lesson 1–2)*

A Reflexive

B Transitive

C Symmetric

D Commutative

Extra Practice   See p. 703.

What You'll Learn
You'll learn to graph linear relations.

Why It's Important
Earth Science
Geologists can use the graph of a relation to predict the time at which Old Faithful will erupt.
See Example 8.

A geyser is a hot spring that throws hot water and steam into the air at intervals. One of the most famous geysers is Old Faithful, located at Yellowstone National Park. In 1938, a park ranger discovered a relationship that can be used to predict Old Faithful's eruptions.

Predicting Old Faithful	
Length of the Eruption (min)	Time Until Next Eruption (min)
1.5	48
2.0	55
2.5	62
3.0	69
3.5	76
4.0	83
4.5	90
5.0	97

The equation $y = 14x + 27$ can be used to predict the eruptions.

y = time until next eruption x = length of an eruption

The solution set is {(1.5, 48), (2.0, 55), (2.5, 62), (3.0, 69), (3.5, 76), (4.0, 83), (4.5, 90), (5.0, 97)}. When you graph the ordered pairs, the points appear to lie on a line. The line contains an infinite number of points. The ordered pair for every point on the line is a solution of the equation $y = 14x + 27$.

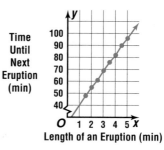

An equation whose graph is a straight line is called a **linear equation**.

Linear Equation in Standard Form
A linear equation is an equation that can be written in the form $Ax + By = C$, where A, B, and C are any numbers, and A and B are not both zero. $Ax + By = C$ is called the *standard form* if A, B, and C are integers.

Linear equations can contain one or two variables.

Linear equations
$3x + 4y = 6$
$y = -2$
$8b = -2a - 5$

Nonlinear equations
$2x + 4y = 3z$
$2x + 2xy = 3$
$2x^2 = 7$

Determine whether each equation is a linear equation. Explain. If an equation is linear, identify A, B, and C when written in standard form.

1 $3x = 5 + y$

First, rewrite the equation so that both variables are on the same side of the equation.

$$3x = 5 + y$$
$$3x - y = 5 \qquad \textit{Subtract y from each side.}$$

The equation is now in the form $Ax + By = C$, where $A = 3$, $B = -1$, and $C = 5$. Therefore, this is a linear equation.

2 $-2x + 3xy = 6$

Since the term $3xy$ has two variables, the equation cannot be written in the form $Ax + By = C$. So, this is not a linear equation.

3 $x = -2$

This equation can be written as $x + 0y = -2$. Therefore, it is a linear equation in the form $Ax + By = C$, where $A = 1$, $B = 0$, and $C = -2$.

Your Turn

a. $8 + y = 2x$ **b.** $-5z = 4x + 2y$ **c.** $y^2 = 4$

To graph a linear equation, make a table of ordered pairs that are solutions. Then graph the ordered pairs and connect the points with a line.

Examples

Graph each equation.

4 $y = 2x - 3$

Select several values for the domain and make a table. Then graph the ordered pairs and connect them to draw the line.

Usually, values of x such as 0 and integers near 0 are chosen so that calculations are simple.

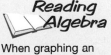

Reading Algebra

When graphing an equation, use arrows to show that the graph continues. Also, label the line with the equation.

x	$2x - 3$	y	(x, y)
-2	$2(-2) - 3$	-7	$(-2, -7)$
-1	$2(-1) - 3$	-5	$(-1, -5)$
0	$2(0) - 3$	-3	$(0, -3)$
1	$2(1) - 3$	-1	$(1, -1)$
2	$2(2) - 3$	1	$(2, 1)$

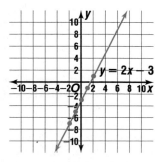

(continued on the next page)

5 $2x + 4y = 8$

In order to find values for y more easily, solve the equation for y.

$2x + 4y = 8$

$\quad\quad 4y = 8 - 2x$ *Subtract 2x from each side.*

$\quad\quad\quad y = \dfrac{8 - 2x}{4}$ *Divide each side by 4.*

Now make a table and draw the graph.

x	$\dfrac{8 - 2x}{4}$	y	(x, y)
-4	$\dfrac{8 - 2(-4)}{4}$	4	$(-4, 4)$
-2	$\dfrac{8 - 2(-2)}{4}$	3	$(-2, 3)$
0	$\dfrac{8 - 2(0)}{4}$	2	$(0, 2)$
2	$\dfrac{8 - 2(2)}{4}$	1	$(2, 1)$
4	$\dfrac{8 - 2(4)}{4}$	0	$(4, 0)$

Your Turn

d. $y = 2x - 1$ **e.** $3x + 2y = 4$

Sometimes, the graph of an equation is a vertical or horizontal line.

Example

6 **Graph $y = 3$. Describe the graph.**

In standard form, this equation is written as $0x + y = 3$. So, for any value of x, $y = 3$. For example, if $x = 0$, $y = 3$; if $x = 1$, $y = 3$; if $x = 3$, $y = 3$. By graphing the ordered pairs $(0, 3)$, $(1, 3)$, and $(3, 3)$, you find that the graph of $y = 3$ is a horizontal line.

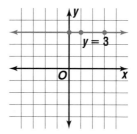

Your Turn

f. Graph the equation $x = -2$. Describe the graph.

The graph of an equation can also pass through the origin.

Graph $y = 2x$.

Make a table and draw the graph.

x	2x	y	(x, y)
−2	2(−2)	−4	(−2, −4)
−1	2(−1)	−2	(−1, −2)
0	2(0)	0	(0, 0)
1	2(1)	2	(1, 2)
2	2(2)	4	(2, 4)

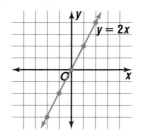

Write $y = 2x$ in standard form.

$$y = 2x$$
$$0 = 2x - y \quad \textit{Subtract y from each side.}$$
$$2x - y = 0 \quad \textit{Symmetric Property of Equality}$$

In any equation where $C = 0$, the graph passes through the origin.

Your Turn

g. Graph the equation $y = -3x$.

Graphs of linear equations can be used to solve real-life problems.

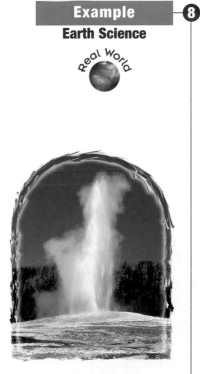

Old Faithful

Refer to the application at the beginning of the lesson. Suppose Old Faithful erupts at 9:46 A.M., and the eruption lasts 3.4 minutes. At about what time will the next eruption occur?

Explore   You need to find the time at which the next eruption will occur.

Plan   Use the graph of $y = 14x + 27$ to solve the problem.

Solve   The graph shows that if an eruption lasts 3.4 minutes, the next eruption will occur in about 75 minutes.

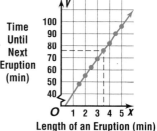

Length of an Eruption (min)

$$9 \text{ h } 46 \text{ min} \rightarrow 9 \text{ h } 46 \text{ min}$$
$$+ \quad 75 \text{ min} \rightarrow \underline{1 \text{ h } 15 \text{ min}}$$
$$10 \text{ h } 61 \text{ min} \rightarrow 11 \text{ h } 1 \text{ min}$$

The next eruption should take place at 11:01 A.M.

Examine   You can verify the answer by using the equation.
$$y = 14x + 27$$
$$y \stackrel{?}{=} 14(3.4) + 27 \quad \textit{Replace x with 3.4.}$$
$$y = 74.6$$

The answer makes sense because 74.6 minutes added to 9:46 is about 11:01.

Check for Understanding

Study the lesson. Then complete the following.

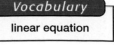

Vocabulary

linear equation

1. **Describe** the graph of $x = -8$.

2. **Explain** how you know that the graph of $y = -2x$ passes through the origin.

Guided Practice

Determine whether each equation is a linear equation. Explain. If an equation is linear, identify A, B, and C. *(Examples 1–3)*

3. $xy = 3$ 4. $3y + 4x = 9$ 5. $2x = 8$

Graph each equation. *(Examples 4–7)*

6. $y = 3x + 2$ 7. $y = -\frac{1}{2}x$ 8. $x = -5$

9. $-2 = 4x + y$ 10. $3x + 2y = 6$ 11. $5y = -2x + 10$

12. **Plants** The number of blooms on a cactus is related to the number of days of sun it gets in a month. This relation is given by the equation $y = 7x - 1$, where x is the number of days of sun and y is the number of blooms. *(Example 8)*

   a. Graph the equation.

   b. Use the graph to determine the number of blooms on a cactus if it gets two days of sun a month.

Exercises

• • • • • • • • • • • • • • • • • • • •

Practice

Determine whether each equation is a linear equation. Explain. If an equation is linear, identify A, B, and C.

13. $xy = 4$ 14. $6 + 2xy = 4$ 15. $2y - 6x = 0$

16. $4x = 8y$ 17. $y = x$ 18. $x + 3xy = 5$

19. $y = x^2 + 8$ 20. $2 - 3x = 8$ 21. $4x + x + y = 9$

Graph each equation.

22. $x = y$ 23. $y = 3x$ 24. $-2x = y$

25. $y = -3$ 26. $x = -3$ 27. $2 = x$

28. $y = 3x + 2$ 29. $2x - 5 = y$ 30. $y = 4 - 2x$

31. $-1 = y + 4x$ 32. $2x + y = -2$ 33. $3 = -3x + y$

34. $4x + 2y = 8$ 35. $3y - 2x = 6$ 36. $-2y = 3x + 1$

Each table lists coordinates of points on a linear graph. Copy and complete each table. (*Hint:* Graph the ordered pairs.)

37.

x	y
0	?
1	-1
2	0
3	1
4	?

38.

x	y
-4	-2
-2	?
0	0
2	1
4	?

39.

x	y
-2	-4
-1	?
0	2
1	?
2	8

Applications and Problem Solving

40. Business Bongo's Clown Service charges a $50 fee for a clown to perform at a birthday party plus an additional $2 per guest.

   a. Write an equation that represents this situation. Let x represent the number of guests and let y represent the cost of the clown's services.

   b. Graph the equation.

   c. Serena's parents can afford to pay $68 for a clown to perform at Serena's party. How many people can attend Serena's party?

41. Geometry Complementary angles are two angles whose sum is 90°. Suppose two complementary angles measure $(2x + 2y)°$ and $(x + 4y)°$.

   a. Write an equation for the sum of the measures of these two angles.

   b. Determine five ordered pairs that satisfy this equation.

   c. Graph the equation.

42. Critical Thinking Graph the equations $y = 2x$, $y = 2x + 1$, $y = 2x + 2$, and $y = 2x + 3$ on the same coordinate plane. Then describe the graphs.

Mixed Review

43. Solve $y = x - 5$ if the domain is $\{-3, 4, 5, 7, 9\}$. *(Lesson 6–2)*

44. Express $\{(-1, -1), (0, 5), (1, -2), (3, 4)\}$ as a table and as a graph. *(Lesson 6–1)*

Solve each proportion. *(Lesson 5–1)*

45. $\dfrac{3}{4} = \dfrac{x}{8}$ **46.** $\dfrac{a}{45} = \dfrac{3}{15}$ **47.** $\dfrac{y}{3} = \dfrac{y + 4}{6}$

48. Statistics Find the range of the data: $\{17, 24, 16, 25, 38, 57\}$. *(Lesson 3–3)*

49. Open-Ended Test Practice The temperature outside is 67°F. Suppose it rises an average of 2°F every hour for h hours to reach a high of 79°F. Write an equation to represent this situation. *(Lesson 1–1)*

Quiz 1   Lessons 6–1 through 6–3

Express each relation as a table and as a graph. Then determine the domain and the range. *(Lesson 6–1)*

1. $\{(1, -1), (3, 5), (0, -2), (3, 3)\}$ **2.** $\{(-0.5, 3), (1.5, 0), (2.5, 4), (3.5, 2.5)\}$

3. Which ordered pairs are solutions to the equation $2y + 4 = 4x$? *(Lesson 6–2)*

   a. $(-1, 3)$ **b.** $(0, -2)$ **c.** $(-2, -6)$ **d.** $(3, -4)$

4. Solve $y = 4x - 3$ if the domain is $\{-2, -1, 2, 3, 5\}$. Graph the solution set. *(Lesson 6–2)*

5. Anne is saving money to buy a $2500 used car. She started with a $500 gift and has been saving $80 per week since then. The equation $y = 500 + 80x$ represents her total savings y after any number of weeks x. *(Lesson 6–3)*

   a. Graph the equation.

   b. Use the graph to estimate when Anne will have enough money.

   c. Verify your estimate using the equation.

Extra Practice   See p. 704.

Lesson 6–3 Graphing Linear Relations   **255**

What You'll Learn

You'll learn to determine whether a given relation is a function.

Why It's Important

Business Functions can be used to determine the cost of renting a car. *See Exercise 48.*

The deepest mine in the world is the Western Deep Levels Gold Mine in South Africa. The graph shows how the temperature of the walls of the mine varies with their depth.

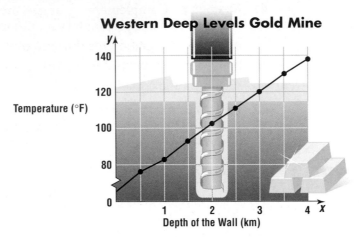

Western Deep Levels Gold Mine

Temperature (°F)

Depth of the Wall (km)

Notice that for each value of *x*, there is exactly one value of *y*. This type of relation is called a **function**.

Functions	A function is a relation in which each member of the domain is paired with *exactly* one member of the range.

Examples

Determine whether each relation is a function. Explain your answer.

1 {(5, 2), (3, 5), (−2, 3), (−5, 1)}

Since each element of the domain is paired with exactly one element of the range, this relation is a function.

2

x	2	3	5	−1	4
y	3	0	0	−2	1

The table represents a function since, for each element of the domain, there is *only one* corresponding element in the range. *It does not matter if two elements of the domain are paired with the same element in the range.*

3

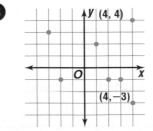

The graph represents a relation that is not a function. Look at the points with the ordered pairs (4, −3) and (4, 4). The element 4 in the domain is paired with both −3 and 4 in the range.

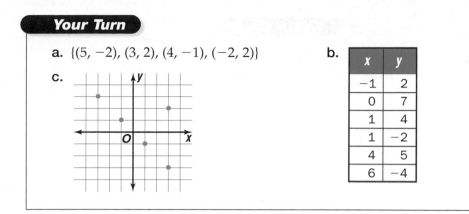

Your Turn

a. $\{(5, -2), (3, 2), (4, -1), (-2, 2)\}$

b.

x	y
−1	2
0	7
1	4
1	−2
4	5
6	−4

c.

To determine whether an equation is a function, you can perform a test on the graph of the equation. Consider the graph shown.

Place a pencil at the left of the graph to represent a vertical line. Move the pencil to the right across the graph.

For each value of x, this vertical line passes through no more than one point on the graph. So, the equation is a function.

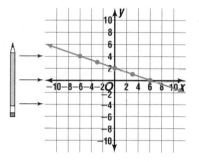

This test is called the **vertical line test**.

Vertical Line Test for a Function	If any vertical line passes through no more than one point of the graph of a relation, then the relation is a function.

Examples

Use the vertical line test to determine whether each relation is a function.

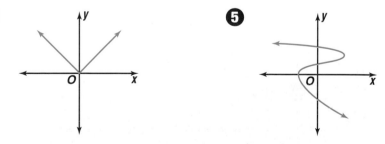

❹   ❺

The relation in Example 4 is a function since any vertical line passes through no more than one point of the graph of the relation. The relation in Example 5 is not a function since there is a vertical line passing through more than one point.

(continued on the next page)

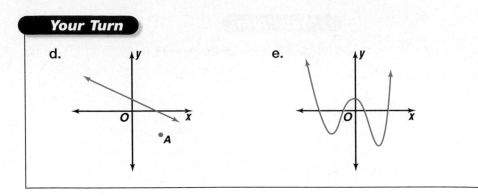

Your Turn

d.

e.

Equations that are functions can be written in a form called **functional notation**. For example, consider the equation $y = 3x + 5$.

equation	**functional notation**
$y = 3x + 5$	$f(x) = 3x + 5$

Reading Algebra

Read the symbol $f(x)$ as *f at x* or *f of x*. Letters other than *f* are also used for names of functions.

In a function, x represents the elements of the domain, and $f(x)$ represents the elements of the range. For example $f(2)$ is the element in the range that corresponds to the member 2 in the domain. We say that $f(2)$ is the **functional value** of f for $x = 2$.

You can find $f(2)$ as shown at the right. The ordered pair (2, 11) is a solution of the function of f.

$f(x) = 3x + 5$
$f(2) = 3 \cdot 2 + 5$ *Replace x with 2.*
$f(2) = 6 + 5 \text{ or } 11$

Examples

If $f(x) = 2x + 3$, find each value.

6 $f(4)$

$f(x) = 2x + 3$
$f(4) = 2(4) + 3$ *Replace x with 4.*
$\quad = 8 + 3$ *Multiply.*
$\quad = 11$ *Add.*

7 $f\left(-\dfrac{1}{2}\right)$

$f(x) = 2x + 3$
$f\left(-\dfrac{1}{2}\right) = 2\left(-\dfrac{1}{2}\right) + 3$ *Replace x with $-\dfrac{1}{2}$.*
$\qquad = -1 + 3$ *Multiply.*
$\qquad = 2$ *Add.*

8 $f(6a)$

$f(x) = 2x + 3$
$f(6a) = 2(6a) + 3$ *Replace x with 6a.*
$\quad = 12a + 3$ *Multiply.*

Your Turn

f. If $g(x) = 3x - 7$, find $g(-2)$, $g(1.5)$, and $g(2h)$.

Functions are often used to solve real-life problems.

9 Anthropologists use the length of certain bones of a human skeleton to estimate the height of the living person. One of these bones is the femur, which extends from the hip to the knee. To estimate the height in centimeters of a female with a femur of length x, the function $h(x) = 61.41 + 2.32x$ can be used. What was the height of a female whose femur measured 46 centimeters?

$h(x) = 61.41 + 2.32x$
$h(46) = 61.41 + 2.32(46)$ *Replace x with 46.*
$\quad\quad = 168.13$

The woman was about 168 centimeters tall.

Check for Understanding

Communicating Mathematics

Study the lesson. Then complete the following.

1. **Compare and contrast** relations and functions.
2. **Explain** why $f(3) = -1$ if $f(x) = 2x - 7$.
3. **You Decide** Lisa says that the graph of every straight line represents a function. Zina says that there are some graphs of straight lines that do not represent a function. Who is correct? Explain your reasoning.

Vocabulary

function
vertical line test
functional notation
functional value

Guided Practice

Determine whether each relation is a function. *(Examples 1–3)*

4. $\{(2, 3), (3, 5), (4, 6), (2, 5), (1, 4)\}$
5. $\{(-1, 6), (1, 4), (-2, 5), (2, 6)\}$

6.

x	y
−2	4
0	3
5	2
2	4

7.

8. Use the vertical line test to determine whether the relation at the right is a function. *(Examples 4 & 5)*

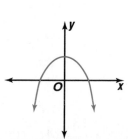

If $f(x) = 4x - 3$, find each value. *(Examples 6–8)*

9. $f(-3)$ **10.** $f\left(\frac{1}{4}\right)$ **11.** $f(b)$ **12.** $f(-2a)$

13. Safety The time in seconds that a traffic light remains yellow is given by the function $t(s) = 0.05s + 1$, where s represents the speed limit. How long will a light remain yellow if the speed limit is 45 miles per hour? *(Example 9)*

Exercises · · · · · · · · · · · · · · · · · · · ·

Practice

Determine whether each relation is a function.

14. $\{(3, 5), (4, 3), (-2, 1), (-1, 4)\}$ **15.** $\{(-2, 3), (-3, 2), (5, 2), (7, 2)\}$

16. $\{(2, -3), (3, -2), (5, -2), (2, -5)\}$ **17.** $\{(0, 2), (-1, 3), (2, 3), (-1, 2)\}$

18. $\{(-3, 5), (-2, 4), (-3, 6), (-2, 3)\}$ **19.** $\{(1, 2), (2, 1), (-1, 2), (-2, 1)\}$

20.

x	y
−2	5
3	−2
4	5
5	5
6	1

21.

x	y
0	1
0	0
−4	0
2	6
0	−4

22.

x	y
−3	4
−2	0
1	6
−2	1
5	2

23.  **24.** **25.**

Use the vertical line test to determine whether each relation is a function.

26. **27.** **28.**

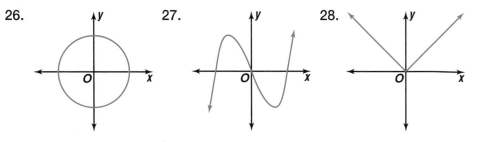

If $f(x) = 2x - 5$ and $g(x) = -3x + 4$, find each value.

29. $f(2)$ **30.** $g(4)$ **31.** $g(-3)$ **32.** $f(-7)$

33. $g(0.2)$ **34.** $f(0.6)$ **35.** $g(1.9)$ **36.** $f(-2.6)$

37. $g\left(\frac{1}{3}\right)$ **38.** $f\left(-\frac{1}{2}\right)$ **39.** $f\left(\frac{3}{4}\right)$ **40.** $g\left(-\frac{5}{6}\right)$

41. $g(h)$ **42.** $f(-2d)$ **43.** $g(-2a)$ **44.** $f(4.5c)$

45. Find $f(-4)$ if $f(x) = 5x + 8$.

46. What is the value of $m(3) + n(-5)$ if $m(x) = 2x$ and $n(x) = 3x - 1$?

47. **Health**   A person's normal systolic blood pressure p depends on the person's age a. To determine the normal systolic blood pressure for an individual, you can use the equation $p = 0.5a + 110$.

a. Write this relation in functional notation.

b. Find the normal systolic blood pressure for a person who is 36 years old.

c. Graph the relation. Describe the graph.

48. **Business**   The cost of a one-day car rental from City-Wide Rentals is given by the function $C(x) = 0.18x + 35$, where x is the number of miles that the car is driven. Suppose Mrs. Burton drove a distance of 95 miles and back in one day. What is the cost of the car rental?

49. **Critical Thinking**   Consider the functions $h(x) = 2x - 5$ and $g(x) = 3x + 2$.

a. For what value of x will $h(x)$ and $g(x)$ be equal?

b. Graph $h(x)$ and $g(x)$. Describe what you observe.

Mixed Review

50. Graph $y = -2x - 6$.   *(Lesson 6–3)*

51. **Geometry**   The equation $A = \frac{1}{2}bh$ can be used to find the area of a triangle. Suppose a triangle has an area of 28 square meters. Find the possible heights of the triangle given the domain values $\{2, 4, 7\}$.   *(Lesson 6–2)*

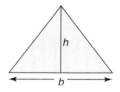

Use the percent proportion to answer each question.   *(Lesson 5–3)*

52. What is 145% of 86?

53. What percent of 60 is 18?

54. **Life Science**   A scale model of a sequoia tree is 4 feet tall. If the actual sequoia tree is 311 feet tall, what is the scale of the model? *(Lesson 5–2)*

55. Evaluate $\frac{x}{5}$ if $x = \frac{3}{4}$.   *(Lesson 4–3)*

56. **Animals**   The fastest swimming bird is the gentoo penguin of Antarctica. It moves through the water at about 17 miles per hour. Traveling at this rate, how far can the gentoo penguin travel in 45 minutes?   *(Lesson 4–1)*

57. What is the value of w if $w - 13.8 = -12.6$?   *(Lesson 3–6)*

58. **Standardized Test Practice**   Find the only solution of $3x + (-5) = 10$ if the replacement set is $\{-5, 3, 5, 8\}$.   *(Lesson 3–4)*

A -5 B 3   C 5   D 8

Gentoo Penguin

Extra Practice   See p. 704.

How Many CENTIMETERS in a Foot?

Materials

- grid paper
- meterstick
- red and blue pens or pencils
- almanac

Functions

Many real-world situations represent functions. Is there a relationship between foot length and shoe size? Does a relationship exist between the month of the year and the average temperature? Let's find out!

Investigate

1. Ask eight to ten females and eight to ten males to participate in your data research.

   a. Make two tables with the headings shown. Use one table to record data for the females and the other to record the data for the males.

Name	Foot Length (cm)	Shoe Size	(Foot Length, Shoe Size)

   b. Measure the standing foot length in centimeters of each participant without shoes. Ask each person what size shoe they typically wear. Record the results in your table.

   c. Write an ordered pair (foot length, shoe size) for each person. Then graph the ordered pairs on a coordinate plane. Use a red pencil to graph the ordered pairs that represent the female data. Use a blue pencil to graph the ordered pairs that represent the male data. Connect each set of ordered pairs with the respective color.

   d. Is either graph a function? Why or why not? Describe any relationship that exists between foot length and shoe size.

   e. Save your tables to use later in the investigation.

2. Choose a city in the United States. Research temperature data for that city. Find the average high and low temperatures for each month. It may be helpful to use an almanac or the Internet.

a. Make a table like the one shown. Record the temperature data for the selected city.

Month	Average High Temperature (°F)	Average Low Temperature (°F)
January		
February		
⋮		

b. Write two sets of ordered pairs. The first set should contain the ordered pairs (month, high temperature). Assign January the number 1, February the number 2, and so on. The second set should contain the ordered pairs (month, low temperature).

c. Graph both sets of data on one coordinate plane. Connect the ordered pairs in the first set using a red pencil. Use a blue pencil to connect the ordered pairs in the second set.

d. Is either graph a function? Explain. Describe any relationship that exists between the month and the average temperature. Are there any similarities between these two graphs?

Extending the Investigation

In this extension, you will continue to investigate functions.

- Select another city in the United States. Make a table of average high and low temperatures for each month. Then graph the data and determine if the graph is a function. Describe any relationships that exist between the graphs.

- Select a city in Hawaii. Make a table of average high and low temperatures for each month. Then graph the data. How does the graph of the temperature data for Hawaii compare to the graph of the temperature data in Exercise 2?

Presenting Your Investigation

Here are some ideas to help you present your conclusions to the class.

- Make a brochure showing the results of your research. Be sure to include the tables and graphs. List any relationships you discovered.

- Discuss any similarities or differences in the graphs.

- Look in an almanac or on the Internet for real-world data whose graph would be a function. Make a poster presenting your data, tables, and graphs.

*inter*NET CONNECTION **Investigation** For more information on temperatures, visit: www.algconcepts.glencoe.com

Math In the Workplace

What You'll Learn

You'll learn to solve problems involving direct variations.

Why It's Important

Electronics
Electricians can use direct variation to determine the weight of copper wire.
See Exercise 32.

Do you know that your weight would vary from planet to planet? The table and graph below show the relation between the weight of an object on Earth and its weight on Mars.

Weight on Earth (lb)	Weight on Mars (lb)
0	0
25	10
50	20
75	30
100	40
125	50
150	60

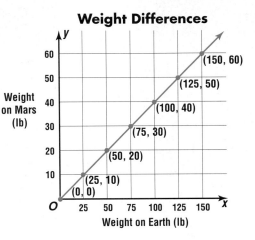

Weight Differences

The weight of an object on Mars depends *directly* on its weight on Earth. That is, as an object's weight on Earth increases, its weight increases on Mars. The relationship between weight on Earth and weight on Mars is shown by the equation $y = 0.4x$. This equation is called a **direct variation**. We say that *y varies directly as x.*

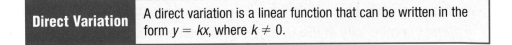

Direct Variation	A direct variation is a linear function that can be written in the form $y = kx$, where $k \neq 0$.

Since the weight on Mars depends on the weight on Earth, the weight on Mars is called the **dependent variable**. The weight on Earth is called the **independent variable**.

As stated, a direct variation can be written in the form $y = kx$. Notice that when $x = 0$, $y = 0$. The graph of a direct variation passes through the origin.

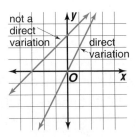

In the equation $y = kx$, k is called the **constant of variation**.

Determine whether the equation is a direct variation.

1 $y = 4x$

Graph the equation.

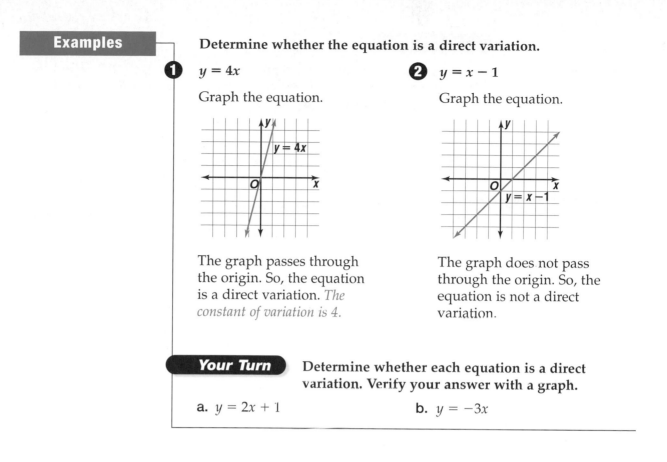

$y = 4x$

2 $y = x - 1$

Graph the equation.

$y = x - 1$

The graph passes through the origin. So, the equation is a direct variation. *The constant of variation is 4.*

The graph does not pass through the origin. So, the equation is not a direct variation.

Your Turn   Determine whether each equation is a direct variation. Verify your answer with a graph.

a. $y = 2x + 1$

b. $y = -3x$

Many real-world problems involve direct variation.

Example

Astronomy Link

Real World

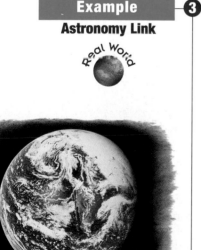

3 The weight of an object on Venus varies directly as its weight on Earth. An object that weighs 80 pounds on Earth weighs 72 pounds on Venus. How much would an object weigh on Venus if its weight on Earth is 90 pounds?

First, find the constant of variation. Let x = weight on Earth and let y = weight on Venus, since the weight on Venus depends on the weight on Earth.

$y = kx$ *Definition of direct variation*
$72 = k \cdot 80$ *Replace y with 72 and x with 80.*
$\dfrac{72}{80} = \dfrac{k \cdot 80}{80}$ *Divide each side by 80.*
$0.9 = k$

Next, use the constant of variation to find y when $x = 90$.

$y = kx$ *Definition of direct variation*
$y = 0.9(90)$ *Replace k with 0.9 and x with 90.*
$y = 81$

So, the object would weigh 81 pounds on Venus.

(continued on the next page)

c. An object that weighs 110 pounds on Earth weighs 18.7 pounds on the moon. How much would an object weigh on the moon if its weight on Earth is 150 pounds?

Direct variations can be used to solve **rate problems**.

Example

Travel Link

Real World

4 The length of a trip varies directly as the amount of gasoline used. Suppose 4 gallons of gasoline were used for a 120-mile trip. At that rate, how many gallons of gasoline are needed for a 345-mile trip?

Let ℓ represent the length of the trip and let g represent the amount of gasoline used for the trip. The statement *the length varies directly as the amount of gasoline* translates into the equation $\ell = kg$ in the same way as *y varies directly as x* translates into $y = kx$. Find the value of k.

$\ell = kg$ *Direct variation*

$120 = k \cdot 4$ *Replace ℓ with 120 and g with 4.*

$\dfrac{120}{4} = \dfrac{4k}{4}$ *Divide each side by 4.*

$30 = k$

Next, find the amount of gasoline needed for a 345-mile trip.

$\ell = kg$

$345 = 30 \cdot g$ *Replace ℓ with 345 and k with 30.*

$\dfrac{345}{30} = \dfrac{30g}{30}$ *Divide each side by 30.*

$11.5 = g$

A 345-mile trip would use 11.5 gallons of gasoline.

Your Turn

d. How many gallons of gasoline would be needed for a 363-mile trip if a 72.6-mile trip used 3 gallons of gasoline?

Direct variations are also related to proportions. Using the table on page 264, many proportions can be formed. Two examples are shown.

weight on Earth → $\dfrac{50}{20} = \dfrac{100}{40}$ ← *weight on Earth*
weight on Mars →   ← *weight on Mars*

weight on Earth → $\dfrac{50}{20} = \dfrac{20}{40}$ ← *weight on Earth*
weight on Mars →   ← *weight on Mars*

Three general forms for proportions like these are $\dfrac{y_1}{y_2} = \dfrac{x_1}{x_2}$, $\dfrac{x_2}{y_2} = \dfrac{x_1}{y_1}$, and $\dfrac{y_1}{x_1} = \dfrac{y_2}{x_2}$. You can use any of these forms to solve direct variation problems.

Example **5** **Suppose y varies directly as x and $y = 36$ when $x = 9$. Find x when $y = 48$.**

Use $\dfrac{y_1}{x_1} = \dfrac{y_2}{x_2}$ to solve the problem.

$\dfrac{36}{9} = \dfrac{48}{x_2}$ *Let $y_1 = 36$, $y_2 = 48$, and $x_1 = 9$.*

$36x_2 = 9(48)$ *Find the cross products.*

$36x_2 = 432$

$\dfrac{36x_2}{36} = \dfrac{432}{36}$ *Divide each side by 36.*

$x_2 = 12$

So, $x = 12$ when $y = 48$.

Your Turn

e. Suppose y varies directly as x and $y = 27$ when $x = 6$. Find x when $y = 45$.

You can also use direct variation to convert measurements.

Example
Measurement Link

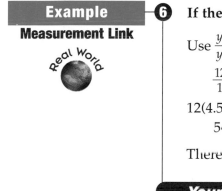
Real World

6 **If there are 12 inches in 1 foot, how many inches are in 4.5 feet?**

Use $\dfrac{y_1}{y_2} = \dfrac{x_1}{x_2}$ to solve the problem.

$\dfrac{12}{1} = \dfrac{x_1}{4.5}$ *Let $y_1 = 12$, $y_2 = 1$, and $x_2 = 4.5$.*

$12(4.5) = x_1$ *Find the cross products.*

$54 = x_1$

There are 54 inches in 4.5 feet.

Your Turn

f. How many pints are in 2.3 gallons if there are 8 pints in 1 gallon?

Check for Understanding

Communicating Mathematics

Study the lesson. Then complete the following.

1. **Explain** how you find the constant of variation in a direct variation.

2. **Give an example** of an equation that is a direct variation.

Math Journal

3. Refer to Example 2. **Write** a few sentences describing the dependent variable and independent variable.

Vocabulary
direct variation
independent variable
dependent variable
constant of variation
rate problems

⏱ Getting Ready | **Find the constant of variation for each direct variation.**

Sample 1: $y = 3x$

Solution: 3

Sample 2: $n = -0.3m$

Solution: -0.3

4. $y = 2x$ **5.** $5c = d$ **6.** $h = -4g$ **7.** $t = \frac{2}{5}r$ **8.** $1.4p = q$

Determine whether each equation is a direct variation. Verify the answer with a graph. *(Examples 1 & 2)*

9. $y = -x$

10. $y = 5x + 5$

Solve. Assume that y varies directly as x. *(Example 5)*

11. If $y = 28$ when $x = 7$, find x when $y = 52$.

12. Find y when $x = 10$ if $y = -8$ when $x = 2$.

13. If there are 3 feet in a yard, how many yards are in 14.1 feet? *(Example 6)*

14. Astronomy The weight of an object on Jupiter varies directly as its weight on Earth. An object that weighs 32.5 pounds on Earth weighs 81.25 pounds on Jupiter. How much would an object weigh on Jupiter if its weight on Earth is 90.2 pounds? *(Examples 3 & 4)*

Exercises • • • •   • • • • • • • • • • • • • •

Practice

Determine whether each equation is a direct variation. Verify the answer with a graph.

15. $y = x + 4$ **16.** $y = -2x$ **17.** $y = -3$

18. $y = -0.5x$ **19.** $x = 2$ **20.** $\frac{y}{x} = 6$

Solve. Assume that y varies directly as x.

21. Find x when $y = 5$ if $y = 3$ when $x = 15$.

22. If $x = 24$ when $y = 8$, find y when $x = 12$.

23. If $y = -7$ when $x = -14$, find x when $y = 10$.

24. Find y when $x = 9$ if $y = -4$ when $x = 6$.

25. If $x = 15$ when $y = 12$, find x when $y = 21$.

26. Find y when $x = 0.5$ if $y = 100$ when $x = 20$.

27. If $y = 12$ when $x = 1\frac{1}{8}$, find y when $x = \frac{1}{4}$.

Solve by using direct variation.

28. How many inches are in 1.25 yards if there are 36 inches in 1 yard?

29. If 1 bushel is 4 pecks, how many bushels are there to 2.4 pecks?

30. How many acres are in 0.75 square mile if there are 640 acres in 1 square mile?

Applications and Problem Solving

31. Consumerism For $1.89, you can buy 75 square feet of plastic wrap. At this rate, how much would 125 square feet of plastic wrap cost?

32. Electronics How much will 25 meters of copper wire weigh if 3.4 meters of the same copper wire weighs 0.442 kilogram?

33. Health The height h of an average person varies directly with their foot length f. Suppose the constant of variation is 7.

  a. Write an equation that represents this situation.

  b. Graph the equation.

  c. Use the graph to find the height of a person whose foot length is 9.5 inches.

34. Critical Thinking Assume that y varies directly as x.

  a. Suppose the value of x is doubled. What is the constant of variation? What happens to the value of y?

  b. Suppose the value of x is tripled. What is the constant of variation? What happens to the value of y?

  c. Suppose the value of y is doubled. What is the constant of variation? What happens to the value of x?

  d. Suppose the value of y is tripled. What is the constant of variation? What happens to the value of x?

Mixed Review

Seal

Sea Lion

35. Find $d(-3)$ if $d(x) = -5d - 4$. *(Lesson 6–4)*

36. Graph $x = 4$. *(Lesson 6–3)*

37. Life Science A full-grown male seal weighs about 550 pounds. This is 50 pounds less than the weight of a full-grown male sea lion. *(Lesson 3–5)*

  a. Write an equation that can be used to find the weight of a full-grown male sea lion.

  b. How much does a full-grown male sea lion weigh?

Find each quotient. *(Lesson 2–6)*

38. $-36 \div 4$ **39.** $-64 \div (-8)$ **40.** $\dfrac{-96}{-16}$

41. Standardized Test Practice Name the coordinates of U, X, and Y. *(Lesson 2–1)*

  A $-5, -2, 3$ **B** $-5, 0, 3$ **C** $-3, 3, 5$ **D** $-4, -2, 0$

Quiz 2   Lessons 6–4 and 6–5

Determine whether each relation is a function. *(Lesson 6–4)*

1. $\{(1, 2), (2, -1), (0, 3), (-1, 4)\}$ **2.** $5x = 6 + y$

3. If $g(x) = 4x + 3$, find $g(-2.6)$. *(Lesson 6–4)*

4. Find x when $y = -8$ if $y = 4$ when $x = 20$. Assume that y varies directly as x. *(Lesson 6–5)*

5. Transportation The length of a trip varies directly as the amount of gasoline used. Julio's car used 5 gallons of gasoline in the first 145 miles of his trip. How much gasoline should he expect to use in the remaining 232 miles of the trip? *(Lesson 6–5)*

What You'll Learn

You'll learn to solve problems involving inverse variations.

Why It's Important

Music Musicians can use inverse variations to find the wavelength of a musical tone. *See Exercise 12.*

Each of the rectangles shown has an area of 12 square units. Notice that as the length increases, the width decreases. As the length decreases, the width increases. However, their product stays the same.

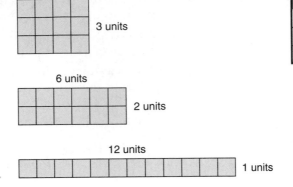

4 units
3 units

6 units
2 units

12 units
1 units

Length	Width	Area
4	3	12
6	2	12
12	1	12

This is an example of an **inverse variation**. We say that y varies inversely as x. This means that as x increases in value, y decreases in value, or as y decreases in value, x increases in value.

Inverse Variation	An inverse variation is described by an equation of the form $xy = k$, where $k \neq 0$.

The equation may also be written as $y = \dfrac{k}{x}$.

The graphs below show how the graph of a direct variation differs from the graph of an inverse variation.

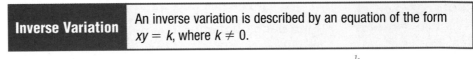

Direct Variation

$y = kx$

Inverse Variation

$xy = k$
or
$y = \dfrac{k}{x}$

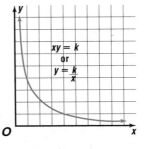

y varies directly as x.
y is directly proportional to x.

The constant of variation is $\dfrac{y}{x}$.

y varies inversely as x.
y is inversely proportional to x.

The constant of variation is xy.

Example

Construction Link

Real World

1 The number of carpenters needed to frame a house varies inversely as the number of days needed to complete the project. Suppose 5 carpenters can frame a certain house in 16 days. How many days will it take 8 carpenters to frame the house? Assume that they all work at the same rate.

Explore   You know that it takes 16 days for 5 carpenters to frame the house. You need to know how many days it will take 8 carpenters to frame the house.

Plan   Solve the problem by using inverse variation.

Solve   Let x = the number of carpenters. Let y = the number of days. First, find the value of k.

$$xy = k \quad \textit{Definition of inverse variation}$$
$$(5)(16) = k \quad \textit{Replace x with 5 and y with 16.}$$
$$80 = k \quad \textit{The constant of variation is 80.}$$

Next, find the number of days for 8 carpenters to frame the house.

$$y = \frac{k}{x} \quad \textit{Divide each side of xy = k by x.}$$

$$y = \frac{80}{8} \quad \textit{Replace k with 80 and x with 8.}$$

$$y = 10$$

A crew of 8 carpenters can frame the house in 10 days.

Your Turn

a. The number of masons needed to build a block basement varies inversely as the number of days needed. If 8 masons can build a block basement in 3 days, how long would it take 9 masons to do it?

You can use a graphing calculator to check the answer to Example 1.

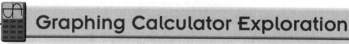

Graphing Calculator Tutorial

See pp. 724–727.

Graphing Calculator Exploration

Step 1   Set the viewing window for x: [0, 10] by 1 and y: [0, 200] by 20.

Step 2   To graph $xy = 80$, rewrite the equation as $y = \frac{80}{x}$. Press [Y=] 80 [÷] [X,T,θ,n] to enter the equation.

(continued on the next page)

Step 3 Press $\boxed{\text{TRACE}}$ to estimate the x- and y-coordinates for any point on the graph. Notice that when $x = 5$, $y = 16$. When x is 8, y is 10. Therefore, the answer given in Example 1 is correct.

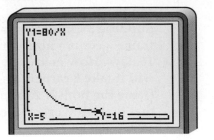

Try These

1. Why doesn't the viewing window allow for negative values to be shown on the graph?

2. Show how to rewrite $xy = 80$ as $y = \dfrac{80}{x}$.

3. Solve Your Turn Exercise a by using a graphing calculator.

Just as with direct variation, you can use a proportion to solve problems involving inverse variation. The proportion $\dfrac{x_1}{x_2} = \dfrac{y_2}{y_1}$ is only one of several that can be formed. *Can you name others?*

Example ❷ **Suppose y varies inversely as x and $y = -6$ when $x = -2$. Find y when $x = 3$.**

$$\dfrac{x_1}{x_2} = \dfrac{y_2}{y_1} \quad \textit{Inverse variation proportion}$$

$$\dfrac{-2}{3} = \dfrac{y_2}{-6} \quad \textit{Let } x_1 = -2, y_1 = -6, \textit{ and } x_2 = 3.$$

$$-2(-6) = 3y_2 \quad \textit{Find the cross products.}$$

$$12 = 3y_2$$

$$4 = y_2$$

Therefore, when $x = 3$, $y = 4$.

Your Turn

b. Suppose y varies inversely as x and $y = 3$ when $x = 12$. Find x when $y = 4$.

Inverse variations can be used to solve rate problems.

Example

Sports Link

Real World

3 In the formula $d = rt$, the time t varies inversely as the rate r. A race car traveling 125 miles per hour completed one lap around a race track in 1.2 minutes. How fast was the car traveling if it completed the next lap in 0.8 minute?

First, solve for the distance d, the constant of variation.

$d = rt$

$d = (125)(1.2)$ *Replace r with 125 and t with 1.2.*

$d = 150$ *The constant of variation is 150.*

Next, find the rate if one lap was completed at 0.8 minute.

$\dfrac{d}{t} = r$ *Divide each side of $d = rt$ by t.*

$\dfrac{150}{0.8} = r$ *Replace d with 150 and t with 0.8.*

$187.5 = r$

The car was traveling 187.5 miles per hour.

Check for Understanding

Communicating Mathematics

Study the lesson. Then complete the following.

Vocabulary

inverse variation

1. **Explain** how to find the constant of variation in an inverse variation.

2. **Complete** the table for the inverse variation $xy = 24$.

x	1	2	3	4	6	8	12	24
y	24	?	?	?	?	?	?	?

Math Journal

3. **Write** a few sentences explaining the difference between inverse variation and direct variation.

Guided Practice

⊕ Getting Ready   **Determine if each equation is an inverse or a direct variation. Find the constant of variation.**

Sample 1: $cd = -3$

Solution: This equation is of the form $xy = k$. So, it is an inverse variation. The constant of variation is -3.

Sample 2: $h = \dfrac{1}{2}g$

Solution: This equation is of the form $y = kx$. So, it is a direct variation. The constant of variation is $\dfrac{1}{2}$.

4. $ab = 4$ **5.** $s - 5r$ **6.** $mn = -7$

7. $y = \dfrac{1}{3}x$ **8.** $y = \dfrac{7}{x}$ **9.** $t = \dfrac{r}{4}$

Solve. Assume that _y_ varies inversely as _x_. *(Example 2)*

10. Suppose $y = 6$ when $x = 2$. Find y when $x = 3$.
11. Find x when $y = 2$ if $y = -4$ when $x = -3$?

12. **Music** The frequency of a musical tone varies inversely as its wavelength. If one tone has a frequency of 440 vibrations per second and a wavelength of 2.4 feet, find the wavelength of a tone that has a frequency of 660 vibrations per second. *(Example 1)*

13. **Travel** The time it takes to travel a certain distance varies inversely to the speed at which you are traveling. Suppose it takes 2.75 hours to drive from one city to another at a rate of 65 miles per hour. How long will the return trip take traveling at 55 miles per hour? *(Example 3)*

Exercises • • • • • • • • • • • • • • • • •

Practice

Solve. Assume that _y_ varies inversely as _x_.

14. Find y when $x = 10$ if $y = 15$ when $x = 4$.
15. Suppose $y = 20$ when $x = 6$. Find x when $y = 12$.
16. If $y = 1.2$ when $x = 3$, find y when $x = 6$.
17. Find y when $x = 2.5$ if $y = 0.15$ when $x = 1.5$.
18. If $y = \frac{2}{3}$ when $x = 6$, find x when $y = 1\frac{1}{3}$.
19. Suppose $y = -\frac{1}{2}$ when $x = -\frac{3}{4}$. Find y when $x = \frac{3}{5}$.

Find the constant of variation. Then write an equation for each statement.

20. y varies inversely as x, and $y = 5$ when $x = 10$.
21. y varies directly as x, and $x = -18$ when $y = 3$.
22. y varies inversely as x, and $y = 4$ when $x = -2$.
23. y varies directly as x, and $x = 7$ when $y = 2.1$.

Match each situation with the equation that represents it. Then tell whether the situation represents a direct variation or an inverse variation.

24. DeShawn purchases a number of pencils for $0.25 each.

    a. $xy = 125$

25. A winning $125 lottery ticket is shared with a number of people.

    b. $y = \dfrac{25}{x}$

26. Twenty-five pieces of candy are shared with several people.

    c. $y = 125x$

27. Each person who has a winning lottery ticket wins $125.

    d. $25x = y$

28. Music The length of a piano string varies inversely as the frequency of its vibrations. A piano string 36 inches long vibrates at a frequency of 480 cycles per second. Find the frequency of a 24-inch string.

29. Geometry The length ℓ of this rectangle varies inversely as its width w.

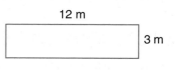

a. Find the constant of variation.

b. What is the width of the rectangle when the length is 15 meters?

c. Write an equation that represents this inverse variation.

d. Draw the graph of the equation.

30. Critical Thinking Assume that y varies inversely as x.

a. Suppose the value of x is doubled. What is the constant of variation? What happens to the value of y?

b. Suppose the value of y is tripled. What is the constant of variation? What happens to the value of x?

Mixed Review

31. Suppose y varies directly as x and $y = 16$ when $x = 12$. Find x when $y = 36$. *(Lesson 6–5)*

32. Is $\{(6, 3), (5, -2), (2, 3), (12, -12)\}$ a function? *(Lesson 6–4)*

Simplify. *(Lesson 3–2)*

33. $-38.9 + 24.2$

34. $12.5 + (-15.6) + 22.7 + (-35.8)$

35. Geography The surface of Lake Michigan is 176 meters above sea level. Its deepest point is 281 meters below sea level. How far below the surface is the deepest part of the lake? *(Lesson 2–4)*

36. Standardized Test Practice Which of the following shows the graph of the relation $\{(-1, 2), (1, 1), (1, 3), (2, 0)\}$? *(Lesson 2–2)*

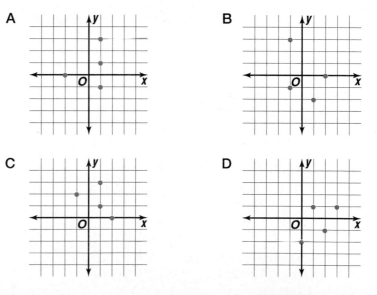

Study Guide and Assessment

Understanding and Using the Vocabulary

After completing this chapter, you should be able to define each term, property, or phrase and give an example or two of each.

*inter*NET
CONNECTION **Review Activities**
For more review activities, visit:
www.algconcepts.glencoe.com

constant of variation *(p. 264)*
dependent variable *(p. 264)*
direct variation *(p. 264)*
domain *(p. 238)*
equation in two variables
   (p. 244)
function *(p. 256)*

functional notation *(p. 258)*
functional value *(p. 258)*
independent variable *(p. 264)*
inverse variation *(p. 270)*
linear equation *(p. 250)*
range *(p. 238)*

rate problems *(p. 266)*
relation *(p. 238)*
solution set *(p. 244)*
vertical line test *(p. 257)*
x-coordinate *(p. 238)*
y-coordinate *(p. 238)*

Choose the correct term or expression to complete each sentence.

1. A (function, **linear equation**) is an equation whose graph is a straight line.
2. The set of all second coordinates of a set of ordered pairs is called the (**range**, domain).
3. Equations that are functions can be written in (**functional notation**, functional value).
4. The domain of $\{(-2, 3), (0, 4), (1, 5)\}$ is ($\{-2, 4, 1\}$, **{−2, 0, 1}**).
5. As x increases in value, y decreases in value in a(n) (direct, **inverse**) variation.
6. Any set of ordered pairs is a (function, **relation**).
7. A direct variation is a linear function that can be written in the form ($xy = k$, **$y = kx$**).
8. In a (relation, **function**), each member of the domain is paired with one member of the range.
9. The equation (**$2 + x = 8$**, $y = 2x + 3$) is an example of *an equation in two variables*.
10. The (**first**, second) coordinate in an ordered pair is called the y-coordinate.

Skills and Concepts

Objectives and Examples

• **Lesson 6–1** Show relations as sets of ordered pairs, as tables, and as graphs.

Express $\{(-1, 0), (0, 3), (1, 2), (2, 1)\}$ as a table and as a graph. Then determine the domain and the range.

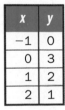

x	y
−1	0
0	3
1	2
2	1

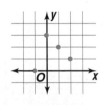

The domain is $\{-1, 0, 1, 2\}$.
The range is $\{0, 3, 2, 1\}$.

Review Exercises

Express each relation as a table and as a graph. Then determine the domain and the range.

11. $\{(-1, 0), (0, 0), (2, 3), (4, 1)\}$
12. $\{(-3, 2), (-2, -1), (0, 4), (3, 3)\}$
13. $\{(-1.5, 2), (1, 3.5), (4.5, -5.5)\}$
14. $\left\{\left(-\frac{1}{2}, 3\right), (1, 2), \left(\frac{1}{2}, 4\right)\right\}$

15. Express the relation as a set of ordered pairs and in a table. Find the domain and range.

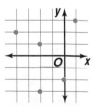

Objectives and Examples

- **Lesson 6–2** Solve linear equations for a given domain.

Solve $y = 2x$ if the domain is $\{-1, 0, 2\}$.

x	2x	y	(x, y)
−1	2(−1)	−2	(−1, −2)
0	2(0)	0	(0, 0)
2	2(2)	4	(2, 4)

The solution is $\{(-1 \ -2), (0, 0), (2, 4)\}$.

- **Lesson 6–3** Graph linear relations.

Graph $y = 2x - 4$.

x	2x − 4	y	(x, y)
0	2(0) − 4	−4	(0, −4)
1	2(1) − 4	−2	(1, −2)
2	2(2) − 4	0	(2, 0)
3	2(3) − 4	2	(3, 2)

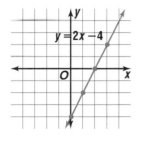

- **Lesson 6–4** Determine whether a given relation is a function.

Determine whether $\{(3, 2), (5, 3), (4, 3), (5, 2)\}$ is a function.

Because there are two values of y for one value of x, 5, the relation is *not* a function.

Review Exercises

Solve each equation if the domain is $\{-1, 0, 2, 4\}$. Graph the solution set.

16. $y = -x$ **17.** $y = \frac{1}{2}x$

18. $y = 3x - 4$ **19.** $3x + 2y = 10$

20. Find the domain of $y = x - 2$ if the range is $\{-3, -1, 4, 6\}$.

21. Find the domain of $y = 2x + 2$ if the range is $\{-2, 0, 2, 4\}$.

Determine whether each equation is a linear equation. Explain. If an equation is linear, identify *A*, *B*, and *C*.

22. $yx = -2$ **23.** $-x = y$

24. $7x + y = -2$ **25.** $3x + 4y = 2z$

Graph each equation.

26. $y = 2x$ **27.** $y = x - 3$

28. $-1 = y$ **29.** $x = 5$

30. $y - -3x \mid 4$ **31.** $4 - 8x - 12$

32. The table lists the coordinates of points on a linear graph. Copy and complete the table.

x	y
?	−2
−2	0
?	2
1	3

Determine whether each relation is a function.

33. $\{(-2, 6), (3, -2), (3, 0), (4, 6)\}$

34. $\{(2, 8), (9, 3), (-3, 7), (4, 3)\}$

35. $\{(-1, 0), (3, 0), (6, 2)\}$

If $g(x) = -3x - 6$, find each value.

36. $g(2)$ **37.** $g(-3.5)$ **38.** $g(-4a)$

Chapter 6 Study Guide and Assessment

Objectives and Examples	Review Exercises

• Lesson 6–5 Solve problems involving direct variation.

Suppose y varies directly as x and $y = 10$ when $x = 15$. Find x when $y = 6$.

Use $\dfrac{y_1}{x_1} = \dfrac{y_2}{x_2}$ to solve the proportion.

$$\dfrac{10}{15} = \dfrac{6}{x_2}$$

$10x_2 = 15(6)$ *Cross products*
$10x_2 = 90$
$\quad x_2 = 9$ So, $x = 9$ when $y = 6$.

Solve. Assume that y varies directly as x.

39. Find y when $x = 7$ if $y = 15$ when $x = 5$.

40. If $y = 12$ when $x = -6$, find y when $x = 2$.

41. Find x when $y = -5$ if $x = -37.5$ when $y = 15$.

Solve by using direct variation.

42. If there are 16 ounces in 1 pound, how many ounces are in 2.8 pounds?

• Lesson 6–6 Solve problems involving inverse variation.

Suppose y varies inversely as x and $y = 10$ when $x = 72$. Find x when $y = 24$.

$$\dfrac{x_1}{x_2} = \dfrac{y_2}{y_1}$$

$$\dfrac{72}{x_2} = \dfrac{24}{10}$$

$720 = 24x_2$ *Cross products*
$\ 30 = x_2$ So, when $y = 24$, $x = 30$.

Solve. Assume that y varies inversely as x.

43. If $x = 3$ when $y = 25$, find x when $y = 15$.

44. Find y when $x = 42$ if $y = 21$ when $x = 56$.

Find the constant of variation. Then write an equation for each statement.

45. y varies inversely as x, and $y = 4$ when $x = 2$.

46. y varies directly as x, and $x = -12$ when $y = 3$.

Applications and Problem Solving

47. Science The table shows how far away a lightning strike is, given the number of seconds between when the lightning is seen and when thunder is heard. Express the relation as a set of ordered pairs. Then find the domain and range. *(Lesson 6–1)*

Time (s)	5	10	15	20	25
Distance (mi)	1	2	3	4	5

48. Sales Jennifer earns a weekly salary of $255 plus a bonus of $0.25 for each CD over 150 that she sells each week. The equation $C(r) = 255 + 0.25(r - 150)$ can be used to determine her total weekly income. How much more will she earn by selling 450 CDs than 200 CDs? *(Lesson 6–2)*

49. Electronics In an electrical transformer, voltage varies directly as the number of turns on the coil. What would be the voltage produced by 66 turns, if 110 volts come from 55 turns? *(Lesson 6–5)*

1. **Write** $y = 2x - 3$ in functional notation.
2. **List** three ways to show a relation.

Express each relation as a table and as a graph. Then determine the domain and range.

3. $\{(-3, -2), (-1, 4), (0, -1), (2, 3), (3, 3)\}$ 4. $\{(-1, 1.5), (2, -4), (3, 2.5), (4, -1.5)\}$

Solve each equation if the domain is $\{-2, -1, 0, 4, 6\}$.

5. $y = \dfrac{1}{2}x$ 6. $y = 4 - x$ 7. $2x - 2y = 6$

Graph each equation.

8. $x = -5$ 9. $y = -3x - 1$ 10. $6x - 2y = 8$

Determine whether each relation is a function.

11. $\{(4, 1), (2, 4), (-1, 2), (-4, 2)\}$ 12. $\{(-3, 2), (0, 4), (1, 5), (3, 0), (-3, -1)\}$

13.

x	y
−1	2
3	−4
−1	0
2	3
1	1

14.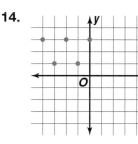

15.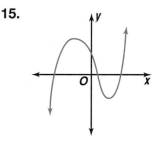

If $f(x) = -3x + 4$ and $g(x) = 2x + 2$, find each value.

16. $f(-1)$ 17. $g(3.5)$ 18. $g\left(\dfrac{1}{4}\right)$ 19. $f(2d)$

Solve. Assume that y varies directly as x.

20. Find y when $x = 12$ if $y = 18$ when $x = 27$.

21. Suppose $y = -84$ when $x = -12$. Find y when $x = -1$.

Solve. Assume that y varies inversely as x.

22. Suppose $x = -16$ when $y = 4$. Find x when $y = 8$.

23. Find y when $x = 10$ if $y = 2.5$ when $x = 19.2$.

24. **Counting**   Refer to the spinners shown.
   a. Suppose each spinner is spun once. Write a relation that shows the different possible outcomes.
   b. Use the relation to find the number of ways to land on an odd number and green.

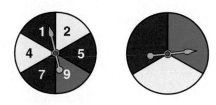

25. **Geometry**   To determine the perimeter P of a rectangle, the equation $P = 2w + 2\ell$, where w is the width and ℓ is the length, can be used. Suppose a rectangle has a perimeter of 64 inches. Find the possible dimensions of the rectangle given the domain values $\{3, 6, 7, 10\}$.

Preparing for Standardized Tests

Probability and Counting Problems

You will need to solve problems involving combinations, permutations, and probability. It will be helpful to familiarize yourself with the terms *combination*, *permutation*, *tree diagram*, and *outcomes*. You will also need to know the definition of probability.

Sketching a picture can help you understand and solve many problems.

$$P(\text{event}) = \frac{\text{number of favorable outcomes}}{\text{total number of possible outcomes}}$$

Proficiency Test Example

How many different five-digit ZIP codes can be made if no digit is repeated in a code?

A 15,120 **B** 30,240
C 60,480 **D** 100,000

Hint Be sure you understand the question before you start to calculate.

Solution A ZIP code consists of five digits. The digits can be 0–9. There are ten possible digits. The problem states that no digit can appear more than once in these ZIP codes. Find the number of possible choices for each of the five positions.

_____ _____ _____ _____ _____

There are 10 possible choices for the first position. There are 9 choices for the second position, because one digit was used for the first position. Following the same reasoning, there are 8 choices for the third position, 7 choices for the fourth position, and 6 choices for the fifth position. Multiply these numbers to find the total number of different codes. Use a calculator.

$$10 \times 9 \times 8 \times 7 \times 6 = 30{,}240$$

So, there are 30,240 ZIP codes. Thus, the answer is B.

SAT Example

Five cards are numbered from 0 to 4. Suppose two cards are selected at random without replacement. What is the probability that the sum of the numbers on the two cards is an even number?

A $\frac{1}{5}$ **B** $\frac{3}{10}$ **C** $\frac{2}{5}$ **D** $\frac{1}{2}$ **E** $\frac{3}{5}$

Hint Decide whether or not order matters. In this problem order doesn't matter. So, it is a combination problem.

Solution First, write the possible outcomes of selecting two cards at random.

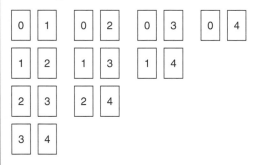

There are 10 possible combinations. There are 4 ways to get a sum that is an even number.

$$P(\text{even}) = \frac{4}{10} \text{ or } \frac{2}{5}$$

The answer is C.

After you work each problem, record your answer on the answer sheet provided or on a sheet of paper.

1. Suppose the spinner shown is spun. What is the probability of landing on an even number?

   A $\frac{5}{8}$ **B** $\frac{1}{2}$

   C $\frac{3}{8}$ **D** $\frac{3}{4}$

2. How many different starting squads of 6 players can be picked from 11 volleyball players?
   A 120   **B** 246   **C** 462   **D** 720

3. Diego has a collection of music CDs. Ten are rock, 5 are jazz, and 8 are country. If he chooses a CD at random, what is the probability that he will *not* choose a jazz CD?

   A $\frac{5}{23}$ **B** $\frac{5}{18}$ **C** $\frac{15}{23}$

   D $\frac{13}{18}$ **E** $\frac{18}{23}$

4. How many different 3-digit security codes can you make using the numbers 1, 2, and 3? You can use the numbers more than once.
   A 6   **B** 9   **C** 18   **D** 27

5. A bag contains only white and blue marbles. The probability of selecting a blue marble is $\frac{1}{5}$. The bag contains 200 marbles. If 100 white marbles are added to the bag, what is the probability of selecting a white marble?

   A $\frac{2}{15}$ **B** $\frac{7}{15}$ **C** $\frac{8}{15}$

   D $\frac{4}{5}$ **E** $\frac{13}{15}$

Test Practice For more test practice questions, visit: www.algconcepts.glencoe.com

6. On Tuesday, the temperature fell 6° in 2 hours. Which of these expresses the temperature change per hour?
   A −3°   **B** −6°   **C** 6°   **D** 12°

7. Tomorrow 352 students will take a test in the cafeteria. If each table seats 12 students and there are already 15 tables in the cafeteria, how many additional tables will be needed?
   A 14   **B** 15   **C** 23   **D** 24

8. Ellen earns $86.80 for working 14 hours. At this rate, how much would she earn if she worked 20 hours?
   A $6.20   **B** $60.76
   C $124.00   **D** $1,215.20

Open-Ended Questions

9. **Grid-In** If $5x - 4 = x - 1$, what is the value of x?

10. The table shows the average monthly low temperatures for Detroit, Michigan.

Month	Temperature (°F)
January	44
February	46
March	54
April	60
May	66
June	72
July	74
August	74
September	70
October	61
November	52
December	45

Source: Excite Travel

Part A Organize the data in a stem-and-leaf plot.

Part B Find the mean, median, and mode. Tell which of these best represents the data.

▶ What You'll Learn in Chapter 7:

- to find the slope of a line given the coordinates of two points on the line *(Lesson 7–1)*,
- to write a linear equation in point-slope form and in slope-intercept form *(Lessons 7–2 and 7–3)*,
- to graph and interpret points on scatter plots *(Lesson 7–4)*,
- to graph linear equations by using the *x*- and *y*-intercepts or the slope and *y*-intercept *(Lesson 7–5)*,
- to explore the effects of changing the slopes and *y*-intercepts of linear functions *(Lesson 7–6)*, **and**
- to write an equation of a line that is parallel or perpendicular to the graph of a given equation and that passes through a given point *(Lesson 7–7)*.

Problem-Solving Workshop

Project

Have you ever heard someone say, "I remember when penny candy actually cost a penny"? The table shows how the value of a dollar has changed. In this project, you will choose ten items and estimate their value in earlier years. *For this project, assume that today's cost and the cost in 1998 are the same.*

Year	Value of a Dollar
1950	$6.76
1958	$5.64
1966	$5.03
1974	$3.31
1982	$1.69
1990	$1.25
1998	$1.00

Working on the Project

Work with a partner and choose a strategy. Use the questions below to help you get started.

- $100 per week today is equivalent to how much money per week in 1958?
- Suppose a hamburger costs $2.80 today. How much would a comparable hamburger have cost in 1950? in 1966?
- How does the value of a dollar change over time? Can this change be described as a linear function?

Strategies

- Look for a pattern.
- Draw a diagram.
- Make a table.
- Work backward.
- Use an equation.
- Make a graph.
- Guess and check.

Technology Tools

- Use a **calculator** to help you estimate the cost of items in different years.
- Use **graphing software** to graph the data points (year, value of a dollar) as ordered pairs in a scatter plot.

*inter*NET
CONNECTION   **Research**   For more information about inflation, visit: www.algconcepts.glencoe.com

Presenting the Project

Write a one-page paper describing your results. Include the following:

- today's costs of the ten items you selected,
- an estimate for the cost or value of each item in 1950 and in another year, and
- a scatter plot showing the value of a dollar as a function of the year.

What You'll Learn

You'll learn to find the slope of a line given the coordinates of two points on the line.

Why It's Important

Construction

Building codes require that certain slopes be used for stairways.
See Exercise 28.

The rows in some movie theaters are so steep that there are no bad seats. We say that the theater has a steep **slope**. Slope is the ratio of the *rise*, or the vertical change, to the *run*, or the horizontal change.

$$\text{slope} = \frac{\text{rise}}{\text{run}} = \frac{9}{10} \quad \begin{array}{l} \textit{vertical change} \\ \textit{horizontal change} \end{array}$$

The slope of the seats in the theater at the right is $\frac{9}{10}$.

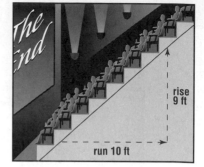
rise 9 ft
run 10 ft

On the graph below, the line passes through points at (4, 0) and (11, 3). The change in *y* or rise is 3 − 0 or 3, while the change in *x* or run is 11 − 4 or 7. Therefore, the slope of this line is $\frac{3}{7}$.

$$\text{slope} = \frac{\text{rise}}{\text{run}} = \frac{\text{change in } y}{\text{change in } x}$$

| **Slope** | **Words:** The slope of a line is the ratio of the change in *y* to the corresponding change in *x*. |
| | **Model:** 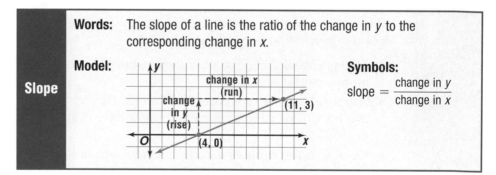   **Symbols:** $\text{slope} = \dfrac{\text{change in } y}{\text{change in } x}$ |

Examples   Determine the slope of each line.

1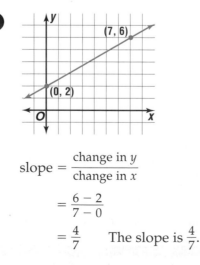

$$\text{slope} = \frac{\text{change in } y}{\text{change in } x}$$

$$= \frac{6 - 2}{7 - 0}$$

$$= \frac{4}{7} \quad \text{The slope is } \frac{4}{7}.$$

2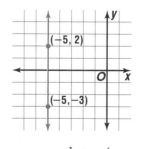

$$\text{slope} = \frac{\text{change in } y}{\text{change in } x}$$

$$= \frac{-3 - 2}{-5 - (-5)}$$

$$= \frac{-5}{0} \quad \text{The slope is undefined. } Why?$$

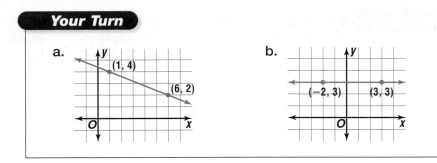

a. (1, 4) (6, 2)

b. (−2, 3) (3, 3)

In Example 1, suppose $2 - 6$ had been used as the change in y and $0 - 7$ had been used as the change in x. Since $\frac{2-6}{0-7}$ is also equal to $\frac{4}{7}$, it does not matter which order is chosen. However, the coordinates of both points must be used in the same order.

In many real-world applications, the slope is the rate of change.

Example

Income Link

Real World

3

The graph at the right shows the hours worked and income for Helena and Steve. Find the slope of each line. To what does the slope refer?

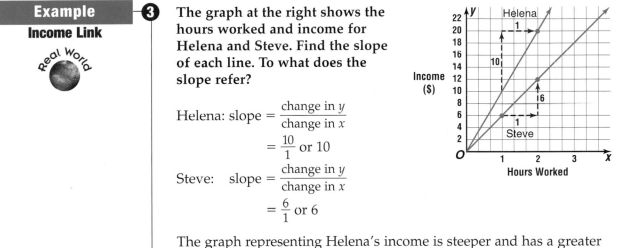

Helena: $\text{slope} = \dfrac{\text{change in } y}{\text{change in } x}$

$= \dfrac{10}{1} \text{ or } 10$

Steve: $\text{slope} = \dfrac{\text{change in } y}{\text{change in } x}$

$= \dfrac{6}{1} \text{ or } 6$

The graph representing Helena's income is steeper and has a greater slope than the graph of Steve's income. This is because Helena makes $10 per hour and Steve makes $6 per hour. The slope refers to each person's pay rate.

Look at the pattern in the graph at the right. Each time x increases 3 units, y decreases 1 unit.

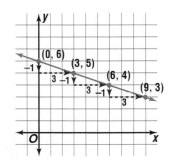

(0, 6) (3, 5) (6, 4) (9, 3)

x	y
0	6
3	5
6	4
9	3

+3, −1 for each step

$\text{slope} = \dfrac{\text{change in } y}{\text{change in } x} = \dfrac{-1}{3} \text{ or } -\dfrac{1}{3}$

Example

4 A line contains the points whose coordinates are listed in the table. Determine the slope of the line.

Each time x decreases 2 units, y increases 3 units.

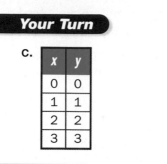

x	y
6	−3
4	0
2	3
0	6

$$\text{slope} = \frac{\text{change in } y}{\text{change in } x} = \frac{3}{-2}$$

The slope of the line containing these points is $-\frac{3}{2}$.

Your Turn

c.

x	y
0	0
1	1
2	2
3	3

d.

x	y
−5	2
−1	3
3	4
7	5

The examples above suggest the following.

Determining Slope Given Two Points	**Words:**	The slope m of a line containing any two points with coordinates (x_1, y_1) and (x_2, y_2) is given by the following formula. $$\text{slope} = \frac{\text{difference of the } y\text{-coordinates}}{\text{difference of the corresponding } x\text{-coordinates}}$$
	Symbols:	$m = \dfrac{y_2 - y_1}{x_2 - x_1}$, where $x_2 \neq x_1$

Examples

Determine the slope of each line.

5 the line through points at $(2, 9)$ and $(6, 9)$

$$m = \frac{y_2 - y_1}{x_2 - x_1}$$

$$= \frac{9 - 9}{6 - 2}$$

$$= \frac{0}{4} \text{ or } 0$$

The slope is 0.

6 the line through points at $(3, -2)$ and $(-4, 7)$

$$m = \frac{y_2 - y_1}{x_2 - x_1}$$

$$= \frac{7 - (-2)}{-4 - 3}$$

$$= \frac{9}{-7} \text{ or } -\frac{9}{7}$$

The slope is $-\frac{9}{7}$.

Your Turn

e. the line through points at $(2, 5)$ and $(3, 9)$

f. the line through points at $(-8, 1)$ and $(4, 1)$

The slopes of lines can be summarized as follows.

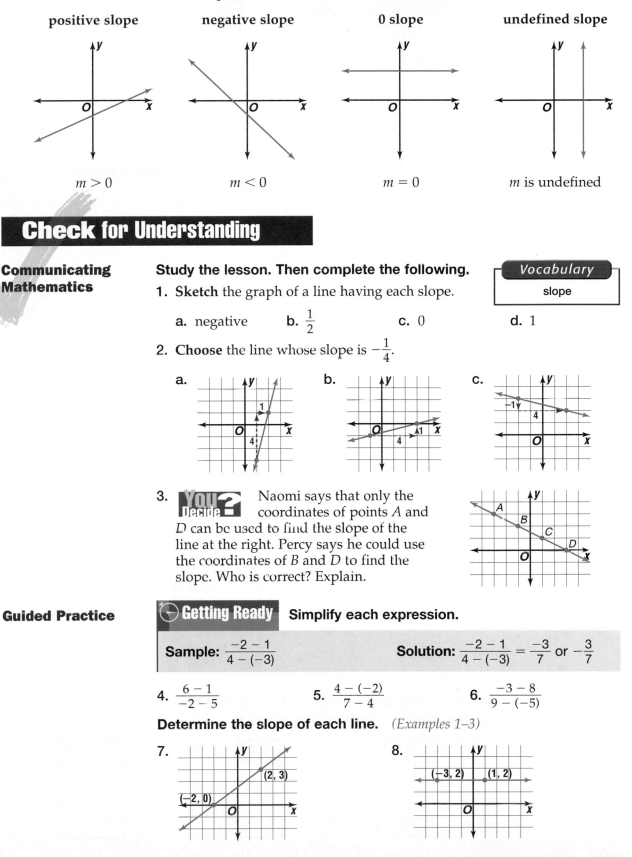

positive slope   negative slope   0 slope   undefined slope

$m > 0$ $m < 0$ $m = 0$ m is undefined

Check for Understanding

Communicating Mathematics

Study the lesson. Then complete the following.

1. **Sketch** the graph of a line having each slope.

   a. negative   **b.** $\frac{1}{2}$ **c.** 0   **d.** 1

Vocabulary

slope

2. **Choose** the line whose slope is $-\frac{1}{4}$.

   a.   **b.**   **c.**

3. **You Decide** Naomi says that only the coordinates of points A and D can be used to find the slope of the line at the right. Percy says he could use the coordinates of B and D to find the slope. Who is correct? Explain.

Guided Practice

Getting Ready **Simplify each expression.**

Sample: $\frac{-2 - 1}{4 - (-3)}$ **Solution:** $\frac{-2 - 1}{4 - (-3)} = \frac{-3}{7}$ or $-\frac{3}{7}$

4. $\frac{6 - 1}{-2 - 5}$ 5. $\frac{4 - (-2)}{7 - 4}$ 6. $\frac{-3 - 8}{9 - (-5)}$

Determine the slope of each line. *(Examples 1–3)*

7.   8.

9. Determine the slope of the line passing through the points whose coordinates are listed in the table. *(Example 4)*

x	−2	−1	0	1
y	−8	−4	0	4

Determine the slope of each line. *(Examples 5–6)*

10. the line through points at (1, 0) and (−2, 9)

11. the line through points at (10, −3) and (5, 3)

12. Entertainment The Cyclone roller coaster was built in 1927 on New York's Coney Island. It has the steepest first drop of any wooden coaster in the world. It drops about 5 feet for every 3 feet of horizontal change. What is the slope of the first drop? *(Example 1)*

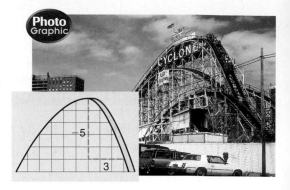

Exercises

Practice

Determine the slope of each line.

13.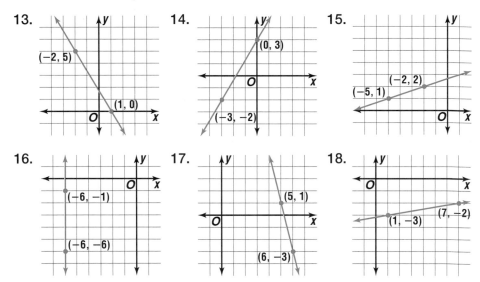

14.

15.

16.

17.

18.

Determine the slope of the line passing through the points whose coordinates are listed in each table.

19.

x	y
0	12
1	10
2	8
3	6

20.

x	y
0	7
2	6
4	5
6	4

21.

x	y
−2	−7
1	−6
4	−5
7	−4

Determine the slope of each line.

22. the line through points at (2, 5) and (6, 9)

23. the line through points at (2, 3) and (−3, 5)

24. the line through points at (4, −3) and (4, 5)

25. Find the slope of a line that passes through points at (−8, 9) and (0, 6).

26. What is the slope of a line that passes through $A(−8, 2)$ and $B(5, 8)$?

Applications and Problem Solving

27. **Driving**   Some roads in the Rocky Mountains have a rise of 7 feet for every 100 horizontal feet. What is the slope of such roads?

28. **Construction**   Building codes regulate the steepness of stairs. Some homes must have steps that are at least 11 inches wide for each 9 inches that they rise.

    a. What is the slope of the stairs?

    b. Describe how changing the width or the height affects the steepness of the stairs.

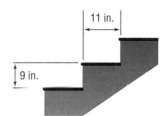

29. **Money**   The linear function $f(x) = 0.25x$ represents the value of x quarters.

    a. Explain how changes in x affect $f(x)$. What does this tell you about the slope of the graph of the equation?

    b. Describe a reasonable domain and range for the function.

Number of Quarters x	Value of Quarters $f(x)$
1	0.25
2	0.50
3	0.75
4	1.00

30. **Critical Thinking**   Determine whether the table represents a function that is linear or nonlinear. Explain.

x	1	3	5	7
y	2	4	6	16

Mixed Review

31. Assume that y varies inversely as x. Suppose $y = 4$ when $x = 6$. Find y when $x = 2$.   *(Lesson 6–6)*

Determine whether each relation is a function.   *(Lesson 6–4)*

32. {(4, 8), (5, 5), (−1, 3), (0, 6)}

33. {(7, 1), (−4, 3), (0, 1), (−4, 0)}

34. **Communication**   The coastline of New Zealand is 8161 miles long. This is about 86% of the length of a cable under the ocean that connects New Zealand and Canada. What is the approximate length of the cable?   *(Lesson 5–3)*

35. Solve $9x = 4x − 10$.   *(Lesson 4–6)*

36. **Standardized Test Practice**   Solve $−5 − 0.8y = 7$.   *(Lesson 4–5)*

    A 15   B −15   C 9.6   D −2.5

New Zealand

Extra Practice   See p. 705.

What You'll Learn

You'll learn to write a linear equation in point-slope form given the coordinates of a point on the line and the slope of the line.

Why It's Important

Population Point-slope equations are useful in making predictions about the population.
See Exercise 40.

The graph shows the growth in the number of sport-utility vehicles in the United States. What is an equation of the line? *You will solve this problem in Exercise 38.*

You can write an equation of a line if you know its slope and the coordinates of one point on the line.

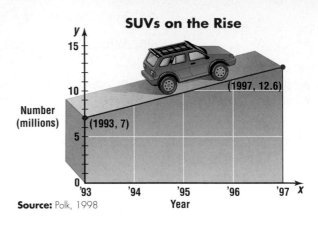

SUVs on the Rise

(1997, 12.6)

(1993, 7)

Number (millions)

Source: Polk, 1998

Year

$$\frac{y_2 - y_1}{x_2 - x_1} = m \qquad \text{\textit{Slope of a line}}$$

$$\frac{y - y_1}{x - x_1} = m \qquad \text{\textit{Replace }} (x_2, y_2) \text{ \textit{with }} (x, y).$$

$$\left(\frac{y - y_1}{x - x_1}\right)(x - x_1) = m(x - x_1) \qquad \text{\textit{Multiply each side by }} (x - x_1).$$

$$y - y_1 = m(x - x_1)$$

This equation is said to be in **point-slope form**. Recall that it is called a *linear equation* because its graph is a straight line. *The graph of a nonlinear equation is not a straight line.*

Point-Slope Form	**Words:**	For a nonvertical line through the point at (x_1, y_1) with slope m, the point-slope form of a linear equation is $y - y_1 = m(x - x_1)$.
	Numbers:	$y - 6 = 4(x - 1)$ $(x_1, y_1) = (1, 6)$ and $m = 4$

The equation of a vertical line through the point at (x_1, y_1) is $x = x_1$.

Examples

Write the point-slope form of an equation for each line passing through the given point and having the given slope.

1 $(4, -5)$, $m = \frac{3}{4}$

$$y - y_1 = m(x - x_1) \qquad \text{\textit{Point-Slope Form}}$$

$$y - (-5) = \frac{3}{4}(x - 4) \qquad \text{\textit{Replace }} x_1 \text{ \textit{with }} 4, y_1$$
$$\text{\textit{with }} -5, \text{ \textit{and }} m \text{ \textit{with }} \frac{3}{4}.$$

$$y + 5 = \frac{3}{4}(x - 4)$$

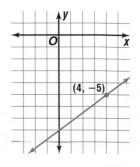

(4, −5)

An equation of the line is $y + 5 = \frac{3}{4}(x - 4)$.

Check: Look at the graph. Choose some other point on the line and determine whether it is a solution of $y + 5 = \frac{3}{4}(x - 4)$. Try $(0, -8)$.

$$y + 5 = \frac{3}{4}(x - 4)$$

$$-8 + 5 \stackrel{?}{=} \frac{3}{4}(0 - 4) \quad \text{Replace } x \text{ with } 0 \text{ and } y \text{ with } -8.$$

$$-3 = -3 \quad \checkmark$$

2 $(-7, 2), m = 0$

$$\begin{array}{ll} y - y_1 = m(x - x_1) & \text{Point-Slope Form} \\ y - 2 = 0[x - (-7)] & \text{Replace } x_1 \text{ with } -7, y_1 \text{ with } 2, \text{ and } m \text{ with } 0. \\ y - 2 = 0 & \\ y - 2 + 2 = 0 + 2 & \text{Add 2 to each side.} \\ y = 2 & \end{array}$$

An equation of the line is $y = 2$.

Your Turn

a. $(-3, 2), m = 2$ **b.** $(5, 4), m = -\frac{2}{3}$

You can also write an equation of a line if you know the coordinates of two points on the line.

Example **3** **Write the point-slope form of an equation of the line at the right.**

First, determine the slope of the line.

$$m = \frac{y_2 - y_1}{x_2 - x_1} = \frac{5 - 1}{-2 - 3} \text{ or } -\frac{4}{5}$$

The slope is $-\frac{4}{5}$. Now use the slope and either point to write an equation.

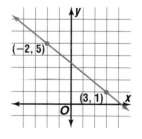

Method 1   Use $(-2, 5)$.

$$\begin{array}{ll} y - y_1 = m(x - x_1) & \text{Point-Slope} \\ & \text{Form} \\ y - 5 = -\frac{4}{5}[x - (-2)] & (x_1, y_1) = \\ & (-2, 5) \\ y - 5 = -\frac{4}{5}(x + 2) & \end{array}$$

Check:

$$y - 5 = -\frac{4}{5}(x + 2)$$

$$1 - 5 \stackrel{?}{=} -\frac{4}{5}(3 + 2) \quad (x, y) = (3, 1)$$

$$-4 \stackrel{?}{=} -\frac{4}{5}(5)$$

$$-4 = -4 \quad \checkmark$$

(continued on the next page)

Method 2   Use (3, 1).

$y - y_1 = m(x - x_1)$ *Point-Slope*
                                     Form

$y - 1 = -\frac{4}{5}(x - 3)$ $(x_1, y_1) =$
                                $(3, 1)$

Check:

$y - 1 = -\frac{4}{5}(x - 3)$

$5 - 1 \overset{?}{=} -\frac{4}{5}(-2 - 3)$ $(x, y) =$
                                   $(-2, 5)$

$4 \overset{?}{=} -\frac{4}{5}(-5)$

$4 = 4$ ✓

Both $y - 5 = -\frac{4}{5}(x + 2)$ and $y - 1 = -\frac{4}{5}(x - 3)$ are point-slope forms of an equation for the line passing through points at (3, 1) and (−2, 5).

Your Turn

Write the point-slope form of an equation for each line.

c.

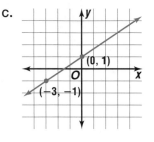

d. the line passing through points at (1, 4) and (3, −5)

In the 1988 Olympics, the women's winning long jump was about 24.7 feet. It is predicted that by 2020 the record will be about 26.3 feet. Write the point-slope form of an equation for the line in the graph at the right.

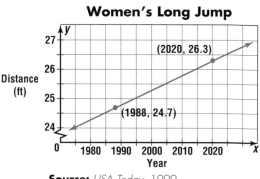

Women's Long Jump

Source: *USA Today, 1999*

First, determine the slope of the line.

$m = \dfrac{y_2 - y_1}{x_2 - x_1} = \dfrac{26.3 - 24.7}{2020 - 1988}$ $(x_1, y_1) = (1988, 24.7), (x_2, y_2) = (2020, 26.3)$

$\qquad\qquad = \dfrac{1.6}{32}$ or $\dfrac{1}{20}$ The slope is $\dfrac{1}{20}$.

Now use the slope and either point to write an equation.

   $y - y_1 = m(x - x_1)$ *Point-Slope Form*

$y - 24.7 = \dfrac{1}{20}(x - 1988)$ *Replace (x_1, y_1) with (1988, 24.7) and m with $\dfrac{1}{20}$.*

An equation of the line is $y - 24.7 = \dfrac{1}{20}(x - 1988)$.

Check by substituting (2020, 26.3) into the equation.

Check for Understanding

Communicating Mathematics

Study the lesson. Then complete the following.

Vocabulary

point-slope form

1. **Tell** what information is needed to write a linear equation in point-slope form.

2. **Sketch** a line whose slope is 2. Then use a point on the line to write an equation of the line in point-slope form.

Guided Practice

⊖ **Getting Ready** **State the slope and the coordinates of a point on the graph of the equation.**

Sample: $y - 2 = 3(x - 5)$

Solution: $y - y_1 = m(x - x_1)$

$y - 2 = 3(x - 5)$

$m = 3, (5, 2)$

3. $y - 6 = 2(x - 1)$ 4. $y + 4 = \frac{2}{5}(x - 8)$ 5. $y - (-5) = -3(x + 9)$

Write the point-slope form of an equation for each line passing through the given point and having the given slope. *(Examples 1 & 2)*

6. $(5, 6), m = 2$

7. $(-2, 4), m = \frac{3}{4}$

8. $(3, -4), m = -5$

9. $(-8, 1), m = 0$

Write the point-slope form of an equation for each line.
(Example 3)

10.

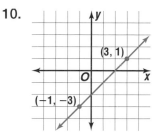

11.

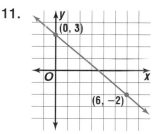

12. the line through points at $(5, 3)$ and $(-2, 1)$

13. the line through points at $(-2, 5)$ and $(4, -7)$

14. **Physical Science**   The table shows the growth of the Pacific giant kelp plant.   *(Example 4)*

   a. Write the point-slope form of an equation that represents the growth.

   b. How much will the plant grow in 7 days?

   c. Would the plant show faster or slower growth if the slope of the equation in part a was 30?

Time (days)	Growth (cm)
x	**y**
2	90
3	135

Lesson 7–2 Writing Equations in Point-Slope Form   **293**

Practice

Write the point-slope form of an equation for each line passing through the given point and having the given slope.

15. $(-1, 4), m = -3$ **16.** $(7, 2), m = 5$ **17.** $(9, 1), m = 6$

18. $(-3, 6), m = \dfrac{2}{3}$ **19.** $(-2, -2), m = \dfrac{4}{5}$ **20.** $(-1, -5), m = -7$

21. $(4, -2),$
m is undefined.   **22.** $(0, 6), m = \dfrac{3}{5}$ **23.** $\left(\dfrac{1}{2}, -4\right), m = -\dfrac{5}{2}$

Write the point-slope form of an equation for each line.

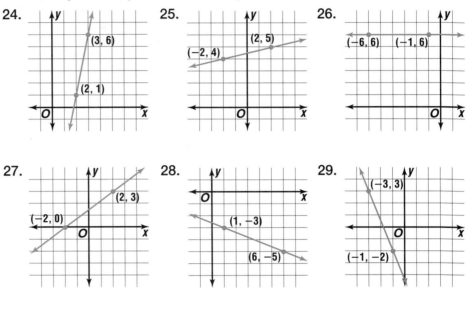

24.   (3, 6)   (2, 1)
25.   (2, 5)   (−2, 4)
26.   (−6, 6)   (−1, 6)

27.   (2, 3)   (−2, 0)
28.   (1, −3)   (6, −5)
29.   (−3, 3)   (−1, −2)

30. the line through points at $(2, 1)$ and $(3, 8)$

31. the line through points at $(4, 5)$ and $(0, 3)$

32. the line through points at $(4, -2)$ and $(0, -1)$

33. the line through points at $(1, -4)$ and $(4, 5)$

34. the line through points at $(6, 9)$ and $(-2, 9)$

35. the line through points at $(-3, 7)$ and $(5, 2)$

36. Write an equation in point-slope form of a line that has a slope of -1.8 and passes through the point at $(-5, 11)$.

37. What is the point-slope form of an equation of the line containing the points $Q(3, 6)$ and $R(-4, 7)$?

Applications and Problem Solving

Real World

38. Transportation Refer to the application at the beginning of the lesson.

   a. Write the point-slope form of an equation for the graph of the line.

   b. Use the equation in part a to predict the number of sport-utility vehicles that will be in the United States in the year 2004.

39. Savings Lynda adds $5 to her savings account each week, as shown in the graph at the right.

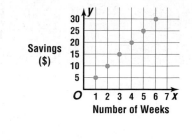

a. Write the point-slope form of an equation of the line through the points that represent her savings.

b. Describe a reasonable domain and range for the equation.

Data Update For the latest information on U.S. population, visit: www.algconcepts. glencoe.com

40. Population The graph shows the percent of U.S. residents who are born outside the United States.

Residents Not Born in U.S.A.

a. Write the point-slope form of an equation for the line passing through the points.

b. Use the graph to estimate the percent of residents that were born outside the United States in 1995. Use the equation from part a to check your estimate.

c. Predict the percent of residents that will be born outside the United States in 2005.

41. Critical Thinking Find the coordinates of a point through which the graph of $y - 11 = \frac{3}{4}(x - 8)$ passes, if the y-coordinate is twice the x-coordinate.

Mixed Review

42. Construction A ramp installed to give handicapped people access to a building has a 3-foot rise and a 36-foot run. What is the slope of the ramp? *(Lesson 7–1)*

Find the domain of each equation if the range is {0, 2, 4}.
(Lesson 6–2)

43. $y = 2x$ **44.** $y = 4 + x$ **45.** $y = 2x - 2$

Find the percent of increase or decrease. *(Lesson 5–5)*

46. original: 15
new: 18

47. original: 6
new: 24

48. Use the percent equation to find 45% of 160. *(Lesson 5–4)*

49. Find $(-1.2)(-7)$. *(Lesson 4–1)*

50. Open-Ended Test Practice List six numbers whose median is 12. *(Lesson 3–3)*

Writing Equations in Slope-Intercept Form

What You'll Learn

You'll learn to write a linear equation in slope-intercept form given the slope and y-intercept.

Why It's Important

Catering Caterers can use linear equations to represent the total cost. *See Exercise 47.*

A California inventor designed the *Skycar*, a car that can fly! It runs on regular gas, reaches speeds of 390 miles per hour, and can climb to 9144 meters. The graph at the right represents its landing from 50 meters off the ground.

The line crosses the y-axis at (0, 50). The number 50 is called the **y-intercept** of the equation. The line crosses the x-axis at (25, 0). Thus, 25 is the **x-intercept** of the equation.

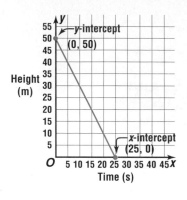

In Lesson 7–2, you learned how to write an equation in point-slope form by using the slope and a point on the line, and two points on the line. You can also write an equation of a line if you know the slope and y-intercept. Consider the graph below, which crosses the y-axis at (0, *b*).

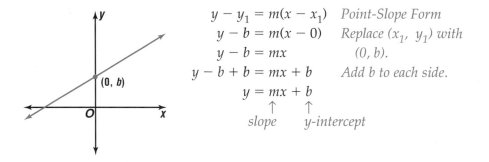

$$y - y_1 = m(x - x_1) \quad \textit{Point-Slope Form}$$
$$y - b = m(x - 0) \quad \textit{Replace } (x_1, y_1) \textit{ with}$$
$$y - b = mx \quad\quad\quad (0, b).$$
$$y - b + b = mx + b \quad \textit{Add b to each side.}$$
$$y = mx + b$$
$$\uparrow \quad\quad \uparrow$$
$$\textit{slope} \quad \textit{y-intercept}$$

Slope-Intercept Form	**Words:** Given the slope *m* and y-intercept *b* of a line, the slope-intercept form of an equation of the line is $y = mx + b$.	**Model:**

Examples

Write an equation in slope-intercept form of each line with the given slope and y-intercept.

1 $m = -4, b = 5$

$y = mx + b \quad$ *Slope-Intercept Form*

$y = -4x + 5 \quad$ *Replace m with −4 and b with 5.*

An equation of the line is $y = -4x + 5$.

2 $m = \frac{1}{3}, b = -6$

$y = mx + b$ *Slope-Intercept Form*

$y = \frac{1}{3}x + (-6)$ *Replace m with $\frac{1}{3}$ and b with -6.*

$y = \frac{1}{3}x - 6$

An equation of the line is $y = \frac{1}{3}x - 6$.

3 $m = 0, b = 3$

$y = mx + b$ *Slope-Intercept Form*

$y = 0x + 3$ *Replace m with 0 and b with 3.*

$y = 3$

The equation of the line is $y = 3$. Remember that a line with a slope of 0 is a horizontal line.

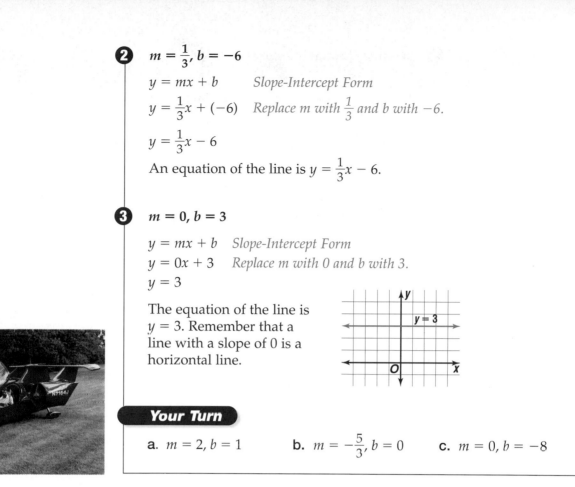

Your Turn

a. $m = 2, b = 1$ **b.** $m = -\frac{5}{3}, b = 0$ **c.** $m = 0, b = -8$

Now you can use the methods in Lesson 7–2 to write equations in slope-intercept form.

Examples

Write an equation of the line in slope-intercept form for each situation.

4 slope 1 and passes through the point at (2, 5)

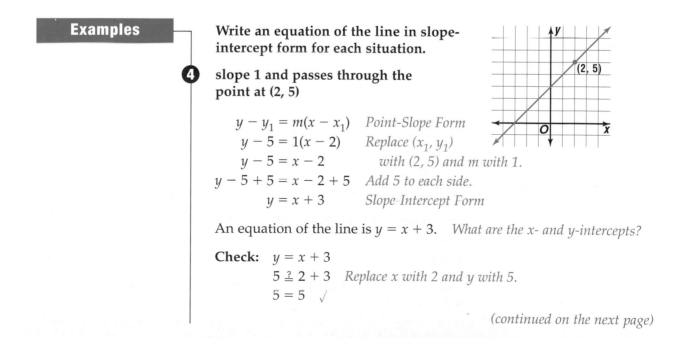

$y - y_1 = m(x - x_1)$ *Point-Slope Form*

$y - 5 = 1(x - 2)$ *Replace (x_1, y_1)*

$y - 5 = x - 2$ *with (2, 5) and m with 1.*

$y - 5 + 5 = x - 2 + 5$ *Add 5 to each side.*

$y = x + 3$ *Slope-Intercept Form*

An equation of the line is $y = x + 3$.   *What are the x- and y-intercepts?*

Check: $y = x + 3$

$5 \overset{?}{=} 2 + 3$ *Replace x with 2 and y with 5.*

$5 = 5$ ✓

(continued on the next page)

⑤ passing through points at $(-4, 4)$ and $(2, 1)$

First, determine the slope of the line.

$$m = \frac{y_2 - y_1}{x_2 - x_1} = \frac{1 - 4}{2 - (-4)} = \frac{-3}{6} \text{ or } -\frac{1}{2}$$

Now substitute the known values into the point-slope form.

$y - y_1 = m(x - x_1)$ *Point-Slope Form*

$y - 4 = -\frac{1}{2}[x - (-4)]$ *Replace (x_1, y_1) with $(-4, 4)$ and m with $-\frac{1}{2}$.*

$y - 4 = -\frac{1}{2}x - 2$ *Distributive Property*

Then write in slope-intercept form.

$y - 4 + 4 = -\frac{1}{2}x - 2 + 4$ *Add 4 to each side.*

$y = -\frac{1}{2}x + 2$ *Slope-Intercept Form*

An equation of the line is $y = -\frac{1}{2}x + 2$. You can see from the graph that the y-intercept is 2. *You can also check by substituting the coordinates of one of the points into the equation.*

Your Turn

Write an equation in slope-intercept form of each line.

d. the line whose slope is $\frac{3}{4}$ and passes through the point at $(8, -2)$

e. the line passing through points at $(2, 4)$ and $(0, 5)$

---

⑥ **Refer to the application at the beginning of the lesson. Write an equation of the line in slope-intercept form.**

The y-intercept of the line is 50. Determine the slope.

$$m = \frac{y_2 - y_1}{x_2 - x_1} = \frac{0 - 50}{25 - 0} \text{ or } -2$$

Now substitute these values into the slope-intercept form.

$y = mx + b$ *Slope-Intercept Form*

$y = -2x + 50$ *Replace m with -2 and b with 50.*

An equation of the line is $y = -2x + 50$. *Check by substituting $(0, 50)$ and $(25, 0)$ into the equation.*

Check for Understanding

Study the lesson. Then complete the following.

Vocabulary
y-intercept
x-intercept
slope-intercept form

1. **Sketch** a line that has a y-intercept of 4 and a negative slope.

2. **Determine** whether the graph of $y = -3x + 8$ passes through the point at $(-5, 23)$. Explain how you found your answer.

3. **You Decide?** Andrew says that every line has an x-intercept. Jacquie disagrees. Who is correct? Give an example to support your answer.

Guided Practice

Write an equation in slope-intercept form of the line with each slope and y-intercept. *(Examples 1–3)*

4. $m = \dfrac{3}{2}, b = 4$

5. $m = \dfrac{1}{4}, b = -3$

6. $m = 0, b = 12$

Write an equation in slope-intercept form of the line having the given slope and passing through the given point. *(Example 4)*

7. $m = 4, (2, 1)$

8. $m = \dfrac{3}{5}, (0, 5)$

9. $m = -\dfrac{1}{2}, (4, -5)$

Write an equation in slope-intercept form of the line passing through each pair of points. *(Example 5)*

10. $(2, 3)$ and $(1, 0)$

11. $(2, -4)$ and $(5, 2)$

12. $(-4, 9)$ and $(5, 9)$

13. **Recycling** Yoshi collected 100 pounds of cans to recycle. He plans to collect an additional 25 pounds each week. The graph shows the amount of cans he plans to collect. *(Example 6)*

    a. Write an equation in slope-intercept form of the line.

    b. What does the slope represent?

    c. Use the equation to predict the total amount of cans Yoshi will have collected after 12 weeks.

Cans (lb)

(4, 200)

(0, 100)

Number of Weeks

Exercises

Practice

Write an equation in slope-intercept form of the line with each slope and y-intercept.

14. $m = 3, b = 1$

15. $m = -2, b = 5$

16. $m = 4, b = -2$

17. $m = \dfrac{3}{2}, b = 7$

18. $m = -\dfrac{1}{4}, b = -8$

19. $m = -\dfrac{2}{5}, b = 3$

20. $m = -3, b = 0$

21. $m = 0, b = 4$

22. $m = -1.6, b = -12$

Write an equation in slope-intercept form of the line having the given slope and passing through the given point.

23. $m = 3, (5, -2)$ **24.** $m = -\frac{1}{4}, (0, 8)$ **25.** $m = 5, (0, 0)$

26. $m = 0, (5, 6)$ **27.** $m = -2, (4, 5)$ **28.** $m = 3, (4, -3)$

29. $m = -5, (6, -4)$ **30.** $m = -\frac{3}{7}, (1, 1)$ **31.** $m = \frac{1}{2}, (-3, -6)$

Write an equation in slope-intercept form of the line passing through each pair of points.

32. $(3, 5)$ and $(2, 1)$ **33.** $(4, 0)$ and $(0, 5)$ **34.** $(4, -3)$ and $(2, 1)$

35. $(3, 0)$ and $(2, -4)$ **36.** $(1, 6)$ and $(-2, 6)$ **37.** $(9, -3)$ and $(-1, -3)$

38. $(7, -2)$ and $(14, -4)$ **39.** $(1, 0)$ and $(6, -4)$ **40.** $(-1, 8)$ and $(3, 6)$

Write an equation in slope-intercept form for each line.

41.   **42.**   **43.**

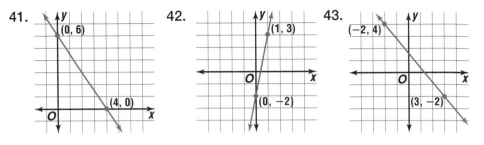

44. Write an equation in slope-intercept form of a line with slope $-\frac{2}{3}$ and y-intercept the same as the line whose equation is $y = 4x - 7$.

45. Write an equation in slope-intercept form of a line with a y-intercept of 11 and slope the same as the line whose equation is $y = -3x + 9$.

Applications and Problem Solving

46. Work For babysitting, Nicole charges a flat fee of $3, plus $5 per hour. The graph at the right shows how much Nicole earns babysitting.

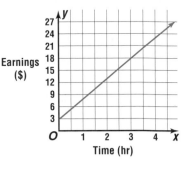

   a. Write an equation in slope-intercept form of the line.

   b. What do you think the slope and y-intercept represent?

   c. Use the equation to find how much money she will make if she babysits 5 hours.

47. Catering A caterer charges $120 to cater a party for 15 people and $200 for 25 people. Assume that the cost y is a linear function of the number of people x.

   a. Write an equation in slope-intercept form for this function.

   b. Explain what the slope represents.

   c. How much would a party for 40 people cost?

48. Critical Thinking   A line contains points at (5, 5) and (9, 1). Write a convincing argument showing that the *x*-intercept of the line is 10. Check by sketching the line.

Mixed Review   **Write the point-slope form of an equation for each line passing through the given point and having the given slope.**   *(Lesson 7–2)*

49. (4, 5), $m = -2$ **50.** (6, −2), $m = \frac{1}{2}$

Express each relation as a table and as a graph. Then determine the domain and the range.   *(Lesson 6–1)*

51. {(−3, 1), (−1, 5), (1, 3), (2, 6)}   **52.** {(−3, 0), (−1, 3), (0, −3), (3, 2)}

53. Measurement   Refer to the application at the beginning of the lesson. Convert 9144 meters to kilometers.   *(Lesson 5–1)*

Find each quotient.   *(Lesson 4–3)*

54. $7.2 \div 72$ **55.** $6 \div (-1.2)$ **56.** $-0.5 \div (-0.2)$

57. Standardized Test Practice   Solve $|c - 3| + 6 = 17$.   *(Lesson 3–7)*
   A {−8, 14}   **B** {8, 14}   **C** {−14, 8}   **D** {−14, 8}

Quiz 1   Lessons 7–1 through 7–3

▶ **Determine the slope of each line.**   *(Lesson 7–1)*

1.

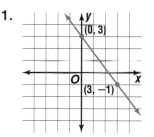

2. the line through points at (0, −2) and (2, 2)

Write the point-slope form of an equation for each line passing through the given point and having the given slope.   *(Lesson 7–2)*

3. (5, 4), $m = -2$ **4.** (−1, 6), $m = \frac{8}{5}$

5. Education   In order to "curve" a set of test scores, a teacher uses the equation $y = 2.5x + 10$, where *y* is the curved test score and *x* is the number of problems answered correctly.   *(Lesson 7–3)*
   a. Find the test score of a student who answers 32 problems correctly.
   b. Explain what the slope and *y*-intercept mean in the equation.

Lesson 7–3 Writing Equations in Slope-Intercept Form   **301**

Math In the Workplace

What You'll Learn
You'll learn to graph and interpret points on scatter plots.

Why It's Important
Insurance Insurance companies can use scatter plots to determine policy rates. *See Exercise 17.*

Look Back

Independent and Dependent Variables: Lesson 6–5

The Caribbean islands have many different species of birds.

To determine if there is a relationship between area and number of bird species, we can display the data points in a graph called a **scatter plot**. In a scatter plot, two sets of data are plotted as ordered pairs in the coordinate plane. The set of all the first coordinates is the domain, and the set of all the second coordinates is the range.

(area, number of species)

x-coordinate   y-coordinate

For example, the point with the box around it is at (95, 236). *The area is the independent quantity, and the number of species is the dependent quantity.*

You can use the scatter plot to draw conclusions and make predictions about the data.

Birds in the Caribbean		
Country	**Area (sq mi)**	**Number of Species**
Aruba	69	236
Anguilla	34	151
Barbados	166	144
Bermuda	21	350
Bonaire	95	236
Cayman Islands	93	176
Grenada	133	117
St. Vincent and the Grenadines	150	113
U.S. Virgin Islands	132	206

Source: *Thayer Birder's Diary, 1998*

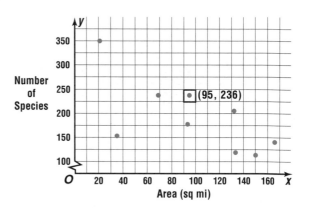

- The country with the least area has 350 species of birds.

- In general, as area increases, the number of species of birds appears to decrease. This shows a relationship between area and the number of species. That is, the number of species is a function of area. *The country of Anguilla is an exception to this generalization because it has fewer bird species than Aruba, which is a larger country.*

- The domain is the set of all *x*-coordinates: {21, 34, 69, 93, 95, 132, 133, 150, 166}. The range is the set of all corresponding *y*-coordinates: {350, 151, 236, 176, 236, 206, 117, 113, 144}.

- Based on the data in the scatter plot, you could predict that a Caribbean country with an area of 50 square miles would have fewer than 350 species of birds.

Often, when data are displayed in a scatter plot, the purpose is to determine whether there is a pattern, trend, or relationship between the variables.

- Data points that appear to go uphill show a relationship that is *positive*.
- Data points that appear to go downhill show a relationship that is *negative*.

Positive Slope

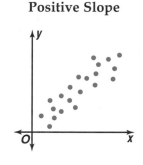

As *x* increases,
y increases.
positive relationship

Negative Slope

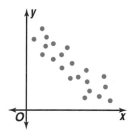

As *x* increases,
y decreases.
negative relationship

No Pattern

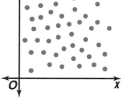

no obvious
pattern;
no relationship

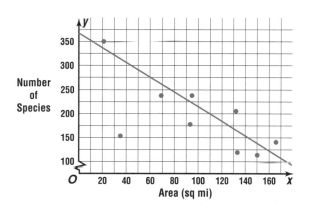

In the opening application, suppose a line were drawn that was close to most of the data points. The line would have a negative slope. That is, as the areas of the countries increase, the numbers of species decrease. Thus, area and numbers of species would have a negative relationship.

Example ❶

Word Processing Link

Real World

The scatter plot at the right shows the word processing speeds of 12 students and the number of weeks they have studied word processing. Determine whether the scatter plot shows a *positive* relationship, *negative* relationship, or *no* relationship. If there is a relationship, describe it.

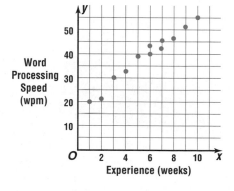

The plot indicates that as the number of weeks of experience increases, the word processing speed increases. Thus, there is a positive relationship between experience and word processing speed.

Determine whether each scatter plot shows a *positive* relationship, *negative* relationship, or *no* relationship. If there is a relationship, describe it.

a.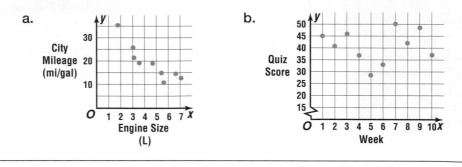

b.

Organizing data in a scatter plot makes it easier to observe patterns.

Example ❷

School Link

Real World

The table shows the amount of time students in math class spent studying and the scores they received on the test.

Study Time (min)	Test Score	Study Time (min)	Test Score
60	92	30	77
55	79	95	94
10	65	60	83
75	87	15	68
120	98	35	79
90	95	70	78
110	75	25	97
45	73	115	100

A. **Make a scatter plot of the data.**

Let the horizontal axis represent study time, and let the vertical axis represent test scores. Then plot the data.

B. **Does the scatter plot show a relationship between study time and test scores? Explain.**

In general, the scatter plot seems to show that a higher grade is directly related to the amount of time spent studying. There is a positive relationship.

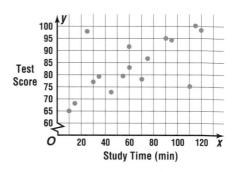

C. **Describe the independent and dependent variables. Then state the domain and the range.**

Because test scores depend on the amount of study time, test scores is the dependent variable, and study time is the independent variable. The domain is the set of all study times and the range is the set of all test scores.

c. Make a scatter plot of the data. Determine whether there is a relationship between annual rainfall and size of pumpkins. Explain.

Annual Rainfall (in.)	43	35	38	42	30	29	44	40	41
Weight of Pumpkin (lb)	204	195	198	200	190	190	206	202	200

Check for Understanding

Communicating Mathematics

Study the lesson. Then complete the following.

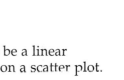
Vocabulary
scatter plot

1. **Describe** how you can use a scatter plot to display two sets of related data. Then explain what the domain, range, independent variable, and dependent variable are.

2. **Explain** how you know whether there appears to be a linear relationship between the two sets of data plotted on a scatter plot. Include the concept of slope in your explanation.

Math Journal

3. **Describe** situations in which data graphed in a scatter plot would have a positive relationship, negative relationship, and no relationship. Tell what the labels on the horizontal and vertical axes for each scatter plot would be.

Guided Practice

4. Determine whether the scatter plot has a *positive* relationship, *negative* relationship, or *no* relationship. If there is a relationship, describe it. (*Example 1*)

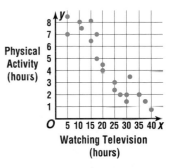

5. **Weather** The average maximum and minimum monthly temperatures in July of ten cities are given in the table at the right. (*Example 2*)

   a. Make a scatter plot of the data.

   b. Does the scatter plot show a linear relationship between maximum and minimum temperatures? Explain.

City	Minimum (°F)	Maximum (°F)
Atlanta, GA	70	88
Lexington, KY	66	86
Galveston, TX	79	87
Columbus, OH	63	84
Jackson, MS	71	92
Albany, NY	60	84
Los Angeles, CA	65	84
Miami, FL	76	89
Portland, OR	57	80
Chicago, IL	63	84

Source: *The World Almanac, 1999*

6. **Health** The scatter plot at the right shows data that compare a person's weekly exercise to that person's resting heart rate. Does there appear to be any relationship between hours of exercise and resting heart rate? If so, describe it. *(Example 2)*

Exercises • • • • • • • • • • • • • • • • • • •

Practice

Determine whether each scatter plot has a *positive* relationship, *negative* relationship, or *no* relationship. If there is a relationship, describe it.

7.

8.

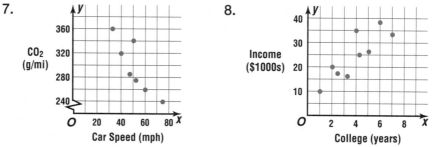

Make a scatter plot of each set of data. Then describe any trend shown in the scatter plot.

9.

Dallas Stars Hockey Team, Selected Players			
Goals	Assists	Goals	Assists
34	47	3	9
28	27	2	6
12	33	20	32
13	18	14	34
12	17	17	17
1	3	9	21
13	14	6	23
4	11	6	14
6	11	4	12

Source: *Infoplease.com, 1999*

10.

Breakfast Sweets	
Sugar (g)	Calories
30	260
36	430
13	420
19	270
15	530
18	430
47	420
52	580
23	520

Source: *Nutrition Action Healthletter, 1996*

11. Emission of CO_2 from car exhaust is harmful to the environment. Refer to the scatter plot in Exercise 7. What recommendation would you make about speed limits based on the data? Explain.

Determine whether a scatter plot of the data for the following would show a *positive*, *negative*, or *no* relationship between the variables.

12. playing time and points scored

13. your height and month of birth

14. size of household and pounds of garbage produced

15. age of car and its value

16. Insurance The table shows data on the age of drivers and the number of traffic accidents they have for each 100,000 drivers of that age.

Age	Number of Accidents	Age	Number of Accidents
16	168	23	120
17	181	24	117
18	190	25	92
19	195	26	80
20	145	27	69
21	126	28	70
22	114	29	75

   a. Make a scatter plot of the data. Determine if there is a *positive*, *negative*, or *no* relationship. If there is a relationship, describe it.

   b. Based on the scatter plot, what recommendation would you make to insurance companies about how to determine insurance rates for different age drivers?

17. Physical Science The table below shows the elevation and the average annual precipitation for places around the world.

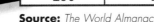

Elevation (ft)	Average Yearly Precipitation (in.)	Elevation (ft)	Average Yearly Precipitation (in.)
20	49	138	21
66	59	164	47
118	37	79	33
82	45	108	27
79	50	125	59
200	32	10	46
23	44	82	27
118	55	171	21
190	23		

Source: *The World Almanac*

   a. Make a scatter plot of the data.

   b. Does there seem to be a relationship? If so, describe it.

   c. Describe the slope of a line passing through the points.

18. Critical Thinking Refer to the scatter plot in Exercise 6. Would the relationship between the variables change if you relabeled the axes so that resting heart rate was on the horizontal axis? Explain.

19. Write an equation in slope-intercept form of the line passing through points at $(-2, -8)$ and $(1, 7)$. *(Lesson 7–3)*

20. Graph $y = x + 4$. *(Lesson 6–3)*

21. Architecture The Chrysler Building in New York is approximately 300 meters high. If a model of the building is 60 centimeters high, find the scale of the model. *(Lesson 5–2)*

22. Solve $-3(5 + x) = 12$. Check your solution. *(Lesson 4–7)*

23. Standardized Test Practice The area of the rectangle is 185.9 square centimeters. Find the length. *(Lesson 4–4)*

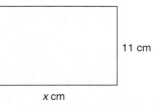

   A 22 cm   **B** 163.9 cm

   C 174.9 cm   **D** 16.9 cm

Chapter 7

Humidity Heats Things Up

Materials

 grid paper

straightedge

graphing
calculator
or graphing
software

Best-Fit Lines and Prediction

In the summertime, have you ever noticed that it feels hotter on days that
are humid? The *heat index* describes what the temperature feels like to
the body when there is humidity.

Investigate

1. The table shows the heat index for various air temperatures and
   percents of relative humidity. For example, at 100°F and 30% relative
   humidity, the heat index is 104°F. So, it feels like 104°F.

Air Temperature	Heat Index at Relative Humidities										
	0%	5%	10%	15%	20%	25%	30%	35%	40%	45%	50%
70°F	64	65	65	65	66	66	67	67	68	68	69
75°F	69	69	70	71	72	72	73	73	74	74	75
80°F	73	74	75	76	77	77	78	79	79	80	81
85°F	78	79	80	81	82	83	84	85	86	87	88
90°F	83	84	85	86	87	88	90	91	93	95	96
95°F	87	88	90	91	93	94	96	98	101	104	107
100°F	91	93	95	97	99	101	**104**	107	110	115	120

   a. Look at the row for 75°F. Write ordered pairs of the form (percent
      humidity, heat index) for the entire row. For example, the first
      ordered pair is (0, 69).

   b. Make a scatter plot of the data.

   c. Does the scatter plot show any relationship between humidity and
      heat index? If so, describe it.

   d. Describe the domain and the range using words and using numbers.

2. When there is a relationship between two sets of data, you can draw a
   best-fit line through the points, as you did in Lesson 7–4. This line is
   close to most of the data points. Use a straightedge to draw a best-fit
   line on the scatter plot you made in Exercise 1.

3. You can use the best-fit line to write a linear equation that describes a relationship between humidity and heat index.

   a. Select two points on the line that you drew in Exercise 2. Use the points to determine the slope of the line. *The points may or may not be points in the scatter plot.*

   b. Write an equation of the line in slope-intercept form.

   c. Describe the independent and dependent variables.

   d. Use your equation to predict the heat index at 75°F when the relative humidity is 60%.

Extending the Investigation

In this extension, you will use a graphing calculator or graphing software to explore lines of best fit.

- Make a scatter plot and draw best-fit lines for 85°, 95°, and one other air temperature of your choice.

- Write an equation in slope-intercept form for each best-fit line.

- Use your equations to predict the heat index for 85°, 95°, and the other air temperature that you chose when the humidity is 60%. Explain whether you think your predictions are reasonable.

- Use a graphing calculator or graphing software to find linear equations for the heat index data. Compare your equations for the best-fit lines to the equations given by the calculator or software.

- What can you say about the relationship between humidity and heat index by examining the slope of the equations for the lines?

- How do the graphs of the best-fit lines compare for the different air temperatures? Explain how the change in temperature affects heat index.

Presenting Your Conclusions

Here are some ideas to help you present your conclusions to the class.

- Make a poster displaying the heat index table, your scatter plots, best-fit lines, and equations. Research heat index and include a written description of a heat index.

- Research windchill factor. Find a table of the values for wind chill. Choose several wind speeds. Draw scatter plots and best-fit lines, and write equations for the data. If you prefer, use a graphing calculator or graphing software to find best-fit lines.

interNET
CONNECTION **Investigation** For more information on windchill factor, visit: www.algconcepts.glencoe.com

Math In the Workplace

What You'll Learn

You'll learn to graph linear equations by using the *x*- and *y*-intercepts or the slope and *y*-intercept.

Why It's Important

Rates Linear graphs are helpful in showing phone costs.
See Exercise 38.

In this problem, the domain and range values are always positive because the values represent time.

During the school year, Takara practices tennis 8 hours each week.

Let x = the number of hours each week she practices serves.
Let y = the number of hours each week she practices returns.

The total amount of time she practices tennis each week can be represented by the equation $x + y = 8$. A simple method of graphing a linear equation is using the points where the line crosses the *x*-axis and the *y*-axis.

To find the *x*-intercept, let $y = 0$.
$$x + y = 8$$
$$x + 0 = 8 \quad y = 0$$
$$x = 8$$

To find the *y*-intercept, let $x = 0$.
$$x + y = 8$$
$$0 + y = 8 \quad x = 0$$
$$y = 8$$

The *x*-intercept is 8, and the *y*-intercept is 8. This means that the graph intersects the *x*-axis at (8, 0) and the *y*-axis at (0, 8). Graph these ordered pairs. Then draw the line that passes through these points. Each point on the graph represents how Takara could spend her 8 hours practicing.

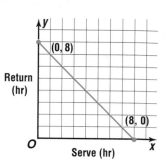

$$(3, 5)$$
↗   ↖
3 hours practicing serves   *5 hours practicing returns*

Examples

Determine the *x*-intercept and *y*-intercept of the graph of each equation. Then graph the equation.

1 $5y - x = 10$

To find the *x*-intercept, let $y = 0$.
$$5y - x = 10$$
$$5(0) - x = 10 \quad \text{Replace } y \text{ with } 0.$$
$$-x = 10$$
$$\frac{-x}{-1} = \frac{10}{-1} \quad \text{Divide each side by } -1.$$
$$x = -10$$

To find the *y*-intercept, let $x = 0$.
$$5y - x = 10$$
$$5y - 0 = 10 \quad \text{Replace } x \text{ with } 0.$$
$$5y = 10$$
$$\frac{5y}{5} = \frac{10}{5} \quad \text{Divide each side by } 5.$$
$$y = 2$$

The *x*-intercept is -10, and the *y*-intercept is 2. This means that the graph intersects the *x*-axis at $(-10, 0)$ and the *y*-axis at $(0, 2)$. Graph these ordered pairs. Then draw the line that passes through these points.

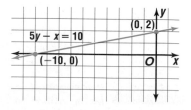

2 $2x - 4y = 8$

To find the x-intercept, let $y = 0$.

$2x - 4y = 8$

$2x - 4(0) = 8$ *Replace y with 0.*

$2x = 8$

$\dfrac{2x}{2} = \dfrac{8}{2}$ *Divide each side by 2.*

$x = 4$

To find the y-intercept, let $x = 0$.

$2x - 4y = 8$

$2(0) - 4y = 8$ *Replace x with 0.*

$-4y = 8$

$\dfrac{-4y}{-4} = \dfrac{8}{-4}$ *Divide each side by −4.*

$y = -2$

The x-intercept is 4, and the y-intercept is −2. This means that the graph intersects the x-axis at (4, 0) and the y-axis at (0, −2). Graph these ordered pairs. Then draw the line that passes through these points.

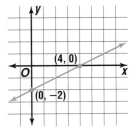

Check: Look at the graph. Choose some other point on the line and determine whether it is a solution of $2x - 4y = 8$. Try (2, −1).

$2x - 4y = 8$

$2(2) - 4(-1) \stackrel{?}{=} 8$ *Replace x with 2 and y with −1.*

$4 - (-4) \stackrel{?}{=} 8$

$8 = 8$ ✓

Your Turn

a. $x + y = 2$ **b.** $3x + y = 3$ **c.** $4x - 5y = 20$

You can easily determine the slope and y-intercept of the graph of an equation in slope-intercept form.

3 **To mail a letter in 2000, it cost $0.33 for the first ounce and $0.22 for each additional ounce. This can be represented by $y = 0.33 + 0.22x$. Determine the slope and y-intercept of the graph of the equation.**

$y = mx + b$ *Slope-Intercept Form*

$y = 0.22x + 0.33$

The slope is 0.22, and the y-intercept is 0.33. So the slope represents the cost per ounce after the first ounce, and the y-intercept represents the cost of the first ounce of mail.

Example ----4

Determine the slope and y-intercept of the graph of $10 + 5y = 2x$.

Write the equation in slope-intercept form to find the slope and y-intercept.

$$10 + 5y = 2x$$
$$10 + 5y - 10 = 2x - 10 \quad \textit{Subtract 10 from each side.}$$
$$5y = 2x - 10$$
$$\frac{5y}{5} = \frac{2x - 10}{5} \quad \textit{Divide each side by 5.}$$
$$y = \frac{2}{5}x - 2$$

The slope is $\frac{2}{5}$, and the y-intercept is -2.

Your Turn

d. $y = 5x + 9$ **e.** $4x + 3y = 6$

You can also graph a linear equation by using the slope and y-intercept.

Examples

Graph each equation by using the slope and y-intercept.

⑤ $y = \frac{2}{3}x - 5$

$$y = mx + b \quad \textit{Slope-Intercept Form}$$
$${\downarrow} {\downarrow}$$
$$y = \frac{2}{3}x + (-5)$$

The slope is $\frac{2}{3}$, and the y-intercept is -5.
Graph the point at $(0, -5)$. Then go up 2
units and right 3 units. This will be the
point at $(3, -3)$. Then draw the line
through points at $(0, -5)$ and $(3, -3)$.

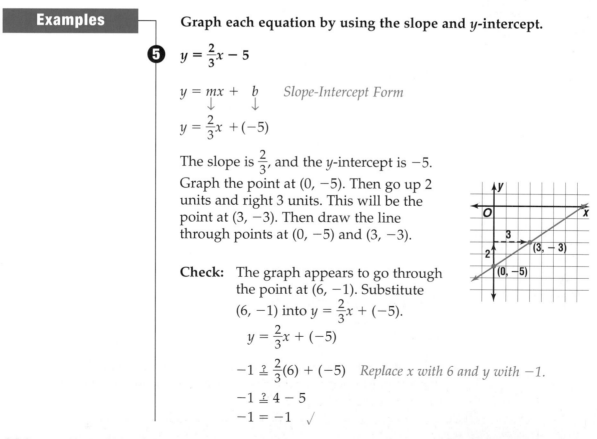

Check: The graph appears to go through
the point at $(6, -1)$. Substitute
$(6, -1)$ into $y = \frac{2}{3}x + (-5)$.

$$y = \frac{2}{3}x + (-5)$$
$$-1 \stackrel{?}{=} \frac{2}{3}(6) + (-5) \quad \textit{Replace x with 6 and y with } -1.$$
$$-1 \stackrel{?}{=} 4 - 5$$
$$-1 = -1 \quad \checkmark$$

6 $3x + 2y = 6$

First, write the equation in slope-intercept form.

$$3x + 2y = 6$$
$$3x + 2y - 3x = 6 - 3x \quad \text{\textit{Subtract 3x from each side.}}$$
$$2y = 6 - 3x$$
$$\frac{2y}{2} = \frac{6 - 3x}{2} \quad \text{\textit{Divide each side by 2.}}$$
$$y = -\frac{3}{2}x + 3 \quad \text{The slope is } -\frac{3}{2}, \text{ and the } y\text{-intercept is 3.}$$

Graph the point at (0, 3). Then go up 3 units and left 2 units. This will be the point at $(-2, 6)$. Then draw the line through (0, 3) and $(-2, 6)$. *Check by substituting the coordinates of another point that appears to lie on the line, such as (2, 0).*

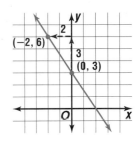

Your Turn

f. $y = \frac{1}{2}x + 3$

g. $x + 4y = -8$

The graph of a horizontal line has a slope of 0 and no x-intercept. The graph of a vertical line has an undefined slope and no y-intercept.

Examples

Graph each equation.

7 $y = 4$

$$y = mx + b$$
$$\downarrow \qquad \downarrow$$
$$y = 0x + 4 \quad \text{\textit{slope = 0, y-intercept = 4}}$$

No matter what the value of x, $y = 4$. So, all ordered pairs are of the form $(x, 4)$. Some examples are $(0, 4)$ and $(-3, 4)$.

8 $x = -2$

slope: undefined, y-intercept: none

No matter what the value of y, $x = -2$. So, all ordered pairs are of the form $(-2, y)$. Some examples are $(-2, -1)$ and $(-2, 3)$.

Your Turn

h. $y = -1$

i. $x = 3$

Communicating Mathematics

Study the lesson. Then complete the following.

1. **Explain** how many points are needed to draw the graph of a linear equation.

2. **You Decide?** Gwen says that $y = 8$ and $x = 8$ are both functions. Darnell says that only one of the equations is a function. Who is correct? Explain.

Guided Practice

Determine the *x*-intercept and *y*-intercept of the graph of each equation. Then graph the equation. *(Examples 1 & 2)*

3. $x + y = -3$

4. $x + 4y = 4$

5. $5x + 2y = 10$

6. $2x - 6y = 12$

Determine the slope and *y*-intercept of the graph of each equation. Then graph the equation. *(Examples 3–8)*

7. $y = x + 1$

8. $y = -4x$

9. $y = \frac{1}{4}x - 3$

10. $y = -2$

11. **Temperature** The equation $F = \frac{9}{5}C + 32$ gives the temperature in degrees Fahrenheit if you know the degrees Celsius. What are the slope and *y*-intercept of the graph of the equation? *(Example 3)*

Exercises

Practice

Determine the *x*-intercept and *y*-intercept of the graph of each equation. Then graph the equation.

12. $x + y = 4$

13. $x + y = -5$

14. $x + 2y = 4$

15. $x - 3y = 3$

16. $5x + y = 5$

17. $2x + y = -6$

18. $4x + 5y = 20$

19. $-3x + 4y = 12$

20. $6x - 3y = 6$

21. $7x - 2y = 14$

22. $2x + 5y = -10$

23. $x + \frac{1}{2}y = 4$

Determine the slope and *y*-intercept of the graph of each equation. Then graph the equation.

24. $y = x + 3$

25. $y = -x + 2$

26. $y = 2x + 1$

27. $y = 3x - 1$

28. $x = 6$

29. $y = 4$

30. $y = \frac{1}{2}x + 3$

31. $y = \frac{3}{4}x - 4$

32. $y = -\frac{3}{2}x + 5$

33. $-2x + y = 3$

34. $x + 2y = 4$

35. $3x + 5y = 10$

36. a. Graph $3x - 2y = 4$.

b. What is the slope of the line?

c. Where does the line intersect the x-axis?

37. a. Graph a line that goes through points at $(-4, 5)$ and $(2, -2)$.

b. Write an equation of the line in slope-intercept form.

c. Name the slope and y-intercept.

Applications and Problem Solving

38. Rates   A long-distance phone company charges $5 per month plus $0.10 per minute.

a. Write a linear equation to represent the total monthly cost $f(x)$ as a function of the number of minutes x.

b. Graph the equation.

c. Explain what the slope and y-intercept represent.

39. Physical Science   The weight of a bucket of water is a linear function of the depth of the water. When there are 4 centimeters of water in the bucket, it weighs 2 pounds. When there are 12 centimeters of water in the bucket, it weighs 4 pounds.

a. Write an equation for the weight of the bucket $f(x)$ as a function of the depth of the water x.

b. Graph the equation.

c. Estimate how much a bucket of water will weigh if there are 18 centimeters of water in it.

40. Critical Thinking   Explain how the pattern in the arithmetic sequence 2, 2.5, 3, 3.5, 4, 4.5, 5, . . . is related to the graph at the right.

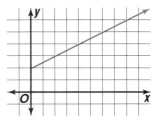

Mixed Review

41. Sketch a scatter plot that shows a positive relationship between two sets of related data.   *(Lesson 7–4)*

42. Entertainment   It costs $60 to rent jet skis for 2.5 hours. At that rate, how much will it cost to rent jet skis for 6 hours?   *(Lesson 6–5)*

Find the probability of each outcome if a die is rolled.   *(Lesson 5–6)*

43. a 4

44. an even number

45. a number greater than 6

Solve each equation. Check your solution.   *(Lesson 3–6)*

46. $-6 + x = 20$

47. $12 = b - 7$

48. $0 = k + 1.5$

49. Standardized Test Practice   Write the fractions in order from least to greatest.   *(Lesson 3–1)*

A $\dfrac{3}{14}, \dfrac{5}{23}, \dfrac{9}{23}$ **B** $\dfrac{5}{23}, \dfrac{9}{23}, \dfrac{3}{14}$ **C** $\dfrac{9}{23}, \dfrac{5}{23}, \dfrac{3}{14}$ **D** $\dfrac{3}{14}, \dfrac{9}{23}, \dfrac{5}{23}$

What You'll Learn

You'll learn to explore the effects of changing the slopes and *y*-intercepts of linear functions.

Why It's Important

Business Families of graphs can display different fees.
See Exercise 10.

The graph representing a cheetah's speed is much steeper than the graph representing a spider's speed or an elephant's speed. This is because the cheetah can travel a greater distance for each unit of time.

What do the graphs have in common? They have the same *y*-intercept, 0. They are called a **family of graphs** because they have at least one characteristic in common that makes them different from other groups of graphs.

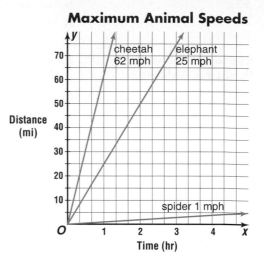

Maximum Animal Speeds

Families of linear graphs often fall into two categories—those with the same slope or those with the same *x*- or *y*-intercept.

Family of Graphs	**Family of Graphs**	**Not a Family of Graphs**
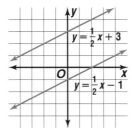	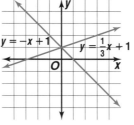	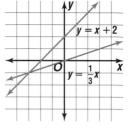
same slope	same *y*-intercept	different slopes and different intercepts

---

Examples

Graph each pair of equations. Describe any similarities or differences. Explain why they are a family of graphs.

1 $y = 3x + 4$
$y = 3x - 2$

The graphs have *y*-intercepts of 4 and −2, respectively.

They are a family of graphs because the slope of each line is 3.

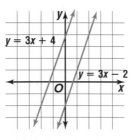

2 $y = x + 3$

$y = -\dfrac{1}{2}x + 3$

Each graph has a different slope.
Each graph has a y-intercept of 3.
Thus, they are a family of graphs.

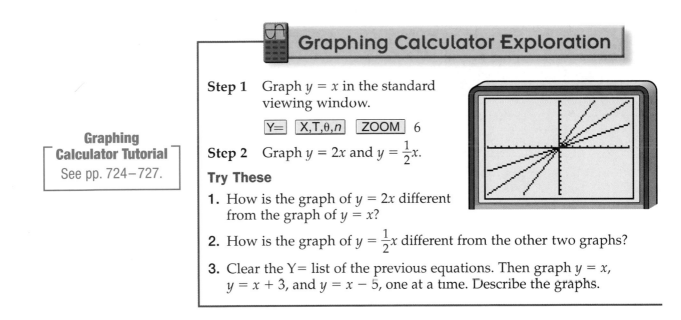

$y = -\dfrac{1}{2}x + 3$

$y = x + 3$

Your Turn

a. $y = 2x - 1$
$y = 2x + 5$

b. $y = x + 1$
$y = 3x + 1$

A graphing calculator is a good tool for exploring families of graphs.

Graphing Calculator Exploration

Graphing Calculator Tutorial
See pp. 724–727.

Step 1 Graph $y = x$ in the standard viewing window.

$\boxed{Y=}$ $\boxed{X,T,\theta,n}$ $\boxed{ZOOM}$ 6

Step 2 Graph $y = 2x$ and $y = \dfrac{1}{2}x$.

Try These

1. How is the graph of $y = 2x$ different from the graph of $y = x$?

2. How is the graph of $y = \dfrac{1}{2}x$ different from the other two graphs?

3. Clear the Y= list of the previous equations. Then graph $y = x$, $y = x + 3$, and $y = x - 5$, one at a time. Describe the graphs.

You can compare graphs of lines by looking at their equations.

Example

Business Link

Real World

3 Matthew and Juan are starting their own pet care business. Juan wants to charge $5 an hour. Matthew thinks they should charge $3 an hour. Suppose x represents the number of hours. Then $y = 5x$ and $y = 3x$ represent how much they would charge, respectively. Compare and contrast the graphs of the equations.

The equations have the same y-intercept, but the graph of $y = 5x$ is steeper. This is because its slope, which represents $5 per hour, is greater than the slope of the graph of $y = 3x$.

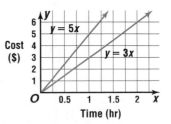

$y = 5x$

$y = 3x$

Cost ($)

Time (hr)

(continued on the next page)

Compare and contrast the graphs of the equations. Verify by graphing the equations.

c. $y = -3x + 4$
$ y = -x + 4$

d. $y = \frac{2}{3}x + 3$

$ y = \frac{2}{3}x - 1$

A **parent graph** is the simplest of the graphs in a family. Let's summarize how changing the m or b in $y = mx + b$ affects the graph of the equation.

Parent: $y = x$ **Parent: $y = -x$**   **Parent: $y = 2x$**

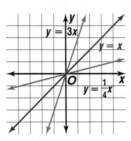 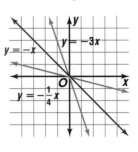 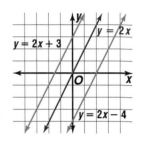

As the value of m increases, the line gets steeper.

As the value of m decreases, the line gets steeper.

As the value of b increases, the graph shifts up on the y-axis. As the value of b decreases, the graph shifts down on the y-axis.

You can change a graph by changing the slope or y-intercept.

Examples

Change $y = -\frac{1}{2}x + 3$ so that the graph of the new equation fits each description.

4 **same y-intercept, steeper negative slope**

The y-intercept is 3, and the slope is $-\frac{1}{2}$. The new equation will also have a y-intercept of 3. In order for the slope to be steeper and still be negative, its value must be less than $-\frac{1}{2}$, such as -2. The new equation is $y = -2x + 3$.

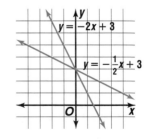

5 **same slope, y-intercept is shifted up 4 units**

The slope of the new equation will be $-\frac{1}{2}$. Since the current y-intercept is 3, the new y-intercept will be $3 + 4$ or 7. The new equation is $y = -\frac{1}{2}x + 7$. *Check by graphing.*

Your Turn Change $y = 2x + 1$ so that the graph of the new equation fits each description.

e. same slope, shifted down 1 unit

f. same y-intercept, less steep positive slope

Check for Understanding

Communicating Mathematics

Study the lesson. Then complete the following.

1. **Describe** how the graph of each equation is different from the graph of $y = \frac{1}{4}x + 1$.

Vocabulary
family of graphs
parent graph

a. $y = \frac{1}{4}x - 3$

b. $y = \frac{1}{8}x + 1$

c. $y = 2x + 1$

d. $y = \frac{1}{4}x + 4$

2. **Sketch** a family of graphs. Identify the parent graph and explain how the graphs are similar.

Math Journal

3. **a. Explain** the connection between a rate, such as miles per hour or dollars per hour, and slope.

b. Describe how changing the rate affects the slope of the graph.

c. Include some examples and sketches of graphs.

Guided Practice

Graph each pair of equations. Describe any similarities or differences and explain why they are a family of graphs. *(Examples 1 & 2)*

4. $y = 3x + 3$

$y - 3x - 2$

5. $y = \frac{1}{2}x - 4$

$y = \frac{3}{2}x - 4$

Compare and contrast the graphs of each pair of equations. Verify by graphing the equations. *(Example 3)*

6. $2x + 1 = y$
$2x = y$

7. $y = -\frac{1}{4}x + 3$

$y = -x + 3$

Change $y = x - 3$ so that the graph of the new equation fits each description. *(Examples 4 & 5)*

8. same y-intercept, steeper positive slope

9. same slope, shifted up 3 units

10. Business Refer to Example 3. Suppose Juan and Matthew both agree to charge \$3 an hour, but Juan wants to charge an additional \$5 fee per visit. Then $y = 3x + 5$ represents how much Juan would charge. *(Example 2)*

a. Graph $y = 3x$ and $y = 3x + 5$.

b. Explain how the graphs are similar and how they are different.

c. What does the slope represent?

Practice

Graph each pair of equations. Describe any similarities or differences. Explain why they are a family of graphs.

11. $y = 5x$
$y = 2x$

12. $y = 3x + 4$
$y = x + 4$

13. $y = 4x$
$y = 4x - 2$

14. $-\frac{2}{3}x + 1 = y$
$\frac{1}{4}x + 1 = y$

15. $y = -\frac{1}{2}x$
$y = -3x$

16. $y = -2x + 2$
$y = -2x - 3$

Compare and contrast the graphs of each pair of equations. Verify by graphing the equations.

17. $y = -x$
$y = x$

18. $\frac{1}{5}x = y$
$5x = y$

19. $y = 3x + 1$
$y = 3x - 8$

20. $y = -\frac{1}{3}x + 4$
$y = -\frac{1}{3}x$

21. $y = -2x - 2$
$y = -\frac{3}{4}x - 2$

22. $y = \frac{1}{2}x - 1$
$x - 3 = 2y$

Change $y = -\frac{4}{5}x + 6$ so that the graph of the new equation fits each description.

23. y-intercept is 0, same slope

24. positive slope, same y-intercept

25. shifted down 4 units, same slope

26. shifted up 2 units, same slope

27. steeper negative slope, same y-intercept

28. less steep negative slope, same y-intercept

29. Write an equation in slope-intercept form of a line passing through points at $(-3, -5)$ and $(3, 7)$. Then write an equation that has the same slope but a different y-intercept.

30. Write an equation of a line whose graph lies between the graphs of $y = 3x + 4$ and $y = 3x + 2$.

Applications and Problem Solving

31. Shipping Most cargo is transported by barge, train, or truck. The graph represents the distances traveled for the same amount of fuel.

a. Are these graphs a family of graphs? Explain.

b. What does the slope represent?

c. Which form of transportation gets the most miles per gallon? Explain.

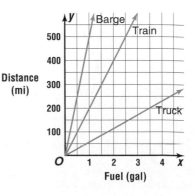

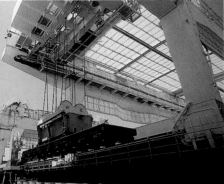

32. Population

> The U.S. Latino population ... is growing at a rate five times faster than that of the general population, according to the Census Bureau.

Source: *USA TODAY*, 1998

Compare and contrast the graphs that represent the growth of the Latino population and the growth of the general population.

33. Critical Thinking Given $A(0, 7)$, $B(2, -3)$, $C(-4, 6)$, and $D(0, 7)$, determine whether $\overline{AB}$ and $\overline{CD}$ are a family of graphs. Explain how you know.

Mixed Review

Determine the slope and *y*-intercept of the graph of each equation. *(Lesson 7–5)*

34. $y = 5x + 1$ **35.** $y = -2x$ **36.** $y = \frac{3}{4}x - 7$

37. Jewelry Necklace charms can be gold or silver, boy-shaped or girl-shaped, and contain a different birthstone for each month of the year. How many different charms are possible? *(Lesson 4–2)*

38. Find $-14 + (-3)$. *(Lesson 3–2)*

39. Standardized Test Practice State the property shown by $5(x + 2) = 5x + 10$. *(Lesson 1–4)*

    A Commutative ($\times$)   **B** Identity
    C Distributive   **D** Associative ($+$)

Quiz 2   Lessons 7–4 through 7–6

1. Animals Determine whether a scatter plot showing the temperature of a glass of water at a given air temperature has a *positive* relationship, *negative* relationship, or *no* relationship. If there is a relationship, describe it. *(Lesson 7–4)*

Determine the *x*-intercept and *y*-intercept of the graph of each equation. Then graph the equation. *(Lesson 7–5)*

2. $2x + y = 4$ **3.** $x + 4y = 4$ **4.** $2x - 3y = -12$

5. Videos The graph shows how much two different video stores pay for movies. *(Lesson 7–6)*

    a. Describe any similarities or differences and explain why they are a family of graphs.

    b. Explain what the slope represents.

    c. Which store pays less per movie?

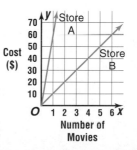

7-7 Parallel and Perpendicular Lines

What You'll Learn

You'll learn to write an equation of a line that is parallel or perpendicular to the graph of a given equation and that passes through a given point.

Why It's Important

Surveying Surveyors use parallel and perpendicular lines to plan construction. *See Exercise 39.*

McKenna is buying a new music system for her car. For Plan A, she makes a down payment of $50 and pays $40 per month. For Plan B, she pays only $40 per month. Suppose x represents the number of months. Then y represents the amount paid, and $y = 40x + 50$ and $y = 40x$ represent Plans A and B, respectively.

The graphs of the equations are a family of graphs because they have the same slope. Because $40x$ is never equal to $40x + 50$, the amount paid for each music system will never be the same in any given month and the graphs will never intersect. These lines are **parallel**.

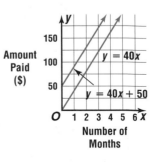

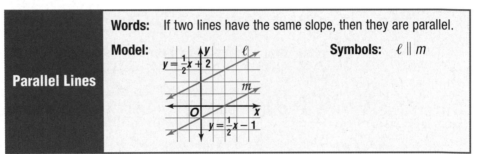

Parallel Lines	**Words:** If two lines have the same slope, then they are parallel.	
	Model:	**Symbols:** $\ell \parallel m$

All vertical lines are parallel.

Example

1 **Determine whether the graphs of the equations are parallel.**

$$y = -\frac{3}{4}x - 2$$

$$4y = -3x + 12$$

First, determine the slopes of the lines. Write each equation in slope-intercept form.

$y = -\frac{3}{4}x - 2$ *Slope-Intercept Form*

$4y = -3x + 12$

$\dfrac{4y}{4} = \dfrac{-3x + 12}{4}$ *Divide each side by 4.*

The slope is $-\frac{3}{4}$.

$y = -\frac{3}{4}x + 3$ The slope is $-\frac{3}{4}$.

The slopes are the same, so the lines are parallel. *Check by graphing.*

Your Turn

a. $y = 2x$
$7 = 2x - y$

b. $y = -3x + 3$
$2y = 6x - 5$

A *parallelogram* is a four-sided figure with two sets of parallel sides.

2 **Determine whether figure *ABCD* is a parallelogram.**

Explore To be a parallelogram, $\overline{AB}$ and $\overline{DC}$ must be parallel. Also, $\overline{AD}$ and $\overline{BC}$ must be parallel.

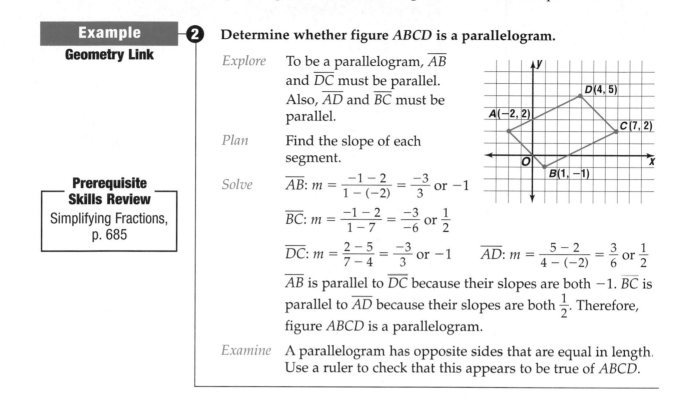

Plan Find the slope of each segment.

Solve $\overline{AB}$: $m = \dfrac{-1-2}{1-(-2)} = \dfrac{-3}{3}$ or -1

$\overline{BC}$: $m = \dfrac{-1-2}{1-7} = \dfrac{-3}{-6}$ or $\dfrac{1}{2}$

$\overline{DC}$: $m = \dfrac{2-5}{7-4} = \dfrac{-3}{3}$ or -1 $\overline{AD}$: $m = \dfrac{5-2}{4-(-2)} = \dfrac{3}{6}$ or $\dfrac{1}{2}$

$\overline{AB}$ is parallel to $\overline{DC}$ because their slopes are both -1. $\overline{BC}$ is parallel to $\overline{AD}$ because their slopes are both $\dfrac{1}{2}$. Therefore, figure *ABCD* is a parallelogram.

Examine A parallelogram has opposite sides that are equal in length. Use a ruler to check that this appears to be true of *ABCD*.

If you know the slope of a line, you can use that information to write an equation of a line that is parallel to it.

Example

3 **Write an equation in slope-intercept form of the line that is parallel to the graph of $y = -4x + 8$ and passes through the point at (1, 3).**

The slope of the given line is -4. So, the slope of the new line will also be -4. Find the new equation by using the point-slope form.

$y - y_1 = m(x - x_1)$ *Point-Slope Form*
$y - 3 = -4(x - 1)$ *Replace (x_1, y_1) with (1, 3) and m with -4.*
$y - 3 = -4x + 4$ *Distributive Property*
$y - 3 + 3 = -4x + 4 + 3$ *Add 3 to each side.*
$y = -4x + 7$

An equation whose graph is parallel to the graph of $4x + y = 8$ and passes through the point at (1, 3) is $y = -4x + 7$. *Check by substituting (1, 3) into $y = -4x + 7$ or by graphing.*

Your Turn

Write an equation in slope-intercept form of the line that is parallel to the graph of each equation and passes through the given point.

c. $y = 6x - 4$; (2, 3)   **d.** $3x + 2y = 9$; (2, 0)

Materials:  grid paper 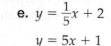 protractor

Step 1 Draw line ℓ through the origin and $P(5, 3)$.

Step 2 Use a protractor to rotate the line $90°$. Label the new line ℓ'.

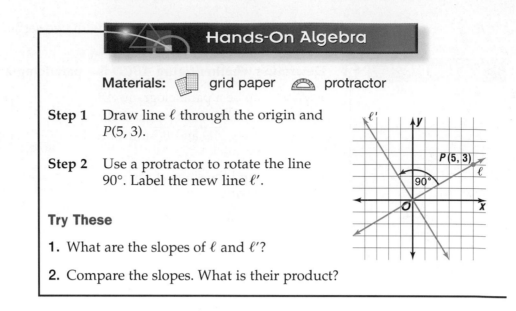

Try These

1. What are the slopes of ℓ and ℓ'?

2. Compare the slopes. What is their product?

The results of the Hands-On Algebra activity lead to the following definition of **perpendicular lines**.

Perpendicular Lines	**Words:** If the product of the slopes of two lines is -1, then the lines are perpendicular.
	Model: 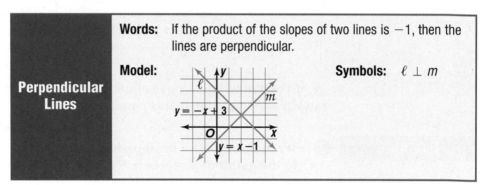 $y = -x + 3$ $y = x - 1$ $\qquad$ **Symbols:** $\ell \perp m$

In a plane, vertical lines are perpendicular to horizontal lines.

Example

4 Determine whether the graphs of the equations are perpendicular.

$$y = \frac{2}{3}x + 1$$

$$y = -\frac{3}{2}x + 2$$

The graphs are perpendicular because the product of their slopes is $\frac{2}{3} \cdot \left(-\frac{3}{2}\right)$ or -1.

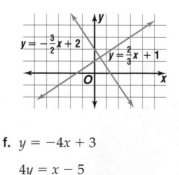

Your Turn

e. $y = \frac{1}{5}x + 2$

$\quad y = 5x + 1$

f. $y = -4x + 3$

$\quad 4y = x - 5$

Example ● **5** Write an equation in slope-intercept form of the line that is perpendicular to the graph of $y = \frac{1}{3}x - 2$ and passes through the point at $(-4, 2)$.

The slope is $\frac{1}{3}$. A line perpendicular to the graph of $y = \frac{1}{3}x - 2$ has slope -3. Find the new equation by using the point-slope form.

$$y - y_1 = m(x - x_1) \qquad \textit{Point-Slope Form}$$
$$y - 2 = -3[x - (-4)] \qquad \textit{Replace } (x_1, y_1) \textit{ with } (-2, 4) \textit{ and m with } -3.$$
$$y - 2 = -3x - 12 \qquad \textit{Distributive Property}$$
$$y - 2 + 2 = -3x - 12 + 2 \qquad \textit{Add 2 to each side.}$$
$$y = -3x - 10$$

The new equation is $y = -3x - 10$. *Check by substituting $(-4, 2)$ into the equation or by graphing.*

Your Turn Write an equation in slope-intercept form of the line that is perpendicular to the graph of each equation and passes through the given point.

g. $y = 2x + 6$; $(0, 0)$ **h.** $2x + 3y = 2$; $(3, 0)$

Check for Understanding

Communicating Mathematics

Study the lesson. Then complete the following.

1. **Compare and contrast** the slopes of lines that are parallel and the slopes of lines that are perpendicular.

Vocabulary

parallel lines
perpendicular lines

2. **Choose** the graph that is parallel to the graph of $3x + 4y = -12$. Explain.

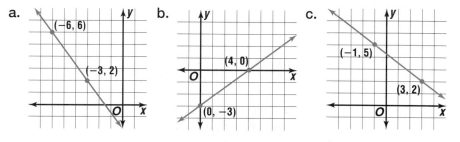

a. $(-6, 6)$ $(-3, 2)$

b. $(4, 0)$ $(0, -3)$

c. $(-1, 5)$ $(3, 2)$

Guided Practice

● **Getting Ready** State the slopes of the lines parallel to and perpendicular to the graph of each equation.

Sample: $y = -3x + 6$

Solution: The slope is -3. Lines parallel to the graph have a slope of -3. Lines perpendicular to the graph have a slope of $\frac{1}{3}$.

3. $y = 4x + 2$ **4.** $5x + y = 8$ **5.** $2x - 3y = 7$

Determine whether the graphs of each pair of equations are _parallel_, _perpendicular_, or _neither_. *(Examples 1 & 4)*

6. $y = 4x - 9$
$y = 4x - 11$

7. $y = -\dfrac{3}{2}x + 5$
$y = 2x + 1$

8. $y = 5x + 3$
$x + 5y = 2$

9. Geometry Determine $m\angle STQ$ if the slope of $\overline{RQ}$ is $\dfrac{2}{5}$ and the slope of $\overline{SP}$ is $-\dfrac{5}{2}$. Explain. *(Example 4)*

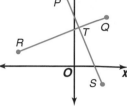

Exercise 9

Write an equation in slope-intercept form of the line that is parallel to the graph of each equation and passes through the given point. *(Example 3)*

10. $y = x + 5;\ (7, 2)$

11. $y - \dfrac{3}{4}x = 4;\ (2, 0)$

Write an equation in slope-intercept form of the line that is perpendicular to the graph of each equation and passes through the given point. *(Example 5)*

12. $y = 2x + 3;\ (3, -4)$

13. $3x + 8y = 4;\ (0, 4)$

14. Exercise Joey and Cortez are both riding their bikes at 12 miles per hour. However, Cortez started 5 miles ahead of Joey. The equations $y = 12x$ and $y = 12x + 5$ represent the positions of Joey and Cortez, respectively. *(Example 1)*

a. If they ride for 3 hours, will Joey catch up to Cortez? Explain your answer in terms of slopes of the graphs of the equations.

b. Are the graphs a family of graphs? Explain.

Exercises •

Practice

Determine whether the graphs of each pair of equations are _parallel_, _perpendicular_, or _neither_.

15. $y = \dfrac{3}{5}x + 2$
$y = -\dfrac{5}{3}x - 1$

16. $y = 5x + 4$
$y = 5x + 8$

17. $x + 3y = 7$
$y = \dfrac{1}{3}x - 9$

18. $y = -2x + 6$
$y = 2x$

19. $y = 2x + 3$
$x + 2y = 4$

20. $y = \dfrac{2}{3}x$
$3x + 2y = 0$

21. $3y = 8$
$y = 3$

22. $-\dfrac{1}{4}x + 5 = 0$
$2y - 9 = 0$

23. $7x + 3y = 4$
$3x - 7y = 1$

Write an equation in slope-intercept form of the line that is parallel to the graph of each equation and passes through the given point.

24. $y = 4x + 5$; $(2, -3)$ **25.** $y = -\frac{1}{2}x - 3$; $(0, 0)$ **26.** $y = 4$; $(2, 5)$

27. $4x + y = 1$; $(-2, 1)$ **28.** $3x + y = 3$; $(3, 5)$ **29.** $x = 3$; $(2, 6)$

Write an equation in slope-intercept form of the line that is perpendicular to the graph of each equation and passes through the given point.

30. $y = 2x + 3$; $(3, -4)$ **31.** $y = -4x + 5$; $(1, 1)$ **32.** $2x - 5y = 3$; $(-2, 7)$

33. $y = 2x$; $(2, 0)$ **34.** $x = 6$; $(4, 2)$ **35.** $3x - 5y = 2$; $(-1, 0)$

Determine whether $\overline{AB}$ and $\overline{CD}$ are *parallel*, *perpendicular*, or *neither*.

36. $A(4, 3)$, $B(5, 2)$, $C(8, 3)$, $D(9, 2)$

37. $A(-2, 3)$, $B(4, 0)$, $C(2, 5)$, $D(-1, 11)$

38. Write an equation in slope-intercept form of each line.

   a. passes through the point at $(4, -2)$ and is parallel to the graph of $5x - 2y = 6$

   b. perpendicular to the line through points at $(1, 2)$ and $(8, 6)$ and passes through the point at $(0, 5)$

Applications and Problem Solving

Real World

39. Surveying Determine whether roads $\overrightarrow{AB}$ and $\overrightarrow{BC}$ are perpendicular in the survey at the right. Explain.

40. Geometry Determine whether quadrilateral *JKLM* is a rectangle if its vertices are $J(0, 5)$, $K(9, 2)$, $L(7, -4)$, and $M(-2, -1)$. Explain how you know.

41. Critical Thinking Suppose the line through points at $(-2, 2)$ and $(x, 5)$ is parallel to the graph of $2x - 4y = 9$. What is x?

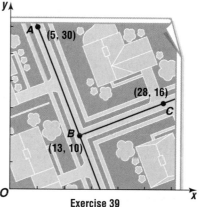

Exercise 39

Mixed Review

Compare and contrast the graphs of each pair of equations. Verify by graphing the equations. *(Lesson 7–6)*

42. $y = 4x$
$y = 4x - 3$

43. $y = -3x + 2$
$y = 3x + 2$

44. $y = x + 2$
$7y = -2x + 14$

A card is drawn from a standard deck of cards. Determine whether each event is *mutually exclusive* or *inclusive*. *(Lesson 5–7)*

45. $P(\text{queen or ace})$ **46.** $P(\text{8 or red card})$ **47.** $P(\text{heart or spade})$

48. Solve $6 = q - 5$. *(Lesson 3–5)*

49. Solve $g = 15.6 - 2.75$. *(Lesson 3–4)*

50. Standardized Test Practice In which quadrant does the graph of $G(-15, 4)$ lie? *(Lesson 2–2)*

   A I   **B** II   **C** III   **D** IV

Extra Practice   See p. 707.

Study Guide and Assessment

Understanding and Using the Vocabulary

After completing this chapter, you should be able to define each term, property, or phrase and give an example or two of each.

*inter*NET
CONNECTION   **Review Activities**
For more review activities, visit:
www.algconcepts.glencoe.com

best-fit line *(p. 308)*
family of graphs *(p. 316)*
parallel lines *(p. 322)*
parent graph *(p. 318)*

perpendicular lines *(p. 324)*
point-slope form *(p. 290)*
scatter plot *(p. 302)*
slope *(p. 284)*

slope-intercept form *(p. 296)*
x-intercept *(p. 296)*
y-intercept *(p. 296)*

State whether each sentence is *true* or *false*. If false, replace the underlined word(s) to make a true statement.

1. In the equation $y = mx + b$, b is the <u>*x*-intercept</u> of the line.
2. A line that appears to go downhill from left to right has a <u>negative</u> slope.
3. <u>Perpendicular</u> lines have the same slopes.
4. The graph of a <u>linear equation</u> is a straight line.
5. You can write an equation of a line if you know the slope and a <u>point on the line</u>.
6. The <u>point-slope form</u> of a linear equation is $y - y_1 = m(x - x_1)$.
7. The graph of a line having a slope of 0 is a <u>vertical</u> line.
8. To determine the *y*-intercept of the graph of an equation, let <u>y</u> = 0 and solve.
9. A <u>scatter plot</u> is a graph in which data are plotted as ordered pairs.
10. Data points that appear to go uphill in a scatter plot show a <u>positive</u> relationship.

Skills and Concepts

Objectives and Examples

- **Lesson 7–1** Find the slope of a line given the coordinates of two points on the line.

Determine the slope of the line through points at $(-3, 1)$ and $(7, 4)$.

$$m = \frac{y_2 - y_1}{x_2 - x_1}$$
$$= \frac{4 - 1}{7 - (-3)} \quad \begin{array}{l}(x_1, y_1) = (-3, 1), \\ (x_2, y_2) = (7, 4)\end{array}$$
$$= \frac{3}{10}$$

The slope is $\frac{3}{10}$.

Review Exercises

Determine the slope of each line.

11.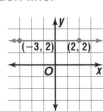
   (0, 1) (3, 3)

12.
   (−3, 2) (2, 2)

13. the line through points at $(6, 0)$ and $(2, 8)$

14. the line through points at $(-4, 9)$ and $(-4, 5)$

Objectives and Examples	Review Exercises

• Lesson 7–2 Write a linear equation in point-slope form given the coordinates of a point on the line and the slope of the line.

$$y - y_1 = m(x - x_1) \quad \textit{Point-Slope Form}$$
$$y - 8 = 2(x - 1) \quad (x_1, y_1) = (1, 8) \text{ and } m = 2$$

Write the point-slope form of an equation for each line passing through the given point and having the given slope.

15. $(2, 5), m = 3$ **16.** $(-1, 6), m = \dfrac{1}{4}$

17. $(0, 3), m = -5$ **18.** $(4, 8), m = 0$

19. Write an equation in point-slope form of a line passing through points at $(10, 4)$ and $(-8, 7)$.

• Lesson 7–3 Write a linear equation in slope-intercept form given the slope and y-intercept.

$$y = mx + b \quad \textit{Slope-Intercept Form}$$
$$y = 4x + 12 \quad m = 4 \text{ and } b = 12$$
slope y-intercept

Write an equation in slope-intercept form of the line with each slope and y-intercept.

20. $m = 1, b = 4$ **21.** $m = 3, b = -9$

22. $m = \dfrac{1}{2}, b = 5$ **23.** $m = 2, b = 0$

24. $m = 0, b = 7$ **25.** $m = 11, b = -6$

• Lesson 7–4 Graph and interpret points on scatter plots.

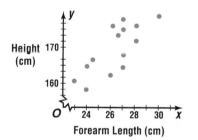

There is a positive relationship between forearm length and height of students. As forearm length increases, height increases.

Determine whether each scatter plot has a *positive* relationship, *negative* relationship, or *no* relationship. If there is a relationship, describe it.

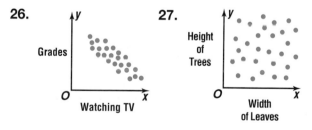

26. **27.**

• Lesson 7–5 Graph linear equations by using the x- and y-intercepts or the slope and y-intercept.

$$\text{slope} = \frac{2}{3}$$

$$y\text{-intercept} = 2$$

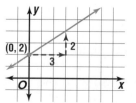

Determine the x-intercept and y-intercept of the graph of each equation. Then graph.

28. $x + 3y = 6$ **29.** $2x - y = 4$

Determine the slope and y-intercept of the graph of each equation. Then graph.

30. $y = x + 5$ **31.** $y = -2x + 3$

Objectives and Examples

Review Exercises

• **Lesson 7–6** Explore the effects of changing the slopes and y-intercepts of linear functions.

The graphs have the same y-intercept and different negative slopes. They are a family of graphs.

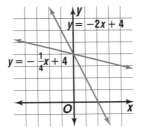

Compare and contrast the graphs of each pair of equations. Verify by graphing the equations.

32. $y = x + 4$
$y = 3x + 4$

33. $y = \frac{1}{3}x - 1$
$y = x - 1$

34. $y = 2x$
$y = 2x + 5$

35. $\frac{1}{2}x + 3 = y$
$\frac{1}{2}x - 4 = y$

• **Lesson 7–7** Write an equation of a line that is parallel or perpendicular to the graph of a given equation and that passes through a given point.

ℓ is parallel to a because their slopes are each $\frac{1}{2}$.

ℓ is perpendicular to b because the product of their slopes is -1.

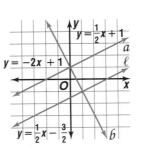

Determine whether the graphs of each pair of equations are *parallel*, *perpendicular*, or *neither*.

36. $y = 2x + 8$
$y = 2x - 4$

37. $y = -3x$
$y = 3x + 1$

38. $y = \frac{1}{7}x - 1$
$y = -7x - 1$

39. $\frac{3}{4}x - 5 = y$
$y = \frac{3}{4}x + 6$

Applications and Problem Solving

40. Building Paved areas are usually slightly inclined so that puddles do not form. The pavement at the right rises 1 inch for every 25 inches of horizontal change. What is the slope? *(Lesson 7–1)*

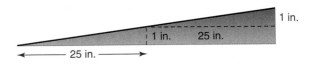

41. Employment The table shows earnings for Employees A and B. *(Lesson 7–6)*

a. Sketch a graph of each employee's earnings by plotting points with the coordinates (hours, earnings).

b. Suppose Employee C earns more money per hour than Employee A and less money per hour than Employee B. How would a graph of this employee's earnings compare with the graphs in part a?

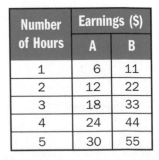

Number of Hours	Earnings ($)	
	A	B
1	6	11
2	12	22
3	18	33
4	24	44
5	30	55

1. **Describe** how changing each of the following affects the graph of a linear function.
   a. slope　　　　　　　　　　　　　**b.** y-intercept

2. **Compare and contrast** the graphs of $y = 4x - 1$, $y = x - 1$, and $y = 3x - 1$. Verify by graphing the equations.

Determine the slope of each line.

3.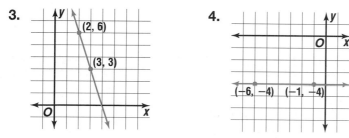

4.

5. the line passing through points at $(1, -2)$ and $(6, 0)$

Write the point-slope form of an equation for each line passing through the given point and having the given slope.

6. $(5, 6)$, $m = 3$　　　　　7. $(-3, 1)$, $m = -2$　　　　　8. $(-4, 8)$, $m = 0$

Write an equation in slope-intercept form of the line passing through each pair of points.

9. $(1, 5)$ and $(2, 8)$　　　　　10. $(3, 1)$ and $(-7, 11)$　　　　　11. $(-4, 0)$ and $(2, 3)$

12. **Music** A disc jockey notices that as the music gets faster, more people start dancing. Would a scatter plot showing speed of music and number of dancers have a *positive* relationship, *negative* relationship, or *no* relationship?

Graph each equation.

13. $x + y = 2$　　　　　14. $x + 3y = 3$　　　　　15. $2x - 4y = 12$

16. $y = 2x - 5$　　　　　17. $y = \frac{1}{3}x + 2$　　　　　18. $y = -1$

19. Are the graphs of the equations *parallel*, *perpendicular*, or *neither*?
$$y = -4x + 9$$
$$y = \frac{1}{4}x - 6$$

20. **Construction** The *pitch* of a roof describes its steepness. Suppose a roof is 40 feet wide and its pitch is $\frac{1}{2}$. Find x, its height above the rafters at its peak. $\left(Hint: \text{Use } \frac{\text{rise}}{\text{run}}.\right)$

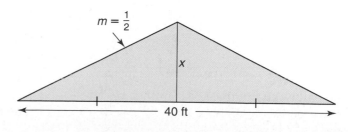

Algebra Word Problems

All standardized tests involve writing and solving equations from realistic settings. You'll need to translate words into equations. You'll need to find patterns in number tables and write equations to represent the patterns.

THE PRINCETON REVIEW

When a question includes a variable, you can "plug-in" a number for the variable that fits the problem.

Proficiency Test Example

Ami dropped a rubber ball from several heights and measured how high the ball bounced. Her results are shown in the table. How high would the ball bounce if it were dropped from a height of 36 centimeters?

Height of Drop (cm)	Bounce Height (cm)
40	30
60	45
80	60
100	75

A 19 cm   **B** 24 cm   **C** 27 cm   **D** 30 cm

Hint Look for a pattern in the table.

Solution The bounce is 30 when the drop height is 40. The bounce is 75 when the drop height is 100. So, the bounce is $\frac{3}{4}$ of the drop height. Check that this pattern is true for each of the other numbers in the table.

Let b represent the bounce height and let d represent the drop height.

$b = \frac{3}{4}d$ *Write an equation.*

$= \frac{3}{4}(36)$ *Replace d with 36.*

$= \frac{108}{4}$ or 27

The ball would bounce 27 centimeters. The answer is C.

ACT Example

After N chocolate bars are divided equally among 6 children, 3 bars remain. How many would remain if $(N + 4)$ chocolate bars were divided equally among 6 children?

A 0   **B** 1   **C** 2   **D** 3   **E** 4

Hint Sketch a diagram to understand the problem. Use the "plug-in" strategy.

Solution One way to solve this kind of problem is to choose a numeric value for N and "plug it in." Suppose each of the 6 children got 2 chocolate bars, and 3 bars remain. Then $N = 6(2) + 3 = 15$.

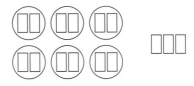

Now, use the value of 15 for N.

$$N + 4 = 15 + 4 \text{ or } 19 \text{ bars}$$

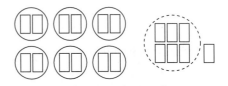

Divide 19 bars among 6 children.

$$19 = 18 + 1 \text{ or } 3(6) + 1$$

The remainder is 1. The answer is B.

After you work each problem, record your answer on the answer sheet provided or on a sheet of paper.

1. Refer to the table and the equation in the first example on the opposite page. From what height should Ami drop the ball so that the bounce height is 105 centimeters?

   A 110 cm   **B** 140 cm
   C 180 cm   **D** 200 cm

2. A cable TV company charges $21.95 per month for basic service. Each premium channel selected costs an additional $5.95 per month. If x represents the number of premium channels selected, which expression can be used to find the monthly cost of cable service?

   A $5.95 + 21.95x$ **B** $21.95 + 5.95x$
   C $21.95 + 5.95 + x$ **D** $21.95 - 5.95x$

3. Steve ran a 12-mile race at an average speed of 8 miles per hour. If Adam ran the same race at an average speed of 6 miles per hour, how many minutes longer than Steve did Adam take to complete the race?

   A 9   **B** 12   **C** 16   **D** 24   **E** 30

4. The bar graph shows the number of people who watch prime time television each day of the week. About how many more people watch on Monday than on Saturday?

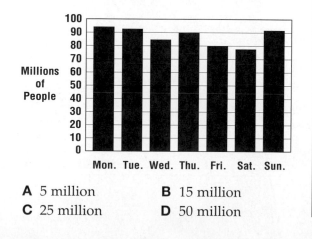

   A 5 million   **B** 15 million
   C 25 million   **D** 50 million

5. Kim earns $4 per hour baby-sitting. If x represents the number of hours Kim baby-sits, which expression could be used to find the amount she earns?

   A $x + 4$ **B** $x + 5$ **C** $4x$ **D** $5x$

6. If Nathan is $\frac{1}{4}$ as old as his father and the sum of their ages is 60, then how old is Nathan?

   A 8   **B** 12   **C** 15   **D** 20   **E** 48

7. There are 5 different pizza toppings. How many different 3-topping pizzas are possible?

   A 10   **B** 15   **C** 60   **D** 125

8. A group of 5 adults and 3 children sees a play. A child's ticket costs $6.25, and an adult's ticket costs $9.75. Which equation can be used to find c, the amount of change from $100 after paying for the group's tickets?

   A $5(6.25) + 3(9.75) = 100 - c$
   B $5(9.75) + 3(6.25) = c$
   C $5(9.75) + 3(6.25) + c = 100$
   D $5(9.75) + 3(6.25) = 100 + c$

Open-Ended Questions

9. **Grid-In** What number increased by two is equal to two less than twice the number?

10. The fare charged by a taxi driver is a $3 fixed charge plus $0.35 per mile. Beth pays $10 for a ride of m miles.

    Part A Write an equation that can be used to find m. Show your work.

    Part B Use the equation in Part A to find how many miles Beth rode. Show your work.

CHAPTER 8 Powers and Roots

▶ What You'll Learn in Chapter 8:

- to use powers in expressions (*Lesson 8–1*),
- to multiply and divide powers (*Lesson 8–2*),
- to simplify expressions containing negative exponents (*Lesson 8–3*),
- to express numbers in scientific notation (*Lesson 8–4*),
- to simplify radicals (*Lesson 8–5*),
- to estimate square roots (*Lesson 8–6*), and
- to use the Pythagorean Theorem to solve problems (*Lesson 8–7*).

Problem-Solving Workshop

Project

The *Super Sweepstakes Contest*, a fictitious contest, offers its winners a choice of prizes. The winner can choose to receive $1 million each year for 25 years or $1 the first year, $2 the second year, $4 the third year, $8 the fourth year, and so on, for 25 years. At the end of 25 years, which prize awards the most money?

Working on the Project

Work with a partner and choose a strategy to help analyze the information. Develop a plan. Here are some questions to help you get started.

- For each prize, how much money is awarded each year for the first five years?
- What is the total amount of money awarded at the end of five years?

Technology Tools

- Use a **spreadsheet** to perform the calculations.
- Use a **graphing calculator** or **graphing software** to show the data in a graph.
- Use a **word processor** to write a report about your solution.

 Research For more information about linear and exponential functions, visit: www.algconcepts.glencoe.com

> ### ► Strategies
>
> **Look for a pattern.**
>
> **Draw a diagram.**
>
> **Make a table.**
>
> **Work backward.**
>
> **Use an equation.**
>
> **Make a graph.**
>
> **Guess and check.**

Presenting the Project

Write a report that includes any spreadsheets or graphs that you have made. As part of your project, answer these questions.

- At the end of 25 years, what is the difference in the total amount of money awarded?
- Suppose you are the prize winner. Which prize would you choose? Explain your reasoning.
- Suppose you are an accountant that works for the *Super Sweepstakes Contest*. Which prize would you like the winner to choose? Explain your reasoning.

8-1 Powers and Exponents

Math
In the Workplace

What You'll Learn
You'll learn to use powers in expressions.

Why It's Important
Landscaping
Landscape architects use the formula $A = \pi r^2$ to find the area of a circular region. The 2 in the formula is an exponent. *See Exercise 43.*

Ancient mathematicians called numbers like 1, 4, 9, and 16 **perfect squares** because they could be represented by a square array of dots.

| 1 | 4 | 9 | 16 |

Perfect squares are the product of a number and itself. For example, 16 is a perfect square because $16 = 4 \times 4$. The expression 4×4 can be written using exponents. An **exponent** tells how many times a number, called the **base**, is used as a factor. Numbers that are expressed using exponents are called **powers**. The expression 4×4 can be written as 4^2.

$$base \rightarrow 4^2 \leftarrow exponent$$

Symbols	Words	Meaning
4^1	4 to the first power	4
4^2	4 to the second power or 4 squared	$4 \cdot 4$
4^3	4 to the third power or 4 cubed	$4 \cdot 4 \cdot 4$
4^4	4 to the fourth power	$4 \cdot 4 \cdot 4 \cdot 4$
4^n	4 to the nth power	$\underbrace{4 \cdot 4 \cdot 4 \cdot \ldots \cdot 4}_{n \text{ factors}}$

Examples

Write each expression using exponents.

1 $2 \cdot 2 \cdot 2 \cdot 2 \cdot 2$

The base is 2. It is a factor 5 times.
$2 \cdot 2 \cdot 2 \cdot 2 \cdot 2 = 2^5$

2 $m \cdot m \cdot m \cdot m$

The base is m. It is a factor 4 times.
$m \cdot m \cdot m \cdot m = m^4$

3 7

The base is 7. It is a factor 1 time.
$7 = 7^1$

Reading Algebra
When no exponent is shown, it is understood to be 1. For example, $10 = 10^1$.

Your Turn

a. $4 \cdot 4 \cdot 4 \cdot 4$ **b.** $x \cdot x \cdot x$ **c.** 10

Example **4** Write $(2)(2)(2)(-5)(-5)$ using exponents.

Use the Associative Property to group the factors with like bases.
$(2)(2)(2)(-5)(-5) = [(2)(2)(2)][(-5)(-5)]$
$= (2)^3(-5)^2$

Your Turn

d. Write $(-1)(-1)(-1)(-1)(3)(3)$ using exponents.

You can use the definition of exponent to write a power as a multiplication expression.

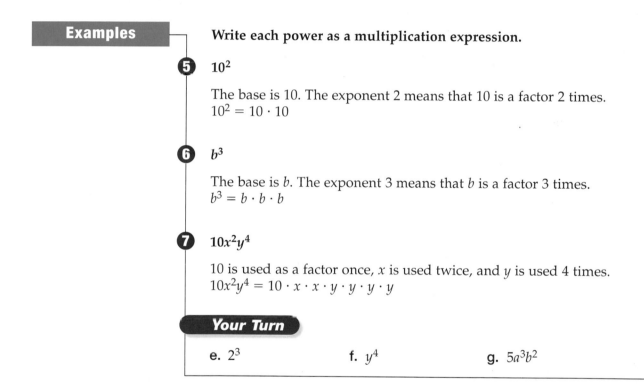

Examples Write each power as a multiplication expression.

5 10^2

The base is 10. The exponent 2 means that 10 is a factor 2 times.
$10^2 = 10 \cdot 10$

6 b^3

The base is b. The exponent 3 means that b is a factor 3 times.
$b^3 = b \cdot b \cdot b$

7 $10x^2y^4$

10 is used as a factor once, x is used twice, and y is used 4 times.
$10x^2y^4 = 10 \cdot x \cdot x \cdot y \cdot y \cdot y \cdot y$

Your Turn

e. 2^3 **f.** y^4 **g.** $5a^3b^2$

You can also use the definition of exponent to evaluate expressions.

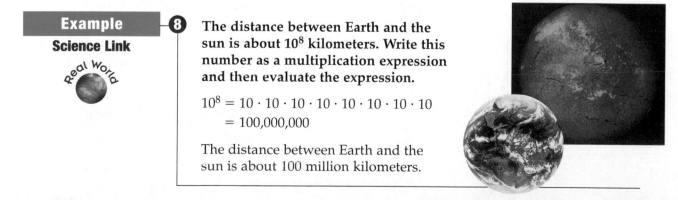

Example **8** The distance between Earth and the
Science Link sun is about 10^8 kilometers. Write this
number as a multiplication expression
and then evaluate the expression.

$10^8 = 10 \cdot 10 \cdot 10 \cdot 10 \cdot 10 \cdot 10 \cdot 10 \cdot 10$
$= 100,000,000$

The distance between Earth and the
sun is about 100 million kilometers.

Exponents are included in the rules for order of operations.

Look Back

Order of
Operations:
Lesson 1–2

Order of Operations	1. Do all operations within grouping symbols first; start with the innermost grouping symbols.
	2. Evaluate all powers in order from left to right.
	3. Next do all multiplications and divisions from left to right.
	4. Then do all additions and subtractions from left to right.

In any expression, an exponent goes with the number or the quantity in parentheses that immediately precedes it.

$2 \cdot 5^3$ means $2 \cdot 5 \cdot 5 \cdot 5$ *The exponent 3 goes with the 5.*

$(2 \cdot 5)^3$ means $(2 \cdot 5)(2 \cdot 5)(2 \cdot 5)$ *The exponent goes with $(2 \cdot 5)$.*

Examples

Evaluate each expression.

9 $4m^3$ if $m = 2$

$4m^3 = 4(2)^3$ *Replace m with 2.*
$\quad\;\; = 4(8)$ *Evaluate the power: $2 \cdot 2 \cdot 2 = 8$.*
$\quad\;\; = 32$ *Multiply.*

10 $3x + y^2$ if $x = -2$ and $y = -3$

$3x + y^2 = 3(-2) + (-3)^2$ *Replace x with -2 and y with -3.*
$\qquad\;\; = 3(-2) + (9)$ *$(-3)^2 = (-3)(-3)$ or 9*
$\qquad\;\; = (-6) + (9)$ *Multiply.*
$\qquad\;\; = 3$ *Add.*

Your Turn

h. $3a^3$ if $a = -2$ **i.** $-5(m + n)^2$ if $m = 4$ and $n = 2$

Squaring a number is related to finding the area of squares. In the following activity, you will investigate perimeters and areas of squares.

Graphing Calculator Exploration

Graphing Calculator Tutorial
See pp. 724–727.

The pattern at the right is made up of perfect squares.

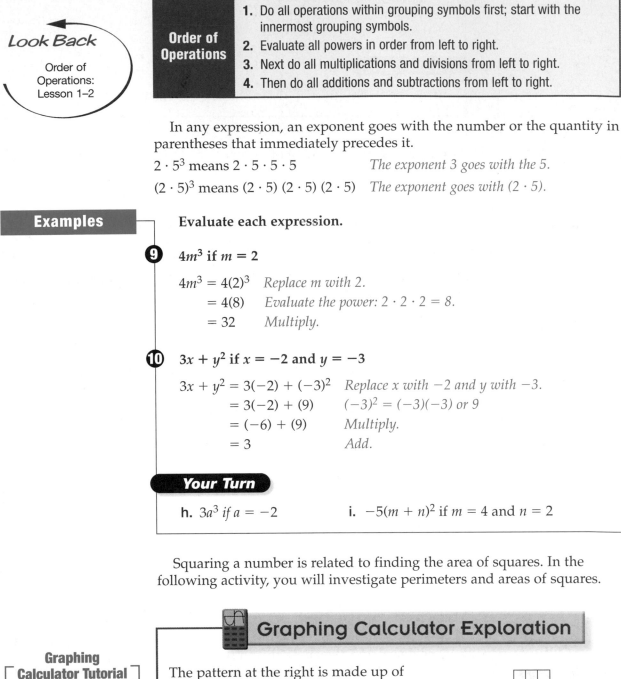

Step 1 Draw the next three squares in the pattern.

Step 2 Find the perimeter and area of each square. Organize your data in a table.

Length of Side	Perimeter	Area

Try These

1. Write an equation that shows the relationship between the length of a side of a square, x, and its perimeter, y.
2. Write an equation that shows the relationship between the length of a side of a square, x, and its area, y.
3. Graph the two functions. Compare and contrast the graphs.
4. When is the value of the perimeter greater than the value of the area? When is the value of the perimeter equal to the area?
5. If the length of a square is doubled, how does its perimeter change? How does its area change?

Check for Understanding

Communicating Mathematics

Study the lesson. Then complete the following.

1. **Write** a definition of *perfect square*.
2. **Explain** what the 2 represents in 10^2.
3. **You Decide** Jan thinks that $(6n)^3$ is equal to $6n^3$. Becky thinks they are not equal. Who is correct? Explain your reasoning.

Vocabulary
perfect squares
exponent
base
powers

Guided Practice

Write each expression using exponents. *(Examples 1–3)*

4. $9 \cdot 9 \cdot 9 \cdot 9$
5. $a \cdot a \cdot a \cdot a \cdot a$
6. 3

Write each power as a multiplication expression. *(Examples 5–7)*

7. 12^4
8. x^5
9. $m^4 n^3$

Evaluate each expression if $a = 3$, $b = -2$, and $c = 4$. *(Examples 8–10)*

10. c^3
11. $2a^4$
12. $3a^2 b$

13. **Number Theory** The prime factorization of 360 is $2 \cdot 2 \cdot 2 \cdot 3 \cdot 3 \cdot 5$. Write the prime factorization using exponents. *(Example 4)*

Exercises

Practice

Write each expression using exponents.

14. $10 \cdot 10 \cdot 10$
15. $(-2)(-2)(-2)(-2)$
16. 6
17. 7 cubed
18. $4 \cdot 4 \cdot 4 \cdot 6 \cdot 6$
19. $2 \cdot 3 \cdot 5 \cdot 2 \cdot 3 \cdot 3$
20. $a \cdot a \cdot a \cdot a \cdot b \cdot b$
21. $3 \cdot x \cdot x \cdot y \cdot y$
22. $(-5)(m)(m)(m)(n)$

Write each power as a multiplication expression.

23. 3^5 **24.** $(-2)^2$ **25.** $2^4 \cdot 3^2$ **26.** $2 \cdot 3^5$

27. y^3 **28.** x^2y^2 **29.** $6ab^4$ **30.** $-2y^4$

Evaluate each expression if $x = -2$, $y = 3$, $z = -1$, and $w = 0.5$.

31. x^5 **32.** $4y^2$ **33.** $-2x^6$ **34.** $x^2 - y^2$

35. $3(y^2 + z)$ **36.** $-2(x^3 + 1)$ **37.** $2w^2$ **38.** wx^3y

39. Find the value of $x^2 + 2x + 1$ if $x = -3$.

40. Which is greater, 2^5 or 5^2?

Applications and Problem Solving

41. Number Theory   The prime factorization of a number is $2 \cdot 3^5$. Find the number.

42. Geometry   Write an expression using exponents that represents the total number of unit cubes in the large cube. Then evaluate the expression.

Exercise 42

43. Landscaping   Landscape architects use the formula $A = \pi r^2$ to find the area of circular flower beds. In the formula, $\pi \approx 3.14$, and r is the radius of the circle.

    a. Estimate the area of a circular flower bed with a radius of 8 feet.

    b. About how many bags of mulch will the landscape architect need if each bag covers about 10 square feet?

44. Critical Thinking   Suppose you raise a negative integer to a positive power. When is the result negative? When is the result positive?

Mixed Review

45. Write an equation of the line that is parallel to the graph of $y = -2x + 3$ and that passes through the point at $(1, 4)$.   *(Lesson 7–7)*

Graph each equation using the slope and y-intercept.   *(Lesson 7–6)*

46. $y = 3x - 2$ **47.** $y = x + 2$ **48.** $-x + 2y = 8$

Solve each problem.   *(Lesson 5–4)*

49. Find 26% of 120.   **50.** 17 is 40% of what number?

51. 9 is what percent of 18?   **52.** 98% of 40 is what number?

53. Standardized Test Practice
A mechanic charges an initial fee of $40 plus $30 for each hour she works. Which equation could be used to find the cost c of a repair job that lasts h hours? *(Lesson 4–5)*

    A $c = 30 + 40h$

    B $c = 40 + 30h$

    C $c = 40 - 30h$

    D $c = 30 - 40h$

Extra Practice   See p. 707.

8-2 Multiplying and Dividing Powers

What You'll Learn

You'll learn to multiply and divide powers.

Why It's Important

Movie Industry
The intensity of sound is measured using *decibels*, a unit that is based on powers of ten. *See page 346.*

Suppose you hear some good news and want to pass it along to two of your friends by e-mail. The next day, each of your friends passes the news along to two of their friends. This pattern continues for several more days.

The table and graph show the number of friends who hear the news on each day. This situation is represented by powers of 2.

Number of Friends

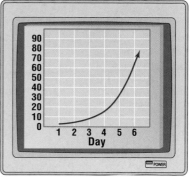

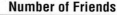

Day	1	2	3	4	5	6
Number	2	4	8	16	32	64
Power of 2	2^1	2^2	2^3	2^4	2^5	2^6

You can use powers of 2 to help find a rule for multiplying powers. What do you notice about the exponents in the following products?

Numerical Products	$4 \cdot 2 = 8$	$4 \cdot 8 = 32$	$8 \cdot 8 = 64$
Powers	$2^2 \cdot 2^1 = 2^3$	$2^2 \cdot 2^3 = 2^5$	$2^3 \cdot 2^3 = 2^6$

These examples suggest that you can multiply powers with the same base by adding the exponents. Think about $a^2 \cdot a^3$.

$a^2 \cdot a^3 = (a \cdot a)(a \cdot a \cdot a)$ *a^2 has two factors; a^3 has three factors.*

$\quad = a \cdot a \cdot a \cdot a \cdot a$ *Substitution Property*

$\quad = a^5$ *The product has $2 + 3$ or 5 factors.*

Product of Powers	**Words:**	You can multiply powers with the same base by adding the exponents.
	Numbers:	$3^3 \cdot 3^2 = 3^{3 + 2}$ or 3^5
	Symbols:	$a^m \cdot a^n = a^{m + n}$

Simplify each expression.

1 $4^3 \cdot 4^5$

$4^3 \cdot 4^5 = 4^{3+5}$ *To multiply powers that have the same base, write*
$ = 4^8$ *the common base, then add the exponents.*

2 $x^3 \cdot x^4$

$x^3 \cdot x^4 = x^{3+4}$ *To multiply powers that have the same base, write*
$ = x^7$ *the common base, then add the exponents.*

> ### Reading Algebra
>
> An expression is *simplified* when each base appears only once and all of the fractions are in simplest form.

3 $(4y^2)(3y)$

$(4y^2)(3y) = (4 \cdot 3)(y^2 \cdot y)$ *Use the Commutative and Associative Properties.*
$ = 12y^{2+1}$ $y = y^1$
$ = 12y^3$

4 $(a^3b^2)(a^2b^4)$

$(a^3b^2)(a^2b^4) = (a^3 \cdot a^2)(b^2 \cdot b^4)$
$ = a^{3+2} \cdot b^{2+4}$
$ = a^5b^6$

Your Turn

a. $10^4 \cdot 10^2$ **b.** $y^4 \cdot y^2$ **c.** $(-3x^2)(5x)$ **d.** $(x^5y^2)(x^4y^6)$

You can use powers of 2 to help find a rule for dividing powers. Study each quotient. What do you notice about the exponents?

Numerical Quotients	$16 \div 8 = 2$	$32 \div 4 = 8$	$64 \div 2 = 32$
Powers	$2^4 \div 2^3 = 2^1$	$2^5 \div 2^2 = 2^3$	$2^6 \div 2^1 = 2^5$

These examples suggest that you can divide powers with the same base by subtracting the exponents. Think about $a^5 \div a^2$. Remember that you can write a division expression as a fraction.

$\dfrac{a^5}{a^2} = \dfrac{a \cdot a \cdot a \cdot a \cdot a}{a \cdot a}$ a^5 *has five factors;* a^2 *has two factors.*

$\phantom{\dfrac{a^5}{a^2}} = \dfrac{\overset{1}{\cancel{a}} \cdot \overset{1}{\cancel{a}} \cdot a \cdot a \cdot a}{\underset{1}{\cancel{a}} \cdot \underset{1}{\cancel{a}}}$ *Notice that* $\dfrac{a \cdot a}{a \cdot a} = 1.$

$\phantom{\dfrac{a^5}{a^2}} = a \cdot a \cdot a$ *The quotient has* $5 - 2$ *or 3 factors.*

$\phantom{\dfrac{a^5}{a^2}} = a^3$

	Words:	You can divide powers with the same base by subtracting the exponents.
Quotient of Powers	**Numbers:**	$\dfrac{5^7}{5^4} = 5^{7-4}$ or 5^3
	Symbols:	$\dfrac{a^m}{a^n} = a^{m-n}$ *The value of a cannot be zero.*

Examples

Simplify each expression.

5 $\dfrac{4^3}{4^2}$

$\dfrac{4^3}{4^2} = 4^{3-2}$ *To divide powers that have the same base, write the common base. Then subtract the exponents.*

$= 4^1$ or 4

6 $\dfrac{x^6}{x^4}$

$\dfrac{x^6}{x^4} = x^{6-4}$ *Write the common base. Then subtract the exponents.*

$= x^2$

7 $\dfrac{8m^4n^5}{2m^3n^2}$

$\dfrac{8m^4n^5}{2m^3n^2} = \left(\dfrac{8}{2}\right)\left(\dfrac{m^4}{m^3}\right)\left(\dfrac{n^5}{n^2}\right)$ *Group the powers that have the same base.*

$= 4m^{4-3}n^{5-2}$ $\quad 8 \div 2 = 4$

$= 4m^1n^3$

$= 4mn^3$

Your Turn

e. $\dfrac{10^5}{10^2}$ **f.** $\dfrac{y^5}{y^4}$ **g.** $\dfrac{u^4b^3}{ab^2}$ **h.** $\dfrac{-30m^5n^2}{10m^3n}$

A special case results when you divide a power by itself. Consider the following two ways to simplify $\dfrac{b^4}{b^4}$, where $b \neq 0$.

Method 1   Definition of Power

$\dfrac{b^4}{b^4} = \dfrac{\overset{1}{\cancel{b}} \cdot \overset{1}{\cancel{b}} \cdot \overset{1}{\cancel{b}} \cdot \overset{1}{\cancel{b}}}{\underset{1}{\cancel{b}} \cdot \underset{1}{\cancel{b}} \cdot \underset{1}{\cancel{b}} \cdot \underset{1}{\cancel{b}}}$

$= 1$

Method 2   Quotient of Powers

$\dfrac{b^4}{b^4} = b^{4-4}$

$= b^0$

Since $\dfrac{b^4}{b^4}$ cannot have two different values, you can conclude that $b^0 = 1$. In general, any nonzero number raised to the zero power is equal to 1.

8 Simplify $\dfrac{x^4y^2}{xy^2}$.

$$\dfrac{x^4y^2}{xy^2} = \left(\dfrac{x^4}{x^1}\right)\left(\dfrac{y^2}{y^2}\right)$$

$$= x^{4-1}y^{2-2}$$

$$= x^3y^0 \qquad y^0 = 1$$

$$= x^3 \cdot 1 \text{ or } x^3$$

Your Turn

i. $\dfrac{a^3b^4}{a^3b}$ j. $\dfrac{10x^4y^3}{5x^4y^2}$ k. $\dfrac{m^{10}n^5}{m^{10}n^5}$

Check for Understanding

Communicating Mathematics

Study the lesson. Then complete the following.

1. **Explain** why $a^4 \cdot a^7$ can be simplified but $a^4 \cdot b^7$ cannot.
2. **You Decide** Tonia says $10^3 \times 10^2 = 100^5$, but Emilio says $10^3 \times 10^2 = 10^5$. Who is correct? Explain your reasoning.

Guided Practice

Simplify each expression.

3. $y^2 \cdot y^5$ 4. $m^5(m)$ 5. $(t^2)(t^2)(t)$ *(Examples 1 & 2)*

6. $(x^3y)(xy^3)$ 7. $(3a^2)(4a^3)$ 8. $(-5x^3)(4x^4)$ *(Examples 3 & 4)*

9. $\dfrac{n^8}{n^5}$ 10. $\dfrac{b^6c^5}{b^3c^2}$ 11. $\dfrac{xy^7}{y^4}$ *(Examples 5 & 6)*

12. $\dfrac{12x^5}{4x^4}$ 13. $\dfrac{ab^5c}{ac}$ 14. $\dfrac{22a^4b^5c^7}{-11abc^2}$ *(Examples 7 & 8)*

15. **Measurement** There are 10^1 millimeters in 1 centimeter and 10^2 centimeters in 1 meter. How many millimeters are in 1 meter? Write your answer as a power and then evaluate the expression. *(Example 1)*

Exercises • • • • • • • • • • • • • • • • • • •

Practice

Simplify each expression.

16. $2^6 \cdot 2^8$ 17. $5^3 \cdot 5$ 18. $y^7 \cdot y^7$ 19. $d \cdot d^5$

20. $(b^4)(b^2)$ 21. $(a^2b)(ab^4)$ 22. $(m^3n)(mn^2)$ 23. $(r^3t^4)(r^4t^4)$

24. $(2xy)(3x^2y^2)$ 25. $(-5a)(-3a)$ 26. $(-10x^3y)(2x^2)$ 27. $(4x^2y^3)(2xy^2)$

28. $(-8x^3y)(2x^4)$ 29. $m^4(m^3b^2)$ 30. $(ab)(ac)(bc)$ 31. $(m^2n)(am)(an^2)$

32. $\dfrac{10^9}{10^3}$ 33. $\dfrac{9^6}{9^5}$ 34. $\dfrac{w^7}{w^3}$ 35. $\dfrac{k^9}{k^4}$

36. $\dfrac{a^4b^8}{ab^2}$ **37.** $\dfrac{24a^3b^6}{-2a^2b^2}$ **38.** $\dfrac{24x^2y^7z^3}{-6x^2y^3z}$ **39.** $\dfrac{-40mn^2}{-10mn^2}$

40. $\dfrac{3}{4}a(12b^2)$ **41.** $\dfrac{5}{6}c(18a^3)$ **42.** $\left(\dfrac{1}{2}a^2\right)\!\left(6ab^2\right)$ **43.** $x^0(2x^3)$

44. Evaluate 5^0.

45. Find the product of $2x$ and $-8x$.

Applications and Problem Solving

46. Geometry The measure of the length of a rectangle is $5x$ and the measure of the width is $3x$. Find the measure of the area.

47. Earth Science At a distance of 10^7 meters from Earth, a satellite can see almost all of our planet. At a distance of 10^{13} meters, a satellite can see all of our solar system. How many times as great is a distance of 10^{13} meters than 10^7 meters?

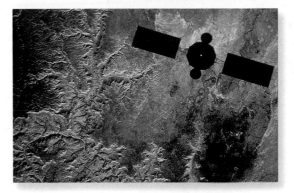

48. Manufacturing The Pizza Parlor uses square boxes to package their pizzas. The drawing shows that a pizza with radius r just fits inside the box. Write an expression for the area of the bottom of the box.

49. Critical Thinking Study the following pattern.

$$(5^2)^3 = (5^2)(5^2)(5^2) \text{ or } 5^6$$

Simplify each expression.

  a. $(10^3)^4$ **b.** $(4^5)^3$ **c.** $(x^2)^4$

  d. Write a rule for finding the *power of a power*.

Mixed Review

Evaluate each expression if $x = -1$, $y = 2$, and $z = -3$. *(Lesson 8–1)*

50. z^3 **51.** $3x^4$ **52.** $5xy^3z$ **53.** $3(x^2 + y^2)$

54. Write an equation of the line that is perpendicular to the graph of $y = 3x + 5$ that passes through the point at $(0, 0)$. *(Lesson 7–7)*

Solve each proportion. *(Lesson 5–1)*

55. $\dfrac{96}{6} = \dfrac{152}{x}$ **56.** $\dfrac{9}{m} = \dfrac{15}{10}$ **57.** $\dfrac{8.6}{25.8} = \dfrac{1}{n}$ **58.** $\dfrac{3}{7} = \dfrac{2.1}{d}$

59. Standardized Test Practice Choose the expression that has a value of 28. *(Lesson 1–2)*

  A $4 + 3 \cdot 4$ **B** $75 \div 3 - 2$ **C** $8 \cdot 4 - 8$ **D** $(5 + 3) \cdot 7 \div 2$

Extra Practice   See p. 708.

Broadcast Technician

Do you dream of being in the movies? If you don't make it onto the big screen, you may find a career behind the scenes as a sound mixer. Sound mixers are broadcast technicians who develop movie sound tracks. Using a process called *dubbing*, they sit at sound consoles and fade in and fade out each sound by regulating its volume. All of the sounds for each scene are blended on a master sound track.

Sound intensity is measured in *decibels*. The decibel scale is based on powers of ten. The softest audible sound is represented by 10^0. The chart lists several common sounds and their intensity as compared to the softest audible sound.

Sound	Decibels	Intensity
jet airplane	140	10^{14}
rock band	120	10^{12}
motorcycle	110	10^{11}
circular saw	100	10^{10}
busy traffic	80	10^8
vacuum cleaner	70	10^7
noisy office	60	10^6
talking	40	10^4
whispering	20	10^2
breathing	10	10^1
softest sound	0	10^0

Find how many times as intense the first sound is as the second.

1. vacuum cleaner, noisy office
2. motorcycle, busy traffic
3. rock band, talking
4. jet airplane, whispering

5. Find two sounds, one of which is 10^2 times as intense as the second.

FAST FACTS   About Broadcast Technicians

Working Conditions
- usually work indoors in pleasant conditions
- work a 40-hour week, but some overtime required to meet deadlines
- evening, weekend, and holiday work

Education
- high school math, physics, and electronics
- postsecondary training in engineering or electronics at a technical school or community college

Job Outlook

Expected Growth in Employment

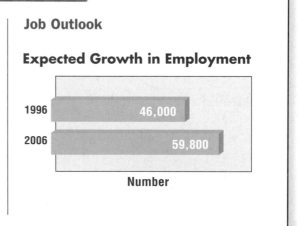

| 1996 | 46,000 |
| 2006 | 59,800 |

Number

***inter*NET**
CONNECTION

Career Data For up-to-date information about a career as a broadcast technician, visit: www.algconcepts.glencoe.com

Math In the Workplace

What You'll Learn
You'll learn to simplify expressions containing negative exponents.

Why It's Important
Electronics Electric current is measured using the prefixes *micro*, which is 10^{-6}, and *milli*, which is 10^{-3}. *See Exercise 14.*

When you look at a rainbow, you are seeing only part of the electromagnetic spectrum. When you listen to the radio or have an X ray, you are also using part of this spectrum. The figure below shows the electromagnetic wave spectrum.

Wavelength (centimeters)

$10^{-10}10^{-9}10^{-8}10^{-7}10^{-6}10^{-5}$ | $10^{-4}10^{-3}10^{-2}10^{-1}1$ | 10^1 10^2 10^3 10^4 10^5 10^6 10^7 10^8 10^9 10^{10}

Gamma rays | X rays | Ultra-violet | Infrared | Radar microwaves | TV FM | Short-waves | C B | Longwaves

Visible light waves

Radio waves

Violet ▮▮▮▮▮ Red

Notice that the wavelengths are written as powers of ten. However, some of the exponents are negative integers. Study the pattern at the right to find the value of 10^{-1} and 10^{-2}. Extending the pattern suggests that $10^{-1} = \frac{1}{10}$ and $10^{-2} = \frac{1}{10^2}$ or $\frac{1}{100}$.

$$
\begin{aligned}
10^3 &= 1000 \\
10^2 &= 100 \\
10^1 &= 10 \\
10^0 &= 1 \\
10^{-1} &= ? \\
10^{-2} &= ?
\end{aligned}
\quad
\begin{aligned}
&\div 10 \\
&\div 10 \\
&\div 10 \\
&\div 10 \\
&\div 10
\end{aligned}
$$

You can use the Quotient of Powers rule and the definition of power to simplify the expression $\frac{x^3}{x^5}$ and write a definition of negative exponents.

Method 1 Quotient of Powers	**Method 2 Definition of Power**
$\frac{x^3}{x^5} = x^{3-5}$ $= x^{-2}$	$\frac{x^3}{x^5} = \frac{\overset{1}{\cancel{x}} \cdot \overset{1}{\cancel{x}} \cdot \overset{1}{\cancel{x}}}{\underset{1}{\cancel{x}} \cdot \underset{1}{\cancel{x}} \cdot \underset{1}{\cancel{x}} \cdot x \cdot x}$ $= \frac{1}{x \cdot x}$ or $\frac{1}{x^2}$

You can conclude that x^{-2} and $\frac{1}{x^2}$ are equal because $\frac{x^3}{x^5}$ cannot have two different values. This and other examples suggest the following definition.

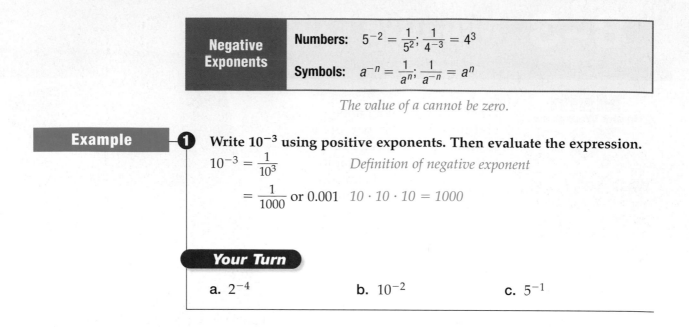

Negative Exponents	Numbers: $5^{-2} = \dfrac{1}{5^2}$; $\dfrac{1}{4^{-3}} = 4^3$
	Symbols: $a^{-n} = \dfrac{1}{a^n}$; $\dfrac{1}{a^{-n}} = a^n$

The value of a cannot be zero.

Example ❶ **Write 10^{-3} using positive exponents. Then evaluate the expression.**

$$10^{-3} = \frac{1}{10^3} \qquad \textit{Definition of negative exponent}$$

$$= \frac{1}{1000} \text{ or } 0.001 \quad \textit{10 · 10 · 10 = 1000}$$

Your Turn

   a. 2^{-4} **b.** 10^{-2} **c.** 5^{-1}

To simplify an expression with a negative exponent, write an equivalent expression that has positive exponents. Also, each base should appear only once and all fractions should be in simplest form.

Examples **Simplify each expression.**

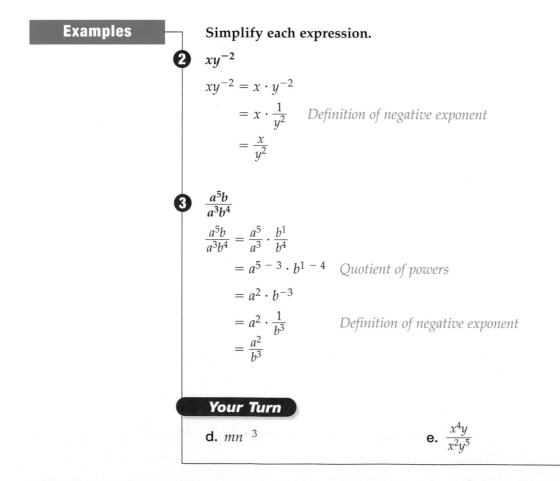

❷ xy^{-2}

$$xy^{-2} = x \cdot y^{-2}$$

$$= x \cdot \frac{1}{y^2} \qquad \textit{Definition of negative exponent}$$

$$= \frac{x}{y^2}$$

❸ $\dfrac{a^5 b}{a^3 b^4}$

$$\frac{a^5 b}{a^3 b^4} = \frac{a^5}{a^3} \cdot \frac{b^1}{b^4}$$

$$= a^{5-3} \cdot b^{1-4} \qquad \textit{Quotient of powers}$$

$$= a^2 \cdot b^{-3}$$

$$= a^2 \cdot \frac{1}{b^3} \qquad \textit{Definition of negative exponent}$$

$$= \frac{a^2}{b^3}$$

Your Turn

   d. mn^{-3} **e.** $\dfrac{x^4 y}{x^2 y^5}$

4 Simplify $\dfrac{-6r^3s^5}{18r^{-7}s^5t^{-2}}$.

$$\dfrac{-6r^3s^5}{18r^{-7}s^5t^{-2}} = \left(\dfrac{-6}{18}\right)\left(\dfrac{r^3}{r^{-7}}\right)\left(\dfrac{s^5}{s^5}\right)\left(\dfrac{1}{t^{-2}}\right)$$

$$= \left(\dfrac{-1}{3}\right)\left(\dfrac{r^3}{r^{-7}}\right)\left(\dfrac{s^5}{s^5}\right)\left(\dfrac{t^0}{t^{-2}}\right) \qquad \dfrac{-6}{18} = -\dfrac{1}{3}, 1 = t^0$$

$$= -\dfrac{1}{3}r^{3-(-7)}s^{5-5}t^{0-(-2)} \qquad \textit{Quotient of powers}$$

$$= -\dfrac{1}{3}r^{10}s^0t^2 \qquad 3-(-7)=10, 0-(-2)=2$$

$$= -\dfrac{r^{10}t^2}{3} \qquad s^0 = 1$$

Your Turn

f. $\dfrac{5a^4b^6}{-25a^{-2}b^6c^{-3}}$

g. $\dfrac{8x^3y^5}{10x^{-3}y^5z}$

Biology Link

Real World

5 The *E. coli* bacteria has a width of 10^{-3} millimeter. The head of a pin has a diameter of 1 millimeter. How many *E. coli* bacteria could fit across the head of a pin?

10^{-3} mm

Photo Graphic

To find how many bacteria, divide 1 by 10^{-3}.

$$\dfrac{1}{10^{-3}} = \dfrac{10^0}{10^{-3}} \qquad 1 = 10^0$$

$$= 10^{0-(-3)}$$

$$= 10^3$$

Since $10^3 = 1000$, about 1000 bacteria could fit across the head of a pin.

Your Turn

h. Refer to the figure on page 347. An ultraviolet wave has a length of 10^{-5} centimeter. An FM radio wave has a length of 10^2 centimeters. How many times as long is the FM wave as the ultraviolet wave?

Check for Understanding

Communicating Mathematics

Study the lesson. Then complete the following.

1. Express 5^{-2} using positive integers. Then evaluate the expression.
2. Evaluate $6t^{-2}$ if $t = 3$.

3. **You Decide?** Booker and Antonio each correctly simplified $a^{-2} \cdot a^3$.

Booker's Method

$a^{-2} \cdot a^3 = a^{-2 + 3}$

$= a^1$ or a

Antonio's Method

$a^{-2} \cdot a^3 = \dfrac{1}{a^2} \cdot a^3$

$= \dfrac{a^3}{a^2}$ or a

a. Which student used the Product of Powers rule?

b. Which student used the definition of negative exponents?

c. Whose method do you prefer? Explain your reasoning.

Guided Practice

Write each expression using positive exponents. Then evaluate the expression. *(Example 1)*

4. 10^{-4} 　　　　　　　　　　　　　**5.** 3^{-3}

Simplify each expression.

6. z^{-1} 　　**7.** n^{-3} 　　**8.** $s^{-2}t^3$ 　　**9.** $p^{-1}q^{-2}r^2$ *(Example 2)*

10. $\dfrac{x^2}{x^3}$ 　　**11.** $\dfrac{a^3}{a^{-4}}$ 　　**12.** $\dfrac{m^4n^{-2}}{m^6n^2}$ 　　**13.** $\dfrac{3a^3bc^5}{27a^4bc^2}$ *(Examples 3–5)*

14. **Electronics** Electric current can be measured in amperes, milliamperes, or microamperes. The prefixes *milli* and *micro* mean 10^{-3} and 10^{-6}, respectively. Express 10^{-3} and 10^{-6} using positive exponents. *(Example 1)*

Exercises · · · · · · · · · · · · · · · · ·

Practice

Write each expression using positive exponents. Then evaluate the expression.

15. 2^{-5} 　　**16.** 10^{-5} 　　**17.** 4^{-1} 　　**18.** 6^{-2}

Simplify each expression.

19. r^{-10} 　　**20.** p^{-6} 　　**21.** $a^4(a^{-2})$ 　　**22.** $x^{-7}(x^4)$

23. $s^{-2}t^4$ 　　**24.** $a^0b^{-1}c^{-2}$ 　　**25.** $15rs^{-2}$ 　　**26.** $10x^{-4}y^{-5}z$

27. $\dfrac{m^2}{m^{-4}}$ 　　**28.** $\dfrac{x^2}{x^3}$ 　　**29.** $\dfrac{k^{-2}}{k^6}$ 　　**30.** $\dfrac{1}{r^{-3}}$

31. $\dfrac{an^3}{n^5}$ 　　**32.** $\dfrac{bm^2}{m^6}$ 　　**33.** $\dfrac{x^3y^{-3}}{x^3y^6}$ 　　**34.** $\dfrac{a^5b^{-3}}{a^7b^3}$

35. $\dfrac{12b^5}{4b^{-4}}$ 　　**36.** $\dfrac{24c^6}{4c^{-2}}$ 　　**37.** $\dfrac{7x^4}{28x}$ 　　**38.** $\dfrac{20y^5}{40y^{-2}}$

39. $\dfrac{4x^3}{28x}$ 　　**40.** $\dfrac{-15r^5s^8}{5r^5s^2}$ 　　**41.** $\dfrac{5ac}{8ab^5c^2}$ 　　**42.** $\dfrac{12c^3d^4f^6}{60cd^6f^3}$

43. Evaluate $4x^{-3}y^2$ if $x = 2$ and $y = 6$.

44. Find the value of $(2b)^{-3}$ if $b = -2$.

45. Which is greater, 2^{-4} or 2^{-6}?

Applications and Problem Solving

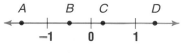

46. Electronics Visible light waves have wavelengths between 10^{-5} centimeter and 10^{-4} centimeter. Express 10^{-5} and 10^{-4} using positive exponents. Then evaluate each expression.

47. Physics Refer to the figure on page 347. Some infrared waves have lengths of 10^{-3} centimeter. Which kind of wave has a length that is 1000 times as long as an infrared wave?

48. Critical Thinking Which point on the number line could be the graph of n^{-2} if n is a positive integer?

$$A \quad\quad B \quad C \quad\quad D$$
$$-1 \quad\; 0 \quad\; 1$$

Mixed Review

Simplify each expression. *(Lesson 8–2)*

49. $(5a^3)(-2a^4)$ **50.** $x^4 \cdot x$ **51.** $\dfrac{n^{10}}{n^4}$ **52.** $\dfrac{15a^3b^6c^2}{-5ab^2c^2}$

Write each expression using exponents. *(Lesson 8–1)*

53. $z \cdot z \cdot z \cdot z$ **54.** $(3)(3)(-2)(-2)(-2)$ **55.** 9

56. Determine the slope of the line passing through points at $(1, -2)$ and $(6, 2)$. *(Lesson 7–1)*

57. Standardized Test Practice The Jaguars softball team played 8 games and scored a total of 96 runs. What was the mean number of runs scored per game? *(Lesson 3–3)*

A 8 **B** 12 **C** 88 **D** 104

Quiz 1   Lessons 8–1 through 8–3

▶ **Write each expression using exponents.** *(Lesson 8–1)*

1. $6 \cdot 6 \cdot 6$ **2.** $x \cdot x \cdot x \cdot x$ **3.** 10 **4.** $(-3)(-3)(-3)$

5. Number Theory The prime factorization of 96 is $2 \cdot 2 \cdot 2 \cdot 2 \cdot 2 \cdot 3$. Write the prime factorization using exponents. *(Lesson 8–1)*

Simplify each expression. *(Lessons 8–2 & 8–3)*

6. $(m^4)(m^6)$ **7.** $\dfrac{x^7}{x^2}$ **8.** $(-3x^2y^3)(-2x^4y^{-1})$ **9.** $\dfrac{3a^2b}{-9a^2b^4}$

10. Biology The length of a *Euglena* protist is about 10^{-2} centimeter. Express 10^{-2} using positive exponents. Then evaluate the expression.

What You'll Learn

You'll learn to express numbers in scientific notation.

Why It's Important

Fiber Optics Laser technicians use scientific notation when they deal with the speed of light, 3×10^8 meters per second.
See Example 7.

One of the hardest things about buying a personal computer may be understanding the abbreviations. For example, a personal computer might have 64 MB of memory, a 2 GB hard drive, and a modem that transmits 56 Kbps.

64 MB → 64 megabytes
2 GB → 2 gigabytes
56 Kbps → 56 kilobytes
                 per second

The prefixes *mega*, *giga*, and *kilo* are metric prefixes. They are used with very large measures. Other prefixes are used with very small measures. The chart shows some metric prefixes.

Prefix	Power of 10	Meaning	Prefix	Power of 10	Meaning
tera	10^{12}	1,000,000,000,000	pico	10^{-12}	0.000000000001
giga	10^{9}	1,000,000,000	nano	10^{-9}	0.000000001
mega	10^{6}	1,000,000	micro	10^{-6}	0.000001
kilo	10^{3}	1000	milli	10^{-3}	0.001

Metric units are used in scientific fields and in industry because calculations are easier with powers of ten. When you multiply a number by a power of ten, the nonzero digits in the original number and the product are the same. The difference is in the position of the decimal point.

$$5 \times 10^2 = 5 \times 100$$
$$= 500 \quad \textit{2 places right}$$

$$4 \times 10^{-1} = 4 \times 0.1$$
$$= 0.4 \quad \textit{1 place left}$$

$$8.23 \times 10^4 = 8.23 \times 10,000$$
$$= 82,300 \quad \textit{4 places right}$$

$$1.23 \times 10^{-3} = 1.23 \times 0.001$$
$$= 0.00123 \quad \textit{3 places left}$$

These and similar examples suggest the following rules for multiplying a number by a power of ten.

Multiplying by Powers of 10	• If the exponent is *positive*, you move the decimal point to the *right*. • If the exponent is *negative*, you move the decimal point to the *left*.

Express each measurement in standard form.

1 **2 megabytes**

2 megabytes = 2×10^6 bytes   *The prefix mega- means 10^6.*

= 2,000,000 bytes   *Move the decimal point 6 places right.*

2 **3.6 nanoseconds**

3.6 nanoseconds = 3.6×10^{-9} seconds   *The prefix nano- means 10^{-9}.*

= 0.0000000036 seconds

Move the decimal point 9 places left.

> **Reading Algebra**
>
> The absolute value of the exponent tells the number of places to move the decimal point.

Your Turn

a. 2 gigabytes   **b.** 3.4 milliseconds

When you deal with very large numbers like 5,800,000 or very small numbers like 0.000076, it is difficult to keep track of the place value. Numbers such as these can be written in **scientific notation**.

Scientific Notation	A number is expressed in scientific notation when it is in the form $a \times 10^n$, where $1 \le a < 10$ and n is an integer.

Follow these steps to write a number in scientific notation.
- First, place the decimal point after the first nonzero digit.
- Then, find the power of ten by counting the decimal places. When the number is greater than one, the exponent of 10 is *positive*. When the number is between zero and one, the exponent of 10 is *negative*.

Examples

Express each number in scientific notation.

3 **5,800,000**

5,800,000 = $5.8 \times 10^?$ *The decimal point moves 6 places.*

= 5.8×10^6 *Since 5,800,000 is greater than one, the exponent is positive.*

4 **0.000076**

0.000076 = $7.6 \times 10^?$ *The decimal point moves 5 places.*

= 7.6×10^{-5} *Since 0.000076 is between zero and one, the exponent is negative.*

Your Turn

c. 3,900,000,000   **d.** 0.0000035

You can use scientific notation to simplify computation.

Evaluate each expression.

5 $400 \times 2{,}000{,}000{,}000$

First express each number in scientific notation. Then use the Associative and Commutative Properties to regroup terms.

$$400 \times 2{,}000{,}000{,}000 = (4 \times 10^2)(2 \times 10^9)$$
$$= (4 \times 2)(10^2 \times 10^9) \quad \textit{Associative and Commutative}$$
$$= 8 \times 10^{11} \qquad\qquad\qquad \textit{Properties}$$
$$= 800{,}000{,}000{,}000$$

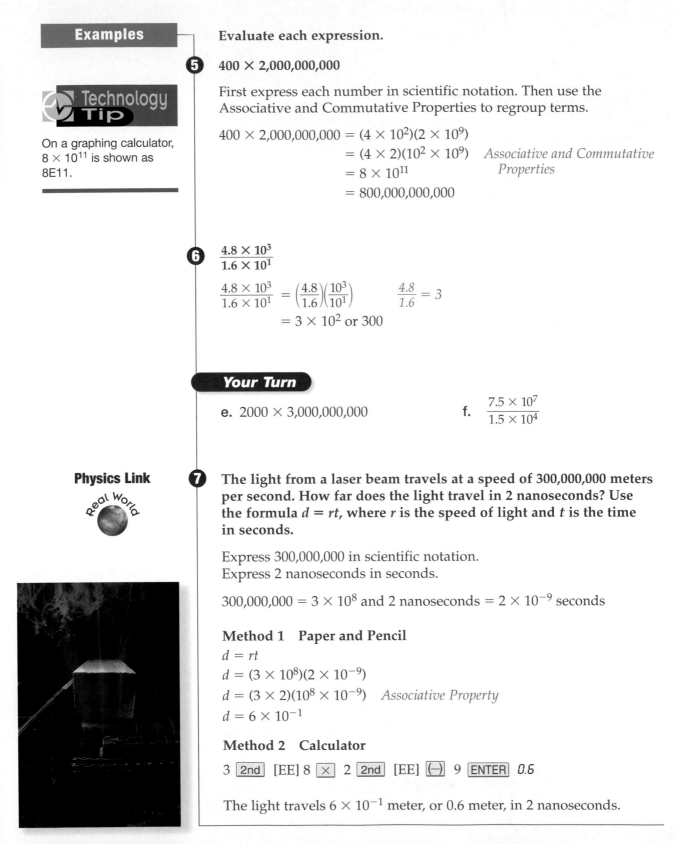

Technology Tip

On a graphing calculator, 8×10^{11} is shown as 8E11.

6 $\dfrac{4.8 \times 10^3}{1.6 \times 10^1}$

$$\frac{4.8 \times 10^3}{1.6 \times 10^1} = \left(\frac{4.8}{1.6}\right)\left(\frac{10^3}{10^1}\right) \qquad \frac{4.8}{1.6} = 3$$
$$= 3 \times 10^2 \text{ or } 300$$

Your Turn

e. $2000 \times 3{,}000{,}000{,}000$

f. $\dfrac{7.5 \times 10^7}{1.5 \times 10^4}$

Physics Link

Real World

7 **The light from a laser beam travels at a speed of 300,000,000 meters per second. How far does the light travel in 2 nanoseconds? Use the formula $d = rt$, where r is the speed of light and t is the time in seconds.**

Express 300,000,000 in scientific notation.
Express 2 nanoseconds in seconds.

$300{,}000{,}000 = 3 \times 10^8$ and $2 \text{ nanoseconds} = 2 \times 10^{-9}$ seconds

Method 1   Paper and Pencil
$d = rt$
$d = (3 \times 10^8)(2 \times 10^{-9})$
$d = (3 \times 2)(10^8 \times 10^{-9}) \quad \textit{Associative Property}$
$d = 6 \times 10^{-1}$

Method 2   Calculator

3 [2nd] [EE] 8 [×] 2 [2nd] [EE] [(−)] 9 [ENTER] *0.6*

The light travels 6×10^{-1} meter, or 0.6 meter, in 2 nanoseconds.

Communicating Mathematics

Study the lesson. Then complete the following.

1. **Tell** whether 23.5×10^3 is expressed in scientific notation. Explain your answer.

2. **Explain** an advantage of expressing very large or very small numbers in scientific notation.

Math Journal

3. **Find** two very large numbers and two very small numbers in a newspaper. Write each number in standard form and in scientific notation.

Guided Practice

⊕ Getting Ready Find each product.

Sample 1: 2×10^3
Solution: 2000 *three places right*

Sample 2: 3.4×10^{-1}
Solution: 0.34 *one place left*

4. 2.45×10^2 5. 6.8×10^1 6. 2×10^6
7. 6.4×10^{-1} 8. 9.23×10^{-2} 9. 3×10^{-4}

Express each measure in standard form. *(Examples 1 & 2)*

10. 5 megaohms

11. 6.5 milliamperes

Express each number in scientific notation. *(Examples 3 & 4)*

12. 9500   13. 56.9   14. 0.0087   15. 0.000023

Evaluate each expression. Express each result in scientific notation and standard form. *(Examples 5–7)*

16. $(2 \times 10^5)(3 \times 10^{-8})$ 17. $\dfrac{5.2 \times 10^5}{2 \times 10^2}$

18. **Health** The length of the virus that causes AIDS is 0.00011 millimeter. Express 0.00011 in scientific notation. *(Example 4)*

Exercises

Practice

Express each measure in standard form.

19. 5.8 billion dollars   20. 4 megahertz   21. 3.9 nanoseconds
22. 82 kilobytes   23. 9 milliamperes   24. 2.3 micrograms

Express each number in scientific notation.

25. 5280   26. 240,000   27. 268.3   28. 25,000,000
29. 0.00032   30. 0.08   31. 0.004296   32. 15.9
33. 0.012   34. 1,000,000   35. 0.000000022   36. 0.0000946

Evaluate each expression. Express each result in scientific notation and standard form.

37. $(3 \times 10^2)(2 \times 10^4)$ 38. $(4 \times 10^2)(1.5 \times 10^6)$
39. $(3 \times 10^{-2})(2.5 \times 10^4)$ 40. $(7.8 \times 10^{-6})(1 \times 10^{-2})$

Evaluate each expression. Express each result in scientific notation and standard form.

41. $\dfrac{6.4 \times 10^5}{3.2 \times 10^2}$

42. $\dfrac{8 \times 10^2}{2 \times 10^{-3}}$

43. $\dfrac{13.2 \times 10^{-6}}{2.4 \times 10^{-1}}$

44. Find the product of (1.2×10^5) and (5×10^{-4}) mentally.

45. Express 5×10^6 in standard form.

Applications and Problem Solving

Real World

Orchid

46. **Biology**   The mass of an orchid seed is 3.5×10^{-6} grams. Express 3.5×10^{-6} in standard form.

47. **Astronomy**   The diameter of Venus is 1.218×10^4 km, the diameter of Earth is 1.276×10^4 km, and the diameter of Mars is 6.76×10^3 km. Rank the planets in order from greatest to least diameter.

48. **Electronics**   Engineering notation is similar to scientific notation. However, in engineering notation, the powers of ten are always multiples of 3, such as 10^3, 10^6, 10^{-9}, and 10^{-12}. For example, 240,000 is expressed as 240×10^3. Express 15,000 ohms in scientific and engineering notation.

49. **Health**   Laboratory technicians look at bacteria through microscopes. A microscope set on 1000× makes an organism appear to be 1000 times larger than its actual size. Most bacteria are between 3×10^{-4} and 2×10^{-3} millimeter in diameter. How large would the bacteria appear under a microscope set on 1000×?

50. **Critical Thinking**   Express each number in scientific notation.
    a. 32×10^5 b. 284×10^3 c. 0.76×10^{-2} d. 0.09×10^{-3}

Mixed Review

Simplify each expression.   *(Lessons 8–2, 8–3)*

51. $y^5(y^{-2})$

52. $15a^{-1}b^3c^{-2}$

53. $\dfrac{30c^4}{-5c^{-2}}$

54. $\dfrac{5r^2s^7}{25r^2s^{10}}$

55. $x^2 \cdot x^3 \cdot x$

56. $\dfrac{2}{3}r(15s^2)$

57. $\dfrac{a^4b^5}{ab^2}$

58. $(-15y^2z)(-2y^2z^0)$

59. $(3x^5y^2)(-2x^{-3}y^4)$

60. **Business**   Julia's wages vary directly as the number of hours she works. If her wages for 5 hours are $34.75, how much will they be for 30 hours?   *(Lesson 6–5)*

61. **Standardized Test Practice**   Choose the graph that represents a function.   *(Lesson 6–4)*

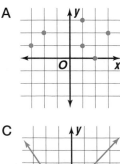

A

B

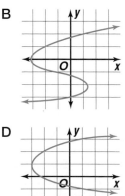

C

D

Extra Practice   See p. 708.

Square Roots

What You'll Learn

You'll learn to simplify radicals by using the Product and Quotient Properties of Square Roots.

Why It's Important

Aviation Pilots can use the formula $d = 1.5\sqrt{h}$ to determine the distance to the horizon. The formula contains a square root symbol. *See Example 6.*

Can you help this character from *Shoe* take his math test?

SHOE

You will find the square root of 225 in Example 3.

In Lesson 8–1, you learned that *squaring* a number means using that number as a factor twice. The opposite of squaring is finding a **square root**. To find a square root of 36, you must find *two equal factors* whose product is 36.

$$6 \times 6 = 36 \quad \rightarrow \quad \text{The square root of 36 is 6.}$$

Square Root	A square root of a number is one of its two equal factors.

The symbol $\sqrt{}$, called a **radical sign**, is used to indicate the square root.

$\sqrt{36} = 6$ *$\sqrt{36}$ indicates the positive square root of 36.*

$-\sqrt{36} = -6$ *$-\sqrt{36}$ indicates the negative square root of 36.*

Examples

Simplify each expression.

1 $\sqrt{49}$
Since $7^2 = 49$, $\sqrt{49} = 7$.

2 $-\sqrt{64}$
Since $8^2 = 64$, $-\sqrt{64} = -8$.

Your Turn

    a. $\sqrt{25}$ **b.** $\sqrt{121}$ **c.** $-\sqrt{25}$ **d.** $-\sqrt{9}$

A **radical expression** is an expression that contains a square root. You can simplify a radical expression like $\sqrt{225}$ by using prime numbers.

A **prime number** is a whole number that has exactly two factors, the number itself and 1. A **composite number** is a whole number that has more than two factors. Every composite number can be written as the product of prime numbers. The tree diagram shows one way to find the prime factors of 225.

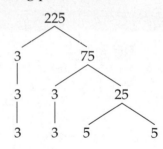

When a number is expressed as a product of prime factors, the expression is called the **prime factorization** of the number. Since 3 and 5 are prime numbers, the prime factorization of 225 is $3 \times 3 \times 5 \times 5$.

To simplify $\sqrt{225}$, use the following property.

Product Property of Square Roots	**Words:** The square root of a product is equal to the product of each square root.
	Numbers: $\sqrt{4 \cdot 9} = \sqrt{4} \cdot \sqrt{9}$
	Symbols: $\sqrt{ab} = \sqrt{a} \cdot \sqrt{b}$ $a \geq 0, b \geq 0$

Examples

Simplify each expression.

3 $\sqrt{225}$

$$\sqrt{225} = \sqrt{3 \cdot 3 \cdot 5 \cdot 5} \quad \textit{Find the prime factorization of 225.}$$
$$= \sqrt{9 \cdot 25} \quad \textit{3} \times \textit{3} = \textit{9, 5} \times \textit{5} = \textit{25}$$
$$= \sqrt{9} \cdot \sqrt{25} \quad \textit{Use the Product Property of Square Roots.}$$
$$= 3 \cdot 5 \text{ or } 15 \quad \textit{Simplify each radical.}$$

4 $\sqrt{576}$

$$\sqrt{576} = \sqrt{2 \cdot 2 \cdot 2 \cdot 2 \cdot 2 \cdot 2 \cdot 3 \cdot 3} \quad \textit{Find the prime factorization of 576.}$$
$$= \sqrt{64 \cdot 9} \quad \textit{2} \cdot \textit{2} \cdot \textit{2} \cdot \textit{2} \cdot \textit{2} \cdot \textit{2} = \textit{64, 3} \cdot \textit{3} = \textit{9}$$
$$= \sqrt{64} \cdot \sqrt{9} \quad \textit{Use the Product Property of Square Roots.}$$
$$= 8 \cdot 3 \text{ or } 24 \quad \textit{Simplify each radical.}$$

Your Turn

e. $\sqrt{144}$ f. $\sqrt{324}$

A similar property for quotients can be used to simplify radicals.

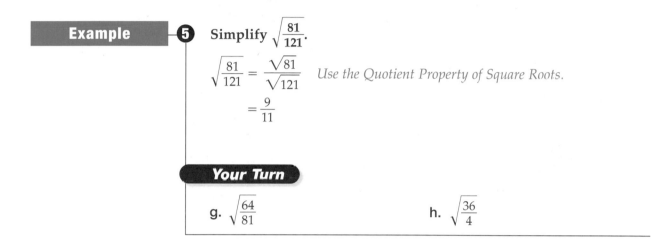

Quotient Property of Square Roots	Words:	The square root of a quotient is equal to the quotient of each square root.
	Numbers:	$\sqrt{\dfrac{4}{9}} = \dfrac{\sqrt{4}}{\sqrt{9}}$
	Symbols:	$\sqrt{\dfrac{a}{b}} = \dfrac{\sqrt{a}}{\sqrt{b}} \quad a \geq 0, b > 0$

Example 5

Simplify $\sqrt{\dfrac{81}{121}}$.

$$\sqrt{\dfrac{81}{121}} = \dfrac{\sqrt{81}}{\sqrt{121}} \qquad \textit{Use the Quotient Property of Square Roots.}$$

$$= \dfrac{9}{11}$$

Your Turn

g. $\sqrt{\dfrac{64}{81}}$

h. $\sqrt{\dfrac{36}{4}}$

You can use a graphing calculator to evaluate square roots. Press 2nd [$\sqrt{\ }$] and then the number to find its positive square root.

Example 6

Aviation Link

Real World

Pilots use the formula $d = 1.5\sqrt{h}$ to determine the distance in miles that an observer can see under ideal conditions. In the formula, d is the distance in miles and h is the height in feet of the plane. If an observer is in a plane that is flying at a height of 3600 feet, how far can he or she see?

$d = 1.5\sqrt{h}$

$d = 1.5 \times \sqrt{3600}$ *Replace h with 3600.*

1.5 2nd [$\sqrt{\ }$] 3600 ENTER   *90*

The observer can see a distance of 90 miles.

Technology Tip

On a graphing calculator, it is not necessary to use the $\times$ key when you multiply square roots.

Communicating Mathematics

Study the lesson. Then complete the following.

1. **Find** the first ten perfect squares.
2. **Write** the symbol for the negative square root of 9.
3. **Explain** why finding a square root and squaring are inverse operations.

Math Journal

Guided Practice

⏱ Getting Ready   Find the prime factorization of each number.

Sample: 81	**Solution:** $81 = 3 \cdot 3 \cdot 3 \cdot 3$ or 3^4

4. 100   5. 121   6. 169   7. 196   8. 256

Simplify.

9. $\sqrt{4}$ 10. $-\sqrt{49}$ 11. $-\sqrt{121}$ *(Examples 1 & 2)*

12. $\sqrt{256}$ 13. $\sqrt{\dfrac{1}{4}}$ 14. $-\sqrt{\dfrac{49}{121}}$ *(Examples 2 & 3)*

15. **Buildings**   A famous 1933 movie about a gorilla helped make the Empire State Building in New York City a popular tourist attraction. If the gorilla's eyes were at a height of 1225 feet, how far could he see in the distance on a clear day? Use the formula $d = 1.5\sqrt{h}$, where d is the distance in miles and h is the height in feet.   *(Example 5)*

Exercises

Practice

Simplify.

16. $-\sqrt{81}$ 17. $\sqrt{100}$ 18. $\sqrt{144}$ 19. $-\sqrt{196}$

20. $-\sqrt{169}$ 21. $\sqrt{529}$ 22. $\sqrt{25}$ 23. $-\sqrt{676}$

24. $\sqrt{441}$ 25. $-\sqrt{484}$ 26. $-\sqrt{1024}$ 27. $\sqrt{289}$

28. $\sqrt{\dfrac{81}{64}}$ 29. $-\sqrt{\dfrac{9}{100}}$ 30. $\sqrt{\dfrac{36}{196}}$ 31. $-\sqrt{\dfrac{25}{400}}$

32. $\sqrt{\dfrac{225}{25}}$ 33. $\sqrt{\dfrac{144}{196}}$ 34. $\sqrt{\dfrac{196}{289}}$ 35. $\sqrt{\dfrac{0.09}{0.16}}$

36. $\sqrt{0.16}$ 37. $-\sqrt{0.0025}$ 38. $\sqrt{0.0036}$ 39. $\sqrt{0.0009}$

40. Find the negative square root of 49.

41. If $x = \sqrt{36}$, what is the value of x?

Applications and Problem Solving

Real World

42. **Geometry**   The area of a square is 25 square inches. Find the length of one of its sides.

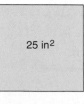
25 in²

43. **Geometry** The formula for the perimeter of a square is $P = 4s$, where s is the length of a side. The area of a square is 169 square meters. Find its perimeter.

44. **Firefighting** The velocity of water discharged from a nozzle is given by the formula $V = 12.14\sqrt{P}$, where V is the velocity in feet per second and P is the pressure at the nozzle in pounds per square inch. Find the velocity of water if the nozzle pressure is 64 pounds per square inch.

45. **Critical Thinking** *True* or *false*: $\sqrt{-36} = -6$. Explain.

Mixed Review

Express each number in scientific notation. *(Lesson 8–4)*

46. 350   47. 63,000   48. 0.023   49. 0.00076

Write each expression using positive exponents. *(Lesson 8–3)*

50. 7^{-2} 51. x^{-4} 52. $ab^{-2}c$ 53. $x^{-2}y^{-3}$

54. **Business** Among high school students ages 15–18, 41% say they are employed. The graph shows the number of hours per week these students work. If you survey 50 high school students who have jobs, predict how many of them work between 11 and 20 hours per week. *(Lesson 5–3)*

High School Workers

Hours per week

10 or less 29%
11–20 40%
21–35 22%
36 or more 9%

Source: Michaels Opinion Research, 1997

55. **Open-Ended Test Practice** Write an equation with variables on both sides in which the solution is −5. *(Lesson 4–6)*

Quiz 2   Lessons 8–4 and 8–5

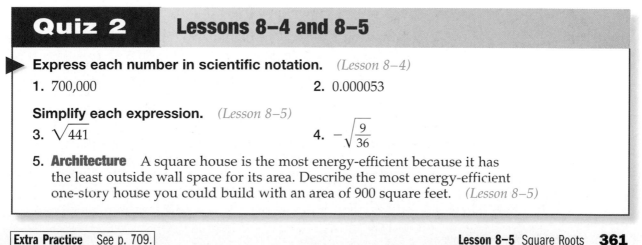

Express each number in scientific notation. *(Lesson 8–4)*

1. 700,000   2. 0.000053

Simplify each expression. *(Lesson 8–5)*

3. $\sqrt{441}$ 4. $-\sqrt{\dfrac{9}{36}}$

5. **Architecture** A square house is the most energy-efficient because it has the least outside wall space for its area. Describe the most energy-efficient one-story house you could build with an area of 900 square feet. *(Lesson 8–5)*

8-6 Estimating Square Roots

Math In the Workplace

What You'll Learn
You'll learn to estimate square roots.

Why It's Important
Law Enforcement
Police officers use the formula $s = \sqrt{30df}$ when they investigate accidents.
See Exercise 38.

Numbers like 25 and 81 are perfect squares because $\sqrt{25} = 5$ and $\sqrt{81} = 9$. But there are many other numbers that are *not* perfect squares. Notice what happens when you find $\sqrt{2}$ and $\sqrt{15}$ with a calculator.

[2nd] [$\sqrt{\ }$] 2 [ENTER] *1.414213562...*

[2nd] [$\sqrt{\ }$] 15 [ENTER] *3.872983346...*

Numbers like $\sqrt{2}$ and $\sqrt{15}$ are not integers or rational numbers because their decimal values do not terminate or repeat. They are **irrational numbers**. You can estimate irrational square roots by using perfect squares.

Hands-On Algebra

Materials: base-ten tiles

You can use base-ten tiles to estimate the square root of 60.

Step 1 Arrange 60 tiles into the largest square possible. The square has 49 tiles, with 11 left over.

Step 2 Add tiles until you have the next larger square. You need to add 4 tiles. This square has 64 tiles.

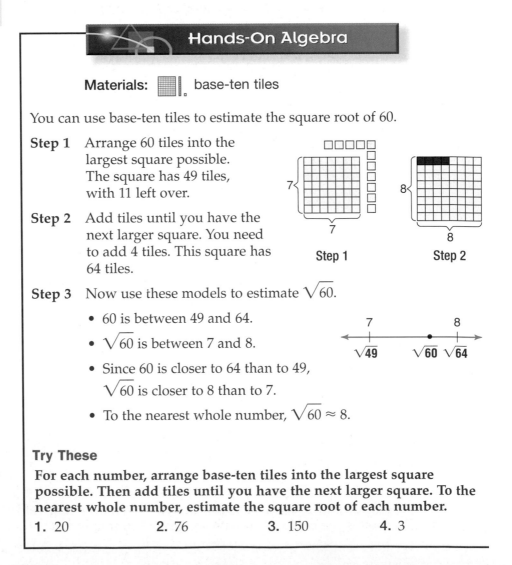

Step 3 Now use these models to estimate $\sqrt{60}$.

- 60 is between 49 and 64.
- $\sqrt{60}$ is between 7 and 8.
- Since 60 is closer to 64 than to 49, $\sqrt{60}$ is closer to 8 than to 7.
- To the nearest whole number, $\sqrt{60} \approx 8$.

Try These

For each number, arrange base-ten tiles into the largest square possible. Then add tiles until you have the next larger square. To the nearest whole number, estimate the square root of each number.

1. 20   **2.** 76   **3.** 150   **4.** 3

Estimate each square root.

1 $\sqrt{22}$

Find the two perfect squares closest to 22. List some perfect squares.
$$1, 4, 9, 16, 25, 36, \ldots$$
22 is between 16 and 25.

$$16 < 22 < 25$$
$$\sqrt{16} < \sqrt{22} < \sqrt{25}$$
$$4 < \sqrt{22} < 5$$

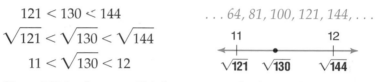

Since 22 is closer to 25 than to 16, the best whole number estimate for $\sqrt{22}$ is 5.

2 $\sqrt{130}$

$$121 < 130 < 144 \qquad \ldots 64, 81, 100, 121, 144, \ldots$$
$$\sqrt{121} < \sqrt{130} < \sqrt{144}$$
$$11 < \sqrt{130} < 12$$

Since 130 is closer to 121 than to 144, the best whole number estimate for $\sqrt{130}$ is 11.

Your Turn

a. $\sqrt{45}$ **b.** $\sqrt{190}$

Gardening Link

3 **A box of fertilizer covers 250 square feet of garden. Find the length in whole feet of the largest square garden that can be fertilized with one box of fertilizer.**

Find the largest perfect square that is less than 250. Then find its square root.

$225 < 250$ $\quad 15^2 = 225$
$\qquad\qquad\quad 16^2 = 256$

The square root of 225 is 15. Therefore, the largest square garden that can be fertilized with one box of fertilizer has a length of 15 feet.

Communicating Mathematics

Study the lesson. Then complete the following.

1. **Explain** why $\sqrt{10}$ is an irrational number.
2. **Graph** $\sqrt{75}$ on a number line.
3. **Write a problem** that can be solved by estimating $\sqrt{200}$.

Guided Practice

⏱ Getting Ready **Find two consecutive perfect squares between which each number lies.**

Sample: 60	**Solution:** 60 is between 49 and 64.

4. 56   5. 85   6. 175   7. 500

Estimate each square root to the nearest whole number.
(Examples 1 & 2)

8. $\sqrt{85}$ 9. $\sqrt{71}$ 10. $\sqrt{149}$ 11. $\sqrt{255}$

12. **Geometry** The area of a square is 200 square inches. Find the length of each side. Round to the nearest whole number. *(Example 3)*

Exercises ● ● ● ● ● ● ● ● ● ● ● ● ● ● ● ● ● ●

Practice

Estimate each square root to the nearest whole number.

13. $\sqrt{3}$ 14. $\sqrt{7}$ 15. $\sqrt{13}$ 16. $\sqrt{19}$ 17. $\sqrt{33}$

18. $\sqrt{56}$ 19. $\sqrt{113}$ 20. $\sqrt{175}$ 21. $\sqrt{410}$ 22. $\sqrt{500}$

23. $\sqrt{575}$ 24. $\sqrt{1000}$ 25. $\sqrt{60.3}$ 26. $\sqrt{94.5}$ 27. $\sqrt{131.4}$

28. $\sqrt{2.314}$ 29. $\sqrt{152.75}$ 30. $\sqrt{189.2}$ 31. $\sqrt{0.08}$ 32. $\sqrt{0.76}$

33. Tell whether 6 is closer to $\sqrt{34}$ or $\sqrt{44}$.
34. Which is closer to $\sqrt{43}$, 6 or 7?

Applications and Problem Solving

Real World

35. **Meteorology** Meteorologists use the formula $t = \sqrt{\dfrac{D^3}{216}}$ to describe violent storms such as hurricanes. In the formula, D is the diameter of the storm in miles, and t is the number of hours it will last. A typical hurricane has a diameter of about 40 miles. Estimate how long the storm will last.

40 mi

Photo Graphic

36. Aviation The British Airways Concorde flies at a height of 60,000 feet. About how far can the pilot see when he or she looks out the window on a clear day? Use the formula $d = 1.5 \sqrt{h}$ and round to the nearest mile.

37. Geometry You can use the formula $A = \sqrt{s(s - a)(s - b)(s - c)}$ to find the area of a triangle given the measures of its sides. In the formula, a, b, and c, are the measures of the sides, A is the area, and s is one-half the perimeter. The figure shows a triangular plot of land that contains a large lake. Estimate the area of the triangle to the nearest square mile.

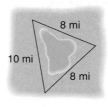

38. Law Enforcement When police officers investigate traffic accidents, they need to determine the speed of the vehicles involved in the accident. They can measure the skid marks and then use the formula $s = \sqrt{30df}$. In the formula, s is the speed in miles per hour, d is the length of the skid marks in feet, and f is the friction factor that depends on road conditions. The table gives some values for f.

	Friction Factor (*f*)	
	concrete	asphalt
wet	0.4	0.5
dry	0.8	1.0

Use the formula to determine the speed in miles per hour for each skid length and road condition. Round to the nearest tenth.

a. 40-foot skid, wet concrete   **b.** 150-foot skid, dry asphalt

c. 350-foot skid, wet asphalt   **d.** 45-foot skid, dry concrete

39. Critical Thinking Find three numbers that have square roots between 3 and 4.

Mixed Review

Find each square root. *(Lesson 8–5)*

40. $\sqrt{0.36}$ **41.** $\sqrt{256}$ **42.** $\sqrt{729}$ **43.** $\sqrt{\dfrac{169}{121}}$

Express each measure in standard form. *(Lesson 8–4)*

44. 2 gigabytes   **45.** 4 nanoseconds   **46.** 6.5 megahertz

Write an equation in slope-intercept form of the line having the given slope that passes through the given point. *(Lesson 7–3)*

47. 5; $(3, -2)$ **48.** $\dfrac{1}{4}$; $(0, 8)$ **49.** -5; $(5, 4)$

50. Travel Tim drove 4 hours at an average speed of 60 miles per hour. How long would it take Tim to drive the same distance at an average speed of 50 miles per hour? *(Lesson 6–6)*

51. Standardized Test Practice Which statement is *not* correct? *(Lesson 3–1)*

A $\dfrac{1}{2} > \dfrac{3}{8}$ **B** $\dfrac{4}{5} < \dfrac{5}{6}$ **C** $\dfrac{8}{11} < \dfrac{7}{13}$ **D** $\dfrac{1}{3} < \dfrac{10}{13}$

Extra Practice See p. 709.

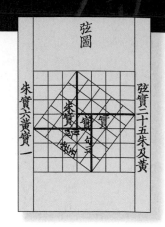

What You'll Learn
You'll learn to use the Pythagorean Theorem to solve problems.

Why It's Important
Carpentry
Carpenters use the Pythagorean Theorem to determine whether the corners of a deck are right angles.
See Example 4.

An ancient Chinese text that was written sometime during 1200–600 B.C. contains a puzzle about a right triangle. The sides of each right triangle have lengths of 3, 4, and 5 units.

The relationship among these lengths forms the basis for one of the most famous theorems in mathematics. It can be illustrated geometrically.

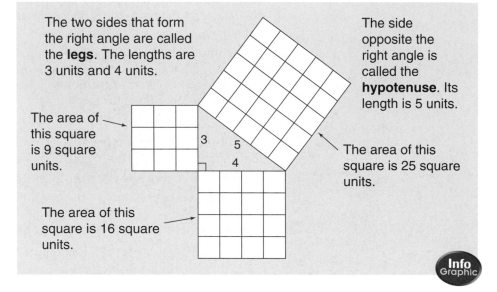

The two sides that form the right angle are called the **legs**. The lengths are 3 units and 4 units.

The side opposite the right angle is called the **hypotenuse**. Its length is 5 units.

The area of this square is 9 square units.

The area of this square is 25 square units.

The area of this square is 16 square units.

The area of the larger square is equal to the total area of the two smaller squares.

$$25 = 9 + 16$$

$$5^2 = 3^2 + 4^2$$

This relationship is true for *any* right triangle and is called the **Pythagorean Theorem**.

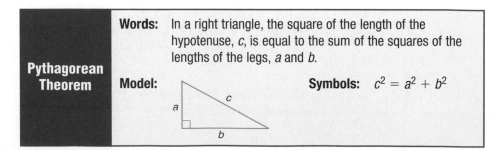

Pythagorean Theorem	**Words:** In a right triangle, the square of the length of the hypotenuse, c, is equal to the sum of the squares of the lengths of the legs, a and b.
	Model: **Symbols:** $c^2 = a^2 + b^2$

Find the length of the hypotenuse of the right triangle.

$$c^2 = a^2 + b^2 \quad \textit{Pythagorean Theorem}$$
$$c^2 = 15^2 + 8^2 \quad \textit{Replace a with 15 and b with 8.}$$
$$c^2 = 225 + 64$$
$$c^2 = 289$$
$$c = \sqrt{289} \quad \textit{Find the square root of each side.}$$
$$c = 17$$

The length of the hypotenuse is 17 feet.

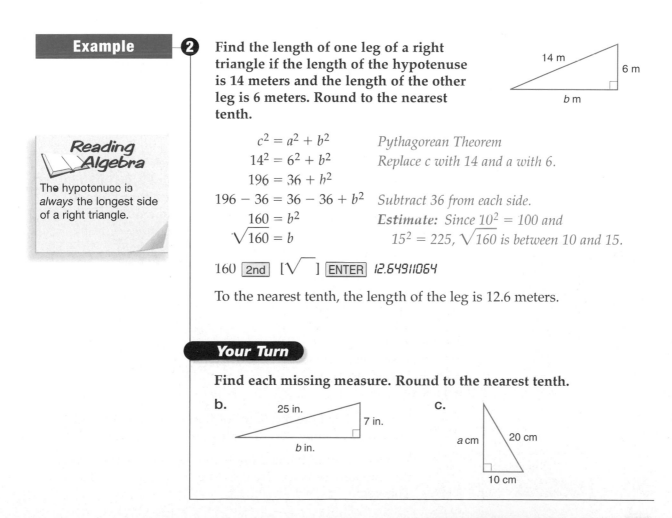

15 ft c ft 8 ft

Your Turn

a. Find the length of the hypotenuse of a right triangle if the lengths of the legs are 6 meters and 8 meters.

You can also use the Pythagorean Theorem to find the length of a leg of a right triangle.

Find the length of one leg of a right triangle if the length of the hypotenuse is 14 meters and the length of the other leg is 6 meters. Round to the nearest tenth.

14 m 6 m *b* m

$$c^2 = a^2 + b^2 \quad \textit{Pythagorean Theorem}$$
$$14^2 = 6^2 + b^2 \quad \textit{Replace c with 14 and a with 6.}$$
$$196 = 36 + b^2$$
$$196 - 36 = 36 - 36 + b^2 \quad \textit{Subtract 36 from each side.}$$
$$160 = b^2 \qquad \textit{Estimate: Since } 10^2 = 100 \text{ and}$$
$$\sqrt{160} = b \qquad\qquad 15^2 = 225, \sqrt{160} \textit{ is between 10 and 15.}$$

160 [2nd] [√] [ENTER] *12.64911064*

To the nearest tenth, the length of the leg is 12.6 meters.

> **Reading Algebra**
>
> The hypotenuse is *always* the longest side of a right triangle.

Your Turn

Find each missing measure. Round to the nearest tenth.

b.   25 in.   7 in.   *b* in.

c.   *a* cm   20 cm   10 cm

You can use the **converse** of the Pythagorean Theorem to test whether a triangle is a right triangle.

Converse of the Pythagorean Theorem	If c is the measure of the longest side of a triangle and $c^2 = a^2 + b^2$, then the triangle is a right triangle.

Examples

3 The measures of the three sides of a triangle are 5, 7, and 9. Determine whether this triangle is a right triangle.

$c^2 = a^2 + b^2$ *Pythagorean Theorem*
$9^2 \stackrel{?}{=} 5^2 + 7^2$ *Replace c with 9, a with 5, and b with 7.*
$81 \stackrel{?}{=} 25 + 49$
$81 \neq 74$

Since $c^2 \neq a^2 + b^2$, the triangle is *not* a right triangle.

Your Turn

The measures of three sides of a triangle are given. Determine whether each triangle is a right triangle.

d. 20, 21, 28   **e.** 37, 12, 34

Carpentry Link

Real World

4 A carpenter checks to be sure the corners of a deck are square by using a 3-4-5 system. He or she measures along one side in 3-foot units and along the adjacent side in the same number of 4-foot units. If the measure of the hypotenuse is the same number of 5-foot units, the corner of the deck is square. *If a corner is square, it is formed by a right angle. So, the triangle is a right triangle.*

When laying out a deck, a carpenter measures along one side a distance of 9 feet and along the adjacent side a distance of 12 feet. The measure of the hypotenuse is 15 feet. Is the corner of the deck square?

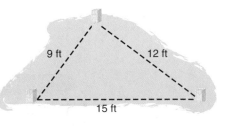

9 ft   12 ft

15 ft

Explore   You know that the measures of the sides are 9, 12, and 15. You want to show that the sides form a right triangle.

Plan   The measures of the legs are 9 and 12, and the measure of the hypotenuse is 15. Use these numbers in the Pythagorean Theorem.

Solve $c^2 = a^2 + b^2$ *Pythagorean Theorem*

$15^2 \stackrel{?}{=} 9^2 + 12^2$ *Replace c with 15, a with 9, and b with 12.*

$225 \stackrel{?}{=} 81 + 144$

$225 = 225$

Since $c^2 = a^2 + b^2$, the triangle is a right triangle and the corner is square.

Examine   The measures of the sides are 9, 12, and 15. They are multiples of a 3, 4, 5 triangle. The answer is reasonable.

Check for Understanding

Communicating Mathematics

Study the lesson. Then complete the following.

1. **Draw** a right triangle and label the right angle, the hypotenuse, and the legs.

2. **Explain** how you know whether a triangle is a right triangle if you know the lengths of the three sides.

> **Vocabulary**
> hypotenuse
> leg
> Pythagorean Theorem
> converse

Guided Practice

⏱ **Getting Ready**   **Determine whether each sentence is *true* or *false*.**

Sample: $4^2 + 5^2 = 6^2$ **Solution:** $16 + 25 \neq 36$; false

3. $9^2 + 10^2 = 11^2$ 4. $7^2 + 24^2 = 25^2$ 5. $3^2 + 12^2 = 20^2$

If *c* is the measure of the hypotenuse and *a* and *b* are the measures of the legs, find each missing measure. Round to the nearest tenth if necessary.

6.
6 ft c ft   8 ft

7.   *(Example 1)*
a cm   15 cm   5 cm

8. $a = 30$, $c = 34$, $b = ?$ 9. $a = 7$, $b = 4$, $c = ?$ *(Example 2)*

The lengths of three sides of a triangle are given. Determine whether each triangle is a right triangle. *(Example 3)*

10. 9 ft, 16 ft, 20 ft   11. 9 mm, 40 mm, 41 mm

12. **Construction**   A builder is laying out the foundation for a house. The measure along one side is 12 feet, the adjacent side is 16 feet, and the hypotenuse is 21 feet. Determine whether the corner is a right angle. *(Example 4)*

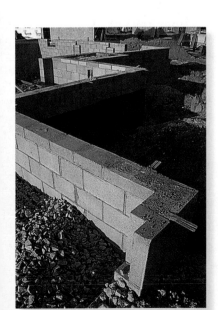

Practice

If *c* is the measure of the hypotenuse and *a* and *b* are the measures of the legs, find each missing measure. Round to the nearest tenth if necessary.

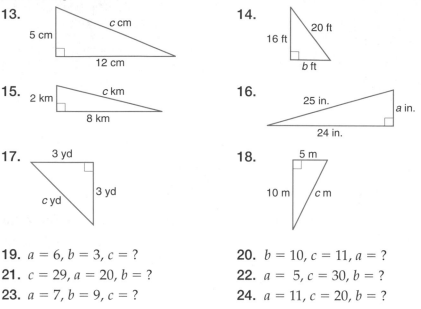

13.

5 cm
c cm
12 cm

14.

20 ft
16 ft
b ft

15.

2 km
c km
8 km

16.

25 in.
a in.
24 in.

17.

3 yd
3 yd
c yd

18.

5 m
10 m
c m

19. $a = 6, b = 3, c = ?$

20. $b = 10, c = 11, a = ?$

21. $c = 29, a = 20, b = ?$

22. $a = 5, c = 30, b = ?$

23. $a = 7, b = 9, c = ?$

24. $a = 11, c = 20, b = ?$

The lengths of three sides of a triangle are given. Determine whether each triangle is a right triangle.

25. 11 in., 12 in., 16 in.

26. 11 cm, 60 cm, 61 cm

27. 6 ft, 8 ft, 9 ft

28. 6 mi, 7 mi, 12 mi

29. 45 m, 60 m, 75 m

30. 1 mm, 1 mm, 2 mm

31. Is a triangle with measures 30, 40, and 50 a right triangle?

32. Find the length of the hypotenuse of a right triangle if the lengths of the legs are 6 miles and 11 miles. Round to the nearest tenth.

Applications and Problem Solving

Real World

33. **Geometry**   Find the length of the diagonal of a rectangle whose length is 8 meters and whose width is 5 meters. Round to the nearest tenth.

34. **Carpentry**   The rise, the run, and the rafter of a pitched roof form a right triangle. Find the length of the rafter that is needed if the rise of the roof is 6 feet and the run is 12 feet. Round your answer to the nearest tenth.

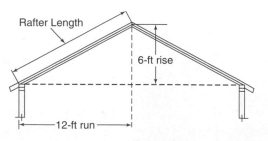

Rafter Length
6-ft rise
12-ft run

35. Architecture   Architects often use arches when they design windows and doors. The Early English, or pointed, arch is based on an *equilateral triangle*. An equilateral triangle has three sides of equal measure.

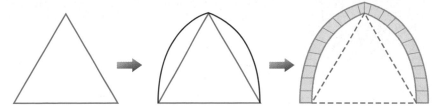

In the figure at the right, the height *h* separates the equilateral triangle into two right triangles of equal size. Find the height of the arch. Round to the nearest tenth.

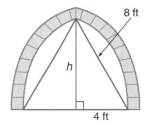

36. Critical Thinking   The figure shows three squares. The two smaller squares each have an area of 1 square unit.

  a. Find the area of the third square.

  b. Draw a square on rectangular dot paper with an area of 5 square units.

Mixed Review

Estimate each square root to the nearest whole number.   *(Lesson 8–6)*

37. 65   **38.** 95   **39.** 200   **40.** 500

Simplify.   *(Lesson 8–5)*

41. $\sqrt{225}$ **42.** $-\sqrt{49}$ **43.** $\sqrt{100}$ **44.** $-\sqrt{\dfrac{9}{25}}$

45. Finance   To save for a new bicycle, Katie begins a savings plan. Her plan can be described by the equation $s = 5w + 100$, where s represents the total savings in dollars and w represents the number of weeks since the start of the savings plan.   *(Lesson 7–3)*

  a. How much money had Katie already saved when the plan started?

  b. How much does Katie save each week?

46. Write the point-slope form of an equation for the line that passes through the point at $(2, -4)$ and has a slope of -1.   *(Lesson 7–2)*

47. Standardized Test Practice   Which set is the domain of the relation shown on the graph?   *(Lesson 6–1)*

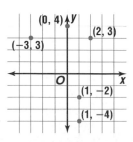

  A $\{-3, 0, 2\}$

  B $\{-4, -2, 3, 4\}$

  C neither A nor B

  D both A and B

Extra Practice   See p. 709.

Must Be TV

Materials

- uncooked spaghetti
- ruler
- grid paper
- scissors
- calculator

Using the Pythagorean Theorem

Look at an advertisement for an electronics store and you will find television sets that have 10-inch, 32-inch, and 36-inch screens. Did you ever wonder what these numbers represent? They represent the hypotenuse of a right triangle.

A television set is shaped like a rectangle with length ℓ and width w. The diagonal is a line segment from one corner to the opposite corner, which separates the screen into two right triangles of equal size. The diagonal is the hypotenuse of the right triangles. So, a 10-inch television set has a screen in which the diagonal is 10 inches.

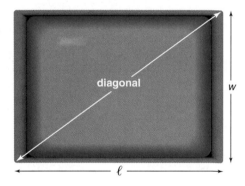

Investigate

1. Use a piece of uncooked spaghetti and grid paper to find several rectangles that have a 10-inch diagonal.

   a. Break the spaghetti so that one piece measures 10 inches.

   b. Place the spaghetti diagonally on the grid paper as shown below.

   c. Using the ends of the spaghetti as endpoints of the diagonal, draw horizontal and vertical lines from the endpoints to complete the rectangle.

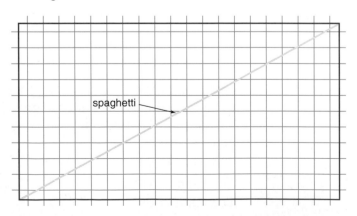

d. Repeat Steps 1b and 1c four more times using the same piece of spaghetti and different pieces of grid paper. For each rectangle, change the angle at which you place the spaghetti on the grid paper.

e. Cut out the rectangles. Choose the rectangle that looks most like a television screen. Describe how this rectangle is different from the others.

2. Television manufacturers have set a standard ratio for the size of a television screen. The ratio of the height to the width is 3:4. For which of your rectangles is the ratio of the height to the width closest to 3:4?

Extending the Investigation

In this extension, you will predict the dimensions of a 19-inch television screen.

- Use a calculator and the Pythagorean Theorem to find the measures of five different rectangles that have a 19-inch diagonal.

- Find the areas of each rectangle and the ratios of each height to width.

- Analyze the data and make a prediction about the dimensions of a 19-inch television screen.

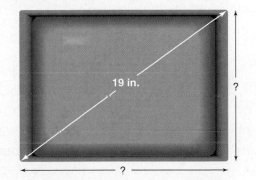

Presenting Your Investigation

Here are some ideas to help you present your conclusions to the class.

- Make a poster that shows scale drawings of your rectangles.

- Write a report in which you explain how you made your prediction. Be sure to include the terms *ratio* and *Pythagorean Theorem* in your report. Then compare your prediction to the dimensions of an actual 19-inch television.

*inter**NET*** **Investigation** For more information on ratios,
CONNECTION visit: www.algconcepts.glencoe.com

Understanding and Using the Vocabulary

After completing this chapter, you should be able to define each term, property, or phrase and give an example or two of each.

*inter*NET
CONNECTION **Review Activities**
For more review activities, visit:
www.algconcepts.glencoe.com

Algebra

base *(p. 336)*
composite number *(p. 358)*
exponent *(p. 336)*
irrational numbers *(p. 362)*
negative exponent *(p. 348)*
perfect square *(p. 336)*
power *(p. 336)*

prime factorization *(p. 358)*
prime number *(p. 358)*
radical expression *(p. 358)*
radical sign *(p. 357)*
scientific notation *(p. 353)*
square root *(p. 357)*

Geometry

hypotenuse *(p. 366)*
leg *(p. 366)*
Pythagorean Theorem *(p. 366)*

Complete each sentence using a term from the vocabulary list.

1. When a number is expressed as a product of factors that are all prime, the expression is called the ___?___ of the number.

2. The symbol $\sqrt{}$ is called a ___?___ .

3. The square roots of numbers such as 2, 3, and 7 are ___?___ .

4. The ___?___ tells the relationship among the measures of the legs of a right triangle and the hypotenuse.

5. In the expression 2^3, the 3 is called the ___?___ . It tells how many times 2 is used as a factor.

6. The ___?___ of a number is one of its two equal factors.

7. The longest side of a right triangle is called the ___?___ .

8. When you express 5,000,000 as 5×10^6, you are using ___?___ .

9. Numbers like 25, 36, and 100 are called ___?___ .

10. Numbers that are expressed using exponents are called ___?___ .

Skills and Concepts

Objectives and Examples	**Review Exercises**

• **Lesson 8–1**   Use powers in expressions.

Write $n \cdot n \cdot n \cdot n$ using exponents.
$n \cdot n \cdot n \cdot n = n^4$

Write $2^3 \cdot 3^2$ as a multiplication expression.
$(2^3)(3^2) = (2)(2)(2)(3)(3)$

Write each expression using exponents.

11. $9 \cdot 9 \cdot 9 \cdot 9 \cdot 9$ 12. 5 squared

13. $2 \cdot 2 \cdot 3 \cdot 3 \cdot 3$ 14. $(-2)(x)(x)(y)(y)$

Write each power as a multiplication expression.

15. 12^3 16. $(-2)^5$ 17. y^2

18. a^3b^2 19. $5x^2y^3z^2$ 20. $-5m^2$

Objectives and Examples	Review Exercises

• Lesson 8–2 Multiply and divide powers.

$$(2y^4)(5y^5) = (2 \cdot 5)(y^4 \cdot y^5)$$
$$= (10)(y^{4+5})$$
$$= 10y^9$$

$$\frac{x^7 y^4}{x^3 y^3} = \left(\frac{x^7}{x^3}\right)\left(\frac{y^4}{y^3}\right)$$
$$= (x^{7-3})(y^{4-3})$$
$$= x^4 y$$

Simplify each expression.

21. $b^2 \cdot b^5$ **22.** $y^3 \cdot y^3 \cdot y$

23. $(a^2 b)(a^2 b^2)$ **24.** $(3ab)(-4a^2 b^3)$

25. $\dfrac{y^{10}}{y^6}$ **26.** $\dfrac{a^2 b^4}{a^3 b^2}$

27. $\dfrac{42x^7}{14x^4}$ **28.** $\dfrac{-15x^3 y^5 z^4}{5x^2 y^2 z}$

• Lesson 8–3 Simplify expressions containing negative exponents.

$$a^2 b^{-3} = a^2 \cdot \frac{1}{b^3}$$
$$= \frac{a^2}{b^3}$$

Write each expression using positive exponents. Then evaluate the expression.

29. 2^{-3} **30.** 10^{-2}

Simplify each expression.

31. x^{-4} **32.** $m^{-2} n^5$

33. $\dfrac{y^3}{y^{-4}}$ **34.** $\dfrac{rs^2}{s^3}$

35. $\dfrac{-25a^5 b^6}{5u^5 b^3}$ **36.** $\dfrac{27b^{-2}}{14b^{-3}}$

• Lesson 8–4 Express numbers in scientific notation.

$$3600 = 3.6 \times 10^3$$

$$0.023 = 2.3 \times 10^{-2}$$

Express each measure in standard form.

37. 1.5 nanoseconds   **38.** 5 kilobytes

Express each number in scientific notation.

39. 240,000   **40.** 0.000314

41. 4,880,000,000   **42.** 0.000015

Evaluate each expression.

43. $(2 \times 10^5)(3 \times 10^6)$ **44.** $\dfrac{8 \times 10^{-3}}{2 \times 10^4}$

• Lesson 8–5 Simplify radicals.

$$\sqrt{400} = \sqrt{2 \cdot 2 \cdot 2 \cdot 2 \cdot 5 \cdot 5}$$
$$= \sqrt{16 \cdot 25}$$
$$= \sqrt{16} \cdot \sqrt{25}$$
$$= 4 \cdot 5 \text{ or } 20$$

Simplify.

45. $\sqrt{121}$ **46.** $-\sqrt{324}$

47. $\sqrt{\dfrac{4}{81}}$ **48.** $-\sqrt{\dfrac{100}{225}}$

Objectives and Examples

- **Lesson 8–6** Estimate square roots.

  Estimate $\sqrt{150}$.

  $$144 < \quad 150 \quad < 169 \quad \ldots 100, 121, 144,$$
  $$169, \ldots$$

  $$\sqrt{144} < \sqrt{150} < \sqrt{169}$$
  $$12 < \sqrt{150} < 13 \quad \textit{150 is closer to 144}$$
  $$\textit{than to 169.}$$

  The best whole number estimate for $\sqrt{150}$
  is 12.

Review Exercises

Estimate each square root to the nearest whole number.

49. $\sqrt{19}$

50. $\sqrt{108}$

51. $\sqrt{200}$

52. $\sqrt{125.52}$

53. Which is closer to $\sqrt{50}$, 7 or 8?

- **Lesson 8–7** Use the Pythagorean Theorem to solve problems.

  Find the length of the missing side.
  $$c^2 = a^2 + b^2$$
  $$25^2 = 15^2 + b^2$$
  $$625 = 225 + b^2$$
  $$625 - 225 = 225 - 225 + b^2$$
  $$400 = b^2$$
  $$20 = b$$

  15 ft · 25 ft · b ft

If c is the measure of the hypotenuse and a and b are the measures of the legs, find each missing measure. Round to the nearest tenth if necessary.

54.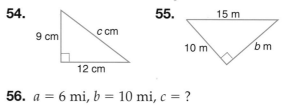

9 cm · c cm · 12 cm

55. 15 m · 10 m · b m

56. $a = 6$ mi, $b = 10$ mi, $c = ?$

57. $b = 6$, $c = 12$, $a = ?$

Applications and Problem Solving

58. Biology *E. coli* is a fast-growing bacteria that splits into two identical cells every 15 minutes. If you start with one *E. coli* bacterium, how many bacteria will there be at the end of 2 hours? *(Lesson 8–1)*

59. Geometry The area of a square is 90 square meters. Estimate the length of its side to the nearest whole number. *(Lesson 8–6)*

60. Construction Park managers want to construct a road from the park entrance directly to a campsite. Use the figure below to find the length of the proposed road. *(Lesson 8–7)*

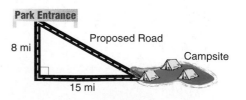

Park Entrance · Proposed Road · Campsite · 8 mi · 15 mi

CHAPTER 8 Test

1. **Write** a multiplication problem whose product is $12x^5$.
2. **Explain** why 3 is the square root of 9.
3. **Draw** a figure that can be used to estimate $\sqrt{96}$.

Evaluate each expression if $x = 2$, $y = 3$, $z = -1$, and $w = -2$.

4. $3x^5$
5. $5w^3y$
6. $12y^{-2}$
7. z^0

Simplify each expression.

8. $(a^3b^2)(a^5b)$
9. $\dfrac{xy^3}{y^5}$
10. m^{-10}
11. $\dfrac{-3r^2s^8}{12r^2s^5}$

Express each number in scientific notation.

12. 0.00000125
13. 36,000,000,000

Evaluate each expression. Express each result in scientific notation and standard form.

14. $(8.2 \times 10^{-5})(1 \times 10^{-3})$
15. $\dfrac{6 \times 10^3}{2 \times 10^{-2}}$

Estimate each square root to the nearest whole number.

16. $\sqrt{5}$
17. $\sqrt{20}$
18. $\sqrt{63}$
19. $\sqrt{395}$

Find each missing measure. Round to the nearest tenth if necessary.

20.
21.
22.

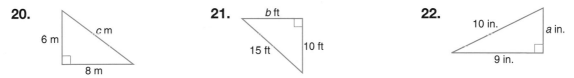

23. **Chemistry** The diameter of an atom is about 10^{-8} centimeter and the diameter of its nucleus is about 10^{-13} centimeter. How many times as large is the diameter of the atom as the diameter of its nucleus?

24. **Sports** Gymnastic routines are performed on a 40-foot by 40-foot square mat. Gymnasts usually use the diagonal of the mat because it gives them more distance to complete their routine. To the nearest whole number, how much longer is the diagonal of the mat than one of its sides?

25. **Sequences** The first term of the sequence $1, \dfrac{1}{2}, \dfrac{1}{4}, \dfrac{1}{8}, \ldots$ can be written as 2^0. Write the next three terms using negative exponents.

Pythagorean Theorem Problems

All standardized tests contain several problems that you can solve using the Pythagorean Theorem. The **Pythagorean Theorem** states that in a right triangle, the sum of the squares of the measures of the legs equals the square of the measure of the hypotenuse.

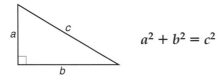

$$a^2 + b^2 = c^2$$

THE PRINCETON REVIEW

Test-Taking Tip

The 3-4-5 right triangle and its multiples, like 6-8-10 and 9-12-15, occur frequently on standardized tests. Other commonly used Pythagorean triples include 5-12-13 and 7-24-25.

Proficiency Test Example

Use the information in the figure below. Find *BC* to the nearest centimeter.

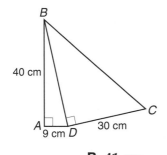

A 31 cm   **B** 41 cm
C 51 cm   **D** 80 cm

Hint   Use the Pythagorean Theorem twice.

Solution   The figure is made up of two right triangles, △*ABD* and △*BDC*. $\overline{BD}$ is the hypotenuse of △*ABD*. $\overline{BD}$ is also a leg of △*BDC*.

First, find *BD*. Then use *BD* to find *BC*.

$(BD)^2 = 9^2 + 40^2$ $(BC)^2 = 30^2 + 41^2$
$(BD)^2 = 81 + 1600$ $(BC)^2 = 900 + 1681$
$(BD)^2 = 1681$ $(BC)^2 = 2581$
    $BD = 41$ $BC \approx 50.80$

The answer is C.

SAT Example

A 25-foot ladder is placed against a vertical wall of a building with the bottom of the ladder standing on concrete 7 feet from the base of the building. If the top of the ladder slips down 4 feet, then the bottom of the ladder will slide out how many feet?

A 5 ft   **B** 6 ft
C 7 ft   **D** 8 ft

Hint   Start by drawing a diagram.

Solution   The ladder placed against the wall forms a 7-24-25 right triangle. After the ladder slips down 4 feet, the new right triangle has sides that are multiples of a 3-4-5 right triangle, 15-20-25.

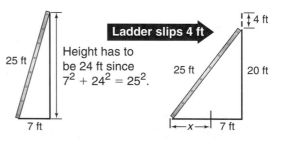

Height has to be 24 ft since $7^2 + 24^2 = 25^2$.

The bottom of the ladder will be 15 feet from the wall. This means the bottom of the ladder will slide out $15 - 7$ or 8 feet.

The answer is D.

After you work each problem, record your answer on the answer sheet provided or on a sheet of paper.

1. Diego bought 3 fish. Every month, the number of fish doubles. The formula $f = 3 \cdot 2^m$ represents the number of fish he will have after m months. How many fish will he have after 4 months if he keeps all of them and none of them dies?

   A 12   **B** 24
   C 48   **D** 1296

2. Which number is greater than 2.8?

   A 175%   **B** $\sqrt{12}$
   C -3.5 **D** 5.6×10^{-1}

3. A wire from the top of a 25-foot flagpole is attached to a point 20 feet from the base of the flagpole. To the nearest whole number, find the length of the wire.

   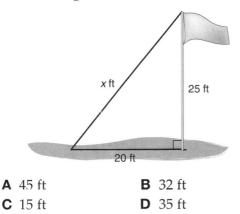

   A 45 ft   **B** 32 ft
   C 15 ft   **D** 35 ft

4. Which of the following are measures of three sides of a right triangle?

   A 4, 7, 8   **B** 10, 15, 20
   C 3, 7, 9   **D** 9, 12, 15

5. Which equation does *not* have 6 as a solution?

   A $2(x - 6) = 0$ **B** $4x + 3 = 27$
   C $2x + 1 = 14$ **D** $2x + x = 18$

6. In the figure, $\triangle ABC$ and $\triangle ACD$ are right triangles. If $AB = 20$, $BC = 15$, and $AD = 7$, then $CD = $ —

   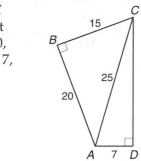

   A 22.
   B 23.
   C 24.
   D 25.

7. A farmer wants to put a fence around a square garden that has a total area of 1600 square feet. If fencing costs \$1.50 per foot, how much will it cost to fence the garden? To solve this problem, begin by—

   A dividing 1600 by \$1.50.
   B multiplying 1600 by \$1.50 and taking the square root of the product.
   C dividing 1600 by 4 and multiplying by \$1.50.
   D finding the square root of 1600 and multiplying by 4.

8. If $x = 3$, $y = -2$, $z = -1$, and $w = 5$, then the value of $9xy^2z^{10}w^0$ is—

   A -590.   **B** -108.
   C 1.   **D** 108.

Open-Ended Questions

9. **Grid-In** On the first test of the semester, Jennifer scored a 60. On the last test of the semester, she scored a 75. By what percent did Jennifer's score improve?

10. The table shows the squares of six whole numbers. Describe one way that the digits in the squares are related to the digits in the original number.

$15^2 = 225$	$55^2 = 3025$	$135^2 = 18,225$
$25^2 = 625$	$95^2 = 9025$	$175^2 = 30,625$

*inter***NET** CONNECTION   **Test Practice** For additional test practice questions, visit:
www.algconcepts.glencoe.com

▶ What You'll Learn in Chapter 9:

- to identify and classify polynomials and find their degree *(Lesson 9–1)*,
- to add and subtract polynomials *(Lesson 9–2)*, **and**
- to multiply polynomials *(Lessons 9–3, 9–4, and 9–5)*.

Problem-Solving Workshop

Project

Congratulations! You have been selected to design the box for a new caramel corn snack. The manufacturers would like you to present three possible box designs. Each box must be a rectangular prism with a surface area of 150 square inches. Make a pattern for each of the three box designs. In a one-page paper, explain which box design you think will be best for the new product.

Working on the Project

Work with a partner and choose a strategy. Here are some suggestions to help you get started.

- Cut apart an empty cereal box along its edges to make a pattern for the surface area of a rectangular prism.

- Find the area of each of the six faces and the surface area of the prism.

- Suppose you don't know the measurements of the prism. Let ℓ represent the length, w represent the width, and h represent the height of the prism. Write a polynomial expression for the surface area of the prism.

> ### ▶ Strategies
> **Look for a pattern.**
> **Draw a diagram.**
> **Make a table.**
> **Work backward.**
> **Use an equation.**
> **Make a graph.**
> **Guess and check.**

Technology Tools

- Use a **calculator** to find the surface area of your patterns.
- Use **drawing software** to make patterns for your boxes.

*inter*NET **Research** For more information about packaging, visit:
CONNECTION www.algconcepts.glencoe.com

Presenting the Project

Prepare a portfolio of your box designs. Make sure your portfolio contains the following information:

- your calculations of the surface area of each box design,
- a one-page paper promoting the box design you think will be best, and
- a name for the new product and a logo or design for the box.

Math
In the Workplace

What You'll Learn

You'll learn to identify and classify polynomials and find their degree.

Why It's Important

Medicine Doctors can use polynomials to study the heart. *See Exercise 57.*

A cube is a solid figure in which all of the faces are squares. Suppose you wanted to paint the cube shown below. You would need to find the surface area of the cube to determine how much paint to buy.

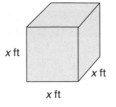

x ft
x ft
x ft

The area of each face of the cube is $x \cdot x$ or x^2. There are six faces to paint.

$$x^2 + x^2 + x^2 + x^2 + x^2 + x^2 = 6x^2$$

So, the surface area of the cube is $6x^2$ square feet.

The expression $6x^2$ is called a **monomial**. A monomial is a number, a variable, or a product of numbers and variables that have only positive exponents. A monomial cannot have a variable as an exponent.

Monomials		Not Monomials	
-4	a number	2^x	has a variable as an exponent
y	a variable	$x^2 + 3$	includes addition
a^2	the product of variables	$5a^{-2}$	includes a negative exponent
$\frac{1}{2}x^2y$	the product of numbers and variables	$\frac{3}{x}$	includes division

Examples

Determine whether each expression is a monomial. Explain why or why not.

1 $-6ab$

$-6ab$ is a monomial. It is the product of a number and variables.

2 $m^2 - 4$

$m^2 - 4$ is *not* a monomial because it includes subtraction.

Your Turn

a. 10 **b.** $5z^{-3}$ **c.** $\dfrac{6}{x}$ **d.** x^2

The sum of one or more monomials is called a **polynomial**. For example, $x^3 + x^2 + 3x + 2$ is a polynomial. The terms of the polynomial are x^3, x^2, $3x$, and 2. *Recall that to subtract, you add the opposite.*

Special names are given to polynomials with two or three terms. A polynomial with two terms is a **binomial**. A polynomial with three terms is a **trinomial**. Here are some examples.

Binomial	Trinomial
$x + 2$	$a + b + c$
$5c - 4$	$x^2 + 5x - 7$
$4w^2 - w$	$3a^2 + 5ab + 2b^2$

Examples

State whether each expression is a polynomial. If it is a polynomial, identify it as a *monomial*, *binomial*, or *trinomial*.

❸ $2m - 7$

The expression $2m - 7$ can be written as $2m + (-7)$. So, it is a polynomial. Since it can be written as the sum of two monomials, $2m$ and   7, it is a binomial.

❹ $x^2 + 3x - 4 - 5$

The expression $x^2 + 3x - 4 - 5$ can be written as $x^2 + 3x + (-9)$. So, it is a polynomial. Since it can be written as the sum of three monomials, it is a trinomial.

❺ $\dfrac{5}{2x} - 3$

The expression $\dfrac{5}{2x} - 3$ is not a polynomial since $\dfrac{5}{2x}$ is not a monomial.

Your Turn

e. $5a - 9 + 3$ **f.** $4m^{-2} + 2$ **g.** $3y^2 - 6 + 7y$

The terms of a polynomial are usually arranged so that the powers of one variable are in descending or ascending order.

Polynomial	Descending Order	Ascending Order	
$2x + x^2 + 1$	$x^2 + 2x + 1$	$1 + 2x + x^2$	
$3y^2 + 5y^3 + y$	$5y^3 + 3y^2 + y$	$y + 3y^2 + 5y^3$	
$x^2 + y^2 + 3xy$	$x^2 + 3xy + y^2$	$y^2 + 3xy + x^2$	*in terms of x*
$2xy + y^2 + x^2$	$y^2 + 2xy + x^2$	$x^2 + 2xy + y^2$	*in terms of y*

The **degree** of a monomial is the sum of the exponents of the variables.

Look Back

Zero Power:
Lesson 8–2

Monomial	Degree
$-3x^2$	2
$5pq^2$	$1 + 2 = 3$
2	0

$p = p^1$
$2 = 2x^0$

To find the degree of a polynomial, you must find the degree of each term. The degree of the polynomial is the greatest of the degrees of its terms.

Polynomial	Terms	Degree of the Terms	Degree of the Polynomial
$2n + 7$	$2n, 7$	1, 0	1
$3x^2 + 5x$	$3x^2, 5x$	2, 1	2
$a^6 + 2a^3 + 1$	$a^6, 2a^3, 1$	6, 3, 0	6
$5x^4 - 4a^2b^6 + 3x$	$5x^4, -4a^2b^6, 3x$	4, 8, 1	8

Examples

Find the degree of each polynomial.

6 $5a^2 + 3$

Term	Degree
$5a^2$	2
3	0

$3 = 3x^0$

So, the degree of $5a^2 + 3$ is 2.

7 $6x^2 - 4x^2y - 3xy$

Term	Degree
$6x^2$	2
$-4x^2y$	2 + 1 or 3
$-3xy$	1 + 1 or 2

$y = y^1$

So, the degree of $6x^2 - 4x^2y - 3xy$ is 3.

Your Turn

h. $3x^2 - 7x$

i. $8m^3 - 2m^2n^2 + 5$

Polynomials can be used to represent many real-world situations.

Example

Science Link

Real World

8 The expression $14x^3 - 17x^2 - 16x + 34$ can be used to estimate the number of eggs that a certain type of female moth can produce. In the expression, x represents the width of the abdomen in millimeters. About how many eggs would you expect this type of moth to produce if her abdomen measures 3 millimeters?

Explore	You know the measurement of the abdomen. You need to determine the number of eggs that the moth could produce.

Plan   In the expression $14x^3 - 17x^2 - 16x + 34$, replace x with 3 to determine the number of eggs that the moth could produce.

Solve
$$14x^3 - 17x^2 - 16x + 34 = 14(3)^3 - 17(3)^2 - 16(3) + 34$$
$$= 14(27) - 17(9) - 16(3) + 34$$
$$= 378 - 153 - 48 + 34$$
$$= 211$$

Examine   Check your computations by adding in a different order.
$$378 - 153 - 48 + 34 = 378 + 34 + (-153) + (-48)$$
$$= 412 + (-201)$$
$$= 211 \ \checkmark$$

The moth could produce about 211 eggs.

Check for Understanding

Communicating Mathematics

Study the lesson. Then complete the following.

1. **Write** an expression that is *not* a monomial. Explain why it is *not* a monomial.
2. **Explain** how to determine the degree of a polynomial.
3. **Arrange** the terms of the polynomial $3x^2 + 5xy + 2y^2$ so that the powers of x are in ascending order.

Math Journal

4. **Write** a definition for monomial, polynomial, binomial, and trinomial. Give three examples of each.

> **Vocabulary**
> monomial
> polynomial
> binomial
> trinomial
> degree

Guided Practice

Determine whether each expression is a monomial. Explain why or why not. *(Examples 1 & 2)*

5. $7xy$

6. $a^2 + 5$

State whether each expression is a polynomial. If it is a polynomial, identify it as a *monomial*, *binomial*, or *trinomial*. *(Examples 3–5)*

7. $2mn + 3$

8. 0

9. $\frac{4}{d} - d^2$

10. $h^2 - 3h + 8 + 2$

Find the degree of each polynomial. *(Examples 6 & 7)*

11. $25x$

12. 5

13. $15m^2 + 4n$

14. $-6y^3z - 4y^2z$

15. Geometry To find the number of diagonals d in an n-sided figure, you can use the formula $d = 0.5n^2 - 1.5n$. *(Example 8)*

a. Find the number of diagonals in an octagon.

b. How many diagonals are in a hexagon?

Exercises • • • • • • • • • • • • • • • • • •

Practice

Determine whether each expression is a monomial. Explain why or why not.

16. $5z$

17. $7a + 2$

18. $\dfrac{5}{x}$

19. $8x^2y$

20. y^{-3}

21. $-10a^2b^2c$

State whether each expression is a polynomial. If it is a polynomial, identify it as a *monomial*, *binomial*, or *trinomial*.

22. y^3

23. $2y^2 + 5y - 7$

24. $a^3 - 3a$

25. $2r + 3s^{-3} - t$

26. $17 + 12x^3 - 3x^4$

27. $2m^2 + 5 - 1$

28. $-4j + 2k^2 - 3$

29. $x^2 - \dfrac{1}{2}x$

30. $\dfrac{x}{y}$

31. $-12a^2b^3cd^5e^3$

32. $2.5w^5 - v^3w^2$

33. $3x + 4x$

Find the degree of each polynomial.

34. 1

35. $8x^2$

36. $-14x^3$

37. $5p^2 + 3p + 5$

38. $4s^2 - 6t^8$

39. $p - 4q + 5r$

40. $25a + a^2$

41. $a^3 + a^2 + a$

42. $15g^3h^4 - 10gh^5$

43. $10v^4w^2 + vw^5$

44. $3rs^4 - 2r^2 + 7$

45. $5cd^4 - 2bcd$

Arrange the terms of each polynomial so that the powers of *x* are in descending order.

46. $2x^2 + x^4 - 6$

47. $x^2 - x + 25 - x^5$

48. $3x - 4x^2y + 5 - 3x^5$

49. $6w^3x + 3wx^2 - 10x^6 + 5x^7$

Arrange the terms of each polynomial so that the powers of *x* are in ascending order.

50. $2 + x^4 + x^5 + x^2$

51. $5x^3 - x^2 + 7 + 2x$

52. $a^3x - 6a^3x^3 + 0.5x^5 + 4x^2$

53. $3x^3y + 5xy^4 - x^2y^3 + y^6$

54. *True* or *false*: Every monomial is a polynomial.

55. *True* or *false*: Every polynomial is a monomial.

Applications and Problem Solving

56. **Geometry** The surface area of a right cylinder is given by the polynomial $2\pi rh + 2\pi r^2$, where r is the radius of the base of the cylinder and h is the height of the cylinder.

    a. Classify the polynomial.

    b. Find the surface area of a right circular cylinder with a height of 5 feet and a radius of 3 feet. Round to the nearest tenth.

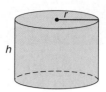

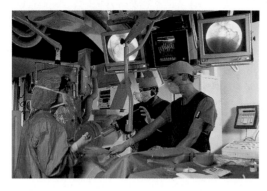

57. **Medicine** Doctors can study the heart of a potential heart attack patient by injecting a dye in a vein near the heart. In a healthy heart, the amount of dye in the bloodstream after t seconds is given by the expression $-0.006t^4 + 1.79t - 0.53t^2 + 0.14t^3$.

    a. Arrange the terms of the polynomial so that the powers of t are in descending order.

    b. Find the degree of the polynomial.

58. **Critical Thinking** A numeral in base 10 can be written in polynomial form. For example, $3247 = 3(10)^3 + 2(10)^2 + 4(10) + 7(10)^0$.

    a. Write the year of your birth in polynomial form.

    b. Suppose 2356 is a numeral in base a. Write 2356 in polynomial form.

    c. What is the degree of the polynomial?

Mixed Review

The lengths of three sides of a triangle are given. Determine whether each triangle is a right triangle. *(Lesson 8–7)*

59. 5 cm, 12 cm, 13 cm   60. 2 in., 3 in., 4 in.   61. 8 ft, 15 ft, 17 ft

Estimate each square root to the nearest whole number. *(Lesson 8–6)*

62. $\sqrt{20}$ 63. $\sqrt{50}$ 64. $\sqrt{75}$ 65. $\sqrt{120}$

66. **Population** In 2050, Earth's population is expected to be 8,900,000,000. Write this number in scientific notation. *(Lesson 8–4)*

Simplify each expression. *(Lesson 8–3)*

67. xy^{-3} 68. $\dfrac{k^{-3}}{k^4}$

69. **Standardized Test Practice** The graph of $y = 3x + 6$ is shown at the right. Which is the slope of a line parallel to the graph of $y = 3x + 6$? *(Lesson 7–7)*

    A -3 B 3

    C $-\dfrac{1}{3}$ D $\dfrac{1}{6}$

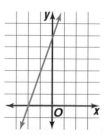

Extra Practice See p. 710.

Adding and Subtracting Polynomials

What You'll Learn

You'll learn to add and subtract polynomials.

Why It's Important

Framing Addition of polynomials can be used to find the size of a picture.
See Exercise 43.

You can use algebra tiles to model polynomials.

Here is a model of the polynomial $2x^2 + 3x + 1$.

Red tiles are used to represent -1, $-x$, and $-x^2$. Here is the model of the polynomial $-x^2 - 2x - 3$.

Once you know how to model polynomials using algebra tiles, you can use algebra tiles to add and subtract polynomials.

Hands-On Algebra
Algebra Tiles

Materials: algebra tiles

Add $x^2 + 2x - 3$ and $x^2 - x + 4$.

Step 1 Model each polynomial.

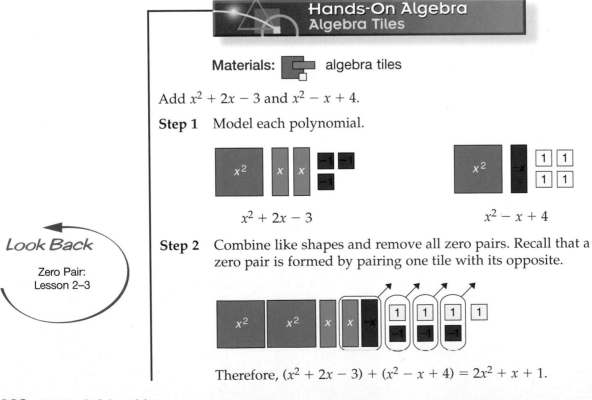

$x^2 + 2x - 3$ $x^2 - x + 4$

Look Back

Zero Pair: Lesson 2–3

Step 2 Combine like shapes and remove all zero pairs. Recall that a zero pair is formed by pairing one tile with its opposite.

Therefore, $(x^2 + 2x - 3) + (x^2 - x + 4) = 2x^2 + x + 1$.

Try These

Use algebra tiles to find each sum.

1. $(3x + 5) + (2x - 3)$ **2.** $(-2x + 3) + (4x - 3)$

3. $(2x^2 + 2x - 4) + (x^2 + 3x + 7)$ **4.** $(3x^2 + x - 4) + (x^2 - 2x - 1)$

5. $(x^2 - 1) + (2x + 3)$ **6.** $(2x^2 + 3) + (x^2 - 2x - 1)$

7. Write a rule for adding polynomials without models.

You can add polynomials by grouping the like terms together and then finding their sum.

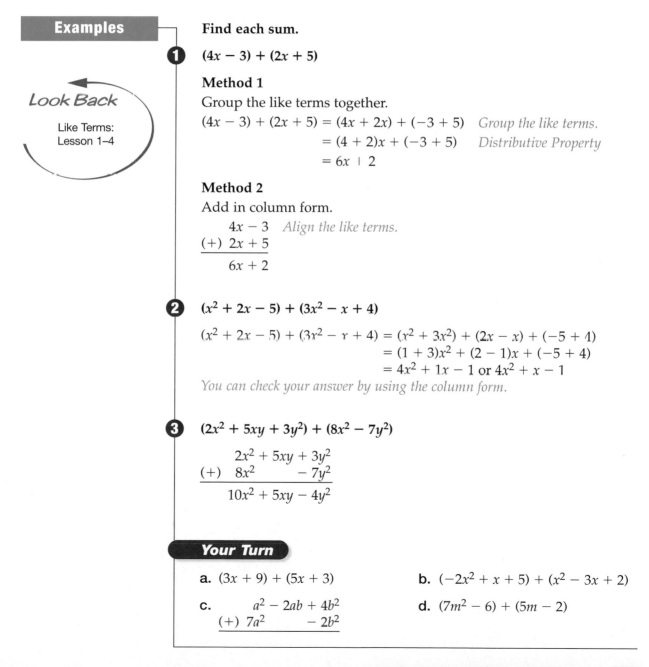

Examples

Look Back

Like Terms:
Lesson 1–4

Find each sum.

1 $(4x - 3) + (2x + 5)$

Method 1
Group the like terms together.
$(4x - 3) + (2x + 5) = (4x + 2x) + (-3 + 5)$ *Group the like terms.*
$= (4 + 2)x + (-3 + 5)$ *Distributive Property*
$= 6x + 2$

Method 2
Add in column form.
$$\begin{array}{r} 4x - 3 \\ (+)\ 2x + 5 \\ \hline 6x + 2 \end{array}$$ *Align the like terms.*

2 $(x^2 + 2x - 5) + (3x^2 - x + 4)$

$(x^2 + 2x - 5) + (3x^2 - x + 4) = (x^2 + 3x^2) + (2x - x) + (-5 + 4)$
$= (1 + 3)x^2 + (2 - 1)x + (-5 + 4)$
$= 4x^2 + 1x - 1 \text{ or } 4x^2 + x - 1$
You can check your answer by using the column form.

3 $(2x^2 + 5xy + 3y^2) + (8x^2 - 7y^2)$

$$\begin{array}{r} 2x^2 + 5xy + 3y^2 \\ (+)\ \ 8x^2 \qquad\ - 7y^2 \\ \hline 10x^2 + 5xy - 4y^2 \end{array}$$

Your Turn

a. $(3x + 9) + (5x + 3)$ **b.** $(-2x^2 + x + 5) + (x^2 - 3x + 2)$

c. $\begin{array}{r} a^2 - 2ab + 4b^2 \\ (+)\ 7a^2 \qquad\ - 2b^2 \end{array}$ **d.** $(7m^2 - 6) + (5m - 2)$

Recall that you can subtract an integer by adding its additive inverse or opposite.

$$2 - 3 = 2 + (-3) \quad \textit{The additive inverse of 3 is } -3.$$

$$5 - (-4) = 5 + 4 \quad \textit{The additive inverse of } -4 \textit{ is } 4.$$

Similarly, you can subtract a polynomial by adding its additive inverse. To find the additive inverse of a polynomial, replace each term with its additive inverse.

Polynomial	Additive Inverse
$a + 2$	$-a - 2$
$x^2 + 3x - 1$	$-x^2 - 3x + 1$
$2x^2 - 5xy + y^2$	$-2x^2 + 5xy - y^2$

$-(a + 2) = -a - 2$

$-(x^2 + 3x - 1) = -x^2 - 3x + 1$

$-(2x^2 - 5xy + y^2) = -2x^2 + 5xy - y^2$

Examples

Find each difference.

4 $(6x + 5) - (3x + 1)$

Method 1

Find the additive inverse of $3x + 1$. Then group the like terms together and add.

The additive inverse of $3x + 1$ is $-(3x + 1)$ or $-3x - 1$.

$$\begin{aligned}(6x + 5) - (3x + 1) &= (6x + 5) + (-3x - 1) \\ &= (6x - 3x) + (5 - 1) \quad \textit{Group the like terms.} \\ &= (6 - 3)x + (5 - 1) \quad \textit{Distributive Property} \\ &= 3x + 4\end{aligned}$$

Method 2

Arrange like terms in column form.

$$\begin{array}{r} 6x + 5 \\ (-)\ 3x + 1 \\ \hline \end{array} \qquad \textit{Add the additive inverse.} \qquad \begin{array}{r} 6x + 5 \\ (+)\ -3x - 1 \\ \hline 3x + 4 \end{array}$$

5 $(2y^2 - 3y + 5) - (y^2 + 2y + 8)$

The additive inverse of $y^2 + 2y + 8$ is $-(y^2 + 2y + 8)$ or $-y^2 - 2y - 8$.

$$\begin{aligned}(2y^2 - 3y + 5) - (y^2 + 2y + 8) &= (2y^2 - 3y + 5) + (-y^2 - 2y - 8) \\ &= (2y^2 - 1y^2) + (-3y - 2y) + (5 - 8) \\ &= (2 - 1)y^2 + (-3 - 2)y + (5 - 8) \\ &= 1y^2 + (-5y) + (-3) \textit{ or } y^2 - 5y - 3\end{aligned}$$

To check this result, add $y^2 - 5y - 3$ and $y^2 + 2y + 8$. Their sum should be $2y^2 - 3y + 5$.

6 $(3x^2 + 5) - (-4x + 2x^2 + 3)$

Reorder the terms so that the powers of x are in descending order.
$-4x + 2x^2 + 3 = 2x^2 - 4x + 3$

Then subtract.

$$
\begin{array}{r}
3x^2 \qquad + 5 \\
(-)\ 2x^2 - 4x + 3 \\
\hline
\end{array}
$$

Add the additive inverse.

$$
\begin{array}{r}
3x^2 \qquad + 5 \\
(+)\ -2x^2 + 4x - 3 \\
\hline
x^2 + 4x + 2
\end{array}
$$

Your Turn

e. $(3x - 2) - (5x - 4)$

f. $(10x^2 + 8x - 6) - (3x^2 + 2x - 9)$

g.
$$
\begin{array}{r}
6m^2 \qquad + 7 \\
(-)\ -2m^2 + 2m - 3 \\
\hline
\end{array}
$$

h. $(5x^2 - 4x) - (2 - 3x)$

Polynomials can be used to represent measures of geometric figures.

| **Example** |
| **Geometry Link** |

7 The measure of the perimeter of triangle *ABC* is $7x + 2y$. Find the measure of the third side of the triangle.

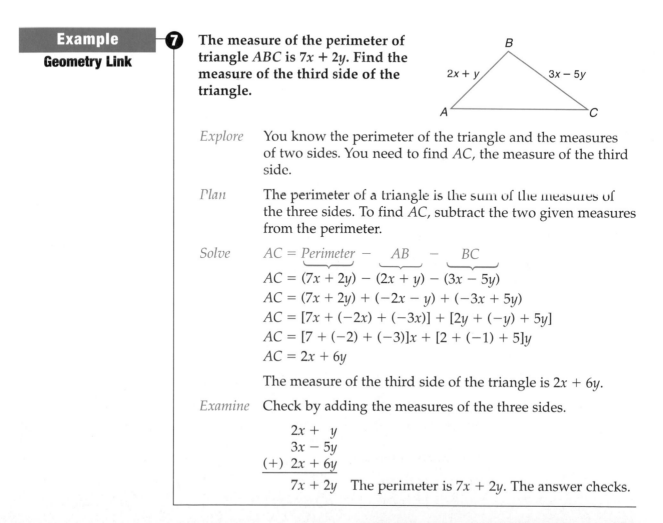

Explore   You know the perimeter of the triangle and the measures of two sides. You need to find *AC*, the measure of the third side.

Plan   The perimeter of a triangle is the sum of the measures of the three sides. To find *AC*, subtract the two given measures from the perimeter.

Solve $AC = \underbrace{Perimeter} - \underbrace{AB} - \underbrace{BC}$
$AC = (7x + 2y) - (2x + y) - (3x - 5y)$
$AC = (7x + 2y) + (-2x - y) + (-3x + 5y)$
$AC = [7x + (-2x) + (-3x)] + [2y + (-y) + 5y]$
$AC = [7 + (-2) + (-3)]x + [2 + (-1) + 5]y$
$AC = 2x + 6y$

The measure of the third side of the triangle is $2x + 6y$.

Examine   Check by adding the measures of the three sides.

$$
\begin{array}{r}
2x + \ y \\
3x - 5y \\
(+)\ 2x + 6y \\
\hline
7x + 2y
\end{array}
$$
The perimeter is $7x + 2y$. The answer checks.

Check for Understanding

Study the lesson. Then complete the following.

1. **Describe** the first step you take when you add or subtract polynomials in column form.
2. **Explain** how addition and subtraction of polynomials are related.

Guided Practice

🕐 **Getting Ready** **Find the additive inverse of each polynomial.**

Sample 1: $6y - 3z$	**Sample 2:** $-a^2 + 2ab - 3b^2$
Solution: $-6y + 3z$	**Solution:** $a^2 - 2ab + 3b^2$

3. $2a + 9b$

4. $-4m + 6n$

5. $x^2 + 8x + 5$

6. $-3h^2 - 2h - 3$

7. $4xy^2 + 6x^2y - y^3$

8. $-2c^3 + c^2 - 3c$

Find each sum. *(Examples 1–3)*

9.
$$\begin{array}{r} 6y - 5 \\ (+)\ 2y + 7 \\ \hline \end{array}$$

10.
$$\begin{array}{r} x^2 - 6x + 5 \\ (+)\ x^2 + 4x - 7 \\ \hline \end{array}$$

11. $(2x^2 + 4x + 5) + (x^2 - 5x - 3)$

12. $(2x^2 - 5x + 4) + (3x^2 - 1)$

Find each difference. *(Examples 4–6)*

13.
$$\begin{array}{r} 3x + 4 \\ (-)\ x + 2 \\ \hline \end{array}$$

14.
$$\begin{array}{r} 6m^2 - 5m + 3 \\ (-)\ 5m^2 + 2m - 7 \\ \hline \end{array}$$

15. $(5x^2 + 4x - 1) - (4x^2 + x + 2)$

16. $(5x^2 - 4x) - (-3x + 2)$

17. **Geometry** The measure of the perimeter of triangle *DEF* is $12x^2 - 7x + 9$. Find the measure of the third side of the triangle. *(Example 7)*

$3x^2 + 2x - 1$

D $8x^2 - 8x + 5$ F

Exercises

• • • • • • • • • • • • • • • • • •

Practice

Find each sum.

18.
$$\begin{array}{r} 2x + 3 \\ (+)\ x - 1 \\ \hline \end{array}$$

19.
$$\begin{array}{r} 2x^2 - 5x + 4 \\ (+)\ 2x^2 + 8x - 1 \\ \hline \end{array}$$

20.
$$\begin{array}{r} 2x^2 + 3xy - 4y^2 \\ (+)\ 4x^2\ \ \ \ \ \ \ + 4y^2 \\ \hline \end{array}$$

21. $(12x + 8y) + (2x - 7y)$

22. $(-7y + 3x) + (4x + 3y)$

23. $(n^2 + 5n + 3) + (2n^2 + 8n + 8)$

24. $(5x^2 - 7x + 9) + (3x^2 + 4x - 6)$

25. $(5n^2 - 4) + (2n^2 - 8n + 9)$

26. $(-2x^2 + 3xy - 3y^2) + (2x^2 - 5xy)$

Find each difference.

27.
$$\begin{array}{r} 9x + 5 \\ (-)\ 4x - 3 \\ \hline \end{array}$$

28.
$$\begin{array}{r} 5a^2 + 7a + 9 \\ (-)\ 3a^2 + 4a + 1 \\ \hline \end{array}$$

29.
$$\begin{array}{r} 3x^2 + 4x - 1 \\ (-)\ 4x^2 +\ \ x + 2 \\ \hline \end{array}$$

Math
In the Workplace

What You'll Learn
You'll learn to multiply a polynomial by a monomial.

Why It's Important
Recreation You can use monomials and polynomials to solve problems involving recreation.
See Exercise 65.

Suppose you have a square whose length and width are x feet. If you increase the length by 3 feet, what is the area of the new figure?

You can model this problem by using algebra tiles. The figures show how to make a rectangle whose length is $x + 3$ feet and whose width is x feet.

The area of any rectangle is the product of its length and its width. The area can also be found by adding the areas of the tiles.

Formula
$A = \ell w$
$A = (x + 3)x$ or $x(x + 3)$
$A = x^2 + 3x$

Algebra Tiles
$A = x^2 + x + x + x$
$A = x^2 + 3x$

Since the areas are equal, $x(x + 3) = x^2 + 3x$. The new area is $x^2 + 3x$ square feet.

The example above shows how the Distributive Property can be used to multiply a polynomial by a monomial.

Multiplying a Polynomial by a Monomial	**Words:**	To multiply a polynomial by a monomial, use the Distributive Property.
	Symbols:	$a(b + c) = ab + ac$
	Model:	

$$
\begin{array}{c|c}
b & c \\
\hline
ab & ac
\end{array}
$$

Examples

Find each product.

1 $y(y + 5)$
$y(y + 5) = y(y) + y(5)$
$\qquad\quad = y^2 + 5y$

Look Back

Multiplying Powers:
Lesson 8–2

2 $b(2b^2 + 3)$
$b(2b^2 + 3) = b(2b^2) + b(3)$
$\qquad\qquad\quad = 2b^3 + 3b$

30. $(3x - 4) - (5x + 2)$ **31.** $(9x - 4y) - (12x - 5y)$
32. $(2x^2 + 5x + 3) - (x^2 - 2x + 3)$ **33.** $(3a^2 + 2a - 5) - (2a^2 - a - 4)$
34. $(3x^2 + 5x + 4) - (-1 + x^2)$ **35.** $(5x^2 - 4xy) - (2y^2 - 3xy)$

Find each sum or difference.
36. $(6x^3 - 10) - (2x^3 + 5x + 7)$ **37.** $(2pq + 3qr - 4pr) + (5pr - 3qr)$
38. $(3 + 2a - a^2) + (a^2 + 8a + 5)$ **39.** $(-x^2y - 4x^2y^2) - (x^2y + 3x^2y^2)$

40. Find the sum of $(x^2 + x + 5)$, $(3x^2 - 4x - 2)$, and $(2x^2 + 2x - 1)$.
41. What is $x^2 + 6x$ minus $3x^2 + 7$?

42. Geometry   The measure of each side of a triangle is $3x + 1$. Find its perimeter.

43. Framing   The standard measurement for a custom-made picture frame is the *united inch*. It refers to the sum of the length and the width of a picture to be framed. Suppose a picture's length is 4 inches longer than its width. Then the width is w and the length would be $w + 4$.

$\longleftarrow\quad w + 4 \quad\longrightarrow$

w

a. Write a polynomial that represents the size of the picture in united inches.

b. If the width is 15 inches, use the polynomial from part a to find the size of the picture in united inches.

44. Critical Thinking   One polynomial is subtracted from a second polynomial and the difference is $3m^2 + m - 5$. What is the difference when the second polynomial is subtracted from the first?

Mixed Review

Find the degree of each polynomial.   *(Lesson 9–1)*
45. $y^2 + 3y + 2$ **46.** $n^6 + 3n^3 + 2$ **47.** $3x^2 + 2x^2y$

48. Sailing   A rope from the top of a mast on a sailboat extends to a point 6 feet from the base of the mast. Suppose the rope is 24 feet long. To the nearest foot, how high is the mast?   *(Lesson 8–7)*

Simplify each expression.   *(Lesson 8–2)*
49. $(x^2y)(x^3y^2)$ **50.** $(4a^2)(2a^4)$ **51.** $(-2y)(-5xy)$

52. Standardized Test Practice   Which relation is a function?
(Lesson 6–4)
    A $\{(3, 1), (4, 2), (2, 3), (1, 2)\}$ **B** $\{(1, 1), (-1, 5), (3, 3), (-1, 2)\}$
    C $\{(-1, 3), (1, 4), (3, 2), (1, 2)\}$ **D** $\{(-2, 1), (4, 2), (3, 3), (3, 2)\}$

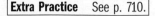

3 $-2n(7 - 5n^2)$

$$-2n(7 - 5n^2) = -2n(7) + (-2n)(-5n^2)$$
$$= -14n + 10n^3$$

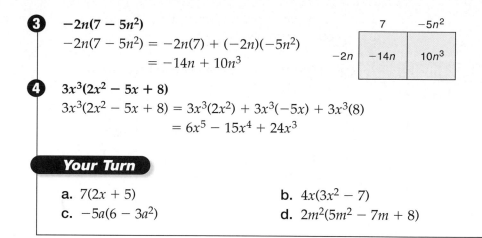

4 $3x^3(2x^2 - 5x + 8)$

$$3x^3(2x^2 - 5x + 8) = 3x^3(2x^2) + 3x^3(-5x) + 3x^3(8)$$
$$= 6x^5 - 15x^4 + 24x^3$$

Your Turn

a. $7(2x + 5)$

b. $4x(3x^2 - 7)$

c. $-5a(6 - 3a^2)$

d. $2m^2(5m^2 - 7m + 8)$

Many equations contain polynomials that must be multiplied.

Examples

Solve each equation.

5 $11(y - 3) + 5 = 2(y + 22)$

$$11(y - 3) + 5 = 2(y + 22)$$

$11y - 33 + 5 = 2y + 44$	*Distributive Property*
$11y - 28 = 2y + 44$	*Combine like terms: $-33 + 5 = -28$.*
$11y - 28 - 2y = 2y + 44 - 2y$	*Subtract 2y from each side.*
$9y - 28 = 44$	
$9y - 28 + 28 = 44 + 28$	*Add 28 to each side.*
$9y = 72$	
$\dfrac{9y}{9} = \dfrac{72}{9}$	*Divide each side by 9.*
$y = 8$	The solution is 8.

Look Back

Solving Equations with Grouping Symbols: Lesson 4–7

6 $w(w + 12) = w(w + 14) + 12$

$$w(w + 12) = w(w + 14) + 12$$

$w^2 + 12w = w^2 + 14w + 12$	*Distributive Property*
$w^2 + 12w - w^2 = w^2 + 14w + 12 - w^2$	*Subtract w^2 from each side.*
$12w = 14w + 12$	
$12w - 14w = 14w + 12 - 14w$	*Subtract 14w from each side.*
$-2w = 12$	
$\dfrac{-2w}{-2} = \dfrac{12}{-2}$	*Divide each side by -2.*
$w = -6$	The solution is -6.

Your Turn

e. $2(5x - 12) = 6(-2x + 3) + 2$

f. $a(a + 2) + 3a = a(a - 3) + 8$

You can apply multiplication of a polynomial by a monomial to problems involving area.

7 **Find the area of the shaded region in simplest form.**

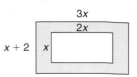

Subtract the area of the smaller rectangle from the area of the larger rectangle.

area of larger rectangle: $3x(x + 2)$ $A = \ell w$
area of smaller rectangle: $2x(x)$
area of shaded region: $3x(x + 2) - 2x(x)$

$A = 3x(x + 2) - 2x(x)$
$A = 3x(x) + 3x(2) - 2x(x)$ *Distributive Property*
$A = 3x^2 + 6x - 2x^2$
$A = 1x^2 + 6x$ or $x^2 + 6x$ *Combine like terms:* $3x^2 - 2x^2 = 1x^2$.

The area of the shaded region is $x^2 + 6x$.

Check for Understanding

Study the lesson. Then complete the following.

1. **Name** the property used to express $3n(2n - 6)$ as $6n^2 - 18n$.

2. **Write** an expression for the area of the rectangle at the right in the following two ways.

   a. a product of a monomial and a polynomial

   b. a sum of monomials

3. Consuelo says that $2x(3x + 4) = 6x^2 + 8x$ is a true statement. Shawn says that $2x(3x + 4) = 6x^2 + 4$ is a true statement. Who is correct? Explain.

Guided Practice

Find each product. *(Examples 1–4)*

4. $2(y + 5)$ 5. $x(x + 2)$ 6. $3a(a - 1)$

7. $-4(x - 2)$ 8. $-3n(5 - 2n)$ 9. $4z(z^2 - 2z)$

10. $2(2d^2 + 3d + 8)$ 11. $3(8y^2 + 3y - 5)$ 12. $-2a(4a^2 - 3a - 1)$

Solve each equation. *(Examples 5 & 6)*

13. $7(x + 2) = 42$ 14. $4(y - 8) + 10 = 2y + 12$

15. $-3(2a - 4) + 9 = 3(a + 1)$ 16. $x(x + 3) + 5x = x(x + 5) + 9$

17. **Geometry** Find the area of the shaded region in simplest form. *(Example 7)*

Exercises

Practice

Find each product.

18. $2(x + 6)$ **19.** $-3(y + 3)$ **20.** $7(2a + 3)$

21. $x(x - 5)$ **22.** $n(n + 4)$ **23.** $z(3z - 2)$

24. $2m(m + 4)$ **25.** $4x(2x - 3)$ **26.** $5y(y + 1)$

27. $-2a(a - 2)$ **28.** $-3x(x - 5)$ **29.** $-5y(6 - 2y)$

30. $3s(4s^2 - 7)$ **31.** $5d(d^2 + 3)$ **32.** $-7p(-3p - 6)$

33. $-2a(5a^2 - 7a + 2)$ **34.** $4x(-2x + 7x^2)$ **35.** $5n(8n^3 + 7n^2 - 3n)$

36. $-3y(6 - 9y + 4y^2)$ **37.** $7(-2a^2 + 5a - 11)$ **38.** $-5x^2(3x^2 - 8x - 12)$

39. $1.2(c^2 - 10)$ **40.** $0.1(4x^2 - 7x)$ **41.** $0.25x(4x - 8)$

42. $\frac{1}{2}(2x^2 - 6x)$ **43.** $\frac{2}{3}(6y^2 - 9y + 3)$ **44.** $\frac{1}{4}x(12x^2 + 8x)$

Solve each equation.

45. $8(x - 3) = 16$ **46.** $4(y - 3) = -8$

47. $40 = 5(a + 10)$ **48.** $2(x + 4) + 9 = 5x - 1$

49. $6(a + 2) - 5 = 2a + 3$ **50.** $8x + 14 = 3(x - 2) + 10$

51. $2(5a - 13) = 6(-2a + 3)$ **52.** $3(2y - 4) = -2(2y - 9)$

53. $5(n + 7) = 3(-2n + 1) - 1$ **54.** $b(b + 12) = b(b + 14) + 12$

55. $c(c + 8) - c(c + 3) - 23 = 3c + 11$

56. $m(m - 8) + 3m = -2 + m(m - 9)$

57. What is the product of $4x$ and $x^2 - 2x + 1$?

58. Multiply $-2n$ and $3 - 5n$.

Simplify.

59. $2u(u^2 - 5u + 4) - 6(a^3 + 3a - 11)$

60. $5t^2(t + 2) - 5t(4t^2 - 3t + 6) + 3(t^2 - 6)$

61. $3n^2(n - 4) + 6n(3n^2 + n - 7) - 4(n - 7)$

Applications and Problem Solving

62. Sports The length of a billiards table is twice as long as its width. Suppose the width of a billiards table is $4x - 1$. What is the length of the table?

Geometry Find the area of the shaded region for each figure.

63.

64.

65. Recreation On the Caribbean island of Trinidad, children play a form of hopscotch called Jumby. The pattern for this game is shown at the right. Suppose the length of each rectangle is $y + 5$ units long and y units wide.

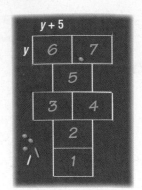

a. Write an expression in simplest form for the area of the pattern.

b. If y represents 10 inches, find the area of the pattern.

66. Critical Thinking Write three different multiplication problems whose product is $6a^2 + 8a$.

Mixed Review

Find each sum or difference. *(Lesson 9–2)*

67. $(2x + 1) + (5x - 3)$

68. $(3a - 2) - (a + 4)$

69. $(x^2 + 2x - 3) + (2x^2 - 4x)$

70. $(y^2 + 8y) - (2y^2 + 5)$

71. Ecology The deer population of the Kaibab Plateau in Arizona from 1905 until 1930 can be estimated by the polynomial $-0.13x^5 + 3.13x^4 + 4000$, where x is the number of years after 1900. Find the degree of the polynomial. *(Lesson 9–1)*

Simplify. *(Lesson 8–5)*

72. $\sqrt{81}$

73. $-\sqrt{121}$

74. $\sqrt{256}$

75. $\sqrt{\dfrac{9}{100}}$

Find the odds of each outcome if a die is rolled. *(Lesson 5–6)*

76. a number less than 5

77. an odd number

78. Standardized Test Practice Which equation is equivalent to $12 + (x - 7) = 21$? *(Lesson 4–7)*

A $12x - 7 = 21$

B $12x - 84 = 21$

C $x + 5 = 21$

D $19 + x = 21$

Quiz 1 — Lessons 9–1 through 9–3

1. Find the degree of $3y^4 + 2y^3 - y^2 + 5y - 3$. *(Lesson 9–1)*

2. Arrange the terms of $2xy + x^2 - y^2$ so that the powers of x are in descending order. *(Lesson 9–1)*

Find each sum or difference. *(Lesson 9–2)*

3. $(4x^2 + 3x) - (6x^2 - 5x + 2)$

4. $(2w^2 - 4w - 12) + (15 - 3w^2 + 2w)$

5. Sports In the National Football League, the length of the playing field is 40 feet longer than twice its width. *(Lesson 9–3)*

a. Express the width and length of the playing field as polynomials.

b. Find the polynomial that represents the area of the playing field.

Extra Practice See p. 710.

Math In the Workplace

What You'll Learn
You'll learn to multiply two binomials.

Why It's Important
Packaging   The FOIL method can be used to find the dimensions of a cereal box. See Exercise 55.

Katie has a square herb garden in which she grows parsley. She wants to increase the length by 2 feet and the width by 1 foot so she can grow sage, rosemary, and thyme. A plan for the garden is shown at the right. Find two different expressions for the area of the new garden.

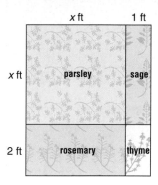

The area of the garden is the product of its length and width.

Formula

$$A = \ell w$$
$$= (x + 2)(x + 1)$$

You can also find the area by adding the areas of the smaller regions.

Sum of Regions

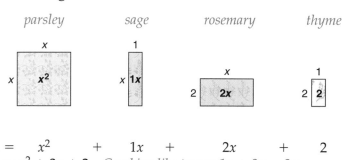

$$A = \quad x^2 \quad + \quad 1x \quad + \quad 2x \quad + \quad 2$$
$$= x^2 + 3x + 2 \quad \textit{Combine like terms: } 1x + 2x = 3x.$$

Since the areas are equal, $(x + 2)(x + 1) = x^2 + 3x + 2$.

The multiplication can also be shown in the model below. Notice that the Distributive Property is used twice.

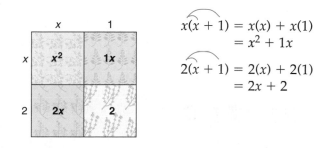

$$x(x + 1) = x(x) + x(1)$$
$$= x^2 + 1x$$

$$2(x + 1) = 2(x) + 2(1)$$
$$= 2x + 2$$

So, $(x + 2)(x + 1) = x^2 + 1x + 2x + 2$ or $x^2 + 3x + 2$.

Materials: 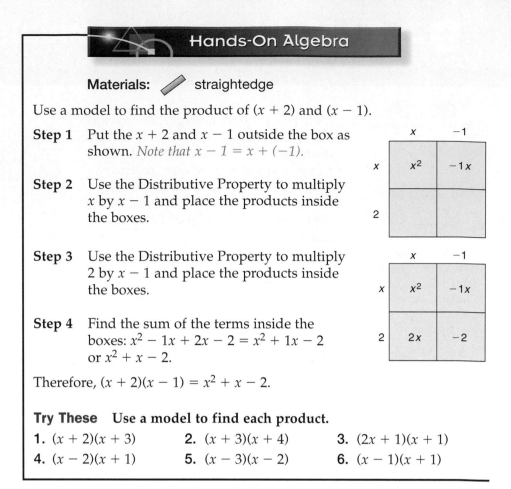 straightedge

Use a model to find the product of $(x + 2)$ and $(x - 1)$.

Step 1 Put the $x + 2$ and $x - 1$ outside the box as shown. *Note that $x - 1 = x + (-1)$.*

Step 2 Use the Distributive Property to multiply x by $x - 1$ and place the products inside the boxes.

Step 3 Use the Distributive Property to multiply 2 by $x - 1$ and place the products inside the boxes.

Step 4 Find the sum of the terms inside the boxes: $x^2 - 1x + 2x - 2 = x^2 + 1x - 2$ or $x^2 + x - 2$.

Therefore, $(x + 2)(x - 1) = x^2 + x - 2$.

Try These **Use a model to find each product.**

1. $(x + 2)(x + 3)$ **2.** $(x + 3)(x + 4)$ **3.** $(2x + 1)(x + 1)$

4. $(x - 2)(x + 1)$ **5.** $(x - 3)(x - 2)$ **6.** $(x - 1)(x + 1)$

You can also use the Distributive Property to multiply binomials.

Examples

Find each product.

❶ $(x + 3)(x - 4)$

$$\begin{aligned}(x + 3)(x - 4) &= x(x - 4) + 3(x - 4) &&\text{Distributive Property}\\ &= x(x) + x(-4) + 3(x) + 3(-4) &&\text{Distributive Property}\\ &= x^2 - 4x + 3x - 12 &&\text{Simplify.}\\ &= x^2 - 1x - 12 \text{ or } x^2 - x - 12 &&\text{Combine like terms.}\end{aligned}$$

❷ $(2y - 1)(y - 3)$

$$\begin{aligned}(2y - 1)(y - 3)\\ &= 2y(y - 3) + (-1)(y - 3) &&\text{Distributive Property}\\ &= 2y(y) + 2y(-3) + (-1)(y) + (-1)(-3) &&\text{Distributive Property}\\ &= 2y^2 - 6y - 1y + 3 &&\text{Simplify.}\\ &= 2y^2 - 7y + 3 &&\text{Combine like terms.}\end{aligned}$$

Your Turn

a. $(y + 4)(y - 2)$ **b.** $(m - 3)(m + 1)$

c. $(x - 5)(x - 2)$ **d.** $(2a - 3)(a - 2)$

Two binomials can always be multiplied using the Distributive Property. However, the following shortcut can also be used. It is called the **FOIL method**.

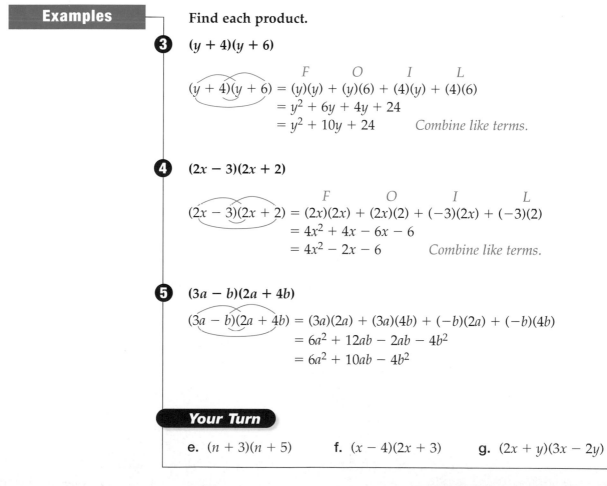

$$(3x + 1)(x + 2) = (3x)(x) \quad + \quad (3x)(2) \quad + \quad (1)(1x) \quad + \quad (1)(2)$$

F	O	I	L
product of FIRST terms	product of OUTER terms	product of INNER terms	product of LAST terms

$$= 3x^2 + 6x + 1x + 2$$
$$= 3x^2 + 7x + 2$$

FOIL Method for Multiplying Two Binomials	To multiply two binomials, find the sum of the products of **F** the First terms, **O** the Outer terms, **I** the Inner terms, and **L** the Last terms.

Examples

Find each product.

3 $(y + 4)(y + 6)$

$$\begin{aligned} & \qquad\qquad\quad F \quad\quad O \quad\quad I \quad\quad L \\ (y + 4)(y + 6) &= (y)(y) + (y)(6) + (4)(y) + (4)(6) \\ &= y^2 + 6y + 4y + 24 \\ &= y^2 + 10y + 24 \qquad \textit{Combine like terms.} \end{aligned}$$

4 $(2x - 3)(2x + 2)$

$$\begin{aligned} & \qquad\qquad\qquad\quad F \quad\quad\quad O \quad\quad\quad I \quad\quad\quad L \\ (2x - 3)(2x + 2) &= (2x)(2x) + (2x)(2) + (-3)(2x) + (-3)(2) \\ &= 4x^2 + 4x - 6x - 6 \\ &= 4x^2 - 2x - 6 \qquad \textit{Combine like terms.} \end{aligned}$$

5 $(3a - b)(2a + 4b)$

$$\begin{aligned} (3a - b)(2a + 4b) &= (3a)(2a) + (3a)(4b) + (-b)(2a) + (-b)(4b) \\ &= 6a^2 + 12ab - 2ab - 4b^2 \\ &= 6a^2 + 10ab - 4b^2 \end{aligned}$$

Your Turn

e. $(n + 3)(n + 5)$ **f.** $(x - 4)(2x + 3)$ **g.** $(2x + y)(3x - 2y)$

Sometimes it is not possible to simplify the product of two binomials.

Example ⑥ Find the product of $x^2 - 4$ and $x + 3$.

$$(x^2 - 4)(x + 3) = (x^2)(x) + (x^2)(3) + (-4)(x) + (-4)(3)$$
$$= x^3 + 3x^2 - 4x - 12 \quad \textit{There are no like terms.}$$

Your Turn

h. Find the product of $x - 2$ and $x^2 + 3$.

You can use the FOIL method to solve problems involving volume.

Example ⑦

Geometry Link

The volume V of a rectangular prism is equal to the area of the base B times the height h. Express the volume of the prism as a polynomial. Use $V = Bh$.

$x - 1$ units
x units
$x + 6$ units

First, find the area of the base. The base is a rectangle.
$B = \ell w$
$B = (x + 6)(x)$ *Replace ℓ with $x + 6$ and w with x.*
$B = x^2 + 6x$ *Use the Distributive Property.*

To find the volume, multiply the area of the base by the height.
$V = Bh$
$V = (x^2 + 6x)(x - 1)$ *Replace B with $x^2 + 6x$ and h with $x - 1$.*
$V = (x^2)(x) + (x^2)(-1) + (6x)(x) + (6x)(-1)$ *Use the FOIL method.*
$V = x^3 - x^2 + 6x^2 - 6x$
$V = x^3 + 5x^2 - 6x$

The volume of the prism is $x^3 + 5x^2 - 6x$ cubic units.

Check for Understanding

Communicating Mathematics

Study the lesson. Then complete the following.

1. Draw a model that represents the product of $x - 2$ and $x + 3$. Then find the product.

2. Write the multiplication problem represented by each model.

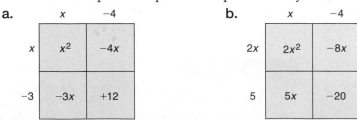

a.

	x	-4
x	x^2	$-4x$
-3	$-3x$	$+12$

b.

	x	-4
$2x$	$2x^2$	$-8x$
5	$5x$	-20

Math Journal

3. Compare and contrast the procedure for multiplying two binomials and the procedure for multiplying a binomial and a monomial.

Guided Practice

4. $(a - 5)(a - 3)$ **5.** $(x + 1)(x + 5)$ **6.** $(g + 1)(g - 9)$
7. $(2m + 3)(3m + 2)$ **8.** $(2j + 1)(j - 3)$ **9.** $(y + 5)(2y + 4)$

Find each product. Use the Distributive Property or the FOIL method.
(*Examples 1–5*)

10. $(x + 2)(x + 4)$ **11.** $(w - 7)(w + 5)$
12. $(a - 3)(a - 6)$ **13.** $(2y + 3)(y - 1)$
14. $(3x - 4)(2x - 3)$ **15.** $(2y + 7)(3y - 5)$
16. $(x + 2y)(x + 3y)$ **17.** $(2a - 5b)(a + 2b)$
18. $(y^2 + 3)(y - 2)$ **19.** $(m^3 - 2m)(m + 3)$

20. Geometry Express the volume of the rectangular prism as a polynomial. Use the formula $V = Bh$, where B is the area of the base and h is the height of the prism. (*Example 6*)

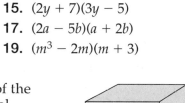
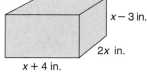

$x - 3$ in.

$2x$ in.

$x + 4$ in.

Exercises

Practice

Find each product. Use the Distributive Property or the FOIL method.

21. $(x + 4)(x + 8)$ **22.** $(r + 7)(r + 2)$ **23.** $(a - 3)(a + 7)$
24. $(x - 3)(x - 7)$ **25.** $(n - 11)(n - 5)$ **26.** $(y - 4)(y + 15)$
27. $(z + 6)(z - 4)$ **28.** $(s - 11)(s + 5)$ **29.** $(x - 4)(3x + 2)$
30. $(2a + 5)(a - 7)$ **31.** $(z - 4)(3z - 5)$ **32.** $(y - 3)(2y - 5)$
33. $(5n + 2)(n - 3)$ **34.** $(x + 7)(2x - 3)$ **35.** $(3a + 1)(3a + 1)$
36. $(2x + 5)(5x + 3)$ **37.** $(4h - 3)(3h + 2)$ **38.** $(7a - 1)(2a - 3)$
39. $(2x + 7)(x - 3)$ **40.** $(8m + 2n)(6m + 5n)$ **41.** $(2y - 4z)(3y - 6z)$
42. $(3a - b)(2a + b)$ **43.** $(2x + 5y)(3x - y)$ **44.** $(5n - 2p)(5n + 2p)$
45. $(x^2 + 1)(x - 2)$ **46.** $(y + 3)(y^2 - 4)$ **47.** $(x^2 + 3)(3x^2 - 1)$
48. $(2y^2 + 1)(y + 1)$ **49.** $(x^2 + 2x)(x - 3)$ **50.** $(x + a)(x + b)$

51. Find the product of $(x + 3)$ and $(x - 3)$.
52. What is the product of x, $(2x + 1)$, and $(x - 3)$?

53. **Geometry**   Find the area of each shaded region in simplest form.

   a.

   $x + 2$

   x

   $x + 1$ | x

   b.

   $2x + 1$

   $x - 1$

   $2x$ | $x - 1$

54. **Number Theory**   Use the FOIL method to find the product of 16 and
   38. (*Hint*: Write 16 as $10 + 6$ and 38 as $30 + 8$.)

55. **Packaging**   A cereal box has a length of $2x$ inches, a width of $x - 2$
   inches, and a height of $2x + 5$ inches.
   a. Express the volume of the package as a polynomial.
   b. Find the volume of the box if $x = 5$.

56. **Critical Thinking**   Use the Distributive Property to find the product of
   $x + 2$ and $x^2 - 3x + 2$.

Mixed Review

57. **Clocks**   Before mechanical clocks were invented, candles were used
   to keep track of time. One formula that was used was $c = 2(5 - t)$,
   where c is the height of the candle in inches and t is the time in hours
   that the candle burns.   (*Lesson 9–3*)
   a. Use the Distributive Property to multiply 2 and $5 - t$.
   b. Find the height of the candle when $t = 2$.

58. Find the sum of $x^2 + 5x$ and $-3x - 7$.   (*Lesson 9–2*)

Write each expression using exponents.   (*Lesson 8–1*)

59. $2 \cdot 2 \cdot 2 \cdot x \cdot x \cdot x$ 60. 10   61. $(3)(3)(-2)(-2)(-2)$

Graph each equation.   (*Lesson 7–5*)

62. $y = 2x + 3$ 63. $y = -x + 2$ 64. $4x + 5y = 20$

Data Update   For the
latest information on
breakfast foods, visit:
www.algconcepts.
glencoe.com

65. **Food**   The graph shows
   the favorite breakfast
   food for students ages
   9–12. Suppose you
   interview 200 students
   who are in this age
   range. How many
   would you expect to
   like cereal for breakfast?
   (*Lesson 5–4*)

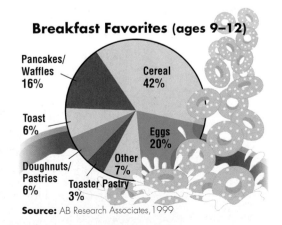

Breakfast Favorites (ages 9–12)

Pancakes/
Waffles
16%

Cereal
42%

Toast
6%

Eggs
20%

Other
7%

Doughnuts/
Pastries
6%

Toaster Pastry
3%

Source: AB Research Associates, 1999

66. **Open-Ended Test Practice**   Draw and label a rectangle whose
   perimeter is 20 centimeters.   (*Lesson 1–5*)

Extra Practice   See p. 711.

Math
In the Workplace

What You'll Learn

You'll learn to develop and use the patterns for $(a + b)^2$, $(a - b)^2$, and $(a + b)(a - b)$.

Why It's Important

Biology Geneticists use a technique similar to finding $(a + b)^2$ to predict the characteristics of a population. *See Example 5.*

Which appears to be greater, the height of the Gateway Arch or its width? It appears to be taller than it is wide. However, in reality, the width and height are equal. They both measure 630 feet.

Sometimes, incorrect conclusions are made from our observations. The same is true in mathematics. An expression like $(x + 1)^2$ is often assumed to be equal to $x^2 + 1^2$. But the model below shows that $(x + 1)^2$ is *not* equal to $x^2 + 1^2$.

Gateway Arch,
St. Louis, Missouri

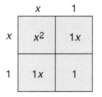

$$(x + 1)^2 = x^2 + 1x + 1x + 1$$
$$= x^2 + 2x + 1$$

twice the product of 1 and x

You can use a similar model to find $(x - 1)^2$. $x - 1 = x + (-1)$

$$(x - 1)^2 = x^2 - 1x - 1x + 1$$
$$= x^2 - 2x + 1$$

twice the product of −1 and x

The square of a sum and the square of a difference can be found by using the following rules.

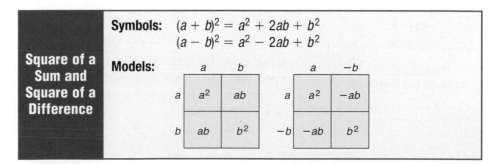

Square of a Sum and Square of a Difference	**Symbols:** $(a + b)^2 = a^2 + 2ab + b^2$ $(a - b)^2 = a^2 - 2ab + b^2$ **Models:**

Find each product.

1 $(x + 4)^2$

$(a + b)^2 = a^2 + 2ab + b^2$ *Square of a Sum*
$(x + 4)^2 = x^2 + 2(x)(4) + 4^2$ *Replace a with x and b with 4.*
$\quad\quad\quad = x^2 + 8x + 16$

2 $(4m + n)^2$

$(a + b)^2 = a^2 + 2ab + b^2$ *Square of a Sum*
$(4m + n)^2 = (4m)^2 + 2(4m)(n) + n^2$ *Replace a with 4m and b with n.*
$\quad\quad\quad\quad = 16m^2 + 8mn + n^2$

3 $(w - 5)^2$

$(a - b)^2 = a^2 - 2ab + b^2$ *Square of a Difference*
$(w - 5)^2 = w^2 - 2(w)(5) + 5^2$ *Replace a with w and b with 5.*
$\quad\quad\quad\quad = w^2 - 10w + 25$

4 $(3p - 2q)^2$

$(a - b)^2 = a^2 - 2ab + b^2$ *Square of a Difference*
$(3p - 2q)^2 = (3p)^2 - 2(3p)(2q) + (2q)^2$ *Replace a with 3p and b with 2q.*
$\quad\quad\quad\quad = 9p^2 - 12pq + 4q^2$

Your Turn

a. $(y + 3)^2$ b. $(3g + h)^2$
c. $(d - 2)^2$ d. $(4x - 3y)^2$

Biologists use a technique that is similar to squaring a sum to find the characteristics of offspring.

Example

Biology Link

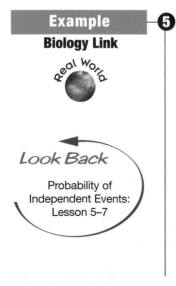

Real World

Look Back

Probability of Independent Events: Lesson 5–7

5 In a certain population, a parent has a 10% chance of passing the gene for brown eyes to its offspring. If an offspring receives one eye-color gene from its mother and one from its father, what is the probability that an offspring will receive at least one gene for brown eyes?

There is a 10% chance of passing the gene for brown eyes. Therefore, there is a 90% chance of *not* passing the gene.

Use the model at the right to show all possible combinations. In the model, B represents the gene for brown eyes and b represents the gene for *not* brown eyes. *Note that the percents are written as decimals.*

		Father	
		B = 0.1	b = 0.9
		BB	**Bb**
	B = 0.1	(0.1)(0.1) = 0.01	(0.1)(0.9) = 0.09
Mother		**bB**	**bb**
	b = 0.9	(0.9)(0.1) = 0.09	(0.9)(0.9) = 0.81

Three of the four small squares in the model contain a B. Add their probabilities.

$0.01 + 0.09 + 0.09 = 0.19$

So, the probability of an offspring receiving at least one gene for brown eyes is 0.19 or 19%.

You can use the FOIL method to find product of the sum and difference of the same two terms. Consider $(x + 3)(x - 3)$.

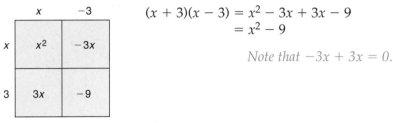

$$(x + 3)(x - 3) = x^2 - 3x + 3x - 9$$
$$= x^2 - 9$$

Note that $-3x + 3x = 0$.

The product of a sum and a difference can be found by using the following rule. Note that the product is the difference of the squares of the terms.

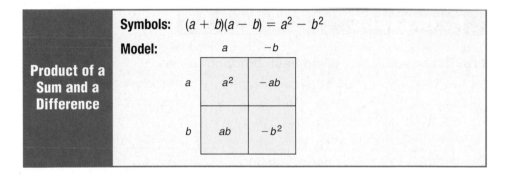

Product of a Sum and a Difference

Symbols: $(a + b)(a - b) = a^2 - b^2$

Model:

	a	$-b$
a	a^2	$-ab$
b	ab	$-b^2$

Examples

Find each product.

6 $(y + 2)(y - 2)$

$(a + b)(a - b) = a^2 - b^2$ *Product of a Sum and a Difference*
$(y + 2)(y - 2) = y^2 - (2)^2$ *Replace a with y and b with 2.*
$= y^2 - 4$

7 $(2r + s)(2r - s)$

$(a + b)(a - b) = a^2 - b^2$
$(2r + s)(2r - s) = (2r)^2 - s^2$ *Product of a Sum and a Difference*
$= 4r^2 - s^2$ *Replace a with 2r and b with s.*

Your Turn

e. $(x + y)(x - y)$ f. $(5m - 6n)(5m + 6n)$

Check for Understanding

Communicating Mathematics

Study the lesson. Then complete the following.

1. **Match** each description with a product.
   a. square of a sum
   b. square of a difference
   c. product of a sum and difference

   i. $(r + t)(r - t)$
   ii. $(z + 1)^2$
   iii. $(6 - a)^2$

2. **You Decide?** Jessica says that the product of two binomials is always a trinomial. Hector disagrees. Who is correct? Explain your reasoning.

Guided Practice

Find each product. *(Examples 1–4, 6, & 7)*

3. $(y + 2)^2$
4. $(2a + 4b)^2$
5. $(m - 9)^2$
6. $(j - 7k)^2$
7. $(x + 7)(x - 7)$
8. $(5r + 9s)(5r - 9s)$

9. **Biology** In a certain population, a parent has a 5% chance of passing the gene for cystic fibrosis to its offspring. What is the probability that an offspring will *not* receive the gene for cystic fibrosis from either of its parents? *(Example 5)*

Exercises •

Practice

Find each product.

10. $(r + 2)^2$
11. $(x + 5)^2$
12. $(a + 2b)^2$
13. $(w - 8)^2$
14. $(2x - y)^2$
15. $(m - 3n)^2$
16. $(p + 2)(p - 2)$
17. $(x + 4y)(x - 4y)$
18. $(2x - 3)(2x + 3)$
19. $(5 + k)^2$
20. $(3x - 2y)^2$
21. $(y + 3z)(y - 3z)$
22. $(6 - 2m)^2$
23. $(3x + 5)(3x - 5)$
24. $(5 + 2p)(5 + 2p)$
25. $(4 + 2x)^2$
26. $(5a - 2b)(5a + 2b)$
27. $(a - 3b)(a - 3b)$
28. $y(y + 1)(y - 1)$
29. $x(x + 1)^2$
30. $(x + 2)(x - 2)(x - 3)$

31. Find the product of $x - 2y$ and $x + 2y$.
32. What is the product of n, $n + 1$, and $n + 1$?

Applications and Problem Solving

33. **Number Theory** Explain how to find the product of 39 and 41 mentally. (*Hint*: Write 39 as $40 - 1$ and 41 as $40 + 1$.)

34. **Photography** Kareem cut off a 1-inch strip all around a square photograph so that it would fit into a square picture frame. He removed a total of 20 square inches.
    a. Draw and label a diagram that describes this situation.
    b. Write an equation that could be used to find the dimensions of the photograph.
    c. Find the original dimensions of the photograph.

35. Geometry The area of a triangle is given by the expression $\frac{1}{2}bh$, where b represents the length of the base and h represents the height. Suppose a right triangle has a base that measures $x - 3$ units and a height of $x + 3$ units.

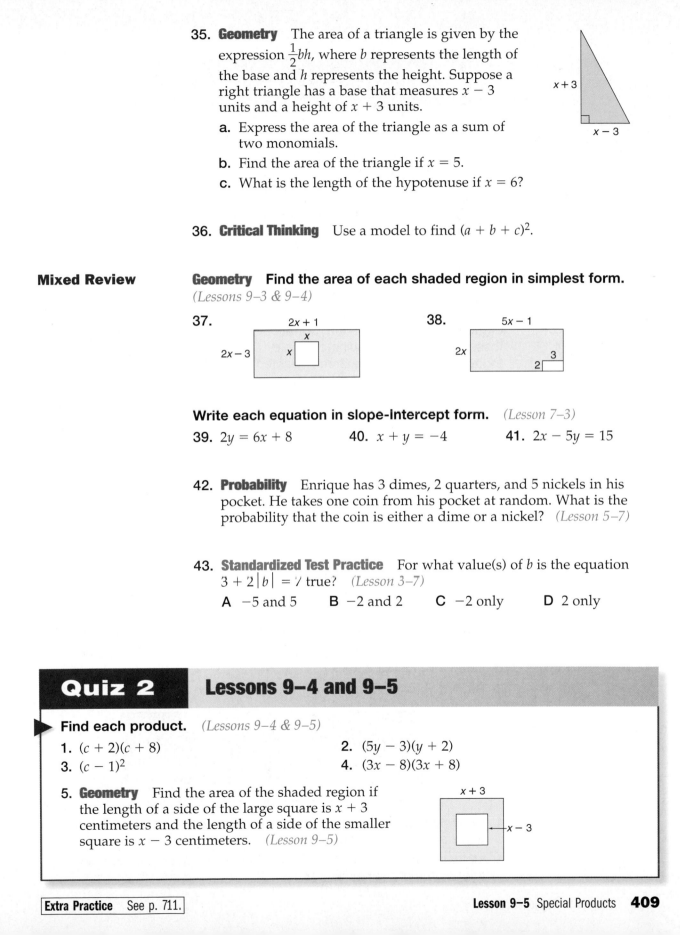

$x + 3$

$x - 3$

a. Express the area of the triangle as a sum of two monomials.

b. Find the area of the triangle if $x = 5$.

c. What is the length of the hypotenuse if $x = 6$?

36. Critical Thinking Use a model to find $(a + b + c)^2$.

Mixed Review

Geometry Find the area of each shaded region in simplest form. (*Lessons 9–3 & 9–4*)

37.
$2x + 1$
x
$2x - 3$
x

38.
$5x - 1$
$2x$
3
2

Write each equation in slope-intercept form. (*Lesson 7–3*)

39. $2y = 6x + 8$ **40.** $x + y = -4$ **41.** $2x - 5y = 15$

42. Probability Enrique has 3 dimes, 2 quarters, and 5 nickels in his pocket. He takes one coin from his pocket at random. What is the probability that the coin is either a dime or a nickel? (*Lesson 5–7*)

43. Standardized Test Practice For what value(s) of b is the equation $3 + 2|b| = 7$ true? (*Lesson 3–7*)

A -5 and 5   B -2 and 2   C -2 only   D 2 only

Quiz 2   Lessons 9–4 and 9–5

Find each product. (*Lessons 9–4 & 9–5*)

1. $(c + 2)(c + 8)$ **2.** $(5y - 3)(y + 2)$

3. $(c - 1)^2$ **4.** $(3x - 8)(3x + 8)$

5. Geometry Find the area of the shaded region if the length of a side of the large square is $x + 3$ centimeters and the length of a side of the smaller square is $x - 3$ centimeters. (*Lesson 9–5*)

$x + 3$

$x - 3$

It's Greek to Me!

Areas and Ratio

Materials

- ruler
- calculator
- grid paper

Early Greek mathematicians enjoyed solving problems that required a lot of thought and investigation. Some of these problems still fascinate people today. For example, the Greeks wanted to construct a cube whose volume was twice the volume of a given cube. First, let's look at similar problems. What happens to the area of a square when you double the length of the sides of that square? Can you construct a square whose area is twice as great as a given square?

Temple of Poseidon

Investigate

Draw squares with different lengths on grid paper and find the ratios of their areas. Use a table or spreadsheet to record the lengths, areas, and ratios.

1. Draw a 3-centimeter by 3-centimeter square. Label it A. Find the area.

2. Draw a second square that has a side length twice that of square A. Label it B. Find the area.

3. Record your information in a table like the one below.

Square A		Square B		Ratios	
Length of Side	Area	Length of Side	Area	Length B / Length A	Area B / Area A
3	9	6	36	2:1	4:1

4. Draw three more pairs of squares and label them A and B. Find the areas and ratios for each pair of squares.

5. Make a conjecture about the ratio of areas of squares if the ratio of the lengths of their sides is 3:1.

Extending the Investigation

In this extension, you will investigate the ratios of the areas of other squares. You will also try to find the length of the sides of two squares so that the area of one square is twice the other.

- Draw pairs of squares so that the ratio of the length of one square to the length of the second square is 3:1. Make a conjecture about the ratio of their areas.

- Draw pairs of squares so that the ratio of the length of one square to the length of the second square is not 2:1 or 3:1. Find the ratio of the lengths of their sides. Make a conjecture about the ratio of their areas.

- Draw pairs of squares so that the lengths of their sides differ by 1. For example, one might have a length of 4 units and the second a length of 5 units. Find the ratio of the lengths of their sides. Make a conjecture about the ratios of their areas.

- Draw pairs of squares so that the lengths of their sides differ by a number greater than 1. Find the ratio of the lengths of their sides. Make a conjecture about the ratios of their areas.

- Draw a 2-centimeter by 2-centimeter square. Then sketch a square whose area is twice as great as the area of the first square. Find the length of its side.

Presenting Your Conclusions

Here are some ideas to help you present your conclusions to the class.

- Make a bulletin board showing the results of this investigation.
- Write a report about your conjectures. Include diagrams or computations that help to explain your findings.
- Make a video showing the ratios that you discovered. Present your conjectures in a creative way.

interNET CONNECTION **Investigation** For more information on three classical problems in Greek mathematics, visit: www.algconcepts.glencoe.com

Study Guide and Assessment

Understanding and Using the Vocabulary

After completing this chapter, you should be able to define each term, property, or phrase and give an example or two of each.

interNET
CONNECTION **Review Activities**
For more review activities, visit:
www.algconcepts.glencoe.com

binomial *(p. 383)*
degree *(p. 384)*

FOIL method *(p. 401)*
monomial *(p. 382)*

polynomial *(p. 383)*
trinomial *(p. 383)*

Choose the letter of the term that best matches each statement or phrase. Each letter is used once.

1. $(x - y)^2$
2. a polynomial with two terms
3. a polynomial with three terms
4. a monomial or a sum of monomials
5. the sum of the exponents of the variables
6. a number, a variable, or a product of numbers and variables
7. Subtract polynomials by adding this.
8. Add polynomials by grouping these together.
9. Use this to multiply any two binomials.
10. Use this to multiply a polynomial by a monomial.

a. additive inverse
b. binomial
c. degree of a polynomial
d. Distributive Property
e. FOIL method
f. like terms
g. monomial
h. polynomial
i. square of a difference
j. trinomial

Skills and Concepts

Objectives and Examples

• **Lesson 9–1** Identify and classify polynomials and find their degree.

Identify $3x^3 - 2x^2$ as either a monomial, a binomial, or a trinomial.

$3x^3 - 2x^2$ can be written as a sum of two monomials, $3x^3$ and $-2x^2$. It is a binomial.

Review Exercises

State whether each expression is a polynomial. If it is a polynomial, identify it as either a *monomial*, *binomial*, or *trinomial*.

11. $7m$
12. $5g^{-2}$
13. $2x^2 + 3x - 4$
14. $3y + 5y + 8$

Find the degree of each polynomial.

15. 7
16. $-4m^5$
17. $10x^2 - x^3y$
18. $2abc + 9a^5b - 4bc^3$
19. Arrange $3d^3 - d^2 + 7cd$ so that the powers of d are in ascending order.

Objectives and Examples

- **Lesson 9–2** Add and subtract polynomials.

$(2x + 4) + (5x − 7)$
$= (2x + 5x) + (4 − 7)$
$= (2 + 5)x + (4 − 7)$
$= 7x − 3$

$(7s^2 + 3s − 4) − (3s^2 − 2s + 5)$

The additive inverse of $(3s^2 − 2s + 5)$
is $(−3s^2 + 2s − 5)$.

$(7s^2 + 3s − 4) − (3s^2 − 2s + 5)$
$= (7s^2 + 3s − 4) + (−3s^2 + 2s − 5)$
$= (7s^2 − 3s^2) + (3s + 2s) + (−4 − 5)$
$= (7 − 3)s^2 + (3 + 2)s + (−4 − 5)$
$= 4s^2 + 5s − 9$

Review Exercises

Find each sum or difference.

20. $\begin{aligned} &8x − 7 \\ (+)\ &3x + 5 \end{aligned}$

21. $\begin{aligned} &6x^2 \qquad + 4 \\ (+)\ &3x^2 − 3x + 2 \end{aligned}$

22. $\begin{aligned} &9m^2 + 6m − 6 \\ (−)\ &7m^2 − 3m + 3 \end{aligned}$

23. $\begin{aligned} &15s^2 \qquad − 6 \\ (−)\ &8s^2 + 6s − 4 \end{aligned}$

24. $(17x − 3y) + (−2x + 5y)$
25. $(10g^2 + 5g) − (6g^2 − 3g)$
26. $(14x^2 + 3x − 6) − (8 − 2x)$
27. $(−3s^2t^2 − 4st^2) + (−7s^2t^2 + 6st^2)$
28. What is $5n^2 + 3$ minus $2n^2 − 4$?

- **Lesson 9–3** Multiply a polynomial by a monomial.

$2x^2(3x^3 + 2x^2 − x)$
$= 2x^2(3x^3) + 2x^2(2x^2) + 2x^2(−x)$
$= 6x^5 + 4x^4 − 2x^3$

Find each product.

29. $3(x − 5)$
30. $n(n + 3)$

31. $2m(m^2 + 4m)$
32. $x^4(3x^2 − 2x)$

33. $5(2x^2 + 3x − 2)$
34. $−3m^2(m^2 − 2m + 1)$

Solve each equation.

35. $x(x + 8) = x(x + 11) + 9$
36. $9(w − 4) + 10 = 2w + 16$

- **Lesson 9–4** Multiply two binomials.

Find the product of $x + 4$ and $x − 3$.

$\overset{F\quad\ \ L}{(x + 4)(x − 3)}$

$= (x)(x) + (x)(−3) + (4)(x) + (4)(−3)$
$= x^2 − 3x + 4x − 12$
$= x^2 + x − 12$

Find each product.

37. $(y − 2)(y + 6)$
38. $(m + 5)(m + 7)$

39. $(x − 1)(x − 3)$
40. $(a + 11)(a − 4)$

41. $(2m − 1)(m − 4)$
42. $(x + 4)(3x − 2)$

43. $(4y + 2)(3y + 1)$
44. $(6x − 3)(2x + 5)$

45. Find the product of $5x + 4$ and $2x − 3$.

Objectives and Examples	**Review Exercises**

- **Lesson 9–5** Develop and use the patterns for $(a + b)^2$, $(a - b)^2$, and $(a + b)(a - b)$.

$(x + 3)^2 = x^2 + 2(x)(3) + 3^2$ *Square of*
$ = x^2 + 6x + 9$ *a sum*

$(y - 5)^2 = y^2 - 2(y)(5) + 5^2$ *Square of*
$ = y^2 - 10y + 25$ *a difference*

$(s + 4)(s - 4) = s^2 - 4^2$ *Product of a sum*
$ = s^2 - 16$ *and difference*

Find each product.

46. $(w + 6)^2$ **47.** $(2x + 3)^2$

48. $(8 + g)^2$ **49.** $(x - 2)^2$

50. $(3y - 1)^2$ **51.** $(5m - 3n)^2$

52. $(y + 4)(y - 4)$ **53.** $(2a + 3)(2a - 3)$

54. $(2p - 3q)(2p + 3q)$ **55.** $(5a + 1)(5a + 1)$

Applications and Problem Solving

56. Recreation Bocce is a game similar to lawn bowling, but it is played on a rectangular dirt court. If the length of the rectangle is $12x$ and the width is $2x + 2$, find the area of the court in terms of x. Write your answer as a polynomial in simplest form. *(Lesson 9–3)*

57. Geometry The area of a triangle is given by $\frac{1}{2}bh$, where b is the length of the base of the triangle and h is the measure of the height of the triangle. In triangle ABC, $b = 3x + 8$ and $h = 7x - 4$. Write the polynomial representing the measure of the area of triangle ABC. *(Lesson 9–4)*

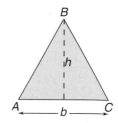

58. Gardening A rectangular garden is 10 feet longer than it is wide. It is surrounded by a brick walkway 3 feet wide, as shown at the right. Suppose the total area of the walkway is 396 square feet. *(Lesson 9–4)*

a. Find the area of the small rectangle.

b. What are the dimensions of the garden?

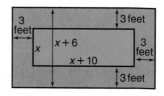

59. Quilting Kirsten is making a quilt to enter in the arts festival. The diagram gives the dimensions of each block of the quilt. *(Lesson 9–5)*

a. Find the area of each block.

b. The completed quilt is a square of 36 blocks. Write a product of binomials that could be used to determine the area of the entire quilt.

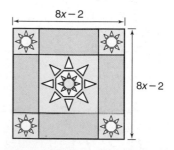

1. **Explain** how to use the FOIL method to multiply two binomials.

2. **Explain** how to subtract polynomials.

3. 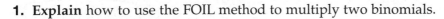 Mark says that $-4wx^{-2}y^3$ is a monomial of degree 2. Linda disagrees. Who is correct? Explain your reasoning.

State whether each expression is a polynomial. If it is a polynomial, identify it as either a _monomial_, _binomial_, or _trinomial_, and state its degree.

4. $4x^3 + 3x^2$
5. $12r^{-2} - 3r + 6$
6. m^0
7. $5w^3yz^2$

Find each sum or difference.

8. $\begin{array}{r} 3x^2 - 5x + 1 \\ (+)\ 5x^2 + 7x + 8 \\ \hline \end{array}$

9. $\begin{array}{r} 8z^2 - 5z \\ (-)\ 6z^2 - 3z + 4 \\ \hline \end{array}$

10. $(y^3 + 6y^2 + 4y) - (2y^3 - 7y)$

11. $(-9h^2 - 5g) + (2g + 7h^2)$

Find each product.

12. $3x(2x^2 + 4x - 5)$
13. $5t^2(-2t + 3t^3 + 4)$
14. $(x + 2)(x + 3)$
15. $(y - 4)(y - 5)$
16. $(x - 2)(x + 5)$
17. $(3n + 4)(2n + 3)$
18. $(x + 5)^2$
19. $(2w - 3)^2$
20. $(7r - 4)(7r + 4)$
21. $(m - 3n)(m + 3n)$

Solve.

22. $2(y - 3) + 9 = 5y - 6$

23. $x(x + 4) + 5x = x(x - 1) + 20$

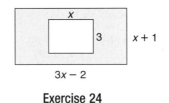

24. **Geometry** Find the area of the shaded region.

Exercise 24

25. **Family Life** Mrs. Douglas wants her five children, Aaron, Briana, Casey, Danielle, and Eddie, to spend a total of 35 hours on chores this week. Aaron, the oldest, works twice as many hours as the others. Danielle has earned an hour off from chores by getting an A on her algebra test. If x is the number of hours Danielle will work, then Briana, Casey, and Eddie will each work $x + 1$ hours, and Aaron will work twice as long, $2x + 2$ hours.

   a. Write an equation to represent the number of hours the children will work.

   b. Use the equation to determine the number of hours each child should work on chores this week.

Percent Problems

Standardized tests contain several types of percent problems. They may be numeric problems, word problems, or data analysis problems.

It will be helpful to familiarize yourself with common fractions, decimals, and percents.

THE PRINCETON REVIEW

A percent is a fraction whose denominator is 100.

$$0.01 = \frac{1}{100} = 1\% \qquad 0.1 = \frac{1}{10} = 10\% \qquad 0.333\ldots = \frac{1}{3} \approx 33.3\%$$

$$0.25 = \frac{1}{4} = 25\% \qquad 0.5 = \frac{1}{2} = 50\% \qquad 0.75 = \frac{3}{4} = 75\%$$

Proficiency Test Example

The table shows recorded music sales. What percent of total sales in 1996 was on compact disc (CD)? Round to the nearest percent.

Sales of Recorded Music (millions)

Year	Format		
	CD	Cassette	Album
1988	149.7	450.1	72.4
1992	407.5	366.4	2.3
1996	778.9	225.3	2.9

Hint Write a ratio, and then write the fraction as a percent.

Solution First, find the total sales in 1996.

$$778.9 + 225.3 + 2.9 = 1007.1$$

Next, write the ratio of compact disc sales to total sales.

$$\frac{778.9}{1007.1}$$

Then write this fraction as a decimal and as a percent. Use your calculator.

$$\frac{778.9}{1007.1} = 0.7734 \text{ or } 77\% \text{ to the nearest percent}$$

Compact disc sales were about 77% of total recorded music sales in 1996.

ACT Example

A bus company charges $5 each way to shuttle passengers between the hotel and a shopping mall. On a given day, the bus company has a total capacity of 250 people on the way to the mall and back. If the bus company runs at 90% of capacity, how much money would it take in that day?

A $1147.50 **B** $1250 **C** $2250
D $2500 **E** $2625

Hint Avoid partial answers. Be sure you answer the question that is asked.

Solution First determine how much money the bus company would make if it ran at total capacity. The 250 passengers would pay $10 each, because the charge is $5 each way.

$$250(\$10) = \$2500$$

Notice that this total amount is answer choice D, but it is a wrong answer. The question asks for the amount when the bus runs at 90% of capacity.

The word *of* is a clue to multiply. Find 90% of $2500.

$$(90\%)(\$2500) = (0.90)(2500) \text{ or } 2250$$

The bus company would take in $2250. Therefore, the answer is C.

After you work each problem, record your answer on the answer sheet provided or on a sheet of paper.

1. Dr. Martinez uses two rooms in her home as an office. The areas of the rooms are 150 square feet and 135 square feet. If the total area of her home is 2000 square feet, what percent of the total area is office space?

   A $6\frac{3}{4}\%$ **B** $7\frac{1}{2}\%$

   C $14\frac{1}{4}\%$ **D** $85\frac{3}{4}\%$

2. A $9.95 calendar is marked down 40%. Before tax, how much is saved on the purchase of one calendar?

   A $1.99   **B** $3.98

   C $4.00   **D** $5.97

3. On a 16-question quiz, Tom answered 2 questions incorrectly. If each question is worth the same number of points, his point total is what percent of the total points?

   A 12.5%   **B** 16%   **C** 85%

   D 87.5%   **E** 94%

4. An increase in prices has made the cost of remodeling an office 12% more than the original cost. The original cost was $7145. What is the best estimate of the new cost?

   A $700   **B** $7700

   C $10,000   **D** $70,000

5. The graph shows the attendance at a park. Predict the attendance in the year 2005.

   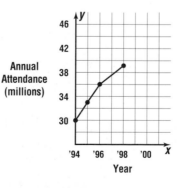

   A 38 million   **B** 48 million

   C 60 million   **D** 100 billion

6. A pair of shoes that regularly sold for $44 is now on sale for $33. What is the percent of discount?

   A 11%   **B** 25%   **C** 75%

   D $33\frac{1}{3}\%$ **E** $66\frac{2}{3}\%$

7. There are 60 students in the band. How many play a percussion instrument?

   Instruments in Bay City Band

   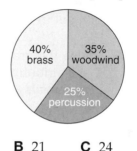

   A 15   **B** 21   **C** 24   **D** 25

8. Which expression could be used to find the value of y?

   A $2x + 1$ **B** $1 - 3x$

   C $3x - 1$ **D** $3x + 1$

x	y
1	2
2	5
3	8
4	11

Open-Ended Questions

9. **Grid-In** A CD player is on sale for $250. If there is a 6% sales tax, what is the total cost?

10. The total land area of a state is 23,159,000 acres. Of that, 13,513,000 acres are cropland, 1,866,000 acres are pastureland, and 3,626,000 acres are forest.

    Part A To the nearest percent, what percent of the state's area is cropland?

    Part B What percent of the state's area is *not* cropland, pastureland, or forest?

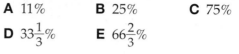

Test Practice For additional test practice questions, visit: www.algconcepts.glencoe.com

▶ What You'll Learn in Chapter 10:

- to find the greatest common factor of a set of numbers or monomials *(Lesson 10–1)*,
- to use the greatest common factor and the Distributive Property to factor polynomials *(Lesson 10–2)*, and
- to factor trinomials and the differences of squares *(Lessons 10–3, 10–4, and 10–5)*.

Problem-Solving Workshop

Project

In 1964, there were about 7600 shopping malls in the United States. By 1997, this number had increased to 42,874. In this project, you will design a new shopping mall for a city near you. Your mall should occupy at least 750,000 square feet. Include a scale drawing of your mall and a one-page proposal promoting your design.

Working on the Project

Work with a partner and choose a strategy to help analyze and complete this project. Develop a plan. Brainstorm with your partner on possible dimensions for the mall. Here are some questions to help you get started.

- Suppose the mall had three levels and each level was shaped like a square. What would be the length of one side of the building?
- Suppose the mall had two levels, each shaped like a rectangle. What are two possible dimensions for the building?

Strategies

Look for a pattern.

Draw a diagram.

Make a table.

Work backward.

Use an equation.

Make a graph.

Guess and check.

Technology Tools

- Use a **calculator** to find possible dimensions for the mall.
- Use a **word processor** to write your proposal.
- Use **drawing software** to make your scale drawing.

interNET CONNECTION **Research** For more information about shopping malls, visit: www.algconcepts.glencoe.com

Presenting the Project

Prepare a portfolio of your scale drawings. Write a one-page proposal that highlights the features of your mall. Make sure that your drawings and proposal include:

- your calculations for the total number of square feet occupied by the mall,
- labels for all dimensions in your scale drawings, and
- the scale that you used for your drawings.

KIRARA BASSO

Math In the Workplace

What You'll Learn

You'll learn to find the greatest common factor of a set of numbers or monomials.

Why It's Important

Crafts Quilters use greatest common factors when they cut fabric. See Exercise 57.

For many centuries, mathematicians have competed to see who could find the largest prime number. As of June 1, 1999, the largest known prime number was $2^{6972593} - 1$. This number has 2,098,960 digits. If it were printed in a newspaper, it would fill 66 pages!

Recall that when two or more numbers are multiplied, each number is a *factor* of the product. For example, 12 can be expressed as the product of different pairs of whole numbers. Factors can be shown geometrically.

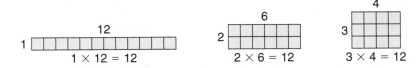

The whole numbers 1, 12, 2, 6, 3, and 4 are the factors of 12.

Some whole numbers have exactly two factors, the number itself and 1. Recall that these numbers are called *prime numbers*. Whole numbers that have more than two factors, such as 12, are called *composite numbers*.

Prime Numbers Less Than 20	Composite Numbers Less Than 20	Neither Prime nor Composite
2, 3, 5, 7, 11, 13, 17, 19	4, 6, 8, 9, 10, 12, 14, 15, 16, 18	0, 1

Examples

Find the factors of each number. Then classify each number as *prime* or *composite*.

1 **72**

To find the factors of 72, list all pairs of whole numbers whose product is 72.

$1 \times 72 \qquad 2 \times 36 \qquad 3 \times 24 \qquad 4 \times 18 \qquad 6 \times 12 \qquad 8 \times 9$

The factors of 72 are 1, 2, 3, 4, 6, 8, 9, 12, 18, 24, 36, and 72. Since 72 has more than two factors, it is a composite number.

*inter***NET** CONNECTION

Data Update For the latest information on prime numbers, visit: www.algconcepts. glencoe.com

2 **37**

There is only one pair of whole numbers whose product is 37.

1×37

The factors of 37 are 1 and 37. Therefore, 37 is a prime number.

Your Turn

a. 25   **b.** 23   **c.** 79   **d.** 51

You can use a graphing calculator to investigate factor patterns.

Graphing Calculator Exploration

Graphing Calculator Tutorial
See pp. 724–727.

The table below shows the numbers 2 through 12 and their factors arranged according to the number of factors.

2 Factors	3 Factors	4 Factors	5 Factors	6 Factors
2: 1, 2 **3:** 1, 3 **5:** 1, 5 **7:** 1, 7 **11:** 1, 11	**4:** 1, 2, 4 **9:** 1, 3, 9	**6:** 1, 2, 3, 6 **8:** 1, 2, 4, 8 **10:** 1, 2, 5, 10	none	**12:** 1, 2, 3, 4, 6, 12

Step 1: Copy the table above.

Step 2: Use the graphing calculator program below to find the factors of the numbers 13 through 20.

```
PROGRAM:FACTOR
:Input "ENTER NUMBER", N
:For (D, 1, N)
:If iPart (N/D) = (N/D)
:Disp D
:END
```

Try These

1. Place the numbers 13 through 20 in the correct column of the table.

2. Predict a number from 21 through 100 that can be placed in each column. Check your prediction by using the calculator program.

3. Explain the pattern in each column.

Since $4 \cdot 3 = 12$, 4 is a factor of 12. However, it is not a prime factor of 12 because 4 is not a prime number. Recall that when a number is expressed as a product of prime factors, the expression is called the *prime factorization* of the number.

You can use a *factor tree* to find the prime factorization of 12. Two different factor trees are shown.

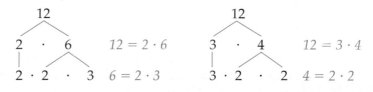

All of the factors in the last row are prime numbers. The factors are in a different order, but the result is the same. Except for the order of the factors, there is only one prime factorization of a number. Thus, the prime factorization of 12 is $2 \cdot 2 \cdot 3$ or $2^2 \cdot 3$.

You can use prime factorization to factor monomials. A monomial is in *factored form* when it is expressed as the product of prime numbers and variables and no variable has an exponent greater than 1.

Look Back

Monomials:
Lesson 9–1

Examples

Factor each monomial.

3 $12a^2b$

$12a^2b = 2 \cdot 2 \cdot 3 \cdot a \cdot a \cdot b$ $12 = 2 \cdot 2 \cdot 3, a^2 = a \cdot a$

4 $100mn^3$

$100mn^3 = 2 \cdot 2 \cdot 5 \cdot 5 \cdot m \cdot n \cdot n \cdot n$ $100 = 2 \cdot 2 \cdot 5 \cdot 5, n^3 = n \cdot n \cdot n$

5 $-25x^2$

To factor a negative integer, first express it as the product of a whole number and -1. Then find the prime factorization.

$-25x^2 = -1 \cdot 25x^2$ $-25 = -1 \cdot 25$

$\qquad\quad = -1 \cdot 5 \cdot 5 \cdot x \cdot x$ $25 = 5 \cdot 5$

Your Turn

e. $15ab^2$ **f.** $84yz^2$ **g.** $-36b^3$

Two or more numbers may have some common prime factors. Consider the prime factorization of 36 and 42.

$$36 = \boxed{2} \cdot 2 \cdot \boxed{3} \cdot 3 \qquad \textit{Line up the common factors.}$$
$$42 = \boxed{2} \cdot \quad \boxed{3} \cdot \quad 7$$

The integers 36 and 42 have 2 and 3 as common prime factors. The product of these prime factors, $2 \cdot 3$ or 6, is called the **greatest common factor (GCF)** of 36 and 42. The GCF is the greatest number that is a factor of both original numbers.

Greatest Common Factor	The greatest common factor of two or more integers is the product of the prime factors common to the integers.

The GCF of two or more monomials is the product of their common factors when each monomial is expressed in factored form.

Find the GCF of each set of numbers or monomials.

6 **24, 60, and 72**

$24 = 2 \cdot 2 \cdot 2 \cdot 3$ *Find the prime factorization of each number.*
$60 = 2 \cdot 2 \cdot \quad 3 \cdot 5$ *Line up as many factors as possible.*
$72 = 2 \cdot 2 \cdot 2 \cdot 3 \cdot 3$ *Circle the common factors.*

The GCF of 24, 60, and 72 is $2 \cdot 2 \cdot 3$ or 12.

7 **15 and 8**

$15 = 3 \cdot 5$
$8 = \qquad 2 \cdot 2 \cdot 2$

There are no common prime factors. The only common factor is 1. So, the GCF of 15 and 8 is 1.

Reading Algebra

Two numbers or monomials whose greatest common factor is 1 are called *relatively prime*. So, 15 and 8 are relatively prime.

8 **$15a^2b$ and $18ab$**

$15a^2b = \qquad 3 \cdot 5 \cdot a \cdot a \cdot b$
$18ab = 2 \cdot 3 \cdot 3 \cdot \quad a \cdot \quad b$

The GCF of $15a^2b$ and $18ab$ is $3 \cdot a \cdot b$ or $3ab$.

Your Turn

h. 75, 100, and 150   **i.** $5a$ and $8b$ **j.** $24ab^2c$ and $60a^2bc$

Knowing the factors of a number can help you with geometry.

Example

Geometry Link

9 **The area of a rectangle is 18 square inches. Find the length and width so that the rectangle has the least perimeter. Assume that the length and width are both whole numbers.**

Explore   You know that the area of the rectangle is 18 square inches. You want to find the length and width so that the rectangle has the least perimeter.

Plan   Find the factors of 18 and draw rectangles with each length and width. Then find each perimeter.

Solve

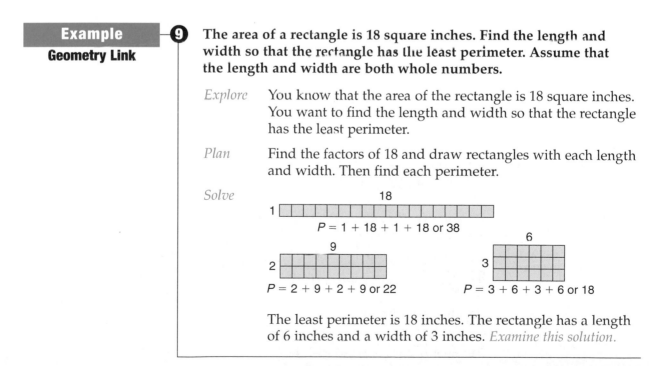

The least perimeter is 18 inches. The rectangle has a length of 6 inches and a width of 3 inches. *Examine this solution.*

Communicating Mathematics

Study the lesson. Then complete the following.

Vocabulary

greatest common factor (GCF)

1. **List** the prime numbers between 20 and 50.
2. **Name** two numbers whose GCF is 4.
3. **Explain** how to find the GCF of $8x^2$ and $16x$.
4.  Jennifer believes that $2 \cdot 3 \cdot 4 \cdot 5$ is the prime factorization of 120, but Arturo disagrees. Who is correct? Explain.

Guided Practice

Getting Ready Find the prime factorization of each number.

Sample 1: 28	**Sample 2:** 60
Solution: $28 = 2 \cdot 2 \cdot 7$	**Solution:** $60 = 2 \cdot 2 \cdot 3 \cdot 5$

5. 21
6. 72
7. 51
8. 150
9. 108
10. 110

Find the factors of each number. Then classify each number as **prime** or **composite**. *(Examples 1 & 2)*

11. 42
12. 47

Factor each monomial. *(Examples 3–5)*

13. $24x^2y$
14. $-16ab^2c$

Find the GCF of each set of numbers or monomials. *(Examples 6–8)*

15. 15, 70
16. 16, 24, 28
17. 20, 21
18. $2x, 5y$
19. $7y^2, 14y^3$
20. $-12ab, 4a^2b^3$

21. **Geometry** The area of a rectangle is 72 square centimeters. Find the length and width so that the rectangle has the greatest perimeter. Assume that the length and width are both whole numbers. *(Example 9)*

Exercises

Practice

Find the factors of each number. Then classify each number as **prime** or **composite**.

22. 19
23. 20
24. 61
25. 45
26. 49
27. 91

Factor each monomial.

28. $20x^2$
29. $-15a^2b$
30. $-24c^3$
31. $50m^2n^2$
32. $44r^2s$
33. $90yz^2$

Find the GCF of each set of numbers or monomials.

34. 24, 40   **35.** 12, 8   **36.** 17, 21

37. 18, 36   **38.** 20, 30   **39.** 45, 72

40. $3x^2$, $3x$ **41.** $18y^2$, $3y$ **42.** $-5ab$, $6b^2$

43. -18, $45mn$ **44.** $24a^2$, $60ab$ **45.** $9x^2y$, $10m^2n$

46. 6, 8, 12   **47.** 20, 21, 25   **48.** 18, 30, 54

49. $5m^2$, $15n^2$, $25mn$ **50.** $6ax^2$, $18ay^2$, $9az^3$ **51.** $15r^2$, $35s^2$, $70rs$

52. What is the greatest prime number less than 90?

53. Find the greatest common factor of $5x^2$, $5y^2$, and $10xy$.

54. *Twin primes* are prime numbers that differ by 2, such as 5 and 7. Find two other sets of twin primes that are between 25 and 45.

Applications and Problem Solving

Real World

55. Crafts Ashley wants to make a quilt from two different kinds of fabric. One is 60 inches wide, and the other is 48 inches wide. What are the dimensions of the largest squares she can cut from both fabrics so there is no wasted fabric?

56. Math History In 1880, English mathematician John Venn (1834–1923) developed a way to show how sets of numbers are related. The *Venn diagram* shows the prime factors of 12 and 28. The common factors are in the overlapping circles, and the GCF of 12 and 28 is $2 \cdot 2$ or 4.

Prime Factors of 12 and 18

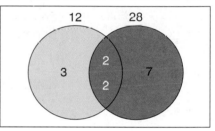

a. Draw a Venn diagram showing the prime factors of 36 and 45.

b. Find the GCF of 36 and 45.

57. Critical Thinking Explain why 2 is the only even prime number.

Mixed Review

Find each product. *(Lessons 9–3, 9–4, and 9–5)*

58. $(x + 3)(x - 3)$ **59.** $(2y + 1)(2y + 1)$ **60.** $(3a - 2)^2$

61. $(z + 4)(z + 3)$ **62.** $(x - 5)(x - 4)$ **63.** $(2n + 1)(n + 4)$

64. $3(x - 5)$ **65.** $2a(3 - a^2)$ **66.** $4x^2y(3x - 2y)$

67. Transportation In 2025, there are expected to be 817,000,000 cars in use worldwide. Write 817,000,000 in scientific notation. *(Lesson 8–4)*

68. Standardized Test Practice Which graph below is *not* the graph of a function? *(Lesson 6–4)*

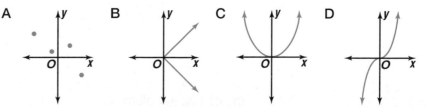

Extra Practice   See p. 711.

A **Puzzling** Perimeter **Problem**

Materials

algebra tiles

Perimeter and Area

You know that the perimeter of a rectangle is the distance around the outside of the figure. It is measured in units like inches, centimeters, or feet. The area of a rectangle is the number of square units needed to cover the surface. It is measured in units like square inches or square centimeters. Are there any rectangles in which the measure of the perimeter is equal to the measure of the area? Let's investigate.

Investigate

1. Use the 1-tiles from a set of algebra tiles.

   a. Make a table like the one below with 21 rows.

Area of Rectangle	Possible Dimensions of Rectangle	Perimeter of Rectangle
1		
2		

A 1-by-2 rectangle and a 2-by-1 rectangle have the same perimeter. So, they are listed only once in the table.

   b. Select one tile. The measure of its area is 1. The dimension of this rectangle is 1 by 1. Find the measure of the perimeter of the rectangle and enter it in column 3.

   c. Select two tiles. You can form a 1-by-2 rectangle with the tiles. This rectangle has an area of 2. Write 1 by 2 in column 2. What is the perimeter of this rectangle? Enter it in column 3.

   d. Repeat this process using three tiles.

   e. When you select four tiles, you have two options: a 1-by-4 rectangle or a 2-by-2 rectangle.

   Write 1 by 4 and 2 by 2 in column 2 as the dimensions of the rectangles. Find the perimeters of the two rectangles and enter the results in column 3.

Your table should look like this.

Area of Rectangle	Possible Dimensions of Rectangle	Perimeters of Rectangle
1	1 by 1	4
2	1 by 2	6
3	1 by 3	8
4	1 by 4, 2 by 2	10, 8

2. Use tiles to form all possible rectangles with areas from 5 through 20. Write the dimensions in column 2 and the perimeters in column 3.

3. Analyze the data. Which rectangle(s) has the same numerical values for perimeter and area?

4. Make a conjecture about whether there are other rectangles that have the same numerical values for perimeter and area.

Extending the Investigation

In this extension, you will continue to investigate the area and perimeter of rectangles. Extend your table through at least 30 squares.

- Study the perimeters for only the rectangles whose dimensions are 1 by n. Make an ordered list of the perimeters. Describe the pattern of the perimeters.

- Study the perimeters for only the rectangles whose dimensions are 2 by n, but not 2 by 2. Make an ordered list of the perimeters. Describe the pattern of the perimeters.

- Study the perimeters for only the rectangles whose dimensions are 3 by n, but not 3 by 3. Make an ordered list of the perimeters. Describe the pattern of the perimeters.

- Study the perimeters for rectangles that are also squares. Make an ordered list of these perimeters. Describe the pattern of the perimeters.

Presenting Your Conclusions

Here are some ideas to help you present your conclusions to the class.

- Make a brochure describing your findings. Include figures and tables to illustrate your results.

- Make a video showing the patterns you discovered. You may want to have students in your class take on the roles of rectangles with various dimensions.

10-2 Factoring Using the Distributive Property

Math
In the Workplace

What You'll Learn

You'll learn to use the GCF and the Distributive Property to factor polynomials.

Why It's Important

Marine Biology You can find the height a dolphin jumps out of the water by evaluating an expression that is written in factored form.
See Exercise 46.

Pentominoes are shapes consisting of five squares. There are twelve different possibilities, as shown below.

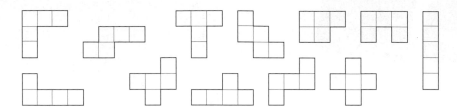

There are twelve shapes, each having an area of 5 square units. So, the total area is 60 square units. It is possible to combine these shapes into a rectangle whose dimensions are 3 units by 20 units.

Since $3 \cdot 20 = 60$, 3 and 20 are factors of 60.

Sometimes, you know the product and are asked to find the factors. This process is called **factoring**.

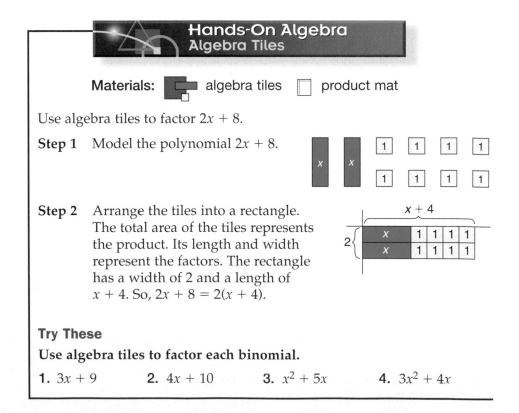

Hands-On Algebra
Algebra Tiles

Materials: algebra tiles ☐ product mat

Use algebra tiles to factor $2x + 8$.

Step 1 Model the polynomial $2x + 8$.

Step 2 Arrange the tiles into a rectangle. The total area of the tiles represents the product. Its length and width represent the factors. The rectangle has a width of 2 and a length of $x + 4$. So, $2x + 8 = 2(x + 4)$.

Try These

Use algebra tiles to factor each binomial.

1. $3x + 9$ **2.** $4x + 10$ **3.** $x^2 + 5x$ **4.** $3x^2 + 4x$

Look Back

Distributive
Property:
Lesson 1–4

In Chapter 9, you used the Distributive Property to multiply a polynomial by a monomial.

$$2y(4y + 5) = 2y(4y) + 2y(5)$$
$$= 8y^2 + 10y$$

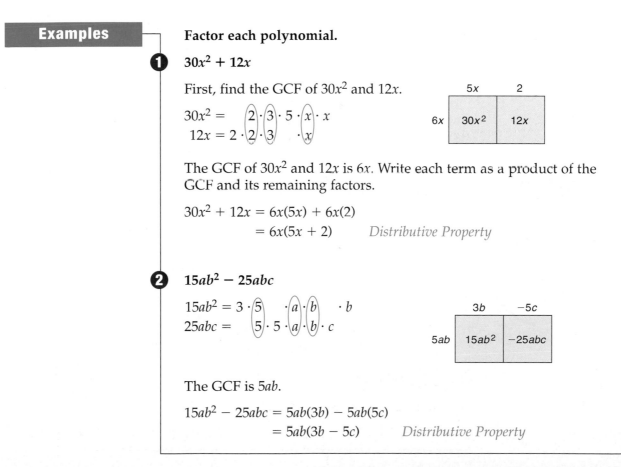

You can reverse this process to express a polynomial in *factored form*. A polynomial is in factored form when it is expressed as the product of polynomials. To factor $8y^2 + 10y$, find the greatest common factor of $8y^2$ and $10y$.

$$8y^2 = 2 \cdot 2 \cdot 2 \cdot y \cdot y$$
$$10y = 2 \cdot 5 \cdot y$$

The GCF of $8y^2$ and $10y$ is $2y$. Write each term as a product of the GCF and its remaining factors. Then use the Distributive Property.

$$8y^2 + 10y = 2y(4y) + 2y(5)$$
$$= 2y(4y + 5) \qquad \textit{Distributive Property}$$

$8y^2 + 10y$ written in factored form is $2y(4y + 5)$.

Examples

Factor each polynomial.

① $30x^2 + 12x$

First, find the GCF of $30x^2$ and $12x$.

$$30x^2 = 2 \cdot 3 \cdot 5 \cdot x \cdot x$$
$$12x = 2 \cdot 2 \cdot 3 \cdot x$$

The GCF of $30x^2$ and $12x$ is $6x$. Write each term as a product of the GCF and its remaining factors.

$$30x^2 + 12x = 6x(5x) + 6x(2)$$
$$= 6x(5x + 2) \qquad \textit{Distributive Property}$$

② $15ab^2 - 25abc$

$$15ab^2 = 3 \cdot 5 \cdot a \cdot b \cdot b$$
$$25abc = 5 \cdot 5 \cdot a \cdot b \cdot c$$

The GCF is $5ab$.

$$15ab^2 - 25abc = 5ab(3b) - 5ab(5c)$$
$$= 5ab(3b - 5c) \qquad \textit{Distributive Property}$$

Factor each polynomial.

❸ $18x^2y + 12xy^2 + 6xy$

$18x^2y = \boxed{2} \quad \cdot \boxed{3} \cdot 3 \cdot \boxed{x} \cdot x \cdot \boxed{y}$
$12xy^2 = \boxed{2} \cdot 2 \cdot \boxed{3} \quad \cdot \boxed{x} \quad \cdot \boxed{y} \cdot y$
$6xy = \boxed{2} \quad \cdot \boxed{3} \quad \cdot \boxed{x} \quad \cdot \boxed{y}$

	$3x$	$2y$	1
$6xy$	$18x^2y$	$12xy^2$	$6xy$

The GCF is $6xy$. When $6xy$ is factored from $6xy$ the remaining factor is 1.

$18x^2y + 12xy^2 + 6xy = 6xy(3x) + 6xy(2y) + 6xy(1)$
$\qquad\qquad\qquad\qquad = 6xy(3x + 2y + 1)$ *Distributive Property*

❹ $7x^2 + 9yz$

$7x^2 = 7 \qquad \cdot x \cdot x$
$9yz = \quad 3 \cdot 3 \qquad \cdot y \cdot z$

There are no common factors of $7x^2$ and $9yz$ other than 1. Therefore, $7x^2 + 9yz$ cannot be factored using the GCF. It is a prime polynomial.

Reading Algebra

A polynomial that cannot be written as a product of two polynomials with integral coefficients is called a *prime polynomial*.

Your Turn

a. $12n^2 - 8n$

b. $16a^2b + 10ab^2$

c. $20rs^2 - 15r^2s + 5rs$

d. $21x + 5y + 16z$

If you know a product and one of its factors, you can use division to find the other factor. To divide a polynomial by a monomial, divide each term of the polynomial by the monomial.

Example

❺ **Divide $15x^3 + 12x^2$ by $3x$.**

$(15x^3 + 12x^2) \div 3x = \dfrac{15x^3}{3x} + \dfrac{12x^2}{3x}$ \qquad *Divide each term by $3x$.*

$\qquad\qquad\qquad\qquad = \dfrac{\overset{5}{15} \cdot \overset{1}{x} \cdot x \cdot x}{\underset{1}{3} \cdot \underset{1}{x}} + \dfrac{\overset{4}{12} \cdot \overset{1}{x} \cdot x}{\underset{1}{3} \cdot \underset{1}{x}}$ \quad *Simplify.*

$\qquad\qquad\qquad\qquad = 5x^2 + 4x$

Therefore, $(15x^3 + 12x^2) \div 3x = 5x^2 + 4x$.

Look Back

Dividing Powers: Lesson 8–2

Your Turn

Find each quotient.

e. $(9b^2 - 15) \div 3$

f. $(10x^2y^2 + 5xy) \div 5xy$

Factoring a polynomial can help simplify computations.

6 A stone walkway is to be built around a square planter that contains a shade tree.

A. **If the walkway is 2 meters wide, write an expression in factored form that represents the area of the walkway.**

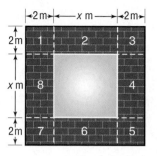

Let x represent the length and width of the planter. You can find the area of the walkway by finding the sum of the areas of the 8 rectangular sections shown in the figure.

The resulting expression can be simplified by first using the Distributive Property to combine like terms and then factoring.

Regions

$$\overbrace{1}\quad\overbrace{2}\quad\overbrace{3}\quad\overbrace{4}\quad\overbrace{5}\quad\overbrace{6}\quad\overbrace{7}\quad\overbrace{8}$$

$A = 2 \cdot 2 + 2 \cdot x + 2 \cdot 2 + 2 \cdot x + 2 \cdot 2 + 2 \cdot x + 2 \cdot 2 + 2 \cdot x$

$A = \quad 4 \quad + \quad 2x \quad + \quad 4 \quad + \quad 2x \quad + \quad 4 \quad + \quad 2x \quad + \quad 4 \quad + \quad 2x$

$A = 16 + 8x$ *$4 + 4 + 4 + 4 = 16$ and $2x + 2x + 2x + 2x = 8x$*

$A = 8(2) + 8(x)$ *The GCF of 16 and $8x$ is 8.*

$A = 8(2 + x)$

B. **If the dimensions of the square planter are 1.5 meters by 1.5 meters, find the area of the walkway.**

$A = 8(2 + x)$

$A = 8(2 + 1.5)$ *Replace x with 1.5.*

$A = 8(3.5)$ or 28

The area of the walkway is 28 square meters.

Check for Understanding

Communicating Mathematics

Study the lesson. Then complete the following.

1. **Illustrate** with algebra tiles or a drawing how to factor $x^2 + 2x$.

2. **Explain** what it means to factor a polynomial.

Math Journal

3. **Write** a few sentences explaining how the Distributive Property is used to factor polynomials. Include at least two examples.

Vocabulary

factoring

⏱ **Getting Ready**   **Find the GCF of the terms in each expression.**

Sample: $9a^2 + 3a$ **Solution:** $9a^2 = 3 \cdot 3 \cdot a$
$3a = \quad 3 \cdot a$ The GCF is $3a$.

4. $8xy + 12mn$ **5.** $x^2 + 2x$ **6.** $5xy + y^2$

7. $18y^2 + 30y$ **8.** $3ab - 2a^2b$ **9.** $15m^2n - 20m^2n$

Factor each polynomial. If the polynomial cannot be factored, write
prime.   *(Examples 1–4)*

10. $3x + 6$ **11.** $2x^2 + 4x$ **12.** $12a^2b + 6a$

13. $7mn - 13yz$ **14.** $3x^2y + 6xy + 9y^2$ **15.** $2a^3b^2 + 8ab + 16a^2b^3$

Find each quotient.   *(Example 5)*

16. $(12m^2 - 15m) \div 3m$ **17.** $(3c^2d + 9cd) \div 3cd$

18. Landscaping   Kiyoshi is planning to build a walkway around her square koi pond. The walkway is 6 feet wide.

   a. If x represents the length and width of the pond, write an expression in factored form that represents the area of the walkway.   *(Example 6)*

   b. If the dimensions of the pond are 8 feet by 8 feet, find the area of the walkway.   *(Example 7)*

Exercises   • • • • • • • • • • • • • • • • • • •

Practice

Factor each polynomial. If the polynomial cannot be factored, write
prime.

19. $9x + 15$ **20.** $6x + 3x^2$ **21.** $8x + 2x^2y$

22. $7a^2b^2 + 3ab^3$ **23.** $3c^2d - 6c^2d^2$ **24.** $7x - 3y$

25. $36mn - 11mn^2$ **26.** $18xy^2 + 24x^2y$ **27.** $19ab + 21xy$

28. $14mn^2 - 2mn$ **29.** $12xy^3 + y^4$ **30.** $3a^2b - 6a^2b^2$

31. $24xy + 18xy^2 - 3y$ **32.** $3x^3y + 9xy + 36xy^2$

33. $x + x^2y^3 + x^3y^2$ **34.** $6x^2 + 9xy + 24x^2y^2$

35. $12axy - 14ay + 20ax$ **36.** $42xyz - 12x^2y^2 + 3x^3y^3$

Find each quotient.

37. $(27x^2 - 21y^2) \div 3$ **38.** $(5abc + c) \div c$

39. $(14ab + 28b) \div 14b$ **40.** $(16x + 24xy) \div 8x$

41. $(4x^2y^2z + 6xz^2) \div 2xz$ **42.** $(3x^2y + 12xyz^2) \div 3xy$

43. Divide $6x^2 + 9$ by 3.

44. What is the GCF of $14abc^2$ and $18c$?

**Applications and
Problem Solving**

45. Geometry   The area of a rectangle is $(16x + 4y)$ square feet. If the width is 4 feet, find the length.

46. **Marine Biology** In a pool at a water park, a dolphin jumps out of the water traveling at 24 feet per second. Its height h, in feet, above the water after t seconds is given by the formula $h = 24t - 16t^2$.

    a. Factor the expression $24t - 16t^2$.

    b. Find the height of a dolphin when $t = 0.75$ second.

47. **Geometry** Write an expression in factored form that represents the area of the shaded region.

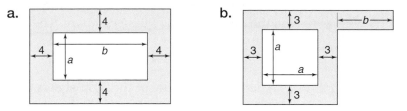

    a.   **b.**

48. **Critical Thinking** The length and width of a rectangle are represented by $2x$ and $9 - 4x$. If x must be an integer, what are the possible measures for the area of this rectangle?

Mixed Review

Classify each number as *prime* or *composite*. *(Lesson 10–1)*

49. 2   **50.** 21   **51.** 49   **52.** 53   **53.** 90

54. **Geometry** The length of a side of a square is $3x + 5$ units. What is the area of the square? *(Lesson 9–5)*

Add or subtract. *(Lesson 9–2)*

55. $(x^2 + 4x - 3) + (2x^2 - 6x - 9)$ **56.** $(2y^2 - 5y + 3) - (5y^2 - 4)$

57. **Open-Ended Test Practice** Write a second degree polynomial. *(Lesson 9–1)*

Quiz 1   Lessons 10–1 and 10–2

1. Find the prime factorization of 24. *(Lesson 10–1)*

Find the GCF of the terms in each expression. Then factor the expression. *(Lessons 10–1, 10–2)*

2. $20s + 40s^2$ **3.** $ax^3 + 7bx^3 + 11cx^3$ **4.** $6x^3 + 12x^2 + 6x$

5. **Construction** A 2-foot wide stone path is to be built along each side of a rectangular flower garden. The length of the garden is twice the width. If the flower garden is bordered on one side by a house, write an expression in factored form to represent the area of the path. *(Lesson 10–2)*

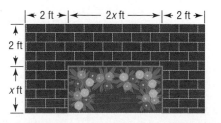

Math
In the Workplace

What You'll Learn

You'll learn to factor trinomials of the form $x^2 + bx + c$.

Why It's Important

Biology Geneticists use Punnett squares, which are similar to the models for factoring trinomials.
See Exercise 51.

In biology, *Punnett squares* are used to show possible ways that traits can be passed from parents to their offspring.

Each parent has two genes for each trait. The letters representing the parent's genes are placed on the outside of the Punnett square. The letters inside the boxes show the possible gene combinations for their offspring.

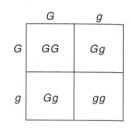

The Punnett square at the right shows the gene combinations for fur color in rabbits.

- G represents the dominant gene for gray fur.
- g represents the recessive gene for white fur.

Notice that the Punnett square is similar to the model for multiplying binomials. The model below shows the product of $(x + 1)$ and $(x + 3)$.

$$(x + 1)(x + 3) = x^2 + 3x + 1x + 3$$
$$= x^2 + 4x + 3$$

In this lesson, you will factor a trinomial into the product of two binomials.

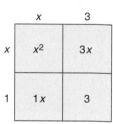

Hands-On Algebra

Materials: 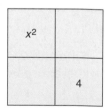 straightedge

Use a model to factor $x^2 + 5x + 4$.

Step 1 Draw a square with four sections. Put the first and last terms into the boxes as shown.

Step 2 Factor x^2 as $x \cdot x$ and place the factors outside the box.

Now, think of factors of 4 to place outside the box.

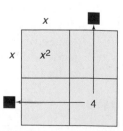

Step 3   The number 4 has two different factor pairs, 2 and 2, and 4 and 1. Try the factor pairs until you find the one that results in a middle term of $5x$.

Try 2 and 2.

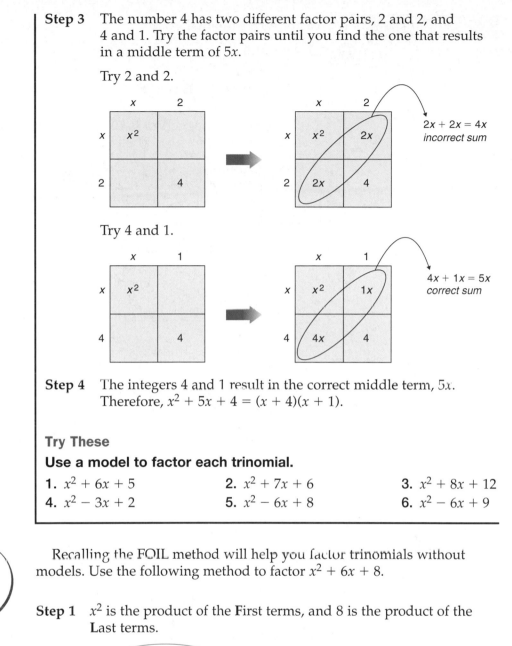

$2x + 2x = 4x$
incorrect sum

Try 4 and 1.

$4x + 1x = 5x$
correct sum

Step 4   The integers 4 and 1 result in the correct middle term, $5x$. Therefore, $x^2 + 5x + 4 = (x + 4)(x + 1)$.

Try These

Use a model to factor each trinomial.

1. $x^2 + 6x + 5$ **2.** $x^2 + 7x + 6$ **3.** $x^2 + 8x + 12$
4. $x^2 - 3x + 2$ **5.** $x^2 - 6x + 8$ **6.** $x^2 - 6x + 9$

Look Back

FOIL Method:
Lesson 9–4

Recalling the FOIL method will help you factor trinomials without models. Use the following method to factor $x^2 + 6x + 8$.

Step 1 x^2 is the product of the First terms, and 8 is the product of the Last terms.

$$x^2 + 6x + 8 = (x + \blacksquare)(x + \blacksquare)$$

Step 2   Try several factor pairs of 8 until the sum of the products of the Outer and Inner terms is $6x$. Check by using FOIL.

Try 1 and 8. $(x + 1)(x + 8) = x^2 + 8x + 1x + 8$
                              $= x^2 + 9x + 8$ *9x is not the*
                                                      correct term.

Try 2 and 4. $(x + 2)(x + 4) = x^2 + 4x + 2x + 8$
                              $= x^2 + 6x + 8$ ✓

Therefore, $x^2 + 6x + 8 = (x + 2)(x + 4)$.

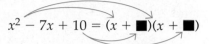

Examples

Factor each trinomial.

1 $x^2 - 7x + 10$

$$x^2 - 7x + 10 = (x + \blacksquare)(x + \blacksquare)$$

Find integers whose product is 10 and whose sum is -7. *Recall that the product of two negative integers is positive.*

Product	Integers	Sum
10	$-1, -10$	$-1 + (-10) = -11$
10	$-2, -5$	$-2 + (-5) = -7$ ✓

Therefore, $x^2 - 7x + 10 = (x - 2)(x - 5)$.

2 $x^2 + 5x - 6$

$$x^2 + 5x - 6 = (x + \blacksquare)(x + \blacksquare)$$

Find integers whose product is -6 and whose sum is 5. *Recall that the product of a positive integer and a negative integer is negative.*

Product	Integers	Sum
-6	$-2, 3$	$-2 + 3 = 1$
-6	$2, -3$	$2 + (-3) = -1$
-6	$-1, 6$	$-1 + 6 = 5$ ✓

You can stop listing factors when you find a pair that works.

Therefore, $x^2 + 5x - 6 = (x - 1)(x + 6)$.

3 $x^2 - 7 - 3x$

First, write the trinomial as $x^2 - 3x - 7$.

$$x^2 - 3x - 7 = (x + \blacksquare)(x + \blacksquare)$$

Find two integers whose product is -7 and whose sum is -3.

Product	Integers	Sum
-7	$-1, 7$	$-1 + 7 = 6$
-7	$1, -7$	$1 + (-7) = -6$

There are no factors of -7 whose sum is -3. Therefore, $x^2 - 3x - 7$ is a prime polynomial.

Your Turn

a. $x^2 + 3x + 2$ **b.** $a^2 + 4a + 3$ **c.** $b^2 + 4b + 4$

d. $y^2 - 7y + 12$ **e.** $n^2 - 5n - 14$ **f.** $m^2 - m + 1$

In the previous lesson, you learned that the terms of a polynomial might have a GCF that can be factored using the Distributive Property. When you factor trinomials, always check for a GCF first.

Example **4** Factor $2x^2 - 20x - 22$.

First, check for a GCF.
$2x^2 - 20x - 22 = 2(x^2 - 10x - 11)$ *The GCF is 2.*

Now, factor $x^2 - 10x - 11$. *Find two integers whose product is -11 and whose sum is -10.*

Product	Integers	Sum
-11	$-1, 11$	$-1 + 11 = 10$
-11	$1, -11$	$1 + (-11) = -10$ ✓

So, $x^2 - 10x - 11 = (x + 1)(x - 11)$.
Therefore, $2x^2 - 20x - 22 = 2(x + 1)(x - 11)$. *Check by using FOIL.*

Your Turn

Factor each polynomial.
g. $3y^2 - 9y - 54$ **h.** $5m^2 + 45m + 100$

The area of a figure can often be expressed as a trinomial.

Example **5**

Gardening Link

Real World

Tammy is planning a rectangular garden in which the width will be 4 feet less than its length. She has decided to put a birdbath within the garden, occupying a space 3 feet by 4 feet. How many square feet are now left for planting? Express the answer in factored form.

Let ℓ = the length of the original rectangle.
Let $\ell - 4$ = the width of the original rectangle.

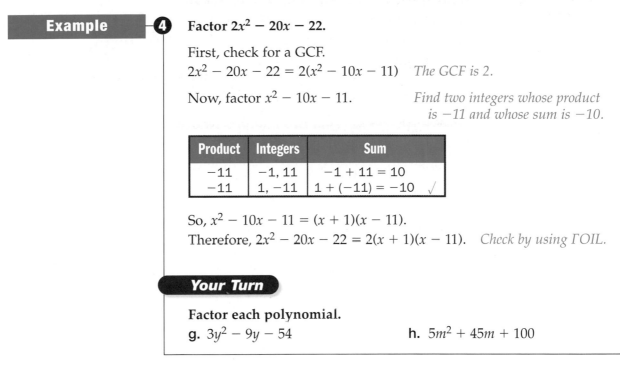

Find the area of the original rectangle.
$A = \ell w$ *Area = length × width*
$A = \ell(\ell - 4)$ *Replace w with $\ell - 4$.*
$A = \ell^2 - 4\ell$ *Distributive Property*

Find the area of the small rectangle.
$A = 4(3)$ or 12

Remaining area = area of original rectangle − area of small rectangle
$$= \quad \ell^2 - 4\ell \quad - \quad 12$$

The remaining area is $\ell^2 - 4\ell - 12$ or $(\ell - 6)(\ell + 2)$.

Check for Understanding

Study the lesson. Then complete the following.

1. **Illustrate** how to factor $x^2 + 7x + 6$ using a model.

2. **Explain** why the trinomial $x^2 + x + 5$ cannot be factored.

3. **Complete** the following sentence.
   When you factor $m^2 - 3m - 10$, you want to find two integers whose
   product is ___?___ and whose sum is ___?___.

Guided Practice

⏱ **Getting Ready** **Find two integers whose product is the first
number and whose sum is the second number.**

Sample: 10, 7   **Solution:** $2 \times 5 = 10$, $2 + 5 = 7$

4. 30, 11   5. 12, −7   6. −10, 3   7. −6, −5   8. −30, −7

**Factor each trinomial. If the trinomial cannot be factored, write
prime.** *(Examples 1–4)*

9. $x^2 + 5x + 6$ 10. $y^2 + 9y + 20$ 11. $a^2 - 5a + 4$

12. $z^2 - 8z + 16$ 13. $x^2 + 3x - 10$ 14. $m^2 - 4m - 21$

15. $w^2 + w + 2$ 16. $3a^2 + 15a + 12$ 17. $2c^2 - 12c - 14$

18. **Geometry**   Find the area of the shaded
    region. Express the area in factored
    form.   *(Example 5)*

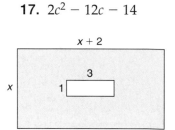

Exercises

• • • • • • • • • • • • • • • • • • •

Practice

**Factor each trinomial. If the trinomial cannot be factored, write
prime.**

19. $b^2 + 5b + 4$ 20. $x^2 + 10x + 25$ 21. $a^2 + 7a + 12$

22. $a^2 + 3a + 5$ 23. $y^2 + 12y + 27$ 24. $z^2 + 13z + 40$

25. $x^2 - 8x + 15$ 26. $a^2 - 4a + 4$ 27. $c^2 - 13c + 36$

28. $d^2 - 11d + 28$ 29. $m^2 - 5m + 1$ 30. $y^2 - 12y + 32$

31. $c^2 + 2c - 3$ 32. $x^2 - 5x - 24$ 33. $r^2 - 3r - 18$

34. $m^2 - 2m - 24$ 35. $n^2 + 13n - 30$ 36. $m^2 + 11m - 12$

37. $x^2 - 17x + 72$ 38. $a^2 - a - 90$ 39. $r^2 + 22r - 48$

40. $4x^2 + 28x + 40$ 41. $3y^2 - 21y + 36$ 42. $z^3 + z^2 - 12z$

43. $m^3 + 3m^2 + 2m$ 44. $3y^3 - 24y^2 + 36y$ 45. $2a^3 + 14a^2 - 16a$

46. Express $x^2 + 24x + 95$ as the product of two binomials.

47. Write a trinomial that cannot be factored.

48. Complete the trinomial $x^2 + 6x +$ ___?___ with a positive integer so
    that the resulting trinomial can be factored.

49. **Geometry** Refer to the figure at
the right.
   a. Express the area of the shaded
      region as a polynomial.
   b. Express the area in factored form.

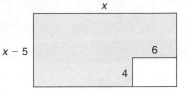

50. **Geometry** The volume of a rectangular prism is $x^3 + 4x^2 + 3x$. Find
    the length, width, and height of the prism if each one can be written
    as a monomial or binomial with integral coefficients. (*Hint*: Use the
    formula $V = \ell wh$.)

51. **Genetics** In guinea pigs, a black coat is a dominant trait over a white
    coat. Let C represent a black coat and c represent a white coat in the
    Punnett squares below. Find the missing genes or gene pair.

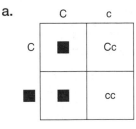

a.

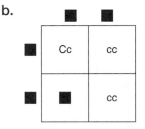

b.

52. **Critical Thinking** Find all values of k so that the trinomial
    $x^2 + kx + 10$ can be factored.

Mixed Review

Find each quotient. (*Lesson 10–2*)

53. $(10x^2 + 25y^2) \div 5$

54. $(2y^2 + 4y) \div y$

55. $(6a^2 + 8ab - 6b^2) \div 2$

56. $(3x^2y^2 + 9x^3y^2z) \div 3x^2y^2$

Find the GCF of each set of numbers or monomials. (*Lesson 10–1*)

57. $12a, 16b$

58. $6a^2b, 9ab^2$

59. $15x, 7y$

60. $15, 60, 75$

61. **Sales** The graph shows the
    value of riding lawn mower
    shipments in 1997 and 2000.
    (*Lesson 7–4*)
   a. Write an equation of the line
      in slope-intercept form.
   b. What does the slope
      represent?
   c. Use the equation to predict
      the value of riding lawn
      mower shipments in 2005.

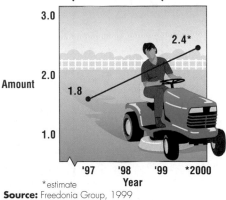

Riding Lawn Mower Shipments
(billion dollars)

*estimate
Source: Freedonia Group, 1999

62. **Standardized Test Practice** Evaluate $8x + 3y$ if $x = 9$ and $y = -2$.
    (*Lesson 2–5*)

   A 66 B 57 C 78 D 11

10-4 Factoring Trinomials: $ax^2 + bx + c$

What You'll Learn

You'll learn to factor trinomials of the form $ax^2 + bx + c$.

Why It's Important

Manufacturing
The volume of a rectangular crate can be expressed in factored form.
See Example 4.

In this lesson, you will learn to factor trinomials in which the coefficient of x^2 is a number other than 1.

Hands-On Algebra

Materials: straightedge

Use a model to factor $2x^2 + 7x + 6$.

Step 1 Draw a square with four sections. Put the first and last terms into the boxes as shown.

Step 2 Factor $2x^2$ as $2x \cdot x$ and place the factors outside the box.

Think of factors of 6 to place outside the box.

Step 3 The number 6 has two different factor pairs, 2 and 3, and 1 and 6. Try the factor pairs until you find the one that results in a middle term of $7x$. First, try 2 and 3. Note that there are two different ways of placing the 2 and 3 outside of the box.

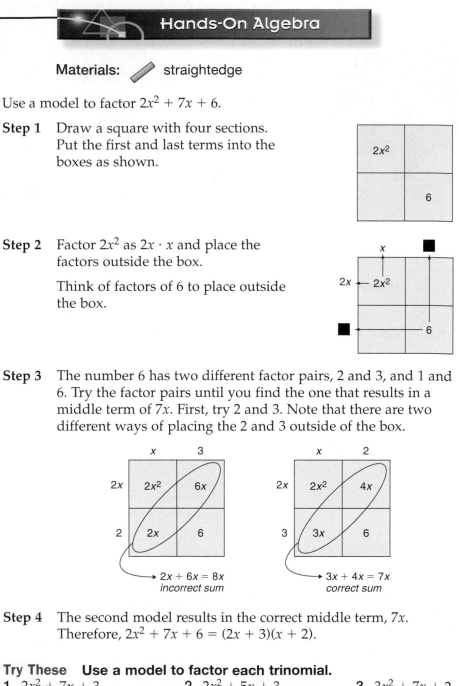

Step 4 The second model results in the correct middle term, $7x$. Therefore, $2x^2 + 7x + 6 = (2x + 3)(x + 2)$.

Try These   Use a model to factor each trinomial.
1. $2x^2 + 7x + 3$
2. $2x^2 + 5x + 3$
3. $3x^2 + 7x + 2$
4. $3x^2 + 8x + 5$
5. $4x^2 + 8x + 3$
6. $4x^2 + 13x + 3$

The FOIL method will help you factor trinomials without models.

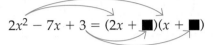
Factor each trinomial.

1 $2x^2 - 7x + 3$

$2x^2$ is the product of the First terms, and 3 is the product of the Last terms.

$$2x^2 - 7x + 3 = (2x + \blacksquare)(x + \blacksquare)$$

The last term, 3, is positive. The sum of the inside and outside terms, -7, is negative. So, both factors of 3 must be negative. Try these factor pairs of 3 until the sum of the products of the **O**uter and **I**nner terms is $-7x$.

Try -3 and -1. $(2x - 3)(x \quad 1) = 2x^2 - 2x - 3x + 3$
$$= 2x^2 - 5x + 3 \quad \textit{-5x is not the correct}$$
$$\textit{middle term.}$$

$$(2x - 1)(x - 3) = 2x^2 - 6x - 1x + 3$$
$$= 2x^2 - 7x + 3 \quad \checkmark$$

Therefore, $2x^2 - 7x + 3 = (2x - 1)(x - 3)$.

2 $3y^2 + 2y - 5$

$3y^2$ is the product of the First terms, and -5 is the product of the Last terms.

$$3y^2 + 2y - 5 = (3y + \blacksquare)(y + \blacksquare)$$

Find integers whose product is -5. Try these factor pairs of -5 until the sum of the products of the **O**uter and **I**nner terms is $2y$.

Try -5 and 1. $(3y - 5)(y + 1) = 3y^2 + 3y - 5y - 5$
$$= 3y^2 - 2y - 5 \quad \textit{-2y is not the correct}$$
$$\textit{middle term.}$$

$$(3y + 1)(y - 5) = 3y^2 - 15y + 1y - 5$$
$$= 3y^2 - 14y - 5 \quad \textit{-14y is not the}$$
$$\textit{correct middle term.}$$

Try 5 and -1. $(3y + 5)(y - 1) = 3y^2 - 3y + 5y - 5$
$$= 3y^2 + 2y - 5 \quad \checkmark$$

Therefore, $3y^2 + 2y - 5 = (3y + 5)(y - 1)$.

Your Turn

a. $2x^2 + 3x + 1$ **b.** $5y^2 + 2y - 3$ **c.** $3z^2 - 8z + 4$

Sometimes the coefficient of x^2 can be factored into more than one pair of integers.

Example — **3** **Factor $4x^2 + 12x + 5$.**

Number	Factor Pairs
4	4 and 1, 2 and 2
5	5 and 1

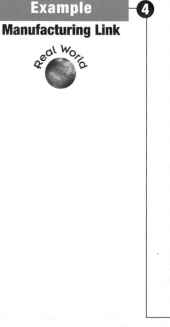

Reading Algebra

When factoring this kind of trinomial, it is important to keep an organized list of the factors.

Try 4 and 1. $(4x + 5)(1x + 1) = 4x^2 + 4x + 5x + 5$
$\qquad\qquad\qquad\qquad\qquad\quad = 4x^2 + 9x + 5$

$\qquad\qquad\quad (4x + 1)(1x + 5) = 4x^2 + 20x + 1x + 5$
$\qquad\qquad\qquad\qquad\qquad\qquad = 4x^2 + 21x + 5$

Try 2 and 2. $(2x + 5)(2x + 1) = 4x^2 + 2x + 10x + 5$
$\qquad\qquad\qquad\qquad\qquad\quad = 4x^2 + 12x + 5$ ✓

Therefore, $4x^2 + 12x + 5 = (2x + 5)(2x + 1)$.

Your Turn

d. $6x^2 + 17x + 5$ **e.** $4x^2 - 8x - 5$

Recall that the first step in factoring any polynomial is to factor out any GCF other than one.

Example — **4**

Manufacturing Link

Real World

The volume of a rectangular shipping crate is $6x^3 - 15x^2 - 36x$. Find possible dimensions for the crate.

The formula for the volume of a rectangular prism is $V = \ell wh$. Find three factors of $6x^3 - 15x^2 - 36x$. First, look for a GCF.

$6x^3 - 15x^2 - 36x = 3x(2x^2 - 5x - 12)$ *The GCF is 3x.*

$3x$ is one factor of $6x^3 - 15x^2 - 36x$. Factor $2x^2 - 5x - 12$ to find the other two factors.

$2x^2 - 5x - 12 = (2x + \blacksquare)(x + \blacksquare)$

The factors of -12 are -3 and 4, 3 and -4, -2 and 6, 2 and -6, -1 and 12, and 1 and -12. Check several combinations; the correct factors are 3 and -4.

$2x^2 - 5x - 12 = (2x + 3)(x - 4)$

So, $6x^3 - 15x^2 - 36x = 3x(2x + 3)(x - 4)$. Therefore, the dimensions can be $3x$, $2x + 3$, and $x - 4$.

Communicating Mathematics

Study the lesson. Then complete the following.

1. **Write** the trinomial and its binomial factors shown by each model.

a.

	$2x$	3
x	$2x^2$	$3x$
-4	$-8x$	-12

b.

	$3x$	-1
$2x$	$6x^2$	$-2x$
-3	$-9x$	3

c.

	x	5
$5x$	$5x^2$	$25x$
3	$3x$	15

2. **You Decide** Jamal factored the trinomial $18k^2 - 24k + 8$ and wrote the answer as $(3k - 2)(6k - 4)$. Jacqui disagrees with Jamal's answer. She says that he did not factor the trinomial completely. Who is correct? Explain your reasoning.

Guided Practice

Factor each trinomial. If the trinomial cannot be factored, write prime. *(Examples 1–4)*

3. $2a^2 + 5a + 3$
4. $3y^2 + 7y + 2$
5. $5x^2 + 13x + 6$
6. $2x^2 + x - 3$
7. $2x^2 + x - 21$
8. $2n^2 - 11n + 7$
9. $10a^2 - 9a + 2$
10. $6y^2 - 11y + 4$
11. $6x^2 + 16x + 10$

12. **Geometry** The measure of the volume of a rectangular prism is $2x^3 + x^2 - 15x$. Find possible dimensions for the prism. *(Example 4)*

Exercises • • • • • • • • • • • • • • • • • •

Practice

Factor each trinomial. If the trinomial cannot be factored, write prime.

13. $2y^2 + 7y + 3$
14. $2x^2 + 11x + 5$
15. $4a^2 + 8a + 3$
16. $2x^2 - 9x - 5$
17. $2q^2 - 9q - 18$
18. $5x^2 - 13x - 6$
19. $7a^2 + 22a + 3$
20. $3y^2 + 7y + 15$
21. $3x^2 + 14x + 8$
22. $2z^2 - 11z + 15$
23. $3x^2 + 14x + 15$
24. $3m^2 + 10m + 8$
25. $3x^2 + 5x + 1$
26. $4x^2 - 8x + 3$
27. $14x^2 + 33x - 5$
28. $6y^2 - 11y + 4$
29. $8m^2 - 10m + 3$
30. $6r^2 + 9r - 42$
31. $6x^2 + 3x - 30$
32. $4x^2 + 10x - 6$
33. $2x^3 + 5x^2 - 12x$
34. $7x - 5 + 6x^2$
35. $11y + 6y^2 - 2$
36. $15x^3 - 11x^2 - 12x$
37. $2a^2 + 5ab - 3b^2$
38. $15x^2 - 13xy + 2y^2$
39. $9k^2 + 30km + 25m^2$

40. Factor $2x^2 + 5x - 25$.

41. What are the factors of the trinomial $6x^3 + 15x^2 - 9x$?

Applications and Problem Solving

42. **Measurement** The volume of a rectangular prism is 60 cubic feet. If the measure of the length, width, and height are consecutive integers, find the dimensions.

43. **Manufacturing** The dimensions of a rectangular piece of metal are shown at the right.

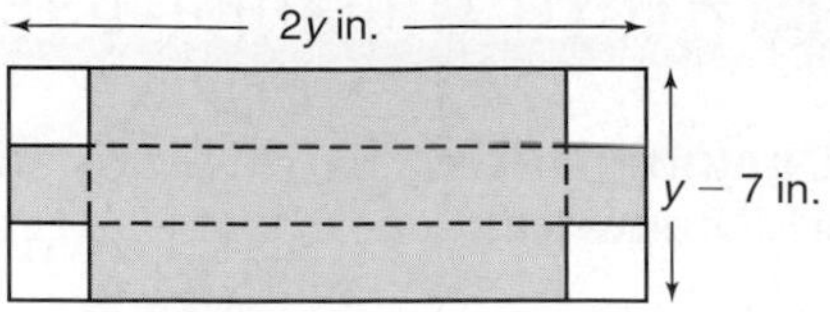

2y in.

y − 7 in.

a. If a 1-inch by 1-inch square is removed from each corner, write an expression that represents the area of the remaining piece of metal. Express the area in factored form.

b. If the metal is folded along the dashed lines, an open box is formed. Write an expression that represents the volume of the box.

c. If $y = 10$ inches, find the area of the metal and the volume of the box.

44. **Critical Thinking** Find all values of k so that the trinomial $4y^2 + ky + 5$ can be factored.

Mixed Review

Factor each polynomial. *(Lessons 10–2 & 10–3)*

45. $x^2 + 14x - 32$

46. $y^2 - 7x + 12$

47. $3a^3 - 15a^2 + 6a$

48. $2n^2 + 2n - 24$

49. **Geometry** Find the length of the diagonal of a rectangle whose length is 24 feet and whose height is 7 feet. *(Lesson 8–7)*

Solve. Assume that *y* varies directly as *x*. *(Lesson 6–5)*

50. If $y = 28$ when $x = 7$, find x when $y = 52$.

51. Find x when $y = 45$, if $y = 27$ when $x = 6$.

52. **Standardized Test Practice** What is the solution of $10 - 3(x + 4) = 16$? *(Lesson 4–7)*

A −6   B $-\dfrac{12}{7}$ C 2   D 6

Quiz 2   Lessons 10–3 and 10–4

▶ **Factor each trinomial.**

1. $x^2 + 3x - 10$

2. $x^2 - 5x - 24$ *(Lesson 10–3)*

3. $2x^2 + 9x + 7$

4. $8x^2 - 16x - 10$ *(Lesson 10–4)*

5. **Geometry** Find the area of the shaded region. Express the area in factored form. *(Lesson 10–3)*

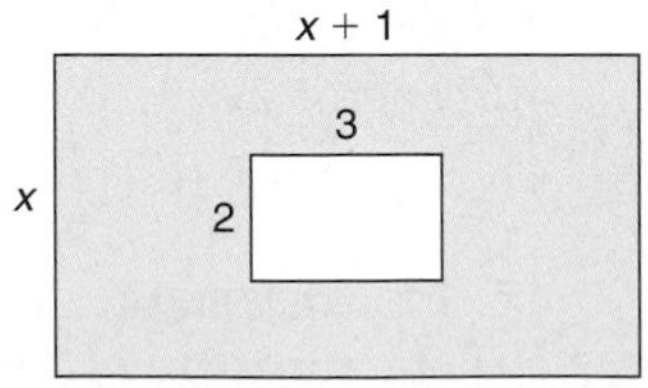

x + 1

3

x

2

Extra Practice   See p. 712.

Math
In the Workplace

What You'll Learn
You'll learn to recognize and factor the differences of squares and perfect square trinomials.

Why It's Important
Manufacturing You can find the area of a washer by using the difference of squares. *See Exercise 52.*

When you studied geometry, you learned to recognize a square as a rectangle in which the length and width are equal. For example, the base of the Arc de Triomphe in Paris, France, is equal to its height. In algebra, you learned to recognize numbers that are perfect squares. They have two equal factors.

In this lesson, you will learn to recognize and factor polynomials that are **perfect square trinomials**. They have two equal binomial factors.

The model shows the product $(x + 3)^2$. You can also use the FOIL method to find the product.

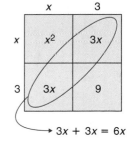

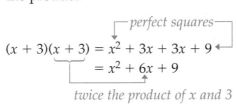

$$(x + 3)(x + 3) = x^2 + 3x + 3x + 9$$
$$= x^2 + 6x + 9$$

perfect squares

twice the product of x and 3

Look Back

Square of a Sum: Lesson 9–5

The square of $(x + 3)$ is the sum of

- the square of the first term of the binomial,
- the square of the last term of the binomial, and
- twice the product of the terms of the binomial.

These observations will help you recognize when a trinomial is a perfect square trinomial. They can be factored as shown.

Factoring Perfect Square Trinomials	**Numbers:**	$x^2 + 6x + 9 = (x + 3)(x + 3)$
		$x^2 - 6x + 9 = (x - 3)(x - 3)$
	Symbols:	$a^2 + 2ab + b^2 = (a + b)(a + b)$
		$a^2 - 2ab + b^2 = (a - b)(a - b)$
	Models:	

	a	b
a	a^2	ab
b	ab	b^2

	a	$-b$
a	a^2	$-ab$
$-b$	$-ab$	b^2

Determine whether each trinomial is a perfect square trinomial. If so, factor it.

① $x^2 + 10x + 25$

To determine whether $x^2 + 10x + 25$ is a perfect square trinomial, answer each question.

- Is the first term a perfect square?   Yes, x^2 is the square of x.
- Is the last term a perfect square?   Yes, 25 is the square of 5.
- Is the middle term twice the product of x and 5?   Yes, $10x = 2(5x)$.

Therefore, $x^2 + 10x + 25$ is a perfect square trinomial.
$x^2 + 10x + 25 = (x + 5)^2$

② $4n^2 - 4n + 1$

- Is the first term a perfect square?   Yes, $4n^2$ is the square of $2n$.
- Is the last term a perfect square?   Yes, 1 is the square of 1 and -1.
- Is the middle term twice the product of $2n$ and -1?   Yes, $2(-2n) = -4n$.

Therefore, $4n^2 - 4n + 1$ is a perfect square trinomial.
$4n^2 - 4n + 1 = (2n - 1)^2$

③ $4p^2 - 12p + 36$

- Is the first term a perfect square?   Yes, $4p^2$ is the square of $2p$.
- Is the last term a perfect square?   Yes, 36 is the square of 6 and -6.
- Is the middle term twice the product of $2p$ and -6?   No, $2(-12p) \neq -12p$.

Therefore, $4p^2 - 12p + 36$ is *not* a perfect square trinomial.

Your Turn

a. $a^2 + 2a + 1$ **b.** $16x^2 + 20x + 25$ **c.** $49x^2 - 14x + 1$

Geometry Link

④ **The area of a square is $x^2 + 18x + 81$. Find the perimeter.**

Factor $x^2 + 18x + 81$ to find the measure of one side of the square.

$x^2 + 18x + 81 = (x + 9)^2$

The measure of one side of the square is $x + 9$. A square has four sides of equal length. So, the perimeter is four times the length of a side.

$4(x + 9) = 4x + 36$ *Distributive Property*

The perimeter of the square is $4x + 36$.

A polynomial like $x^2 - 9$ is called the **difference of squares**. Although this is not a trinomial, it *can* be factored into two binomials. The model shows how to factor $x^2 - 9$.

Look Back

Product of a Sum and a Difference: Lesson 9–5

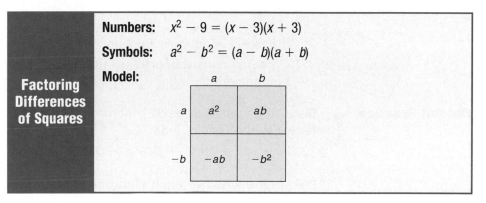

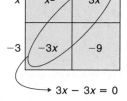

A difference of squares can be factored as shown.

Factoring Differences of Squares	**Numbers:** $x^2 - 9 = (x - 3)(x + 3)$ **Symbols:** $a^2 - b^2 = (a - b)(a + b)$ **Model:**

Examples

Determine whether each binomial is the difference of squares. If so, factor it.

5 $a^2 - 25$

a^2 and 25 are both perfect squares, and $a^2 - 25$ is a difference.

$a^2 - 25 = (a)^2 - (5)^2$ *$a \cdot a = a^2, 5 \cdot 5 = 25$*

$= (a - 5)(a + 5)$ *Use the Difference of Squares.*

6 $y^2 + 100$

y^2 and 100 are both perfect squares. But $y^2 + 100$ is a sum, not a difference. Therefore $y^2 + 100$ is *not* the difference of squares. It is a prime polynomial.

7 $3n^2 - 48$

First, look for a GCF. Then, determine whether the remaining factor is the difference of squares.

$3n^2 - 48 = 3(n^2 - 16)$ *The GCF of $3n^2$ and 48 is 3.*

$= 3[(n)^2 - (4)^2]$ *$n \cdot n = n^2, 4 \cdot 4 = 16$*

$= 3(n - 4)(n + 4)$ *Use the Difference of Squares.*

Your Turn

d. $121 - p^2$ **e.** $25x^3 - 100x$ **f.** $4a^2 + 49$

The following chart summarizes factoring methods.

Factoring Method	Number of Terms		
	Two	Three	Four or more
greatest common factor	✓	✓	✓
difference of squares	✓		
perfect square trinomials		✓	
trinomial with two binomial factors		✓	

Check for Understanding

Communicating Mathematics

Study the lesson. Then complete the following.

1. **State** whether $4c^2 - 7$ can be factored as the difference of squares. Explain.

Math Journal

2. **Copy** the chart shown above into your math journal. Then write a polynomial that can be factored by using each method.

Guided Practice

Determine whether each trinomial is a perfect square trinomial. If so, factor it. *(Examples 1–3)*

3. $y^2 + 14y + 49$ 4. $x^2 - 10x + 100$ 5. $a^2 - 10a + 25$

Determine whether each binomial is the difference of squares. If so, factor it. *(Examples 5–7)*

6. $16x^2 - 25$ 7. $8x^2 - 50y^2$ 8. $49m^2 + 16$

Factor each polynomial. If the polynomial cannot be factored, write *prime*.

9. $3x^2 + 15$ 10. $y^2 + 6y - 9$ 11. $3y^2 + 21y - 24$

12. **Geometry**   The measure of the area of a square is $4x^2 + 20xy + 25y^2$. Find the measure of the perimeter.   *(Example 4)*

Exercises

Practice

Determine whether each trinomial is a perfect square trinomial. If so, factor it.

13. $r^2 + 8r + 16$ 14. $x^2 - 16x + 64$ 15. $a^2 + 2a + 1$
16. $4a^2 + 4a + 1$ 17. $4z^2 - 20z + 25$ 18. $9m^2 + 15m + 25$
19. $9a^2 + 24a + 16$ 20. $d^2 - 22d + 121$ 21. $49 + 14z + z^2$

Determine whether each binomial is the difference of squares. If so, factor it.

22. $x^2 - 16$ 23. $a^2 - 36$ 24. $y^2 - 20$
25. $1 - 9m^2$ 26. $16m^2 - 25n^2$ 27. $y^2 + z^2$
28. $8a^2 - 18$ 29. $2z^2 - 98$ 30. $49 - a^2b^2$

31. Write a polynomial that is the difference of two squares. Then factor it.

32. Is $x^2 + x - 1$ a perfect square trinomial? Explain.

Factor each polynomial. If the polynomial cannot be factored, write *prime*.

33. $5x^2 + 25$ **34.** $a^2 - 16b^2$ **35.** $y^2 - 5y + 6$

36. $m^2 + 8m + 16$ **37.** $2x^2 - 72$ **38.** $3a^2b + 6ab + 9ab^2$

39. $8xy^2 - 13x^2y$ **40.** $2r^2 + 3r + 1$ **41.** $x^2 - 6x - 9$

42. $z^3 + 6z^2 + 9z$ **43.** $20n^2 + 34n + 6$ **44.** $b^2 + 6 - 7b$

45. $8w^2 + 14w - 15$ **46.** $a^3 - 17a^2 + 72a$ **47.** $5x^2 + 15x + 10$

48. $7a^2 - 21a$ **49.** $2x^3 - 32x$ **50.** $2x^2 - 11x - 21$

Applications and Problem Solving

51. Number Theory The difference of two numbers is 2. The difference of their squares is 12. Find the numbers.

52. Manufacturing A metal washer is manufactured by stamping out a circular hole from a metal disk. In the figure, r represents the radius of the metal disk. The radius of the hole is 1 centimeter.

  a. Write an expression in factored form for the area of the washer. (*Hint:* Use $A = \pi r^2$.)

  b. If $r = 2$ centimeters, find the area of the washer to the nearest hundredth.

53. Critical Thinking The area of a square is $81 - 90x + 25x^2$. If x is a positive integer, what is the least possible measure for the square's perimeter?

Mixed Review

Factor each polynomial. (*Lessons 10–3 & 10–4*)

54. $4y^2 + 16y + 15$ **55.** $4x^2 + 11x - 3$

56. $a^3 - 7a^2 + 12a$ **57.** $m^2 - 5m - 14$

Simplify each expression. (*Lesson 8–3*)

58. n^{-2} **59.** $a^5(a^{-3})$ **60.** $\dfrac{1}{r^{-3}}$ **61.** $\dfrac{3c^2d^3f^4}{9c^4d^2f^4}$

62. Describe the change to the graph of $y = 4x$ when $y = 4x - 5$ is graphed. (*Lesson 7–6*)

63. Standardized Test Practice Which graph is the best example of data that exhibit a linear relationship between the variables x and y? (*Lesson 6–3*)

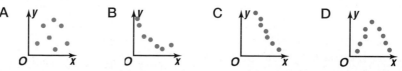

Extra Practice   See p. 713.

Study Guide and Assessment

Understanding and Using the Vocabulary

After completing this chapter, you should be able to define each term, property, or phrase and give an example or two of each.

*inter*NET
CONNECTION   **Review Activities**
For more review activities, visit:
www.algconcepts.glencoe.com

difference of squares *(p. 447)*
factoring *(p. 428)*
greatest common factor (GCF) *(p. 422)*
perfect square trinomials *(p. 445)*
prime polynomial *(p. 430)*

State whether each sentence is *true* or *false*. If false, replace the underlined word or number to make a true sentence.

1. The prime factorization of 12 is $\underline{3 \cdot 4}$.

2. $\underline{3x}$ is the greatest common factor of $6x^2$ and $9x$.

3. When two or more numbers are multiplied, each number is a <u>factor</u> of the product.

4. $2y$ and $(y + 3)$ are factors of $\underline{2y^2 + 3}$.

5. When you factor trinomials, always check for a <u>GCF</u> first.

6. The number <u>51</u> is an example of a prime number.

7. $(x - 3)(x + 3)$ is the factored form of $\underline{x^2 + 9}$.

8. Whole numbers that have more than two factors are called <u>composite numbers</u>.

9. A polynomial is in <u>factored form</u> when it is expressed as the product of polynomials.

10. $4a^2 - b^2$ is an example of a <u>perfect square trinomial</u>.

Skills and Concepts

Objectives and Examples	Review Exercises
• **Lesson 10–1** Find the greatest common factor of a set of numbers or monomials.	**Find the GCF of each set of numbers or monomials.**

Find the GCF of $12x^2y$ and $30xy^2$.
$12x^2y = 2 \cdot \boxed{2} \cdot \boxed{3} \cdot \boxed{x} \cdot x \cdot \boxed{y}$
$30xy^2 = \boxed{2} \cdot \boxed{3} \cdot 5 \cdot \boxed{x} \cdot \boxed{y} \cdot y$

The GCF of $12x^2y$ and $30xy^2$ is
$2 \cdot 3 \cdot x \cdot y$ or $6xy$.

11. $20, 25$

12. $12, 18, 42$

13. $20, 25, 28$

14. $5xy, 10x$

15. $9x^2, 9x$

16. $6a^2b, 18a^2b^2, 9ab^2$

Objectives and Examples

- **Lesson 10–2** Use the GCF and the Distributive Property to factor polynomials.

Factor $12x^2 - 8xy$.

The GCF of $12x^2$ and $8xy$ is $4x$. Write each term as a product of the GCF and its remaining factors.

$$12x^2 - 8xy = 4x(3x) - 4x(2y)$$
$$= 4x(3x - 2y) \quad \textit{Distributive Property}$$

Review Exercises

Factor each polynomial. If the polynomial cannot be factored, write *prime***.**

17. $5x + 30y$
18. $16a^2 + 32b^2$
19. $12ab - 18a^2$
20. $5mn^2 + 10mn$
21. $3xy + 12x^2y^2$

Find each quotient.

22. $(20x^3 + 15x^2) \div 5x$
23. $(40a^2b^2 - 8ab) \div 8ab$

- **Lesson 10–3** Factor trinomials of the form $x^2 + bx + c$.

Factor $x^2 + 3x - 10$.

Find integers whose product is -10 and whose sum is 3.

Product	Integers	Sum
-10	$2, -5$	$2 + (\ 5) - -3$
-10	$-2, 5$	$-2 + 5 = 3$ $\checkmark$

Therefore, $x^2 + 3x - 10 = (x - 2)(x + 5)$.

Factor each trinomial. If the trinomial cannot be factored, write *prime***.**

24. $y^2 + 9y + 14$
25. $x^2 - 8x + 15$
26. $a^2 + 5a - 7$
27. $x^2 - 2x - 8$
28. $y^2 + 7y + 12$
29. $x^2 + 2x - 35$
30. $u^2 - a - 1$
31. $2n^2 - 8n - 24$

- **Lesson 10–4** Factor trinomials of the form $ax^2 + bx + c$.

Factor $2x^2 + 5x + 3$.

$$2x^2 + 5x + 3 = (2x + \blacksquare)(x + \blacksquare)$$

$$(2x + 3)(x + 1) = 2x^2 + 2x + 3x + 3$$
$$= 2x^2 + 5x + 3$$

Therefore, $2x^2 + 5x + 3 = (2x + 3)(x + 1)$.

Factor each trinomial. If the trinomial cannot be factored, write *prime***.**

32. $2z^2 + 7z + 5$
33. $3x^2 + 8x + 5$
34. $3a^2 + 8a + 4$
35. $6a^2 - a - 2$
36. $3x^2 - 7x - 6$
37. $2y^2 - 9y - 18$
38. $2x^2 + 5x + 6$
39. $15a^2 - 20a + 5$

Objectives and Examples	Review Exercises

• Lesson 10–5 Recognize and factor the differences of squares and perfect square trinomials.

Factor $a^2 + 6a + 9$.
$$a^2 + 6a + 9 = a^2 + 2(3a) + 3^2$$
$$= (a + 3)^2$$

Factor $x^2 - 25$.
$$x^2 - 25 = x^2 - 5^2$$
$$= (x + 5)(x - 5)$$

Factor each polynomial. If the polynomial cannot be factored, write *prime*.

40. $y^2 + 8y + 16$

41. $a^2 - 12a + 36$

42. $n^2 + 2n - 1$

43. $25x^2 + 20x + 4$

44. $y^2 - 81$

45. $4x^2 - 9$

46. $a^2 + 49$

47. $12c^2 - 12$

Applications and Problem Solving

48. Physics If a flare is launched into the air, its height h above the ground after t seconds is given by the formula $h = vt - 16t^2$. In the formula, v represents the initial velocity in feet per second. *(Lesson 10–2)*
   a. Factor the expression $vt - 16t^2$.
   b. If the flare is launched with an initial velocity of 144 feet per second, find the height after 2 seconds.

49. Genetics The Punnett square below represents the possible gene combinations for hair length in dogs. H represents long hair, and h represents short hair. Find the missing genes for the parents. *(Lesson 10–3)*

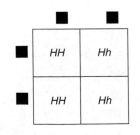

	■	■
■	HH	Hh
■	HH	Hh

50. Geometry The area of a square is $(25x^2 + 30x + 9)$ square units. Find the perimeter. *(Lesson 10–5)*

1. **Explain** what it means to *factor* a polynomial.

2. **Write** two monomials whose GCF is 1.

3. **Write** the trinomial and its binomial factors shown by the model.

4. **List** two different methods of factoring polynomials.

5. **Classify** the number 15 as *prime* or *composite*. Explain your reasoning.

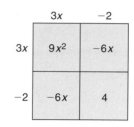

Exercise 3

Factor each monomial.

6. $25x^2y^2$

7. $-15b^3$

8. $24a^2b$

Find the GCF of each set of numbers or monomials.

9. 24, 60

10. $16a^2, 30a^3$

11. $20a^2b, 25a^2b^2$

Factor each polynomial. If the polynomial cannot be factored, write *prime*.

12. $12x^2 + 18x$

13. $3x^2y \quad 12xy^2$

14. $6a^3 + 8a^2 + 2a$

15. $x^2 + 9x + 8$

16. $m^2 - 10m + 24$

17. $y^2 - 3y - 18$

18. $3x^2 + x - 14$

19. $3m^2 + 17m + 10$

20. $2x^2 - 18$

21. $n^2 - 8n - 16$

22. $y^2 + 10y + 25$

23. $25m^2 - 16$

23. $3r^2 + r + 1$

24. $6x^3 + 15x^2 - 9x$

25. **Geometry** Find the area of the shaded region. Express the area in factored form.

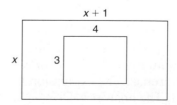

Preparing for Standardized Tests

Function and Graph Problems

All standardized tests include problems with functions and graphs.

You'll need to understand these concepts.

function

table of values

graph of a line

equation of a line

y-intercept, x-intercept

slope

THE
PRINCETON
REVIEW

Slope is the ratio of the change in y to the change in x. A line that slopes upward from left to right has a positive slope.

Proficiency Test Example

Martin is paid by the hour for babysitting. His hourly wage is a fixed amount plus an additional amount for each child. The graph shows his hourly wage for up to 5 children. If x represents the number of children, which expression can be used to find Martin's hourly wage, y?

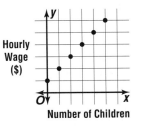

A x **B** $x - 1$ **C** $2x + 1$ **D** $x + 1$

Hint Study the graph. The points lie on a line. Find the y-intercept and the slope.

Solution The fixed amount is represented by the y-intercept. It is the point that represents 0 children. The y-intercept is 1.

The slope of the line shows the amount for each child. Moving left to right, each point is one unit higher than the previous point. So the slope is 1.

Martin's hourly wage is the fixed amount, $1, plus the amount per child, $1, times the number of children, x. The expression is $1x + 1$ or $x + 1$. The answer is D.

SAT Example

What is the equation of a line that is parallel to the line whose equation is $y = \frac{2}{3}x + 5$ and passes through the point at $(-6, 2)$?

A $y = \frac{2}{3}x + 5$ **B** $y = \frac{2}{3}x - 2$ **C** $y = \frac{2}{3}x + 6$

D $y = \frac{2}{3}x - \frac{22}{3}$ **E** $y = -\frac{3}{2}x - 7$

Hint Memorize the slope-intercept and point-slope forms of linear equations.

Solution The equation of the given line is in slope-intercept form. The slope is $\frac{2}{3}$. Parallel lines have the same slope. So, the slope of the parallel line must also be $\frac{2}{3}$. This eliminates answer choice E.

The parallel line must pass through the point at $(-6, 2)$. Write the equation of the line in point-slope form.

$y - 2 = \frac{2}{3}[x - (-6)]$

$y - 2 = \frac{2}{3}(x + 6)$

$y - 2 = \frac{2}{3}x + \frac{2}{3}(6)$ *Distributive Property*

$y - 2 = \frac{2}{3}x + 4$

$y = \frac{2}{3}x + 6$

The answer is C.

After you work each problem, record your answer on the answer sheet provided or on a sheet of paper.

1. Use the function table to find the value of y when $x = 5$.

x	y
0	3
1	17
2	31
3	45

   A 59   **B** 60
   C 73   **D** 75

2. What is the y-intercept of the line determined by the equation $5x + 2 = 7y - 3$?

   A -1 **B** $-\dfrac{1}{7}$ **C** $\dfrac{1}{7}$ **D** $\dfrac{5}{7}$ **E** 5

3. Which expression could be used to find the value of y in the graph?

   A $1 - 2x$ **B** $1 - 3x$
   C $2x + 1$ **D** $3x - 1$

4. The charge to enter a nature preserve is a fixed amount per vehicle plus a fee for each person in it. The table shows some charges. What would the charge be for a vehicle with 8 people?

People	Charge
1	$1.50
2	$2.00
3	$2.50
4	$3.00

   A $3.50   **B** $4.00   **C** $5.00   **D** $6.00

5. At what point does the line MN cross the y-axis?

   A $(-4, 0)$ **B** $(0, -4)$
   C $(-2, 0)$ **D** $(0, -2)$

   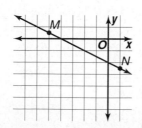

6. The average of two numbers x and y is A. Which of the following is an expression for y?

   A $\dfrac{A + x}{2}$ **B** $\dfrac{A}{2} - x$ **C** $2A - x$
   D $A - x$ **E** $x - A$

7. Write 4^{-4} without using an exponent.

   A 0.00039   **B** 0.0039
   C 0.016   **D** 256

8. Which expression should come next in the pattern $2x, 4x^2, 8x^3, 16x^4, \ldots$?

   A $24x^5$ **B** $32x^5$
   C $24x^6$ **D** $32x^6$

Open-Ended Questions

9. **Grid-In**   The graph of $y = 4x - 2$ is shown. What is the x-intercept?

   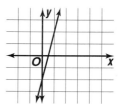

10. The graph shows the distance traveled by an African elephant.

    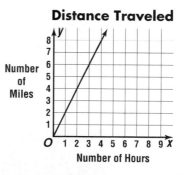

    Distance Traveled

    Part A   What is the slope of the line?
    Part B   Explain what the slope represents.

Quadratic and Exponential Functions

▶ What You'll Learn in Chapter 11:

- to graph quadratic functions *(Lesson 11–1)*,
- to recognize characteristics of families of parabolas *(Lesson 11–2)*,
- to solve quadratic equations by graphing, factoring, completing the square, and using the Quadratic Formula *(Lessons 11–3, 11–4, 11–5, and 11–6)*, **and**
- to graph exponential functions *(Lesson 11–7)*.

Problem-Solving Workshop

Project

Do you want to be a millionaire? In this project, you will make a plan for saving money. Your goal is to reach $1,000,000 at the end of a 40-year period, following the guidelines listed below.

1. The annual growth rate of your investment is 5%.
2. The money is invested for exactly 40 years.
3. Your initial deposit is any amount of your choice.
4. Interest is calculated at the end of each year.
5. Choose an additional amount of money to deposit after interest is calculated each year. This amount is fixed, or the same, for all 40 years.
6. After 40 years, you should have as close to $1,000,000 as possible.

Working on the Project

Work with a partner and choose a strategy to help analyze and solve the problem. Here are some questions to help you get started.

- At the beginning of year 1, you decide to invest $5000. How much money will you have at the end of year 1, including interest?
- At the end of year 1, you invest an additional $2000. How much money will you have at the end of year 2, including interest?

▶ Strategies

Look for a pattern.

Draw a diagram.

Make a table.

Work backward.

Use an equation.

Make a graph.

Guess and check.

Technology Tools

- Use a **spreadsheet** to calculate each year-end balance.
- Use **graphing software** to make a graph of your investment over 40 years.

*inter***NET** **CONNECTION** **Research** For more information about investing, visit: www.algconcepts.glencoe.com

Presenting the Project

Write a one-page paper describing your investment plan. Include:

- successful and unsuccessful strategies used to plan,
- a table or spreadsheet showing each year-end balance, and
- a graph of your savings over the 40-year period.

Math
In the Workplace

What You'll Learn

You'll learn to graph quadratic functions.

Why It's Important

Architecture
Architects graph quadratic functions when designing arches.
See Exercise 15.

The Exchange House in London, England, is supported by a steel arch shaped like a **parabola**. This parabola can be modeled by the **quadratic function** $y = -0.025x^2 + 2x$, where y represents the height of the arch and x represents the horizontal distance from one end of the base in meters. What is the maximum height of the arch? *This problem will be solved in Example 4.*

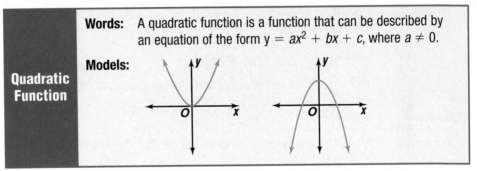

| Quadratic Function | **Words:** A quadratic function is a function that can be described by an equation of the form $y = ax^2 + bx + c$, where $a \neq 0$. |

Graphs of quadratic functions all have the shape of a parabola.

You can use a table of values to graph a quadratic function.

Examples

Graph each quadratic function by making a table of values.

1 $y = x^2 + 1$

First, choose integer values for x. Evaluate the function for each x-value. Graph the resulting coordinate pairs and connect the points with a smooth curve.

Look Back

Graphing Relations:
Lesson 6–3

x	$x^2 + 1$	y	(x, y)
-2	$(-2)^2 + 1$	5	$(-2, 5)$
-1	$(-1)^2 + 1$	2	$(-1, 2)$
0	$0^2 + 1$	1	$(0, 1)$
1	$1^2 + 1$	2	$(1, 2)$
2	$2^2 + 1$	5	$(2, 5)$

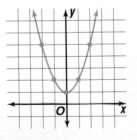

2 $y = -x^2$

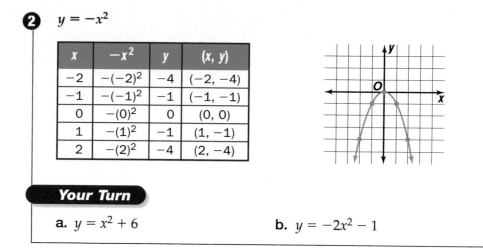

x	$-x^2$	y	(x, y)
-2	$-(-2)^2$	-4	$(-2, -4)$
-1	$-(-1)^2$	-1	$(-1, -1)$
0	$-(0)^2$	0	$(0, 0)$
1	$-(1)^2$	-1	$(1, -1)$
2	$-(2)^2$	-4	$(2, -4)$

Your Turn

a. $y = x^2 + 6$ **b.** $y = -2x^2 - 1$

In Example 1, the *lowest point*, or **minimum**, of the graph of $y = x^2 + 1$ is at $(0, 1)$. Since the coefficient of x^2 is positive, the graph opens *upward*.

In Example 2, the *highest point*, or **maximum**, of the graph of $y = -x^2$ is at $(0, 0)$. Since the coefficient of x^2 is negative, the graph opens *downward*.

The maximum or minimum point of a parabola is called the **vertex**. The vertical line containing the vertex of a parabola is called the **axis of symmetry**. If you picked up the graph of $y = x^2 + 1$ or $y = -x^2$ and folded it along the axis of symmetry, the two halves of each graph would coincide. In both examples, the axis of symmetry is the line $x = 0$.

You can use the rule below to find the equation of the axis of symmetry.

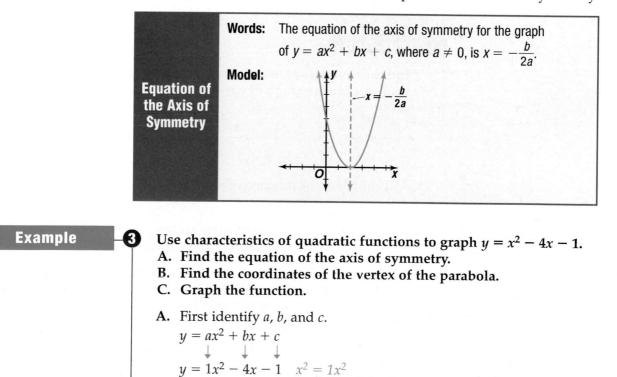

Equation of the Axis of Symmetry

Words: The equation of the axis of symmetry for the graph of $y = ax^2 + bx + c$, where $a \neq 0$, is $x = -\dfrac{b}{2a}$.

Model:

$x = -\dfrac{b}{2a}$

Example

3 Use characteristics of quadratic functions to graph $y = x^2 - 4x - 1$.
   A. Find the equation of the axis of symmetry.
   B. Find the coordinates of the vertex of the parabola.
   C. Graph the function.

   A. First identify a, b, and c.
   $$y = ax^2 + bx + c$$
   $$y = 1x^2 - 4x - 1 \quad x^2 = 1x^2$$
   So, $a = 1$, $b = -4$, and $c = -1$.   *(continued on the next page)*

Now, find the equation of the axis of symmetry.

$$x = -\frac{b}{2a} \quad \textit{Equation of axis of symmetry}$$

$$x = -\frac{-4}{2(1)} \quad a = 1, b = -4$$

$$x = 2 \quad \textit{Simplify.}$$

B. Next, find the vertex. Since the equation of the axis of symmetry is $x = 2$, the x-coordinate of the vertex must be 2. Substitute 2 for x in the equation $y = x^2 - 4x - 1$ to solve for y.

$$y = x^2 - 4x - 1$$
$$= (2)^2 - 4(2) - 1$$
$$= 4 - 8 - 1 \text{ or } -5$$

The point at $(2, -5)$ is the vertex. *This point is a minimum.*

C. Construct a table. Choose some values for x that are less than 2 and some that are greater than 2. This ensures that points on each side of the axis of symmetry are graphed.

x	$x^2 - 4x - 1$	y	(x, y)
0	$0^2 - 4(0) - 1$	−1	(0, −1)
1	$1^2 - 4(1) - 1$	−4	(1, −4)
2	$2^2 - 4(2) - 1$	−5	(2, −5)
3	$3^2 - 4(3) - 1$	−4	(3, −4)
4	$4^2 - 4(4) - 1$	−1	(4, −1)

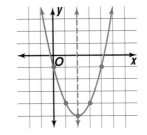

Your Turn

Find the coordinates of the vertex and the equation of the axis of symmetry for the graph of each equation. Then graph the function.

c. $y = x^2 + x$

d. $y = -x^2 + 2x - 3$

Models of quadratic functions can be seen in the real world.

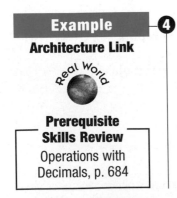

Example

Architecture Link

Real World

Prerequisite Skills Review

Operations with Decimals, p. 684

4 **Refer to the application at the beginning of the lesson. What is the maximum height of the parabolic arch?**

The maximum height of the arch is the y-coordinate of the vertex. Find the equation of the axis of symmetry for $h(x) = -0.025x^2 + 2x$.

$$x = -\frac{b}{2a}$$

$$= -\frac{2}{2(-0.025)} \quad a = -0.025, b = 2$$

$$= -\frac{2}{-0.05} \text{ or } 40 \quad \textit{Simplify.}$$

Next, find the vertex. Since the equation of the axis of symmetry is $x = 40$, the x-coordinate of the vertex must be 40. Substitute 40 for x in the function $h(x) = -0.025x^2 + 2x$. Then solve for h.

$$h(x) = -0.025x^2 + 2x$$
$$h(40) = -0.025(40)^2 + 2(40) \quad \textit{Replace x with 40.}$$
$$= -40 + 80 \text{ or } 40 \quad \textit{Simplify.}$$

The point $(40, 40)$ is the vertex. So, the maximum height is 40 meters.

Check for Understanding

Communicating Mathematics

Study the lesson. Then complete the following.

1. **Compare and contrast** quadratic and linear functions.

2. **Explain** what effect a negative coefficient of x^2 has on the orientation of a parabola.

3. **Name** the point on the graph of a quadratic function that has a unique y-coordinate.

Vocabulary

parabola
quadratic function
minimum
maximum
vertex
axis of symmetry

Guided Practice

⊙ **Getting Ready** **Identify the values for a, b, and c for each quadratic function in the form $y = ax^2 + bx + c$.**

Sample 1: $y = x^2 - 9$

Solution: $a = 1, b = 0, c = -9$

Sample 2: $y = -4x^2 - 9x + 13$

Solution: $a = -4, b = -9,$
$c = 13$

4. $y = 6x^2 - 3x + 1$

5. $y = x^2 + 4$

6. $y = -x^2 - x - 12$

7. $y - 3x^2 + 4x$

Graph each quadratic function by making a table of values.
(Examples 1 & 2)

8. $y = -2x^2$

9. $y = x^2 + 3x$

10. $y = -x^2 - 2x + 5$

Write the equation of the axis of symmetry and the coordinates of the vertex of the graph of each quadratic function. Then graph the function. *(Example 3)*

11. $y = x^2 + 2$

12. $y = x^2 + 8x + 12$

13. $y = -x^2 + 4x + 3$

14. $y = -x^2 + 10x$

15. **Architecture** Mr. Kwan is sketching the windows for a building. Their shape is a parabola modeled by the equation $h = -w^2 + 9$, where h is the height of the window and w is the width in feet. *(Example 4)*

a. Graph the function.

b. Find the maximum height of each window.

c. Find the width of each window at its base.

Practice

Graph each quadratic function by making a table of values.

16. $y = 2x^2$ **17.** $y = 3x^2$ **18.** $y = -5x^2$

19. $y = x^2 + 4$ **20.** $y = 2x^2 - 1$ **21.** $y = -x^2 + 8x$

22. $y = -x^2 + 4x + 1$ **23.** $y = 3x^2 - 6x + 1$ **24.** $y = \frac{1}{2}x^2 - 6x + 5$

Write the equation of the axis of symmetry and the coordinates of the vertex of the graph of each quadratic function. Then graph the function.

25. $y = x^2$ **26.** $y = -2x^2$ **27.** $y = 4x^2$

28. $y = 2x^2 + 3$ **29.** $y = -x^2 + 1$ **30.** $y = x^2 - 6x$

31. $y = x^2 - 4x + 2$ **32.** $y = x^2 - 4x + 10$ **33.** $y = -3x^2 - 6x + 4$

34. $y = -\frac{1}{4}x^2 + x - 2$ **35.** $y = x^2 + 5x$ **36.** $y = 3x^2 - 3x - 2$

Match each function with its graph.

37. $y = x^2 + 2x + 1$ **38.** $y = x^2 - 2x - 1$ **39.** $y = -x^2 + 4x - 3$

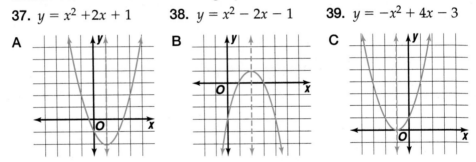

40. Suppose the equation of the axis of symmetry for a quadratic function is $x = -3$ and one of the x-intercepts is -8. What is the other x-intercept?

41. Suppose the points at $(-8, 5)$ and $(6, 5)$ are on the graph of a parabola. What is the equation of the axis of symmetry?

Applications and Problem Solving

Real World

42. Football   Mark wanted to know at which angle he should kick a football for maximum distance. He used a device to simulate kicking a football at a constant velocity, only changing the angle. He recorded his results in the table.

Angle	Distance (ft)
30°	140
35°	152
40°	160
45°	162
50°	160
55°	152

a. Graph the data in the table.

b. Which angle gives the maximum distance?

c. Predict how far the football will go if it is kicked at a 60° angle.

43. **Business** The profit function of a small business can be expressed as $P(x) = -x^2 + 300x$, where x represents the number of employees.
   a. How many employees will yield the maximum profit?
   b. What is the maximum profit?

44. **Critical Thinking** Graph $y = x^2$. Then graph $y = x^2 - 4$ on the same axes. Describe the difference between the graphs.

Factor each polynomial. *(Lesson 10–5)*

Mixed Review

45. $x^2 - 49$ 46. $a^2 + 10a + 25$ 47. $2g^2 - 16g + 32$

48. **Geometry** The volume of a rectangular prism is $3x^3 - 6x^2 - 24x$. Find possible dimensions for the prism. *(Lesson 10–4)*

49. Arrange the terms of the polynomial $-4x^2 + 5 - x^3 - 8x$ so that the powers of x are in descending order. Then state the degree of the polynomial. *(Lesson 9–1)*

Use the map for Exercises 50–51.

50. You can use the letters and numbers on the map to form ordered pairs and name square areas. State the locations of Fancyburg and Northam Parks as ordered pairs. *(Lesson 2–2)*

51. Find the straight-line distance in miles between Fancyburg and Northam Parks. *(Lesson 5–2)*

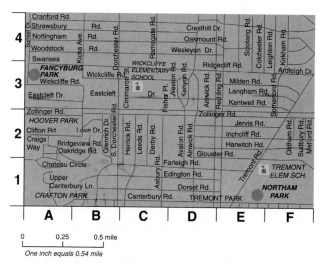

52. **Standardized Test Practice** The line graph shows the amount of soda an average American drinks over a five-year period. During which of the following periods was the greatest change in consumption? *(Lesson 1–7)*
   A 1993–1994   B 1994–1995
   C 1995–1996   D 1996–1997

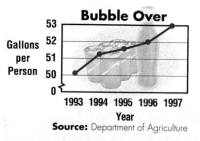

Bubble Over

Source: Department of Agriculture

Extra Practice See p. 713.

Math In the Workplace

What You'll Learn

You'll learn the characteristics of families of parabolas.

Why It's Important

Design Graphic artists use families of parabolas to create the illusion of motion. *See Example 5.*

Have you ever wondered how your favorite animated movie was created? Artists make a series of individual images and play them in a sequence to create the illusion of motion. Animated designs are created by using families of graphs.

Families of linear graphs may share the same slope or y-intercept. Families of parabolas share the same vertex, axis of symmetry, or have the same shape. Graphing calculators make it easy to study families of parabolas.

Examples

Graph each group of equations on the same screen. Compare and contrast the graphs. What conclusions can be drawn?

1 $y = x^2$, $y = 0.2x^2$, $y = 3x^2$

Use the Y= screen to enter functions into the graphing calculator.

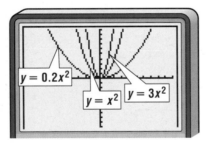

Each graph opens upward and has its vertex at the origin. The graph of $y = 0.2x^2$ is wider than the graph of $y = x^2$. The graph of $y = 3x^2$ is more narrow than the graph of $y = x^2$.

The parent graph for each family is $y = x^2$.

The shape of the parabola narrows as the coefficient of x^2 becomes greater. The shape of the parabola widens as the coefficient of x^2 becomes smaller.

2 $y = x^2$, $y = x^2 - 6$, $y = x^2 + 3$

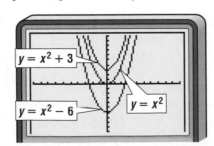

Each graph opens upward and has the same shape as $y = x^2$. Yet, each parabola has a different vertex located along the y-axis.

A constant greater than 0 shifts the graph upward, and a constant less than 0 shifts the graph downward along the y-axis.

3 $y = x^2$, $y = (x + 2)^2$, $y = (x - 4)^2$

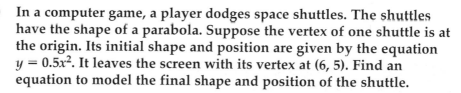

Each graph opens upward and has the same shape as $y = x^2$. However, each parabola has a different vertex located along the x-axis.

Find the number for x that results in 0 inside the parentheses. The graph shifts this number of units to the left or right.

4 $y = x^2$, $y = (x - 7)^2 + 2$

The graph of $y = (x - 7)^2 + 2$ has the same shape as the graph of $y = x^2$. However, it shifts to the right 7 units because a positive 7 will result in zero inside the parentheses. It also shifts upward 2 units because of the 2 outside the parentheses.

Your Turn

a. $y = x^2$, $2x^2$, $4x^2$

b. $y = x^2$, $x^2 - 1$, $x^2 - 8$

c. $y = x^2$, $-x^2$

d. $y = -x^2$, $y = -(x + 2)^2$, $y = -(x + 4)^2$

Sometimes computers are used to generate families of graphs.

Example

Computer Animation Link

Real World

5 **In a computer game, a player dodges space shuttles. The shuttles have the shape of a parabola. Suppose the vertex of one shuttle is at the origin. Its initial shape and position are given by the equation $y = 0.5x^2$. It leaves the screen with its vertex at (6, 5). Find an equation to model the final shape and position of the shuttle.**

The shape of the shuttle remains the same. However, the vertex shifts from (0, 0) to the right 6 units and up 5 units.

Begin with the original equation.

$y = 0.5x^2$

$y = 0.5(x - 6)^2$

If $x = 6$, then $x - 6 = 0$. Shift the vertex to the right 6 units.

$y = 0.5(x - 6)^2 + 5$

The 5 outside the parentheses shifts the entire parabola up 5 units.

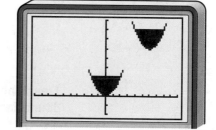

So, the final shape and position of the shuttle can be described by the equation $y = 0.5(x - 6)^2 + 5$.

Communicating Mathematics

Study the lesson. Then complete the following.

1. **Describe** the parabola whose equation is $y = 100x^2$.

2. **You Decide?** Vanessa says that the graphs of $y = (x - 2)^2$ and $y = x^2 - 2$ are the same. Vickie says that they are different. Who is correct? Explain your answer and sketch the graphs.

3. **Match** each equation with its corresponding graph.

$y = -3x^2$ $y = x^2 + 3$ $y = (x - 3)^2$

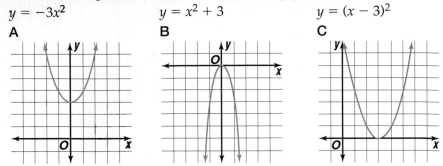

A   B   C

Guided Practice

4. Graph $y = x^2$, $y = 0.5x^2$, and $y = 0.1x^2$ on the same axes. Compare and contrast the graphs. *(Example 1)*

Describe how each graph changes from the parent graph of $y = x^2$. Then name the vertex of each graph. *(Examples 1–4)*

5. $y = 0.7x^2$ 6. $y = x^2 + 10$

7. $y = (x + 4)^2$ 8. $y = (x + 2)^2 - 9$

9. **Computer Animation** Refer to Example 5. Suppose the shuttle is programmed to move to point (4, 3) before leaving the screen. Write the equation that describes its location. *(Example 5)*

Exercises •

Practice

Graph each group of equations on the same screen. Compare and contrast the graphs.

10. $y = -x^2$
    $y = -4x^2$
    $y = -6x^2$

11. $y = (x + 1)^2$
    $y = (x + 2)^2$
    $y = (x + 3)^2$

12. $y = -x^2 - 1$
    $y = -x^2 - 3$
    $y = -x^2 - 5$

Describe how each graph changes from the parent graph of $y = x^2$. Then name the vertex of each graph.

13. $y = 5x^2$ 14. $y = x^2 - 8$ 15. $y = (x - 7)^2$

16. $y = (x - 3)^2$ 17. $y = -2x^2$ 18. $y = -0.9x^2$

19. $y = 2x^2 + 1$ 20. $y = -(x + 4)^2$ 21. $y = 0.4x^2 - 8$

22. $y = -x^2 + 6$ 23. $y = (x + 1)^2 - 5$ 24. $y = [x - (-2)]^2 + 3$

25. What is the equation of the parabola that moves the parent graph $y = x^2$ to the left 8 units?

26. Suppose the parent graph is $y = x^2 + 3$. Write the equation of the parabola that would move it down 5 units.

Applications and Problem Solving

Real World

27. Fireworks In Cincinnati, fireworks are launched from a large barge on the Ohio River for a Labor Day celebration. Their flight can be modeled by the function $h(t) = -4.9(t - 4)^2 + 80$, where the height h is in meters and the time t is in seconds. Suppose that buildings obstruct the fireworks. So, the barge is relocated 30 meters to the east. Write a function to model the path of the fireworks in the new location.

28. Baseball A baseball player hits a pop-up. The height of the ball can be modeled by the function $h(t) = -16(t - 3.5)^2 + 200$, where the height h is in feet and the time t is in seconds. Another pop-up is hit with the same velocity, but from one half a foot higher above the ground.

  a. Write equations to model the height of the ball for each hit.

  b. Graph both equations and find each vertex.

  c. Does the higher height affect its maximum height? Does the higher height affect the time it takes to reach its maximum height?

29. Critical Thinking Graphs of two quadratic functions are shown. The graph of $y = x^2$ is the parent graph. Write the equation for the other graph.

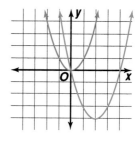

Mixed Review

Find the coordinates of the vertex for each quadratic function. *(Lesson 11–1)*

30. $y = -4x^2 + 8x + 13$ **31.** $y = 2x^2 + 6x + 3$

32. Finance Tracy invested $500, earning an annual interest rate r. The value of her investment at the end of two years can be represented by the polynomial $500r^2 + 1000r + 500$. *(Lesson 10–5)*

  a. Write an expression in factored form for the value of her investment at the end of two years.

  b. If $r = 6\%$, find the value of her investment.

Find each product. *(Lesson 9–4)*

33. $(a + 7)(a - 4)$ **34.** $(3n + 6)(n - 4)$

35. Standardized Test Practice Nine tickets, numbered 3 through 11, are placed in an empty hat. If one ticket is drawn at random from the hat, what is the probability that a prime number will be on the ticket? *(Lesson 5–6)*

  A $\frac{1}{9}$ **B** $\frac{4}{9}$ **C** $\frac{5}{9}$ **D** $\frac{1}{3}$

Extra Practice See p. 713.

11-3 Solving Quadratic Equations by Graphing

What You'll Learn

You'll learn to locate the roots of quadratic equations by graphing the related functions.

Why It's Important

Manufacturing
You can use quadratic equations to solve problems with production and profit. *See Exercise 24.*

The Buckingham Fountain in Chicago has 133 jets through which water flows. The paths of the water streams are in the shape of a parabola. They can be modeled by the quadratic function $h(d) = -2d^2 + 4d + 6$, where $h(d)$ represents the height of a stream of water at any distance d from its jet in feet.

A **quadratic equation** is an equation in which the value of the related quadratic function is 0. For example, if you substitute 0 for $h(d)$ in the function above, the result is the quadratic equation $0 = -2d^2 + 4d + 6$. The solutions of a quadratic equation are called the **roots** of the equation. The roots of a quadratic equation can be found by finding the x-intercepts or **zeros** of the related quadratic function.

Examples

Landmark Link

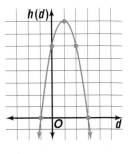

1 **Find the distance from the jet where the water hits the ground by graphing. Use the function $h(d) = -2d^2 + 4d + 6$.**

Explore Graph the parabola to determine where the stream of water will hit the ground.

Plan Find the solution of the equation by looking at the values of d where $h(d)$ is 0.

Solve Make a table of values to graph the related function $h(d) = -2d^2 + 4d + 6$.

d	$-2d^2 + 4d + 6$	$h(d)$
-1	$-2(-1)^2 + 4(-1) + 6$	0
0	$-2(0)^2 + 4(0) + 6$	6
1	$-2(1)^2 + 4(1) + 6$	8
2	$-2(2)^2 + 4(2) + 6$	6
3	$-2(3)^2 + 4(3) + 6$	0

The roots of $0 = -2d^2 + 4d + 6$ are -1 and 3.

Examine Since d represents distance, it cannot be negative. Therefore, $d = -1$ is not a solution, and the only reasonable solution is 3. The water will hit the ground 3 feet from the jet.

2 **Find the roots of $x^2 - 10x + 16 = 0$ by graphing the related function.**

Graph the related function $f(x) = x^2 - 10x + 16$. Before making a table of values, first find the equation of the axis of symmetry. This will make it easier to select x-values for your table.

$$x = -\frac{b}{2a} \qquad \textit{Equation of the axis of symmetry}$$

$$x = -\frac{-10}{2(1)} \text{ or } 5 \quad \textit{a = 1 and b = −10}$$

The equation of the axis of symmetry is $x = 5$. Now, make a table using x-values around 5. Graph each point on a coordinate plane.

x	$x^2 - 10x + 16$	$f(x)$
2	$2^2 - 10(2) + 16$	0
4	$4^2 - 10(4) + 16$	−8
5	$5^2 - 10(5) + 16$	−9
6	$6^2 - 10(6) + 16$	−8
8	$8^2 - 10(8) + 16$	0

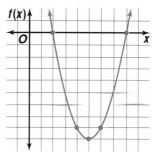

The zeros of the function appear to be 2 and 8. So, the roots are 2 and 8.

Check: Substitute 2 and 8 for x in the equation $x^2 - 10x + 16 = 0$.

$$x^2 - 10x + 16 = 0 \qquad\qquad x^2 - 10x + 16 = 0$$
$$2^2 - 10(2) + 16 \stackrel{?}{=} 0 \qquad\qquad 8^2 - 10(8) + 16 \stackrel{?}{=} 0$$
$$4 - 20 + 16 \stackrel{?}{=} 0 \qquad\qquad 64 - 80 + 16 \stackrel{?}{=} 0$$
$$0 = 0 \ \checkmark \qquad\qquad 0 = 0 \ \checkmark$$

Your Turn

a. Find the roots of $0 = x^2 - 5x + 4$ by graphing the related function.

Sometimes exact roots cannot be found by graphing. In this case, estimate solutions by stating the consecutive integers between which the roots are located.

Example **3** **Estimate the roots of $-x^2 + 2x + 1 = 0$.**

Find the equation of the axis of symmetry.

$$x = -\frac{b}{2a} \qquad \textit{Equation of the axis of symmetry}$$

$$x = -\frac{2}{2(-1)} \text{ or } 1 \quad \textit{a = −1 and b = 2}$$

The equation of the axis of symmetry is $x = 1$. Now, make a table using x-values around 1. Graph each point on a coordinate plane.

(continued on the next page)

x	$-x^2 + 2x + 1 = 0$	$f(x)$
-1	$-(-1)^2 + 2(-1) + 1$	-2
0	$-0^2 + 2(0) + 1$	1
1	$-(1)^2 + 2(1) + 1$	2
2	$-(2)^2 + 2(2) + 1$	1
3	$-(3)^2 + 2(3) + 1$	-2

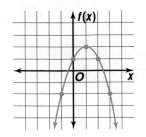

The x-intercepts of the graph are between -1 and 0 and between 2 and 3. So, one root of the equation is between -1 and 0, and the other root is between 2 and 3.

Your Turn

b. Estimate the roots of $y = x^2 - 2x - 9$.

Example

Number Theory Link

4 **Find two numbers whose sum is 4 and whose product is 5.**

Explore Let x = one of the numbers.
Then $4 - x$ = the other number.

Plan Since the product of the two numbers is 5, you know that $5 = x(4 - x)$.

$5 = x(4 - x)$

$5 = 4x - x^2$ *Distributive Property*

$0 = -x^2 + 4x - 5$ *Subtract 5 from each side.*

Solve You can solve $0 = -x^2 + 4x - 5$ by graphing the related function $f(x) = -x^2 + 4x - 5$. The equation of the axis of symmetry is $x = -\dfrac{4}{2(-1)}$ or 2. So, choose x-values around 2 for your table.

x	$-x^2 + 4x - 5$	$f(x)$
0	$-0^2 + 4(0) - 5$	-5
1	$-(1)^2 + 4(1) - 5$	-2
2	$-(2)^2 + 4(2) - 5$	-1
3	$-(3)^2 + 4(3) - 5$	-2
4	$-(4)^2 + 4(4) - 5$	-5

The graph has no x-intercepts since it does not cross the x-axis. This means the equation $x^2 - 4x + 5 = 0$ has no real roots. Thus, it is *not* possible for two numbers to have a sum of 4 and a product of 5. *Examine this solution by testing several pairs of numbers.*

The graphing calculator is a useful tool when making tables in order to solve functions.

Graphing Calculator Tutorial
See pp. 724–727.

Solve $x^2 + x - 2 = 0$ by making a table.

Enter the equation in the ▢Y= screen. Then press ▢2nd [TABLE]. A table with two columns labeled X and Y₁ will appear on your screen.

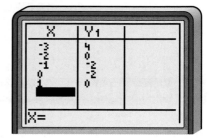

Begin entering values for x. For each x-value entered, a corresponding y-value is calculated using the equation $y = x^2 + x - 2$.

Since $y = 0$ when $x = -2$ and $x = 1$, the roots of the equation are -2 and 1.

Try These

Make a table to find the roots of each equation.

1. $x^2 - 18x + 81 = 0$ **2.** $x^2 - 2x - 15 = 0$

Check for Understanding

Communicating Mathematics

Study the lesson. Then complete the following.

1. **Explain** why finding the x-intercepts of the graph of a quadratic function can be used to solve the related quadratic equation.

2. **Write** the related function for $4 = x^2 - 3x$.

3. **Sketch** a parabola that has 1 root.

> **Vocabulary**
> quadratic equation
> roots
> zeros

Guided Practice

⊙**Getting Ready**   **State the roots of each quadratic equation.**

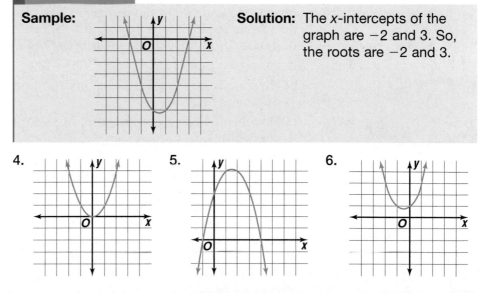

Sample:

Solution: The x-intercepts of the graph are -2 and 3. So, the roots are -2 and 3.

4.   5.   6.

Solve each equation by graphing the related function. If exact roots cannot be found, state the consecutive integers between which the roots are located. *(Examples 1–3)*

7. $x^2 - 5x + 6 = 0$ **8.** $x^2 + 2x - 15 = 0$ **9.** $2x^2 - 3x - 7 = 0$

10. Number Theory Use a quadratic equation to find two numbers whose sum is 4 and whose product is -12. *(Example 4)*

Exercises

Practice

Solve each equation by graphing the related function. If exact roots cannot be found, state the consecutive integers between which the roots are located.

11. $x^2 + 2x + 1 = 0$ **12.** $x^2 - 4x + 3 = 0$ **13.** $x^2 - 5x - 14 = 0$

14. $-x^2 + 6x + 7 = 0$ **15.** $x^2 - 10x + 25 = 0$ **16.** $x^2 + 3x - 2 = 0$

17. $x^2 + 4x - 2 = 0$ **18.** $x^2 - 2x + 2 = 0$ **19.** $-2x^2 + 3x + 4 = 0$

Use a quadratic equation to determine the two numbers that satisfy each situation.

20. Their sum is 18 and their product is 81.

21. Their difference is 4 and their product is 32.

Use the given roots and vertex of a quadratic equation to graph the related quadratic function.

22. roots: $-2, 4$
vertex: $(1, 9)$

23. roots: $-7, -3$
vertex: $(-5, -4)$

Applications and Problem Solving

24. Business Mr. Jamison owns a manufacturing company that produces key rings. Last year, he collected data about the number of key rings produced per day and the corresponding profit. He then modeled the data using the function $P(k) = -2k^2 + 12k - 10$, where the profit P is in thousands of dollars and the number of key rings k is in thousands.

a. Graph the profit function.

b. How many key rings must be produced per day so that there is no profit and no loss?

c. How many key rings must be produced for the maximum profit?

d. What is the maximum profit?

25. Skydiving On March 12, 1999, Adrian Nicholas broke two world records by flying 10 miles in 4 minutes 55 seconds without a plane. He jumped from an airplane at 35,000 feet and did not activate his parachute until he was 500 feet above the ground. If air resistance was ignored, how long would Nicholas free-fall? Use the formula $h(t) = -16t^2 + h_0$, where the time t is in seconds and the initial height h_0 is in feet.

26. **Geometry** Mrs. Parker has 200 feet of fencing to use for enclosing a rectangular running yard at her dog kennel.

   a. Make a drawing of the enclosed yard. Then write a formula for the perimeter and solve for the width.

   b. Write a quadratic equation for the area A of the yard. Use the width from part a.

   c. Show graphically how the area of the yard changes when the length changes. Can the yard have a length of 100 feet? Explain.

27. **Critical Thinking** Suppose the value of a quadratic function is negative when $x = 1$ and positive when $x = 2$. Explain why it is reasonable to assume that the related equation has a root between 1 and 2.

Mixed Review

Describe how each graph changes from its parent graph of $y = x^2$. Then name the vertex. *(Lesson 11–2)*

28. $y = -x^2$ **29.** $y = x^2 + 4$ **30.** $y = (x + 6)^2$

31. If the equation of the axis of symmetry of a quadratic function is $x = 0$, and one of the x-intercepts is -4, what is the other x-intercept? *(Lesson 11–1)*

32. Standardized Test Practice Evaluate $4^x + (x^2)^3$ if $x = 2$. *(Lesson 8–1)*

   A 6   **B** 20   **C** 32   **D** 80

Quiz 1   Lessons 11–1 through 11–3

Write the equation of the axis of symmetry and the coordinates of the vertex for each quadratic function. Then graph the function. *(Lesson 11–1)*

1. $y = x^2 + 6x - 5$ **2.** $y = -2x^2 - 9$ **3.** $y = -3x^2 - 6x + 4$

Describe how each graph changes from its parent graph of $y = x^2$. Then name the vertex of each graph. *(Lesson 11–2)*

4. $y = x^2 + 7$ **5.** $y = (x - 6)^2$ **6.** $y = -0.8x^2$

Solve each equation by graphing the related function. If exact roots cannot be found, state the consecutive integers between which the roots are located. *(Lesson 11–3)*

7. $x^2 - 5x + 6 = 0$ **8.** $x^2 + 6x + 10 = 0$ **9.** $x^2 - 2x - 1 = 0$

10. Diving Anna is doing a back dive from a 10-meter platform. Her path of descent is given by the graph of $h(d) = -d^2 + 2d + 10$, where the height h and distance from the platform d are both in meters. *(Lessons 11–1 & 11–3)*

   a. Graph the function.

   b. At her maximum height, how far away from the platform is she?

   c. About how far away from the platform will she enter the water?

Extra Practice See p. 714.

11-4 Solving Quadratic Equations by Factoring

What You'll Learn

You'll learn to solve quadratic equations by factoring and by using the Zero Product Property.

Why It's Important

Photography You can use quadratic equations to solve problems involving photography.
See Exercise 34.

At an outdoor adventure park, you can jump from a cliff 26 feet high into a pool of water below. The equation $h = -16t^2 + 4t + 26$ describes your height h in feet t seconds after jumping.

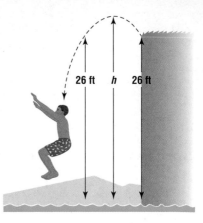

26 ft h 26 ft

Most people would jump up and out to avoid hitting any rocks below. On the way back down, they will pass the height of 26 feet again. To find the time at which this occurs, let $h = 26$ in the equation above and solve for t.
This problem will be solved in Example 2.

$$26 = -16t^2 + 4t + 26$$
$$0 = -16t^2 + 4t \qquad \textit{Subtract 26 from each side.}$$
$$0 = 4t(-4t + 1) \qquad \textit{Factor } -16t^2 - 8t.$$

To solve this equation, find the values of t that make the product $4t(-4t + 1)$ equal to 0.

Consider the following products.

$$3(0) = 0 \qquad 0(-8) = 0 \qquad \left(\frac{1}{2} - \frac{1}{2}\right)0 = 0 \qquad 0(x + 2) = 0$$

Since 0 times any number is equal to 0, *at least one* of the factors in the cases above is zero. This leads us to the **Zero Product Property**.

Zero Product Property	For all numbers a and b, if $ab = 0$, then $a = 0$, $b = 0$ or both a and b equal 0.

We can use this property to solve any equation that is written in the form $ab = 0$.

Example

1 Solve $3x(x - 1) = 0$. Check your solution.

If $3x(x - 1) = 0$, then $3x = 0$ or $x - 1 = 0$.   *Zero Product Property*
$3x = 0$ or $x - 1 = 0$ *Solve each equation.*
   $x = 0$ $x = 1$

Check: Substitute 0 and 1 for x in the original equation.

$$3x(x - 1) = 0 \qquad\qquad \text{or} \qquad\qquad 3x(x - 1) = 0$$
$$3(0)(0 - 1) \stackrel{?}{=} 0 \qquad\qquad\qquad\qquad 3(1)(1 - 1) \stackrel{?}{=} 0$$
$$0(-1) \stackrel{?}{=} 0 \qquad\qquad\qquad\qquad\qquad 3(0) \stackrel{?}{=} 0$$
$$0 = 0 \; \checkmark \qquad\qquad\qquad\qquad\qquad\qquad 0 = 0 \; \checkmark$$

The solutions are 0 and 1.

 **Your Turn** Solve each equation. Check your solution.

a. $z(z - 8) = 0$ **b.** $(a - 4)(4a + 3) = 0$

Example ❷	Refer to the application at the beginning of the lesson. Solve $4t(-4t + 1) = 0$ to find how long it would take you to reach the height of 26 feet again.

Recreation Link

If $4t(-4t + 1) = 0$, then $4t = 0$ or $-4t + 1 = 0$. *Zero Product Property*

$4t = 0$ or $-4t + 1 = 0$ *Solve each equation.*

$t = 0$ $-4t = -1$

 $t = \dfrac{1}{4}$

Check: Graph the equation and find the x-intercepts. We will graph the equation $h = 4t(-4t + 1)$ on a graphing calculator.

Enter the equation in the $\boxed{Y=}$ screen. Press $\boxed{\text{GRAPH}}$. To find the zeros, press $\boxed{\text{2nd}}$ [CALC] $\boxed{\blacktriangledown}$ 2. Use the arrow keys to move the cursor to the left of one of the x-intercepts. Press $\boxed{\text{ENTER}}$. Move the cursor to the right of the x-intercept. Press $\boxed{\text{ENTER}}$ twice. Coordinates of the root appear. Repeat for the other root.

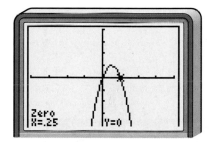

Set viewing window for x: $[-1, 1]$ by 0.25 and y: $[-1, 1]$ by 0.25.

The solutions are 0 and $\dfrac{1}{4}$. The solution 0 represents the beginning of the jump from 26 feet. So, you would return to your original location after a fourth of a second.

You will often need to factor an equation before using the Zero Product Property to solve for the variable.

Example ❸	Solve $x^2 + 4x - 12 = 0$. Check your solution.

$x^2 + 4x - 12 = 0$

$(x - 2)(x + 6) = 0$ *Factor.*

$x - 2 = 0$ or $x + 6 = 0$ *Zero Product Property*

$x = 2$ $x = -6$ *Check this solution.*

The solutions are -6 and 2.

Look Back

Factoring Trinomials: Lessons 10–3 & 10–4

Your Turn

c. Solve $x^2 - 2x = 3$. Check your solution.

The length ℓ of a rectangle is 2 feet more than twice its width w. The area of the rectangle is 144 square feet. Find the measures of the sides.

Explore The formula for the area of a rectangle is $A = \ell w$ and $\ell = 2w + 2$.

Plan

<u>Area</u>	<u>equals</u>	<u>length</u>	<u>times</u>	<u>width.</u>
144	=	$2w + 2$	$\cdot$	w

Solve
$144 = 2w^2 + 2w$ *Distributive Property*
$0 = 2w^2 + 2w - 144$ *Subtract 144 from each side.*
$0 = 2(w^2 + w - 72)$ *Factor out the GCF, 2.*
$0 = w^2 + w - 72$ *Divide each side by 2.*
$0 = (w + 9)(w - 8)$ *Factor.*

$w + 9 = 0$ or $w - 8 = 0$ *Zero Product Property*
     $w = -9$ $w = 8$ *Solve each equation.*

Examine Since width cannot be negative, the width must be equal to 8 feet. Substituting 8 for w, the length of the rectangle is $2(8) + 2$ or 18 feet.

Check for Understanding

Communicating Mathematics

Study the lesson. Then complete the following.

1. **Define** the Zero Product Property in your own words.

2. **Justify** why you should check your answers by substituting values into the original equation rather than into a simplified version.

3. **YOU Decide**   Lashonda said that she should solve $2x^2 - 3x - 35 = 0$ by graphing. Jerome disagreed. He said that she should solve the equation by factoring. Who is correct, and why?

Guided Practice

Solve each equation. Check your solution.

4. $(x + 1)(x - 7) = 0$ 5. $3b(b - 5) = 0$ 6. $(2m - 1)(m + 4) = 0$

7. $x^2 - 6x + 9 = 0$ 8. $x^2 = x$ 9. $7x^2 - 14x = 56$

10. **Geometry** The height of a triangle measures 5 centimeters more than its base. The area of the triangle is 18 square centimeters. Find the measures of the base and height of the triangle.

Exercises   • • • • •   •   •   •   •   •   •   •   •   •   •

Practice

Solve each equation. Check your solution.

11. $3m(m + 2) = 0$ 12. $5z(z + 8) = 0$ 13. $(p + 7)(p - 6) = 0$

14. $(s - 2)(s + 11) = 0$ 15. $(r + 1)(2r - 8) = 0$ 16. $(x - 2)(3x + 4) = 0$

17. $x^2 + 10x + 16 = 0$ 18. $z^2 + 5z - 6 = 0$ 19. $p^2 - 11p + 24 = 0$

20. $n^2 + 9n + 18 = 0$ 21. $m^2 - m - 12 = 0$ 22. $r^2 + 14r = 0$

23. $3w^2 - 9w = 0$ **24.** $15 = x^2 - 2x$ **25.** $2x^2 - 8x + 8 = 0$

26. $2a^2 - 70 = 4a$ **27.** $3b^2 - 12b - 15 = 0$ **28.** $2v^2 - 17v = 9$

For each problem, define a variable. Then use an equation to solve the problem.

29. The length of Luis' house is ten feet longer than it is wide. The area in square feet is 875. Find the dimensions of the house.

30. Find two consecutive even integers whose product is 120.

31. Find two integers whose difference is 3 and whose product is 88.

Applications and Problem Solving

Real World

32. Physics A flare is launched from a life raft with an initial upward velocity of 192 feet per second. How many seconds will it take for the flare to return to the sea? Use the formula $h = 192t - 16t^2$, where the height h of the flare is in feet and the time t is in seconds.

33. Geometry Four corners are cut from a rectangular piece of cardboard that measures 3 feet by 5 feet. The cuts are x feet from the corners as shown in the figure at the right. After the cuts are made, the sides are folded up to form an open box. The area of the bottom of the box is 3 square feet. Find the dimensions of the box.

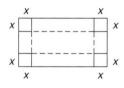

34. Photography A rectangular photograph is 4 inches wide and 6 inches long. The photograph is enlarged by increasing the length and width by an equal amount. If the area of the new photograph is twice as large as the original area, what are the dimensions of the new photograph?

35. Critical Thinking The graph of a quadratic function is shown.

a. What are the zeros of the function?

b. Write an equation for the graph.

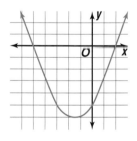

Mixed Review

Solve each equation by graphing the related function. *(Lesson 11–3)*

36. $x^2 - 10x + 21 = 0$ **37.** $x^2 - 2x + 5 = 0$ **38.** $x^2 + 4x = 12$

39. Graph $y = 0.25x^2$, $y = x^2$, and $y = 4x^2$ on the same set of axes. Compare and contrast the graphs. *(Lesson 11–2)*

Factor. *(Lesson 10–3)*

40. $y^2 + 7y + 6$ **41.** $b^2 + b - 42$

42. Standardized Test Practice Which of the following expressions is equivalent to $2^3 \times 4$? *(Lesson 8–2)*

A 8^4 **B** 8^3 **C** 6^4 **D** 2^6 **E** 2^5

What You'll Learn

You'll learn to solve quadratic equations by completing the square.

Why It's Important

Construction You can use quadratic equations to determine the dimensions of a room.
See Exercise 13.

Petunias

Daylilies

Carmen planted daylilies on a square piece of land with an area of 49 square feet. She wants to plant petunias around the daylilies to form a border whose area is 72 square feet. What length should she make the outer square?

Let x = the length of the outer square. Find a relationship between the variable and given numerical information to write and solve an equation.

Total area	*minus*	*the daylily area*	*is*	*the petunia area.*
x^2	$-$	49	$=$	72

$x^2 = 121$ *Add 49 to each side.*

$\sqrt{x^2} = \pm\sqrt{121}$ *Find the square root of each side.*

$x = \pm 11$

Length cannot be negative, so the sides of the outer square measure 11 feet.

Look Back

Perfect Square Trinomials:
Lesson 10–5

We were able to easily solve the problem above because 121 is a perfect square. However, few quadratic expressions are perfect squares. To make *any* quadratic expression a perfect square, a method called **completing the square** may be used.

If given an equation, you would complete the square by writing one side of the equation as a perfect square. This is illustrated below.

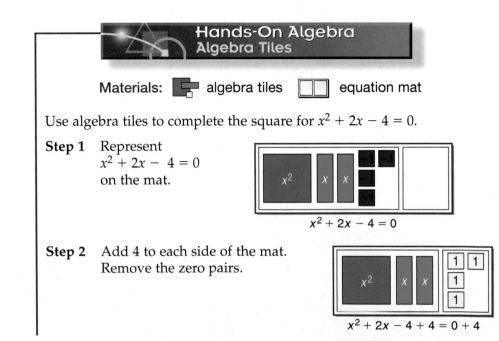

Hands-On Algebra
Algebra Tiles

Materials: ▪ algebra tiles ▫ equation mat

Use algebra tiles to complete the square for $x^2 + 2x - 4 = 0$.

Step 1 Represent
$x^2 + 2x - 4 = 0$
on the mat.

$x^2 + 2x - 4 = 0$

Step 2 Add 4 to each side of the mat. Remove the zero pairs.

$x^2 + 2x - 4 + 4 = 0 + 4$

Step 3 Begin to arrange the x^2 and x-tiles into a square.

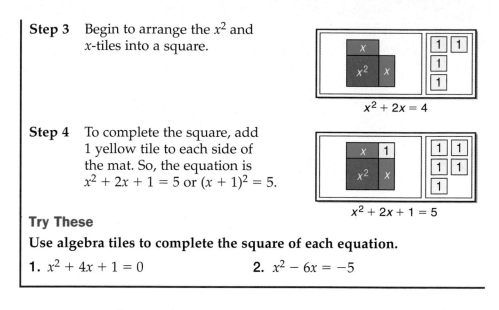

$x^2 + 2x = 4$

Step 4 To complete the square, add 1 yellow tile to each side of the mat. So, the equation is $x^2 + 2x + 1 = 5$ or $(x + 1)^2 = 5$.

$x^2 + 2x + 1 = 5$

Try These

Use algebra tiles to complete the square of each equation.

1. $x^2 + 4x + 1 = 0$ **2.** $x^2 - 6x = -5$

To complete the square for any quadratic expression of the form $x^2 + bx$, you can follow the steps below.

    Step 1 Take $\frac{1}{2}$ of b, the coefficient of x.

    Step 2 Square the result of Step 1.

    Step 3 Add the result of Step 2 to $x^2 + bx$.

Example ➊ **Find the value of c that makes $x^2 + 14x + c$ a perfect square.**

    Step 1 Find one half of 14. $\frac{14}{2} = 7$

    Step 2 Square the result of Step 1. $7^2 = 49$

    Step 3 Add the result of Step 2 to $x^2 + 14x$. $x^2 + 14x + 49$

    Thus, $c = 49$. Notice that $x^2 + 14x + 49 = (x + 7)^2$.

Your Turn

    a. Find the value of c that makes $x^2 - 6x + c$ a perfect square.

Once a perfect square is found, we can solve equations by taking the square root of each side.

Example ➋ **Solve $x^2 + 12x - 13 = 0$ by completing the square.**

Reading Algebra

Read $\pm\sqrt{49}$ as *plus or minus the square root of 49.*

$$x^2 + 12x - 13 = 0 \qquad \textit{x}^2 + 12x - 13 \textit{ is not a perfect square.}$$
$$x^2 + 12x = 13 \qquad \textit{Add 13 to each side.}$$
$$x^2 + 12x + 36 = 13 + 36 \qquad \textit{Since } \left(\tfrac{12}{2}\right)^2 = 36, \textit{ add 36 to each side.}$$
$$(x + 6)^2 = 49 \qquad \textit{Factor } x^2 + 12x + 36.$$
$$\sqrt{(x + 6)^2} = \pm\sqrt{49} \qquad \textit{Take the square root of each side.}$$

(continued on the next page)

$$x + 6 = \pm 7$$
$$x + 6 - 6 = \pm 7 - 6 \quad \textit{Subtract 6 from each side.}$$
$$x = 7 - 6 \quad \text{or} \quad x = -7 - 6 \quad \textit{Simplify each equation.}$$
$$x = 1 \qquad\qquad x = -13$$

The solutions are -13 and 1. *Check the solution by using the graphing calculator or by substituting -13 and 1 for x in the original equation.*

Your Turn

b. Solve $x^2 + 6x - 16 = 0$ by completing the square.

You can complete the square only if the coefficient of the first term is 1. If the coefficient is *not* 1, first divide each term by the coefficient.

Example ③

Construction Link

A school wants to redesign its nurse's station. Because of building restrictions, the maximum length of the rectangular room is 20 meters less than twice its width. Find the dimensions, to the nearest tenth of a meter, for the largest possible nurse's station if its area is to be 60 square meters.

Explore Draw a picture.
Let w = the maximum width.
Then $2w - 20$ = the maximum length.

$2w - 20$ — w

Plan $\underbrace{Area}$ $\underbrace{equals}$ $\underbrace{length}$ $\underbrace{times}$ $\underbrace{width.}$
          60 $=$ $2w - 20$ $\cdot$ w

Solve $60 = w(2w - 20)$
             $60 = 2w^2 - 20w$ *Distributive Property*
             $30 = w^2 - 10w$ *Divide each side by 2.*
$30 + 25 = w^2 - 10w + 25$ $\left(\frac{-10}{2}\right)^2 = 25.$ *Add 25 to each side.*
             $55 = (w - 5)^2$ *Factor $w^2 - 10w + 25$.*
          $\sqrt{55} = \sqrt{(w - 5)^2}$ *Take the square root of each side.*
         $\pm\sqrt{55} = w - 5$
     $5 \pm \sqrt{55} = w - 5 + 5$ *Add 5 to each side.*
     $5 \pm \sqrt{55} = w$

The roots are $5 + \sqrt{55}$ and $5 - \sqrt{55}$. You can use a calculator to find decimal approximations for these numbers.

$5 + \sqrt{55} \approx 12.4$ and $5 - \sqrt{55} \approx -2.4$

The solution of -2.4 is not reasonable since there cannot be negative width. The dimensions of the nurse's station should be about 12.4 meters by $2(12.4) - 20$ or 4.8 meters.

Examine Since $12.4(4.8) = 59.52$, the answer seems reasonable.

Communicating Mathematics

Study the lesson. Then complete the following.

Vocabulary

completing the square

1. **Explain** why completing the square is a good strategy to use in solving the equation $x^2 - 5x - 7 = 0$.

Math Journal

2. **List** the steps you would take to complete the square of the expression $x^2 - 10x$.

Guided Practice

Getting Ready **State whether each trinomial is a perfect square.**

Sample 1: $x^2 - 4x - 4$ **Sample 2:** $x^2 + 6x + 9$

Solution: No, it is not factorable.   **Solution:** Yes, $x^2 + 6x + 9 = (x + 3)^2$.

3. $x^2 - 2x + 1$ 4. $x^2 - 18x - 9$

5. $x^2 + 7x - 14$ 6. $x^2 + 8x + 16$

Find the value of *c* that makes each trinomial a perfect square.
(Example 1)

7. $x^2 + 2x + c$ 8. $z^2 - 18z + c$ 9. $m^2 + 3m + c$

Solve each equation by completing the square. *(Example 2)*

10. $p^2 - 6p = 0$ 11. $r^2 + 7r + 12 = 0$ 12. $x^2 - 4x - 9 = 0$

13. **Construction** The Thompsons' rectangular bathroom measures 6 feet by 9 feet. They want to double the area of the bathroom by increasing the length and width by the same amount. Find the dimensions, to the nearest foot, of the new bathroom. *(Example 3)*

Exercises

Practice

Find the value of *c* that makes each trinomial a perfect square.

14. $z^2 - 10z + c$ 15. $f^2 + 12f + c$ 16. $h^2 - 20h + c$

17. $s^2 - 2s + c$ 18. $q^2 + 14q + c$ 19. $x^2 - 24x + c$

20. $z^2 + 16z + c$ 21. $k^2 + 5k + c$ 22. $r^2 + 9r + c$

Solve each equation by completing the square.

23. $x^2 + 2x - 3 = 0$ 24. $x^2 - 8x + 7 = 0$ 25. $p^2 + 6p = -5$

26. $r^2 - 16r = 0$ 27. $s^2 - 14s = 0$ 28. $x(x + 10) - 1 = 0$

29. $q^2 - 2q = 30$ 30. $4x^2 + 16x + 24 = 0$ 31. $3z^2 - 18z = 30$

32. $m^2 - 2m = 6$ 33. $d^2 + 3d - 10 = 0$ 34. $a^2 - \frac{7}{2}a + \frac{3}{2} = 0$

Applications and Problem Solving

Real World

35. **Physics** Omar set off a rocket in a field. The height h of the rocket in feet can be roughly estimated using the formula $h = -16t^2 + 128t$, where t is the time in seconds. How long will it take the rocket to reach a height of 240 feet?

36. Photography Sheila places a 54 square inch photo behind a 12-inch-by-12-inch piece of matting. The photograph is positioned so that the matting is twice as wide at the top and bottom as the sides.

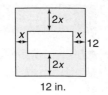

a. Write an equation for the area of the photo in terms of x.

b. Find the dimensions of the photo.

37. Critical Thinking Find the values of b that make $x^2 + bx + 81$ a perfect square.

Mixed Review

Solve each quadratic equation by factoring. *(Lesson 11–4)*

38. $a^2 + 3a - 28 = 0$

39. $x^2 - 7x + 10 = 0$

40. $z^2 + 5z = -4$

41. $2b^2 + 12b + 18 = 0$

42. Rockets Jon is launching rockets in an open field. The path of the rocket can be modeled by the quadratic function $h(t) = -16t^2 + 96t$, where $h(t)$ is the height in feet any time t in seconds. *(Lesson 11–3)*

a. Graph the quadratic function.

b. After how many seconds will the rocket reach a height of 80 feet for the *second* time?

c. After how many seconds will the rocket hit the ground?

d. What is the rocket's maximum height?

43. Geometry The area of a rectangle is represented by $(2ab + 8b^2)$.

a. Factor the expression to find the dimensions of the rectangle. *(Lesson 10–2)*

b. Using the dimensions from part a, write an expression for the perimeter of the rectangle in simplest form. *(Lessons 9–2 & 9–3)*

c. Find the dimensions, area, and perimeter of the rectangle if $a = 5$ inches and $b = 1$ inch. *(Lesson 1–2)*

44. Standardized Test Practice The graph shows how each Ohio tax dollar was divided among programs and services in a recent year. If Maria paid $32 in state taxes on her last paycheck, what amount of her money went towards education? *(Lessons 5–3 & 5–4)*

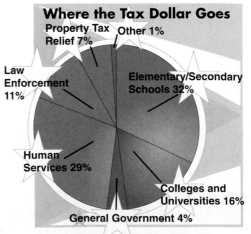

Where the Tax Dollar Goes

Property Tax Relief 7%
Other 1%
Law Enforcement 11%
Elementary/Secondary Schools 32%
Human Services 29%
Colleges and Universities 16%
General Government 4%

Source: Ohio Dept. of Taxation, 1998

A $5.12 **B** $10.24

C $15.36 **D** $24.96

Extra Practice See p. 714.

11-6 The Quadratic Formula

What You'll Learn

You'll learn to solve quadratic equations by using the Quadratic Formula.

Why It's Important

Science You can use the Quadratic Formula to solve problems dealing with physics.
See Exercise 10.

Football punters can use a formula to calculate their "hang time," the total amount of time the ball stays in the air. Manuel is a punter for his high school team. He can kick the ball with an upward velocity of 64 ft/s, and his foot meets the ball 2 feet off the ground. The formula is $h(t) = -16t^2 + 64t + 2$, where $h(t)$ is the ball's height in feet for any time t, in seconds, since the ball was kicked. What is the hang time ? *This problem will be solved in Example 3.*

The table below summarizes the ways you have learned to solve quadratic equations. Although these methods can be used with all quadratic equations, each method is most useful in certain situations.

Method	When Is the Method Useful?
Graphing	Use only to find approximations of solutions.
Factoring	Use when the quadratic is easy to factor.
Completing the Square	Use when the coefficient of x^2 is 1 and all other coefficients are fairly small.

An alternative method is to develop a general formula for solving *any* quadratic equation. Begin with the general form of a quadratic equation, $ax^2 + bx + c = 0$, where $a \neq 0$, and complete the square.

$$ax^2 + bx + c = 0$$

$$x^2 + \frac{b}{a}x + \frac{c}{a} = 0 \qquad \textit{Divide by a so the coefficient of } x^2 \textit{ is 1.}$$

$$x^2 + \frac{b}{a}x = -\frac{c}{a} \qquad \textit{Subtract } \frac{c}{a} \textit{ from each side.}$$

Now complete the square.

$$x^2 + \frac{b}{a}x + \left(\frac{b}{2a}\right)^2 = -\frac{c}{a} + \left(\frac{b}{2a}\right)^2 \qquad \frac{1}{2} \cdot \frac{b}{a} = \frac{b}{2a}$$

$$\left(x + \frac{b}{2a}\right)^2 = -\frac{c}{a} + \frac{b^2}{4a^2} \qquad \textit{Factor the left side of the equation.}$$

$$\left(x + \frac{b}{2a}\right)^2 = \frac{4a}{4a}\left(-\frac{c}{a}\right) + \frac{b^2}{4a^2} \qquad \textit{The common denominator is } 4a^2.$$

$$\left(x + \frac{b}{2a}\right)^2 = \frac{b^2 - 4ac}{4a^2} \qquad \textit{Simplify the right side of the equation.}$$

Finally, take the square root of each side and solve for x.

$$\sqrt{\left(x + \frac{b}{2a}\right)^2} = \pm\sqrt{\frac{b^2 - 4ac}{4a^2}}$$ *Take the square root of each side.*

$$x + \frac{b}{2a} = \frac{\pm\sqrt{b^2 - 4ac}}{2a}$$ *Simplify:* $\sqrt{4a^2} = 2a.$

$$x = \frac{\pm\sqrt{b^2 - 4ac}}{2a} - \frac{b}{2a}$$ *Subtract* $\frac{b}{2a}$ *from each side.*

$$x = \frac{-b \pm \sqrt{b^2 - 4ac}}{2a}$$ *Combine the terms.*

The result is called the **Quadratic Formula** and can be used to solve *any* quadratic equation.

The Quadratic Formula	$x = \dfrac{-b \pm \sqrt{b^2 - 4ac}}{2a}, a \neq 0$

Examples

Use the Quadratic Formula to solve each equation.

1 $x^2 - 4x + 3 = 0$

$$x = \frac{-b \pm \sqrt{b^2 - 4ac}}{2a}$$

$$x = \frac{-(-4) \pm \sqrt{(-4)^2 - 4(1)(3)}}{2(1)}$$ $a = 1, b = -4,$ *and* $c = 3$

$$x = \frac{4 \pm \sqrt{16 - 12}}{2}$$ $(-4)^2 = 16$ *and* $4(1)(3) = 12$

$$x = \frac{4 \pm \sqrt{4}}{2}$$

$$x = \frac{4 \pm 2}{2}$$

$$x = \frac{4 + 2}{2} \quad \text{or} \quad x = \frac{4 - 2}{2}$$

$$x = \frac{6}{2} \text{ or } 3 \qquad x = \frac{2}{2} \text{ or } 1$$

Check: Substitute values into the original equation.

$$x^2 - 4x + 3 = 0 \quad \text{or} \quad x^2 - 4x + 3 = 0$$
$$3^2 - 4(3) + 3 \stackrel{?}{=} 0 \qquad 1^2 - 4(1) + 3 \stackrel{?}{=} 0$$
$$9 - 12 + 3 \stackrel{?}{=} 0 \qquad 1 - 4 + 3 \stackrel{?}{=} 0$$
$$0 = 0 \ \checkmark \qquad\qquad 0 = 0 \ \checkmark$$

The roots are 3 and 1.

2 $-2x^2 + 3x - 5 = 0$

$$x = \frac{-b \pm \sqrt{b^2 - 4ac}}{2a}$$

$$x = \frac{-3 \pm \sqrt{3^2 - 4(-2)(-5)}}{2(-2)} \qquad a = -2, b = 3, \text{ and } c = -5$$

$$x = \frac{-3 \pm \sqrt{9 - 40}}{-4} \qquad 3^2 = 9 \text{ and } 4(-2)(-5) = 40$$

$$x = \frac{-3 \pm \sqrt{-31}}{-4}$$

The square root of a negative number, such as $\sqrt{-31}$, is not a real number. So, there are no real solutions for x.

Check: Use a graphing calculator. It is clear that the graph of $f(x) = -2x^2 + 3x - 5$ never crosses the x-axis. Therefore, there are no real roots for the equation $-2x^2 + 3x - 5 = 0$.

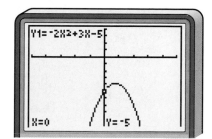

Your Turn

    a. $-3x^2 + 6x + 9 = 0$ **b.** $x^2 + 4x + 2 = 0$

Sometimes when using the Quadratic Formula, the solutions are irrational numbers. A calculator may be useful to estimate their values.

Example

Sports Link

Real World

3 Refer to the application at the beginning of the lesson. What is Manuel's hang time if the ball is not caught?

Since the final height of the ball is 0, solve the equation $0 = -16t^2 + 64t + 2$ by using the Quadratic Formula.

$$t = \frac{-b \pm \sqrt{b^2 - 4ac}}{2a}$$

$$t = \frac{-64 \pm \sqrt{64^2 - 4(-16)(2)}}{2(-16)} \qquad a = -16, b = 64, \text{ and } c = 2$$

$$t = \frac{-64 \pm \sqrt{4096 + 128}}{-32} \qquad 64^2 = 4096 \text{ and } 4(-16)(2) = -128$$

$$t = \frac{-64 \pm \sqrt{4224}}{-32} \qquad 4096 - (-128) = 4096 + 128$$

$$t = \frac{-64 + \sqrt{4224}}{-32} \quad \text{or} \quad t = \frac{-64 - \sqrt{4224}}{-32}$$

$$t \approx -0.031 \qquad\qquad\qquad t \approx 4.031$$

Since time cannot be negative, the only approximate solution is 4.031. The football has a hang time of about 4 seconds.

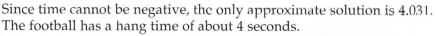

Check for Understanding

Math Journal

Guided Practice

Study the lesson. Then complete the following.

Vocabulary

Quadratic Formula

1. **Describe** a case in which there are no real solutions to a quadratic equation.

2. **List** the steps you take when you use the Quadratic Formula to solve a quadratic equation.

⏱ Getting Ready Find the value of $b^2 - 4ac$ for each equation.

Sample 1: $-3z^2 + 8z + 5 = 0$	**Sample 2:** $6a^2 = 3$
Solution: $8^2 - 4(-3)(5)$	**Solution:** Rewrite $6a^2 = 3$ as
$\quad = 64 + 60$	$6a^2 - 3 = 0$.
$\quad = 124$	$0^2 - 4(6)(-3) = 72$

3. $x^2 + 10x + 13 = 0$ 4. $3x^2 - 9x + 2 = 0$

5. $4x^2 = 12$ 6. $-2x^2 - 7x = -5$

Use the Quadratic Formula to solve each equation. *(Examples 1 & 2)*

7. $z^2 - 25 = 0$ 8. $r^2 - 5r + 7 = 0$ 9. $d^2 + 4d = -3$

10. **Physics** Josefina tosses a ball upward with an initial velocity of 76 feet per second. She tosses it from an initial height of 2 feet. Find the approximate time that the ball hits the ground. Use the function $h(t) = -16t^2 + 76t + 2$. *(Example 3)*

Exercises

• • • • •   • • • • • • • • • • • • • • •

Practice

Use the Quadratic Formula to solve each equation.

11. $x^2 + 7x + 6 = 0$ 12. $r^2 + 10r + 9 = 0$ 13. $-a^2 + 5a - 6 = 0$

14. $-b^2 + 6b + 7 = 0$ 15. $9d^2 - 4d + 3 = 0$ 16. $3v^2 + 5v - 8 = 0$

17. $x^2 - 8x = 0$ 18. $-p^2 + 6p - 5 = 0$ 19. $w^2 + w - 7 = 0$

20. $z^2 + 9z - 2 = 0$ 21. $4g^2 + 8g - 3 = 0$ 22. $-2t^2 + 4t = 3$

23. $2c^2 + 14c = 10$ 24. $5z^2 + 12 = -6z$ 25. $8k^2 - 5k = 7$

Real World

Applications and Problem Solving

26. **Skiing** Brandi, an amateur skier, begins on the "bunny hill," which has a vertical drop of 200 feet. Her speed at the bottom of the hill is given by the equation $v^2 = 64h$, where the velocity v is in feet per second and the height h of the hill is in feet. Assuming there is no friction, approximate Brandi's velocity at the bottom of the hill.

27. **Space** Suppose an astronaut can jump vertically with an initial velocity of 5 m/s. The time that it takes before he touches the ground is given by the equation $0 = 5t - 0.5at^2$. The time t is in seconds and the acceleration due to gravity a is in m/s^2.

   a. If the astronaut jumps on the moon where $a = 1.6$ m/s^2, how long will it take him to reach the ground?

   b. How long will it take him to reach the ground if he jumps on Earth where $a = 9.8$ m/s^2?

28. Critical Thinking The expression $b^2 - 4ac$ is called the *discriminant* of a quadratic equation. It determines the number of real solutions you can expect when solving a quadratic equation.

a. Copy and complete the table below.

Equation	$x^2 - 4x + 1 = 0$	$x^2 + 6x + 11 = 0$	$x^2 - 4x + 4 = 0$
Value of the Discriminant			
Number of *x*-intercepts			
Number of Real Solutions			

b. Use the table above to describe the discriminant of a quadratic equation for each type of real solution.

   i. two solutions   **ii.** 1 solution   **iii.** no solutions

Mixed Review

Find the value of *c* to make a perfect square. *(Lesson 11–5)*

29. $a^2 + 4a + c$ **30.** $x^2 + 8x + c$ **31.** $v^2 - 18v + c$

32. Geometry The area of a circle varies directly as the square of the radius *r*. If the radius of a circle is doubled, how many times as large is the area of the circle? *(Lesson 6–5)*

33. Standardized Test Practice If four times a number is equal to that number decreased by 5, what is the number? *(Lesson 4–6)*

   A -5 **B** $-\dfrac{5}{3}$ **C** $-\dfrac{1}{4}$ **D** $\dfrac{1}{3}$ **E** 4

Quiz 2   Lessons 11–4 through 11–6

Solve each quadratic equation by factoring. *(Lesson 11–4)*

  1. $x^2 + 4x - 12 = 0$ **2.** $x^2 - 8x + 16 = 0$ **3.** $x^2 - 10x = -21$

Solve each quadratic equation by completing the square. *(Lesson 11–5)*

  4. $x^2 + 4x + 3 = 0$ **5.** $x^2 - 6x + 7 = 0$ **6.** $2x^2 - 8x = 4$

Solve each equation by using the Quadratic Formula. *(Lesson 11–6)*

  7. $x^2 + 3x + 3 = 0$ **8.** $x^2 + 8x + 15 = 0$ **9.** $x^2 + 5x + 3 = 0$

10. Construction A company is designing a parabolic arch bridge to span a creek. The height of the bridge is given by the equation $h = -0.02d^2 + 0.8d$, where the height *h* and the distance *d* from one end of the base are in feet. Find the width of the bridge. *(Lesson 11–6)*

Math In the Workplace

Civil Engineer

Civil engineers work in the oldest branch of engineering. They design and supervise the construction of buildings, bridges, roads, and sewer systems. One of the most innovative structures in bridges, the parabolic arch, was first used extensively by the Romans. Civil engineers continue to use the parabolic arch today for its simple beauty.

A parabolic arch supports the Hulme Arch Bridge in England. The Hulme Arch can be modeled by the quadratic function $h(x) = -0.037x^2 + 25$, where $h(x)$ represents the height of the arch at any point x to the left or right of the center. All measures are in meters.

1. Use the Quadratic Formula to find the approximate zeros of the quadratic function $h(x) = -0.037x^2 + 25$ to the nearest whole number.
2. What do the zeros represent on the Hulme Arch?
3. Find the width of the Hulme Arch to the nearest meter.
4. What is the height of the Hulme Arch?
5. Graph the Hulme Arch on a coordinate plane.

FAST FACTS About Civil Engineers

Working Conditions
- usually work in office buildings, industrial plants, or outside on construction sites
- 40-hour work weeks, except around deadlines when more hours are expected

Education
- college degree in engineering, physical science, or mathematics
- creative and detail-oriented personalities are well-suited for the profession

Earnings

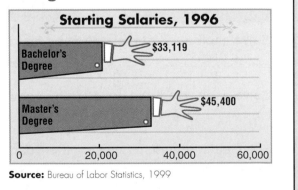

Starting Salaries, 1996

| Bachelor's Degree | $33,119 |
| Master's Degree | $45,400 |

| 0 | 20,000 | 40,000 | 60,000 |

Source: Bureau of Labor Statistics, 1999

*inter*NET CONNECTION   **Career Data**   For the latest information on civil engineers, visit:

www.algconcepts.glencoe.com

11-7 Exponential Functions

What You'll Learn
You'll learn to graph exponential functions.

Why It's Important
Finance You can use exponential functions to study the growth of bank accounts and populations.
See Example 4.

Often, patterns can be used to describe and classify types of functions.

Hands-On Algebra

Materials: ☐ large sheet of paper

Step 1 Fold a large sheet of paper in half. Unfold it and record how many sections are formed by the creases. Refold the paper.

Step 2 Fold the paper in half again. Record how many sections are formed by the creases. Refold the paper.

Step 3 Continue folding in half and recording the number of sections until you can no longer fold the paper.

Try These
1. How many folds could you make?
2. How many sections were formed?

The data collected in the Hands-On Algebra activity could be represented in a table where you can see a pattern. Each fold increases the number of sections by a factor of 2.

Folds	Sections	Pattern
1	2	$2 = 2^1$
2	4	$2(2) = 2^2$
3	8	$2(2)(2) = 2^3$
4	16	$2(2)(2)(2) = 2^4$
5	32	$2(2)(2)(2)(2) = 2^5$
6	64	$2(2)(2)(2)(2)(2) = 2^6$

Let x equal the number of folds and let y equal the number of sections. Then the function $y = 2^x$ represents the number of sections for any number of folds. This is an example of an **exponential function**. An exponential function is of the form $y = a^x$, where $a > 0$ and $a \neq 1$.

You can make a table to help graph exponential functions.

Example ———❶ Graph $y = 2^x$.

x	2^x	y
-1	2^{-1}	0.5
0	2^0	1
1	2^1	2
2	2^2	4
3	2^3	8

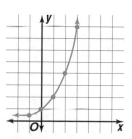

The graph of an exponential function changes little for small x values. However, as the values of x increase, the y values increase quickly.

Exponential functions of the form $y = a^x$ have a y-intercept at 1. If a constant c is added to the function, it shifts the graph up or down c units.

Graph each exponential function. Then state the y-intercept.

2 $y = 5^x$

x	5^x	y
-2	5^{-2}	0.04
-1	5^{-1}	0.2
0	5^0	1
1	5^1	5
2	5^2	25

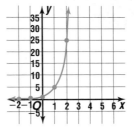

To find the y-intercept, let $x = 0$ and solve for y.

$$y = 5^0 \text{ or } 1$$

In this case, the y-intercept is 1.

3 $y = 2^x + 3$

x	$2^x + 3$	y
-2	$2^{-2} + 3$	3.25
-1	$2^{-1} + 3$	3.5
0	$2^0 + 3$	4
1	$2^1 + 3$	5
2	$2^2 + 3$	7
3	$2^3 + 3$	11

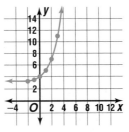

Reading Algebra

Read $2^x + 3$ as *two to the x plus three.*

The y-intercept is 4. *The constant is 3: $1 + 3 = 4$.*

Your Turn Graph each function. Then state the y-intercept.

a. $y = 2.5^x$ **b.** $y = 4^x + 3$

Quantities that increase rapidly have *exponential growth.* For instance, money in the bank may grow at an exponential rate.

4 When Taina was 10 years old, she received a certificate of deposit (CD) for $2000 with an annual interest rate of 5%. After eight years, how much money will she have in the account?

After the first year, the CD is worth the initial deposit plus interest.
$2000 + 2000(0.05) = 2000(1 + 0.05)$ *Distributive Property*
$= 2000(1.05)$

Taina can make a table to look for patterns in the growth of her CD.

Year	Balance	Pattern
0	2000	$2000(1.05)^0$
1	$2000(\mathbf{1.05}) = 2100$	$2000(1.05)^1$
2	$(2000 \cdot \mathbf{1.05})\mathbf{1.05} = 2205.00$	$2000(1.05)^2$
3	$(2000 \cdot \mathbf{1.05} \cdot \mathbf{1.05})\mathbf{1.05} = 2315.25$	$2000(1.05)^3$

Let x represent the number of years. Then the function that represents the balance in Taina's CD is $B(x) = 2000(1.05)^x$. After eight years, it will have a balance of $2954.91.

For exponential data, you can create a best-fit curve that passes through most of the data points. You can then use this curve to make predictions.

Graphing Calculator Tutorial
See pp. 724–727.

Make sure that the STATPLOT is turned on.

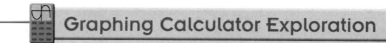

Graphing Calculator Exploration

The city of Manchester, England, went on the World Wide Web in 1993. The table shows the average number of "hits" per week since 1994.

Years After 1990	"Hits" per Week
4	10,000
5	20,000
6	37,500
7	62,500
8	190,000

Step 1 Use the Edit option in the STAT menu to enter the year data in L1 and the visitor data in L2.

Step 2 Find an equation for a best-fit curve.

Enter: [STAT] [▶] 0 [2nd] [L1] [,]
[2nd] [L2] [,] [VARS] [▶] [ENTER] [ENTER] [ENTER]

The screen displays the equation $y = a \cdot b^x$ and gives values for a and b.

Step 3 The exponential equation has been stored in the [Y=] menu. To see a graph of the function in the proper window, press [ZOOM] 9.

Try This

Use the graph of the best-fit curve to predict the number of weekly "hits" for the year 2005.

Check for Understanding

Communicating Mathematics

Study the lesson. Then complete the following.

Vocabulary
exponential function

1. **Describe** the general shape of an exponential function.

2. **Explain** why the y-intercept of an exponential function in the form $y = a^x$ is always 1.

⏱ Getting Ready | Use a calculator to find each value to the nearest hundredth.

Sample 1: 3^{-4}
Solution:
3 ⌃ ⊖ 4 [ENTER] *0.01*

Sample 2: 1.25^5
Solution:
1.25 ⌃ 5 [ENTER] *3.05*

3. 2^8 **4.** 4^{-1} **5.** 1.08^3

Graph each exponential function. Then state the *y*-intercept.
(Examples 1–3)

6. $y = 2^x + 2$ **7.** $y = 2^x - 6$ **8.** $y = 3^x + 5$

9. Finance When Lindsay was born, her parents opened a savings account with $500. The account pays an annual interest of 2%.
(Example 4)

a. Make a table showing the account's growth for 3 years.

b. Write a function that represents the amount of money in Lindsay's account after *x* number of years.

c. How much money will be in Lindsay's account after 18 years?

d. If the initial deposit had been $1000, how much would Lindsay's account be worth after 18 years?

Exercises • • • • • • • • • • • • • • • • • •

Practice

Graph each exponential function. Then state the *y*-intercept.

10. $y = 3^x$ **11.** $y = 4^x$ **12.** $y = 2^x - 5$
13. $y = 2^x + 1$ **14.** $y = 3^x + 2$ **15.** $y = 4^x - 3$
16. $y = 2^{2x}$ **17.** $y = 3^{2x} - 1$ **18.** $y = 2^{0.5x}$

Find the amount of money in a bank account given the following conditions.

19. initial deposit = $5000, annual rate = 3%, time = 2 years
20. initial deposit = $1500, annual rate = 10%, time = 10 years
21. initial deposit = $3000, annual rate = 5.5%, time = 5 years

Applications and Problem Solving

22. Taxes In a recent year, the average tax refund had risen to about $1700. The table shows the nest egg you would make for yourself by investing that refund at a 10% interest rate. Find your savings after 30 years.

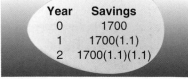

Compounding Tax Refunds

Year	Savings
0	1700
1	1700(1.1)
2	1700(1.1)(1.1)

Source: National Tax Services

23. Finance Nuno bought a new car for $18,000. Once the car is driven off the lot, the car depreciates in value 15% each year. How much will the car be worth in 5 years?

24. **Population** The Atlanta, Georgia, metropolitan area experienced population growth throughout the 1990s. If the population in 1990 was 2,960,000 and the annual rate of increase was 3%, predict the population in the year 2002. Use the function $P(t) = 2,960,000(1.03)^t$, where t is the number of years since 1990.

25. **Money** Suppose on the first day of the year, your parents offered you $500 for the month of January. If you did not take the $500, they would give you a penny on the first day and then continue to double the amount each following day. Which is the better deal?

   a. Copy and complete the table.

   b. Write a function that represents the amount of money you will receive after x days.

   c. Use the equation in part b to find the day when you receive a payment of more than $500.

Day	Cents	Pattern
1	1	
2	$2 \cdot 1$	
3	$2 \cdot 2 \cdot 1$	
4		
5		

26. **Critical Thinking** Make a graph of each exponential function listed below. Then describe the shape of an exponential function when a number less than 1 is raised to the x power compared to a number greater than 1.

   a. $y = 2^x$ b. $y = 1.5^x$ c. $y = 0.5^x$ d. $y = 0.25^x$

Mixed Review

Use the Quadratic Formula to solve each equation. *(Lesson 11–6)*

27. $3p^2 - 12 = 0$

28. $x^2 + 4x = 10$

29. **Flooring** There are 30 square yards in a roll of carpet. If the carpet is unrolled, it forms a rectangle. The length is four yards more than twice its width. *(Lesson 11–5)*

   a. Write a quadratic equation to represent the carpet's area.

   b. Find the dimensions.

Write the point-slope form of an equation for each line passing through the given point and having the given slope. *(Lesson 7–2)*

30. $(4, -3), 2$

31. $(-2, 6), 0$

32. **Open-Ended Test Practice** During a road-trip, Javonte recorded the number of miles he had driven by the end of each hour. He graphed the number of miles driven on the number line shown. *(Lesson 7–1)*

   a. Find the slope of the line between hours 0 and 2.

   b. Find the slope of the line between hours 2 and 3.

   c. Find the slope of the line between hours 3 and 6.

   d. What does the slope represent? (*Hint:* Look at the units.)

Extra Practice   See p. 715.

Sticky Note Sequences

Materials:

 five 3-in. by 3-in. self-adhesive notes

scissors

Geometric Sequences

In the Chapter 3 Investigation, you examined arithmetic sequences. In this investigation, you will look at a different type of sequence.

Investigate

1. Use a self-adhesive note to represent 1 whole unit. Label it "1." Cut a second self-adhesive note in half to represent one half of a unit. Label one of the resulting pieces "$\frac{1}{2}$." Cut the other piece in half again to represent one-fourth of a unit. Label one of the resulting pieces "$\frac{1}{4}$." Continue this process. Arrange the self-adhesive notes as shown below.

a. Make a table like the one shown. Write the numbers 1–10 in column 1. Record the fractions from each note in column 2.

b. Find the ratio of the new portion and the previous portion to complete column 3.

Term Number	Portion of Self-Adhesive Note	New Portion / Previous Portion
1	1	—
2	$\frac{1}{2}$	$\frac{1}{2} \div 1 = ?$
3	$\frac{1}{4}$	$\frac{1}{4} \div \frac{1}{2} = ?$

c. List the numbers in column 2 from greatest to least.

d. The sequence in part c is called a **geometric sequence** because the quotient between any two consecutive terms, called the **common ratio**, is always the same. What is the common ratio?

2. A geometric sequence can be written as a formula in several ways.

   a. Copy and complete a table like the one below for terms 1–10.

Term	Portion of Self-Adhesive Note	Form 1	Form 2	Form 3
1	1	1	1	$1 \cdot \left(\frac{1}{2}\right)^0$
2	$\frac{1}{2}$	$1 \cdot \frac{1}{2}$	$1 \cdot \frac{1}{2}$	$1 \cdot \left(\frac{1}{2}\right)^1$
3	$\frac{1}{4}$	$\frac{1}{2} \cdot \frac{1}{2}$	$1 \cdot \frac{1}{2} \cdot \frac{1}{2}$	$1 \cdot \left(\frac{1}{2}\right)^2$

   b. You can describe Form 1 by using the formula *portion = previous portion times one half*. How could you describe Forms 2 and 3?

3. Some bacteria reproduce exponentially by *fission*. One cell splits to become two cells. There are now two bacteria instead of one.

   a. Copy and complete a table like the one below for terms 1–10. Use 3 self-adhesive notes to represent 3 bacteria. Cut each in half to produce 6 bacteria. Continue this process.

   b. Write a sequence that models the bacteria's growth. Then write a recursive formula representing the sequence.

Term Number	Number of Bacteria	New Number of Bacteria / Previous Number of Bacteria
1	3	—
2	6	$6 \div 3 = ?$

Extending the Investigation

In this extension, you will examine other sequences and their formulas.

1. Graph the sequence in Exercise 1. Label the *x*-axis "term number" and the *y*-axis "portion of self-adhesive note." What type of function would describe the graph?

2. Use the sequence in Exercise 1 to find the sum of the first 10 terms. Suppose the sequence continued for 100 terms. What would you expect the sum to be?

3. Repeat 1 and 2 for Exercise 3.

Presenting Your Investigation

Here are some ideas to help you present your conclusions to the class.

• Make a brochure showing the sequences you wrote in this investigation.

• Describe a situation that results in a sequence similar to those in this investigation.

*inter*NET CONNECTION **Investigation** For more information on geometric sequences, visit: www.algconcepts.glencoe.com

Understanding and Using the Vocabulary

After completing this chapter, you should be able to define each term, property, or phrase and give an example or two of each.

interNET CONNECTION Review Activities
For more review activities, visit:
www.algconcepts.glencoe.com

axis of symmetry *(p. 459)*
completing the square *(p. 478)*
exponential function *(p. 489)*
geometric sequence *(p. 494)*
maximum *(p. 459)*

minimum *(p. 459)*
parabola *(p. 458)*
quadratic equation *(p. 468)*
Quadratic Formula *(p. 484)*
quadratic function *(p. 458)*

roots *(p. 468)*
vertex *(p. 459)*
zeros *(p. 468)*

Choose the letter of the term that best matches each statement or phrase. Each letter is used once.

1. the maximum or minimum point of a parabola
2. Use this to find the exact roots of any quadratic equation.
3. the method of making a quadratic expression into a perfect square
4. the solutions of a quadratic equation
5. the shape of the graph of any quadratic function
6. a vertical line containing the vertex of a parabola
7. $f(x) = a^x, a > 0$ and $a \neq 1$
8. $f(x) = ax^2 + bx + c, a \neq 0$
9. the x-intercepts of a quadratic function
10. the highest point on the graph of a quadratic function

a. exponential function
b. Quadratic Formula
c. completing the square
d. zeros
e. maximum
f. vertex
g. roots
h. axis of symmetry
i. quadratic function
j. parabola

Skills and Concepts

Objectives and Examples

• **Lesson 11–1** Graph quadratic functions.

Write the equation of the axis of symmetry and the coordinates of the vertex of $f(x) = x^2 - 8x + 12$.

$$x = -\frac{-8}{2(1)} \text{ or } 4$$

Since $4^2 - 8(4) + 12 = -4$, the graph has a vertex at $(4, -4)$.

Review Exercises

Write the equation of the axis of symmetry and the coordinates of the vertex of each quadratic function.

11. $y = x^2 - 3$
12. $y = -x^2 + 4x + 5$
13. $y = 3x^2 + 6x - 17$
14. $y = x^2 - 3x - 4$

Objectives and Examples

Review Exercises

• **Lesson 11-1** Graph quadratic functions.

Graph
$f(x) = x^2 - 8x + 12.$

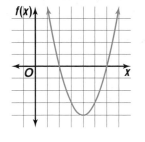

Use the results from Exercises 11–14 and a table to graph each quadratic function.

15. $y = x^2 - 3$

16. $y = -x^2 + 4x + 5$

17. $y = 3x^2 + 6x - 17$

18. $y = x^2 - 3x - 4$

• **Lesson 11-2** Recognize characteristics of families of parabolas.

Describe how the graph of $y = (x + 12)^2 - 5$ changes from the graph of $y = x^2$.

If $x = -12$, then $x + 12 = 0$. The graph shifts 12 units to the left. The -5 outside the parentheses shifts the graph 5 units down. The vertex is at $(-12, -5)$.

Describe how each graph changes from its parent graph of $y = x^2$. Then name the vertex.

19. $y = (x - 8)^2$

20. $y = x^2 - 10$

21. $y = 0.4x^2$

22. $y = (x + 3)^2$

23. $y = (x - 1)^2 + 7$

24. $y = -1.5x^2$

• **Lesson 11-3** Locate the roots of quadratic equations by graphing.

Find the roots of $0 = x^2 - 8x + 12$.

Based on the graph of the related function shown above, the roots are 2 and 6.

Solve each equation by graphing the related function.

25. $x^2 - x - 12 = 0$

26. $x^2 + 10x + 25 = 0$

27. $2x^2 - 5x + 4 = 0$

28. $x^2 + x - 4 = 0$

29. $6x^2 - 13x = 15$

• **Lesson 11-4** Solve quadratic equations by factoring.

Solve $x^2 - 8x - 20 = 0$.

$x^2 - 8x - 20 = 0$
$(x - 10)(x + 2) = 0$
$x - 10 = 0$ or $x + 2 = 0$
$\quad x = 10 \qquad\qquad x = -2$

Solve each equation. Check your solution.

30. $y(y + 11) = 0$

31. $(t + 4)(2t - 10) = 0$

32. $z^2 - 36 = 0$

33. $x^2 + 16x + 64 = 0$

34. $b^2 - 5b - 14 = 0$

35. $4g^2 - 4 = 0$

36. $2x^2 + 10x + 12 = 0$

Objectives and Examples

- **Lesson 11–5** Solve quadratic equations by completing the square.

$x^2 + 6x + 2 = 0$
$\quad x^2 + 6x = -2$ *Subtract 2.*
$x^2 + 6x + 9 = -2 + 9$ *Complete the square.*
$\quad (x + 3)^2 = 7$
$\quad\quad x + 3 = \pm\sqrt{7}$ *Take the square root.*
$\quad\quad\quad x = -3 \pm \sqrt{7}$ *Subtract 3.*

- **Lesson 11–6** Solve equations by using the Quadratic Formula.

Solve $-x^2 + 3x + 40 = 0$.

$$x = \frac{-b \pm \sqrt{b^2 - 4ac}}{2a}$$

$$x = \frac{-3 \pm \sqrt{3^2 - 4(-1)(40)}}{2(-1)}$$

$$x = \frac{-3 \pm \sqrt{169}}{-2}$$

$x = \dfrac{-3 + 13}{-2}$ or $x = \dfrac{-3 - 13}{-2}$

$x = -5$ $x = 8$

- **Lesson 11–7** Graph exponential functions.

Graph $y = 2^x - 1$.

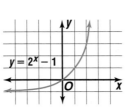

$y = 2^x - 1$

Review Exercises

Find the value of c that makes each trinomial a perfect square.

37. $r^2 + 4r + c$ **38.** $t^2 - 12t + c$

Solve each quadratic equation by completing the square.

39. $t^2 - 4t - 21 = 0$ **40.** $x^2 - 16x + 23 = 0$

Solve each equation by using the Quadratic Formula. Check your solution.

41. $r^2 + 10r + 9 = 0$
42. $-2d^2 + 8d = 0$
43. $3n^2 - 2n - 1 = 0$
44. $s^2 + 2s + 2 = 0$
45. $m^2 + 5m = -3$
46. $-4x^2 + 8x + 3 = 0$

Graph each exponential function. Then state the y-intercept.

47. $y = 2^x$
48. $y = 2.7^x$
49. $y = 3^x + 3$
50. $y = 4^x - 4$

Applications and Problem Solving

51. Finance Game room tickets cost $5. The function $I(x) = -50x^2 + 100x + 6000$ represents the weekly income when x is the number of 50¢ price increases. Graph the function to find the ticket price that results in the maximum income. *(Lesson 11–1)*

52. Tuition A college board voted to increase tuition prices a maximum of 4% annually. If tuition is $6678, in how many years will it exceed $10,000? Use the function $P(t) = 6678(1.04)^t$, where t is the number of years. *(Lesson 11–7)*

For each situation, state whether it is *always* true, *sometimes* true, or *never* true.

1. The graph of a quadratic function opens upward in a "u-shape."
2. You can use the Quadratic Formula to solve quadratic equations.
3. There are two real solutions to any quadratic equation.

Write the equation of axis of symmetry and the coordinates of the vertex of the graph of each quadratic function. Then graph the function.

4. $y = x^2 + 6x + 8$
5. $y = -x^2 + 3x$

Describe how each graph changes from its parent graph of $y = x^2$. Then name the vertex of each graph.

6. $y = x^2 - 10$
7. $y = (x - 9)^2$
8. $y = (x + 7)^2 + 3$

Solve each equation by graphing the related function. If exact roots cannot be found, state the consecutive integers between which the roots are located.

9. $y = x^2 - 8x + 16$
10. $y = x^2 - 5x - 1$

Solve each equation by factoring. Check your solution.

11. $(r - 3)(r - 8) = 0$
12. $s^2 - 5s + 4 = 0$
13. $x^2 + 7x + 10 = 0$

Find the value of c that makes each trinomial a perfect square.

14. $b^2 + 10b + c$
15. $x^2 + 8x + c$
16. $g^2 - 2g + c$

Solve each equation by completing the square.

17. $a^2 + 14a = -45$
18. $v^2 - 6v + 3 = 0$

Solve each equation by using the Quadratic Formula.

19. $2n^2 - n - 3 = 0$
20. $3x^2 + x + 5 = 0$

Graph each exponential function. Then state the y-intercept.

21. $y = 1.5^x$
22. $y = 2^x - 3$

23. **Physics** A plane flying 400 feet over Antarctica drops equipment needed by research scientists below. How long will it take the equipment to touch ground? Use the formula $h = -16t^2 + 400$, where the height h is in feet and the time t is in seconds.

24. **Geometry** The length of a rectangle is 4 inches more than its width. The area of the rectangle is 60 square inches. Find the dimensions of the rectangle.

25. **Finance** A school district has a $64 million budget. The school board has voted to increase the budget by only 1% each year. After how many years will the budget be over $70 million? Use the function $B(t) = 64(1.01)^t$, where t is the time in years.

Polynomial and Factoring Problems

State proficiency tests may include a few polynomial problems. In addition, many questions on the SAT and ACT ask you to factor or evaluate polynomials.

Memorize these polynomial relationships.

1. difference of squares $a^2 - b^2 = (a + b)(a - b)$
2. perfect square trinomials $a^2 + 2ab + b^2 = (a + b)^2$;
                                       $a^2 - 2ab + b^2 = (a - b)^2$

> **THE PRINCETON REVIEW**
>
> Look for factors in all problems that include polynomials. Factoring is often the quick way to find an answer.

Proficiency Test Example

Evaluate $2x^2y - 4y + x^4$ if $x = 2$ and $y = -1$.

A 12   **B** 16   **C** 20   **D** 28

> **Hint** Apply the rules for operations carefully when negative numbers are involved.

Solution Substitute 2 for x and -1 for y. Then simplify the expression.

$$
\begin{aligned}
2x^2y - 4y + x^4 &= 2(2)^2(-1) - 4(-1) + (2)^4 \\
&= 2(4)(-1) - 4(-1) + 16 \\
&= 8(-1) + 4 + 16 \\
&= -8 + 20 \\
&= 12
\end{aligned}
$$

The answer is A.

SAT Example

The SAT contains 15 *quantitative comparison* questions, where you compare two quantities and decide which is greater. The quantities are labeled *Column A* and *Column B*. Choose answer choice

A if the quantity in Column A is greater;

B if the quantity in Column B is greater;

C if the two quantities are equal;

D if the relationship cannot be determined from the information given.

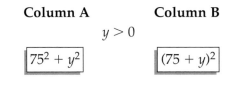

Column A   Column B

$y > 0$

$\boxed{75^2 + y^2}$ $\boxed{(75 + y)^2}$

> **Hint** The given information, y is greater than 0, is true for both columns.

Solution Write the expression in Column B as the trinomial $75^2 + 150y + y^2$.

75^2 and y^2 are equal in both. $150y$ is always greater than 0. Therefore, the quantity in Column B is always greater.

The answer is B.

After you work each problem, record your answer on the answer sheet provided or on a sheet of paper.

1. Evaluate $h^2 - 5h$ if $h = -3$.

   A -24 **B** -6

   C 6 **D** 24

2. A swimming pool is shaped like a rectangular prism. Its volume is 1200 cubic feet. The dimensions of a smaller pool are half the size of the larger pool's dimensions. What is the volume, in cubic feet, of the smaller pool?

   A 150 **B** 300 **C** 600 **D** 1200

3. For all $x \neq -3$, which of the following is equivalent to the expression $\dfrac{x^2 - x - 12}{x + 3}$?

   A $x - 4$

   B $x - 2$

   C $x + 2$

   D $x + 4$

   E $x + 6$

4. If $x = -2$, then find $3x^2 - 5x - 6$.

   A -8 **B** -4 **C** 10 **D** 16

5. The figures below represent a sequence of numbers. What is an expression for the nth term of this sequence?

   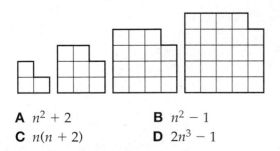

   A $n^2 + 2$ **B** $n^2 - 1$

   C $n(n + 2)$ **D** $2n^3 - 1$

6. Which quadratic equation has roots $\dfrac{1}{2}$ and $\dfrac{1}{3}$?

   A $5x^2 - 5x - 2 = 0$

   B $5x^2 - 5x + 1 = 0$

   C $6x^2 + 5x + 1 = 0$

   D $6x^2 - 6x + 1 = 0$

   E $6x^2 - 5x + 1 = 0$

7. Use the function table to find the value of y when $x = 8$.

   A 9 **B** 10.5

   C 12 **D** 13.5

x	y
1	1.5
2	3.0
3	4.5
4	6.0
5	7.5

Quantitative Comparison

8. **Column A**   **Column B**

   $$a > 0;\ a \neq b$$

   $\dfrac{a^2 - b^2}{a - b}$ b

   A if the quantity in Column A is greater;

   B if the quantity in Column B is greater;

   C if the two quantities are equal;

   D if the relationship cannot be determined from the information given.

Open-Ended Questions

9. **Grid-In**   Solve for x if $\dfrac{x^2 + 7x + 12}{x + 4} = 5$.

10. Consider $2x + 3y = 15$.

    Part A   Make a table of at least six pairs of values that satisfy the equation above.

    Part B   Use your table from Part A to graph the equation.

*inter***NET** **CONNECTION**   **Test Practice**   For additional test practice questions, visit: www.algconcepts.glencoe.com

CHAPTER 12 Inequalities

▶ What You'll Learn in Chapter 12:

- to graph inequalities on a number line *(Lesson 12–1),*
- to solve inequalities involving addition, subtraction, multiplication, and division *(Lessons 12–2 and 12–3),*
- to solve inequalities involving more than one operation *(Lesson 12–4),*
- to solve compound inequalities *(Lesson 12–5),*
- to solve inequalities involving absolute value *(Lesson 12–6),* and
- to graph inequalities in the coordinate plane *(Lesson 12–7).*

Problem-Solving Workshop

Project

Plan a one-week menu of foods that require little or no preparation.
1. Total Calories must be between 1200 and 1800 per day.
2. Calories from fat must be at most 30% of the total Calories. Note that 1 gram of fat has 9 Calories.
3. Total sodium must be less than 2400 milligrams.
4. Total protein must be at least 46 grams.

Working on the Project

Work with a partner to solve the problem.

- Write inequalities to represent the Calories, fat (g), sodium (mg), and protein (g) allowed per day.
- Determine whether the sample menu in the table below satisfies the nutritional guidelines.

Technology Tools

- Use a **spreadsheet** to help write a menu that meets all guidelines.
- Use **publishing software** to illustrate your menu.

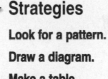

> **Strategies**
>
> Look for a pattern.
> Draw a diagram.
> Make a table.
> Work backward.
> Use an equation.
> Make a graph.
> Guess and check.

Food (one serving)	Calories	Fat (g)	Sodium (mg)	Protein (g)
canned chicken noodle soup	90	2	940	6
granola cereal and skim milk	270	9	20	5
canned tuna in water (2 ounces)	60	0.5	250	13
frozen dinner	450	20	1530	10
white bread (3 slices)	240	3	480	6

interNET CONNECTION **Research** For more information about nutrition, visit: www.algconcepts.glencoe.com

Presenting the Project

Compile a portfolio of your work for this project. Include the menu for each of seven days, the nutritional value of each food, and a list of inequalities to show that all guidelines were met.

Math In the Workplace

What You'll Learn
You'll learn to graph inequalities on a number line.

Why It's Important
Postal Service
Inequalities can be used to describe postal regulations.
See Exercise 12.

In recent years, several states have considered raising the legal driving age or placing restrictions on beginning drivers. The state of New York has considered restricting the time of day when teens can drive. However, New York's minimum driving age remains at 16, as it does in many states.

You can express New York's minimum driving age with an inequality. If *d* represents the ages of people old enough to drive, then $d \geq 16$.

$$\underbrace{\textit{The ages of all drivers}}_{d} \quad \underbrace{\textit{are greater than or equal to}}_{\geq} \quad \underbrace{\textit{16 years.}}_{16}$$

$d \geq 16$ *is the same as* $16 \leq d$.

Verbal phrases like *greater than* or *less than* are used to describe inequalities. The chart below lists other phrases that indicate inequalities.

Look Back

Inequalities:
Lesson 3–1

Inequalities			
<	≤	>	≥
• less than • fewer than	• less than or equal to • at most • no more than • a maximum of	• greater than • more than	• greater than or equal to • at least • no less than • a minimum of

Example

1 **Write an inequality to describe people who are *not* of legal driving age in New York.**

Let *d* represent people who are *not* of legal driving age.
Then *d* is *less than* 16, or $d < 16$.

Your Turn

a. Lisa can carry no more than 50 newspapers on her paper route. Express the number of papers that Lisa can carry as an inequality.

Not only can inequalities be expressed through words and symbols, but they can also be graphed.

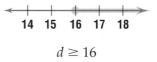

Reading Algebra

On a graph, a circle means that the number is *not* included. A bullet means that the number *is* included.

Consider the inequality $d \geq 16$.

Step 1 Graph a bullet at 16, since d can *equal* 16.

Step 2 Graph all numbers *greater than* 16 by drawing a line and an arrow to the right.

$$d \geq 16$$

The graph shows all values that are *greater than or equal to* 16.

Now suppose we want to graph all numbers that are *less than* 16. We want to include all numbers up to 16, but not including 16.

Step 1 Graph a circle at 16, since d *does not equal* 16.

Step 2 Graph all numbers *less than* 16 by drawing a line and an arrow to the left.

$$d < 16$$

The graph shows all values that are *less than* 16.

Examples

Graph each inequality on a number line.

❷ $x > -4$

Since x cannot equal -4, graph a circle at -4 and shade to the right.

❸ $n \geq 1.5$

Since n can equal 1.5, graph a bullet at 1.5 and shade to the right.

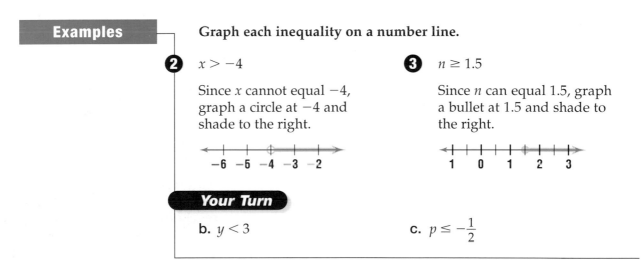

Your Turn

b. $y < 3$

c. $p \leq -\dfrac{1}{2}$

You can also write inequalities given their graphs.

Examples

Write an inequality for each graph.

❹

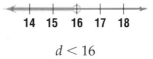

Locate where the graph begins. This graph begins at 1, but 1 is not included. Also note that the arrow is to the left. The graph describes values that are less than 1.

So, $x < 1$.

(continued on the next page)

5

$$\begin{array}{c|c|c|c|c|c} & | & \bullet & | & | & | \\ \hline 0 & \frac{1}{4} & \frac{2}{4} & \frac{3}{4} & 1 \end{array}$$

Locate where the graph begins. This graph begins at $\frac{1}{4}$ and includes $\frac{1}{4}$. Note that the arrow is to the right. The graph describes values greater than or equal to $\frac{1}{4}$.

So, $x \geq \frac{1}{4}$.

Your Turn

d.

$$\begin{array}{c|c|c|c|c} & | & | & \bullet & | & | \\ \hline -4 & -3 & -2 & -1 & 0 \end{array}$$

e.

$$\begin{array}{c|c|c|c|c} & | & | & \bullet & | & | \\ \hline 0 & \frac{1}{3} & \frac{2}{3} & 1 & 1\frac{1}{3} \end{array}$$

Inequalities are commonly used in the real world.

Example

Sports Link

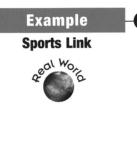

6 **To play junior league soccer, you must be at least 14 years of age. Write an inequality to express this information. Then graph the inequality on a number line.**

Let a represent the ages of people who can play junior league soccer. Then write an inequality using $\geq$ since the soccer players have to be greater than or equal to 14 years of age.

$a \geq 14$

To graph the inequality, first graph a bullet at 14. Then include all ages *greater than* 14 by drawing a line and an arrow to the right.

$$\begin{array}{c|c|c|c|c} & | & | & \bullet & | & | \\ \hline 12 & 13 & 14 & 15 & 16 \end{array}$$

Check for Understanding

Communicating Mathematics

Study the lesson. Then complete the following.

1. **Match** each inequality symbol with its description.

   i. less than or equal to   **a.** $\geq$

   ii. at least   **b.** $\leq$

   iii. fewer than   **c.** $>$

   iv. greater than   **d.** $<$

2. **Describe** the graph of $x \geq 12$.

3. **Explain** the difference between the graphs of $x \leq 4$ and $x < 4$.

4. **You Decide?** Soto says that $x \geq 7$ is the same as $7 \leq x$. Darrell says that it is not. Who is correct? Explain.

Guided Practice

Write an inequality to describe each number. *(Example 1)*

5. a number less than 14

6. a number no less than 0

Graph each inequality on a number line. *(Examples 2 & 3)*

7. $x > 6$

8. $x \leq 2$

9. $5 > x$

Write an inequality for each graph. *(Examples 4 & 5)*

10.
```
 ← + + + ● + + + →
   6  7  8  9  10
```

11.
```
 ← + + + ○ + + →
  −3 −2 −1  0  1
```

12. **Mail** All letters mailed with one first-class stamp can weigh no more than 1 ounce. Write an inequality to express this information. Then graph the inequality. *(Example 6)*

Exercises • • • • • • • • • • • • • • • • • • •

Practice

Write an inequality to describe each number.

13. a number greater than 4

14. a number less than or equal to 13

15. a number less than 9

16. a maximum number of 7

17. a number that is at least -8

18. a number more than -6

Graph each inequality on a number line.

19. $x < 7$

20. $y > 3$

21. $b \leq 0$

22. $a \leq 5$

23. $x > -4$

24. $1 \leq z$

25. $x \geq 2.5$

26. $d < 1.9$

27. $x < -3.4$

28. $g > \dfrac{3}{4}$

29. $1\dfrac{1}{2} \geq x$

30. $h \leq -\dfrac{1}{3}$

Write an inequality for each graph.

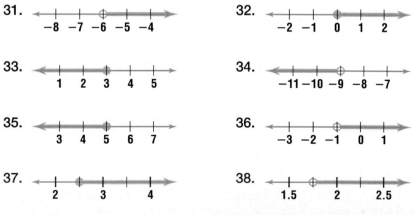

31.
```
 ← + + ○ + + →
  −8 −7 −6 −5 −4
```

32.
```
 ← + + ● + + →
  −2 −1  0  1  2
```

33.
```
 ← + + ● + + →
   1  2  3  4  5
```

34.
```
 ← + + ○ + + →
 −11 −10 −9 −8 −7
```

35.
```
 ← + + ● + + →
   3  4  5  6  7
```

36.
```
 ← + + ○ + + →
  −3 −2 −1  0  1
```

37.
```
 ← + ● + + →
   2    3    4
```

38.
```
 ← + ○ + + →
  1.5   2   2.5
```

Applications and Problem Solving

Real World

For Exercises 39–41, write an inequality to represent each situation. Then graph the inequality on a number line.

39. **Conservation**   At a fish farm, you must return any fish that you catch if it weighs less than 3 pounds.

40. **Entertainment**   A roller coaster requires a person to be at least 42 inches tall to ride it.

41. **Safety**   The elevators in an office building have been approved to hold a maximum of 3600 pounds.

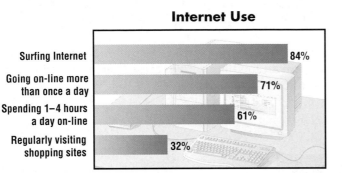

Internet Use

Surfing Internet — 84%

Going on-line more than once a day — 71%

Spending 1–4 hours a day on-line — 61%

Regularly visiting shopping sites — 32%

Source: *USA TODAY*

42. **Critical Thinking**   The survey results of 100 college students concerning Internet use are shown in the graph. Write an inequality to represent the number of students that may spend at least 5 hours on-line each day.

Mixed Review

43. **Finance**   Mr. Johnson invested $10,000 at an annual interest rate of 6%. Graph the function $B(t) = 10,000(1.06)^t$, where t is the time in years. How many years will it take before the balance is over $15,000? *(Lesson 11–7)*

Solve each equation by using the Quadratic Formula.   *(Lesson 11–6)*

44. $t^2 - 3t - 40 = 0$ 45. $3a^2 + a + 1 = 0$ 46. $x^2 - 6x + 7 = 0$

47. Find the prime factorization of 24. How many distinct prime factors are there?   *(Lesson 10–1)*

Simplify each expression.

48. $-\sqrt{16}$ *(Lesson 8–5)*

49. $\dfrac{-18x^3y}{3x}$ *(Lesson 8–2)*

50. **Standardized Test Practice**   If two lines are perpendicular to each other, which of the following statements could be true?   *(Lesson 7–7)*

   I.   The slopes of both lines are equal.

   II.   One slope is positive, and the other slope is negative.

   III.   The slope of one line is the reciprocal of the slope of the other line.

   A I only   **B** II only   **C** III only

   D I and II   **E** II and III

Extra Practice   See p. 715.

Solving Addition and Subtraction Inequalities

Math
In the Workplace

What You'll Learn

You'll learn to solve inequalities involving addition and subtraction.

Why It's Important

Sports You can use inequalities to set goals in competitions. *See Example 2.*

The Iditarod, nicknamed "The Last Great Race on Earth," is a dogsled race in Alaska. The 12- to 16-dog teams cover over 1150 miles in subzero weather. The fastest time recorded for finishing the race is 218 hours (9 days, 2 hours).

Suppose it takes a team 73 hours to reach the first checkpoint of the Iditarod and 98 hours to reach the second. How much time can be spent on the last leg of the race to beat the record time?

Let t represent the time for the last leg of the race. Write an inequality.

The sum of the times	*must be less than*	*the record time.*
$73 + 98 + t$	$<$	218

$$171 + t < 218$$

If this were an equation, we would subtract 171 from each side to solve for t. The same procedure can be used with inequalities, as explained by the properties below. *This problem will be solved in Example 2.*

These properties are also true for $\leq$ and $\geq$.

Addition and Subtraction Properties for Inequalities	**Words:**	For any inequality, if the same quantity is added or subtracted to each side, the resulting inequality is true.
	Symbols:	For all numbers a, b, and c,
		1. if $a > b$, then $a + c > b + c$ and $a - c > b - c$.
		2. if $a < b$, then $a + c < b + c$ and $a - c < b - c$.
	Numbers:	$-5 < 1$ $2 > -4$
		$-5 + 2 < 1 + 2$ $2 - 3 > -4 - 3$
		$-3 < 3$ $-1 > -7$

Example

1   **Solve $x + 14 \geq 5$. Check your solution.**

$$x + 14 \geq 5$$
$$x + 14 - 14 \geq 5 - 14 \quad \text{\textit{Subtract 14 from each side.}}$$
$$x \geq -9$$

(continued on the next page)

Check: Substitute a number less than -9, the number -9, and a number greater than -9 into the inequality.

Let $x = -10$.
$$x + 4 \geq 5$$
$$-10 + 14 \overset{?}{\geq} 5$$
$$4 \geq 5 \quad \text{false}$$

Let $x = -9$.
$$x + 14 \geq 5$$
$$-9 + 14 \overset{?}{\geq} 5$$
$$5 \geq 5 \quad \text{true}$$

Let $x = 0$.
$$x + 14 \geq 5$$
$$0 + 14 \overset{?}{\geq} 5$$
$$14 \geq 5 \quad \text{true}$$

The solution is {all numbers greater than or equal to -9}.

Your Turn Solve each inequality. Check your solution.

a. $x + 2 < 7$ **b.** $x - 6 \geq 12$

A more concise way to express the solution to an inequality is to use **set-builder notation**. The solution in Example 1 in set-builder notation is $\{x \mid x \geq -9\}$.

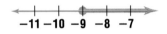

$\{x \qquad\qquad \mid \qquad\qquad x \geq -9\}$

The set of all numbers x   such that   x is greater than or equal to -9.

In Lesson 12–1, you learned that you can show the solution to an inequality on a line graph. The solution, $\{x \mid x \geq -9\}$, is shown below.

$$\xleftarrow{\qquad} \underset{-11 \; -10 \; -9 \; -8 \; -7}{+\!\!+\!\!\bullet\!\!+\!\!+} \xrightarrow{\qquad}$$

Examples

Sports Link

Real World

2 Refer to the application at the beginning of the lesson. Solve $171 + t < 218$ to find the time needed to finish the last leg of the Iditarod and beat the record.

$$171 + t < 218$$
$$171 + t - 171 < 218 - 171 \quad \textit{Subtract 171 from each side.}$$
$$t < 47 \qquad\qquad \textit{This means all numbers less than 47.}$$

The solution can be written as $\{t \mid t < 47\}$. So any time less than 47 hours will beat the record.

3 Solve $7y + 4 > 8y - 12$. Graph the solution.

$$7y + 4 > 8y - 12$$
$$7y + 4 - 7y > 8y - 12 - 7y \quad \textit{Subtract 7y from each side.}$$
$$4 > y - 12$$
$$4 + 12 > y - 12 + 12 \quad \textit{Add 12 to each side.}$$
$$16 > y$$

Since $16 > y$ is the same as $y < 16$, the solution is $\{y \mid y < 16\}$.

The graph of the solution has a circle at 16, since 16 is not included. The arrow points to the left.

$$\xleftarrow{\qquad} \underset{14 \quad 15 \quad 16 \quad 17 \quad 18}{+\!\!+\!\!\diamond\!\!+\!\!+} \xrightarrow{\qquad}$$

Your Turn Solve each inequality. Graph the solution.

c. $5y - 3 > 6y - 9$ **d.** $3r + 7 \leq 2r + 4$

You can also use inequalities to describe some geometry concepts.

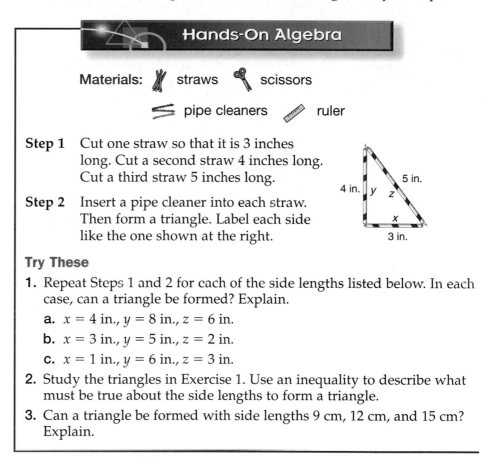

Hands-On Algebra

Materials: straws   scissors   pipe cleaners   ruler

Step 1   Cut one straw so that it is 3 inches long. Cut a second straw 4 inches long. Cut a third straw 5 inches long.

Step 2   Insert a pipe cleaner into each straw. Then form a triangle. Label each side like the one shown at the right.

Try These

1. Repeat Steps 1 and 2 for each of the side lengths listed below. In each case, can a triangle be formed? Explain.

   a. $x = 4$ in., $y = 8$ in., $z = 6$ in.

   b. $x = 3$ in., $y = 5$ in., $z = 2$ in.

   c. $x = 1$ in., $y = 6$ in., $z = 3$ in.

2. Study the triangles in Exercise 1. Use an inequality to describe what must be true about the side lengths to form a triangle.

3. Can a triangle be formed with side lengths 9 cm, 12 cm, and 15 cm? Explain.

Check for Understanding

Communicating Mathematics

Study the lesson. Then complete the following.

1. **Compare and contrast** solving inequalities by using addition and subtraction with solving equations by using addition and subtraction.

2. **Write** two inequalities that each have $\{y \mid y > 4\}$ as their solution.

3. **List** three situations in which solving an inequality may be helpful.

Math Journal

> **Vocabulary**
>
> set-builder notation

Guided Practice

 **Getting Ready** Write an inequality for each statement.

Sample: Five more than a number is greater than sixteen.
Solution: $n + 5 > 16$

4. A number minus three is greater than or equal to ten.

5. The sum of 8 and a number is at most 12.

Solve each inequality. Check your solution. *(Examples 1 & 2)*

6. $x + 9 < 12$

7. $v - 12 \geq 5$

8. $a - 6 \leq -14$

9. $h - 7 \leq 14$

10. $4.6 + x > 2.1$

11. $y - \dfrac{2}{3} < 1\dfrac{1}{6}$

Solve each inequality. Graph the solution. *(Example 3)*

12. $z - 12 \geq 2z + 4$

13. $4x - 1 > 5x$

interNET
CONNECTION

Data Update For the latest prices on computer hardware, visit: www.algconcepts. glencoe.com

14. Budgeting Antonio can spend no more than $1000 on a new computer system. The hard drive he wants costs $220. The monitor costs $300. How much money does Antonio have to spend on other components? *(Example 2)*

Exercises • • • • • • • • • • • • • • • • • • •

Practice

Solve each inequality. Check your solution.

15. $n + 6 > 9$

16. $x - 7 > -3$

17. $-3 + b < -8$

18. $g + 12 \leq 5$

19. $r - 8 < 11$

20. $x + 6 \geq 14$

21. $w - 9 > 13$

22. $4 + p \leq 1$

23. $t - 5 \leq -5$

24. $x + 3 \geq 19$

25. $-2 < c + 2$

26. $11 \leq -4 + m$

27. $d + 1.4 < 6.8$

28. $-3 + x > 11.9$

29. $-0.2 \geq 0.3 + z$

30. $s - (-2) > \dfrac{3}{4}$

31. $3\dfrac{3}{8} + v \leq 5\dfrac{7}{8}$

32. $\dfrac{1}{2} > x - \dfrac{2}{3}$

Solve each inequality. Graph the solution.

33. $7 > 8g - 7g - 6$

34. $5n < 4n + 10$

35. $3x + 12 \geq 2x - 4$

36. $6a \leq 5(a - 1)$

37. $-(-t + 9) \geq 0$

38. $3(v - 5) > 2(v - 2)$

Write an inequality for each statement. Then solve.

39. Eight less than a number is not greater than 12.

40. The sum of 5, 10, and a number is more than 25.

Applications and Problem Solving

41. Fundraising Westfield High School is having a raffle to raise money for Habitat for Humanity. Any homeroom that sells at least 150 tickets will use three days of school to help build a home. Ms. Martinez' homeroom is keeping a table of the number of tickets sold each day. How many more tickets do they need to sell to use three days of school to help build a home?

Day	Tickets
1	32
2	19
3	?
4	?
5	?

42. Wrestling   Carlos weighs 175 pounds. At least how much weight must he lose to wrestle in the 171-pound weight class?

43. Academics   Alissa must earn 475 out of 550 points to receive a grade of B. So far, she has earned 244 test points, 82 quiz points, and all 50 homework points. How many points must she score on her final exam to earn at least a B in the class?

44. Volunteerism   Each summer, 70 men bicycle in the Journey of Hope to raise money for people with disabilities. They begin their journey in San Francisco, California, and end in Washington, D.C. The map shows the miles they cycle in California. It takes them at most 285 miles to cycle to Nevada. How many miles is it from Jackson, California, to their first stop in Nevada?

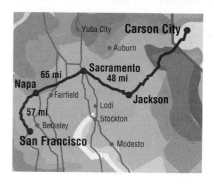

45. Critical Thinking   Is it possible for the solution set of an inequality to be the empty set (∅)? If so, give an example.

Mixed Review

Write an inequality to describe each number.   *(Lesson 12–1)*

46. a number no more than 1

47. a number less than −8

48. a number greater than 3

49. a minimum number of 5

50. The graph of which of the following equations is shown at the right: $y = 2^x$, $y = 2^x + 1$, $y = 2^x - 1$, or $y = -2^x$?   *(Lesson 11–7)*

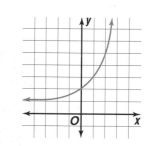

51. Factor the trinomial $3x^2 - 13x - 10$.   *(Lesson 10–4)*

52. Write $3x + y = -8$ in slope-intercept form.   *(Lesson 7–3)*

53. If y varies inversely as x and $y = 8$ when $x = 24$, find y when $x = 6$.   *(Lesson 6–6)*

54. Standardized Test Practice   Suppose $**x** = 4x - 2$. If $**x** = 10$, then what is the value of x?   *(Lesson 4–4)*

    A 2　　　　　　**B** 3　　　　　　**C** 38　　　　　　**D** 40

Extra Practice   See p. 716.

What You'll Learn

You'll learn to solve inequalities involving multiplication and division.

Why It's Important

Savings You can use inequalities when you are trying to budget your money.
See Exercise 37.

In 1999, Hurricane Floyd caused the largest U.S. peacetime evacuation. Many people trying to leave Charleston, South Carolina, were able to drive only 5 miles per hour. They needed to be at least 100 miles inland to avoid the dangers of the hurricane. How long would it take them to reach safety?

Recall that rate times time equals distance or $rt = d$. Since people needed to drive a distance *greater than or equal to* 100 miles, use an inequality. *This inequality will be solved in Example 1.*

$$\underbrace{\text{Rate}}\quad \underbrace{\text{times}}\quad \underbrace{\text{time}}\quad \underbrace{\text{is greater than or equal to}}\quad \underbrace{\text{distance.}}$$

$$5 \quad\cdot\quad t \quad\quad \geq \quad\quad 100$$

$$5t \geq 100$$

If we were solving $5t = 100$, we would divide each side by 5. Will this work when solving inequalities? Consider the inequality $12 > 4$.

Divide by 4.

$$12 > 4$$
$$\frac{12}{4} \overset{?}{>} \frac{4}{4}$$
$$3 > 1 \quad \text{true}$$

Divide by −4.

$$12 > 4$$
$$\frac{12}{-4} \overset{?}{>} \frac{4}{-4}$$
$$-3 > -1 \quad \text{false}$$

These examples lead us to the Division Property for Inequalities.

This property is also true for ≤ and ≥ .

<table>
<tr><td rowspan="4">Division Property for Inequalities</td><td>Words:</td><td>If you divide each side of an inequality by a positive number, the inequality remains true. If you divide each side of an inequality by a negative number, the inequality symbol must be reversed for the inequality to remain true.</td></tr>
<tr><td>Symbols:</td><td>For all numbers a, b, and c,

1. if c is positive and $a < b$, then $\frac{a}{c} < \frac{b}{c}$, and

if c is positive and $a > b$, then $\frac{a}{c} > \frac{b}{c}$.

2. if c is negative and $a < b$, then $\frac{a}{c} > \frac{b}{c}$, and

if c is negative and $a > b$, then $\frac{a}{c} < \frac{b}{c}$.</td></tr>
<tr><td>Numbers:</td><td>If $9 > 6$, then $\frac{9}{3} > \frac{6}{3}$ or $3 > 2$.

If $9 > 6$, then $\frac{9}{-3} < \frac{6}{-3}$ or $-3 < -2$.</td></tr>
</table>

1 **Refer to the application at the beginning of the lesson. How long would it take people to reach safety?**

$5t \geq 100$

$\dfrac{5t}{5} \geq \dfrac{100}{5}$ *Divide each side by 5.*

$t \geq 20$ *Keep the symbol facing the same direction.*

It would take at least 20 hours to drive at least 100 miles.

Prerequisite Skills Review
Operations with Decimals, p. 684

2 **Solve $-10x \leq 25.6$. Check your solution.**

$-10x \leq 25.6$

$\dfrac{-10x}{-10} \geq \dfrac{25.6}{-10}$ *Divide each side by –10 and reverse the symbol.*

$x \geq -2.56$

Check: Substitute -2.56 and a number greater than -2.56, such as 0, into the inequality.

Let $x = -2.56$.   Let $x = 0$.

$-10x \leq 25.6$ $-10x \leq 25.6$

$-10(-2.56) \stackrel{?}{\leq} 25.6$ $-10(0) \stackrel{?}{\leq} 25.6$

$25.6 \leq 25.6$ true $0 \leq 25.6$ true

The solution set is $\{x \mid x \geq -2.56\}$.

Your Turn

Solve each inequality. Check your solution.

a. $8x > 40$ **b.** $-x \leq 4.7$

Up to this point, we have shown how to solve inequalities by dividing. We can also solve inequalities by multiplying. Use the Multiplication Property for Inequalities.

This property is also true for $\leq$ and $\geq$.

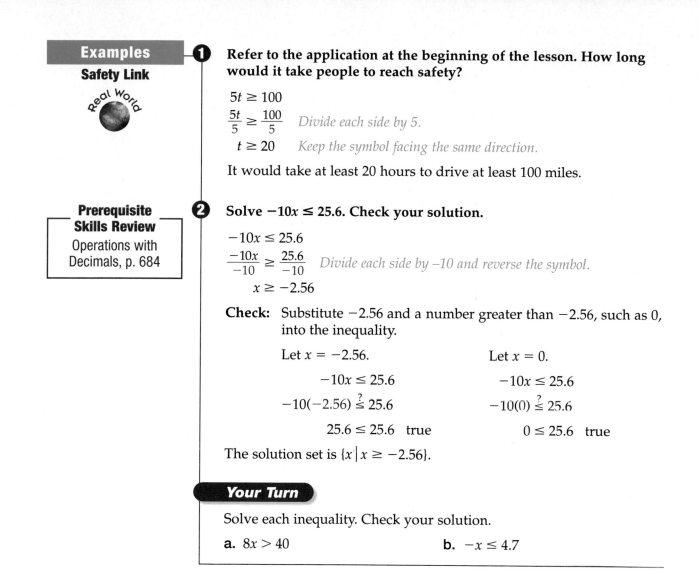

Multiplication Property for Inequalities	**Words:**	If you multiply each side of an inequality by a positive number, the inequality remains true. If you multiply each side of an inequality by a negative number, the inequality symbol must be reversed for the inequality to remain true.
	Symbols:	For all numbers a, b, and c, **1.** if c is positive and $a < b$, then $ac < bc$, and if c is positive and $a > b$, then $ac > bc$. **2.** if c is negative and $a < b$, then $ac > bc$, and if c is negative and $a > b$, then $ac < bc$.
	Numbers:	If $3 > -7$, then $3(2) > -7(2)$ or $6 > -14$. If $3 > -7$, then $3(-2) < -7(-2)$ or $-6 < 14$.

Solve $-\frac{x}{3} < -6$. **Check your solution.**

$$-\frac{x}{3} < -6$$

$$-3\left(-\frac{x}{3}\right) > -3(-6) \quad \textit{Multiply each side by –3 and reverse the symbol.}$$

$$x > 18$$

Check: Substitute 18 and a number greater than 18, such as 21, into the inequality.

Let $x = 18$.

$$-\frac{x}{3} < -6$$

$$-\frac{18}{3} \overset{?}{<} -6$$

$$-6 < -6 \quad \text{false}$$

Let $x = 21$.

$$-\frac{x}{3} < -6$$

$$-\frac{21}{3} \overset{?}{<} -6$$

$$-7 < -6 \quad \text{true}$$

The solution is $\{x \mid x > 18\}$.

Your Turn

c. $\dfrac{x}{-4} \le 8$

d. $\dfrac{2}{3}x > 22$

Check for Understanding

Communicating Mathematics

Study the lesson. Then complete the following.

1. **Compare and contrast** the Division Property for Inequalities and the Multiplication Property for Inequalities.
2. **Explain** why 5 is *not* a solution of the inequality $4x > 20$.
3. **Graph** the solution of $-2x \ge -8$ on a number line.

Guided Practice

Solve each inequality. Check your solution.

4. $2z > 12$ 5. $-3x \le 27$ 6. $-7b < 14$ (*Examples 1 & 2*)

7. $\dfrac{x}{4} \ge 5$ 8. $-\dfrac{r}{6} < 6$ 9. $\dfrac{2}{5}a \le -12$ (*Example 3*)

10. **Time**   Cherise drives to the restaurant where she works by different routes, but the trip is at most 10 miles. Considering the stops from traffic lights, she thinks her average driving speed is about 40 miles per hour. How much time does it take Cherise to get to work? (*Example 1*)

Exercises

Practice

Solve each inequality. Check your solution.

11. $-6h \le 12$ **12.** $-9n < 18$ **13.** $8d \le -24$

14. $\frac{a}{6} > 5$ **15.** $\frac{g}{3} > -12$ **16.** $-\frac{b}{8} \ge 5$

17. $7x \ge 49$ **18.** $-4y \le -40$ **19.** $3z > -9$

20. $-\frac{p}{2} > -1$ **21.** $\frac{c}{9} > -6$ **22.** $-\frac{a}{8} \le -4$

23. $3k > 5$ **24.** $-2t \le 11$ **25.** $-3 > -6w$

26. $\frac{3}{4}x < 3$ **27.** $5 \le \frac{5}{6}y$ **28.** $-\frac{2}{5}v \le 20$

29. $5.3v \ge 10.6$ **30.** $4.1x < -6.15$ **31.** $-28 \le -0.1s$

32. $-\frac{h}{3.8} \ge 2$ **33.** $\frac{n}{10.5} < 10$ **34.** $-\frac{x}{0.5} < -7$

Applications and Problem Solving

Real World

35. Geometry An *acute angle* has a measure less than 90°. If the measure of an acute angle is $2x$, what is the value of x?

$2x°$

36. Production The ink cartridge that Bill just bought for his printer can print up to 900 pages of text. Suppose Bill prints handbooks that are 32 pages each. How many complete handbooks can he print with this cartridge?

37. Budgeting Jenny mows lawns to earn money. She wants to earn at least $200 to buy a new stereo system. If she charges $12 a lawn, at least how many does she need to mow?

38. Critical Thinking Use a counterexample to show that if $x < y$, then $x^2 < y^2$ is not always true.

Mixed Review

Solve each inequality. Check your solution. *(Lesson 12–2)*

39. $z + 1 \le 5$

40. $-3 \ge b + 11$

41. Contests At a beach museum in San Pedro, California, more than 600 people built a life-size sand sculpture of a whale. Use an inequality to represent the number of people who built the sculpture. *(Lesson 12–1)*

42. Solve $x^2 + 5x + 4 = 0$ by factoring. *(Lesson 11–3)*

43. Find the product of $2v - 1$ and $2v + 1$. *(Lesson 9–5)*

44. Earthquakes On September 2, 1999, the state of Illinois experienced an earthquake tremor. It measured 3.5 on the Richter scale, releasing about 1.6×10^7 Joules of energy. The largest earthquake ever recorded in Illinois measured 5.5 on the Richter scale, releasing about 5.7×10^{11} Joules of energy. How many times stronger was the earthquake than the tremor? *(Lesson 8–4)*

45. Probability There are 20 students in a class. Each student's name is written on a separate slip of paper and placed in a box. A name is randomly drawn to determine who will read the daily announcements. Then the slip of paper is replaced in the box. What is the probability that the same name is drawn two days in a row? *(Lesson 5–7)*

46. Standardized Test Practice If the figure at the right is a square, then what is the value of x? *(Lesson 4–4)*

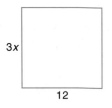
$3x$

12

A 3   B 4

C 12   D 36

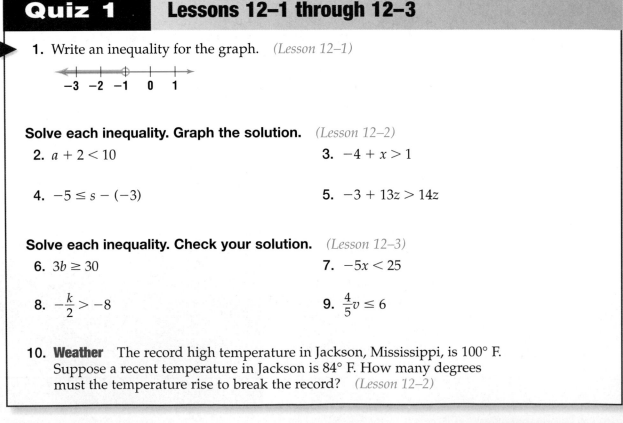

Quiz 1   Lessons 12–1 through 12–3

1. Write an inequality for the graph. *(Lesson 12–1)*

$$-3 \quad -2 \quad -1 \quad 0 \quad 1$$

Solve each inequality. Graph the solution. *(Lesson 12–2)*

2. $a + 2 < 10$

3. $-4 + x > 1$

4. $-5 \leq s - (-3)$

5. $-3 + 13z > 14z$

Solve each inequality. Check your solution. *(Lesson 12–3)*

6. $3b \geq 30$

7. $-5x < 25$

8. $-\dfrac{k}{2} > -8$

9. $\dfrac{4}{5}v \leq 6$

10. Weather The record high temperature in Jackson, Mississippi, is 100° F. Suppose a recent temperature in Jackson is 84° F. How many degrees must the temperature rise to break the record? *(Lesson 12–2)*

Extra Practice See p. 716.

Math In the Workplace

What You'll Learn

You'll learn to solve inequalities involving more than one operation.

Why It's Important

School You can determine what score is needed to receive a certain class grade. See Example 5.

Rafael is going to visit his aunt in Mexico. His aunt told him that during the time he'll be there, temperatures are usually warmer than 30° Celsius. Rafael wants to know this temperature in Fahrenheit so he can pack the right clothes.

The relationship between degrees Fahrenheit F and degrees Celsius C is given by the formula $\frac{5}{9}(F - 32) = C$.

Since Rafael's aunt said that it would be warmer than 30° C, we can use the inequality below to solve for F. *This inequality will be solved in Example 2.*

$$\underbrace{\frac{5}{9}(F - 32)}_{\text{The temperature}} \quad \underbrace{>}_{\text{is warmer than}} \quad \underbrace{30}_{30° C.}$$

This inequality involves more than one operation. So, the best strategy to use in solving the inequality is to undo the operations in reverse order. In other words, work backward just as you did in solving equations with more than one operation.

Example 1

Look Back

Solving Multi-Step Equations: Lesson 4–5

Solve $9 + 3x < 27$. Check your solution.

$$9 + 3x < 27$$
$$9 + 3x - 9 < 27 - 9 \quad \textit{Subtract 9 from each side.}$$
$$3x < 18$$
$$\frac{3x}{3} < \frac{18}{3} \quad \textit{Divide each side by 3.}$$
$$x < 6$$

Check: Substitute 6 and a number less than 6 into the inequality.

Let $x = 6$.
$$9 + 3x < 27$$
$$9 + 3(6) \overset{?}{<} 27$$
$$9 + 18 \overset{?}{<} 27$$
$$27 < 27 \quad \text{false}$$

Let $x = 0$.
$$9 + 3x < 27$$
$$9 + 3(0) \overset{?}{<} 27$$
$$9 + 0 \overset{?}{<} 27$$
$$9 < 27 \quad \text{true}$$

The solution is $\{x \mid x < 6\}$.

Your Turn Solve each inequality. Check your solution.

a. $4 + 2x \leq 12$
b. $8x - 5 \geq 11$

2 Refer to the application at the beginning of the lesson. Solve the inequality for *F*, degrees Fahrenheit.

$$\frac{5}{9}(F - 32) > 30$$

$$\frac{9}{5} \cdot \frac{5}{9}(F - 32) > \frac{9}{5} \cdot 30 \qquad \textit{Multiply each side by } \frac{9}{5}, \textit{ the reciprocal of } \frac{5}{9}.$$

$$F - 32 > 54$$

$$F - 32 + 32 > 54 + 32 \qquad \textit{Add 32 to each side.}$$

$$F > 86 \qquad \textit{Check your solution.}$$

Therefore, Rafael can expect it to be warmer than 86° F in Mexico.

Solving inequalities is similar to solving equations. The *only* exception is that with inequalities, you must reverse the inequality symbol if you multiply or divide by a negative number.

Example

3 Solve $-4x + 3 \geq 23 + 6x$. Check your solution.

$$-4x + 3 \geq 23 + 6x$$

$$-4x + 3 - 6x \geq 23 + 6x - 6x \qquad \textit{Subtract 6x from each side.}$$

$$-10x + 3 \geq 23$$

$$-10x + 3 - 3 \geq 23 - 3 \qquad \textit{Subtract 3 from each side.}$$

$$-10x \geq 20$$

$$\frac{-10x}{-10} \leq \frac{20}{-10} \qquad \textit{Divide each side by } -10 \textit{ and reverse the symbol.}$$

$$x \leq -2$$

The solution is $\{x \,|\, x \leq -2\}$. *Check your solution.*

Your Turn Solve each inequality. Check your solution.

c. $10 - 5x < 25$ **d.** $-3x + 1 > -17$

To solve inequalities that contain grouping symbols, you may use the Distributive Property first.

Example

4 Solve $8 \leq -2(x - 5)$. Check your solution.

$$8 \leq -2(x - 5)$$

$$8 \leq -2x + 10 \qquad \textit{Distributive Property}$$

$$8 - 10 \leq -2x + 10 - 10 \qquad \textit{Subtract 10 from each side.}$$

$$-2 \leq -2x$$

$$\frac{-2}{-2} \geq \frac{-2x}{-2} \qquad \textit{Divide each side by } -2 \textit{ and reverse the symbol.}$$

$$1 \geq x$$

The solution is $\{x \,|\, x \leq 1\}$. *Check your solution.*

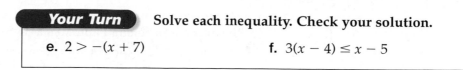

Your Turn   Solve each inequality. Check your solution.

e. $2 > -(x + 7)$

f. $3(x - 4) \leq x - 5$

You can use a graphing calculator to solve multi-step inequalities.

Graphing
Calculator Tutorial
See pp. 724–727.

Graphing Calculator Exploration

Solve $-5 - 8x \geq 43$ by using a graphing calculator.

Step 1   Clear the $\boxed{Y=}$ list to enter the inequality $-5 - 8x \geq 43$. (The $\geq$ symbol is item 4 on the TEST menu.)

Step 2   Press $\boxed{ZOOM}$ 6. Use the $\boxed{TRACE}$ and arrow keys to to move the cursor along the graph. You should see a line above the x-axis for values of x that are less than or equal to -6.

This represents the solution $\{x \mid x \leq -6\}$.

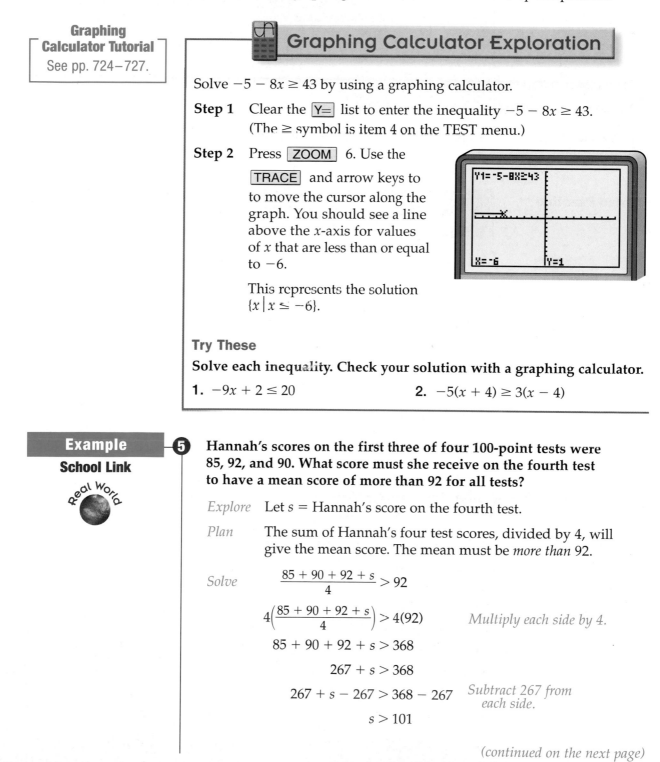

Try These

Solve each inequality. Check your solution with a graphing calculator.

1. $-9x + 2 \leq 20$

2. $-5(x + 4) \geq 3(x - 4)$

Example 5

School Link

Real World

Hannah's scores on the first three of four 100-point tests were 85, 92, and 90. What score must she receive on the fourth test to have a mean score of more than 92 for all tests?

Explore   Let s = Hannah's score on the fourth test.

Plan   The sum of Hannah's four test scores, divided by 4, will give the mean score. The mean must be *more than* 92.

Solve
$$\frac{85 + 90 + 92 + s}{4} > 92$$

$$4\left(\frac{85 + 90 + 92 + s}{4}\right) > 4(92) \qquad \textit{Multiply each side by 4.}$$

$$85 + 90 + 92 + s > 368$$

$$267 + s > 368$$

$$267 + s - 267 > 368 - 267 \qquad \textit{Subtract 267 from each side.}$$

$$s > 101$$

(continued on the next page)

Examine   Substitute a number greater than 101, such as 102, into
the original problem. Hannah's average would be
$\frac{85 + 90 + 92 + 102}{4}$ or 92.25. Since 92.25 > 92 is a true
statement, the solution is correct. Hannah must score
more than 101 points out of a 100-point test. Without
extra credit, this is not possible. So, Hannah cannot have
a mean score over 92.

Check for Understanding

**Communicating
Mathematics**

1. **Name** the operations used to solve $5 - 2x < -9$.
2. **Write** an inequality requiring more than one operation to solve.

Guided Practice

⊕ Getting Ready   **State which operation you would perform first to
solve the inequality.**

Sample: $12x + 4 > 20$	**Solution:** Subtract 4 from each side.

3. $-15z + 7 \geq -10$
4. $24 < 8b - 3$
5. $4.5a - (-3.1) > 8.2$
6. $\frac{n + 3}{9} \leq -11$

Solve each inequality. Check your solution.   *(Examples 1–4)*

7. $2y + 4 > 12$
8. $8 - 3h \leq 20$
9. $5 - 2x > 7$
10. $7z - 4 \geq 10$
11. $10 - 5w < w + 22$
12. $3(n - 4) > 2(n + 6)$

13. **School**   Kira wants her average math grade to be at least 90. Her test
scores are 88, 93, and 87. What score does she need on her fourth test
to ensure an average score of at least 90?   *(Example 5)*

Exercises   • • • • • • • • • • • • • • • • • • •

Practice

Solve each inequality. Check your solution.

14. $4t - 8 > 16$
15. $3b + 9 < 45$
16. $2 - 3n \leq 17$
17. $-20 \geq 8 - 7x$
18. $5 \leq 7c - 2$
19. $-6g - 1 < -13$
20. $-12 + 11r \leq 54$
21. $8 - 4v \geq 6$
22. $7 + 0.1a > 9$
23. $0.3m - 2.1 \leq -3.0$
24. $\frac{s}{4} - 6 < -11$
25. $\frac{11 - 6d}{5} > 1$
26. $3h - 5 \geq 2h + 4$
27. $5x + 3 \leq 2x + 9$
28. $6j - 9 > j + 6$
29. $2(7 - 2y) > 10$
30. $-\frac{1}{6}(z + 2) \leq -1$
31. $\frac{2}{3}(b + 1) < \frac{1}{2}(b + 5)$

Write and solve an inequality for each situation.

32. The sum of twice a number and 17 is no greater than 41.

33. Five times the sum of a number and 6 is less than 35.

34. Two thirds of a number decreased by 7 is at least 9.

Applications and Problem Solving

35. **Employment** Sophie is a receptionist at a hair salon. She earns $7.00 an hour, plus 10% of any hair products she sells. Suppose she works 22 hours a week. How much money in hair products must she sell to earn at least $180?

36. **Recreation** The admission fee to a state fair is $5.00. Each ride costs an additional $1.50.

    a. Suppose Pilar does not want to spend more than $20. How many rides can she go on?

    b. The fair has a special admission price for $14, which includes unlimited rides. For how many rides is this a better deal than paying for each ride separately?

37. **Finance** At a bank, an advertisement reads, "In one year, your earnings will be greater than your original investment plus 6% of the investment." Suppose Diego invests $1400. How much money can he expect to have at the end of the year?

38. **Critical Thinking** Would the solution of $x^2 > 4$ be $x > 2$? Justify your answer.

Mixed Review

Solve each inequality. Check your solution. *(Lesson 12–3)*

39. $5p > 35$

40. $-24 \le -8v$

41. $-\dfrac{x}{4} \ge 10$

42. $\dfrac{2}{5}z < -6$

43. **Budgeting** Haley earned $36 babysitting. She plans to buy her two favorite books that cost $8.25 each. With the rest of the money, she plans to go to dinner and see a movie with friends. At most, how much money can she spend on a movie and dinner? *(Lesson 12–2)*

44. Find the value of c that makes $x^2 - 6x + c$ a perfect trinomial square. *(Lesson 11–5)*

45. **Standardized Test Practice** If $x \ne 5$, then $\dfrac{x^2 - 3x - 10}{x - 5}$ is equivalent to which of the following? *(Lesson 10–3)*

    A $x + 2$

    B $x - 3$

    C $x - 5$

    D $x + 10$

What You'll Learn

You'll learn to solve compound inequalities.

Why It's Important

Nutrition
Pharmacists use inequalities to write prescriptions.
See Example 2.

Lamar is buying vitamins for his dog. The daily dose for the vitamins is based on the dog's weight. Lamar's dog weighs 32 pounds. Since 32 is greater than 25, but less than or equal to 50, he will give his dog 2 tablets.

Daily Dose	Dog's Weight (pounds)
1 tablet	25 or less
2 tablets	greater than 25 and less than or equal to 50
3 tablets	more than 50

Another way to write this information is to use an inequality. Let w represent the weight that requires 2 tablets.

Weight is greater than 25.
$w > 25$ *and*

Weight is less than or equal to 50.
$w \leq 50$

These two inequalities form a **compound inequality**. The compound inequality $w > 25$ and $w \leq 50$ can be written without using the word *and*.

Method 1 $25 < w \leq 50$

This can be read as 25 is less than w, which is less than or equal to 50.

Method 2 $50 \geq w > 25$

This can be read as 50 is greater than or equal to w, which is greater than 25.

Note that in each, both inequality symbols are facing the same direction.

Example ① **Write $x \geq 2$ and $x < 7$ as a compound inequality without using *and*.**

$x \geq 2$ and $x < 7$ can be written as $2 \leq x < 7$ or as $7 > x \geq 2$.

Your Turn

a. $x < 10$ and $x \geq -4$ **b.** $x \leq 6$ and $x \geq 2$

A compound inequality using *and* is true if and only if *both* inequalities are true. Thus, the graph of a compound inequality using *and* is the **intersection** of the graphs of the two inequalities.

Consider the inequality $-2 < x < 3$. It can be written using *and*: $x > -2$ and $x < 3$. To graph, follow the steps below.

Step 1   Graph $x > -2$.

Step 2   Graph $x < 3$.

Step 3   Find the intersection of the graphs.

The solution, shown by the graph of the intersection, is $\{x \mid -2 < x < 3\}$.

Example

Pet Care Link

Real World

2 Refer to the application at the beginning of the lesson. Graph the solution of $25 < w \le 50$.

Rewrite the compound inequality using *and*.
$25 < w \le 50$ is the same as $w > 25$ and $w \le 50$.

Step 1 Graph $w > 25$.

```
 ←┼──┼──┼──┼──◆──┼──┼──┼──┼──┼──┼──→
  10 15 20 25 30 35 40 45 50 55 60
```

Step 2 Graph $w \le 50$.

```
 ←┼──┼──┼──┼──┼──┼──┼──┼──◆──┼──┼──→
  10 15 20 25 30 35 40 45 50 55 60
```

> **Reading Algebra**
>
> Most of the time, $<$ or $\le$ symbols are used with compound inequalities.

Step 3 Find their intersection.

```
 ←┼──┼──┼──◇━━━━━━━━━━━━◆──┼──┼──→
  10 15 20 25 30 35 40 45 50 55 60
```

The solution is $\{w \mid 25 < w \le 50\}$.

Your Turn

c. Graph the solution of $3 \le x \le 5$.

Often, you must solve a compound inequality before graphing it.

Example

3 Solve $4 < x + 3 \le 12$. Graph the solution.

Step 1 Rewrite the compound inequality using *and*.

$$4 < x + 3 \le 12$$

$$x + 3 > 4 \qquad \text{and} \qquad x + 3 \le 12$$

Step 2 Solve each inequality.

$$x + 3 > 4 \qquad \text{and} \qquad x + 3 \le 12$$
$$x + 3 - 3 > 4 - 3 \qquad\qquad x + 3 - 3 \le 12 - 3$$
$$x > 1 \qquad\qquad\qquad x \le 9$$

Step 3 Rewrite the inequality as $1 < x \le 9$.

The solution is $\{x \mid 1 < x \le 9\}$.
The graph of the solution is shown at the right.

```
 ←┼──◇━━━━━━━━━━━━━━━━◆──┼──→
  0  1  2  3  4  5  6  7  8  9 10
```

Your Turn

d. Solve $-2 < x - 4 < 2$. Graph the solution.

Another type of compound inequality uses the word *or*. This type of inequality is true if one or more of the inequalities is true. The graph of a compound inequality using *or* is the **union** of the graphs of the two inequalities.

Example ─④ **Graph the solution of $x > 0$ or $x \le -1$.**

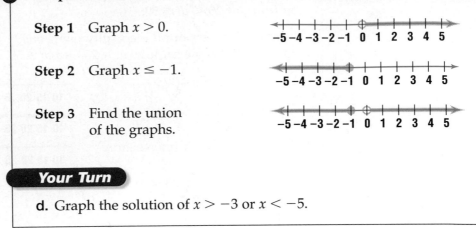

Step 1 Graph $x > 0$.

Step 2 Graph $x \le -1$.

Step 3 Find the union of the graphs.

Your Turn

d. Graph the solution of $x > -3$ or $x < -5$.

Sometimes you must solve compound inequalities containing the word *or* before you are able to graph the solution.

Example ─⑤ **Solve $3x \ge 15$ or $-2x < 4$. Graph the solution.**

$$3x \ge 15 \qquad\qquad \text{or} \qquad\qquad -2x < 4$$
$$\frac{3x}{3} \ge \frac{15}{3} \qquad\qquad\qquad\qquad \frac{-2x}{-2} > \frac{4}{-2}$$
$$x \ge 5 \qquad\qquad\qquad\qquad\qquad x > -2$$

Now graph the solution.

Step 1 Graph $x \ge 5$.

Step 2 Graph $x > -2$

Step 3 Find the union of the graphs.

The last graph shows the solution $\{x \mid x > -2\}$.

Your Turn

e. Solve $-6x > 18$ or $x - 2 < 1$. Graph the solution.

Check for Understanding

Communicating Mathematics

1. **Define** compound inequality in your own words.

2. **Write** a compound inequality for *x is greater than 3 and less than or equal to 5.*

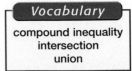

Vocabulary

compound inequality
intersection
union

State whether to find the intersection or union of the two inequalities to graph the solution.

Sample 1: $-8 \leq x + 4 < 12$ | **Solution:** Find the intersection since the inequalities can be joined by the word *and*.

Sample 2: $x > -9$ or $x < 5$ | **Solution:** Find the union since the inequalities are joined by the word *or*.

3. $x > -10$ and $x < 3$ 4. $x < 7$ or $x < 4$
5. $x - 5 < 2$ or $x \leq 15$ 6. $-13 < 9x \leq -11$

Write each compound inequality without using *and*. *(Example 1)*

7. $\ell > 5$ and $\ell < 10$ 8. $b < 3$ and $b \geq -2$

Graph the solution of each compound inequality. *(Examples 2 & 4)*

9. $y < 5$ and $y \geq 0$ 10. $x > 3$ or $x < -7$

Solve each compound inequality. Graph the solution. *(Examples 3 & 5)*

11. $10 > c + 2 > 5$ 12. $2 > s + 4 \geq -3$
13. $0 \leq 2v \leq 8$ 14. $-5x > 10$ or $7x < 28$
15. $j + 6 > 6$ or $-4j \geq 4$ 16. $-1 + r < -4$ or $-1 + r > 3$

17. **Construction** Odyssey of the Mind competitions encourage students to use creativity in solving difficult problems. One year, students had to construct a balsa-wood structure. The structure needed to be at least 9 inches tall and no more than 11.5 inches tall. Write a compound inequality describing the height of the structure. Graph the solution. *(Example 2)*

Exercises   • • • • • • • • • • • • • • • • • •

Practice

Write each compound inequality without using *and*.

18. $b > 0$ and $b < 5$ 19. $h > -8$ and $h < 8$
20. $y \leq 4$ and $y \geq -1$ 21. $g \geq 2$ and $g \leq 5$
22. $-2 < r$ and $1 > r$ 23. $6 < x$ and $x \leq 8$

Graph the solution of each compound inequality.

24. $x > 0$ or $x \leq -4$ 25. $z \leq 2$ and $z > -2$
26. $k < 7$ and $k > 5$ 27. $y \geq 16$ and $y \leq 21$
28. $b > 4$ or $b < 0$ 29. $m \leq 10$ or $m < 6$

Solve each compound inequality. Graph the solution.

30. $2 \leq a + 3 < 7$ 31. $9 \geq x + 1 \geq 5$
32. $-16 < 8s < 16$ 33. $9 \geq 3w > 0$
34. $-6 > r - 2 > -10$ 35. $2 \leq h + 5 \leq 8$

Solve each compound inequality. Graph the solution.

36. $-5 < y - 1 < -4$

37. $16 < -4c < 20$

38. $6 \geq x - (-5) \geq 0$

39. $z + 3 > 7$ or $z - 5 \leq -12$

40. $v - 8 > 12$ or $v + 4 < 20$

41. $r \leq -1$ or $-8r < 0$

42. $3j > 18$ or $j - 3 \geq 5$

43. $p - 5 > -3$ or $-2p \leq 2$

44. $c - 2.4 \geq 7.6$ or $c - 8.8 \leq 0$

45. $d + 2.1 > 3.6$ or $d - 4 > 0.5$

46. $8x \geq 4$ or $x + \dfrac{3}{4} < 1$

47. $\dfrac{w}{3} \leq -2$ or $-w < 2$

Write a compound inequality for each solution shown below.

48.
$$\begin{array}{c}\text{-5 -4 -3 -2 -1 0 1 2 3 4 5}\end{array}$$

49.
$$\begin{array}{c}\text{-5 -4 -3 -2 -1 0 1 2 3 4 5}\end{array}$$

50.
$$\begin{array}{c}\text{-8 -7 -6 -5 -4 -3 -2 -1 0 1 2}\end{array}$$

51.
$$\begin{array}{c}\text{-4 -3 -2 -1 0 1 2 3 4 5 6}\end{array}$$

Solve each compound inequality.

52. $2 > 3y + 2 > -13$

53. $4x - 7 \leq 5$ or $-(x - 8) < -1$

54. $2x - 1 > -5$ or $-3(x + 1) > 6$

55. $10 \leq 2(k - 6) \leq 14$

Applications and Problem Solving

56. Taxes Matthew Brooks is single and has a part-time job while attending college. Last year, he paid \$649 in federal income tax. Write an inequality for his taxable income. Use the table that describes the different tax brackets.

If Form 1040A, line 24, is —		And you are —			
At least	But less than	Single	Married filing jointly	Married filing separately	Head of a household
			Your tax is —		
4,200	4,250	634	634	634	634
4,250	4,300	641	641	641	641
4,300	4,350	649	649	649	649
4,350	4,400	656	656	656	656
4,400	4,450	664	664	664	664
4,450	4,500	671	671	671	671
4,500	4,550	679	679	679	679
4,550	4,600	686	686	686	686

Source: Ohio Department of Taxation, 1998

57. Cooking A box of macaroni and cheese lists two sets of directions for cooking. It says to heat the macaroni for 11 to 13 minutes on the stove or 12 to 14 minutes in the microwave.

a. Write an inequality that represents possible heating times.

b. Graph the solution.

58. Geometry To construct any triangle, the sum of the lengths of two sides must be greater than the length of the third. Suppose that two sides of a triangle have lengths of 4 inches and 12 inches. What are the possible values for the length of the third side? Express your answer as a compound inequality.

59. Chemistry Soil pH is measured on a scale from 0 to 14. The pH level describes the soil as shown below.

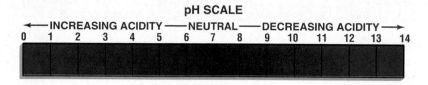

pH SCALE

←INCREASING ACIDITY——NEUTRAL——DECREASING ACIDITY→
0 1 2 3 4 5 6 7 8 9 10 11 12 13 14

Most plants grow best if the soil pH is between 6.5 and 7.2. Mr. Cohen took samples of soil from his garden and found pH values of 6.3, 6.4, and 6.7. What range of values must the fourth sample have if the soil pH is the best for growing plants? (*Hint:* Find the mean soil pH.)

60. Critical Thinking Graph each compound inequality. Then state the solution.

a. $x > 4$ or $x \le 4$ **b.** $2 > 3z + 2 > 14$

Mixed Review

Solve each inequality. Check your solution. (*Lessons 12–3 & 12–4*)

61. $2y + 4 < 4$ **62.** $-3n - 8 > 22$

63. $-2 \le 0.6x - 5$ **64.** $\frac{2}{3}p - 3 \le 7$

65. $-10b < -60$ **66.** $3r \ge -12$

67. Welding Maxwell is welding two pieces of iron together. During this process, the iron melts and begins to boil. Small droplets erupt and follow the paths of parabolic arcs. They can be modeled by the quadratic function $h(d) = -d^2 + 4d + 30$, where $h(d)$ represents the height of the arc above the ground at any horizontal distance d from the two pieces of iron. All measures are in inches. (*Lesson 11–1*)

a. Graph the function.

b. How high above the pieces of iron do the iron droplets jump?

68. Find the GCF of $8x^2$, $2x$, and $4xy$. (*Lesson 10–1*)

69. Simplify $3(b - 6)$. (*Lesson 9–3*)

70. Simplify $(c^2 + 5) - (c^2 - 8c + 1)$. (*Lesson 9–2*)

71. Standardized Test Practice Emily drove 8 miles in 12 minutes. At this rate, how many miles will she drive in 1 hour? (*Lesson 5–1*)

A 4 mi   B 20 mi   C 40 mi

D 56 mi   E 96 mi

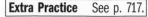

12-6 Solving Inequalities Involving Absolute Value

Math
In the Workplace

What You'll Learn
You'll learn to solve inequalities involving absolute value.

Why It's Important
Manufacturing
Employees use inequalities involving absolute value to determine tolerances.
See Example 3.

Ms. Gibson is a bank teller who handles thousands of dollars every day. At the end of each day, she is required to balance her drawer and be within $2.00 of the expected balance. If Ms. Gibson's expected balance is $8758.20, for what range of values is there an acceptable amount of money in her drawer?

If b represents the actual balance of the drawer, then you can write the following inequality to represent acceptable balances.

$$\underbrace{\text{The difference of b and \$8758.20}}_{|b - 8758.20|} \quad \underbrace{\text{is less than}}_{<} \quad \underbrace{\$2.00}_{2}$$

We use absolute value because it does not make a difference whether the actual balance is less than or greater than the expected balance. *You will solve this problem in Exercise 42.*

The three types of open sentences that can involve absolute value are listed below. Note that in each case, n is nonnegative since the absolute value of a number can only equal 0 or a positive number.

Look Back

Absolute Value:
Lesson 3–7

$$|x| = n \qquad\qquad |x| < n \qquad\qquad |x| > n$$

You have already studied equations involving absolute value. Inequalities involving absolute value are similar. Consider the graphs and solutions of the two open sentences below.

$|x| = 3$

The distance from 0 is 3.
So, $x = 3$ or $x = -3$.

$|x| < 3$

The distance from 0 is less than 3.
So, $x > -3$ and $x < 3$, or $-3 < x < 3$.

For both equations and inequalities involving absolute value, there are two cases to consider.

Case 1 The value within the absolute value symbols is positive.

Case 2 The value within the absolute value symbols is negative.

Solve $|x - 5| \leq 2$. Graph the solution.

Case 1 $x - 5$ is positive.

$$x - 5 \leq 2$$
$$x - 5 + 5 \leq 2 + 5 \quad \textit{Add 5.}$$
$$x \leq 7$$

Case 2 $x - 5$ is negative.

$$-(x - 5) \leq 2$$
$$-(x - 5)(-1) \geq 2(-1) \quad \textit{Multiply by } -1$$
$$\textit{and reverse}$$
$$x - 5 \geq -2 \quad \textit{the symbol.}$$
$$5 + 5 \geq -2 + 5 \quad \textit{Add 5.}$$
$$x \geq 3$$

So, the solution is $\{x \,|\, 3 \leq x \leq 7\}$.

The solution makes sense since 3 and 7 are at most 2 units from 5.

Your Turn

a. Solve $|x - 7| < 4$. Graph the solution.

As in Example 1, when solving an inequality involving absolute value and the symbols $<$ or $\leq$, the solution can be written as an inequality using *and*. However, when solving an inequality involving absolute value and the symbols $>$ or $\geq$, the solution can be written as an inequality using *or*.

Example ━② **Solve $|6x| > 18$. Graph the solution.**

Case 1 $6x$ is positive.

$$6x > 18$$
$$\frac{6x}{6} > \frac{18}{6} \quad \textit{Divide by 6.}$$
$$x > 3$$

Case 2 $6x$ is negative.

$$-6x > 18$$
$$\frac{-6x}{-6} < \frac{18}{-6} \quad \textit{Divide by } -6 \textit{ and}$$
$$\textit{reverse the symbol.}$$
$$x < -3$$

So, the solution is $\{x \,|\, x < -3 \text{ or } x > 3\}$.

Your Turn

b. Solve $|x - 2| \geq 3$. Graph the solution.

Inequalities involving absolute value are often used to indicate *tolerance*. Tolerance is the amount of error or uncertainty that is allowed when taking measurements.

Example	**3**	When producing $\frac{1}{2}$-inch bolts for bicycle parts, the tolerance is

Measurement Link

Real World

When producing $\frac{1}{2}$-inch bolts for bicycle parts, the tolerance is 0.005 inch. What is the range of acceptable bolt measures?

Explore   The difference in the actual size of the bolt and its expected size has to be less than or equal to 0.005 inch.

Plan   Let m = the actual measure of the bolt.

Then, $\left| m - \dfrac{1}{2} \right| \le 0.005$.

Solve $|m - 0.5| \le 0.005$ *Write $\frac{1}{2}$ as 0.5.*

Case 1 $m - 0.5$ is positive.	**Case 2** $m - 0.5$ is negative.
$m - 0.5 \le 0.005$	$-(m - 0.5) \le 0.005$
$m - 0.5 + 0.5 \le 0.005 + 0.5$	$-(m - 0.5)(-1) \ge 0.005(-1)$
$m \le 0.505$	$m - 0.5 \ge -0.005$
	$m - 0.5 + 0.5 \ge -0.005 + 0.5$
	$m \ge 0.495$

The solution is $\{m \mid 0.495 \le m \le 0.505\}$.

Examine   The bolt must measure from 0.495 inch to 0.505 inch, inclusive. To check the solution, choose a value for m within this range and one outside of this range. Substitute them into the original problem. Which value results in a true inequality?

Check for Understanding

Communicating Mathematics

1. **Compare and contrast** the graphs of the solutions for $|x| < 7$ and $|x| > 7$.

2. **Graph** the solutions for $|x - 2| > 1$ and $|x - 2| < 1$. For which inequality could you take the *intersection* of the graphs of two inequalities to find the solution? For which inequality could you take the *union* of the graphs of two inequalities to find the solution?

3. **You Decide** Madison says that the solution for $|x| \le 0$ is the same as the solution for $|x| \ge 0$. Mia says it is not. Who is correct? Explain.

Guided Practice

Getting Ready **Write two inequalities to describe the solution.**

Sample 1: $|x| \le 5$	**Sample 2:** $|x| > 1$
Solution: $x \le 5$ and $x \ge -5$	**Solution:** $x > 1$ or $x < -1$

4. $|x| < 10$ 5. $|x| \le 3$

6. $|x| \ge 2$ 7. $|x| > 8$

Solve each inequality. Graph the solution. *(Examples 1 & 2)*

8. $|n - 4| < 5$

9. $|x - 2| < 6$

10. $|3j| \leq 12$

11. $|t - 5| \geq 3$

12. $|2y| > 2$

13. $|s + 4| \geq 3$

14. **Measurement** Refer to Example 3. What are the possible measures for the bolt if the tolerance is 0.05 inch? Does a lesser or greater tolerance ensure more accurate measurements? Explain. *(Example 3)*

Exercises •

Practice

Solve each inequality. Graph the solution.

15. $|m + 1| < 5$

16. $|3v| < 15$

17. $|z + 7| \leq 2$

18. $|x + 3| \leq 8$

19. $|p - 1| < 2$

20. $|r + 4| < 4$

21. $|7t| \leq 14$

22. $|a - 3| \leq 4$

23. $|k + 2| < 3.5$

24. $|5n| \geq 30$

25. $|y + 3| \geq 6$

26. $|z - 1| \geq 1$

27. $|9x| > 18$

28. $|w + 2| > 5$

29. $|r - 4| \geq 1$

30. $|a - 8| \geq 3$

31. $|h - 3| > 9$

32. $|d + 9| > 0.2$

Write an inequality involving absolute value for each statement. Do not solve.

33. Quincy's golf score s was within 4 strokes of his average score of 90.

34. The measure m of a board used to build a cabinet must be within $\frac{1}{4}$ inch of 46 inches, inclusive, to fit properly.

35. The cruise control of a car set at 65 mph should keep the speed s within 3 mph, inclusive, of 65 mph.

For each graph, write an inequality involving absolute value.

36.
```
 <-+--o--+--+--+--+--+--+--+--o--+->
  -5-4-3-2-1  0  1  2  3  4  5
```
37.
```
 <-+--+--+--+--+--+--+--+--o--+--+->
  -5-4-3-2-1  0  1  2  3  4  5
```

Solve each inequality. Graph the solution.

38. $|2x - 11| \geq 7$

39. $|3x - 12| < 12$

40. $4|x + 3| \leq 8$

41. $10|x + 1| > 90$

Applications and Problem Solving

For Exercises 42–43, write an inequality involving absolute value and solve.

42. **Finance** Refer to the application at the beginning of the lesson. What are acceptable balances for Ms. Gibson's drawer?

43. **Chemistry** For a chemistry project, Marvin must pour 3.25 milliliters of liquid into a beaker. If he does not pour within 0.05, inclusive, of 3.25 milliters, the results will be inaccurate. How many milliliters of liquid can Marvin use?

44. Critical Thinking The *percent of error* is the ratio of the greatest possible error (tolerance) to a measurement. You can find the percent of error using this formula.

$$\text{percent of error} = \left| \frac{\text{greatest possible error}}{\text{measurement}} \right| \cdot 100$$

One rating system for in-line skate bearings is based on tolerances. The table shows tolerances for the outside diameter of bearings measuring 22 millimeters.

Rating	Tolerance (mm)
1	0.010
3	0.008
5	0.005

a. Find the percent of error for each rating to the nearest hundredth.

b. What can you conclude about the rating system?

Mixed Review

Graph the solution of each compound inequality. *(Lesson 12–5)*

45. $m < -7$ or $m \geq 0$

46. $x \geq -2$ and $x \leq 5$

47. $y > -5$ and $y < 0$

48. $r > 2$ or $r < -2$

49. Solve $1 \geq 4y + 5$. *(Lesson 12–4)*

50. Sales Grant bought a new sweater for $44.52. This included 6% sales tax. What was the cost of the sweater before tax? *(Lesson 5–5)*

51. Solve $|x - 2| = 4$. *(Lesson 3–7)*

52. Open-Ended Test Practice Write three fractions whose sum is $1\frac{5}{8}$. *(Lesson 3–2)*

Quiz 2   Lessons 12–4 through 12–6

Solve each inequality. *(Lesson 12–4)*

1. $3x + 8 < 11$

2. $5 - 6n > -19$

3. $9d - 4 \geq 8 - d$

Solve each compound inequality. Graph the solution. *(Lesson 12–5)*

4. $1 + x < -4$ or $1 + x < 4$

5. $-2 \leq n + 3 < 4$

6. $-6 < -2f < 10$

Solve each inequality. Graph the solution. *(Lesson 12–6)*

7. $|y - 7| < 2$

8. $|a + 8| \geq 1$

9. $|5x| < 20$

10. Travel Before the meter begins ticking, the charge for a taxi is $1.60. For each mile driven, there is an additional charge of 80 cents. What is the greatest distance you could travel in this taxi if you do not want to pay more than $10.00? *(Lesson 12–4)*

Extra Practice   See p. 717.

What You'll Learn
You'll learn to graph inequalities on the coordinate plane.

Why It's Important
Budgeting By graphing inequalities, you can solve problems where there are many solutions.
See Example 3.

Mr. Wheat is planning to take his youth group to a music festival. Lawn tickets cost $20, and pavilion tickets cost $30. If he plans to spend at most $300, how many of each ticket can Mr. Wheat purchase?

Let x represent the number of lawn tickets. Let y represent the number of pavilion tickets. Then the inequality below represents the solution.

The cost of lawn tickets	*plus*	*the cost of pavilion tickets*	*is at most*	*$300.*
$20x$	$+$	$30y$	$\leq$	300

The inequality is written in two variables. It is similar to an equation written in two variables. An easy way to show the solution of an inequality is to graph it in the coordinate plane. *This problem will be solved in Example 3.*

The solution set of an inequality in two variables contains many ordered pairs. The graph of these ordered pairs fills an area on the coordinate plane called a **half-plane**. The graph of an equation defines the **boundary** or edge for each half-plane. Use these steps to graph $y > 3$.

Reading Algebra

The related equation for $y > 3$ is $y = 3$.

Step 1 Determine the boundary by graphing the related equation, $y = 3$.

Step 2 Draw a *dashed* line since the boundary is *not* part of the graph.

Step 3 Determine which half-plane is the solution. To do this, substitute a point from each half-plane into the inequality. Find which point results in a true statement.

Test the point at $(5, 8)$.
$y > 3$
$8 > 3$ *Replace y with 8.*
true

Test the point at $(-3, 1)$.
$y > 3$
$1 > 3$ *Replace y with 1.*
false

The half-plane that contains $(5, 8)$ is the solution. Shade that half-plane. Any point in the shaded region is a solution of the inequality $y > 3$.

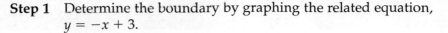

Example 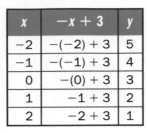 **1** **Graph $y > -x + 3$.**

Step 1 Determine the boundary by graphing the related equation, $y = -x + 3$.

x	$-x + 3$	y
−2	−(−2) + 3	5
−1	−(−1) + 3	4
0	−(0) + 3	3
1	−1 + 3	2
2	−2 + 3	1

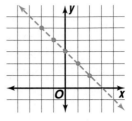

Step 2 Draw a *dashed* line since the boundary is not included.

Step 3 Test any point to find which half-plane is the solution. Use (0, 0) since it is the easiest point to use in calculations.

$y > -x + 3$
$0 > -(0) + 3$ *x = 0, y = 0*
$0 > 3$ false

Since (0, 0) does *not* result in a true inequality, the half-plane containing (0, 0) is *not* the solution. Thus, shade the other half-plane.

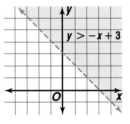

Your Turn

a. Graph $y < x - 7$.

When graphing inequalities, the boundary line is not always dashed. Consider the graph of $y \geq 3$. Since the inequality means $y > 3$ or $y = 3$, the boundary is part of the solution. This is indicated by graphing a solid line.

Example **2** **Graph $4x + y \leq 12$.**

To make a table or graph for the boundary line, solve the inequality for y in terms of x.

$\qquad 4x + y \leq 12$
$4x + y - 4x \leq 12 - 4x$ *Subtract 4x from each side.*
$\qquad\qquad y \leq -4x + 12$ *Rewrite 12 − 4x as −4x + 12.*

Step 1 Determine the boundary by graphing $y = -4x + 12$.

Step 2 Draw a *solid* line since the boundary is included.

Step 3 Test $(0, 0)$ to find which half-plane contains the solution.

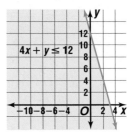

$$4x + y \le 12$$
$$4(0) + 0 \le 12 \quad x = 0, y = 0$$
$$0 \le 12 \quad \text{true}$$

The half-plane that contains $(0, 0)$ should be in the solution, which is indicated by the shaded region.

Your Turn

b. Graph $-2x + y \le 10$.

When solving real-life inequalities, the domain and range of the inequality are often restricted to nonnegative numbers or whole numbers.

Example

Budgeting Link

Real World

❸ **Refer to the application at the beginning of the lesson. How many lawn and pavilion tickets can Mr. Wheat purchase?**

First, solve for y in terms of x.

$$20x + 30y \le 300$$
$$20x + 30y - 20x \le 300 \quad 20x \quad \textit{Subtract 20x from each side.}$$
$$30y \le -20x + 300$$
$$\frac{30y}{30} \le \frac{-20x + 300}{30} \quad \textit{Divide each side by 30.}$$
$$y \le \frac{-20x}{30} + \frac{300}{30}$$
$$y \le -\frac{2}{3}x + 10$$

Step 1 Determine the boundary by graphing $y = -\frac{2}{3}x + 10$.

Step 2 Draw a *solid* line since the boundary is included.

Step 3 Test $(0, 0)$ to find which half-plane contains the solution.

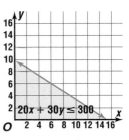

(continued on the next page)

Mr. Wheat cannot buy a negative number of tickets, nor can he buy portions of tickets. The solution is positive ordered pairs that are whole numbers beneath or on the graph of the line $y = -\frac{2}{3}x + 10$.

One solution is (12, 2). This represents 12 lawn tickets and 2 pavilion tickets costing $300.

Check for Understanding

Communicating Mathematics

Math Journal

1. **Explain** how to determine whether the boundary is a solid line or dashed line when graphing inequalities in two variables.

2. **Describe** how you could check whether a point is part of the solution of an inequality.

Guided Practice

⏱ **Getting Ready** Test (0, 0) to find which half-plane is the solution of each inequality.

Sample: $3x + y > 4$

Solution: $3x + y > 4$

$3(0) + 0 > 4$

$0 > 4$ false

Shade the half-plane *not* containing the point at (0, 0).

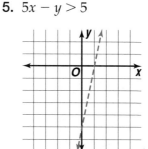

3. $x < -2$ 4. $y \leq 1$ 5. $5x - y > 5$

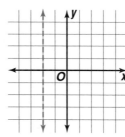
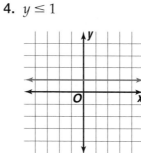

Graph each inequality. *(Examples 1 & 2)*

6. $y \leq -1$ 7. $y \geq x - 7$ 8. $y < 3x + 1$

9. $2x + y < 0$ 10. $-4x + y > -8$ 11. $x + 3y \geq 9$

12. **Sales** Tickets for the winter dance are $5 for singles and $8 for couples. To cover the deejay, photographer, and decoration expenses, a minimum of $1000 must be made from ticket sales. Write and graph an inequality that describes the number of singles' and couples' tickets that must be sold. Name at least one solution. *(Example 3)*

Exercises • • • • • • • • • • • • • • • • • •

Practice

Graph each inequality.

13. $y \leq 7$

14. $x \geq 5$

15. $y < x - 6$

16. $y < -x + 3$

17. $x + y > 8$

18. $y > x$

19. $-y \leq x$

20. $y \leq 2x$

21. $y < -3x + 4$

22. $-x + y < -5$

23. $x + 2y < 10$

24. $2x + y \leq 6$

25. $-3x + y \geq -1$

26. $3x - 2y > -12$

27. $y \leq 2(2x + 1)$

28. $3y > 3(4x - 3)$

29. $2(x + y) \leq 14$

30. $-3(5x - y) > 0$

For Exercises 31–32, write an inequality and graph the solution.

31. The sum of two numbers is greater than four.

32. Twice a number is less than or equal to another number.

Applications and Problem Solving

33. Animals Amara and Toshi have set a goal to find homes for more than twelve pets through the Humane Society.

   a. Write and graph an inequality to determine how many homes each girl must find to reach the goal.

   b. List three of the solutions.

   c. Describe the limitations on the solution set.

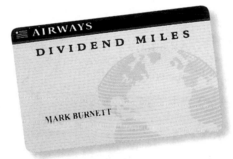

34. Airlines For each mile that Mr. Burnett flies, he earns one point toward a free flight. His credit card company is associated with the airline, and he earns one-half point for each dollar charged. Suppose he needs at least 20,000 points for a free flight. Show with a graph the miles that Mr. Burnett must fly and the money he must charge to get a free flight.

35. Critical Thinking Graph the intersection of the solutions of $4x + 2y \geq 8$ and $y < x$.

Mixed Review

Solve each inequality. Graph the solution. *(Lessons 12–5 & 12–6)*

36. $|t + 4| \geq 3$

37. $|h - 7| < 2$

38. $-1 < p + 1 \leq 6$

39. $-5b > 10$ or $b + 4 > 5$

40. Travel Ben and Pam left the park at the same time. Ben traveled north at 45 miles per hour. Pam traveled east at 60 miles per hour. After 1 hour, how far apart are Ben and Pam? *(Lesson 8–7)*

41. Write 848.3 in scientific notation. *(Lesson 8–4)*

42. Standardized Test Practice Suppose you toss 2 coins at the same time. What is the probability that both land heads up? *(Lesson 5–6)*

   A 0   **B** 0.25   **C** 0.50   **D** 0.75   **E** 1

Extra Practice   See p. 717.

Investigation

Parabolas and Pavilions

Materials

 grid paper

 ruler

yellow and blue colored pencils

Quadratic Inequalities

In this investigation, you will learn how to solve and graph **quadratic inequalities**.

Investigate

1. Graph the quadratic equation $y = x^2 + 2x - 3$ on a piece of graph paper.

   a. With a yellow colored pencil, shade the region inside the parabola.

   b. Make a table like the one below. Fill in column 1 with five points that appear in the yellow region, such as (0, 0). Compare the value of the y-coordinate with the value of $y = x^2 + 2x - 3$ when it is evaluated at the x-coordinate. When the quadratic equation is evaluated at 0, the result is -3. The y-coordinate, 0, is greater than -3. Place the correct inequality symbol in column 3 for each of the other four points that fall in the yellow region.

Point	y-coordinate	$<$ or $>$	$y = x^2 + 2x - 3$
(0, 0)	0	$>$	$y = (0)^2 + 2(0) - 3$ or -3

   c. Write a quadratic inequality that compares the points of the yellow region with the quadratic equation $y = x^2 + 2x - 3$.

   d. With a blue colored pencil, shade the region outside the parabola. Repeat parts b–c for the blue region.

2. Graph the quadratic equation $y = -x^2 - 2x + 3$ on a separate piece of graph paper.

   a. Shade the region inside the parabola yellow. Then shade the region outside the parabola blue. Make tables similar to those in Step 1. Write inequalities describing the yellow and blue regions.

   b. Suppose you wanted to include the values on the boundary line of a quadratic inequality. Explain how you would write the inequality to include the boundary line.

   c. Now suppose you did *not* want to include the boundary line of a quadratic inequality. Explain how you would draw the graph to show that the boundary line was *not* included in the solution.

3. Spring Town is building a community center, the Spring Town Pavilion. The building is to be 20 feet longer than it is wide and to have an area greater than or equal to 1500 square feet.

a. Let w represent the width of the building. Then $w + 20$ represents the length. The area of the building, $w(w + 20)$, must be greater than or equal to 1500 square feet. That is, $w^2 + 20w - 1500 \geq 0$. Graph $f(w) = w^2 + 20w - 1500$.

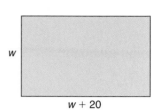

b. Find the values for w that satisfy the inequality in part a. To do this, choose values for w both inside and outside of your parabola.

c. What restrictions are placed on your solution for w since it represents width?

d. Use your graph to find two possible dimensions for the Spring Town Pavilion.

Extending the Investigation

In this extension, you will continue to investigate quadratic inequalities.

Graph each quadratic inequality.

1. $y < x^2 + 4x - 8$ **2.** $y \geq -x^2 + 2x + 15$ **3.** $y \leq -3x^2 + 3$

4. The Spring Town Pavilion is to have a deck for small outdoor gatherings. The length of the deck is to be 6 feet more than the width. The area of the deck is at least one-fifth of the area of the community center building. Write and graph a quadratic inequality that describes this problem. Then give three possible dimensions for the new deck.

Presenting Your Investigation

Here are some ideas to help you present your conclusions to the class.

• Make a poster that describes how to solve quadratic inequalities. Include at least two different inequalities. Show how the graphs are shaded to represent the solutions.

• Write and solve a problem similar to that in Step 3. Make a brochure that describes the problem, the graph of its solution, and a list of possible solutions

 Investigation For more information on graphing quadratic inequalities, visit: www.algconcepts.glencoe.com

Understanding and Using the Vocabulary

After completing this chapter, you should be able to define each term, property, or phrase and give an example or two of each.

*inter*NET
CONNECTION **Review Activities**
For more review activities, visit:
www.algconcepts.glencoe.com

boundary *(p. 535)*
compound inequality *(p. 524)*
half-plane *(p. 535)*

intersection *(p. 524)*
quadratic inequalities *(p. 540)*

set-builder notation *(p. 510)*
union *(p. 525)*

Complete each sentence using a term from the vocabulary list.

1. The solution of a compound inequality using *or* can be found by the ___?___ of the graphs of the two inequalities.

2. Graph the related equation of an inequality to find the ___?___ of the half-plane.

3. Use a test point from each ___?___ to find the solution of an inequality in two variables.

4. An inequality of the form $x < y < z$ is called a(n) ___?___ .

5. The ___?___ for the graph of an inequality in two variables will either be a dashed or solid line.

6. The ___?___ of two graphs is the area where they overlap.

7. A(n) ___?___ is an area on the coordinate plane representing the solution for an inequality in two variables.

8. $2x + 3 < 5$ or $x \geq -1$ is an example of a(n) ___?___ .

9. A solution written in the form $\{x \mid x \leq 3\}$ is written in ___?___ .

10. The solution of a compound inequality using *and* can be found by the ___?___ of the graphs of the two inequalities.

Skills and Concepts

Objectives and Examples	Review Exercises

• **Lesson 12–1** Graph inequalities on a number line.

Graph $x < 5$ on a number line.

The graph begins at 5, but 5 is not included. The arrow is to the left. The graph describes values that are less than 5.

Graph each inequality on a number line.

11. $x > -3$ 12. $z \leq 2$

13. $3.5 > x$ 14. $a \geq -1\frac{1}{2}$

Write an inequality for each graph.

15.
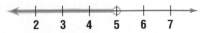
$-7 \quad -6 \quad -5 \quad -4 \quad -3$

16.
$5 \quad 6 \quad 7 \quad 8 \quad 9$

$2 \quad 3 \quad 4 \quad 5 \quad 6 \quad 7$

Objectives and Examples	Review Exercises

• Lesson 12–2 Solve inequalities involving addition and subtraction.

Solve $x - 6 > 2$.
$$x - 6 > 2$$
$x - 6 + 6 > 2 + 6$ *Add 6 to each side.*
$$x > 8$$

The solution is $\{x \mid x > 8\}$.

Solve each inequality. Check your solution.

17. $x + 3 > 7$

18. $a - 4 \geq 2$

19. $\frac{2}{3} \leq y - \frac{1}{2}$

20. $12x - 11x + 5 < 3$

21. $3(x + 1) \geq 4x$

• Lesson 12–3 Solve inequalities involving multiplication and division.

Solve $-4x < -8$.
$$-4x < -8$$
$\dfrac{-4x}{-4} > \dfrac{-8}{-4}$ *Divide each side by -4 and reverse the symbol.*
$$x > 2$$

The solution is $\{x \mid x > 2\}$.

Solve each inequality. Check your solution.

22. $3y \geq 12$

23. $-6n > 30$

24. $0.2w \leq -1.8$

25. $\frac{x}{5} > 3$ **26.** $-\frac{t}{2} \geq -14$

27. $\frac{3}{4}w \leq 6$ **28.** $\frac{h}{2.3} < -7$

• Lesson 12–4 Solve inequalities involving more than one operation.

Solve $2(x + 1) < 4 - 3x$.
$2(x + 1) < 4 - 3x$
$\;\;2x + 2 < 4 - 3x$ *Distributive Property*
$\;\;5x + 2 < 4$ *Add 3x to each side.*
$\;\;\;\;\;\;5x < 2$ *Subtract 2 from each side.*
$\;\;\;\;\;\;\;x < \frac{2}{5}$ *Divide each side by 5.*

The solution is $\left\{x \mid x < \frac{2}{5}\right\}$.

Solve each inequality. Check your solution.

29. $2x + 6 \geq 14$ **30.** $9 < -0.2y - 1$

31. $\frac{2}{3}t - 4 \leq 2$ **32.** $3(4 + n) < 21$

Write and solve an inequality.

33. Four times a number decreased by 3 is greater than 25.

34. Seven minus two times a number is no less than nine.

• Lesson 12–5 Solve compound inequalities and graph the solution.

Solve $3 < x + 2 \leq 8$.

$\;\;\;\;x + 2 > 3$ and $x + 2 \leq 8$
$x + 2 - 2 > 3 - 2$ $x + 2 - 2 \leq 8 - 2$
$\;\;\;\;\;\;x > 1$ $x \leq 6$

The solution is $\{x \mid 1 < x \leq 6\}$.

Solve each compound inequality. Graph the solution.

35. $-4 \leq y + 3 < 2$

36. $2 \leq 3 - t$ or $3 - t > 5$

37. $3a \geq 6$ or $-5 - a \geq -6$

38. $9 < 2x + 1 < 13$

Objectives and Examples

- **Lesson 12–6** Solve inequalities involving absolute value and graph the solution.

Solve $|x + 1| < 1$.

Case 1 $x + 1$ is positive.
$x + 1 < 1$
   $x < 0$ *Subtract 1 from each side.*

Case 2 $x + 1$ is negative.
   $-(x + 1) < 1$
$-(x + 1)(-1) > 1(-1)$ *Multiply by −1 and reverse the symbol.*
   $x + 1 > -1$
      $x > -2$ *Subtract 1 from each side.*

The solution is $\{x \mid -2 < x < 0\}$.

Review Exercises

Solve each inequality. Graph the solution.

39. $|t - 1| \le 5$

40. $|a + 7| < -2$

41. $|x + 3| < 1$

42. $|s + 4| \ge 3$

43. $|y - 2| > 0$

Write an inequality involving absolute value for each statement. Do not solve.

44. Bianca's guess g was within $6 of the actual value of $25.

45. The difference between Greg's score s on his final exam and his mean grade of 80 is more than 5 points.

- **Lesson 12–7** Graph inequalities in the coordinate plane.

Graph $2x + y < 5$.

First, solve for y.
   $2x + y < 5$
$2x + y - 2x < 5 - 2x$
      $y < -2x + 5$

Graph the related function $y = -2x + 5$ as a dashed line and shade.

Graph each inequality.

46. $y \ge x - 1$

47. $y < -2x + 4$

48. $\dfrac{y}{2} > x - 2$

49. $x - y \ge 3$

Applications and Problem Solving

50. Shipping An empty book crate weighs 30 pounds. The weight of a book is 1.5 pounds. For shipping, the crate can weigh no more than 60 pounds. What is the acceptable number of books that can be packed in a crate? *(Lesson 12–4)*

51. Geometry An obtuse angle measures more than 90° but less than 180°. If the measure of an obtuse angle is $3x$, what are the possible values for x? *(Lesson 12–5)*

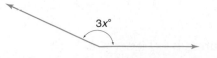

$3x°$

1. **List** at least three verbal phrases that are used to describe inequalities.

2. If you multiply or divide each side of an inequality by a negative number, what must happen to the symbol for the inequality to remain true?

3. Before solving an inequality involving absolute value, which two cases must you consider?

Write an inequality for each graph.

4.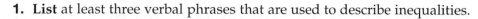

   1 2 3 4 5

5.
   −4 −3 −2 −1 0

Solve each inequality. Check your solution.

6. $2 + x \geq 12$

7. $5t + 6 \leq 4t - 3$

8. $8 < -4t$

9. $-0.2x > -6$

10. $\dfrac{t}{4} > 1$

11. $-\dfrac{2}{5}m \leq 10$

12. $-3r - 1 \geq -16$

13. $7x - 12 < 30$

14. $2(h - 3) > 6$

15. $8(1 - 2z) \leq 25 + z$

Solve each inequality. Graph the solution.

16. $x + 1 > -2$ and $x + 1 < 6$

17. $4 < 3j - 2 \leq 7$

18. $2n + 5 \geq 15$ or $2n + 5 \leq 3$

19. $-6c > -24$ or $c + 0.25 < 1.3$

20. $|x + 3| \geq 4$

21. $|4b| \leq 16$

Graph each inequality.

22. $y \geq 5x - 6$

23. $4x - 2y > -6$

24. **Car Rental** Justine is renting a car that costs $32 a day with free unlimited mileage. Since she is under the age of 25, it costs her an additional $10 per day. Justine does not want to pay any more than $200 on car rental costs. For what number of days can she rent a car?

25. **Manufacturing** Ball bearings are used to connect moving parts and minimize friction. Ball bearings for an automobile will work properly only if their diameter is within 0.01 inch, inclusive, of 5 inches. Write and solve an inequality to represent the range of acceptable diameters for these ball bearings.

Preparing for Standardized Tests

Angle, Triangle, and Quadrilateral Problems

You are already aware of some geometry concepts that you need to know for standardized tests. Be sure that you also know the meanings and definitions of the following terms.

> If there is no figure or diagram, draw one yourself.

right   obtuse   acute   triangle   quadrilateral

equilateral   isosceles   similar ($\approx$)   congruent ($\cong$)   collinear

reflection   translation   rotation   dilation

Proficiency Test Example

The triangles below are similar. Find the length of side *KL*.

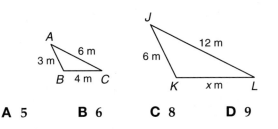

A 5   **B** 6   **C** 8   **D** 9

> **Hint** The measures of corresponding sides of similar polygons are proportional.

Solution Find the corresponding sides of the two triangles. Side *AB* of $\triangle ABC$ corresponds to side *JK* of $\triangle JKL$. Side *BC* of $\triangle ABC$ corresponds to side *KL* of $\triangle JKL$. Using these two pairs of corresponding sides, write a proportion.

$$\frac{AB}{JK} = \frac{BC}{KL}$$

$\dfrac{3}{6} = \dfrac{4}{x}$ *Substitute side measures.*

$3x = 4(6)$ *Cross multiply.*

$3x = 24$

$x = 8$ *Divide each side by 3.*

Side *KL* measures 8 meters. The answer is C.

ACT Example

In the figure below, $\overline{ON}$ is congruent to $\overline{LN}$, $m\angle LON = 30$, and $m\angle LMN = 40$. What is $m\angle NLM$?

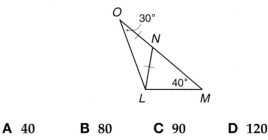

A 40   **B** 80   **C** 90   **D** 120

> **Hint** Examine *all* of the triangles in the figure.

Solution Label the unknown angles as 1 and 2. Find $m\angle 2$.

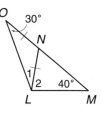

Since $\overline{ON} \cong \overline{LN}$, $\triangle ONL$ is isosceles and the base angles are equal. So, $m\angle 1 = 30$. Since the sum of the angle measures in any triangle is 180, find the sum of the angle measures in $\triangle OML$.

$180 = 30 + 40 + (30 + m\angle 2)$

$180 = 100 + m\angle 2$

$80 = m\angle 2$ *Subtract 100 from each side.*

The answer is B.

After you work each problem, record your answer on the answer sheet provided or on a sheet of paper.

1. The rectangles are similar. Find the value of x.

A 3   B 7

C 14   D 15

21 in.   7 in.

x in.   1 in.

2. $\triangle ABC$ is isosceles. What is $m\angle C$?

A 45   B 50

C 80   D 100

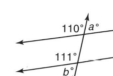

3. What is the sum of a, b, and c?

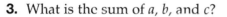

A 180   B 240

C 270   D 360

E It cannot be determined from the information given.

4. The formula for the area of a trapezoid is $A = \frac{1}{2}h(b_1 + b_2)$, where b_1 and b_2 are the lengths of the bases and h is the height. If the area of trapezoid $QRST$ is 72 square centimeters, then what is the height?

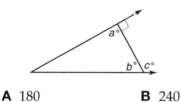

R   6 cm   S

h

Q   12 cm   T

A 1 cm   B 2 cm   C 4 cm   D 8 cm

5. Find the x-intercept of $y = -\frac{2}{3}x + 4$.

A 6   B -6 C 4   D 0

6. Simplify the expression $\dfrac{(-5)(4)\,|-6|}{-3}$.

A -120 B -40 C 40   D 120

7. The square of a number is 255 greater than twice the number. What is the number?

A 15 or -17 B 17 or -15

C 31 or -33 D 33 or -31

Quantitative Comparison

8.

110° $a°$

111°

$b°$

Column A	Column B
a	b

A if the quantity in Column A is greater;

B if the quantity in Column B is greater;

C if the two quantities are equal;

D if the relationship cannot be determined from the information given.

Open-Ended Questions

9. Grid-In In the figure, what is the sum of a, b, and c?

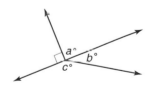

10. Draw a triangle that fits each description. If a drawing is not possible, explain why.

Part A

i. an obtuse equilateral triangle

ii. an acute equilateral triangle

Part B

i. an obtuse isosceles triangle

ii. an acute isosceles triangle

Systems of Equations and Inequalities

▶ What You'll Learn in Chapter 13:

- to solve systems of equations by graphing *(Lesson 13–1),*
- to determine whether a system of equations has one solution, no solution, or infinitely many solutions by graphing *(Lesson 13–2),*
- to solve systems of equations by substitution or elimination *(Lessons 13–3, 13–4, and 13–5),*
- to solve quadratic-linear systems of equations *(Lesson 13–6),* and
- to solve systems of inequalities by graphing *(Lesson 13–7).*

Problem-Solving Workshop

Project

What system of inequalities can be used to describe the design at the right?

Make your own design and write a system of inequalities that describes it. Exchange designs with another pair of students and write a system of inequalities that describes their design.

Working on the Project

Work with a partner and choose a strategy to help analyze and complete this project. Here are some suggestions to help you get started.

- Write an equation for the line that passes through points at $(-4, 0)$ and $(-4, 5)$. Write an inequality for the shaded area to the right of this line.
- Suppose your design has several colors. How could you describe the design so that another person could draw your design exactly without seeing the original?

Strategies

Look for a pattern.

Draw a diagram.

Make a table.

Work backward.

Use an equation.

Make a graph.

Guess and check.

Technology Tools

- Use **drawing software** to make your design.
- Use a **graphing calculator** to find equations for the lines in the given design.

interNET
CONNECTION   **Research** For more information about graphs of equations and inequalities, visit: www.algconcepts.glencoe.com

Presenting the Project
PORTFOLIO

Your portfolio should contain the following items:

- a system of inequalities that describes the design above,
- a sketch of your design on grid paper, and
- a system of inequalities that describes the design.

What You'll Learn
You'll learn to solve systems of equations by graphing.

Why It's Important
Fund-raising
Systems of equations can be used to determine how many items need to be sold in order to make a profit.
See Example 3.

Look Back
Graphing Linear Equations: Lesson 7–5

Highland High School Mathematics Club is planning a fund-raiser. They plan to sell mascot beanbags for a profit of $4 each and caps for a profit of $5 each. The ten members each plan to sell 5 items. If they can make a profit of $230 selling the 50 items, the club can buy a new printer. How many of each item should they try to sell?

x = the number of beanbags $\qquad$ y = the number of caps

You can write two equations to represent this situation.

$x + y = 50$ $\qquad \leftarrow$ *the number of items*
$4x + 5y = 230$ $\quad \leftarrow$ *the total profit*

Together, these equations are called a **system of equations**. The solution to this system is the ordered pair that satisfies both equations. *This problem will be solved in Example 3.*

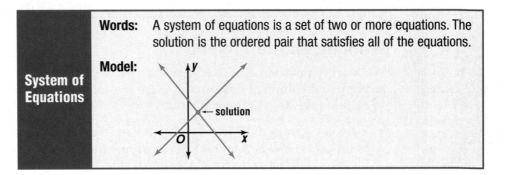

System of Equations

Words: A system of equations is a set of two or more equations. The solution is the ordered pair that satisfies all of the equations.

Model:

One method for solving a system of equations is to graph the equations on the same coordinate plane. The coordinates of the point at which the graphs intersect are the solution.

Examples

Solve each system of equations by graphing.

1 $y = 2x$
$y = -x + 3$

The graphs appear to intersect at (1, 2). Check this estimate by substituting the coordinates into each equation.

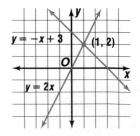

Check:

$y = 2x$ $\qquad\qquad\qquad\qquad$ $y = -x + 3$
$2 \stackrel{?}{=} 2(1)$ $\quad$ *Replace x with 1* $\qquad$ $2 \stackrel{?}{=} -1 + 3$ $\quad$ *Replace x with 1*
$2 = 2$ $\checkmark$ $\quad$ *and y with 2.* $\qquad\quad$ $2 = 2$ $\checkmark$ $\qquad$ *and y with 2.*

The solution of the system of equations is (1, 2).

② $x - y = -2$
$x + y = 4$

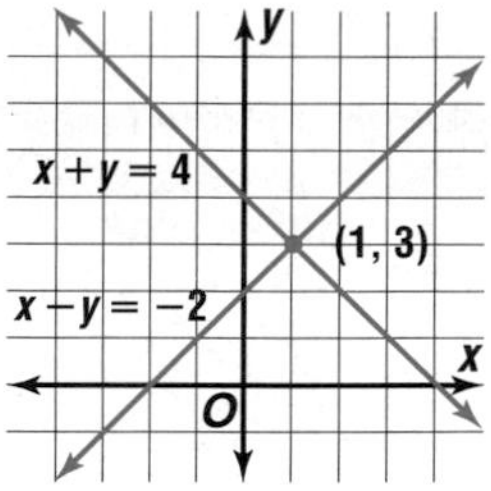

The graphs appear to intersect at
(1, 3). *Check this estimate.*

The solution of the system of equations is (1, 3).

Your Turn

a. $y = 2x$
$y = -x + 6$

b. $x + y = 1$
$2x + y = 4$

A graphing calculator can be used to solve systems of equations or to check solutions.

**Graphing
Calculator Tutorial**
See pp. 724−727.

Graphing Calculator Exploration

You can use a graphing calculator to solve systems of equations.

$3x + y = 1$
$-x + 2y = 16$

Step 1 Solve each equation for y.

$3x + y = 1 \rightarrow y = -3x + 1$

$-x + 2y = 16 \rightarrow y = \frac{1}{2}x + 8$

Step 2 Graph the equations in the standard viewing window.

Press $\boxed{\text{Y=}}$ to enter $-3x + 1$ as Y_1. Scroll down to enter $\frac{1}{2}x + 8$ as Y_2. Then press $\boxed{\text{ZOOM}}$ 6.

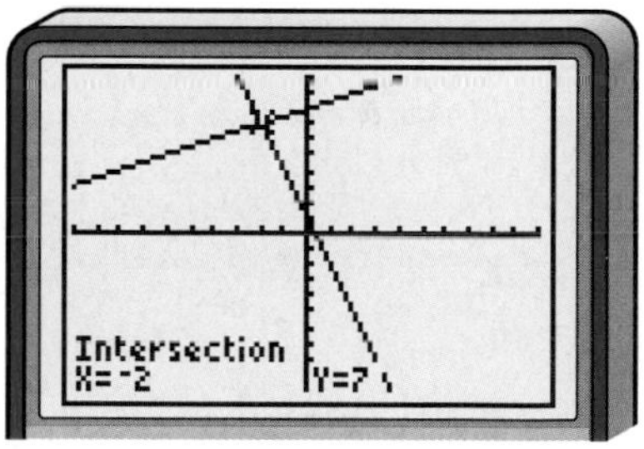

Step 3 Use the INTERSECT feature to find the intersection point.

Enter: $\boxed{\text{2nd}}$ [CALC] 5 $\boxed{\text{ENTER}}$ $\boxed{\text{ENTER}}$ $\boxed{\text{ENTER}}$

The solution is $(-2, 7)$. *Check by substituting the coordinates into the equations.*

Try These
Use a graphing calculator to solve each system of equations.

1. $y = x + 7$
$y = -x + 9$

2. $y = -3x$
$4x + y = 2$

3. $2x - y = 5$
$x + y = 16$

Systems of equations can model many real-world applications.

Example **3**

Sales Link

Real World

Refer to the application at the beginning of the lesson. How many mascot beanbags and caps should the math club try to sell if they sell a total of 50 items and want to make a $230 profit?

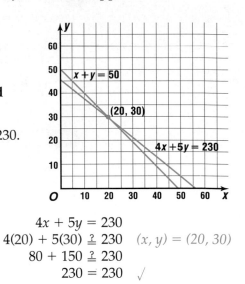

Graph $x + y = 50$ and $4x + 5y = 230$. The graphs appear to intersect at $(20, 30)$. Check this estimate.

Check:

$x + y = 50$

$20 + 30 \stackrel{?}{=} 50$ $(x, y) = (20, 30)$

$50 = 50$ $\checkmark$

$4x + 5y = 230$

$4(20) + 5(30) \stackrel{?}{=} 230$ $(x, y) = (20, 30)$

$80 + 150 \stackrel{?}{=} 230$

$230 = 230$ $\checkmark$

They should try to sell 20 mascot beanbags and 30 caps.

Check for Understanding

Communicating Mathematics

Study the lesson. Then complete the following.

1. **Describe** the solution of a system of equations.

2. **Sketch** a system of linear equations that has $(2, 3)$ as its solution.

3. **Determine** the solution of the system of equations represented by each pair of lines.

   a. ℓ and m

   b. m and n

   c. n and ℓ

> **Vocabulary**
>
> system of equations

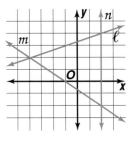

Guided Practice

Solve each system of equations by graphing.   *(Examples 1–3)*

4. $y = x - 3$
$y = -x + 3$

5. $x = -4$
$y = \frac{1}{2}x + 3$

6. $x + y = 6$
$y = 2$

7. $x + y = -2$
$x - y = 4$

8. **Shopping**   Randall would like to buy 10 Valentine's bouquets. The standard bouquet costs $7 and the deluxe one costs $12. He can only afford to spend $100.   *(Example 3)*

   a. Write a system of equations for the number of standard bouquets x and the number of deluxe bouquets y that he can buy.

   b. Use a graphing calculator to find the number of each type of bouquet he can buy.

Exercises

Practice

Solve each system of equations by graphing.

9. $x = 5$
   $y = -4$

10. $x = 4$
    $y = x - 5$

11. $y = x + 2$
    $y = -x + 2$

12. $y = -2x + 3$
    $y = -\frac{1}{2}x$

13. $y = 6$
    $y = \frac{4}{3}x + 2$

14. $y = 2x - 7$
    $y = \frac{3}{2}x - 6$

15. $y = 2$
    $x + y = 7$

16. $x + y = 3$
    $x - y = 1$

17. $x - y = 1$
    $y = -2x - 7$

18. $2x - y = 4$
    $x + y = -4$

19. $x + y = -3$
    $\frac{1}{2}x + y = 0$

20. $5x + 4y = 10$
    $2x + y = 1$

21. Find the solution of the system $y = x + 4$ and $3x + 2y = 18$.

22. What is the solution of the system $2x + 10y = 0$ and $x + y = 4$?

Applications and Problem Solving

![Real World]

23. **Submarines**   Two submarines began dives in the same vertical position and were trying to meet at a designated point. If one submarine was on a course approximated by the equation $x + 4y = -14$ and the other was on a course approximated by the equation $x + 3y = -8$, at what location would they meet? Write the coordinates of the point.

24. **Geometry**   The graphs of the equations $y = x + 2$, $3x + y = 6$, and $y = 5x + 6$ contain the sides of a triangle.
    a. Graph the equations.
    b. Find the coordinates of the vertices of the triangle.

25. **Critical Thinking**   Use your knowledge of slope and y-intercepts to write a system of equations with a solution of $(0, 4)$.

Mixed Review

26. Which ordered pair, $(-2, 0)$, $(-1, 2)$, $(0, -1)$, or $(2, -2)$, is a solution of $3x + 4y \geq 2$?   *(Lesson 12–7)*

Solve each inequality. Then graph the solution set.   *(Lesson 12–6)*

27. $|x - 3| < 2$

28. $|a + 5| \geq 3$

29. $|2m + 2| > 6$

30. **School**   Everyone in Mr. McClain's algebra class is at least 15 years old. Write an inequality to express this information and then make a graph of the inequality.   *(Lesson 12–1)*

31. Solve $2x^2 - 7x + 6 = 0$ by factoring.   *(Lesson 11–4)*

32. **Standardized Test Practice**   Which graph represents the function $y = x^2 + 2x - 3$?   *(Lesson 11–1)*

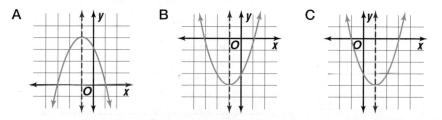

13-2 Solutions of Systems of Equations

Math
In the Workplace

What You'll Learn

You'll learn to determine whether a system of equations has one solution, no solution, or infinitely many solutions by graphing.

Why It's Important

Trains Train engineers must know if the tracks they are running on intersect another track.
See Example 6.

Trains in North America transport both people and materials. The track systems used for trains consist of three types: parallel, intersecting, or the same tracks. Train engineers must know which tracks intersect and which tracks run parallel. They must also know whether two trains are running on the same track.

In the same way, graphs of systems of linear equations may be intersecting lines, parallel lines, or the same line. Systems of equations can be described by the number of solutions they have.

Systems of Equations

consistent
at least one solution

inconsistent
no solution

independent
exactly one solution

dependent
infinitely many solutions

The different possibilities for the graphs of two linear equations are summarized in the following table.

Graph	Description of Graph	Slopes and Intercepts of Lines	Number of Solutions	Type of System
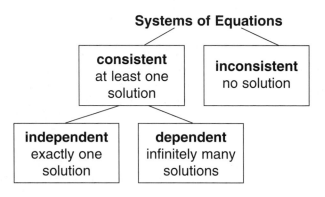	intersecting lines	different slopes	1	consistent and independent
	same line	same slope, same intercepts	infinitely many	consistent and dependent

Graph	Description of Graph	Slopes and Intercepts of Lines	Number of Solutions	Type of System
 $y = -2x + 4$ $y = -2x$	parallel lines	same slope, different intercepts	0	inconsistent

Examples

State whether each system is *consistent and independent, consistent and dependent,* or *inconsistent.*

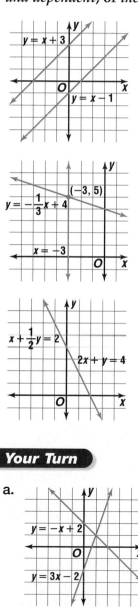

①

$y = x + 3$

$y = x - 1$

The graphs appear to be parallel lines. Since they do not intersect, there is no solution. This system is inconsistent.

②

$y = -\frac{1}{3}x + 4$

$(-3, 5)$

$x = -3$

The graphs appear to intersect at the point at $(-3, 5)$. Because there is one solution, this system of equations is consistent and independent.

③

$x + \frac{1}{2}y = 2$

$2x + y = 4$

Each equation has the same graph. Because any ordered pair on the graph will satisfy both equations, there are infinitely many solutions. The system is consistent and dependent.

Your Turn

a.
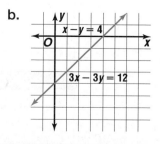

$y = -x + 2$

$y = 3x - 2$

b.

$x - y = 4$

$3x - 3y = 12$

You can determine the number of solutions to a system of equations by graphing.

Examples

Look Back

Graphing Linear Equations: Lesson 7–5

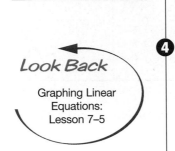

Determine whether each system of equations has *one* solution, *no* solution, or *infinitely many* solutions by graphing. If the system has one solution, name it.

4 $y = x + 2$
$y = -3x - 6$

The graphs appear to intersect at $(-2, 0)$. Therefore, this system of equations has one solution, $(-2, 0)$. Check that $(-2, 0)$ is a solution to each equation.

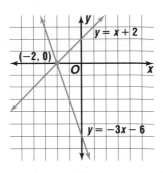

Check:

$y = x + 2$
$0 \stackrel{?}{=} -2 + 2$ *Replace x with -2*
$0 = 0$ ✓   *and y with 0.*

$y = -3x - 6$
$0 \stackrel{?}{=} -3(-2) - 6$ *Replace x with -2*
$0 = 0$ ✓   *and y with 0.*

The solution of the system of equations is $(-2, 0)$.

5 $2x + y = 4$
$2x + y = 6$

Write each equation in slope-intercept form.

$2x + y = 4 \quad \rightarrow \quad y = -2x + 4$
$2x + y = 6 \quad \rightarrow \quad y = -2x + 6$

The graphs have the same slope and different y-intercepts. The system of equations has no solution.

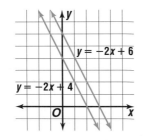

Your Turn

c. $y = x + 3$
   $y = -2x + 3$

d. $2x + y = 6$
   $4x + 2y = 12$

e. $3x - y = 3$
   $3x - y = 0$

Example

Transportation Link

Real World

6 The system of equations below represents the tracks of two trains. Do the tracks intersect, run parallel, or are the trains running on the same track? Explain.

$x + 2y = 4$
$3x + 6y = 12$

$x + 2y = 4$ $3x + 6y = 12$
                       $3(x + 2y) = 3(4)$
                          $x + 2y = 4$ *Divide each side by 3.*

One equation is the multiple of the other. Each equation has the same graph and there are infinitely many solutions. Therefore, the trains are running on the same track.

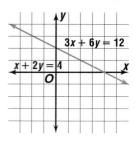

Check for Understanding

Communicating Mathematics

Study the lesson. Then complete the following.

1. **Describe** the possible graphs of a system of two linear equations.

2. **State** the number of solutions for each system of equations described below.

    a. One equation is a multiple of the other.

    b. The equations have the same *y*-intercept and different slopes.

    c. The equations have the same slope and different *y*-intercepts.

Math Journal

3. **Create** memory devices or other ways to remember the definitions of the terms in the Vocabulary box. Describe them in your journal.

> **Vocabulary**
> consistent
> independent
> dependent
> inconsistent

Guided Practice

State whether each system is *consistent and independent*, *consistent and dependent*, or *inconsistent*. *(Examples 1–3)*

4.

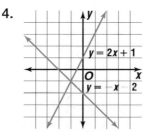

5.

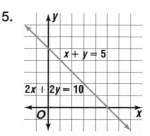

Determine whether each system of equations has *one* solution, *no* solution, or *infinitely many* solutions by graphing. If the system has one solution, name it. *(Examples 4 & 5)*

6. $y = x$
$y = x + 5$

7. $y = -x$
$y = 3x - 4$

8. $2x + y = -3$
$6x + 3y = -9$

9. Animals A dog sees a cat 60 feet away and starts running after it at 50 feet per second. At the same time, the cat runs away at 30 feet per second. This situation can be represented by the following system of equations and the graph at the right. *(Example 6)*

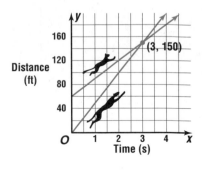

dog: $y = 50x$ cat: $y = 30x + 60$

a. Is this system of equations *consistent and independent*, *consistent and dependent*, or *inconsistent*? Explain.

b. Explain what the point at (3, 150) represents.

Exercises •

Practice

State whether each system is *consistent and independent*, *consistent and dependent*, or *inconsistent*.

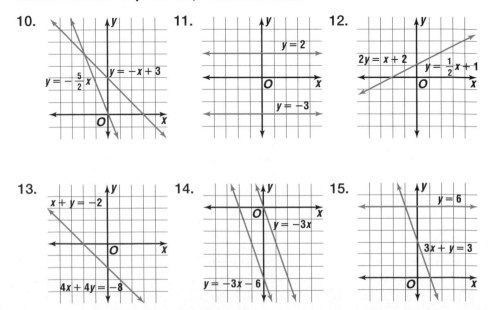

10. $y = -\frac{5}{2}x$ $y = -x + 3$

11. $y = 2$ $y = -3$

12. $2y = x + 2$ $y = \frac{1}{2}x + 1$

13. $x + y = -2$ $4x + 4y = -8$

14. $y = -3x$ $y = -3x - 6$

15. $y = 6$ $3x + y = 3$

Determine whether each system of equations has *one* solution, *no* solution, or *infinitely many* solutions by graphing. If the system has one solution, name it.

16. $y = 4x - 6$
$y = 4x - 1$

17. $y = 3x - 2$
$4y = 12x - 8$

18. $y = \frac{1}{2}x$
$y = -2x + 5$

19. $y = -x + 3$
$y = \frac{1}{5}x - 3$

20. $x = 3$
$2x - 3y = 0$

21. $3x - 2y = -6$
$3x - 2y = 6$

22. $x + y = 5$
$2x + 2y = 8$

23. $6x + y = -3$
$-x + y = 4$

24. $x - 4y = -4$
$\frac{1}{2}x - 2y = -2$

25. Does the system $x - y = 4$ and $x - 3y = 2$ have *one* solution, *no* solution, or *infinitely many* solutions? If the system has one solution, name it.

26. Without graphing, determine whether the system $x - 3y = 11$ and $2x - 6y = -5$ has *one* solution, *no* solution, or *infinitely many* solutions. Explain how you know.

Applications and Problem Solving

Real World

27. Animals Refer to Exercise 9. Suppose the dog is chasing another dog whose distance y can be represented by the equation $y = 50x + 20$. What is the solution? Explain what this means in terms of the dogs.

28. Ballooning A hot air balloon is 10 meters above the ground rising at a rate of 15 meters per minute. Another balloon is 150 meters above the ground descending at a rate of 20 meters per minute.
   a. Write a system of equations to represent the balloons.
   b. What is the solution of the system of equations?
   c. Explain what the solution means.

29. Critical Thinking Write an equation of a line in slope-intercept form that, together with the equation $x + 3y = 9$, forms a system that is inconsistent.

Mixed Review

30. What is the solution of the system $y = x + 4$ and $y = -3x - 4$? *(Lesson 13–1)*

31. Graph $y \geq 2x + 1$. *(Lesson 12–7)*

32. Number Theory Find two numbers if one number is 6 less than the other and whose product is 7. *(Lesson 11–5)*

Factor each polynomial. If the polynomial cannot be factored, write *prime*. *(Lesson 10–2)*

33. $4x - 8$

34. $13x + 2m$

35. $3a^2b^2 + 6ab - 9a$

36. Open-Ended Test Practice Write a square root whose best whole number estimate is 12. *(Lesson 8–6)*

Extra Practice   See p. 718.

Math In the Workplace

What You'll Learn
You'll learn to solve systems of equations by the substitution method.

Why It's Important
Metallurgy Systems of equations can be used to make metal alloys.
See Example 6.

Anita and Tionna were both driving from college to their hometown. Anita left first and traveled an average speed of 55 miles per hour. Tionna left an hour later and traveled an average speed of 65 miles per hour in the same direction. How long did it take Tionna to catch up with Anita? How many miles did she drive? *You will solve this problem in Exercise 31.*

You can solve this problem by letting y represent the distance traveled and x represent the time since Tionna started driving. Then write a system of equations.

Anita: $y = 55x + 55$ *Since Anita left an hour earlier, she had driven 55 miles before Tionna started the trip.*

Tionna: $y = 65x$ *Note that they will both have traveled the same distance, y, when they meet.*

The exact coordinates of the point where the lines intersect cannot be easily determined from a graph. The solution of this system can be found by using an algebraic method called **substitution**.

Hands-On Algebra
Algebra Tiles

Materials: algebra tiles   equation mat

Use a model to solve the system $y = x + 1$ and $3x + y = 9$.

Step 1   Let a green tile represent x. Then, since $y = x + 1$, 1 green tile and 1 yellow tile represent y.

Step 2   Represent $3x + y = 9$ on the equation mat. On one side of the mat, place three green tiles for $3x$ and 1 green tile and 1 yellow tile for y. On the other side, place 9 yellow tiles.

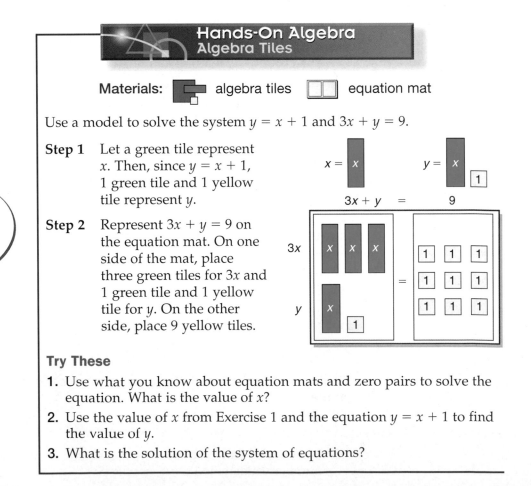

Look Back
Solving Equations with Algebra Tiles: Lesson 3–5

Try These

1. Use what you know about equation mats and zero pairs to solve the equation. What is the value of x?
2. Use the value of x from Exercise 1 and the equation $y = x + 1$ to find the value of y.
3. What is the solution of the system of equations?

Use substitution to solve each system of equations.

1 $y = 2x$

$3x + y = 5$

The first equation tells you that y is equal to $2x$. So, substitute $2x$ for y in the second equation. Then solve for x.

$$3x + y = 5$$
$$3x + 2x = 5 \quad \textit{Replace y with 2x.}$$
$$5x = 5$$
$$\frac{5x}{5} = \frac{5}{5} \quad \textit{Divide each side by 5.}$$
$$x = 1$$

Now substitute 1 for x in either equation and solve for y. *Choose the equation that is easier for you to solve.*

$$y = 2x$$
$$y = 2(1) \text{ or } 2 \quad \textit{Replace x with 1.}$$

The solution of this system of equations is (1, 2). You can see from the graph that the solution is correct. *You can also check by substituting (1, 2) into each of the original equations.*

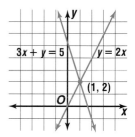

2 $x + y = 1$

$x = y + 6$

Substitute $y + 6$ for x in the first equation. Then solve for y.

$$x + y = 1$$
$$y + 6 + y = 1 \quad \textit{Replace x with y + 6.}$$
$$2y + 6 = 1$$
$$2y + 6 - 6 = 1 - 6 \quad \textit{Subtract 6 from each side.}$$
$$2y = -5$$
$$\frac{2y}{2} = \frac{-5}{2} \quad \textit{Divide each side by 2.}$$
$$y = -\frac{5}{2}$$

Now substitute $-\frac{5}{2}$ for y in either equation and solve for x.

$$x = y + 6$$
$$x = -\frac{5}{2} + 6 \quad \textit{Replace y with } -\frac{5}{2}.$$
$$x = \frac{7}{2}$$

The solution of this system of equations is $\left(\frac{7}{2}, -\frac{5}{2}\right)$. *Check by replacing (x, y) with $\left(\frac{7}{2}, -\frac{5}{2}\right)$ in each equation.*

$x - 3y = 3$
$2x - y = 11$

Solve the first equation for x since the coefficient of x is 1.

$x - 3y = 3 \quad \rightarrow \quad x = 3 + 3y$

Next, find the value of y by substituting $3 + 3y$ for x in the second equation.

$2x - y = 11$
$2(3 + 3y) - y = 11$
$6 + 6y - y = 11$
$6 + 5y = 11$
$6 + 5y - 6 = 11 - 6$
$5y = 5$
$\dfrac{5y}{5} = \dfrac{5}{5}$
$y = 1$

Now substitute 1 for y in either equation and solve for x.

$x - 3y = 3$
$x - 3(1) = 3$ *Replace y with 1.*
$x - 3 = 3$
$x - 3 + 3 = 3 + 3$
$x = 6$

The solution is (6, 1).

Your Turn

a. $x = 1 + 6y$
$\quad x + 2y = 9$

b. $x + y = 3$
$\quad -2x - 7y = 4$

In Lesson 13–2, you learned how to tell whether a system has one solution, no solution, or infinitely many solutions by looking at the graph. You can also determine this information algebraically.

Use substitution to solve each system of equations.

❹ $y = 4x + 1$
$4x - y = 7$

Find the value of x by substituting $4x + 1$ for y in the second equation.

$4x - y = 7$
$4x - (4x + 1) = 7$ *Replace y with 4x + 1.*
$4x - 4x - 1 = 7$ *Distributive Property*
$-1 = 7$

The statement $-1 = 7$ is false. This means that there are no ordered pairs that are solutions to both equations. Compare the slope-intercept forms of the equations, $y = 4x + 1$ and $y = 4x - 7$. Notice that the graphs of these equations have the same slope but different y-intercepts. Thus, the lines are parallel, and the system has no solution.

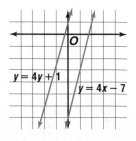

$y = 4y + 1$

$y = 4x - 7$

5 $x = 3 - 2y$
$2x + 4y = 6$

$$2x + 4y = 6$$
$$2(3 - 2y) + 4y = 6 \quad \textit{Replace x with } 3 - 2y.$$
$$6 - 4y + 4y = 6 \quad \textit{Distributive Property}$$
$$6 = 6$$

The statement $6 = 6$ is true. This means that an ordered pair for a point on either line is a solution to both equations. The system has infinitely many solutions.

Your Turn

c. $y = 2x + 1$
$4x - 2y = -9$

d. $x - 6y = 5$
$2x = 12y + 10$

Systems of equations can be used to solve mixture problems.

Example

Metals Link

Real World

6 A metal alloy is 25% copper. Another metal alloy is 50% copper. How much of each alloy should be used to make 1000 grams of a metal alloy that is 45% copper?

Explore   Let a = the number of grams of the 25% copper alloy.
Let b = the number of grams of the 50% copper alloy.

Prerequisite Skills Review
Decimals and Percents, p. 689

	25% Copper	50% Copper	45% Copper
Total Grams	a	b	1000
Grams of Copper	$0.25a$	$0.50b$	$0.45(1000)$

Plan   Write two equations to represent the information.

$$a + b = 1000 \qquad\qquad \leftarrow \textit{total grams}$$
$$0.25a + 0.50b = 0.45(1000) \quad \leftarrow \textit{grams of copper}$$

Solve   Use substitution to solve this system. Since $a + b = 1000$, $a = 1000 - b$.

$$0.25a + 0.50b = 0.45(1000)$$
$$0.25(1000 - b) + 0.50b = 0.45(1000) \quad \textit{Replace a with } 1000 - b.$$
$$250 - 0.25b + 0.50b = 450 \quad \textit{Distributive Property}$$
$$250 + 0.25b = 450$$
$$250 + 0.25b - 250 = 450 - 250 \quad \textit{Subtract 250.}$$
$$0.25b = 200$$
$$\frac{0.25b}{0.25} = \frac{200}{0.25} \quad \textit{Divide each side by 0.25.}$$
$$b = 800$$

(continued on the next page)

Now substitute 800 for b in either equation and solve for a.

$$a + b = 1000$$
$$a + 800 = 1000 \qquad \textit{Replace b with 800.}$$
$$a + 800 - 800 = 1000 - 800 \qquad \textit{Subtract 800 from each side.}$$
$$a = 200$$

So, 200 grams of the 25% copper alloy and 800 grams of the 50% copper alloy should be used.

Examine Check by substituting (200, 800) into the original equations. The solution is correct.

Check for Understanding

Communicating Mathematics

Study the lesson. Then complete the following.

1. **Explain** when you might choose to use substitution rather than graphing to solve a system of equations.

2. **State** what you would conclude if the solution of a system of linear equations yields the equation $4 = 4$.

3.  Faith and Todd are using substitution to solve the system $x + 3y = 8$ and $4x - y = 9$. Faith says that the first step is to solve for x in $x + 3y = 8$. Todd disagrees. He says to solve for y in $4x - y = 9$. Who is right and why?

> **Vocabulary**
>
> substitution

Guided Practice

Use substitution to solve each system of equations. *(Examples 1–5)*

4. $y = x - 4$
   $3x + 2y = 2$

5. $y = 3x$
   $7x - y = 16$

6. $x = 2y$
   $4x + 2y = 15$

7. $3x - 7y = 12$
   $x - 2y = 4$

8. $x = 2y + 5$
   $3x - 6y = 15$

9. $4y = -3x + 8$
   $3x + 4y = 6$

10. **Mixtures** MX Labs needs to make 500 gallons of a 34% acid solution. The only solutions available are 25% acid and 50% acid. How many gallons of each solution should be mixed to make the 34% solution? *(Example 6)*

Exercises

Practice

Use substitution to solve each system of equations.

11. $y = x$
    $3x + y = 4$

12. $x = 2y$
    $x + y = 3$

13. $y = 3x$
    $x + 2y = -21$

14. $2y = x$
    $x - y = 10$

15. $y = 3x - 8$
    $3x - y = 12$

16. $y = x + 7$
    $x + y = 1$

17. $y = \frac{1}{2}x + 3$
    $y - 5 = x$

18. $x = 7 - y$
    $2x - y = 8$

19. $x - 3y = -9$
    $5x - 2y = 7$

20. $x + y = 0$
$4x + 4y = 0$

21. $x - 2y = -2$
$5x - 4y = 2$

22. $2x - y = 6$
$3x - 5y = 9$

23. $x - y = 12$
$3x - y = 16$

24. $4x - 3y = -6$
$x + 5y = 10$

25. $x - 3y = 3$
$2x + 9y = 11$

26. $4x - y = -3$
$y + 2 = x$

27. $x - 6y = 5$
$2x - 12y = 10$

28. $x - 5y = 11$
$3x + y = 7$

29. Use substitution to solve $2x - y = -4$ and $-3x + y = -9$.

30. What is the solution of the system $x - 2y = 5$ and $3x - 5y = 8$?

Applications and Problem Solving

31. Driving   Refer to the application at the beginning of the lesson.
  a. After how many hours did Tionna catch up with Anita?
  b. How many miles did she drive?

32. Exercise   Laura and Ji-Yong were jogging on a 10-mile path. Laura had a 2-mile head start on Ji-Yong. If Laura ran at an average rate of 5 miles per hour and Ji-Yong ran at an average rate of 8 miles per hour, how long would it take for Ji-Yong to catch up with Laura?

33. Critical Thinking   Suppose you have three equations A, B, and C. Suppose a system containing any two of the equations is consistent and independent. Must the three equations together be consistent? Give an example or a counterexample to support your answer.

Mixed Review

34. Determine whether the system $y = -2x$ and $2y + 4x = 0$ has *one* solution, *no* solution, or *infinitely many* solutions by graphing. If the system has one solution, name it.   *(Lesson 13–2)*

35. Solve $x - y = -5$ and $x + y = 3$ by graphing.   *(Lesson 13–1)*

Write a compound inequality for each solution set shown.
(Lesson 12–5)

36.
$$-5\ -4\ -3\ -2\ -1\ \ 0\ \ 1\ \ 2\ \ 3\ \ 4\ \ 5$$

37.
$$-5\ -4\ -3\ -2\ -1\ \ 0\ \ 1\ \ 2\ \ 3\ \ 4\ \ 5$$

38. Solve $-5 + x > 12.8$.   *(Lesson 12–2)*

39. Open-Ended Test Practice   Mr. Drew is a salesperson. The formula for his daily income is $t(s) = 0.15s + 25$, where s is his total sales for the day. Suppose Mr. Drew earns \$94 on Monday. What were his total sales for that day?   *(Lesson 6–4)*

Quiz 1   Lessons 13–1 through 13–3

Determine whether each system of equations has *one* solution, *no* solution, or *infinitely many* solutions by graphing. If the system has one solution, name it.
(Lessons 13–1 & 13–2)

1. $y = x - 5$
$y = -\dfrac{1}{4}x$

2. $y = 2x + 5$
$y = x + 3$

3. $x + y = 3$
$x + y = 1$

4. $y = -2x + 3$
$2y = -4x + 6$

5. Sales   Ms. Williams mixed nuts that cost \$3.90 per pound with nuts that cost \$4.30 per pound to obtain a mixture of 50 pounds of nuts worth \$4.20 per pound. How many pounds of each type of nut did she use?   *(Lesson 13–3)*

13-4 Elimination Using Addition and Subtraction

Math In the Workplace

What You'll Learn

You'll learn to solve systems of equations by the elimination method using addition and subtraction.

Why It's Important

Entertainment
Solving systems of equations using elimination could be used to determine entertainment costs. *See Exercise 35.*

In 1974, the Kartchner Caverns were discovered near Tucson, Arizona. In 1999, visitors were allowed to tour the caverns and see the stalactites and stalagmites. Three adults and 10 students in one van paid $102 for a tour of the caverns, and 3 adults and 7 students in a second van paid $84 for the tour. When they arrived back at school, the bookkeeper needed to record how much each individual admission had cost, but everybody had forgotten the prices by that time. How can the bookkeeper figure out the prices for adult admission and for student admission?

a = the price for an adult s = the price for a student

You can write two equations to represent this situation.

$3a + 10s = 102$ ← *total cost for the first van*
$3a + 7s = 84$ ← *total cost for the second van*

This problem will be solved in Example 2.

Another algebraic method for solving systems of equations is called **elimination**. You can eliminate one of the variables by adding or subtracting the equations.

| **Example** | **①** | Use elimination to solve the system of equations. |

$2x + y = 3$
$x + y = 1$

$$2x + y = 3 \quad \textit{Write the equations in column form.}$$
$$\underline{(-)\, x + y = 1} \quad \textit{Subtract the equations to eliminate the y terms.}$$
$$x + 0 = 2$$
$$x = 2 \quad \text{The value of } x \text{ is 2.}$$

Now substitute in either equation to find the value of y. *Choose the equation that is easier for you to solve.*

$$x + y = 1$$
$$2 + y = 1 \qquad \textit{Replace x with 2.}$$
$$2 + y - 2 = 1 - 2 \quad \textit{Subtract 2 from each side.}$$
$$y = -1 \qquad \text{The value of } y \text{ is } -1.$$

The solution of the system of equations is $(2, -1)$.

Check: $2x + y = 3$

$2(2) + (-1) \stackrel{?}{=} 3$ *Replace (x, y) with (2, −1).*

$4 + (-1) \stackrel{?}{=} 3$

$3 = 3$ $\checkmark$

$x + y = 1$

$2 + (-1) \stackrel{?}{=} 1$ *Replace (x, y) with (2, −1).*

$1 = 1$ $\checkmark$

Your Turn   Use elimination to solve each system of equations.

a. $x + y = 5$
$2x + y = 4$

b. $3x + 5y = -2$
$3x - 2y = -16$

Example **2**

Entertainment Link

Real World

Refer to the application at the beginning of the lesson. Find the admission price for an adult and for a student.

$3a + 10s = 102$ *Write the equations in column form.*

$(-)\ 3a + \ \ 7s = \ \ 84$ *Subtract the equations to eliminate the a terms.*

$\overline{\ \ \ 0\ \ +\ \ 3s\ =\ \ 18}$

$3s = 18$

$\dfrac{3s}{3} = \dfrac{18}{3}$ *Divide each side by 3.*

$s = 6$ The value of *s* is 6.

Now substitute in either equation to find the value of *a*.

$3a + 7s = 84$

$3a + 7(6) = 84$ *Replace s with 6.*

$3a + 42 = 84$

$3a + 42 - 42 = 84 - 42$ *Subtract 42 from each side.*

$3a = 42$

$\dfrac{3a}{3} = \dfrac{42}{3}$ *Divide each side by 3.*

$a = 14$ The value of *a* is 14.

The solution of the system of equations is (14, 6). This means that the cost for adults was $14 and the cost for students was $6.

Check: $3a + 10s = 102$

$3(14) + 10(6) \stackrel{?}{=} 102$ *Replace (a, s) with (14, 6).*

$42 + 60 \stackrel{?}{=} 102$

$102 = 102$ $\checkmark$

$3a + 7s = 84$

$3(14) + 7(6) \stackrel{?}{=} 84$ *Replace (a, s) with (14, 6).*

$42 + 42 \stackrel{?}{=} 84$

$84 = 84$ $\checkmark$

In some systems of equations, the coefficients of terms containing the same variable are additive inverses. For these systems, the elimination method can be applied by adding the equations.

Example 3

Use elimination to solve the system of equations.

$4x - 6y = 10$

$3x + 6y = 4$

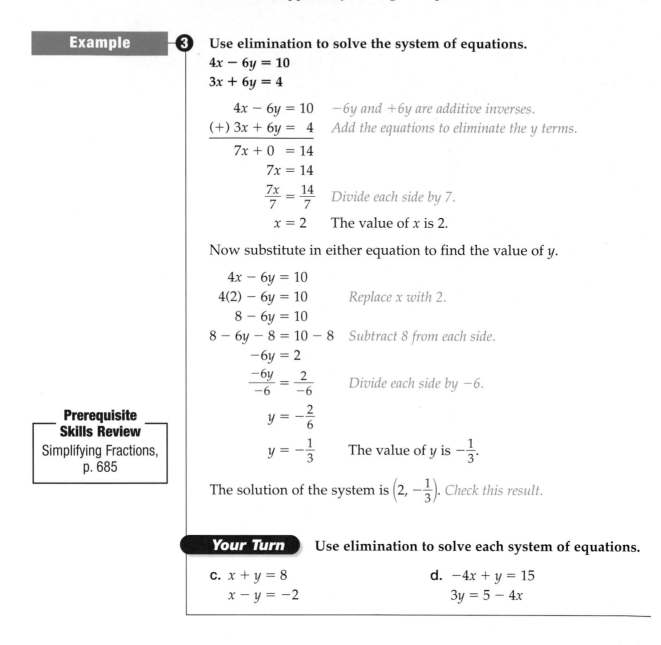

$$\begin{array}{rl} 4x - 6y = 10 & \text{$-6y$ and $+6y$ are additive inverses.} \\ (+)\ 3x + 6y = 4 & \text{Add the equations to eliminate the y terms.} \\ \hline 7x + 0 = 14 & \\ 7x = 14 & \\ \dfrac{7x}{7} = \dfrac{14}{7} & \text{Divide each side by 7.} \\ x = 2 & \text{The value of x is 2.} \end{array}$$

Now substitute in either equation to find the value of y.

$$\begin{array}{rl} 4x - 6y = 10 & \\ 4(2) - 6y = 10 & \text{Replace x with 2.} \\ 8 - 6y = 10 & \\ 8 - 6y - 8 = 10 - 8 & \text{Subtract 8 from each side.} \\ -6y = 2 & \\ \dfrac{-6y}{-6} = \dfrac{2}{-6} & \text{Divide each side by -6.} \\ y = -\dfrac{2}{6} & \\ y = -\dfrac{1}{3} & \text{The value of y is $-\dfrac{1}{3}$.} \end{array}$$

Prerequisite Skills Review
Simplifying Fractions, p. 685

The solution of the system is $\left(2, -\dfrac{1}{3}\right)$. *Check this result.*

Your Turn Use elimination to solve each system of equations.

c. $x + y = 8$
$ x - y = -2$

d. $-4x + y = 15$
$ 3y = 5 - 4x$

Systems of equations can be used to solve **digit problems**.

Example 4

Number Theory Link

The sum of the digits of a two-digit number is 8. If the tens digit is 4 more than the units digit, what is the number?

Let t represent the tens digit and let u represent the units digit.

$t + u = 8$ ← *the sum of the digits*
$t = u + 4$ ← *the relationship between the digits*

Rewrite the second equation so that the t and u are on the same side of the equation.

$$t = u + 4 \rightarrow t - u = 4$$

Then use elimination to solve.

$t + u = 8$	*Write the equations in column form.*
$(+)\ t - u = 4$	*Add the equations to eliminate the u terms.*

$$2t + 0 = 12$$
$$2t = 12$$
$$\frac{2t}{2} = \frac{12}{2} \quad \textit{Divide each side by 2.}$$
$$t = 6 \quad \text{The tens digit is 6.}$$

Now substitute to find the units digit.

$$t + u = 8$$
$$6 + u = 8 \quad \textit{Replace t with 6.}$$
$$6 + u - 6 = 8 - 6 \quad \textit{Subtract 6 from each side.}$$
$$u = 2 \quad \text{The units digit is 2.}$$

Since t is 6 and u is 2, the number is 62. *Check this solution.*

Check for Understanding

Communicating Mathematics

Study the lesson. Then complete the following.

Vocabulary

elimination
digit problems

1. **Explain** when you would use elimination to solve a system of equations.

2. **Describe** the result when you add each pair of equations. What does the result tell you about the system of equations?

   a. $2x - y = 12$
   $-2x + y = 14$

   b. $x + 5y = 3$
   $-x - 5y = -3$

Guided Practice

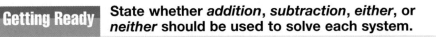

Getting Ready State whether *addition*, *subtraction*, *either*, or *neither* should be used to solve each system.

Sample: $3x + y = 6$
$4x + y = 7$

Solution: $3x + y = 6$
$(-)\ 4x + y = 7$ Subtraction should be used because it
$\overline{\quad -x \quad = -1}$ eliminates the y terms.

3. $x - 2y = 3$
$3x + 2y = 8$

4. $x + y = 5$
$x - y = 9$

5. $4x - 5y = 18$
$2x - 5y = 1$

6. $x + 7y = 3$
$2x + y = 4$

Use elimination to solve each system of equations. *(Examples 1–3)*

7. $x + y = 7$
$x - y = 9$

8. $x + y = 5$
$2x + y = 4$

9. $x + 2y = 6$
$3x - 2y = 2$

10. $-7x + 4y = 6$
$7x + y = 19$

11. $3x - 2y = 10$
$3x - 5y = 1$

12. $4x = 5y - 9$
$4x + 3y = -1$

13. Number Theory   The sum of two numbers is 42. The greater number is three more than twice the other number. *(Example 4)*

    a. Write a system of equations to represent the problem.

    b. What are the numbers?

Exercises • • • • • • • • • • • • • • • • •

Practice

Use elimination to solve each system of equations.

14. $x + y = 3$
$x - y = 3$

15. $x - y = 9$
$x + y = 19$

16. $x + 4y = 6$
$x + 3y = 5$

17. $x + y = 10$
$3x + y = 0$

18. $3x + y = 13$
$2x - y = 2$

19. $9x + 2y = 12$
$7x - 2y = -12$

20. $-3x + 4y = 5$
$3x - 2y = -7$

21. $2x - 3y = 19$
$2x + 3y = 13$

22. $11x - 3y = 10$
$-2x + 3y = 8$

23. $2x - y = -8$
$x - y = 7$

24. $2x - 5y = 9$
$2x - 3y = 11$

25. $3x - 2y = 10$
$2x - 2y = 5$

26. $x + 2y = 8$
$3x = 6 - 2y$

27. $x + y = 7$
$21 - 3x = 3y$

28. $y = 3x - 5$
$x - y = 13$

29. $x = y - 7$
$2x - 5y = -2$

30. $5x - y = -3$
$2y = 10x - 7$

31. $x = 10 - y$
$2x - y = -4$

32. What is the solution of the system $3x + 5y = -16$ and $3x - 2y = -2$?

33. Find the solution of the system $5s + 4t = 12$ and $3s = 4 + 4t$.

Applications and Problem Solving

Real World

34. Spas   The Feel Better Spa has two specials for new members. They can receive 3 facials and 5 manicures for $114 or 3 facials and 2 manicures for $78. What are the prices for facials and manicures?

35. Entertainment   The cost of admission to Water World was $137.50 for 13 children and 2 adults in one party. The admission was $103.50 for 9 children and 2 adults in another party.

    a. Write a system of equations to represent this problem.

    b. How much is admission for children and adults?

36. **Number Theory** The difference between two numbers is 15. The greater number is two less than twice the other number.
   a. Write a system of equations to represent this situation.
   b. Find the numbers.

37. **Number Theory** The sum of a number and twice a second number is 29. The second number is ten less than three times the other number.
   a. Write a system of equations to represent this situation.
   b. Find the numbers.

38. **Critical Thinking** The solution of a system of equations is $(-2, 5)$, and the first equation is $3x + 4y = 14$.
   a. Write a second equation for this system.
   b. Is this the only equation that could be in the system? Explain.

Mixed Review

39. Use substitution to solve the system of equations. *(Lesson 13–3)*

$$y = -3x - 5$$
$$4x + y = 6$$

State whether each system is *consistent and independent*, *consistent and dependent*, or *inconsistent*. *(Lesson 13–2)*

40.   41.   42.

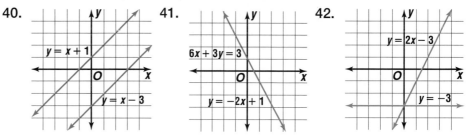

Solve each inequality. *(Lesson 12–4)*

43. $3y < 12$ 44. $6 - 4m \leq 22$ 45. $9 - 3x > 15$

Find each product. *(Lesson 9–4)*

46. $(m + 4)(m + 7)$ 47. $(a - 3)(a - 3)$ 48. $(2x - 2)(3x + 6)$

49. Write the point-slope form of an equation for a line passing through points at $(6, 1)$ and $(7, -4)$. *(Lesson 7–2)*

50. **Standardized Test Practice** In a scale model of a classic car, the front bumper measures 1.5 inches. If the actual bumper measures 3 feet, what is the scale of the model? *(Lesson 5–2)*

   A 1 in. = 16 in.   **B** 1 in. = 18 in.
   C 1 in. = 24 in.   **D** 1 in. = 43 in.

13-5 Elimination Using Multiplication

What You'll Learn

You'll learn to solve systems of equations by the elimination method using multiplication and addition.

Why It's Important

Transportation
Systems of equations can be used to determine the rate of a river's current. *See Example 5.*

You have learned when and how to solve systems of equations by graphing, substitution, and elimination using addition or subtraction. The best times to use these methods are summarized in the table below.

Method	The Best Time to Use	Example
graphing	if the equations are easy to graph, or if you want an estimate because graphing usually does not give an exact solution	$y = x + 4$ $y = -x + 1$
substitution	if one of the variables in either equation has a coefficient of 1 or -1	$y = 2x + 3$ $3x - 5y = 8$
elimination using addition	if the coefficients of one variable are additive inverses	$4x - y = 9$ $-4x + 3y = 6$
elimination using subtraction	if the coefficients of one variable are the same	$2x + y = 4$ $3x + y = 1$

Sometimes neither of the variables in a system of equations can be eliminated by simply adding or subtracting the equations. In this case, another method is to multiply one or both of the equations by some number so that adding or subtracting eliminates one of the variables.

Examples

1 **Use elimination to solve the system of equations.**

$2x + 3y = 9$
$8x - 5y = 19$

Multiply the first equation by -4 so that the x terms are additive inverses.

$2x + 3y = 9$ **Multiply by -4.** $-8x - 12y = -36$
$8x - 5y = 19$ $\underline{(+)\ 8x -\ \ 5y =\ \ \ \ 19}$
$$0 - 17y = -17$$
$$\frac{-17y}{-17} = \frac{-17}{-17}$$
$$y = 1$$

Now find the value of x by replacing y with 1 in either equation.

$2x + 3y = 9$
$2x + 3(1) = 9$ *Replace y with 1.*
$2x + 3 = 9$
$2x + 3 - 3 = 9 - 3$ *Subtract 3 from each side.*
$2x = 6$
$\dfrac{2x}{2} = \dfrac{6}{2}$ *Divide each side by 2.*
$x = 3$

The solution of this system of equations is (3, 1). *Check this solution.*

2 Use elimination to solve the system of equations.
$5x - 6y = 25$
$4x + 2y = 3$

Multiply the second equation by 3 so the y terms are additive inverses.

$5x - 6y = 25$
$4x + 2y = 3$ 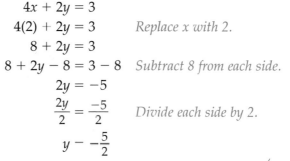 Multiply by 3.

$5x - 6y = 25$
$(+)\ 12x + 6y = \quad 9$

$17x + 0\ = 34$

$\dfrac{17x}{17} = \dfrac{34}{17}$

$x = 2$

Now find the value of y by replacing x with 2 in either equation.

$4x + 2y = 3$
$4(2) + 2y = 3$ *Replace x with 2.*
$8 + 2y = 3$
$8 + 2y - 8 = 3 - 8$ *Subtract 8 from each side.*
$2y = -5$
$\dfrac{2y}{2} = \dfrac{-5}{2}$ *Divide each side by 2.*
$y - -\dfrac{5}{2}$

The solution of the system of equations is $\left(2, -\dfrac{5}{2}\right)$. *Check this solution.*

Your Turn

a. $2x - y = 16$
     $5x + 3y = -4$

b. $4x - 3y = -2$
     $2x + 7y = 16$

c. $2x - 6y = -8$
     $4x - 3y = 11$

Example

Transportation Link

Real World

3 Chris and Alana each have prepaid cards for the Metro train. In May, Chris took the train 15 times during rush hour and 29 non-rush hour times, and it cost $64.80. Alana took the train 30 times during rush hour and 14 non-rush hour times, and it cost $76.80. What are the rush hour and non-rush hour fares?

$r =$ the rush hour fare $n =$ the non-rush hour fare
$15r + 29n = 64.80$ *← Chris' expenses*
$30r + 14n = 76.80$ *← Alana's expenses*

Multiply the first equation by -2 to eliminate the r terms.

Prerequisite Skills Review
Operations with Decimals, p. 684

$15r + 29n = 64.80$
$30r + 14n = 76.80$ Multiply by –2.

$-30r - 58n = -129.60$
$(+)\ 30r + 14n = \quad 76.80$

$0\ - 44n = \ -52.80$

$\dfrac{-44n}{-44} = \dfrac{-52.80}{-44}$

$n = 1.20$

(continued on the next page)

**The Metrorail,
Washington, D.C.**

Now find the value of r by replacing n with 1.20 in either equation.

$$15r + 29n = 64.80$$
$$15r + 29(1.20) = 64.80 \qquad \textit{Replace n with 1.20.}$$
$$15r + 34.80 = 64.80$$
$$15r + 34.80 - 34.80 = 64.80 - 34.80 \quad \textit{Subtract 34.80 from each side.}$$
$$15r = 30$$
$$\frac{15r}{15} = \frac{30}{15} \qquad \textit{Divide each side by 15.}$$
$$r = 2$$

The solution is (2, 1.20). This means that the rush hour fare is $2.00 and the non-rush hour fare is $1.20. Do these fares seem reasonable? *Check this solution by substituting (2, 120) for (r, n) in the original equations.*

Sometimes it is necessary to multiply each equation by a different number and then add in order to eliminate one of the variables.

Example **4** **Use elimination to solve the system of equations.**

$$3x + 4y = -25$$
$$2x - 3y = 6$$

Multiply the first equation by 2 and the second equation by -3 so the x terms are additive inverses.

$$3x + 4y = -25 \quad \boxed{\text{Multiply by 2.}} \qquad 6x + 8y = -50$$

$$2x - 3y = 6 \quad \boxed{\text{Multiply by -3.}} \qquad \underline{(+)-6x + 9y = -18}$$
$$0 + 17y = -68$$
$$\frac{17y}{17} = \frac{-68}{17}$$
$$y = -4$$

Now find the value of x by replacing y with -4 in either equation.

$$3x + 4y = -25$$
$$3x + 4(-4) = -25 \qquad \textit{Replace y with } -4.$$
$$3x - 16 = -25$$
$$3x - 16 + 16 = -25 + 16 \quad \textit{Add 16 to each side.}$$
$$3x = -9$$
$$\frac{3x}{3} = \frac{-9}{3} \qquad \textit{Divide each side by 3.}$$
$$x = -3$$

The solution of the system of equations is $(-3, -4)$. You could also have solved the system of equations by multiplying the first equation by 3 and the second equation by 4. *Why?*

d. $5x + 3y = 12$
$4x - 5y = 17$

e. $4x + 3y = 19$
$3x - 4y = 8$

Systems of equations can be used to solve rate problems.

Example ⑤

Transportation Link

Real World

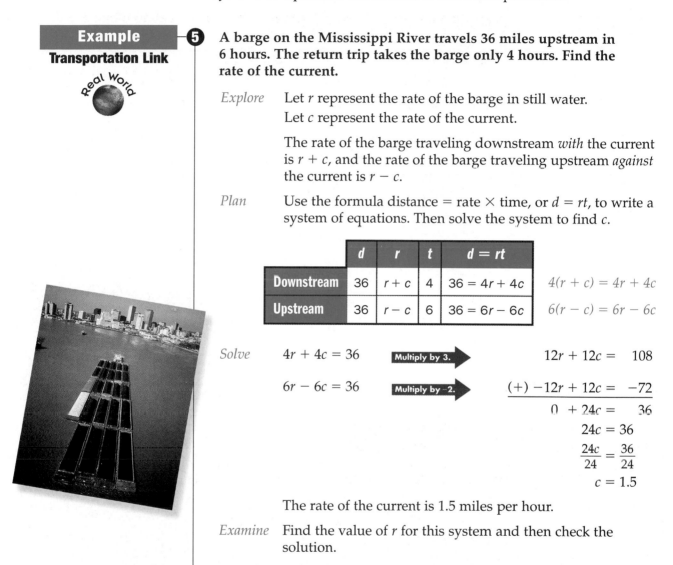

A barge on the Mississippi River travels 36 miles upstream in 6 hours. The return trip takes the barge only 4 hours. Find the rate of the current.

Explore   Let r represent the rate of the barge in still water.
Let c represent the rate of the current.

The rate of the barge traveling downstream *with* the current is $r + c$, and the rate of the barge traveling upstream *against* the current is $r - c$.

Plan   Use the formula distance = rate × time, or $d = rt$, to write a system of equations. Then solve the system to find c.

	d	r	t	$d = rt$	
Downstream	36	$r + c$	4	$36 = 4r + 4c$	$4(r + c) = 4r + 4c$
Upstream	36	$r - c$	6	$36 = 6r - 6c$	$6(r - c) = 6r - 6c$

Solve $4r + 4c = 36$ **Multiply by 3.** → $12r + 12c = \quad 108$

$6r - 6c = 36$ **Multiply by –2.** → $\underline{(+) -12r + 12c = \quad -72}$

$0 + 24c = \quad 36$

$24c = 36$

$\dfrac{24c}{24} = \dfrac{36}{24}$

$c = 1.5$

The rate of the current is 1.5 miles per hour.

Examine   Find the value of r for this system and then check the solution.

Check for Understanding

Communicating Mathematics

Study the lesson. Then complete the following.

1. **Explain** when you would use elimination with multiplication to solve a system of equations.

2. **Write** a system of equations in which you can eliminate the variable x by multiplying one equation by 3 and then adding the equations.

3. **Match** each system of equations with the method listed below that would best solve the system.

  a. $y = 3x$
  $6x - 2y = 9$

  b. $2x + 4y = 5$
  $3x - 2y = -3$

  c. $10x - 2y = 12$
  $-10x + 3y = -13$

  d. $x + 7y = 5$
  $3x + 7y = 1$

  | substitution |
  | elimination (+) |
  | elimination (−) |
  | elimination (×) |

Guided Practice

⏱ Getting Ready Explain the steps you would take to eliminate the variable *y* in each system of equations.

Sample: $4x + 3y = -7$
$3x + y = 1$

Solution: Multiply second equation by −3. Then add.

4. $2x + 6y = 10$
$5x + 3y = 1$

5. $3x - 2y = 3$
$8x + y = 27$

6. $2x + y = 4$
$7x - 4y = 29$

Use elimination to solve each system of equations. *(Examples 1–4)*

7. $x + 2y = 6$
$-4x + 5y = 2$

8. $x - 5y = 0$
$2x - 3y = 7$

9. $5x + 8y = 1$
$2x - 4y = -14$

10. $x + 2y = 5$
$3x + y = 7$

11. $3x - y = -5$
$6x - 2y = 8$

12. $2x - 5y = -10$
$7x - 3y = -6$

13. Entertainment A fishing boat traveled 48 miles upstream in 4 hours. Returning at the same rate, it took 3 hours. *(Example 5)*

  a. Find the rate of the current.

  b. Find the rate of the boat in still water.

Exercises

Practice **Use elimination to solve each system of equations.**

14. $x + 8y = 3$
$2x + 3y = -7$

15. $4x + y = 8$
$x - 7y = 2$

16. $2x + y = 6$
$3x - 7y = 9$

17. $2x + 5y = 13$
$4x - 3y = -13$

18. $-3x + 2y = 10$
$-2x - y = -5$

19. $3x - 2y = 0$
$x + 6y = 5$

20. $-5x + 8y = 21$
$10x + 3y = 15$

21. $6x - 4y = 11$
$2x + 2y = 7$

22. $6x - 5y = -6$
$3x + 10y = 72$

23. $2x - 7y = 9$
$-3x + 4y = 6$

24. $5x + 3y = 4$
$-4x + 5y = -18$

25. $6x - 3y = 7$
$18x - 9y = 21$

26. $7x - 4y = 16$
$2x + 3y = 17$

27. $-5x - 2y = 12$
$11x - 5y = -17$

28. $7x - 3y = -9$
$5x - 4y = -25$

29. What is the solution of the system $9x + 8y = 7$ and $18x - 15y = 14$?

30. Use elimination to find the solution of the system $\frac{1}{3}x - y = -1$ and $\frac{1}{5}x - \frac{2}{5}y = -1$.

Determine the best method to solve each system of equations. Then solve.

31. $x + 2y = 6$
$2x - 4y = -20$

32. $y = 3x - 2$
$x - 5y = -4$

33. $y = \frac{1}{2}x + 6$
$2x - 4y = 8$

Applications and Problem Solving

Real World

34. Traveling A riverboat on the Mississippi River travels 30 miles upstream in 2 hours and 30 minutes. The return trip downstream takes only 2 hours.

   a. Find the rate of the current.

   b. Find the rate of the riverboat in still water.

35. Entertainment The science club purchased tickets for a magic show. They paid $108 for 6 tickets in section A and 10 tickets in section B. The following week, they paid $104 for 4 tickets in section A and 12 tickets in section B.

   a. Write a system of equations to represent the problem.

   b. What are the prices of the tickets in section A and B?

36. Critical Thinking The solution of the system $5x + 6y = -9$ and $10x + 8y = c$ is $(9, b)$. Find values for c and b.

Mixed Review

Use elimination to solve each system of equations. *(Lesson 13–4)*

37. $x - y = 8$
$x + y = 6$

38. $x + y = -4$
$4x + y = 2$

39. $2x + 4y = 7$
$2x - y = 2$

40. Use substitution to solve $x = 5 + 2y$ and $3x - 4y = 3$. *(Lesson 13–3)*

Solve each inequality. *(Lesson 12–3)*

41. $-4m > 16$

42. $\frac{x}{2} < 6$

43. $\frac{3a}{4} \geq -3$

44. Use the Quadratic Formula to solve $n^2 - 5n + 12 = 0$. *(Lesson 11–6)*

45. Sports Suppose a baseball player hits a fly ball into right field above first base. The player hits the ball with his bat at a height of 2 feet above the ground and sends the ball at an upward velocity of 25 meters per second. The height of the ball in t seconds can be approximated by the formula $h = -5t^2 + 25t + 2$. If the ball is not caught, how long before it hits the ground? Round to the nearest second. *(Lesson 11–3)*

46. Standardized Test Practice The graph shows the top turkey producing states. What percent of turkeys does the state of North Carolina produce? Round to the nearest percent. *(Lesson 5–3)*

   A 5%   **B** 19%

   C 33%   **D** 45%

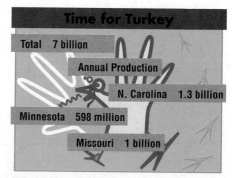

Time for Turkey

Total   7 billion

Annual Production

N. Carolina   1.3 billion

Minnesota   598 million

Missouri   1 billion

Source: U.S. Census Bureau

Matrices

In this investigation, you will use matrices to solve systems of equations. Consider the following problem.

Look Back

Matrices:
Pages 80–81

Kristin and Scott are sales associates at Crazy Computers. On Monday, Kristin sold 4 Model A computers and 1 Model B computer for a total of $6395. Scott sold 2 Model A computers and 3 Model B computers for a total of $7195. What was the price of each computer?

Investigate

You can write a system of equations to solve this problem. You may want to use matrices to solve this system of equations.

1. Write each equation in standard form. Then write the coefficients and constants of the system in a special matrix called an **augmented matrix**. Write the coefficients in the matrix to the left of the dashed line. Write the constants to the right of the dashed line.

$$3x + 5y = 7$$
$$6x - 1y = -8$$
$$\Rightarrow \begin{bmatrix} 3 & 5 & | & 7 \\ 6 & -1 & | & -8 \end{bmatrix}$$

   a. Write $2x - 3y = 5$ and $x + 4y = -7$ as an augmented matrix.

   b. Write $x + 6y = -1$ and $y = 5$ as an augmented matrix.

2. You can use **row operations** to simplify an augmented matrix.

   • Switch any two rows.
   • Replace any row with a nonzero multiple of that row.
   • Replace any row with the sum or difference of that row and a multiple of another row.

   a. Switch rows 1 and 2 of
   $\begin{bmatrix} 0 & -5 & | & 2 \\ 1 & 3 & | & 4 \end{bmatrix}$.

   b. Multiply row 2 of
   $\begin{bmatrix} 1 & -3 & | & 5 \\ 0 & 2 & | & -8 \end{bmatrix}$ by 0.5.

   c. Replace row 2 of $\begin{bmatrix} 1 & -2 & | & -5 \\ -1 & 3 & | & 4 \end{bmatrix}$ with the sum of rows 1 and 2.

   d. In $\begin{bmatrix} 1 & -3 & | & 2 \\ 0 & 3 & | & -5 \end{bmatrix}$, what row operation would you use to get a 0 in the second column of row 1?

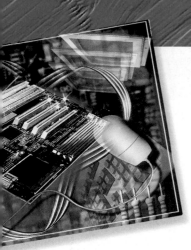

3. The goal of using row operations to solve a system of equations is to get to the **identity matrix**, which is $\begin{bmatrix} 1 & 0 \\ 0 & 1 \end{bmatrix}$. Suppose the solution of a system of equations is $\begin{bmatrix} 1 & 0 & | & -4 \\ 0 & 1 & | & 2 \end{bmatrix}$. What does this matrix represent?

4. Solve the problem. Let x = the cost of the Model A computer and let y = the cost of the Model B computer.

$$4x + 1y = 6395 \quad \leftarrow Kristin's\ sales$$
$$2x + 3y = 7195 \quad \leftarrow Scott's\ sales$$

Step 1 Write the system as an augmented matrix.
$\begin{bmatrix} 4 & 1 & | & 6395 \\ 2 & 3 & | & 7195 \end{bmatrix}$

Step 2 Multiply row 2 by 2.
$2 \times 2 = 4; 2 \times 3 = 6; 2 \times 7195 = 14{,}390$
$\begin{bmatrix} 4 & 1 & | & 6395 \\ 4 & 6 & | & 14{,}390 \end{bmatrix}$

Step 3 Replace row 2 with the difference of row 1 and row 2. $4 - 4 = 0; 1 - 6 = -5;$ $6395 - 14{,}390 = -7995$
$\begin{bmatrix} 4 & 1 & | & 6395 \\ 0 & -5 & | & -7995 \end{bmatrix}$

Step 4 Divide row 2 by -5. $0 \div (-5) = 0;$ $-5 \div (-5) = 1; -7995 \div (-5) = 1599$
$\begin{bmatrix} 4 & 1 & | & 6395 \\ 0 & 1 & | & 1599 \end{bmatrix}$

Step 5 Replace row 1 with the difference of row 1 and row 2. $4 - 0 = 4; 1 - 1 = 0;$ $6395 - 1599 = 4796$
$\begin{bmatrix} 4 & 0 & | & 4796 \\ 0 & 1 & | & 1599 \end{bmatrix}$

Step 6 Divide row 1 by 4.
$4 \div 4 = 1; 0 \div 4 = 0; 4796 \div 4 = 1199$
$\begin{bmatrix} 1 & 0 & | & 1199 \\ 0 & 1 & | & 1599 \end{bmatrix}$

Extending the Investigation

In this extension, you will use matrices as a problem-solving tool.

1. Suppose Scott sells 5 Model TX laser printers and 6 Model DM ink jet printers for a total of $4185. Kristin sells 6 Model TX laser printers and 5 Model DM ink jet printers for a total of $4549. What is the cost of each printer?

2. At Crazy Computers, there are two prices for computer software, Price A and Price B. Suppose Aislyn purchases 3 Price A software and 1 Price B software for a total of $250 and Devin purchases 5 Price A software and 2 Price B software for a total of $450. What are the two prices?

Presenting Your Investigation

Write a real-world problem that can be solved by using matrices and a system of equations. Present your problem and solution on a poster board.

*inter*NET
CONNECTION
Investigation For more information on matrices, visit: www.algconcepts.glencoe.com

13-6 Solving Quadratic-Linear Systems of Equations

What You'll Learn

You'll learn to solve systems of quadratic and linear equations.

Why It's Important

Meteorology
Quadratic-linear systems of equations can be used to determine when airplanes will cross jet streams and experience turbulence. See Exercise 24.

In late November, the jet stream moving across North America could be described by the quadratic equation $y = \frac{1}{4}x^2 - 12$, where Chicago was at the origin. Suppose a plane's route is described by the linear equation $y = -\frac{1}{2}x$. What are the coordinates of the point at which turbulence will occur? *This problem will be solved in Example 7.*

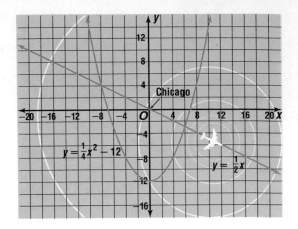

Like a linear system of equations, the solution of a **quadratic-linear** system of equations is the ordered pair that satisfies both equations. A quadratic-linear system can have 0, 1, or 2 solutions, as shown below.

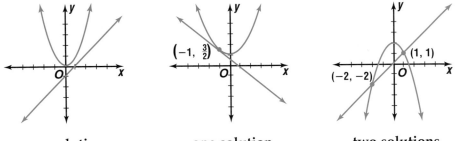

| no solution | one solution | two solutions |

You can solve quadratic-linear systems of equations by using some of the methods you used for solving systems of linear equations. One method is graphing.

Examples

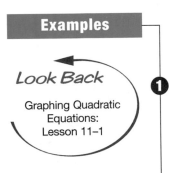

Look Back

Graphing Quadratic Equations: Lesson 11-1

Determine whether each system of equations has *one* solution, *two* solutions, or *no* solution by graphing. If the system has one solution or two solutions, name them.

1 $y = x^2$
$y = x + 2$

The graphs appear to intersect at $(-1, 1)$ and $(2, 4)$. Check this estimate by substituting the coordinates into each equation.

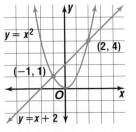

Check: $y = x^2$ $y = x + 2$
 $1 \overset{?}{=} (-1)^2$ $(x, y) = (-1, 1)$ $1 \overset{?}{=} -1 + 2$ $(x, y) = (-1, 1)$
 $1 = 1$ ✓ $1 = 1$ ✓

Check that the ordered pair (2, 4) satisfies both equations.

The solutions of the system of equations are $(-1, 1)$ and $(2, 4)$.

2 $y = 2x^2 + 5$
 $y = -x + 3$

Because the graphs do not
intersect, there is no solution
to this system of equations.

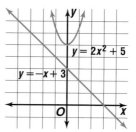

3 $y = -x^2 + 3$
 $y = 2x + 4$

The graphs appear to intersect
at $(-1, 2)$.

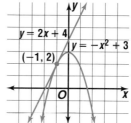

Check: $y = -x^2 + 3$ $y = 2x + 4$
 $2 \overset{?}{=} -(-1)^2 + 3$ $(x, y) = (-1, 2)$ $2 \overset{?}{=} 2(-1) + 4$ $(x, y) = (-1, 2)$
 $2 \overset{?}{=} -(1) + 3$ $2 \overset{?}{=} -2 + 4$
 $2 = 2$ ✓ $2 = 2$ ✓

The solution of the system of equations is $(-1, 2)$.

Your Turn

a. $y = x^2$ **b.** $y = 2x^2 + 1$ **c.** $y = -x^2 + 3$
 $y = x - 2$ $y = 1$ $y = -2x$

Chicago, Illinois

You can also solve quadratic-linear systems of equations by using the
substitution method.

Examples

4 **Use substitution to solve each system of equations.**
$y = -4$
$y = x^2 - 4$
Substitute -4 for y in the second equation. Then solve for x.
 $y = x^2 - 4$
 $-4 = x^2 - 4$ *Replace y with -4.*
$-4 + 4 = x^2 - 4 + 4$ *Add 4 to each side.*
 $0 = x^2$
 $0 = x$ *Take the square root of each side.*

The solution of the system of equations is $(0, -4)$.

(continued on the next page)

Check:

Sketch the graphs of the equations. The parabola and line appear to intersect at $(0, -4)$. The solution is correct.

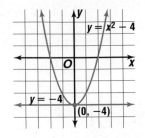

5 $y = x^2$
$y = -3$

Substitute -3 for y in the first equation.

$y = x^2$
$-3 = x^2$ *Replace y with -3.*
$\sqrt{-3} = x$ *Take the square root of each side.*

There is no real solution because the square root of a negative number is not a real number. *Check by graphing.*

6 $y = x^2 - 4x + 6$
$y = -x + 4$

Substitute $-x + 4$ for y in the first equation.

$y = x^2 - 4x + 6$
$-x + 4 = x^2 - 4x + 6$ *Replace y with $-x + 4$.*
$-x + 4 - 4 = x^2 - 4x + 6 - 4$ *Subtract 4 from each side.*
$-x = x^2 - 4x + 2$
$-x + x = x^2 - 4x + 2 + x$ *Add x to each side.*
$0 = x^2 - 3x + 2$
$0 = (x - 2)(x - 1)$ *Factor.*

$0 = x - 2$ or $0 = x - 1$ *Zero Product Property*
$x = 2$　　　　　$x = 1$

Substitute the values of x in either equation to find the corresponding values of y. *Choose the equation that is easier for you to solve.*

$y = -x + 4$　　　　　　　　　　$y = -x + 4$
$y = -2 + 4$ *Replace x with 2.*　　$y = -1 + 4$ *Replace x with 1.*
$y = 2$　　　　　　　　　　　　$y = 3$

The solutions of the system of equations are $(2, 2)$ and $(1, 3)$. The graph shows that the solutions are probably correct. *You can also check by substituting the ordered pairs into the original equations.*

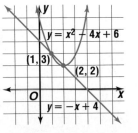

d. $y = x^2 - 12$
$\quad x = 3$

e. $y = -x^2$
$\quad y = x + 2$

f. $y = x^2 - 3x - 6$
$\quad y = 2x$

Example
Meteorology Link

Real World

7 Refer to the application at the beginning of the lesson. What are the coordinates of the point at which turbulence will occur?

Use substitution to solve the system $y = \frac{1}{4}x^2 - 12$ and $y = -\frac{1}{2}x$.

Substitute $-\frac{1}{2}x$ for y in the first equation.

$$y = \frac{1}{4}x^2 - 12$$

$$-\frac{1}{2}x = \frac{1}{4}x^2 - 12 \qquad \textit{Replace } y \textit{ with } -\frac{1}{2}x.$$

$$4\left(-\frac{1}{2}x\right) = 4\left(\frac{1}{4}x^2 - 12\right) \qquad \textit{Multiply each side by 4.}$$

$$-2x = x^2 - 48$$

$$-2x + 2x = x^2 - 48 + 2x \qquad \textit{Add 2x to each side.}$$

$$0 = x^2 + 2x - 48$$

$$0 = (x + 8)(x - 6) \qquad \textit{Factor.}$$

$0 = x + 8 \quad$ or $\quad 0 = x - 6 \qquad \textit{Zero Product Property}$
$x = -8 \qquad\qquad x = 6$

Substitute the values of x to find the corresponding values of y.

$$y = -\frac{1}{2}x \qquad\qquad\qquad y = -\frac{1}{2}x$$

$y = -\frac{1}{2}(-8)$ or $4 \quad$ *Replace x* $y = -\frac{1}{2}(6)$ or -3 *Replace x*
                          with -8.   *with 6.*

The solutions of the system are $(-8, 4)$ and $(6, -3)$. This means that turbulence will occur as the plane passes through points having these coordinates. *Check by substituting the coordinates into each equation and by looking at the graph at the beginning of the lesson.*

interNET
CONNECTION

Data Update For the latest information about the jet stream, visit: www.algconcepts. glencoe.com

Check for Understanding

Communicating Mathematics

Math Journal

Study the lesson. Then complete the following.

1. **State** the solution of the system of equations shown at the right.

2. **Sketch** a system of quadratic-linear equations that has solutions $(-2, 1)$ and $(2, 1)$.

3. **List** the different methods of solving linear and quadratic-linear systems of equations. Describe the situation in which each method is most useful.

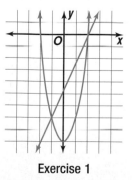

Exercise 1

Solve each system of equations by graphing. *(Examples 1–3)*

4. $y = x^2$
$y = -2x - 1$

5. $y = x^2$
$y = 2x$

Use substitution to solve each system of equations. *(Examples 4–6)*

6. $y = x^2 - 6$
$y = -5x$

7. $y = x^2 + 5$
$y = -3$

8. Business Students in an Algebra I class at Banneker High School are simulating the start-up of a company. The income y can be described by the equation $y = \frac{1}{8}x^2$, where x represents the time in months. The expenses have been growing at a constant rate and can be defined by the equation $y = x$. When will income equal expenses? *(Example 7)*

Exercises

• • • • • • • • • • • • • • • • • • •

Practice

Solve each system of equations by graphing.

9. $y = x^2 - 4$
$y = 2x - 4$

10. $x = 2$
$y = x^2 + 1$

11. $y = 2x + 1$
$y = x^2 + 4$

12. $y = x^2 - 6$
$y = x - 4$

13. $y = -x^2 + 5$
$y = -x + 3$

14. $y = -x^2 + 4$
$y = \frac{1}{2}x + 5$

Use substitution to solve each system of equations.

15. $y = \frac{1}{8}x^2 + 5$
$x = -4$

16. $y = -3x^2$
$y = 2$

17. $y = \frac{1}{2}x^2$
$y = 3x + 8$

18. $y = \frac{1}{2}x^2 - 4$
$y = 3x + 4$

19. $y = x^2 + 3$
$y = -\frac{1}{2}x - 5$

20. $y = x^2 - x - 3$
$y = -1$

21. What is the solution of the system $y = x^2 - 9$ and $y = 3x - 9$?

22. Find the solution of the system $y = -x^2 + 2x - 3$ and $y = -2x + 1$.

Applications and Problem Solving

23. Geometry Four corners are cut from a rectangular piece of cardboard that is 10 feet by 4 feet. The cuts are x feet from the corners, as shown in the figure at the right. After the cuts are made, the sides of the rectangle are folded to form an open box. The area of the bottom of the box is 16 square feet.

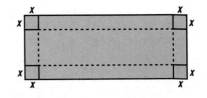

a. Write two equations to represent the area A of the bottom of the box.

b. What are the dimensions of the box?

c. What is the volume of the box?

24. **Meteorology** Refer to the application at the beginning of the lesson. If a plane's route is described by the linear equation $y = x - 14$, will it experience turbulence by the jet stream that is described by the quadratic equation $y = \frac{1}{4}x^2 - 12$? Explain.

25. **Critical Thinking** Suppose the perimeter and area of a square have the same measure. Find the length of the sides of the square. Explain how you can use a system of quadratic-linear equations to find the answer.

Mixed Review

26. **Sports** Diego kayaks 16 miles downstream in 2 hours. It takes him 8 hours to make the return trip. What is the rate of the current? *(Lesson 13–5)*

27. Solve the system $4x + 3y = 5$ and $-3x - 2y = -4$ by using elimination. *(Lesson 13–4)*

28. Describe how the graph of $y = -(x - 3)^2$ changes from its parent graph of $y = x^2$. *(Lesson 11–2)*

Solve each equation. *(Lesson 3–7)*

29. $2 + |x| = 9$

30. $|y| + 4 = 3$

31. $12 = 6 + |w + 2|$

32. **Standardized Test Practice** Which expression shows *3 less than the product of 5 and d*? *(Lesson 1–1)*

    A $3 - 5d$ **B** $5d + 3$ **C** $5d - 3$ **D** $5 + d - 3$

Quiz 2   Lessons 13–4 through 13–6

Use elimination to solve each system of equations. *(Lessons 13–4 & 13–5)*

1. $x + y = 6$
   $2x - y = 6$

2. $2x - 3y = -9$
   $x - 2y = -5$

3. $3x - 5y = 8$
   $4x - 7y = 10$

4. Use substitution to find the solution of the system $y = x^2 + 5$ and $y = 6x - 3$. *(Lesson 13–6)*

5. **Number Theory** The difference between two numbers is 38. The greater number is three times the other number minus two. *(Lesson 13–4)*
   a. Write a system of equations to represent the situation.
   b. What are the numbers?

Extra Practice   See p. 719.   **Lesson 13–6** Solving Quadratic-Linear Systems of Equations

13-7 Graphing Systems of Inequalities

What You'll Learn

You'll learn to solve systems of inequalities by graphing.

Why It's Important

Money Systems of inequalities can be useful in helping you get the most for your money.
See Example 3.

Thomas Edison was a newsboy during the Civil War. One day, he persuaded the editor to give him 300 copies of the paper instead of the usual 100. He went to the train station where people were eager for news of the war. He was able to sell the papers for 10¢ and 25¢ instead of the usual 5¢. If he wanted to earn at least $52, how many papers could he have sold for 10¢ and 25¢?

x = the number of 10¢ papers y = the number of 25¢ papers

The following **system of inequalities** can be used to represent the conditions of this problem.

$x + y \leq 300$ *He can sell as many as 300 papers.*
$0.10x + 0.25y \geq 52$ *He wants to earn at least $52.*

Since both x and y represent the number of papers, neither can be a negative number. Thus, $x \geq 0$ and $y \geq 0$. The solution of the system is the set of all ordered pairs that satisfy both inequalities and lie in the first quadrant. The solution can be determined by graphing each inequality in the same coordinate plane as shown below.

Look Back

Graphing Linear Inequalities in Two Variables: Lesson 12–7

Recall that the graph of each inequality is called a *half-plane*.

- Any point in the yellow region satisfies $x + y \leq 300$.
- Any point in the blue region satisfies $0.10x + 0.25y \geq 52$.
- Any point in the green region satisfies both inequalities. The intersection of the two half-planes represents the solution to the system of inequalities.

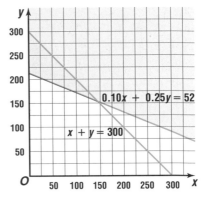

- The graphs of $x + y = 300$ and $0.10x + 0.25y = 52$ are the boundaries of the region and are included in the graph of the system.

This solution is a region that contains the graphs of an infinite number of ordered pairs. An example is (50, 225). This means that Edison could have sold 50 papers at 10¢ and 225 papers at 25¢ to earn at least $52.

Check:

$x + y \leq 300$
$50 + 225 \stackrel{?}{\leq} 300$ $(x, y) = (50, 225)$
$275 \leq 300$ ✓

$0.10x + 0.25y \geq 52$
$0.10(50) + 0.25(225) \stackrel{?}{\geq} 52$ $(x, y) = (50, 225)$
$61.25 \geq 52$ ✓

Remember that the boundary lines of inequalities are only included in the solution if the inequality symbol is greater than or equal to, $\geq$, or less than or equal to, $\leq$.

Examples

Solve each system of inequalities by graphing. If the system does not have a solution, write *no solution*.

1 $y \leq -1$
$y > x - 3$

The solution is the ordered pairs in the intersection of the graphs of $y \leq -1$ and $y > x - 3$. The region is shaded in green at the right. The graphs of $y = -1$ and $y = x - 3$ are the boundaries of this region. The graph of $y = x - 3$ is a dashed line and is not included in the solution of the system. *Choose a point and check the solution.*

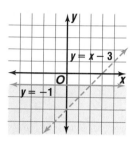

2 $2y < x - 2$
$-3x + 6y \geq 12$

The graphs of $2y = x - 2$ and $-3x + 6y = 12$ are parallel lines. *Check this.*

Because the regions in the solution of $2y < x - 2$ and $-3x + 6y \geq 12$ have no points in common, the system of inequalities has no solution.

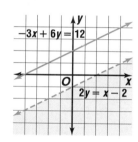

Your Turn

a. $x \geq -1$
   $y < x - 2$

b. $y \geq 4x + 5$
   $x \mid y \geq 3$

Example

Spending Link

Luisa has $96 to spend on gifts for the holidays. She must buy at least 9 gifts. She plans to buy puzzles that cost $8 or $12. How many of each puzzle can she buy?

x = the number of $8 puzzles y = the number of $12 puzzles

The following system of inequalities can be used to represent the conditions of this problem.

$x + y \geq 9$ *She wants to buy at least 9 puzzles.*
$8x + 12y \leq 96$ *The total cost must be no more than $96.*

Because the number of puzzles she can buy cannot be negative, both $x \geq 0$ and $y \geq 0$.

(continued on the next page)

The solutions are all of the ordered pairs in the intersection of the graphs of these inequalities. Only the first quadrant is used because $x \geq 0$ and $y \geq 0$.

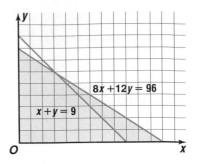

Any point in the region is a possible solution. For example, since the point at (9, 1) is in the region, Luisa could buy 9 puzzles for $8 and 1 for $12. The greatest number of gifts that she could buy would be 12 puzzles for $8 and no puzzles for $12.

A graphing calculator can be helpful in solving systems of inequalities or in checking solutions.

Graphing Calculator Tutorial
See pp. 724–727.

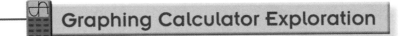

Graphing Calculator Exploration

Use a graphing calculator to solve the system of inequalities.

$y \geq 4x - 3$
$y \leq -2x + 1$

The graphing calculator graphs equations and shades above the first equation entered and below the second equation entered. (Note that inequalities that have > or ≥ are lower boundaries and inequalities that have < or ≤ are upper boundaries.)

Step 1 Press 2nd [DRAW] 7 to choose the SHADE feature.

Step 2 Enter the equation that is the lower boundary of the region to be shaded, 4 X,T,θ,n − 3.

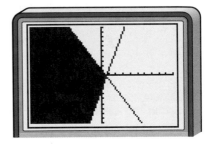

Step 3 Press , and enter the equation that is the upper boundary of the region to be shaded, (−) 2 X,T,θ,n + 1) ENTER.

Try These

Use a graphing calculator to graph each system of inequalities. State one possible solution.

1. $y \geq x - 1$
$y < x + 1$

2. $y \geq x - 2$
$y \leq 3$

3. $y > 2x$
$y \leq -x + 1$

Communicating Mathematics

Study the lesson. Then complete the following.

1. **Compare and contrast** the solution of a system of equations and the solution of a system of inequalities.

2. **Sketch** a system of linear equations that has no solution.

3. **You Decide?** Kyle says that the solution of $y \geq 2x + 2$ and $y \leq -x - 1$ is all of the points in region B. Tarika says that the solution is all of the points in region D. Who is correct? Explain.

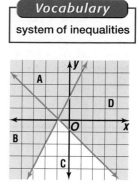

Exercise 3

Guided Practice

Getting Ready State whether each ordered pair is a solution of the system of inequalities $x \geq -2$ and $y \leq 5$.

Sample: $(-3, 5)$ **Solution:** $x \geq -2$
$$-3 \overset{?}{\geq} -2 \quad \textit{Replace x with } -3.$$
$(-3, 5)$ is not a solution.

4. $(6, -1)$ 5. $(8, 7)$ 6. $(0, 0)$

Solve each system of inequalities by graphing. If the system does not have a solution, write **no solution**. *(Examples 1 & 2)*

7. $x \geq -1$
$ y \geq -4$

8. $y \leq 5$
$ y \geq -x + 1$

9. $x + y > 2$
$ y < x + 6$

10. $y \geq \frac{1}{2}x - 2$
$ x - y < 3$

11. **Shopping** Nathaniel must buy two types of cookies for a banquet dinner. He only has $25 to spend and needs at least 6 dozen cookies. The grocery store has small sugar cookies for $3 a dozen and chocolate chip cookies for $4 a dozen. How many of each type cookie can he buy? List three possible solutions. *(Example 3)*

Exercises

Practice

Solve each system of inequalities by graphing. If the system does not have a solution, write **no solution**.

12. $x < 2$
$ y > -1$

13. $y > 0$
$ x \leq 0$

14. $y > 2$
$ y > -x + 2$

15. $y \geq x$
$ y \leq -x$

16. $y \leq 2x$
$ y \geq -3x$

17. $y \geq -3$
$ y < -3x + 1$

18. $x > 2$
$ y > -2x + 3$

19. $y \leq 2x - 1$
$ y \geq 2x + 2$

20. $x + y \geq 2$
$ x + y \leq 6$

21. $y \leq 2x + 2$
$ 2x + y \leq 4$

22. $y + 4 < x$
$ 2y + 4 > -3x$

23. $2y + x < 4$
$ 3x - y > 1$

24. Use the graphs of $y \geq x - 3$ and $y \geq -x - 1$ to determine the solution of the system.

25. Find the solution of the system $3x + 2y \leq -6$ and $y > x + 2$ by graphing.

26. Basketball Nykia must score at least 20 points in the last game of the season to tie the school record for total points in a season. Usually she has no more than 40 opportunities to shoot the ball (both field goals and free throws combined) and makes half of her shots. What combination of shots can Nykia make to score the 20 points? (*Hint*: Field goals are worth 2 points and free throws are worth 1 point.)

27. Communication A long-distance carrier offers three fee plans to their customers.

• 15 cents a minute with no monthly service charge
• 10 cents a minute and a monthly service charge of $5.25
• 5 cents a minute and a monthly service charge of $7.50

Depending on how much people use the phone each month, which plan should they select?

28. Critical Thinking Write an inequality involving absolute value that has no solution.

Determine whether each system of equations has *one* solution, *two* solutions, or *no* solution by graphing. If the system has one or two solutions, name them. *(Lesson 13–6)*

29. $y = x^2$
   $y = 4$

30. $y = x^2 + 3$
   $y = 2x - 1$

31. $y = -x^2 + 2$
   $y = -x + 5$

32. Use elimination to find the solution of the system $5x + 4y = -2$ and $3x - 2y = 1$. *(Lesson 13–5)*

33. Mixtures The Coffee Hut mixes coffee beans that cost $5.25 per pound with coffee beans that cost $6.50 per pound. They need a mixture of 5 pounds of coffee beans that costs $5.50 per pound. How many pounds of each type of coffee bean should they use? *(Lesson 13–3)*

Solve each equation. *(Lesson 4–4)*

34. $-4w = 32$

35. $13 = 2.6a$

36. $\frac{2}{5}y = -8$

37. Standardized Test Practice Find $-15 + (-3) + (-7) + 6$. *(Lesson 2–3)*

   A -19 **B** 1 **C** 11 **D** 16

Extra Practice See p. 720.

Car Dealer

Do you work well with people? Do cars interest you? If so, you may consider a career as a car dealer. Car dealers must keep track of the number of cars they need to sell each week to ensure maximum profit.

Suppose a car dealer receives a profit of $500 for each mid-sized car m sold and $750 for each sport-utility vehicle s sold. The dealer must sell at least two mid-sized cars for each sport-utility vehicle and must earn at least $3500 per week. The table and graph shown represent this situation.

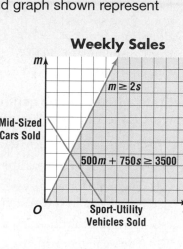

Weekly Sales

Situation	Inequality
At least 2 mid-sized cars must be sold for each sport-utility vehicle.	$m \geq 2s$
Earn a profit of at least $3500.	$500m + 750s \geq 3500$

1. Suppose a car dealer sells 2 sport-utility vehicles. How many mid-sized cars must be sold to earn at least $3500?
2. If a car dealer sells only one sport-utility vehicle, how many mid-sized cars must be sold to meet the goal?

FAST FACTS About Car Dealers

Working Conditions
- usually work in a comfortable environment
- evening, weekend, and holiday work
- work a 40-hour week, with some overtime

Education
- good communication skills and computer skills are helpful
- high school math and business classes
- college degree in business is helpful

Job Outlook

Expected Growth in Earnings

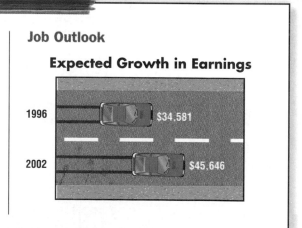

1996 $34,581

2002 $45,646

*inter*NET
CONNECTION
www.algconcepts.glencoe.com

Career Data For the latest information on car dealers, visit:

Understanding and Using the Vocabulary

After completing this chapter, you should be able to define each term, property, or phrase and give an example or two of each.

*inter***NET**
CONNECTION   **Review Activities**
For more review activities, visit:
www.algconcepts.glencoe.com

augmented matrix *(p. 578)*
consistent *(p. 554)*
dependent *(p. 554)*
digit problems *(p. 568)*
elimination *(p. 566)*

identity matrix *(p. 579)*
inconsistent *(p. 554)*
independent *(p. 554)*
quadratic-linear *(p. 580)*

row operations *(p. 578)*
substitution *(p. 560)*
system of equations *(p. 550)*
system of inequalities *(p. 586)*

State whether each sentence is *true* or *false*. If false, replace the underlined word(s) to make a true statement.

1. Inconsistent is the description used for a system of equations that has <u>no solution</u>.
2. A system of inequalities can be solved by <u>graphing</u> the inequalities.
3. The solution to a system of equations is the ordered pair that satisfies <u>two</u> of the equations in the system.
4. A system of linear equations that has <u>at least</u> one solution is called independent.
5. The exact coordinates of the point where two or more lines intersect can be determined by using <u>substitution</u>.
6. A set of two or more equations is called a <u>system of equations</u>.
7. <u>Dependent</u> systems of equations have infinitely many solutions.
8. The description of a system of linear equations depends on the number of <u>solutions</u>.
9. To solve a system of equations by <u>substitution</u>, one of the variables is eliminated by either adding or subtracting the equations.
10. A system of equations that has at least one solution can be described as <u>inconsistent</u>.

Skills and Concepts

Objectives and Examples	**Review Exercises**

• **Lesson 13–1** Solve systems of equations by graphing.

$y = x + 1$
$x + y = 3$

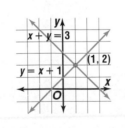

The solution is (1, 2).

Solve each system of equations by graphing.

11. $y = 4x$
    $y = -x + 5$

12. $x = y$
    $y = 2x - 3$

13. $y = \frac{1}{2}x + 3$
    $y = -x$

14. $y = 5$
    $2y = x + 3$

15. $x - 3y = -6$
    $y = -\frac{1}{2}x - 3$

16. $y = 3x + 1$
    $y = -3x + 1$

Objectives and Examples

Review Exercises

• **Lesson 13–2** Determine whether a system of equations has one solution, no solution, or infinitely many solutions by graphing.

$y = x + 5$
$y = -4x$

This system has one solution, $(-1, 4)$.

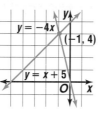

Determine whether each system of equations has *one* solution, *no* solution, or *infinitely many* solutions by graphing. If the system has one solution, name it.

17. $y = -x + 1$
$y = x - 5$

18. $y = 2x + 3$
$y = 2x - 2$

19. $x + y = -2$
$4x + 4y = -8$

20. $y = \frac{1}{2}x - 1$
$y = x$

• **Lesson 13–3** Solve systems of equations by the substitution method.

To solve $y = 3x$ and $2x + y = 10$, substitute $3x$ for y in the second equation.

$2x + 3x = 10$ $y = 3x$
$\quad\; 5x = 10$ $y = 3(2)$ or 6
$\quad\;\; x = 2$ The solution is $(2, 6)$.

Use substitution to solve each system of equations.

21. $y = 2x$
$4x + y = 3$

22. $3y = x$
$4x + 2y = 14$

23. $y = 3x + 1$
$3x - y = 4$

24. $y = x + 1$
$3x - 2y = -12$

25. $2x - 3y = 4$
$4x - 6y = 8$

26. $y - 3 = x$
$2x + y = 4$

• **Lesson 13–4** Solve systems of equations by the elimination method using addition and subtraction.

$\quad\;\; 3x - 4y = \;\;\; 5$ $3x - 4(-2) = 5$
$(-)\; 3x + 2y = -7$ $\quad\; 3x + 8 = 5$
$\overline{\qquad\qquad\qquad}$ $\qquad\;\; 3x = -3$
$\quad\;\;\; -6y = \;\; 12$ $\qquad\quad\; x = -1$
$\qquad\;\;\; y = -2$

The solution is $(-1, -2)$.

Use elimination to solve each system of equations.

27. $2x + y = 2$
$x + y = -2$

28. $x - 2y = 4$
$2x + 2y = 2$

29. $x - y = 5$
$-x + y = 3$

30. $5x + 3y = 11$
$2x + 3y = 2$

31. $4x + 3y = 3$
$-4x - 3y = -3$

32. $y = 3 - 4x$
$x + y = 2$

• **Lesson 13–5** Solve systems of equations by the elimination method using multiplication and addition.

$3x + 2y = 5$
$x + y = 1$ **Multiply by –2.**

$\quad\;\; 3x + 2y = \;\;\; 5$
$(+)\; -2x - 2y = -2$
$\overline{\qquad\qquad\qquad}$
$\qquad\qquad x = \;\;\; 3$

By substituting to find the value of y, you find that the solution is $(3, -2)$.

Use elimination to solve each system of equations.

33. $2x + 3y = 4$
$x + 2y = 2$

34. $4x + 3y = 3$
$-2x + 4y = 4$

35. $3x + 4y = 2$
$2x - y = 1$

36. $x - 3y = 2$
$-3x + 5y = 2$

37. $3x - 2y = 4$
$4x - 3y = -5$

38. $5x + 3y = 4$
$6x + 4y = 3$

Objectives and Examples	Review Exercises

- **Lesson 13–6** Solve systems of quadratic and linear equations.

$y = 2x + 3$
$y = x^2 - 5$

Substitute $x^2 - 5$ for y in the first equation.

$$y = 2x + 3$$
$$x^2 - 5 = 2x + 3$$
$$x^2 - 5 - 2x - 3 = 2x + 3 - 2x - 3$$
$$x^2 - 2x - 8 = 0$$
$$(x + 2)(x - 4) = 0$$

$x + 2 = 0$ or $x - 4 = 0$
$\quad x = -2 \qquad\qquad x = 4$

By finding the corresponding values of y, you find that the solutions are $(-2, -1)$ and $(4, 11)$.

Determine whether each system of equations has *one* solution, *two* solutions, or *no* solution by graphing. If the system has one or two solutions, name them.

39. $y = -x^2$
$\quad\ y = x - 2$

40. $y = x^2 + 2$
$\quad\ y = 6$

41. $y = 2x^2$
$\quad\ y = -x - 4$

42. $y = x^2 - 1$
$\quad\ y = -4x - 5$

Use substitution to solve each system of equations.

43. $y = x^2 + 2$
$\quad\ y = 3x + 6$

44. $y = -x^2 + 4$
$\quad\ y = 4x + 10$

45. Solve the system $y = x^2 - 2x$ and $y = 4x - 9$ by using substitution.

- **Lesson 13–7** Solve systems of inequalities by graphing.

Solve the system of inequalities $x \geq -2$ and $y < x - 3$ by graphing.

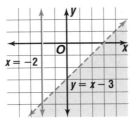

Solve each system of inequalities by graphing.

46. $x < 4$
$\quad\ y \geq -2$

47. $x \geq -3$
$\quad\ y \leq x + 2$

48. $x + y > -4$
$\quad\ x + y \leq 2$

49. $y \geq 2x + 1$
$\quad\ y > -x - 1$

50. Use the graphs of $-6x + 3y > 9$ and $y < 2x - 4$ to determine the solution of the system. If the system does not have a solution, write *no solution*.

Applications and Problem Solving

51. Mixtures Mr. Collins mixed almonds that cost $4.25 per pound with cashews that cost $6.50 per pound. He now has a mixture of 20 pounds of nuts that costs $5.60 per pound. How many pounds of each type of nut did he use? *(Lesson 13–3)*

52. Number Theory The difference between the tens digit and the units digit of a two-digit number is 3. Suppose the tens digit is one less than twice the units digit. What is the number? *(Lesson 13–4)*

53. Geometry The difference between the length and width of a rectangle is 7 feet. Find the dimensions of the rectangle if its perimeter is 50 feet. *(Lesson 13–5)*

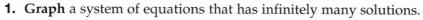

1. **Graph** a system of equations that has infinitely many solutions.

2. **Explain** when you would use elimination with subtraction to solve a system of equations.

Solve each system of equations by graphing.

3. $y = 3$
   $y = x + 4$

4. $x + y = -2$
   $2x - y = -4$

5. $y = -x^2 - 1$
   $y = -5$

State whether each system is *consistent and independent*, *consistent and dependent*, or *inconsistent*.

6.

7.

8. Use graphing to determine whether the system $y = -2x$ and $2x + y = 4$ has *one* solution, *no* solution, or *infinitely many* solutions. If the solution has one solution, name it.

Use substitution to solve each system of equations.

9. $y = 3x$
   $x + y = 4$

10. $x + y = -2$
    $x = y + 10$

11. $y = 5x - 3$
    $10x - 2y = -2$

12. $y = x^2 - 15$
    $y = 2x$

13. $y = 5x + 4$
    $y = x^2 + 5x$

14. $y = 3x + 2$
    $y = x^2 + 6$

Use elimination to solve each system of equations.

15. $x + y = 5$
    $x - y = -9$

16. $4x - 5y = 7$
    $x + 5y = 8$

17. $2x - y = 32$
    $y = 60 - 2x$

18. $x + 3y = -1$
    $2x + 4y = -2$

19. $5x - 2y = 3$
    $15x - 6y = 9$

20. $-5x + 8y = 21$
    $10x + 3y = 15$

Solve each system of inequalities by graphing.

21. $y \leq -3$
    $y > -x - 2$

22. $y < x + 4$
    $y > x - 2$

23. $x \leq 2y$
    $2x + 3y \leq 6$

24. Find two numbers whose sum is 64 and whose difference is 42.

25. **Transportation** Two trains travel toward each other on parallel tracks at the same time from towns 450 miles apart. Suppose one train travels 6 miles per hour faster than the other train. What is the rate of each train if they meet in 5 hours?

Perimeter, Area, and Volume Problems

Standardized tests often include questions on perimeter, area, circumference, and volume. You'll need to know and apply formulas for each of these measurements. Be sure you know these terms.

Use the information given in the figure.

area   circumference   height   perimeter   width
base   diameter   length   radius

Proficiency Test Example	**SAT Example**

Richard plans to increase the floor area of his health club's weight room. The figure below shows the existing floor area with a solid line and the additional floor area with a dotted line. Find the length, in feet, of the new weight room.

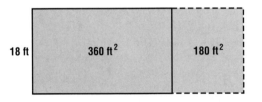

Hint Familiarize yourself with the formulas for the area and perimeter of quadrilaterals.

Solution The existing room has an area of 360 square feet and a width of 18 feet. The addition has an area of 180 square feet and a width of 18 feet.

To find the length of the new weight room, first find the area of the new room.

$$A = 360 + 180 \text{ or } 540 \text{ square feet}$$

Next, use the formula for the area of a rectangle to find the length of the room.

$A = \ell w$
$540 = \ell \cdot 18$ *Replace A with 540 and w with 18.*
$\dfrac{540}{18} = \dfrac{18\ell}{18}$ *Divide each side by 18.*
$30 = \ell$

The length of the new room is 30 feet.

What is the diameter of a circle with a circumference of 5 inches?

A $\dfrac{5}{\pi}$ in.   **B** $\dfrac{10}{\pi}$ in.   **C** 5 in.

D 5π in.   **E** 10π in.

Hint If a geometry problem has no figure, sketch one.

Solution

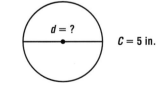

You know the circumference of the circle. You need to find the diameter of the circle.

Use the formula $C = \pi d$, where C is the circumference and d is the diameter to find the circumference of the circle.

First, replace C with 5. Then solve the equation for d.

$C = \pi d$
$5 = \pi d$ *Replace C with 5.*
$\dfrac{5}{\pi} = \dfrac{\pi d}{\pi}$ *Divide each side by π.*
$\dfrac{5}{\pi} = d$

The answer is A.

After you work each problem, record your answer on the answer sheet provided or on a sheet of paper.

1. Quinn's neighborhood is a rectangle 2 miles by 1.5 miles. How many miles does Quinn jog if he jogs around the boundary of his neighborhood?

   A 3   **B** 3.5   **C** 6   **D** 7

2. If $AC = 4$, what is the area of $\triangle ABC$?

   A $\frac{1}{2}$ **B** 2

   C 4   **D** 8

3. Micela is making a poster that has a length of 36 inches. If the maximum perimeter is 96 inches, which inequality could be used to determine the width w of the poster?

   A $96 \geq 2(36) + 2w$ **B** $96 \leq 2(36) + 2w$

   C $96 \geq 36 + 2w$ **D** $96 \geq 36 + w$

4. If the area of a circle is 16 square meters, what is its radius in meters?

   A $\frac{8}{\pi}$ **B** $\frac{16}{\pi}$ **C** $\frac{4\sqrt{\pi}}{\pi}$

   D 12π **E** $144\pi^2$

5. If you double the length and the width of a rectangle, how does its perimeter change?

   A It increases by $1\frac{1}{2}$.

   B It doubles.

   C It quadruples.

   D It does not change.

6. Mr. Tremaine needs to find the area of his backyard so that he can buy the right amount of sod for it. What is the area?

   A 192 ft^2

   B 360 ft^2

   C 456 ft^2

   D 720 ft^2

   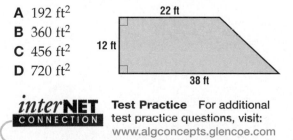

inter*NET* CONNECTION   **Test Practice** For additional test practice questions, visit: www.algconcepts.glencoe.com

7. The table shows the speed of a car and the distance needed to safely stop the car. Which equation represents the data?

   A $y = x + 25$
   B $y = x^2 + 20$
   C $y = x^2 \times 20$
   D $y = x^2 \div 20$
   E $y = x^2 \div 25$

Speed (x)	Distance (y)
20	20
30	45
40	80
50	125

Quantitative Comparison

8.

   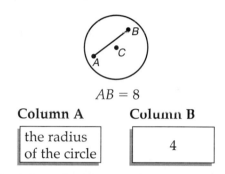

   $AB = 8$

Column A	Column B
the radius of the circle	4

   A if the quantity in Column A is greater;

   B if the quantity in Column B is greater;

   C if the two quantities are equal;

   D if the relationship cannot be determined from the information given.

Open-Ended Problems

9. **Grid-In** Allie sketched her walking route. The curved part is a semicircle with a radius of 10.5 meters. If Allie walks this route 10 times, how many meters does she walk? Round to the nearest meter.

   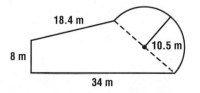

10. Suppose you have 120 feet of fence.

    Part A What is the greatest possible rectangular area you can enclose?

    Part B What are the dimensions of the rectangle that has this area?

▶ What You'll Learn in Chapter 14:

- to describe the relationships among sets of numbers *(Lesson 14–1)*,
- to find the distance between two points in the coordinate plane *(Lesson 14–2)*,
- to simplify, add, and subtract radical expressions *(Lesson 14–3 and 14–4)*, **and**
- to solve simple radical equations *(Lesson 14–5)*.

Problem-Solving Workshop

Project

You are to design three swimming pools that hold a total of 200,000 gallons of water or less. You will need to meet the following requirements.

Pool #1: a square with semicircles on each side, at least 5 feet deep
Pool #2: a rectangle with semicircles on the sides, at least 4 feet deep
Pool #3: a circle, at least 3 feet deep

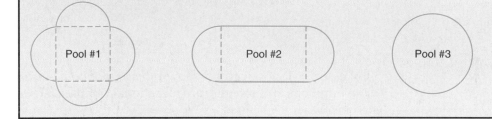

Pool #1   Pool #2   Pool #3

Working on the Project

Work with a partner to solve the problem.

- Suppose a swimming pool with a square shape measures 20 feet on a side. If the pool is 5 feet deep, how many cubic inches of water does it hold? How many gallons? (*Hint:* 1 gal ∼ 231 in³)
- Find the diameter of a circular pool that is 3.5 feet deep and holds 16,000 gallons of water.

Technology Tools

- Use **spreadsheet software** to find the dimensions.
- Use **drawing software** to draw each pool.

*inter***NET** **CONNECTION** **Research** For more information about swimming pools, visit: www.algconcepts.glencoe.com

Presenting the Project

Prepare a brochure of your designs. Include the following information:

- diagrams of the three pools with dimensions labeled,
- the volume of each pool and the total volume of the water, and
- a paragraph describing how you found the dimensions of each pool.

Strategies

> **Strategies**
>
> Look for a pattern.
> Draw a diagram.
> Make a table.
> Work backward.
> Use an equation.
> Make a graph.
> Guess and check.

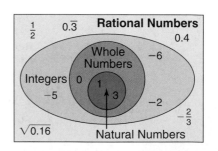

What You'll Learn
You'll learn to describe the relationships among sets of numbers.

Why It's Important
Meteorology
Meteorologists use real numbers when determining the duration of a thunderstorm.
See Exercise 16.

Graphs show data in a way that is easy to read and understand. The graph shows population estimates for Earth. The numbers used in the graph are **real numbers**. All of the numbers you use in everyday life are real numbers.

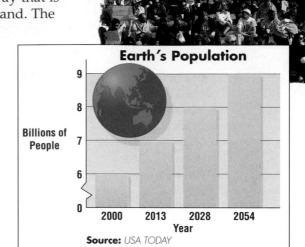

Earth's Population

Billions of People

| | 2000 | 2013 | 2028 | 2054 |

Year

Source: *USA TODAY*

In Lesson 3–1, you learned about *rational numbers*. Natural numbers, whole numbers, and integers are all rational numbers. These sets are listed below.

Natural Numbers: $\{1, 2, 3, 4, \dots\}$

Whole Numbers: $\{0, 1, 2, 3, \dots\}$

Integers: $\{\dots, -2, -1, 0, 1, 2, \dots\}$

Rational Numbers: $\{$all numbers that can be expressed in the form $\frac{a}{b}$, where a and b are integers and $b \neq 0\}$

Recall that repeating or terminating decimals are also rational numbers because they can be expressed as $\frac{a}{b}$, where a and b are integers and $b \neq 0$. The square roots of perfect squares are also rational numbers. For example, $\sqrt{0.16}$ is a rational number since $\sqrt{0.16} = 0.4$. However, $\sqrt{21}$ is irrational because 21 is not a perfect square.

Look Back

Venn Diagram:
Lesson 2–1

The Venn diagram shows the relationship among the rational numbers. For example, the whole numbers are a subset of the integers.

Rational Numbers

$\frac{1}{2}$ $0.\overline{3}$ 0.4

Whole Numbers -6

Integers   0   1   3 -2

-5 $-\frac{2}{3}$

$\sqrt{0.16}$ Natural Numbers

In Lesson 8–6, you learned about *irrational numbers*. A few examples of irrational numbers are shown below.

$0.1010010001\dots$ π $0.153768\dots$ $-\sqrt{23}$

The set of rational numbers and the set of irrational numbers together form the set of real numbers.

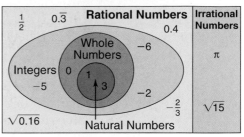

Real Numbers

Rational Numbers	**Irrational Numbers**

$\frac{1}{2}$ $0.\overline{3}$ 0.4 -6 π

Whole Numbers

Integers 0 1 3 -2 $\sqrt{15}$

-5 $-\frac{2}{3}$

$\sqrt{0.16}$ Natural Numbers

Examples

Name the set or sets of numbers to which each real number belongs.

① −5   This number is an integer and a rational number.

② $\frac{12}{3}$ Since $\frac{12}{3} = 4$, this number is a natural number, a whole number, an integer, and a rational number.

> *Reading Algebra*
>
> Natural numbers are also called *counting numbers*.

③ $\sqrt{95}$ $\sqrt{95} = 9.746794345 \ldots$ It is not the square root of a perfect square. So, it is irrational.

④ $0.\overline{7}$ This repeating decimal is a rational number since it is equivalent to $\frac{7}{9}$. *7 ÷ 9 = 0.7777777 . . .*

⑤ $-\sqrt{4}$ Since $-\sqrt{4} = -2$, this number is an integer and a rational number.

Your Turn   Name the set or sets of numbers to which each real number belongs. Let N = natural numbers, W = whole numbers, Z = integers, Q = rational numbers, and I = irrational numbers.

a. 3.141592 . . .   **b.** $-\sqrt{9}$ **c.** 0.1666666 . . .   **d.** $\frac{20}{4}$ **e.** 0

If you graph all of the rational numbers, you would still have some "holes" in the number line. The irrational numbers "fill in" the number line. The graph of all real numbers is the entire number line without any "holes".

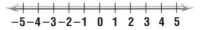

$-5\ -4\ -3\ -2\ -1\ \ 0\ \ 1\ \ 2\ \ 3\ \ 4\ \ 5$

This is illustrated by the Completeness Property.

<table>
<tr><td>**Completeness Property for Points on the Number Line**</td><td>Each real number corresponds to exactly one point on the number line. Each point on the number line corresponds to exactly one real number.</td></tr>
</table>

Every real number can be graphed. You have learned how to graph rational numbers. Irrational numbers can also be graphed. Use a calculator or a table of squares and square roots to find approximate values of square roots that are irrational. These values can be used to approximate the graphs of square roots.

Examples

Find an approximation, to the nearest tenth, for each square root. Then graph the square root on a number line.

6 $\sqrt{3}$

Enter: [2nd] [$\sqrt{}$] 3 [ENTER] *1.732050808*

An approximate value for $\sqrt{3}$ is 1.7.

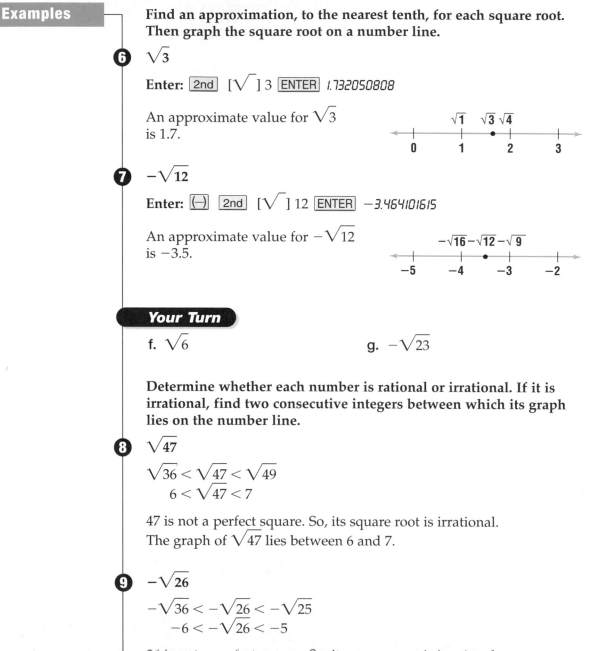

7 $-\sqrt{12}$

Enter: [(−)] [2nd] [$\sqrt{}$] 12 [ENTER] *−3.464101615*

An approximate value for $-\sqrt{12}$ is −3.5.

Your Turn

f. $\sqrt{6}$ g. $-\sqrt{23}$

Determine whether each number is rational or irrational. If it is irrational, find two consecutive integers between which its graph lies on the number line.

8 $\sqrt{47}$

$$\sqrt{36} < \sqrt{47} < \sqrt{49}$$
$$6 < \sqrt{47} < 7$$

47 is not a perfect square. So, its square root is irrational. The graph of $\sqrt{47}$ lies between 6 and 7.

9 $-\sqrt{26}$

$$-\sqrt{36} < -\sqrt{26} < -\sqrt{25}$$
$$-6 < -\sqrt{26} < -5$$

26 is not a perfect square. So, its square root is irrational. The graph of $-\sqrt{26}$ lies between −6 and −5.

Your Turn

h. $\sqrt{64}$

i. $-\sqrt{28}$

You have solved equations that involve rational number solutions. Some equations have solutions that are irrational numbers.

Example

Science Link

Real World

⑩ The time *t* in seconds it takes for a pendulum to complete one full swing (back and forth) is given by the equation $t = 2\pi\sqrt{\dfrac{\ell}{9.8}}$, where ℓ is the length of the pendulum in meters. Suppose a pendulum has a length of 4.9 meters. How long does it take the pendulum to complete one full swing?

$t = 2\pi\sqrt{\dfrac{\ell}{9.8}}$

$t = 2\pi\sqrt{\dfrac{4.9}{9.8}}$ *Replace ℓ with 4.9.*

$t = 2\pi\sqrt{0.5}$

Enter: 2 [2nd] [π] [2nd] [√] 0.5 [ENTER] 4.442882938

The pendulum will complete one full swing in about 4.4 seconds.

Check for Understanding

Communicating Mathematics

Study the lesson. Then complete the following.

1. **Give an example** of a number that is an integer and a rational number.

2. **Give** two counterexamples for the statement *all square roots are irrational numbers*.

3. **Write** a square root that is an irrational number.

Vocabulary

real numbers

Guided Practice

Name the set or sets of numbers to which each real number belongs. Let N = natural numbers, W = whole numbers, Z = integers, Q = rational numbers, and I = irrational numbers. *(Examples 1–5)*

4. $-\sqrt{36}$ 5. $0.3131131113\ldots$ 6. 4   7. $0.\overline{2}$

Find an approximation, to the nearest tenth, for each square root. Then graph the square root on a number line. *(Examples 6 & 7)*

8. $\sqrt{7}$ 9. $-\sqrt{20}$ 10. $\sqrt{32}$ 11. $\sqrt{54}$

Determine whether each number is *rational* or *irrational*. If it is irrational, find two consecutive integers between which its graph lies on the number line. *(Examples 8 & 9)*

12. $\sqrt{63}$ **13.** $-\sqrt{25}$ **14.** $-\sqrt{147}$ **15.** $\sqrt{49}$

16. Meteorology To estimate the amount of time t in hours that a thunderstorm will last, meteorologists use the formula $t = \sqrt{\dfrac{d^3}{216}}$, where d is the diameter of the storm in miles. *(Example 10)*

a. Estimate how long a thunderstorm will last if its diameter is 8.4 miles.

b. Find the diameter of a storm that will last for 1 hour.

Exercises • • • • • • • • • • • • • • • • • •

Practice

Name the set or sets of numbers to which each real number belongs. Let N = natural numbers, W = whole numbers, Z = integers, Q = rational numbers, and I = irrational numbers.

17. $-\dfrac{1}{2}$ **18.** $\sqrt{17}$ **19.** 0 **20.** $\dfrac{15}{3}$

21. -5 **22.** $1.202002\ldots$ **23.** $\sqrt{81}$ **24.** 0.125

25. $-\sqrt{121}$ **26.** $\dfrac{6}{18}$ **27.** $0.834834\ldots$ **28.** $-\dfrac{3}{1}$

Find an approximation, to the nearest tenth, for each square root. Then graph the square root on a number line.

29. $\sqrt{3}$ **30.** $\sqrt{8}$ **31.** $-\sqrt{10}$ **32.** $\sqrt{17}$

33. $-\sqrt{40}$ **34.** $\sqrt{37}$ **35.** $\sqrt{52}$ **36.** $-\sqrt{99}$

37. $\sqrt{108}$ **38.** $-\sqrt{112}$ **39.** $\sqrt{250}$ **40.** $\sqrt{300}$

Determine whether each number is *rational* or *irrational*. If it is irrational, find two consecutive integers between which its graph lies on the number line.

41. $\sqrt{16}$ **42.** $-\sqrt{12}$ **43.** $\sqrt{41}$ **44.** $\sqrt{24}$

45. $-\sqrt{81}$ **46.** $\sqrt{66}$ **47.** $-\sqrt{7}$ **48.** $\sqrt{98}$

49. $\sqrt{125}$ **50.** $\sqrt{144}$ **51.** $-\sqrt{169}$ **52.** $\sqrt{220}$

53. Graph $\sqrt{2}$, π, and $-\sqrt{5}$ on a number line.

54. Write $2.\overline{4}$, 2.41, and $\sqrt{6}$ in order from least to greatest.

Applications and Problem Solving

Real World

55. Science Refer to Example 10. Suppose a pendulum has a length of 18 meters. How long does it take the pendulum to complete one full swing?

56. Geometry The radius of a circle is given by $r = \sqrt{\dfrac{A}{\pi}}$, where A is the area of the circle. Suppose a circle has an area of 25 square centimeters. What is the measure of the radius?

57. Electricity The voltage V in a circuit is given by $V = \sqrt{PR}$, where P is the power in watts and R is the resistance in ohms. A circuit is designed with two resistance settings, 4.6 ohms and 5.2 ohms, and two power settings, 1200 watts and 1500 watts. Which settings can be used so that the voltage of the circuit is between 75 and 85 volts?

58. Geometry Find the area of each rectangle. Round answers to the nearest tenth. (*Hint:* Use the Pythagorean Theorem.)

a.  b.

18 in.   9 in.

22 cm   15 cm

59. Critical Thinking Find all numbers of the form $\sqrt{n}$ such that n is a natural number and the graph of $\sqrt{n}$ lies between each pair of numbers on the number line.

a. 4 and 5   b. 4.25 and 4.5

Mixed Review

60. Find the solution of the system of inequalities $y \le x + 3$ and $y > -x - 4$. (*Lesson 13–7*)

61. What is the solution of the system of equations $y = x^2 + 2$ and $y = 4x - 1$? (*Lesson 13–6*)

62. Graph the solution of $|4n| > 16$ on a number line. (*Lesson 12–6*)

Solve each inequality. (*Lesson 12–2*)

63. $x + 3 < -4$ **64.** $w - 14 \ge -8$ **65.** $3.4 + m \le 1.6$

66. Write an inequality for the graph shown at the right. (*Lesson 12–1*)

−3−2−1 0 1 2 3 4 5 6 7

67. Graph $y = 2^x - 1$. Then state the y-intercept. (*Lesson 11–7*)

Factor each polynomial. If the polynomial cannot be factored, write prime. (*Lesson 10–5*)

68. $x^2 - 25y^2$ **69.** $4a^2 - 36$ **70.** $m^2 + 2m + 4$

71. Find the degree of $5x^3 - 3x^3y - 6xy$. (*Lesson 9–4*)

72. Money Cleavon has $35 to spend on CDs. Suppose CDs cost $13.85 each, including tax. Write an equation that can be used to determine how many CDs Cleavon can buy. (*Lesson 4–4*)

73. Standardized Test Practice What is the value of a if $6 - (-8) = a$? (*Lesson 2–3*)

A 14   B 6   C −14   D −6

What You'll Learn
You'll learn to find the distance between two points in the coordinate plane.

Why It's Important

Engineering
Engineers can use the distance formula to determine the amount of cable needed to install a cable system. *See Exercise 9.*

A coordinate system is superimposed over a map of Washington, D.C. Jessica and Omar walk from the Metro Center to 15th Street and then up 15th Street to McPherson Square. How far is the Metro center from McPherson Square? *This problem will be solved in Example 2.*

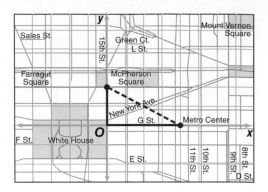

Recall that subtraction is used to find the distance between two points that lie on a vertical or horizontal line.

In the following activity, you will find the distance between two points that do not lie on a horizontal or vertical line.

Look Back

Graphing Ordered Pairs: Lesson 2–2

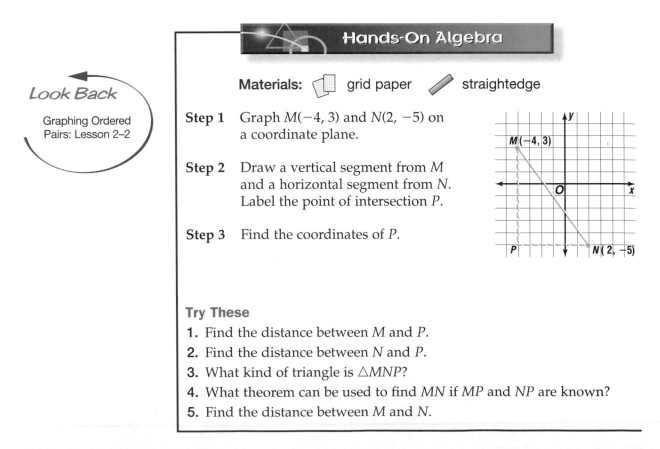

Hands-On Algebra

Materials: grid paper   straightedge

Step 1 Graph $M(-4, 3)$ and $N(2, -5)$ on a coordinate plane.

Step 2 Draw a vertical segment from M and a horizontal segment from N. Label the point of intersection P.

Step 3 Find the coordinates of P.

Try These
1. Find the distance between M and P.
2. Find the distance between N and P.
3. What kind of triangle is $\triangle MNP$?
4. What theorem can be used to find MN if MP and NP are known?
5. Find the distance between M and N.

In the activity, you discovered that $(MN)^2 = (MP)^2 + (NP)^2$. You can solve this equation for MN to find the distance between any two points M and N.

$$(MN)^2 = (MP)^2 + (NP)^2$$
$$\sqrt{(MN)^2} = \sqrt{(MP)^2 + (NP)^2} \quad \textit{Take the positive square root of each side.}$$
$$MN = \sqrt{(MP)^2 + (NP)^2}$$

The **Distance Formula** is a generalization of this formula.

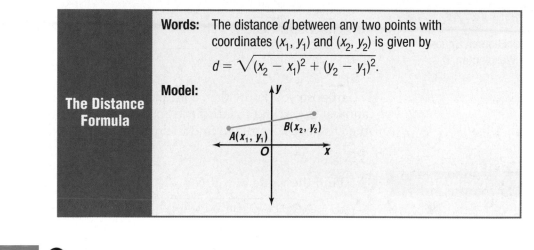

The Distance Formula

Words: The distance d between any two points with coordinates (x_1, y_1) and (x_2, y_2) is given by
$$d = \sqrt{(x_2 - x_1)^2 + (y_2 - y_1)^2}.$$

Model:

$A(x_1, y_1)$ $B(x_2, y_2)$

Example ➊ Find the distance between points $A(1, 2)$ and $B(-3, 7)$.

$d = \sqrt{(x_2 - x_1)^2 + (y_2 - y_1)^2}$ *Distance Formula*

$\quad = \sqrt{(-3 - 1)^2 + (7 - 2)^2}$ $(x_1, y_1) = (1, 2), \text{ and } (x_2, y_2) = (-3, 7)$

$\quad = \sqrt{(-4)^2 + 5^2}$

$\quad = \sqrt{16 + 25}$

$\quad = \sqrt{41} \text{ or about 6.4 units}$

Your Turn

a. Find the distance between points $C(3, 5)$ and $D(6, 4)$.

Example ➋

Map Link

Real World

Refer to the application at the beginning of the lesson. The Metro Center is located 4 blocks east of the intersection of 15ᵗʰ Street and G Street. McPherson Square is located 2 blocks north of the intersection. How far is the Metro Center from McPherson Square?

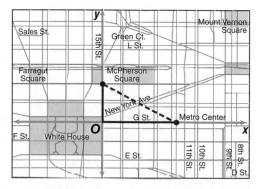

(continued on the next page)

McPherson Square,
Washington, D.C.

Let the Metro Center be represented by (4, 0) and McPherson Square by (0, 2). So, $x_1 = 4$, $y_1 = 0$, $x_2 = 0$, and $y_2 = 2$.

$d = \sqrt{(x_2 - x_1)^2 + (y_2 - y_1)^2}$ *Distance Formula*

$\quad = \sqrt{(0 - 4)^2 + (2 - 0)^2}$ $(x_1, y_1) = (4, 0)$, *and* $(x_2, y_2) = (0, 2)$

$\quad = \sqrt{(-4)^2 + (2)^2}$

$\quad = \sqrt{16 + 4}$

$\quad = \sqrt{20}$ or about 4.5 blocks

The Metro Center is about 4.5 blocks from McPherson Square.

Suppose you know the coordinates of a point, one coordinate of another point, and the distance between the two points. You can use the Distance Formula to find the missing coordinate.

Example ➌ **Find the value of a if $G(4, 7)$ and $H(a, 3)$ are 5 units apart.**

$d = \sqrt{(x_2 - x_1)^2 + (y_2 - y_1)^2}$ *Distance Formula*

$5 = \sqrt{(a - 4)^2 + (3 - 7)^2}$ $(x_1, y_1) = (4, 7)$, *and* $(x_2, y_2) = (a, 3)$

$5 = \sqrt{(a - 4)^2 + (-4)^2}$

$5 = \sqrt{a^2 - 8a + 16 + 16}$ $(a - 4)^2 = a^2 - 8a + 16$ *and* $(-4)^2 = 16$

$5 = \sqrt{a^2 - 8a + 32}$

$5^2 = (\sqrt{a^2 - 8a + 32})^2$ *Square each side.*

$25 = a^2 - 8a + 32$

$0 = a^2 - 8a + 7$ *Subtract 25 from each side.*

$0 = (a - 7)(a - 1)$ *Factor.*

$a - 7 = 0 \quad$ or $\quad a - 1 = 0$ *Zero Product Property.*

$\quad a = 7 \qquad\qquad a = 1$

The value of a is 7 or 1.

Look Back

Zero Product
Property:
Lesson 11–4

Your Turn

b. Find the value of a if $J(a, 5)$ and $K(-7, 3)$ are $\sqrt{29}$ units apart.

Check for Understanding

**Communicating
Mathematics**

Study the lesson. Then complete the following.

1. Name the theorem that is used to derive the Distance Formula.

Vocabulary

Distance Formula

2. **Tell** how to find the distance between the points $A(10, 3)$ and $B(2, 3)$ without using the Distance Formula.

3. **You Decide** Anna says that to find the distance between $A(5, -8)$ and $B(7, -6)$, you should take the square root of $(7 - 5)^2 + [-6 - (-8)]^2$. Nick disagrees. He says to take the square root of $(5 - 7)^2 + [-8 - (-6)]^2$. Who is correct? Explain your reasoning.

Guided Practice

Find the distance between each pair of points. Round to the nearest tenth, if necessary. *(Example 1)*

4. $C(5, -1), D(2, 2)$ 5. $M(6, 8), N(3, 4)$ 6. $R(-3, 8), S(5, 4)$

Find the value of *a* if the points are the indicated distance apart. *(Example 3)*

7. $A(3, -1), B(a, 7); d = 10$ 8. $J(5, a), K(6, 1); d = \sqrt{10}$

9. **Engineering** The MAXTEC Company is installing a fiber optic cable system between two of their office buildings. One of the buildings, MAXTEC West, is located 5 miles west and 2 miles north of the corporate office. The other building, MAXTEC East, is located 4 miles east and 5 miles north of the corporate office. How much cable will be needed to connect MAXTEC West and MAXTEC East? Round to the nearest tenth. *(Example 2)*

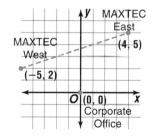

Exercises

Practice

Find the distance between each pair of points. Round to the nearest tenth, if necessary.

10. $X(5, 0), Y(12, 0)$ 11. $A(-6, -4), B(-6, 8)$
12. $M(-2, -5), N(3, 7)$ 13. $P(-4, 0), Q(3, -3)$
14. $V(-3, -4), W(-1, -2)$ 15. $C(7, 2), D(-4, 10)$
16. $E(3, -6), F(9, -2)$ 17. $G(-4, -6), H(-7, -3)$

Find the value of *a* if the points are the indicated distance apart.

18. $A(a, -5), B(-3, -2); d = 5$ 19. $D(-3, a), E(5, 2); d = 17$
20. $Q(7, 2), R(-1, a); d = 10$ 21. $G(7, -3), H(5, a); d = \sqrt{85}$
22. $T(6, -3), U(-3, a); d = \sqrt{130}$ 23. $U(1, -6), V(10, a); d = \sqrt{145}$

24. Find the distance between $J(-9, 5)$ and $K(-4, -2)$.

25. What is the distance between $C(-8, 1)$ and $D(5, 6)$?

26. What is the value of c if $W(1, c)$ and $V(-4, 9)$ are 13 units apart?

27. Suppose $M(b, 9)$ and $N(20, -5)$ are $\sqrt{340}$ units apart. What is the value of b?

28. **Geometry** Triangle MNP has vertices $M(-6, 14)$, $N(2, -1)$, and $P(-2, 2)$. Find the perimeter of $\triangle MNP$. Round to the nearest tenth.

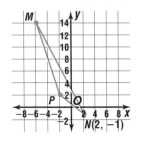

29. **Geometry** An isosceles triangle is a triangle with at least two congruent sides. Determine whether $\triangle CDE$ with vertices $C(1, 6)$, $D(2, 1)$, and $E(4, 1)$ is isosceles. Explain your reasoning.

30. **School** At Dannis State University, Webber Hall is located 35 meters west and 55 meters south of the student union. Packard Hall is located 42 meters east and 30 meters south of the student union.

  a. Draw a diagram on a coordinate grid to represent this situation.

  b. How far is Webber Hall from Packard Hall? Round to the nearest tenth.

31. **Engineering** Refer to Exercise 9. Suppose the MAXTEC Company builds another office building, MAXTEC North, 4 miles east and 2 miles north of the corporate office. How much cable will be needed to connect each of the buildings to the corporate office? Round to the nearest tenth.

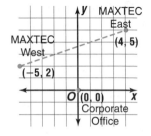

32. **Critical Thinking** If the diagonals of a trapezoid have the same length, then the trapezoid is isosceles. The vertices of trapezoid $ABCD$ are $A(-2, 2)$, $B(10, 6)$, $C(9, 8)$, and $D(0, 5)$. Is the trapezoid isosceles? Explain.

Mixed Review

Find an approximation, to the nearest tenth, for each square root. Then graph the square root on a number line. *(Lesson 14–1)*

33. $\sqrt{12}$ **34.** $-\sqrt{27}$ **35.** $\sqrt{135}$

36. Solve $y \geq -4$ and $y < -2x + 1$ by graphing. *(Lesson 13–7)*

Solve each inequality. Check your solution. *(Lesson 12–3)*

37. $\dfrac{k}{3} \geq 2$ **38.** $\dfrac{-n}{4} > 3$ **39.** $\dfrac{3}{4}b \leq -6$

40. Factor $12m^2 + 7m - 10$. *(Lesson 10–4)*

41. Find the solution of $|3 + w| + 4 = 2$. *(Lesson 3–7)*

42. Standardized Test Practice The table shows the number of pets that have lived at the White House. Which two pets account for exactly half of the pets that have lived at the White House? *(Lesson 1–7)*

A dog and cat   **B** horse and bird
C bird and dog   **D** dog and horse

White House Pets	
dog	23
bird	16
horse	11
cat	10
cow	4
goat	4

Source: *USA TODAY*

Quiz 1   Lessons 14–1 and 14–2

Name the set or sets of numbers to which each real number belongs. Let N = natural numbers, W = whole numbers, Z = integers, Q = rational numbers, and I = irrational numbers. *(Lesson 14–1)*

1. 0

2. $\sqrt{36}$

3. 0.121231234 . . .

4. Find an approximation, to the nearest tenth, for $-\sqrt{35}$. *(Lesson 14–1)*

5. Write $\sqrt{2}$, 1.22, and $1.\overline{2}$ in order from least to greatest. *(Lesson 14–1)*

Find the distance between each pair of points. Round to the nearest tenth, if necessary. *(Lesson 14–2)*

6. $A(2, 0)$, $B(-1, 3)$ **7.** $C(4, 5)$, $D(-3, -2)$ **8.** $E(0, -4)$, $F(8, 7)$

9. What is the value of m if $A(m, 8)$ and $B(3, 4)$ are 5 units apart? *(Lesson 14–2)*

10. Geometry A scalene triangle is a triangle with no congruent sides. Determine whether $\triangle XYZ$ with vertices $X(5, 4)$, $Y(1, 5)$, and $Z(-1, 1)$ is scalene. Explain. *(Lesson 14–2)*

Extra Practice See p. 720.

Lesson 14–2 The Distance Formula **611**

A **Cut-Out** Caper

Materials

 grid paper

✏ ruler

✂ scissors

Midpoints

In Lesson 14–2, you learned how to find the distance between two points in a coordinate plane. In this Investigation, you will learn how to find the **midpoint** of a segment. The midpoint of a line segment is the point on the segment that separates it into two segments of equal length.

Investigate

1. Graph $A(-2, -1)$, $B(6, -1)$, $C(10, 3)$, and $D(2, 3)$ on a coordinate plane.

   a. Connect the points. What type of figure is formed?

   b. Make a table and record the coordinates as shown.

Coordinates of First Point	Coordinates of Second Point	Coordinates of Midpoint
$A(-2, -1)$	$B(6, -1)$	$E(\ ,\ \ )$
$B(6, -1)$	$C(10, 3)$	$F(\ ,\ \ )$
$C(10, 3)$	$D(2, 3)$	$G(\ ,\ \ )$
$D(2, 3)$	$A(-2, -1)$	$H(\ ,\ \ )$

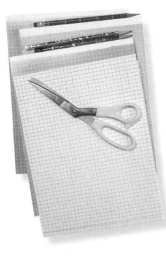

   c. Fold the paper so that points A and B coincide. Crease the paper to mark the midpoint of $\overline{AB}$. Label this point E. Find the coordinates of point E. Why is point E the midpoint of $\overline{AB}$?

   d. Fold the paper so that points B and C coincide. Crease the paper to mark the midpoint of $\overline{BC}$. Label this point F. Find the coordinates of point F. Continue this process with points C and D and points D and A. Label the midpoints G and H, respectively. Record the coordinates of these midpoints in your table.

   e. Draw $\overline{AC}$ and $\overline{BD}$. These are the diagonals of the figure. Find the midpoints of $\overline{AC}$ and $\overline{DB}$. Label the midpoints M and N, respectively. What are the coordinates of M and N? What appears to be true about the midpoints of the diagonals?

   f. Draw $\overline{EG}$ and $\overline{FH}$. What are the coordinates of their midpoints?

   g. Notice the triangles formed on the inside of the figure. Explain any relationships that exist between the triangles. You may cut out the triangles to help see the relationships.

2. Graph $X(-3, -3)$, $Y(3, -3)$, and $Z(3, 5)$ on a coordinate plane.

   a. Connect the points. What type of figure is formed?

   b. Make a table and record the coordinates as shown.

Coordinates of First Point	Coordinates of Second Point	Coordinates of Midpoint
$X(-3, -3)$	$Y(3, -3)$	$U(\ ,\ \ )$
$Y(3, -3)$	$Z(3, 5)$	$V(\ ,\ )$
$Z(3, 5)$	$X(-3, -3)$	$W(\ ,\ )$

   c. Use paper folding to find the midpoints of the segments. Label the midpoints U, V, and W as shown in the table. What are the coordinates of the midpoints? Record the results.

   d. Draw $\overline{YW}$. Compare the measures of $\overline{YW}$, $\overline{XW}$, and $\overline{ZW}$. What appears to be true? Use the Distance Formula to verify your results.

   e. Draw $\overline{WU}$ and $\overline{WV}$. Four triangles are formed. Explain any relationships that exist between the triangles. You may cut out the triangles to help see the relationships.

3. Refer to the tables. Study each pair of first point coordinates, second point coordinates, and the corresponding midpoint coordinates.

   a. Explain how the concept of *mean*, or average, can be used to find the midpoint of a segment.

   b. Write a formula to find the midpoint of two points $X(a, b)$ and $Y(c, d)$.

Extending the Investigation

In this extension, you will continue to investigate midpoints.

- Graph points $A(3, 2)$, $B(-1, -6)$, and $C(-5, -4)$ on a coordinate plane. Connect the points to form a figure. Using the Midpoint Formula you discovered in Exercise 3b, determine the coordinates of the midpoints of each side of the figure. Label the points D, E, and F. Use the Distance Formula to verify that the midpoints separate each segment into two equal lengths.

- Use the Midpoint Formula to determine the midpoint of each pair of points.
  a. $M(4, 3)$, $N(2, 5)$ b. $R(1, 0)$, $S(-3, 2)$ c. $C(3, -6)$, $D(1, -4)$

Presenting Your Investigation

Here are some ideas to help you present your conclusions to the class.

- Make a poster showing the results of your research. Include the tables, graphs, and formula for finding the midpoint of a segment. List any relationships you discovered.

- Discuss any similarities in the triangles formed inside of the figures.

 Investigation For more information on midpoints, visit: www.algconcepts.glencoe.com

What You'll Learn
You'll learn to simplify radical expressions.

Why It's Important
Science Scientists can use radical expressions to find the speed of a river. *See Exercise 41.*

Why does the character from the comic yell *"the square root of sixteen"* when hitting the golf ball?

FoxTrot

FOXTROT©1998 Bill Amend. Reprinted with permission of UNIVERSAL PRESS SYNDICATE. All rights reserved.

The expression $\sqrt{16}$ is a **radical expression**. Since the radicand, 16, is a perfect square, $\sqrt{16} = 4$.

In Lesson 8–5, you learned to simplify radical expressions using the Product Property of Square Roots and prime factorization. You can simplify radical expressions in which the radicand is not a perfect square in a similar manner.

Examples

1 **Simplify $\sqrt{75}$.**

$$\begin{aligned}
\sqrt{75} &= \sqrt{3 \cdot 5 \cdot 5} & &\textit{Prime factorization} \\
&= \sqrt{3 \cdot 25} & &\textit{5} \times \textit{5} = \textit{25} \\
&= \sqrt{3} \cdot \sqrt{25} & &\textit{Product Property of Square Roots} \\
&= \sqrt{3} \cdot 5 \text{ or } 5\sqrt{3} & &\textit{Simplify } \sqrt{25}.
\end{aligned}$$

Your Turn **Simplify each square root. Leave in radical form.**

a. $\sqrt{68}$ **b.** $\sqrt{375}$

Sports Link

*inter*NET
CONNECTION

Data Update For the latest information on softball, visit:
www.algconcepts.
glencoe.com

2 **On a softball field, the distance from second base to home plate is $\sqrt{800}$ meters. Express $\sqrt{800}$ in simplest radical form.**

$$\begin{aligned}
\sqrt{800} &= \sqrt{2 \cdot 2 \cdot 2 \cdot 2 \cdot 2 \cdot 5 \cdot 5} & &\textit{Prime factorization} \\
&= \sqrt{16 \cdot 2 \cdot 25} & &\textit{4} \times \textit{4} = \textit{16, 5} \times \textit{5} = \textit{25} \\
&= \sqrt{16} \cdot \sqrt{2} \cdot \sqrt{25} & &\textit{Product Property of Square Roots} \\
&= 4 \cdot \sqrt{2} \cdot 5 & &\textit{Simplify } \sqrt{16} \textit{ and } \sqrt{25}. \\
&= 20\sqrt{2}
\end{aligned}$$

The distance is $20\sqrt{2}$ meters.

The Product Property can also be used to multiply square roots.

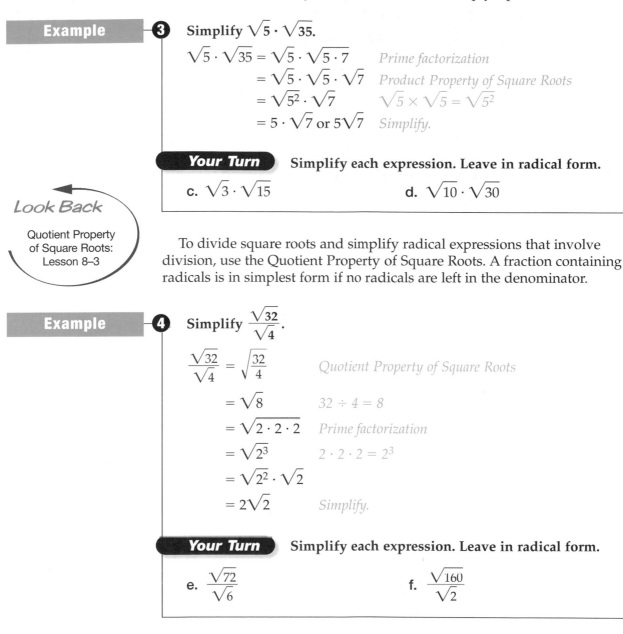

Example ❸ Simplify $\sqrt{5} \cdot \sqrt{35}$.

$$\begin{aligned}
\sqrt{5} \cdot \sqrt{35} &= \sqrt{5} \cdot \sqrt{5 \cdot 7} && \textit{Prime factorization} \\
&= \sqrt{5} \cdot \sqrt{5} \cdot \sqrt{7} && \textit{Product Property of Square Roots} \\
&= \sqrt{5^2} \cdot \sqrt{7} && \sqrt{5} \times \sqrt{5} = \sqrt{5^2} \\
&= 5 \cdot \sqrt{7} \text{ or } 5\sqrt{7} && \textit{Simplify.}
\end{aligned}$$

Your Turn Simplify each expression. Leave in radical form.

c. $\sqrt{3} \cdot \sqrt{15}$ d. $\sqrt{10} \cdot \sqrt{30}$

Look Back

Quotient Property
of Square Roots:
Lesson 8–3

To divide square roots and simplify radical expressions that involve division, use the Quotient Property of Square Roots. A fraction containing radicals is in simplest form if no radicals are left in the denominator.

Example ❹ Simplify $\dfrac{\sqrt{32}}{\sqrt{4}}$.

$$\begin{aligned}
\frac{\sqrt{32}}{\sqrt{4}} &= \sqrt{\frac{32}{4}} && \textit{Quotient Property of Square Roots} \\
&= \sqrt{8} && 32 \div 4 = 8 \\
&= \sqrt{2 \cdot 2 \cdot 2} && \textit{Prime factorization} \\
&= \sqrt{2^3} && 2 \cdot 2 \cdot 2 = 2^3 \\
&= \sqrt{2^2} \cdot \sqrt{2} \\
&= 2\sqrt{2} && \textit{Simplify.}
\end{aligned}$$

Your Turn Simplify each expression. Leave in radical form.

e. $\dfrac{\sqrt{72}}{\sqrt{6}}$ f. $\dfrac{\sqrt{160}}{\sqrt{2}}$

To eliminate radicals from the denominator of a fraction, you can use a method for simplifying radical expressions called **rationalizing the denominator**.

Example ❺ Simplify $\dfrac{\sqrt{5}}{\sqrt{10}}$.

$$\begin{aligned}
\frac{\sqrt{5}}{\sqrt{10}} &= \frac{\sqrt{5}}{\sqrt{10}} \cdot \frac{\sqrt{10}}{\sqrt{10}} && \frac{\sqrt{10}}{\sqrt{10}} = 1 \\
&= \frac{\sqrt{5 \cdot 10}}{\sqrt{10 \cdot 10}}
\end{aligned}$$

(continued on the next page)

$$= \frac{\sqrt{50}}{\sqrt{100}} \qquad \textit{Simplify.}$$

$$= \frac{\sqrt{25 \cdot 2}}{10}$$

$$= \frac{\sqrt{25} \cdot \sqrt{2}}{10} \qquad \textit{Product Property of Square Roots}$$

$$= \frac{5 \cdot \sqrt{2}}{10} \text{ or } \frac{\sqrt{2}}{2} \qquad \textit{Simplify.}$$

Your Turn   **Simplify each expression. Leave in radical form.**

g. $\dfrac{\sqrt{6}}{\sqrt{8}}$ h. $\dfrac{\sqrt{32}}{\sqrt{3}}$

Binomials of the form $a\sqrt{b} + c\sqrt{d}$ and $a\sqrt{b} - c\sqrt{d}$ are **conjugates** of each other because their product is a rational number.

Look Back

Product of a Sum and a Difference: Lesson 9–5

$(6 + \sqrt{3})(6 - \sqrt{3}) = 6^2 - (\sqrt{3})^2$ *Use the pattern $(a + b)(a - b) = a^2 - b^2$ to simplify the product.*

$$= 36 - 3$$
$$= 33$$

Conjugates are useful for simplifying radical expressions because their product is always a rational number.

Example ⑥   **Simplify $\dfrac{6}{3 - \sqrt{2}}$.**

To rationalize the denominator, multiply both the numerator and denominator by $3 + \sqrt{2}$, which is the conjugate of $3 - \sqrt{2}$.

$$\frac{6}{3 - \sqrt{2}} = \frac{6}{3 - \sqrt{2}} \cdot \frac{3 + \sqrt{2}}{3 + \sqrt{2}} \qquad \textit{Notice that } \frac{3 + \sqrt{2}}{3 + \sqrt{2}} = 1.$$

$$= \frac{6(3) + 6\sqrt{2}}{3^2 - (\sqrt{2})^2} \qquad \begin{array}{l}\textit{Distributive Property}\\ (a - b)(a + b) = a^2 - b^2\end{array}$$

$$= \frac{18 + 6\sqrt{2}}{9 - 2} \qquad \textit{Simplify.}$$

$$= \frac{18 + 6\sqrt{2}}{7}$$

Your Turn   **Simplify each expression. Leave in radical form.**

i. $\dfrac{3}{3 - \sqrt{5}}$ j. $\dfrac{4}{5 + \sqrt{6}}$

Radical expressions are in simplest form if the following conditions are met.

	A radical expression is in simplest form when the following three conditions have been met.
Simplified Form for Radicals	1. No radicands have perfect square factors other than 1.
	2. No radicands contain fractions.
	3. No radicals appear in the denominator of a fraction.

Consider the expression $\sqrt{x^2}$. It appears that $\sqrt{x^2} = x$. However, if $x = -3$, then $\sqrt{(-3)^2}$ is 3, not -3. For radical expressions like $\sqrt{x^2}$, use absolute value to ensure nonnegative results. The results of simplifying a few radical expressions are listed below.

$$\sqrt{x^2} = |x| \quad \sqrt{x^3} = x\sqrt{x} \quad \sqrt{x^4} = x^2 \quad \sqrt{x^5} = x^2\sqrt{x} \quad \sqrt{x^6} = |x^3|$$

For $\sqrt{x^3}$, absolute value is not necessary. If x were negative, then x^3 would be negative, and $\sqrt{x^3}$ is not a real number. *Why is absolute value not used for $\sqrt{x^4}$?*

Example

7 **Simplify $\sqrt{98ab^2c^4}$. Use absolute value symbols if necessary.**

$\sqrt{98ab^2c^4}$

$\quad = \sqrt{2 \cdot 7 \cdot 7 \cdot a \cdot b^2 \cdot c^4}$ *Prime factorization*

$\quad = \sqrt{2 \cdot 49 \cdot a \cdot b^2 \cdot c^4}$ *7 × 7 = 49*

$\quad = \sqrt{2} \cdot \sqrt{49} \cdot \sqrt{a} \cdot \sqrt{b^2} \cdot \sqrt{c^4}$ *Product Property of Square Roots*

$\quad = \sqrt{2} \cdot 7 \cdot \sqrt{a} \cdot |b| \cdot c^2$ *Simplify.*

$\quad = 7|b|c^2\sqrt{2a}$ *The absolute value of b ensures a nonnegative result.*

Your Turn

Simplify each expression. Use absolute value symbols if necessary.

k. $\sqrt{63ab^2}$ **l.** $\sqrt{200x^2y^3}$

Check for Understanding

Communicating Mathematics

Study the lesson. Then complete the following.

1. **Explain** why absolute values are sometimes needed when simplifying radical expressions containing variables.

2. **Explain** how you can show whether $8 + \sqrt{2}$ and $8 - \sqrt{2}$ are conjugates.

Vocabulary

radical expression
rationalizing the denominator
conjugates

3. 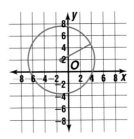 LaToya says that, in simplest form, the expression $\dfrac{2}{3+\sqrt{5}}$ is written as $\dfrac{6-2\sqrt{5}}{4}$. Greg disagrees. He says it should be written as $\dfrac{3-\sqrt{5}}{2}$. Who is correct? Explain your reasoning.

Guided Practice

⊘ Getting Ready

State the conjugate of each expression. Then multiply the expression by its conjugate.

Sample: $1 + \sqrt{7}$ **Solution:** The conjugate is $1 - \sqrt{7}$.

$$(1 + \sqrt{7})(1 - \sqrt{7}) = 1^2 - (\sqrt{7})^2$$
$$= 1 - 7 \text{ or } -6$$

4. $5 - \sqrt{6}$ **5.** $2 + \sqrt{8}$ **6.** $2\sqrt{7} - 3\sqrt{2}$

Simplify each expression. Leave in radical form. *(Examples 1–6)*

7. $\sqrt{75}$ **8.** $\sqrt{96}$ **9.** $\sqrt{6} \cdot \sqrt{15}$

10. $\dfrac{\sqrt{36}}{\sqrt{3}}$ **11.** $\dfrac{\sqrt{3}}{\sqrt{7}}$ **12.** $\dfrac{1}{6 - \sqrt{3}}$

Simplify each expression. Use absolute value symbols if necessary. *(Example 7)*

13. $\sqrt{36x^2y}$ **14.** $\sqrt{50m^4n^5}$

15. Geometry The radius of the circle is $\sqrt{32}$ units long. Express the length of the radius in simplest form. *(Example 2)*

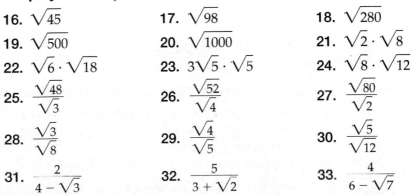

Exercises

Practice

Simplify each expression. Leave in radical form.

16. $\sqrt{45}$ **17.** $\sqrt{98}$ **18.** $\sqrt{280}$

19. $\sqrt{500}$ **20.** $\sqrt{1000}$ **21.** $\sqrt{2} \cdot \sqrt{8}$

22. $\sqrt{6} \cdot \sqrt{18}$ **23.** $3\sqrt{5} \cdot \sqrt{5}$ **24.** $\sqrt{8} \cdot \sqrt{12}$

25. $\dfrac{\sqrt{48}}{\sqrt{3}}$ **26.** $\dfrac{\sqrt{52}}{\sqrt{4}}$ **27.** $\dfrac{\sqrt{80}}{\sqrt{2}}$

28. $\dfrac{\sqrt{3}}{\sqrt{8}}$ **29.** $\dfrac{\sqrt{4}}{\sqrt{5}}$ **30.** $\dfrac{\sqrt{5}}{\sqrt{12}}$

31. $\dfrac{2}{4 - \sqrt{3}}$ **32.** $\dfrac{5}{3 + \sqrt{2}}$ **33.** $\dfrac{4}{6 - \sqrt{7}}$

Simplify each expression. Use absolute value symbols if necessary.

34. $\sqrt{16gh^4}$ **35.** $\sqrt{40m^2}$ **36.** $\sqrt{47c^6d}$

37. $\sqrt{54a^2b^3}$ **38.** $\sqrt{125rst}$ **39.** $\sqrt{36x^3y^4z^5}$

40. Quilting The quilt pattern *Sky Rocket* is shown at the right. Suppose the legs of the indicated triangle each measure 8 centimeters. Use the Pythagorean Theorem to find the measure of the hypotenuse. Express the answer as a radical in simplest form.

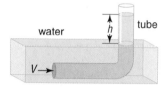

Sky Rocket

41. Science An L-shaped tube like the one shown can be used to measure the speed V in miles per hour of water in a river. By using the formula $V = \sqrt{2.5h}$, where h is the height in inches of the column of water above the surface, the speed of the water can be found.

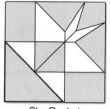

water   tube

$V \rightarrow$

a. Suppose the tube is placed in a river and the height of the column of water is 4.8 inches. What is the speed of the water?

b. What would h be if the speed is exactly 5 miles per hour?

42. Critical Thinking Is the sentence $\sqrt{a \cdot b} = \sqrt{a} \cdot \sqrt{b}$ true for negative values of a and b? Explain your reasoning.

Mixed Review

43. Point J is located at $(-1, 5)$, and point K is located at $(m, 2)$. What is the value of m if the points are $\sqrt{18}$ units apart? *(Lesson 14–2)*

Name the set or sets of numbers to which each real number belongs. Let N = natural numbers, W = whole numbers, Z = integers, Q = rational numbers and I = irrational numbers. *(Lesson 14–1)*

44. 0.75 **45.** $-\sqrt{144}$ **46.** $\dfrac{20}{4}$

47. Use substitution to solve $y = 2x + 3$ and $y - 2x = -5$. *(Lesson 13–3)*

48. Solve $15 > a - 3 > 10$. Graph the solution. *(Lesson 12–5)*

49. Tell whether 9 is closer to $\sqrt{79}$ or $\sqrt{89}$. *(Lesson 8–6)*

50. Standardized Test Practice Two dice are rolled. Find the probability that a number less than 5 is rolled on one die and an odd number is rolled on the other die. *(Lesson 5–7)*

A $\dfrac{1}{2}$ **B** $\dfrac{2}{3}$ **C** $\dfrac{1}{3}$ **D** $\dfrac{3}{4}$

14-4 Adding and Subtracting Radical Expressions

Math
In the Workplace

What You'll Learn
You'll learn to add and subtract radical expressions.

Why It's Important
Hobbies Knowing how to add radical expressions can help you find the perimeter of a sail on a sailboat. *See Exercise 37.*

Look Back

Adding and Subtracting Monomials: Lesson 9–2

To find the exact perimeter of quadrilateral *ABCD*, you will need to add radical expressions.

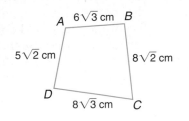

Radical expressions with the same radicands can be added or subtracted in the same way that monomials are added or subtracted.

Monomials

$$5x + 3x = (5 + 3)x$$
$$= 8x$$

$$8y - 2y = (8 - 2)y$$
$$= 6y$$

Radical Expressions

$$5\sqrt{2} + 3\sqrt{2} = (5 + 3)\sqrt{2}$$
$$= 8\sqrt{2}$$

$$8\sqrt{3} - 2\sqrt{3} = (8 - 2)\sqrt{3}$$
$$= 6\sqrt{3}$$

Notice that the Distributive Property was used to simplify each radical expression.

Examples

Simplify each expression.

1 $6\sqrt{7} - 2\sqrt{7}$

$6\sqrt{7} - 2\sqrt{7} = (6 - 2)\sqrt{7}$ *Distributive Property*
$= 4\sqrt{7}$

2 $5\sqrt{5} + 3\sqrt{5} - 18\sqrt{5}$

$5\sqrt{5} + 3\sqrt{5} - 18\sqrt{5} = (5 + 3 - 18)\sqrt{5}$ *Distributive Property*
$= -10\sqrt{5}$

Your Turn

a. $8\sqrt{6} + 3\sqrt{6}$

b. $5\sqrt{2} - 12\sqrt{2}$

c. $4\sqrt{3} + 7\sqrt{3} - 2\sqrt{3}$

d. $3\sqrt{13} - 2\sqrt{13} - 6\sqrt{13}$

3 **Refer to the beginning of the lesson. Find the exact perimeter of quadrilateral** *ABCD*.

$P = 6\sqrt{3} + 8\sqrt{2} + 8\sqrt{3} + 5\sqrt{2}$ *Like terms: $6\sqrt{3}$ and $8\sqrt{3}$;*
$8\sqrt{2}$ and $5\sqrt{2}$

$= 6\sqrt{3} + 8\sqrt{3} + 8\sqrt{2} + 5\sqrt{2}$ *Commutative Property*

$= (6 + 8)\sqrt{3} + (8 + 5)\sqrt{2}$ *Distributive Property*

$= 14\sqrt{3} + 13\sqrt{2}$

The exact perimeter of quadrilateral *ABCD* is $14\sqrt{3} + 13\sqrt{2}$ centimeters.

In Example 3, the expression $14\sqrt{3} + 13\sqrt{2}$ cannot be simplified further for the following reasons.

- The radicands are different.
- There are no common factors.
- Each radicand is in simplest form.

If the radicals in a radical expression are not in simplest form, simplify them first. Then use the Distributive Property wherever possible to further simplify the expression.

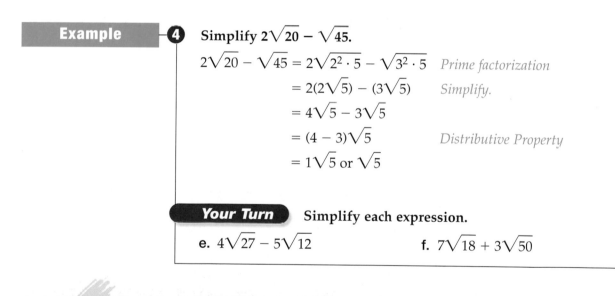

Example

4 **Simplify** $2\sqrt{20} - \sqrt{45}$.

$2\sqrt{20} - \sqrt{45} = 2\sqrt{2^2 \cdot 5} - \sqrt{3^2 \cdot 5}$ *Prime factorization*

$= 2(2\sqrt{5}) - (3\sqrt{5})$ *Simplify.*

$= 4\sqrt{5} - 3\sqrt{5}$

$= (4 - 3)\sqrt{5}$ *Distributive Property*

$= 1\sqrt{5}$ or $\sqrt{5}$

Your Turn **Simplify each expression.**

e. $4\sqrt{27} - 5\sqrt{12}$ **f.** $7\sqrt{18} + 3\sqrt{50}$

Check for Understanding

Communicating Mathematics

Study the lesson. Then complete the following.

1. **Describe** in your own words how to add radical expressions.

2. **Explain** why you should simplify each radical in a radical expression before adding or subtracting.

Math Journal

3. **Explain** how you use the Distributive Property to simplify the sum or difference of like radicals.

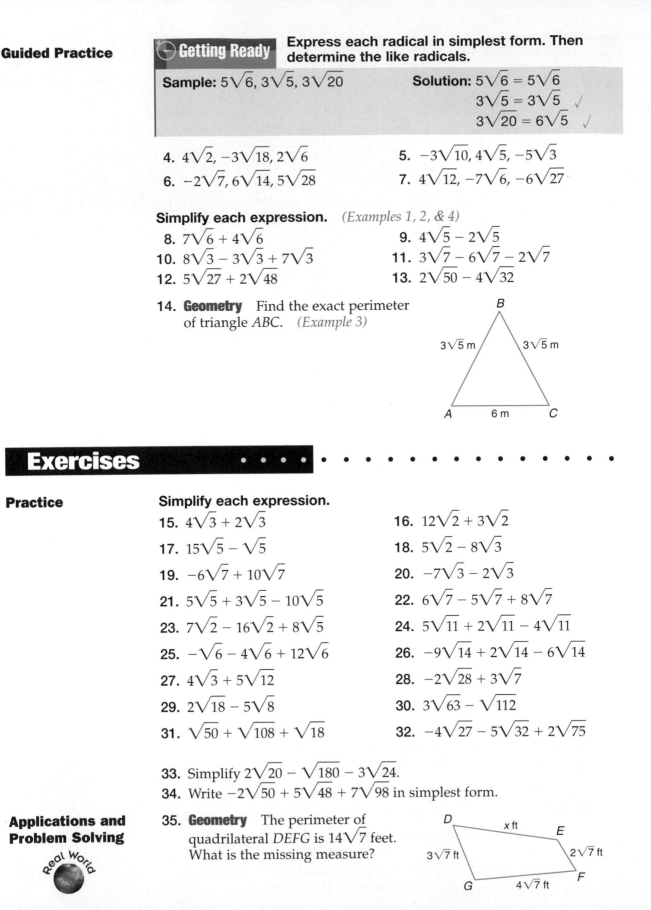

⏱ **Getting Ready** | **Express each radical in simplest form. Then determine the like radicals.**

Sample: $5\sqrt{6}$, $3\sqrt{5}$, $3\sqrt{20}$　　　　**Solution:** $5\sqrt{6} = 5\sqrt{6}$
　　　　　　　　　　　　　　　　　　　　　　　$3\sqrt{5} = 3\sqrt{5}$ ✓
　　　　　　　　　　　　　　　　　　　　　　　$3\sqrt{20} = 6\sqrt{5}$ ✓

4. $4\sqrt{2}$, $-3\sqrt{18}$, $2\sqrt{6}$ 　　　　　**5.** $-3\sqrt{10}$, $4\sqrt{5}$, $-5\sqrt{3}$

6. $-2\sqrt{7}$, $6\sqrt{14}$, $5\sqrt{28}$ 　　　　　**7.** $4\sqrt{12}$, $-7\sqrt{6}$, $-6\sqrt{27}$

Simplify each expression. *(Examples 1, 2, & 4)*

8. $7\sqrt{6} + 4\sqrt{6}$ 　　　　　　　　　**9.** $4\sqrt{5} - 2\sqrt{5}$

10. $8\sqrt{3} - 3\sqrt{3} + 7\sqrt{3}$ 　　　　　**11.** $3\sqrt{7} - 6\sqrt{7} - 2\sqrt{7}$

12. $5\sqrt{27} + 2\sqrt{48}$ 　　　　　　　**13.** $2\sqrt{50} - 4\sqrt{32}$

14. Geometry Find the exact perimeter of triangle *ABC*. *(Example 3)*

B
$3\sqrt{5}$ m 　 $3\sqrt{5}$ m
A 　 6 m 　 C

Exercises

Practice

Simplify each expression.

15. $4\sqrt{3} + 2\sqrt{3}$ 　　　　　　　**16.** $12\sqrt{2} + 3\sqrt{2}$

17. $15\sqrt{5} - \sqrt{5}$ 　　　　　　　**18.** $5\sqrt{2} - 8\sqrt{3}$

19. $-6\sqrt{7} + 10\sqrt{7}$ 　　　　　**20.** $-7\sqrt{3} - 2\sqrt{3}$

21. $5\sqrt{5} + 3\sqrt{5} - 10\sqrt{5}$ 　　　**22.** $6\sqrt{7} - 5\sqrt{7} + 8\sqrt{7}$

23. $7\sqrt{2} - 16\sqrt{2} + 8\sqrt{5}$ 　　　**24.** $5\sqrt{11} + 2\sqrt{11} - 4\sqrt{11}$

25. $-\sqrt{6} - 4\sqrt{6} + 12\sqrt{6}$ 　　　**26.** $-9\sqrt{14} + 2\sqrt{14} - 6\sqrt{14}$

27. $4\sqrt{3} + 5\sqrt{12}$ 　　　　　　**28.** $-2\sqrt{28} + 3\sqrt{7}$

29. $2\sqrt{18} - 5\sqrt{8}$ 　　　　　　**30.** $3\sqrt{63} - \sqrt{112}$

31. $\sqrt{50} + \sqrt{108} + \sqrt{18}$ 　　　**32.** $-4\sqrt{27} - 5\sqrt{32} + 2\sqrt{75}$

33. Simplify $2\sqrt{20} - \sqrt{180} - 3\sqrt{24}$.

34. Write $-2\sqrt{50} + 5\sqrt{48} + 7\sqrt{98}$ in simplest form.

Applications and Problem Solving

🌎 *Real World*

35. Geometry The perimeter of quadrilateral *DEFG* is $14\sqrt{7}$ feet. What is the missing measure?

D 　 x ft 　 E
$3\sqrt{7}$ ft 　　　　　 $2\sqrt{7}$ ft
G 　 $4\sqrt{7}$ ft 　 F

Close to Home

COUNT OFF BY THE SQUARE ROOT OF 7 !

UHH....

36. **Media** Refer to the comic. Suppose the first team member says "$\sqrt{7}$," the second says "$2\sqrt{7}$," and so on.
   a. Write an expression that could be used to determine the response of the fourth student in line.
   b. Determine the response of the fourth person in line.
   c. Suppose the coach asks the students to count off by $\sqrt{5}$. Find the response of the last person in line.

37. **Hobbies** Ling is making a model of a sailboat. The dimensions of one of the sails are shown in the diagram.
   a. Find the missing measure. (*Hint:* Use the Pythagorean Theorem.)
   b. Determine the exact perimeter of the sail.

6 in.

x in.

$2\sqrt{3}$ in.

38. **Critical Thinking** Is the set of irrational numbers closed under addition? Explain your reasoning.

Mixed Review

39. Simplify $\sqrt{45mn^3p^2}$. (*Lesson 14–3*)

Find the distance between each pair of points. Round to the nearest tenth, if necessary. (*Lesson 14–2*)

40. $C(-3, 5)$, $D(0, 8)$

41. $J(4, -2)$, $K(-6, -1)$

42. **Geometry** The base of a triangle measures 3 feet more than its height. The area of the triangle is 20 square feet. Write and solve a quadratic equation to find the height and base of the triangle. (*Lesson 11–4*)

Factor each monomial. (*Lesson 10–1*)

43. $-26a^3b^2$

44. $36xy^4z$

45. Find the product of $a + 2c$ and $a - 2c$. (*Lesson 9–5*)

46. **Standardized Test Practice** The scatter plot shows the relationship between the cost of a stereo system and the rating it received by a consumer group. Which of the following best describes the relationship? (*Lesson 7–4*)

   A positive
   B dependent
   C no pattern
   D negative

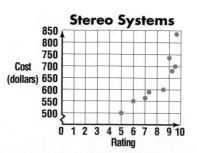

Stereo Systems

Cost (dollars)

14-5 Solving Radical Equations

What You'll Learn

You'll learn to solve simple radical equations in which only one radical contains a variable.

Why It's Important

Engineering
Engineers use radical equations to determine the velocity of roller coasters.
See Example 5.

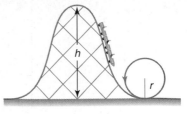

The speed of a roller coaster as it travels through a loop depends on the height of the hill from which the coaster has just descended. The equation $s = 8\sqrt{h - 2r}$ gives the speed s in feet per second where h is the height of the hill and r is the radius of the loop.

Suppose the owner of an amusement park wants to design a roller coaster that will travel at a speed of 40 feet per second as it goes through a loop with a radius of 30 feet. How high should the hill be? *This problem will be solved in Example 5.*

Equations like $s = 8\sqrt{h - 2r}$ that contain radicals with variables in the radicand are called **radical equations**. To solve these equations, first isolate the radical on one side of the equation. Then square each side of the equation to eliminate the radical.

Examples

Solve each equation. Check your solution.

❶ $\sqrt{x} + 5 = 8$

$$\sqrt{x} + 5 = 8$$
$$\sqrt{x} + 5 - 5 = 8 - 5 \quad \textit{Subtract 5 from each side.}$$
$$\sqrt{x} = 3$$
$$(\sqrt{x})^2 = 3^2 \quad \textit{Square each side.}$$
$$x = 9$$

Check: $\sqrt{x} + 5 = 8$
$$\sqrt{9} + 5 \stackrel{?}{=} 8 \quad \textit{Replace x with 9.}$$
$$3 + 5 \stackrel{?}{=} 8$$
$$8 = 8 \quad \checkmark$$

The solution is 9.

❷ $\sqrt{m + 3} + 2 = 6$

$$\sqrt{m + 3} + 2 = 6$$
$$\sqrt{m + 3} + 2 - 2 = 6 - 2 \quad \textit{Subtract 2 from each side.}$$
$$\sqrt{m + 3} = 4$$
$$(\sqrt{m + 3})^2 = 4^2 \quad \textit{Square each side.}$$
$$m + 3 = 16$$
$$m + 3 - 3 = 16 - 3 \quad \textit{Subtract 3 from each side.}$$
$$m = 13 \quad \textit{Check this result.}$$

The solution is 13.

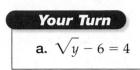

a. $\sqrt{y} - 6 = 4$

b. $\sqrt{a-1} + 5 = 7$

You can use a graphing calculator to solve radical equations.

Graphing Calculator Tutorial

See pp. 724–727.

Graphing Calculator Exploration

Find the solution of $\sqrt{x} + 5 = 8$.

Step 1 Set the viewing window for x: [0, 15] by 1 and y: [0, 15] by 1.

Step 2 Press $\boxed{Y=}$. Enter the left side of the equation, $\sqrt{x} + 5$, as Y_1. Enter the right side of the equation, 8, as Y_2.

Step 3 Press $\boxed{\text{GRAPH}}$ to graph the equations.

Step 4 The solution of the equation is the intersection point of the two lines. Use the INTERSECT feature to find the intersection point.

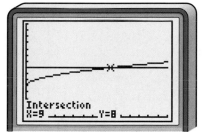

Enter: $\boxed{\text{2nd}}$ [CALC] 5

$\boxed{\text{ENTER}}$ $\boxed{\text{ENTER}}$ $\boxed{\text{ENTER}}$

The graph intersects at the point (9, 8). The solution is the x-coordinate, 9.

Try These

1. Solve the equation in Example 2 by using a graphing calculator.
2. Use a graphing calculator to find the solution of Your Turn Exercise b.
3. What changes must be made to the graphing calculator before finding the solution of Your Turn Exercise a?

Squaring each side of an equation may produce results that do not satisfy the *original* equation. So, you must check all solutions when you solve radical equations.

Examples

Solve each equation. Check your solution.

❸ $\sqrt{n+2} = n - 4$

$$\sqrt{n+2} = n - 4$$
$$(\sqrt{n+2})^2 = (n-4)^2 \qquad \text{\textit{Square each side.}}$$
$$n + 2 = n^2 - 8n + 16$$
$$n - n + 2 - 2 = n^2 - 8n + 16 - n - 2 \qquad \text{\textit{Subtract n and 2 from each side.}}$$
$$0 = n^2 - 9n + 14$$
$$0 = (n-7)(n-2) \qquad \text{\textit{Factor.}}$$

(continued on the next page)

$$n - 7 = 0 \quad \text{or} \quad n - 2 = 0 \quad \textit{Use the Zero Product Property.}$$
$$n = 7 \qquad\qquad n = 2$$

Check:
$$\sqrt{n + 2} = n - 4 \qquad\qquad \sqrt{n + 2} = n - 4$$
$$\sqrt{7 + 2} \stackrel{?}{=} 7 - 4 \qquad\qquad \sqrt{2 + 2} \stackrel{?}{=} 2 - 4$$
$$\sqrt{9} \stackrel{?}{=} 3 \qquad\qquad\qquad \sqrt{4} \stackrel{?}{=} -2$$
$$3 = 3 \quad \checkmark \qquad\qquad\qquad 2 \neq -2$$

Since 2 does not satisfy the original equation, 7 is the only solution.

❹ $\sqrt{3h - 5} + 5 = h$

$$\sqrt{3h - 5} + 5 = h$$
$$\sqrt{3h - 5} + 5 - 5 = h - 5 \qquad\qquad \textit{Subtract 5 from each side.}$$
$$\sqrt{3h - 5} = h - 5$$
$$(\sqrt{3h - 5})^2 = (h - 5)^2 \qquad\qquad \textit{Square each side.}$$
$$3h - 5 = h^2 - 10h + 25$$
$$3h - 3h - 5 + 5 = h^2 - 10h + 25 - 3h + 5 \quad \textit{Add} -3h \textit{ and 5 to each side.}$$
$$0 = h^2 - 13h + 30$$
$$0 = (h - 10)(h - 3) \qquad\qquad \textit{Factor.}$$
$$h - 10 = 0 \quad \text{or} \quad h - 3 = 0 \qquad \textit{Use the Zero Product}$$
$$h = 10 \qquad\qquad h = 3 \qquad\qquad \textit{Property.}$$

Check:
$$\sqrt{3h - 5} + 5 = h \qquad\qquad \sqrt{3h - 5} + 5 = h$$
$$\sqrt{3(10) - 5} + 5 \stackrel{?}{=} 10 \qquad\qquad \sqrt{3(3) - 5} + 5 \stackrel{?}{=} 3$$
$$\sqrt{30 - 5} + 5 \stackrel{?}{=} 10 \qquad\qquad \sqrt{9 - 5} + 5 \stackrel{?}{=} 3$$
$$\sqrt{25} + 5 \stackrel{?}{=} 10 \qquad\qquad\qquad \sqrt{4} + 5 \stackrel{?}{=} 3$$
$$5 + 5 \stackrel{?}{=} 10 \qquad\qquad\qquad 2 + 5 \stackrel{?}{=} 3$$
$$10 = 10 \quad \checkmark \qquad\qquad\qquad 7 \neq 3$$

Since 3 does not satisfy the original equation, 10 is the only solution.

Your Turn

c. $\sqrt{d + 1} = d - 1$ 　　　　　　**d.** $\sqrt{3x - 14} + x = 6$

Radical equations are used in many real-life situations.

Example

Engineering Link

Real World

❺ **Refer to the application at the beginning of the lesson. How high should the hill be?**

Explore 　　You know the speed of the coaster and the radius of the loop. You need to know the height of the hill.

Plan   Use the equation $s = 8\sqrt{h - 2r}$ to find the height of the hill.

Solve

$$s = 8\sqrt{h - 2r}$$

$$40 = 8\sqrt{h - 2(30)} \quad \textit{Replace s with 40 and r with 30.}$$

$$40 = 8\sqrt{h - 60}$$

$$\frac{40}{8} = \frac{8\sqrt{h - 60}}{8} \quad \textit{Divide each side by 8.}$$

$$5 = \sqrt{h - 60}$$

$$5^2 = (\sqrt{h - 60})^2 \quad \textit{Square each side.}$$

$$25 = h - 60$$

$$25 + 60 = h - 60 + 60 \quad \textit{Add 60 to each side.}$$

$$85 = h$$

The height of the hill is 85 feet.

Examine   Check the result.

$$s = 8\sqrt{h - 2r}$$

$$40 \overset{?}{=} 8\sqrt{85 - 2(30)} \quad \textit{Replace s with 40, h with 85, and}$$
$$\textit{r with 30.}$$

$$40 \overset{?}{=} 8\sqrt{85 - 60}$$

$$40 \overset{?}{=} 8\sqrt{25}$$

$$40 \overset{?}{=} 8 \cdot 5$$

$$40 = 40 \quad \checkmark$$

The answer is correct.

Check for Understanding

**Communicating
Mathematics**

Study the lesson. Then complete the following.

1. **Tell** the first step you should take when solving
a radical equation.

2. **Explain** why it is important to check radical equations after finding
possible solutions.

Math Journal

3. **Write** an example of a radical equation. Then write the steps for
solving the equation.

Vocabulary
radical equation

Guided Practice

⏲ **Getting Ready**   **Square each side of the following equations.**

Sample: $\sqrt{y - 1} = 4$ **Solution:** $\sqrt{y - 1} = 4$
$$(\sqrt{y - 1})^2 = 4^2$$
$$y - 1 = 16$$

4. $\sqrt{x} = 7$ 5. $\sqrt{a + 5} = 2$ 6. $9 = \sqrt{2c - 3}$

Solve each equation. Check your solution. *(Examples 1–4)*

7. $\sqrt{x} - 4 = 7$

8. $\sqrt{a - 2} - 5 = 3$

9. $\sqrt{m + 5} + 1 = m$

10. $\sqrt{1 + 2x} = x + 1$

11. **Engineering** Refer to the application at the beginning of the lesson. Suppose the owner of an amusement park wants to design a roller coaster that will travel through a loop at a speed of 56 feet per second after descending a 120-foot hill. What should the radius of the loop be? *(Example 5)*

Exercises • • • • • • • • • • • • • • • • • •

Practice

Solve each equation. Check your solution.

12. $\sqrt{x} = 3$

13. $\sqrt{a} = -4$

14. $7 = \sqrt{7m}$

15. $\sqrt{-3d} = 6$

16. $\sqrt{y} + 5 = 0$

17. $0 = \sqrt{2c} - 2$

18. $\sqrt{m - 4} = 6$

19. $\sqrt{w + 6} = 9$

20. $3 = \sqrt{4n + 1}$

21. $11 = \sqrt{2z - 5}$

22. $\sqrt{8h + 1} - 5 = 0$

23. $2 = \sqrt{3b - 5} + 6$

24. $\sqrt{x + 6} = x$

25. $p = \sqrt{5p - 6}$

26. $\sqrt{8 - b} = b - 2$

27. $\sqrt{k - 2} + 4 = k$

28. $t - 1 = \sqrt{2t + 6}$

29. $5 + \sqrt{3j - 5} = j$

30. Solve $\sqrt{\dfrac{a}{4}} = 6$.

31. Find the solution of $\sqrt{\dfrac{5x}{7}} - 8 = 2$.

Applications and Problem Solving

Real World

32. **Science** The speed of sound S, in meters per second, near Earth's surface can be determined using the formula $S = 20\sqrt{t + 273}$, where t is the surface temperature in degrees Celsius. Suppose a racing team has designed a car that can travel 340 meters per second, in hopes of breaking the sound barrier. At what temperature would the speed of sound be 340 meters per second?

33. **Science** The formula $t = \sqrt{\dfrac{2s}{g}}$ can be used to determine the time t, in seconds, it takes an object initially at rest to fall s meters. In this formula, g is the acceleration due to gravity in meters per second squared.

   a. Suppose, on the moon, a rock falls 7.2 meters in 3 seconds. What is the acceleration due to gravity on the moon?

   b. Suppose on Earth a rock falls 78.4 meters in 4 seconds. What is the acceleration due to gravity on Earth?

34. Critical Thinking Find two numbers such that the square root of their sum is 5 and the square root of their product is 12.

Mixed Review

Simplify each expression. *(Lesson 14–4)*

35. $-7\sqrt{5} + 2\sqrt{5}$ **36.** $4\sqrt{7} - 3\sqrt{28}$ **37.** $-\sqrt{54} - \sqrt{18} + \sqrt{24}$

Simplify each expression. Leave in radical form. *(Lesson 14–3)*

38. $\sqrt{5} \cdot \sqrt{12}$ **39.** $\dfrac{\sqrt{15}}{\sqrt{3}}$ **40.** $\dfrac{5}{4 - \sqrt{7}}$

41. Find the solution of $y = \dfrac{1}{2}x - 4$ and $y = -3$ by graphing.
(Lesson 13–1)

Write an inequality for each graph. *(Lesson 12–1)*

42.    **43.**

44. Animals There are two field mice living in a barn. Suppose the number of mice triples every 4 months. How many mice will there be after 3 years if none of them have died? *(Lesson 11–7)*

45. Open-Ended Test Practice Suppose y varies inversely as x and $y = 18$ when $x = 15$. Find y when $x = 12$. *(Lesson 6–6)*

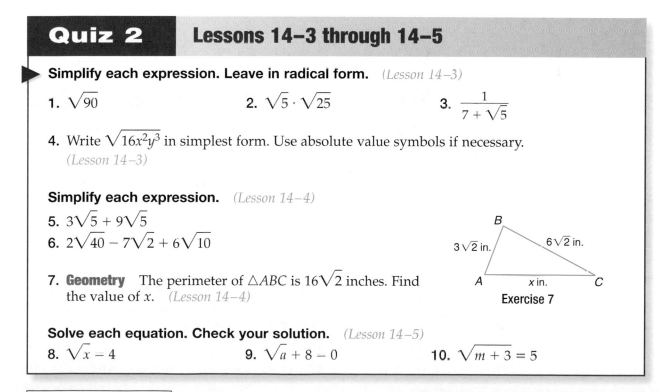

Quiz 2   Lessons 14–3 through 14–5

Simplify each expression. Leave in radical form. *(Lesson 14–3)*

1. $\sqrt{90}$ **2.** $\sqrt{5} \cdot \sqrt{25}$ **3.** $\dfrac{1}{7 + \sqrt{5}}$

4. Write $\sqrt{16x^2y^3}$ in simplest form. Use absolute value symbols if necessary.
(Lesson 14–3)

Simplify each expression. *(Lesson 14–4)*

5. $3\sqrt{5} + 9\sqrt{5}$
6. $2\sqrt{40} - 7\sqrt{2} + 6\sqrt{10}$

7. Geometry The perimeter of $\triangle ABC$ is $16\sqrt{2}$ inches. Find the value of x. *(Lesson 14–4)*

$3\sqrt{2}$ in. $6\sqrt{2}$ in. x in.   Exercise 7

Solve each equation. Check your solution. *(Lesson 14–5)*

8. $\sqrt{x} - 4$ **9.** $\sqrt{a} + 8 - 0$ **10.** $\sqrt{m + 3} = 5$

Understanding and Using the Vocabulary

*inter*NET
CONNECTION **Review Activities**
For more review activities, visit:
www.algconcepts.glencoe.com

After completing this chapter, you should be able to define each term, property, or phrase and give an example or two of each.

conjugates *(p. 616)*
Distance Formula *(p. 607)*

radical equations *(p. 624)*
real numbers *(p. 600)*

rationalizing the
   denominator *(p. 615)*

Choose the correct term to complete each sentence.

1. To find the distance d between any two points (x_1, y_1) and (x_2, y_2), you can use the formula $(\sqrt{(x_2 - x_1)^2 - (y_2 - y_1)^2},\ \sqrt{(x_2 - x_1)^2 + (y_2 - y_1)^2})$.

2. The product of conjugates is (sometimes, always) a rational number.

3. Rationalizing the denominator is a way of eliminating (radicals, perfect squares) from the denominator of a fraction.

4. Natural numbers, whole numbers, and integers are all (irrational, rational) numbers.

5. The expression $(\sqrt{7ab},\ \sqrt{8mn})$ is in simplest form.

6. Radical equations (sometimes, always) have more than one solution.

7. The number 0 is a (whole, natural) number.

8. Binomials of the form $a\sqrt{b} + c\sqrt{d}$ and $a\sqrt{b} - c\sqrt{d}$ are (conjugates, radical equations).

9. The equation $\sqrt{x} + 4 = 0$ has (one, no) solution.

10. The set of (integers, real numbers) contains the sets of rational and irrational numbers.

Skills and Concepts

Objectives	Examples
• **Lesson 14–1** Describe the relationships among sets of numbers.	Name the set or sets of numbers to which each real number belongs. Let N = natural numbers, W = whole numbers, Z = integers, Q = rational numbers, and I = irrational numbers.

$\sqrt{16} = 4$; so $\sqrt{16}$ is a natural number, a whole number, an integer, and a rational number.

The number -3 is an integer and a rational number.

2.151151115 . . . is not the square root of a perfect square. So, it is an irrational number.

11. $-\dfrac{1}{4}$ 12. π

13. $-\sqrt{25}$ 14. $\dfrac{24}{6}$

Find an approximation, to the nearest tenth, for each square root. Then graph the square root on a number line.

15. $\sqrt{5}$ 16. $-\sqrt{15}$

17. $\sqrt{111}$ 18. $-\sqrt{260}$

Chapter 14 Study Guide and Assessment

Objectives and Examples	Review Exercises

• Lesson 14-2 Find the distance between two points in the coordinate plane.

Find the distance between points $A(3, -2)$ and $B(5, 8)$.

$$d = \sqrt{(x_2 - x_1)^2 + (y_2 - y_1)^2}$$
$$= \sqrt{(5 - 3)^2 + (8 - (-2))^2}$$
$$= \sqrt{(2)^2 + (10)^2}$$
$$= \sqrt{4 + 100}$$
$$= \sqrt{104}$$
$$\approx 10.2$$

Find the distance between each pair of points. Round to the nearest tenth, if necessary.

19. $P(1, 2), Q(4, 6)$
20. $J(-6, -3), K(1, 0)$
21. $X(4, -1), Y(-2, 5)$
22. $R(-9, -5), S(-2, 19)$

Find the value of a if the points are the indicated distance apart.

23. $G(a, 2), H(-3, 5); d = \sqrt{13}$
24. $C(4, -7), D(-6, a); d = \sqrt{116}$

• Lesson 14-3 Simplify radical expressions.

$$\sqrt{288} = \sqrt{12^2 \cdot 2}$$
$$= 12\sqrt{2}$$

$$\sqrt{32x^3y^2} = \sqrt{32x^3y^2}$$
$$= \sqrt{2 \cdot 2 \cdot 2 \cdot 2 \cdot 2 \cdot x^3 \cdot y^2}$$
$$= \sqrt{2 \cdot 16 \cdot x^3 \cdot y^2}$$
$$= \sqrt{2} \cdot \sqrt{16} \cdot \sqrt{x^3} \cdot \sqrt{y^2}$$
$$= \sqrt{2} \cdot 4 \cdot x \cdot \sqrt{x} \cdot |y|$$
$$= 4x|y|\sqrt{2x}$$

Simplify each expression. Leave in radical form.

25. $\sqrt{50}$ **26.** $\sqrt{3} \cdot \sqrt{6}$

27. $\dfrac{\sqrt{54}}{\sqrt{2}}$ **28.** $\dfrac{\sqrt{7}}{\sqrt{13}}$

29. $\dfrac{1}{4 - \sqrt{5}}$ **30.** $\dfrac{3}{13 - \sqrt{2}}$

Simplify each expression. Use absolute value symbols if necessary.

31. $\sqrt{300a^2bc^4}$ **32.** $\sqrt{121m^5n^4p^3}$

• Lesson 14-4 Add and subtract radical expressions.

$$8\sqrt{7} - 6\sqrt{7} = (8 - 6)\sqrt{7}$$
$$= 2\sqrt{7}$$

$$13\sqrt{2} + 4\sqrt{2} - 7\sqrt{2} = (13 + 4 - 7)\sqrt{2}$$
$$= 10\sqrt{2}$$

$$\sqrt{54} + \sqrt{96} = \sqrt{3^2 \cdot 6} + \sqrt{4^2 \cdot 6}$$
$$= 3\sqrt{6} + 4\sqrt{6}$$
$$= (3 + 4)\sqrt{6} \text{ or } 7\sqrt{6}$$

Simplify each expression.

33. $2\sqrt{5} + 5\sqrt{5}$
34. $3\sqrt{8} - 6\sqrt{8}$
35. $-5\sqrt{6} + 14\sqrt{6} - 9\sqrt{6}$
36. $12\sqrt{3} - 14\sqrt{3} + 6\sqrt{3}$
37. $7\sqrt{2} + 5\sqrt{18}$
38. $-5\sqrt{40} + 10\sqrt{10}$

39. Simplify $2\sqrt{45} - 5\sqrt{80}$.
40. Write $2\sqrt{45} + 3\sqrt{75} - \sqrt{50}$ in simplest form.

Objectives and Examples	Review Exercises

• Lesson 14–5 Solve simple radical equations in which only one radical contains a variable.

Solve $\sqrt{x+4} - 8 = -5$.

$$\sqrt{x+4} - 8 = -5$$
$$\sqrt{x+4} - 8 + 8 = -5 + 8$$
$$\sqrt{x+4} = 3$$
$$(\sqrt{x+4})^2 = 3^2$$
$$x + 4 = 9$$
$$x + 4 - 4 = 9 - 4$$
$$x = 5$$

The solution is 5.

Solve each equation. Check your solution.

41. $\sqrt{y} = 5$ **42.** $\sqrt{m} = -3$

43. $\sqrt{9h} = 9$ **44.** $\sqrt{-4a} = 8$

45. $7 = \sqrt{a} + 10$ **46.** $\sqrt{n-12} = 15$

47. $t - 1 = \sqrt{3t + 7}$ **48.** $\sqrt{4x - 3} = x$

49. Solve $\sqrt{\dfrac{g}{6}} = 5$.

50. Find the solution of $\sqrt{\dfrac{3x}{4}} + 2 = 5$.

Applications and Problem Solving

51. Weather The graph shows record low temperatures. Name the set or sets of numbers to which the temperatures belong. *(Lesson 14–1)*

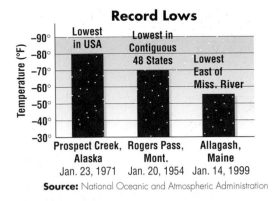

Record Lows

Source: National Oceanic and Atmospheric Administration

52. Geometry Quadrilateral *EFGH* is a square with vertices *E*(2, 5), *F*(6, 1), *G*(2, −3), and *H*(−2, 1). *(Lesson 14–2)*

  a. Draw quadrilateral *EFGH* on a coordinate plane.

  b. Determine the perimeter of the figure. Round to the nearest tenth.

53. Gems The weight of the largest pearl ever found in a giant clam is about $\sqrt{200}$ pounds. Express the weight in simplest radical form. *(Lesson 14–3)*

54. Geometry Triangle *XYZ* has a perimeter of $28\sqrt{5}$ inches. What is the measure of side *YZ*? *(Lesson 14–4)*

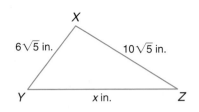

55. Nature The equation $S = 3\sqrt{d}$ gives the speed *S* of a tidal wave in meters per second if *d* is the depth of the water in meters. *(Lesson 14–5)*

  a. Suppose a tidal wave is traveling 186 meters per second. What is the depth of the water?

  b. What is the depth of the water if a tidal wave is traveling 153 meters per second?

1. **Name** two examples of each of the following.
   a. integer   **b.** irrational number   **c.** rational number

2. **Write** a pair of conjugates.

Name the set or sets of numbers to which each real number belongs.
Let N = natural numbers, W = whole numbers, Z = integers,
Q = rational numbers, and I = irrational numbers.

3. $\dfrac{5}{3}$ 4. 0   5. $-0.45275563\ldots$ 6. $-\sqrt{81}$

7. Graph $\sqrt{5}$, $-\sqrt{3}$, and 4.5 on a number line.

8. Write 3.22, -3.22, $\sqrt{11}$, and $3.\overline{2}$, in order from least to greatest.

Find the distance between each pair of points. Round to the nearest tenth, when necessary.

9. $G(7, 4)$, $H(-2, 4)$ 10. $X(3, -3)$, $Y(-9, 2)$ 11. $M(1, -1)$, $N(-5, 1)$

12. Suppose $A(5, m)$ and $B(8, 1)$ are 5 units apart. What is the value of m?

Simplify each expression. Leave in radical form.

13. $\sqrt{80}$ 14. $\sqrt{3} \cdot \sqrt{27}$ 15. $4\sqrt{3} \cdot \sqrt{6}$

16. $\dfrac{\sqrt{96}}{\sqrt{8}}$ 17. $\dfrac{\sqrt{7}}{\sqrt{11}}$ 18. $\dfrac{7}{7 + \sqrt{5}}$

19. $-4\sqrt{7} + 6\sqrt{7} - 12\sqrt{7}$ 20. $5\sqrt{18} - 2\sqrt{50}$

Simplify each expression. Use absolute value symbols if necessary.

21. $\sqrt{25a^3b^2}$ 22. $\sqrt{56m^4np}$

Solve each equation. Check your solution.

23. $\sqrt{b} + 8 = 5$ 24. $\sqrt{4x - 3} = 6 - x$

25. **Science** The distance d in miles a person can see on any planet is given by the formula $d = \sqrt{\dfrac{rh}{2640}}$, where r is the radius of the planet in miles and h is the height of the person in feet. Suppose a person 6 feet tall is standing on Mars. If the radius of the planet is 2109 miles, how far could the person see? Round to the nearest tenth.

Systems of Equations Problems

On standardized tests, you will often need to translate problems into systems of equations and inequalities. You will also need to solve systems of equations using substitution or addition methods.

THE PRINCETON REVIEW

To solve systems of equations on standardized tests, try adding them first. This technique may work because the systems often are of the form $x + y = 6$ and $x - y = 4$.

Proficiency Test Example

Hector has 20 coins. They are all quarters and nickels. The total value is $2.20. Which system of equations can be used to determine the number of quarters q and the number of nickels n?

A $q + n = 20$
     $0.30qn = 2.20$

B $q + n = 20$
     $0.25q + 0.05n = 2.20$

C $q + n = 20$
     $0.05q + 0.25n = 2.20$

D $q + n = 20$
     $0.25q + 0.05n = 20$

Hint   First write the equations in words and then translate them into symbols.

Solution   The number of quarters plus the number of nickels is 20.

$$q + n = 20$$

The value of the quarters plus the value of the nickels equals the total value. The value of the quarters is 25 cents times the number of quarters or $0.25q$. The value of the nickels is 5 cents times the number of nickels or $0.05n$.

$$0.25q + 0.05n = 2.20$$

The answer is B.

SAT Example

If $2x + 3y = 20$ and $3x + 2y = 40$, what is the value of $x + y$?

Hint   Look for ways to solve a problem without lengthy calculations.

Solution   Read carefully. You must find the value of $x + y$, not the individual values of x and y.

Add the two equations.

$$\begin{array}{r} 2x + 3y = 20 \\ + \; 3x + 2y = 40 \\ \hline 5x + 5y = 60 \end{array}$$

Notice that the coefficients of x and y are both 5. Divide each side of the equation by 5. Simplify each term.

$$\frac{5x + 5y}{5} = \frac{60}{5}$$

$$\frac{\overset{1}{\cancel{5}}(x + y)}{\underset{1}{\cancel{5}}} = \frac{\overset{12}{\cancel{60}}}{\underset{1}{\cancel{5}}}$$

$$x + y = 12$$

The value of $x + y$ is 12. The answer is 12.

After you work each problem, record your answer on the answer sheet provided or on a sheet of paper.

1. Jared's total score on the SAT was 1340. His math score m was 400 points less than twice his verbal score v. Which system of equations will help determine his scores?

A $m + v = 1340$
$m = 2v - 400$

B $m + v = 1340$
$m = 400 - 2v$

C $m + v = 1340$
$400m = 2v$

D $m - v = 1340$
$m = 2v - 400$

2. The flag for Monroe High School has an area of 120 square feet. A new flag will have a width $1\frac{1}{2}$ times the width of the old flag. The length remains the same. What is the area, in square feet, of the new flag?

A 160   **B** 180   **C** 240   **D** 360

3. At what point do the lines with equations $y = 2x - 2$ and $7x - 3y = 11$ intersect?

A $(5, 8)$ **B** $(8, 5)$ **C** $\left(\frac{5}{8}, -1\right)$

D $\left(\frac{5}{8}, 1\right)$ **E** $\left(\frac{25}{16}, \frac{9}{8}\right)$

4. Triangle LMN is similar to triangle PQR. What is PQ?

A 4.4 in.
B 7 in.
C 10 in.
D 11.2 in.

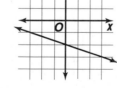

5. Which equation represents the line graphed?

A $y = -\frac{1}{3}x - 2$

B $y = -\frac{1}{3}x + 2$

C $y = \frac{1}{3}x - 2$ **D** $y = \frac{1}{3}x + 2$

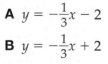

Test Practice For additional test practice questions, visit: www.algconcepts.glencoe.com

6. Gina has x marbles and Dawn has y marbles. Together they have q marbles. If Gina gives Dawn 3 of her marbles, then they will have an equal number. How many marbles does Dawn have?

A $\frac{q - 6}{2}$ **B** $\frac{q - 3}{2}$ **C** $\frac{q + 6}{2}$

D $q - 3$ **E** $q + 3$

7. What is the slope of a line parallel to the line graphed?

A -3 **B** $-\frac{1}{3}$

C $\frac{1}{3}$ **D** 3

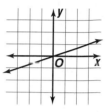

Quantitative Comparison

8. $9^n - 8^n = 1^n$

Column A	Column B
1	n

A if the quantity in Column A is greater;
B if the quantity in Column B is greater;
C if the two quantities are equal;
D if the relationship cannot be determined from the information given.

Open-Ended Questions

9. **Grid-In** $\angle ABC$ is a right angle. $\overrightarrow{BC}$ bisects $\angle EBD$. If $m\angle ABD$ is 130, then what is x?

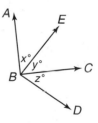

10. Talent show tickets are $3 for adults and $2 for students. Ticket sales of at least $600 are needed to cover costs. At least 5 times as many students as adults attend.

Part A Write a system of inequalities to find the number of adult and student tickets that need to be sold. Let x represent the number of adult tickets and let y represent the number of student tickets.

Part B Graph the system of inequalities. Give one example of adult and student ticket sales that would solve the system.

CHAPTER 15

Rational Expressions and Equations

▶ What You'll Learn in Chapter 15:

- to simplify rational expressions (*Lesson 15–1*),
- to add, subtract, multiply, and divide rational expressions (*Lessons 15–2, 15–4, and 15–5*),
- to divide polynomials by binomials (*Lesson 15–3*), **and**
- to solve equations containing rational expressions (*Lesson 15–6*).

Problem-Solving Workshop

Project

Penguins may waddle slowly on land, but the gentoo penguin of Antarctica is thought to be the fastest swimming bird in the world. You can use the formula $d = rt$ to model the distance d, rate r, and time t of the penguin. Suppose the penguin swims a distance of 20 miles. The equation that relates t and r is $20 = rt$.

In this project, you will graph $20 = rt$ and describe the graph. You will then choose four other values for d and graph their equations.

Working on the Project

Work with a partner and choose a strategy. Here are some suggestions to help you get started.

- Choose several values of r and solve the equation for t. Write the results as ordered pairs (r, t).
- Graph the ordered pairs on a coordinate plane.
- Research the average distance that four other animals could travel in one hour.

Technology Tools

- Use **spreadsheet software** or a **graphing calculator** to prepare your graphs.
- Use **presentation software** to prepare and give your presentation.

 Research For more information about animal speeds, visit: www.algconcepts.glencoe.com

> ### Strategies
>
> Look for a pattern.
>
> Draw a diagram.
>
> Make a table.
>
> Work backward.
>
> Use an equation.
>
> Make a graph.
>
> Guess and check.

Presenting the Project

Prepare a presentation of your findings for your animals. Make sure that your presentation includes:

- a table of ordered pairs and a graph for each value of d,
- an explanation of your findings, including a comparison of the graphs, and
- a discussion of how these graphs differ from the graph of an equation such as $d = 20t$. Use the terms *direct variation* and *inverse variation* in your discussion.

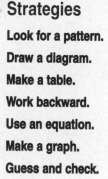

Simplifying Rational Expressions

What You'll Learn

You'll learn to simplify rational expressions.

Why It's Important

Photography
Photographers use a rational expression to find the amount of light that falls on an object.
See Exercise 63.

Would you like to invest $100 so that it grows to $200? You can use the equation $y = \dfrac{72}{x}$ to estimate how long it will take to double your investment. In the equation, y is the number of years, and x is the annual interest rate expressed as a percent.

The expression $\dfrac{72}{x}$ is a **rational expression**.

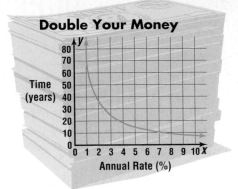

Double Your Money

Time (years) — Annual Rate (%)

Rational Expression	A rational expression is an algebraic fraction whose numerator and denominator are polynomials.

Study the graph of $y = \dfrac{72}{x}$ shown above. As the value of x decreases, the value of y increases. But, the value of x can never be equal to zero.

Zero cannot be the denominator of a fraction because division by zero is undefined. So, any value assigned to a variable that results in a denominator of zero must be excluded from the domain of the variable. These values are called **excluded values**.

You can use the graph of the related **rational function** of a rational expression to investigate excluded values of the variable.

Graphing Calculator Tutorial
See pp. 724–727.

Graphing Calculator Exploration

The screen shows the graph of the rational function $y = \dfrac{1}{x}$. The graph is made up of two branches. One branch is in the first quadrant, and the other branch is in the third quadrant.

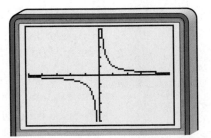

Try These

1. Explain why the graph has two branches.
2. Describe what happens to the graph as the value of x approaches 0.

3. Explain why the two branches of the graph do not touch each other. How does this relate to the excluded value of x?

4. Graph each function. State the excluded value(s) of x.

   a. $y = \dfrac{1}{x - 1}$ **b.** $y = \dfrac{1}{x - 2}$ **c.** $y = \dfrac{1}{x + 2}$

   d. $y = \dfrac{1}{x(x - 1)}$ **e.** $y = \dfrac{1}{(x - 1)(x + 2)}$ **f.** $y = \dfrac{1}{(x + 3)(x + 4)}$

5. **Make a conjecture** about how to find the excluded values of a rational expression without using a calculator.

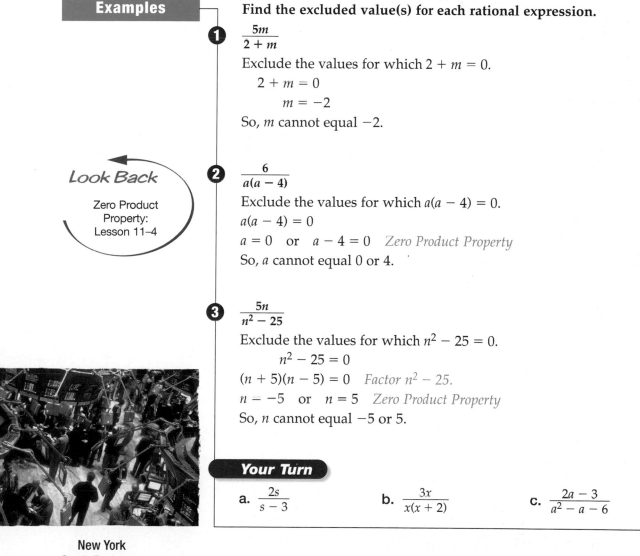

Examples

Find the excluded value(s) for each rational expression.

1 $\dfrac{5m}{2 + m}$

Exclude the values for which $2 + m = 0$.

$$2 + m = 0$$
$$m = -2$$

So, m cannot equal -2.

Look Back

Zero Product Property: Lesson 11–4

2 $\dfrac{6}{a(a - 4)}$

Exclude the values for which $a(a - 4) = 0$.

$a(a - 4) = 0$

$a = 0$ or $a - 4 = 0$ *Zero Product Property*

So, a cannot equal 0 or 4.

3 $\dfrac{5n}{n^2 - 25}$

Exclude the values for which $n^2 - 25 = 0$.

$$n^2 - 25 = 0$$
$(n + 5)(n - 5) = 0$ *Factor $n^2 - 25$.*
$n = -5$ or $n = 5$ *Zero Product Property*

So, n cannot equal -5 or 5.

Your Turn

a. $\dfrac{2s}{s - 3}$ **b.** $\dfrac{3x}{x(x + 2)}$ **c.** $\dfrac{2a - 3}{a^2 - a - 6}$

New York
Stock Exchange

Recall that you can simplify a fraction by using the following steps.

• First, factor the numerator and denominator.
• Then, divide the numerator and denominator by the greatest common factor.

Example
History Link

4 In the 1960 presidential election, more than 60% of the registered voters cast ballots. No presidential election since 1960 has had a greater voter turnout. Express 60% as a fraction in simplest form.

$$60\% = \frac{60}{100}$$

$$= \frac{2 \cdot 2 \cdot 3 \cdot 5}{2 \cdot 2 \cdot 5 \cdot 5} \qquad \textit{Factor 60 and 100.}$$

$$= \frac{\overset{1}{\cancel{2}} \cdot \overset{1}{\cancel{2}} \cdot 3 \cdot \overset{1}{\cancel{5}}}{\underset{1}{\cancel{2}} \cdot \underset{1}{\cancel{2}} \cdot 5 \cdot \underset{1}{\cancel{5}}} \text{ or } \frac{3}{5} \qquad \textit{The GCF of 60 and 100 is } 2 \cdot 2 \cdot 5.$$

Your Turn Simplify each fraction.

d. $\dfrac{9}{24}$ **e.** $\dfrac{15}{20}$ **f.** $\dfrac{12}{30}$

You can use the same procedure to simplify rational expressions that have polynomials in the numerator and denominator. To *simplify* means that the numerator and denominator have no factors in common, except 1.

Examples

From this point on, you can assume that all values that give a denominator of zero are excluded.

5 $\dfrac{8a^2b}{12ab^3}$

$$\frac{8a^2b}{12ab^3} = \frac{2 \cdot 2 \cdot 2 \cdot a \cdot a \cdot b}{2 \cdot 2 \cdot 3 \cdot a \cdot b \cdot b \cdot b} \qquad \textit{Note that } a \neq 0 \text{ and } b \neq 0.$$

$$= \frac{\overset{1}{\cancel{2}} \cdot \overset{1}{\cancel{2}} \cdot 2 \cdot \overset{1}{\cancel{a}} \cdot a \cdot \overset{1}{\cancel{b}}}{\underset{1}{\cancel{2}} \cdot \underset{1}{\cancel{2}} \cdot 3 \cdot \underset{1}{\cancel{a}} \cdot \underset{1}{\cancel{b}} \cdot b \cdot b} \text{ or } \frac{2a}{3b^2} \qquad \textit{The GCF is 4ab.}$$

6 $\dfrac{2x-2}{5x-5}$

$$\frac{2x-2}{5x-5} = \frac{2(x-1)}{5(x-1)} \qquad \textit{Factor } 2x-2 \text{ and } 5x-5.$$

$$= \frac{2(\cancel{x-1})}{5(\cancel{x-1})} \text{ or } \frac{2}{5} \qquad \textit{The GCF is } (x-1).$$

Look Back

Factoring Trinomials: Lesson 10–3

7 $\dfrac{a(a+1)}{a^2+3a+2}$

$$\frac{a(a+1)}{a^2+3a+2} = \frac{a(a+1)}{(a+2)(a+1)} \qquad \textit{Factor } a^2+3a+2.$$

$$= \frac{a(\cancel{a+1})}{(a+2)(\cancel{a+1})} \text{ or } \frac{a}{a+2} \qquad \textit{The GCF is } (a+1).$$

8 $\dfrac{x^2 - 9}{x^2 - x - 6}$

$$\dfrac{x^2 - 9}{x^2 - x - 6} = \dfrac{(x - 3)(x + 3)}{(x + 2)(x - 3)} \qquad \textit{Factor } x^2 - 9 \textit{ and } x^2 - x - 6.$$

$$= \dfrac{\overset{1}{\cancel{(x - 3)}}(x + 3)}{(x + 2)\cancel{(x - 3)}_1} \text{ or } \dfrac{x + 3}{x + 2} \qquad \textit{The GCF is } (x - 3).$$

9 $\dfrac{4 - 2x}{x^2 - 4x + 4}$

$$\dfrac{4 - 2x}{x^2 - 4x + 4} = \dfrac{2(2 - x)}{(x - 2)(x - 2)} \qquad \textit{Factor } 4 - 2x \textit{ and } x^2 - 4x + 4.$$

$$= \dfrac{2(-1)(x - 2)}{(x - 2)(x - 2)} \qquad \textit{Factor } -1 \textit{ from } (2 - x).$$

$$= \dfrac{2(-1)\overset{1}{\cancel{(x - 2)}}}{(x - 2)\cancel{(x - 2)}_1} \text{ or } \dfrac{-2}{x - 2} \qquad \textit{The GCF is } (x - 2).$$

Your Turn

g. $\dfrac{14y^3z}{8y^2z^4}$ **h.** $\dfrac{-12a^2b^2}{18a^5}$ **i.** $\dfrac{3y - 9}{4y - 12}$

j. $\dfrac{x(x + 3)}{x^2 + 6x + 9}$ **k.** $\dfrac{a^2 - 2a - 8}{a^2 + 3a + 2}$ **l.** $\dfrac{x^2 - 4x}{16 - x^2}$

Check for Understanding

Communicating Mathematics

Study the lesson. Then complete the following.

1. **Explain** why $x = 2$ is an excluded value for $\dfrac{x}{x - 2}$.

2. **List** the steps you would use to simplify a rational expression.

3. ![You Decide] Sam and Darnell are trying to simplify $\dfrac{x + 3}{x + 4}$. Sam says the answer is $\dfrac{3}{4}$. Darnell says it is already in simplest form. Who is correct? Explain your reasoning.

Vocabulary

rational expression
excluded values
rational function

Guided Practice

Getting Ready   Factor each expression.

Sample 1: $5y + 15$	**Sample 2:** $x^2 + 5x + 6$
Solution: $5(y + 3)$	**Solution:** $(x + 3)(x + 2)$

4. $4x - 20$ **5.** $16x^2 - 20x$ **6.** $y^2 + 6y + 8$

7. $a^2 - a - 20$ **8.** $n^2 - 3n - 18$ **9.** $x^2 - 10x + 24$

Find the excluded value(s) for each rational expression.
(Examples 1–3)

10. $\dfrac{7}{x-2}$

11. $\dfrac{4a}{a(a+6)}$

12. $\dfrac{2x+3}{x^2-2x-15}$

Simplify each rational expression. *(Examples 4–9)*

13. $\dfrac{8}{20}$

14. $\dfrac{5x}{25x^2}$

15. $\dfrac{12y^2z}{18yz^4}$

16. $\dfrac{x(x+3)}{7(x+3)}$

17. $\dfrac{3y+9}{4y+12}$

18. $\dfrac{x^2-3x}{x^2+x-12}$

19. $\dfrac{x^2+2x-8}{x^2+5x+4}$

20. $\dfrac{x^2+6x+5}{x^2+3x-10}$

21. $\dfrac{25-x^2}{x^2+x-30}$

22. **Sports** The National Hockey League team that wins the first game of the best-of-seven series for the Stanley Cup has about an 80% chance of winning the series. Express 80% as a fraction in simplest form. *(Example 4)* **Source:** NHL, 1997

Exercises

Practice

Find the excluded value(s) for each rational expression.

23. $\dfrac{x}{x+5}$

24. $\dfrac{3a}{2a+4}$

25. $\dfrac{n+6}{n-10}$

26. $\dfrac{5}{a(a+8)}$

27. $\dfrac{x+2}{(x+2)(x-3)}$

28. $\dfrac{5m}{(m-2)(m-3)}$

29. $\dfrac{9x}{x^2+5x}$

30. $\dfrac{x^2+6x+5}{x^2+3x-10}$

31. $\dfrac{y+3}{y^2-16}$

Simplify each rational expression.

32. $\dfrac{10}{16}$

33. $\dfrac{12}{36}$

34. $\dfrac{15}{20}$

35. $\dfrac{4c}{6d}$

36. $\dfrac{8xy}{24x^2}$

37. $\dfrac{-36abc^2}{9ac}$

38. $\dfrac{2(x+1)}{8(x+1)}$

39. $\dfrac{n-3}{5(n-3)}$

40. $\dfrac{x(x+5)}{y(x+5)}$

41. $\dfrac{(x+3)(x-2)}{(x-2)(x+1)}$

42. $\dfrac{(y+6)(y-6)}{(y+6)(y+6)}$

43. $\dfrac{x^2-3x}{2(x-3)}$

44. $\dfrac{a^2-a}{a-1}$

45. $\dfrac{r^2-r-6}{3r-9}$

46. $\dfrac{x^2+3x+2}{x^2+2x+1}$

47. $\dfrac{y^2+7y+12}{y^2-16}$

48. $\dfrac{x-3}{x^2+x-12}$

49. $\dfrac{y^2+4y+4}{y^2+y-2}$

50. $\dfrac{r^2-4r-5}{r^2-2r-15}$

51. $\dfrac{z^2-z-20}{z^2+7z+12}$

52. $\dfrac{6x^2+24x}{x^2+8x+16}$

53. $\dfrac{8m^2-16m}{m^2-4m+4}$

54. $\dfrac{r^2+6r+5}{2r^2-2}$

55. $\dfrac{c^2-c-20}{c^3+10c^2+24c}$

56. $\dfrac{9-3x}{x^2-6x+9}$

57. $\dfrac{x^2-4x}{16-x^2}$

58. $\dfrac{12-4y}{y^2+y-12}$

59. Write $\dfrac{4a}{3a + a^2}$ in simplest form.

60. What are the excluded values of x for $\dfrac{x^2 - 9}{x^2 + 5x + 6}$?

61. Investing Refer to the application at the beginning of the lesson. How long would it take to double your money if it is invested at a rate of 8%?

62. Entertainment The graph shows the number of films that adults see in a theater in a typical month.

   a. What fraction of adults see one movie?

   b. What fraction of adults see three or more movies?

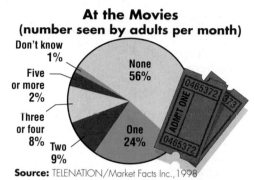

At the Movies
(number seen by adults per month)

Don't know 1%
None 56%
Five or more 2%
Three or four 8%
Two 9%
One 24%

Source: TELENATION/Market Facts Inc., 1998

63. Photography The intensity of light that falls on an object is given by the formula $I = \dfrac{P}{d^2}$. In the formula, d is the distance from the light source, P is the power in lumens, and I is the intensity of light in lumens per square meter. An object being photographed is 3 meters from a 72-lumen light source. Find the intensity.

64. Critical Thinking Write rational expressions that have the following as excluded values.

   a. 0   **b.** −3   **c.** −2 and 7

Mixed Review

Solve each equation. *(Lesson 14–6)*

65. $\sqrt{x} = 4$ **66.** $\sqrt{y} = 2\sqrt{5}$

67. $\sqrt{n - 3} = 4$ **68.** $\sqrt{a} + 4 = 29$

Simplify each expression. *(Lesson 14–5)*

69. $2\sqrt{3} + \sqrt{3} + 4\sqrt{3}$ **70.** $6\sqrt{5} - 3\sqrt{5} + 10\sqrt{5}$

71. $4\sqrt{3} + \sqrt{12}$ **72.** $2\sqrt{50} + 3\sqrt{5}$

73. Find the distance between $J(2, 1)$ and $K(5, 5)$. *(Lesson 14–2)*

74. The sum of two numbers is 9. Their difference is 3. Find the numbers. *(Lesson 13–4)*

75. Standardized Test Practice Which property or properties of inequalities allow(s) you to conclude that if $2y + 8 > 18$, then $y > 5$? *(Lesson 12–4)*

   A Distributive Property   **B** Addition Property

   C Division Property   **D** Addition and Division Properties

Extra Practice   See p. 722.

What You'll Learn

You'll learn to multiply and divide rational expressions.

Why It's Important

Carpentry
Carpenters divide rational expressions to find how many boards of a certain size can be cut from a piece of wood.
See Exercise 42.

In a recent year, about $\frac{2}{5}$ of Americans used the Internet at home. Of those, about $\frac{1}{2}$ used the Internet for e-mail. To find what part of Americans used e-mail at home, multiply $\frac{2}{5}$ and $\frac{1}{2}$.

You can use two methods to multiply rational numbers.

- **Method 1** Multiply numerators and multiply denominators. Then divide each numerator and denominator by the greatest common factor.
- **Method 2** Divide numerators and denominators by any common factors. Then multiply numerators and denominators.

Method 1
Multiply, then simplify.

$$\frac{2}{5} \cdot \frac{1}{2} = \frac{2}{10}$$

$$= \frac{\overset{1}{\cancel{2}}}{\underset{5}{\cancel{10}}} \text{ or } \frac{1}{5} \quad \textit{The GCF is 2.}$$

Method 2
Simplify, then multiply.

$$\frac{2}{5} \cdot \frac{1}{2} = \frac{\overset{1}{\cancel{2}}}{5} \cdot \frac{1}{\underset{1}{\cancel{2}}} \quad \textit{2 is a common factor.}$$

$$= \frac{1}{5}$$

Both methods have the same result. About $\frac{1}{5}$ of Americans used e-mail at home.

You can use the same methods to multiply rational expressions.

Examples

Find each product.

1 $\dfrac{6r^2}{5s^2} \cdot \dfrac{10rs}{6r^3}$

Method 1 $\dfrac{6r^2}{5s^2} \cdot \dfrac{10rs}{6r^3} = \dfrac{60r^3s}{30r^3s^2} \quad \textit{Multiply.}$

$$= \frac{\overset{1}{\cancel{2}} \cdot 2 \cdot \overset{1}{\cancel{3}} \cdot \overset{1}{\cancel{5}} \cdot \overset{1}{\cancel{r}} \cdot \overset{1}{\cancel{r}} \cdot \overset{1}{\cancel{r}} \cdot \overset{1}{\cancel{s}}}{\underset{1}{\cancel{2}} \cdot \underset{1}{\cancel{3}} \cdot \underset{1}{\cancel{5}} \cdot \underset{1}{\cancel{r}} \cdot \underset{1}{\cancel{r}} \cdot \underset{1}{\cancel{r}} \cdot \underset{1}{\cancel{s}} \cdot s} \text{ or } \frac{2}{s} \quad \textit{Simplify.}$$

Method 2 $\dfrac{6r^2}{5s^2} \cdot \dfrac{10rs}{6r^3} = \dfrac{\overset{1}{\cancel{6r^2}}}{\underset{1}{\cancel{5s^2}}\,_s} \cdot \dfrac{\overset{2}{\cancel{10rs}}^{\,1\,1}}{\underset{1\,1}{\cancel{6r^3}}} \quad \textit{Simplify.}$

$$= \frac{2}{s} \quad\quad\quad\quad\quad \textit{Multiply.}$$

2 $\dfrac{3m}{n+2} \cdot \dfrac{n+2}{9m^2}$

$$\dfrac{3m}{n+2} \cdot \dfrac{n+2}{9m^2} = \dfrac{\overset{1}{\cancel{3}}\overset{1}{\cancel{m}}}{\cancel{n+2}} \cdot \dfrac{\overset{1}{\cancel{n+2}}}{\underset{3\,m}{\cancel{9m^2}}} \qquad 3, m, \text{ and } n+2 \text{ are common factors.}$$

$$= \dfrac{1}{3m}$$

3 $\dfrac{y-3}{y+5} \cdot \dfrac{2y^2+10y}{2y-6}$

$$\dfrac{y-3}{y+5} \cdot \dfrac{2y^2+10y}{2y-6} = \dfrac{y-3}{y+5} \cdot \dfrac{2y(y+5)}{2(y-3)} \qquad \textit{Factor } 2y^2+10y \textit{ and } 2y-6.$$

$$= \dfrac{\overset{1}{\cancel{y-3}}}{\cancel{y+5}} \cdot \dfrac{2y\overset{1}{\cancel{(y+5)}}}{\underset{1}{\cancel{2}}\underset{1}{\cancel{(y-3)}}} \qquad 2, y-3, \text{ and } y+5 \text{ are} \atop \text{common factors.}$$

$$= \dfrac{y}{1} \text{ or } y$$

4 $\dfrac{x^2-25}{x^2-3x-10} \cdot \dfrac{x+2}{x}$

$$\dfrac{x^2-25}{x^2-3x-10} \cdot \dfrac{x+2}{x} = \dfrac{(x+5)(x-5)}{(x-5)(x+2)} \cdot \dfrac{x+2}{x} \qquad \textit{Factor } x^2-25 \textit{ and} \atop x^2-3x-10.$$

$$= \dfrac{(x+5)\overset{1}{\cancel{(x-5)}}}{\underset{1}{\cancel{(x-5)}}\underset{1}{\cancel{(x+2)}}} \cdot \dfrac{\overset{1}{\cancel{x+2}}}{x} \qquad x-5 \text{ and } x+2 \text{ are} \atop \text{common factors.}$$

$$= \dfrac{x+5}{x}$$

Your Turn

a. $\dfrac{12x}{5y} \cdot \dfrac{20y^2}{36x^2}$

b. $\dfrac{x+2}{8} \cdot \dfrac{12x}{(x+2)(x-2)}$

c. $\dfrac{n-2}{6} \cdot \dfrac{3}{n^2-2n}$

d. $\dfrac{a^2+6a+9}{a^2-4} \cdot \dfrac{a-2}{a+3}$

Look Back

Reciprocal:
Lesson 4–3

To divide a rational number by any nonzero number, multiply by its reciprocal. You can use the same method to multiply rational expressions.

$$\dfrac{2}{3} \div \dfrac{1}{2} = \dfrac{2}{3} \cdot \dfrac{2}{1} \qquad \textit{The reciprocal of}$$
$$= \dfrac{4}{3} \qquad\qquad \dfrac{1}{2} \textit{ is } \dfrac{2}{1}.$$

$$\dfrac{5}{x} \div \dfrac{y}{z} = \dfrac{5}{x} \cdot \dfrac{z}{y} \qquad \textit{The reciprocal of}$$
$$= \dfrac{5z}{xy} \qquad\qquad \dfrac{y}{z} \textit{ is } \dfrac{z}{y}.$$

5 The Indianapolis 500 is a 500-mile automobile race. Each lap is $2\frac{1}{2}$ miles long. How many laps does a driver complete to race the entire 500 miles?

To find the number of laps, divide 500 by $2\frac{1}{2}$.

$500 \div 2\frac{1}{2} = \dfrac{500}{1} \div \dfrac{5}{2}$ *Express 500 and $2\frac{1}{2}$ as improper fractions.*

$\qquad\qquad = \dfrac{500}{1} \cdot \dfrac{2}{5}$ *The reciprocal of $\frac{5}{2}$ is $\frac{2}{5}$.*

$\qquad\qquad = \dfrac{\overset{100}{\cancel{500}}}{1} \cdot \dfrac{2}{\underset{1}{\cancel{5}}}$ *5 is a common factor.*

$\qquad\qquad = \dfrac{200}{1}$ or 200

A driver completes 200 laps.

Find each quotient.

6 $\dfrac{6x^3}{y} \div \dfrac{2x}{y^2}$

$\dfrac{6x^3}{y} \div \dfrac{2x}{y^2} = \dfrac{6x^3}{y} \cdot \dfrac{y^2}{2x}$ *The reciprocal of $\frac{2x}{y^2}$ is $\frac{y^2}{2x}$.*

$\qquad\qquad = \dfrac{\overset{3\,x^2}{\cancel{6x^3}}}{\underset{1}{\cancel{y}}} \cdot \dfrac{\overset{y}{\cancel{y^2}}}{\underset{1\;1}{\cancel{2x}}}$ *2, x, and y are common factors.*

$\qquad\qquad = \dfrac{3x^2y}{1}$ or $3x^2y$

7 $\dfrac{2m+8}{m+5} \div (m+4)$

$\dfrac{2m+8}{m+5} \div (m+4) = \dfrac{2m+8}{m+5} \cdot \dfrac{1}{m+4}$ *The reciprocal of $(m+4)$ is $\dfrac{1}{m+4}$.*

$\qquad\qquad = \dfrac{2(m+4)}{m+5} \cdot \dfrac{1}{m+4}$ *Factor $2m+8$.*

$\qquad\qquad = \dfrac{2\cancel{(m+4)}^{\,1}}{m+5} \cdot \dfrac{1}{\underset{1}{\cancel{m+4}}}$ *$(m+4)$ is a common factor.*

$\qquad\qquad = \dfrac{2}{m+5}$

Your Turn

e. $\dfrac{12y}{7z^2} \div \dfrac{3xy}{2z}$

f. $\dfrac{x^2+4x+3}{x^2} \div (x+3)$

Sometimes it is necessary to factor -1 from one of the terms.

8 Find $\dfrac{x^2 - 4}{2y} \div \dfrac{2 - x}{6xy}$.

$$\dfrac{x^2 - 4}{2y} \div \dfrac{2 - x}{6xy} = \dfrac{x^2 - 4}{2y} \cdot \dfrac{6xy}{2 - x} \qquad \text{The reciprocal of } \dfrac{2 - x}{6xy} \text{ is } \dfrac{6xy}{2 - x}.$$

$$= \dfrac{(x + 2)(x - 2)}{2y} \cdot \dfrac{6xy}{2 - x} \qquad \text{Factor } x^2 - 4.$$

$$= \dfrac{(x + 2)(x - 2)}{2y} \cdot \dfrac{6xy}{-1(x - 2)} \qquad \text{Factor } -1 \text{ from } 2 - x.$$

$$= \dfrac{(x + 2)(x - 2)}{2y} \cdot \dfrac{\overset{3}{6}xy\overset{1}{}}{-1(x - 2)} \qquad x - 2, y, \text{ and } 2 \text{ are common factors.}$$

$$= \dfrac{3x(x + 2)}{-1} \text{ or } -3x(x + 2)$$

Reading Algebra

Instead of factoring -1 from $2 - x$, you could have factored -1 from $x - 2$ and obtained the same results.

Your Turn

g. Find $\dfrac{x^2 - 9}{8x^3} \div \dfrac{6 - 2x}{4x}$

Check for Understanding

Communicating Mathematics

Study the lesson. Then complete the following.

1. **Identify** and correct the error that was made while finding the quotient.

$$\dfrac{a^2 - 25}{3a} \div \dfrac{a + 5}{15a^2} = \dfrac{3a}{(a + 5)(a - 5)} \cdot \dfrac{a + 5}{15a^2}$$

$$= \dfrac{\overset{1\,1}{3a}}{(a + 5)(a - 5)} \cdot \dfrac{a + 5}{15a^2} \text{ or } \dfrac{1}{5a(a - 5)}$$

Math Journal

2. **Write** a short paragraph explaining which method you prefer when finding the product of rational expressions: to simplify first and then multiply or to multiply first and then simplify. Include examples to support your point of view.

Guided Practice

Getting Ready **Find the reciprocal of each expression.**

Sample 1: $\dfrac{m}{2}$

Solution: The reciprocal is $\dfrac{2}{m}$.

Sample 2: $(a + b)$

Solution: The reciprocal is $\dfrac{1}{a + b}$.

3. $\dfrac{x^2}{4}$

4. y

5. $\dfrac{x - 4}{x + 5}$

6. $y + 6$

Find each product. *(Examples 1–4)*

7. $\dfrac{a^2b}{b^2c} \cdot \dfrac{c}{d}$

8. $\dfrac{5(n-1)}{-3} \cdot \dfrac{9}{n-1}$

9. $\dfrac{2x-10}{x^2+x-12} \cdot \dfrac{x-3}{x-5}$

10. $\dfrac{y^2+3y-10}{2y} \cdot \dfrac{y^2-3y}{y^2-5y+6}$

Find each quotient. *(Examples 6–8)*

11. $\dfrac{12a^3}{bc} \div \dfrac{3a^2}{bc}$

12. $\dfrac{9xy}{5} \div 3x^2y^2$

13. $\dfrac{3x^2+6x}{x} \div \dfrac{2x+4}{x^2}$

14. $\dfrac{x^2-5x+6}{2x^2} \div \dfrac{3-x}{4}$

15. **Sewing** It takes $1\frac{1}{2}$ yards of fabric to make one flag for the school color guard. How many flags can be made from 18 yards of fabric? *(Example 5)*

Exercises • • • • • • • • • • • • • • • • • •

Practice

Find each product.

16. $\dfrac{ab}{ac} \cdot \dfrac{c}{d}$

17. $\dfrac{3x}{2y} \cdot \dfrac{y^2}{6}$

18. $\dfrac{6a^2}{8n^2} \cdot \dfrac{12n}{9a}$

19. $\dfrac{3(a-b)}{a} \cdot \dfrac{a^2}{a-b}$

20. $\dfrac{2(a+2b)}{5} \cdot \dfrac{5}{3(a+2b)}$

21. $\dfrac{7s}{s+2} \cdot \dfrac{2s+4}{21}$

22. $\dfrac{3x+30}{2x} \cdot \dfrac{4x}{4x+40}$

23. $\dfrac{x+3}{x+4} \cdot \dfrac{x}{x^2+7x+12}$

24. $\dfrac{3a-6}{a^2-9} \cdot \dfrac{a+3}{a^2-2a}$

25. $\dfrac{x^2-9}{x+7} \cdot \dfrac{2x+14}{x^2+6x+9}$

26. $\dfrac{x}{x^2+8x+15} \cdot \dfrac{2x+10}{x^2}$

27. $\dfrac{n^2}{n^2-4} \cdot \dfrac{n^2-5n+6}{n^2-3n}$

Find each quotient.

28. $\dfrac{a^2}{b} \div \dfrac{a^2}{b^2}$

29. $\dfrac{7a^2b}{xy} \div \dfrac{7}{6xy}$

30. $2xz \div \dfrac{4xy}{z}$

31. $\dfrac{2a^3}{a+1} \div \dfrac{a^2}{a+1}$

32. $\dfrac{b^2-9}{4b} \div (b-3)$

33. $\dfrac{y^2+8y+16}{y^2} \div (y+4)$

34. $\dfrac{y^2}{y+2} \div \dfrac{y}{y+2}$

35. $\dfrac{m^2+2m+1}{2} \div \dfrac{m+1}{m-1}$

36. $\dfrac{x^2-4x+4}{3x} \div \dfrac{x^2-4}{6}$

37. $\dfrac{x^2+7x+10}{x-1} \div \dfrac{x^2+2x-15}{1-x}$

38. $\dfrac{x^2-16}{16-x^2} \div \dfrac{7}{x}$

39. $\dfrac{a^2-9}{9} \div \dfrac{6-2a}{27a^2}$

40. Find the product of $\dfrac{y-3}{8}$ and $\dfrac{12}{y-3}$.

41. What is the quotient when $\dfrac{x^2}{x^2-y^2}$ is divided by $\dfrac{x^2}{x+y}$?

Applications and Problem Solving

42. Carpentry How many boards, each 2 feet 8 inches long, can be cut from a board 16 feet 6 inches long?

43. Probability Two darts are randomly thrown one at a time. Assume that both hit the target.
   a. What is the probability that both will hit the shaded region?
   b. Find the probability if $x = 3$.

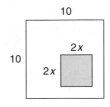

44. Critical Thinking Find two different pairs of rational expressions whose product is $\frac{5x^2}{8y^3}$.

Mixed Review

Simplify. *(Lesson 15–1)*

45. $\dfrac{14a^2b^4c}{2a^3bc}$

46. $\dfrac{m^2 - 16}{m^2 - 8m + 16}$

47. $\dfrac{8 - 4x}{x^2 - 4x + 4}$

48. Cycling You can use the formula $s = 4\sqrt{r}$ to find the greatest speed at which a cyclist can go around a corner safely and not be likely to tip over. In the formula, s is the speed in miles per hour, and r is the radius of the corner in feet. If $s = 8$ miles per hour, find r. *(Lesson 14–5)*

Use elimination to solve each system of equations. *(Lesson 13–5)*

49. $x + y = 4$
$2x - 3y = -7$

50. $x + 3y = -4$
$x - 2y = 6$

51. $3a + 4b = -25$
$2a - 3b = 6$

52. Graph $2x + 3y \leq 12$. *(Lesson 12–7)*

53. Business Paul wants to start a small lawn-care company to earn money in the summer. He estimates that his expenses will be $200 for gasoline and equipment. He plans to charge $25 per lawn. The equation $P = 25n - 200$ can be used to find his profits. In the equation, n represents the number of lawns he cuts. *(Lesson 6–2)*
   a. Determine the ordered pairs that satisfy the equation if the domain is {5, 8, 10}.
   b. Graph the relation.

54. Standardized Test Practice A 60-kilogram mass is 140 centimeters from the fulcrum of a lever. How far from the fulcrum must an 80-kilogram mass be to balance the lever? *(Lesson 4–4)*

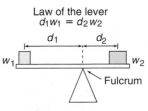

Law of the lever
$d_1 w_1 = d_2 w_2$

   A 186.7 cm   **B** 34.3 cm
   C 120 cm   **D** 105 cm

Math In the Workplace

What You'll Learn

You'll learn to divide polynomials by binomials.

Why It's Important

Aviation You can use polynomials to determine distance, rate, and time. *See Exercise 32.*

In the previous lesson, you learned that some divisions can be performed using factoring.

$$(x^2 + 4x + 3) \div (x + 3) = \frac{x^2 + 4x + 3}{x + 3}$$

$$= \frac{\overset{1}{\cancel{(x + 3)}}(x + 1)}{\underset{1}{\cancel{(x + 3)}}} \quad \textit{Factor the dividend.}$$

$$= x + 1$$

Therefore, $(x^2 + 4x + 3) \div (x + 3) = x + 1$. Since the remainder is 0, the divisor is a factor of the dividend.

You can also use algebra tiles to model the division shown above.

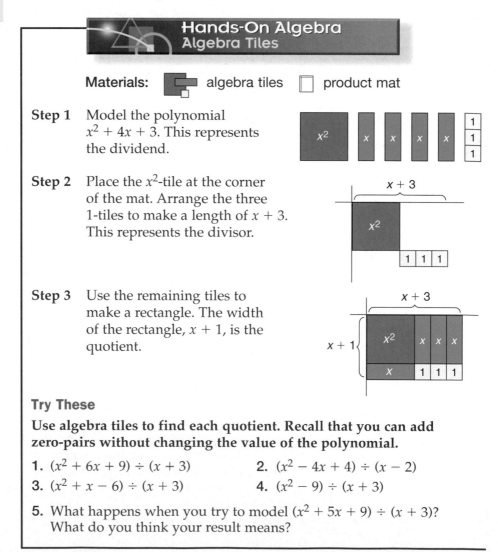

Hands-On Algebra
Algebra Tiles

Materials: algebra tiles   product mat

Step 1 Model the polynomial $x^2 + 4x + 3$. This represents the dividend.

Step 2 Place the x^2-tile at the corner of the mat. Arrange the three 1-tiles to make a length of $x + 3$. This represents the divisor.

Step 3 Use the remaining tiles to make a rectangle. The width of the rectangle, $x + 1$, is the quotient.

Try These

Use algebra tiles to find each quotient. Recall that you can add zero-pairs without changing the value of the polynomial.

1. $(x^2 + 6x + 9) \div (x + 3)$

2. $(x^2 - 4x + 4) \div (x - 2)$

3. $(x^2 + x - 6) \div (x + 3)$

4. $(x^2 - 9) \div (x + 3)$

5. What happens when you try to model $(x^2 + 5x + 9) \div (x + 3)$? What do you think your result means?

You can also divide polynomials using long division. Follow these steps to divide $2x^2 + 7x + 3$ by $2x + 1$.

Step 1   To find the first term of the quotient, divide the first term of the dividend, $2x^2$, by the first term of the divisor, $2x$.

$$
\begin{array}{r}
x \\
2x + 1\overline{)2x^2 + 7x + 3} \\
\underline{(-)\ 2x^2 + 1x} \\
6x
\end{array}
$$

$2x^2 \div 2x = x$
Multiply x and 2x + 1.
Subtract.

Step 2   To find the next term of the quotient, divide the first term of the partial dividend, $6x$, by the first term of the divisor, $2x$.

$$
\begin{array}{r}
x + 3 \\
2x + 1\overline{)2x^2 + 7x + 3} \\
\underline{(\ \ )\ 2x^2 + 1x} \\
6x + 3 \\
\underline{(-)\ 6x + 3} \\
0
\end{array}
$$

Bring down 3; $6x \div 2x = 3$.
Multiply 3 and 2x + 1.
Subtract.

Therefore, $(2x^2 + 7x + 3) \div (2x + 1) = x + 3$.

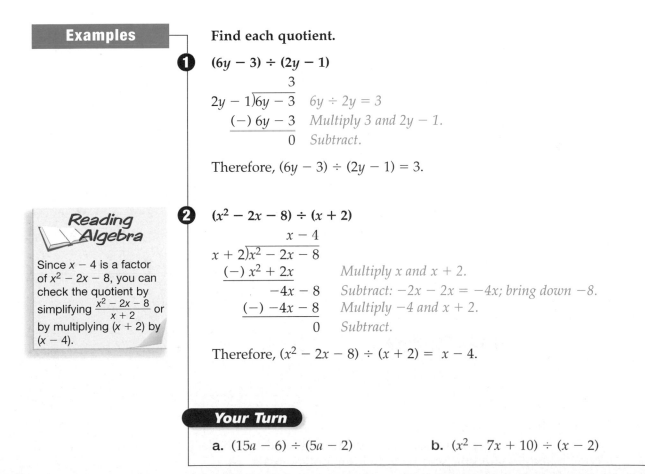

Examples

Find each quotient.

1 $(6y - 3) \div (2y - 1)$

$$
\begin{array}{r}
3 \\
2y - 1\overline{)6y - 3} \\
\underline{(-)\ 6y - 3} \\
0
\end{array}
$$

$6y \div 2y = 3$
Multiply 3 and 2y − 1.
Subtract.

Therefore, $(6y - 3) \div (2y - 1) = 3$.

Reading Algebra

Since $x - 4$ is a factor of $x^2 - 2x - 8$, you can check the quotient by simplifying $\dfrac{x^2 - 2x - 8}{x + 2}$ or by multiplying $(x + 2)$ by $(x - 4)$.

2 $(x^2 - 2x - 8) \div (x + 2)$

$$
\begin{array}{r}
x - 4 \\
x + 2\overline{)x^2 - 2x - 8} \\
\underline{(-)\ x^2 + 2x} \\
-4x - 8 \\
\underline{(-)\ -4x - 8} \\
0
\end{array}
$$

Multiply x and x + 2.
Subtract: $-2x - 2x = -4x$; bring down -8.
Multiply -4 and x + 2.
Subtract.

Therefore, $(x^2 - 2x - 8) \div (x + 2) = x - 4$.

Your Turn

a. $(15a - 6) \div (5a - 2)$ **b.** $(x^2 - 7x + 10) \div (x - 2)$

If the divisor is *not* a factor of the dividend, the remainder will not be 0. The quotient can be expressed as follows.

$$\text{quotient} = \text{partial quotient} + \frac{\text{remainder}}{\text{divisor}}$$

Example ❸ Find $(8y^2 - 2y + 1) \div (2y - 1)$.

$$
\begin{array}{r}
4y + 1 \\
2y - 1 \overline{) 8y^2 - 2y + 1} \\
(-)\ 8y^2 - 4y \\
\hline
2y + 1 \\
(-)\ 2y - 1 \\
\hline
2
\end{array}
$$

Multiply 4y and 2y − 1.
Subtract. Then bring down 1.
Multiply 1 and 2y − 1.
Subtract. The remainder is 2.

The quotient is $4y + 1$ with remainder 2.

So, $(8y^2 - 2y + 1) \div (2y - 1) = 4y + 1 + \dfrac{2}{2y - 1}$.

Your Turn

c. Find $(2a^2 + 7a + 3) \div (a + 2)$.

In an expression like $x^2 - 4$ there is no x term. In such situations, rename the dividend using zero as the coefficient of the missing term.

Example ❹ Find $(x^2 - 4) \div (x + 1)$.

$$
\begin{array}{r}
x - 1 \\
x + 1 \overline{) x^2 + 0x - 4} \\
(-)\ x^2 + 1x \\
\hline
-1x - 4 \\
-1x - 1 \\
\hline
-3
\end{array}
$$

Rename $x^2 − 4$ as $x^2 + 0x − 4$.
Multiply x and x + 1.
Subtract. Then bring down −4.
Multiply −1 and x + 1.
Subtract. The remainder is −3.

Therefore, $(x^2 - 4) \div (x + 1) = x - 1 + \dfrac{-3}{x + 1}$.

Your Turn

d. Find $(a^3 + 8a - 20) \div (a - 2)$.

If you know the area of a rectangle and the length of one side, you can find the width by dividing polynomials.

5 Find the width of a rectangle if its area is $10x^2 + 29x + 21$ square units and its length is $2x + 3$ units.

$2x + 3$

$10x^2 + 29x + 21$

To find the width, divide the area $10x^2 + 29x + 21$ by the length $2x + 3$.

$$
\begin{array}{r}
5x + 7 \\
2x + 3\overline{)10x^2 + 29x + 21} \\
(-)\ 10x^2 + 15x \\
\hline
14x + 21 \\
(-)\ 14x + 21 \\
\hline
0
\end{array}
$$

Multiply 5x and 2x + 3.
Subtract. Then bring down 21.
Multiply 7 and 2x + 3.
The remainder is 0.

Therefore, the width of the rectangle is $5x + 7$ units.
You can check your answer by multiplying (2x + 3) and (5x + 7).

Check for Understanding

Communicating Mathematics

Study the lesson. Then complete the following.

1. **Identify** the dividend, divisor, quotient, and remainder.
   $(3k^2 - 7k - 5) \div (3k + 2) = k - 3 + \dfrac{1}{3k + 2}$

2. **Write** a division problem represented by the model.

3. **Explain** how you know whether the quotient is a factor of the dividend.

$x - 4$

$x + 1 \Big\{$ | x^2 | |
| x | |

Guided Practice

⏱ **Getting Ready** **Find each quotient.**

Sample 1: $x^2 \div x$	**Sample 2:** $4x^2 \div 2x$
Solution: x	**Solution:** $2x$

4. $3y^2 \div y$
5. $6a^2 \div 2a$
6. $10x^3 \div 5x$

Find each quotient. *(Examples 1–4)*

7. $(x^2 + 4x) \div (x + 4)$
8. $(a^2 + 3a + 2) \div (a + 1)$
9. $(2x^2 + 3x - 3) \div (2x - 1)$
10. $(s^3 + 9) \div (s - 3)$

11. **Geometry** Find the length of a rectangle if its area is $2x^2 - 5x - 12$ square inches and its width is $x - 4$ inches. *(Example 5)*

$x - 4$ | $2x^2 - 5x - 12$

Exercises

Practice

Find each quotient.

12. $(12y - 4) \div (3y - 1)$

13. $(x^2 - 3x) \div (x - 3)$

14. $(8a^2 + 6a) \div (4a + 3)$

15. $(6r^3 - 15r^2) \div (2r - 5)$

16. $(a^2 + 6a + 5) \div (a + 5)$

17. $(x^2 + x - 12) \div (x - 3)$

18. $(s^2 + 11s + 18) \div (s + 2)$

19. $(a^2 - 2a - 35) \div (a - 7)$

20. $(c^2 + 12c + 36) \div (c + 9)$

21. $(3t^2 - 10t - 24) \div (3t - 4)$

22. $(2m^2 + 7m + 3) \div (m + 2)$

23. $(2b^2 + 3b - 6) \div (2b - 1)$

24. $(a^3 + 8a - 21) \div (a - 2)$

25. $(x^3 + 27) \div (x + 3)$

26. $(x^3 - 8) \div (x - 2)$

27. $(4x^4 - 2x^2 + x + 1) \div (x - 1)$

28. Find the quotient when $x^2 + 9x + 20$ is divided by $x + 4$.

29. What is the quotient when $2x^2 - 9x + 9$ is divided by $2x - 3$?

Applications and Problem Solving

Real World

30. **Geometry**   The volume of a rectangular prism is $x^3 + 6x^2 + 8x$ cubic feet. If the height of the prism is $x + 4$ feet and the length is $x + 2$ feet, find the width of the prism.

31. **Transportation**   The distance from San Francisco, CA, to New York, NY, is 2807 miles. To the nearest hour, find the number of hours it would take to travel this distance using each method of transportation at the given average speed. Use the formula $d = rt$.

  a. walking, 3 mph

  b. stagecoach, 20 mph

  c. automobile, 55 mph

  d. jumbo jet, 608 mph

32. **Aviation**   The distance d in miles flown by an airplane is given by the polynomial $x^2 + 501x + 500$. Suppose that $x + 500$ represents the speed r of the airplane. Use the formula $d = rt$ to find the following.

  a. Find the polynomial that represents the time t in hours.

  b. If $x = 5$, find the distance, rate, and time for the airplane.

33. **Critical Thinking**   Find the value of k if the remainder is 15 when $x^3 - 7x^2 + 4x + k$ is divided by $x - 2$.

Mixed Review

Find each product or quotient.   *(Lesson 15–2)*

34. $\dfrac{3x + 9}{x} \cdot \dfrac{x^2}{x^2 - 9}$

35. $\dfrac{a^2}{b^2} \div \dfrac{a^2}{b^2}$

36. **Sports**   When a professional football team from the west coast plays a professional football team from the east coast on television on Monday nights, the west coast team wins 64% of the time. Express 64% as a fraction in simplest form.   **Source:** Stanford University Sleep Disorders Clinic, 1998   *(Lesson 15–1)*

Determine whether each system of equations has *one* solution, *no* solution, or *infinitely many* solutions. *(Lesson 13–2)*

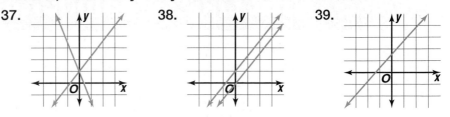

37.   38.   39.

Solve each inequality. *(Lesson 12–4)*

40. $3x - 1 > 14$ **41.** $-7y + 6 \leq 48$

42. Standardized Test Practice Which is the graph of $y = x^2 + 2$?
(Lesson 11–1)

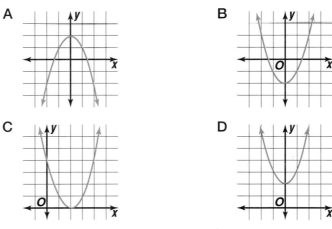

A   B

C   D

Quiz 1   Lessons 15–1 through 15–3

Find the excluded value(s) for each rational expression. Then simplify. *(Lesson 15–1)*

1. $\dfrac{m^2 - 2m}{m - 2}$

2. $\dfrac{x^2 - 4x + 4}{x^2 + 4x - 12}$

Find each product or quotient. *(Lesson 15–2)*

3. $\dfrac{y + 3}{2y + 8} \cdot \dfrac{y^2 + 5y + 4}{y^2 + 4y + 3}$

4. $\dfrac{a^2 b^2}{6a + 18} \div \dfrac{ab}{a^2 - 9}$

5. $\dfrac{6 - 2a}{a^2 + a - 2} \cdot \dfrac{a^2 + 3a + 2}{a - 3}$

6. $\dfrac{x^2}{x^2 - 16} \div \dfrac{x}{x^2 - 16}$

Find each quotient. *(Lesson 15–3)*

7. $(y^2 - 5y) \div (y - 5)$ **8.** $(n^2 + 2n - 10) \div (n - 2)$ **9.** $(x^3 - 1) \div (x - 1)$

10. Fast Food About 90% of Americans eat fast food in a given month.
Of these, 60% use the drive-through window at least once. What percent
of Americans use a fast-food drive-through at least once a month?
Source: Maritz Marketing Research Inc. *(Lesson 15–2)*

Extra Practice   See p. 722.

What You'll Learn

You'll learn to add and subtract rational expressions with like denominators.

Why It's Important

Entertainment
Rational expressions can be combined to determine concert profits.
See Exercise 35.

Why is stadium food so expensive? One reason is that many different people share the earnings, as shown in the diagram at the right. How much more do the team owners receive than the concession companies? To solve the problem, find $\frac{4}{10} - \frac{1}{10}$. You know how to add and subtract fractions with like denominators.

Where Does the Money Go?

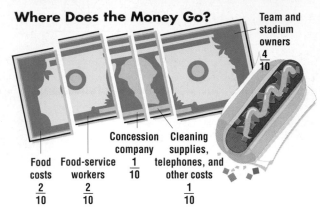

Team and stadium owners $\frac{4}{10}$

Concession company $\frac{1}{10}$

Cleaning supplies, telephones, and other costs $\frac{1}{10}$

Food costs $\frac{2}{10}$ Food-service workers $\frac{2}{10}$

Step 1 Add or subtract the numerators.

Step 2 Write the sum or difference over the common denominator.

$$\frac{4}{10} - \frac{1}{10} = \frac{4-1}{10}$$
$$= \frac{3}{10}$$

So, the team owners receive $\frac{3}{10}$ more of the food sales than the concession companies. You can use this same method to add or subtract rational expressions with like denominators.

Examples

Find each sum or difference.

1 $\dfrac{2}{a} + \dfrac{4}{a}$

$\dfrac{2}{a} + \dfrac{4}{a} = \dfrac{2+4}{a}$ *The common denominator is a.*
Add the numerators.

$= \dfrac{6}{a}$

2 $\dfrac{5x}{11} - \dfrac{2x}{11}$

$\dfrac{5x}{11} - \dfrac{2x}{11} = \dfrac{5x-2x}{11}$ *The common denominator is 11.*
Subtract the numerators.

$= \dfrac{3x}{11}$

Your Turn

a. $\dfrac{4m}{7} + \dfrac{m}{7}$

b. $\dfrac{7}{x} - \dfrac{3}{x}$

When adding or subtracting fractions, the result is not always in simplest form. To find the simplest form, divide the numerator and the denominator by the greatest common factor (GCF).

$$\frac{3}{8} + \frac{1}{8} = \frac{4}{8}$$

$$= \frac{\overset{1}{\cancel{4}}}{\underset{2}{\cancel{8}}} \quad \textit{The GCF of 4}$$
$$\textit{and 8 is 4.}$$

$$= \frac{1}{2}$$

$$\frac{9}{16} - \frac{3}{16} = \frac{6}{16}$$

$$= \frac{\overset{3}{\cancel{6}}}{\underset{8}{\cancel{16}}} \quad \textit{The GCF of 6}$$
$$\textit{and 16 is 2.}$$

$$= \frac{3}{8}$$

A similar process is used to combine rational expressions.

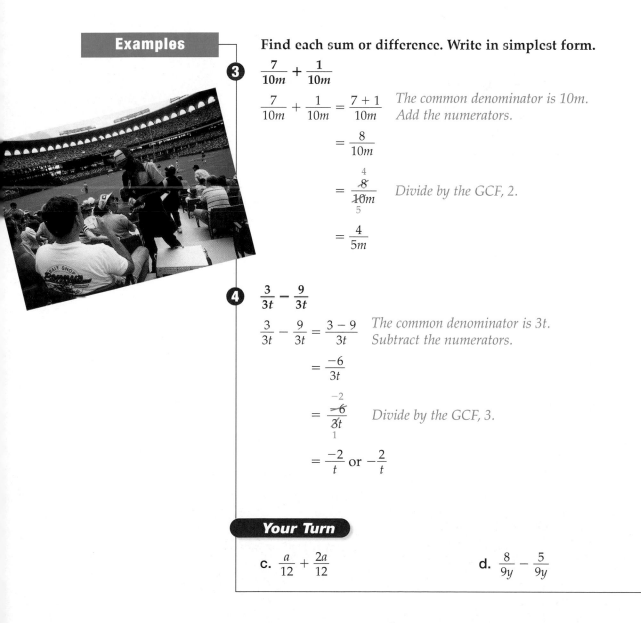

Examples

Find each sum or difference. Write in simplest form.

3 $\dfrac{7}{10m} + \dfrac{1}{10m}$

$$\frac{7}{10m} + \frac{1}{10m} = \frac{7+1}{10m} \quad \textit{The common denominator is 10m.}$$
$$\textit{Add the numerators.}$$

$$= \frac{8}{10m}$$

$$= \frac{\overset{4}{\cancel{8}}}{\underset{5}{\cancel{10m}}} \quad \textit{Divide by the GCF, 2.}$$

$$= \frac{4}{5m}$$

4 $\dfrac{3}{3t} - \dfrac{9}{3t}$

$$\frac{3}{3t} - \frac{9}{3t} = \frac{3-9}{3t} \quad \textit{The common denominator is 3t.}$$
$$\textit{Subtract the numerators.}$$

$$= \frac{-6}{3t}$$

$$= \frac{\overset{-2}{\cancel{-6}}}{\underset{1}{\cancel{3t}}} \quad \textit{Divide by the GCF, 3.}$$

$$= \frac{-2}{t} \text{ or } -\frac{2}{t}$$

Your Turn

c. $\dfrac{a}{12} + \dfrac{2a}{12}$

d. $\dfrac{8}{9y} - \dfrac{5}{9y}$

Sometimes, the denominators of rational expressions are binomials.

Find each sum or difference. Write in simplest form.

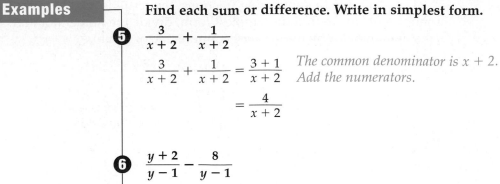

5 $\dfrac{3}{x+2} + \dfrac{1}{x+2}$

$$\dfrac{3}{x+2} + \dfrac{1}{x+2} = \dfrac{3+1}{x+2}$$ *The common denominator is $x + 2$.*
Add the numerators.

$$= \dfrac{4}{x+2}$$

6 $\dfrac{y+2}{y-1} - \dfrac{8}{y-1}$

$$\dfrac{y+2}{y-1} - \dfrac{8}{y-1} = \dfrac{y+2-8}{y-1}$$ *The common denominator is $y - 1$.*
Subtract the numerators.

$$= \dfrac{y-6}{y-1}$$

Your Turn

e. $\dfrac{4m}{2m+3} + \dfrac{5}{2m+3}$ **f.** $\dfrac{5x}{x+2} - \dfrac{2x}{x+2}$

Sometimes you must factor in order to simplify the sum or difference of rational expressions.

Examples

Find each sum or difference. Write in simplest form.

7 $\dfrac{b}{b+4} + \dfrac{b+8}{b+4}$

$$\dfrac{b}{b+4} + \dfrac{b+8}{b+4} = \dfrac{b+(b+8)}{b+4}$$ *The common denominator is $b + 4$.*
Add the numerators.

$$= \dfrac{2b+8}{b+4}$$

$$= \dfrac{2(b+4)}{b+4}$$ *Factor the numerator.*

$$= \dfrac{2(\cancel{b+4})^{\,1}}{\cancel{b+4}_{\,1}}$$ *Divide by the GCF, $b + 4$.*

$$= 2$$

8 $\dfrac{15x}{4x-1} - \dfrac{3x+3}{4x-1}$

$$\dfrac{15x}{4x-1} - \dfrac{3x+3}{4x-1} = \dfrac{15x-(3x+3)}{4x-1}$$ *The common denominator is $4x - 1$.*
Subtract the numerators.

$$= \dfrac{15x-3x-3}{4x-1}$$ *Distributive Property*

$$= \dfrac{12x-3}{4x-1}$$

$$= \frac{3(4x - 1)}{4x - 1} \qquad \textit{Factor the numerator.}$$

$$= \frac{3(\overset{1}{\cancel{4x - 1}})}{\underset{1}{\cancel{4x - 1}}} \qquad \textit{Divide by the GCF, } 4x - 1.$$

$$= 3$$

Your Turn

g. $\dfrac{3r}{r + 5} + \dfrac{15}{r + 5}$ **h.** $\dfrac{m}{m + 3} - \dfrac{3m + 6}{m + 3}$

You can solve some geometry problems by adding rational expressions.

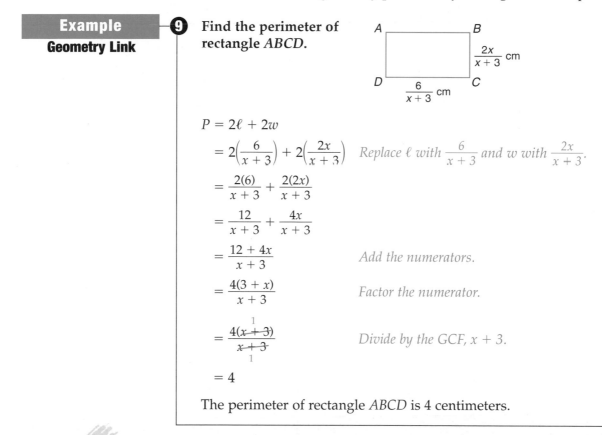

Example
⑨
Geometry Link

Find the perimeter of rectangle $ABCD$.

$P = 2\ell + 2w$

$$= 2\left(\frac{6}{x + 3}\right) + 2\left(\frac{2x}{x + 3}\right) \quad \textit{Replace } \ell \textit{ with } \frac{6}{x + 3} \textit{ and } w \textit{ with } \frac{2x}{x + 3}.$$

$$= \frac{2(6)}{x + 3} + \frac{2(2x)}{x + 3}$$

$$= \frac{12}{x + 3} + \frac{4x}{x + 3}$$

$$= \frac{12 + 4x}{x + 3} \qquad \textit{Add the numerators.}$$

$$= \frac{4(3 + x)}{x + 3} \qquad \textit{Factor the numerator.}$$

$$= \frac{4(\overset{1}{\cancel{x + 3}})}{\underset{1}{\cancel{x + 3}}} \qquad \textit{Divide by the GCF, } x + 3.$$

$$= 4$$

The perimeter of rectangle $ABCD$ is 4 centimeters.

Check for Understanding

Communicating
Mathematics

Study the lesson. Then complete the following.

1. **Explain** how the sum of two fractions with the same denominator can have a sum of zero.

2. **Identify** the mistake in the solution to the problem below. Then write the correct sum.

$$\frac{4x}{2x + 1} + \frac{2}{2x + 1} = \frac{4x + 2}{4x + 2} \text{ or } 1$$

3. **Complete** the table.

a	b	$a + b$	$a - b$	$a \cdot b$	$a \div b$
$\dfrac{2}{t}$	$\dfrac{1}{t}$	$\dfrac{3}{t}$	$\dfrac{1}{t}$	$\dfrac{2}{t^2}$	2
$\dfrac{12}{y}$	$\dfrac{4}{y}$				
$\dfrac{x}{x-1}$	$\dfrac{1}{x-1}$				

Guided Practice

Find each sum or difference. Write in simplest form.

Sample 1: $\dfrac{4}{15} + \dfrac{3}{15}$

Solution: $\dfrac{4}{15} + \dfrac{3}{15} = \dfrac{4 + 3}{15}$

$= \dfrac{7}{15}$

Sample 2: $\dfrac{8}{9} - \dfrac{2}{9}$

Solution: $\dfrac{8}{9} - \dfrac{2}{9} = \dfrac{8 - 2}{9}$

$= \dfrac{6}{9}$ or $\dfrac{2}{3}$

4. $\dfrac{5}{7} + \dfrac{1}{7}$

5. $\dfrac{4}{12} + \dfrac{5}{12}$

6. $\dfrac{6}{10} - \dfrac{1}{10}$

Find each sum or difference. Write in simplest form. *(Examples 1–8)*

7. $\dfrac{5}{x} + \dfrac{2}{x}$

8. $\dfrac{2t}{3} - \dfrac{t}{3}$

9. $\dfrac{7y}{y} - \dfrac{8y}{y}$

10. $\dfrac{a}{12} + \dfrac{2a}{12}$

11. $\dfrac{2x}{x-3} - \dfrac{6}{x-3}$

12. $\dfrac{2c+3}{c-4} - \dfrac{c-2}{c-4}$

13. **Geometry** Find the perimeter of the rectangle. *(Example 9)*

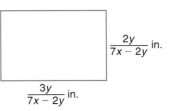

$\dfrac{2y}{7x - 2y}$ in.

$\dfrac{3y}{7x - 2y}$ in.

Exercises • • • • • • • • • • • • • • • • • • •

Practice

Find each sum or difference. Write in simplest form.

14. $\dfrac{7}{m} + \dfrac{4}{m}$

15. $\dfrac{x}{8} + \dfrac{6x}{8}$

16. $\dfrac{7}{15z} - \dfrac{3}{15z}$

17. $\dfrac{5t}{2} - \dfrac{4t}{2}$

18. $\dfrac{8p}{13} - \dfrac{3p}{13}$

19. $\dfrac{1}{6r} + \dfrac{6}{6r}$

20. $\dfrac{a}{2} + \dfrac{a}{2}$

21. $\dfrac{n}{3} + \dfrac{2n}{3}$

22. $\dfrac{5}{3y} - \dfrac{2}{3y}$

23. $\dfrac{5x}{24} - \dfrac{3x}{24}$

24. $\dfrac{8}{27m} + \dfrac{4}{27m}$

25. $\dfrac{5}{2z} + \dfrac{-7}{2z}$

26. $\dfrac{8}{y-2} - \dfrac{6}{y-2}$ **27.** $\dfrac{y}{a+1} - \dfrac{y}{a+1}$ **28.** $\dfrac{3}{x+2} - \dfrac{2}{x+2}$

29. $\dfrac{2x}{x+3} + \dfrac{6}{x+3}$ **30.** $\dfrac{8n+3}{3n+4} - \dfrac{2n-5}{3n+4}$ **31.** $\dfrac{10m-1}{4m-3} - \dfrac{8-2m}{4m-3}$

32. What is $\dfrac{3c-7}{2c-3}$ minus $\dfrac{c-4}{2c-3}$?

33. Find the sum of $\dfrac{x+y}{y-2}$ and $\dfrac{y-x}{y-2}$.

Applications and Problem Solving

34. Geometry Find the perimeter of the isosceles triangle.

$\dfrac{x}{x-2}$ m

$\dfrac{2x-5}{x-2}$ m

35. Entertainment A musical group receives \$15,000 for each concert. The five members receive equal shares of the money. They must pay 0.3 of their earnings for other salaries and expenses.

    a. Write a rational expression that represents how much money each member of the music group actually receives for a concert.

    b. Simplify the expression you wrote in part a to find how much each member of the group receives for a concert.

36. Critical Thinking Which rational expression is not equivalent to the others?

    a. $\dfrac{4}{x-5}$ **b.** $-\dfrac{4}{5-x}$ **c.** $\dfrac{-4}{x-5}$ **d.** $\dfrac{-4}{5-x}$

Mixed Review

Find each quotient. *(Lesson 15–3)*

37. $(a^2 + 9a + 20) \div (a + 5)$ **38.** $(x^2 - 2x - 35) \div (x - 7)$

39. $(2y^2 - 3y - 35) \div (2y + 7)$ **40.** $(t^2 + 12t + 36) \div (t + 9)$

41. Find the product of $\dfrac{5x^2y}{8ab}$ and $\dfrac{12a^2b}{25x}$. *(Lesson 15–2)*

42. Solve the system $y = x^2 - 2$ and $y = -x$ by graphing. *(Lesson 13–6)*

43. Find $(3x + y)^2$. *(Lesson 9–5)*

44. Standardized Test Practice About 25% of all identical twins are "mirror" twins—a reflection of each other. In a group of identical twins, 35 pairs were found to be "mirror" twins. About how many pairs of identical twins would you expect to be in the group? *(Lesson 5–4)*

    A 9   **C** 70

    B 140   **D** 100

Extra Practice See p. 723.

Math
In the Workplace

What You'll Learn

You'll learn to add and subtract rational expressions with unlike denominators.

Why It's Important

Teaching Teachers can use LCM when dividing their classes into groups.
See Example 3.

Mrs. Reimer likes to arrange her students' desks in groups of 4, 6, or 9 because these arrangements work best for group assignments. What is the least number of student desks Mrs. Reimer can have in her classroom?

The least number of desks is given by the **least common multiple (LCM)** of the three numbers. The least common multiple is the least number that is a common multiple of two or more numbers.

multiples of 4: 0, 4, 8, 12, 16, 20, 24, 28, 32, **36**, . . .

multiples of 6: 0, 6, 12, 18, 24, 30, **36**, . . .

multiples of 9: 0, 9, 18, 27, **36**, . . .

Zero is a multiple of every number, but it cannot be the LCM. The LCM of 4, 6, and 9 is 36. Thus, Mrs. Reimer must have at least 36 desks in her classroom. You can also use prime factorization to find the LCM.

$$4 = 2 \cdot 2 \qquad 6 = 2 \cdot 3 \qquad 9 = 3 \cdot 3$$

The prime factors are 2 and 3. The greatest number of times 2 appears is twice (in 4). The greatest number of times 3 appears is twice (in 9). So, the LCM of 4, 6, and 9 is $2 \cdot 2 \cdot 3 \cdot 3$ or 36. This is the same answer you got by listing the multiples.

Examples

Find the LCM for each pair of expressions.

1 $15a^2b, 18a^3$

$15a^2b = 3 \cdot 5 \cdot a \cdot a \cdot b$ *Factor each expression.*
$18a^3 = 2 \cdot 3 \cdot 3 \cdot a \cdot a \cdot a$

$LCM = 2 \cdot 3 \cdot 3 \cdot 5 \cdot a \cdot a \cdot a \cdot b$ *Use each factor the greatest number of*
$= 90a^3b$ *times it appears in either factorization.*

2 $x^2 + x - 6, 2x^2 - 3x - 2$

$x^2 + x - 6 = (x + 3)(x - 2)$ *Factor each expression.*
$2x^2 - 3x - 2 = (2x + 1)(x - 2)$

$LCM = (x + 3)(x - 2)(2x + 1)$ *Use each factor the greatest number of*
   times it appears in either factorization.

Your Turn

a. $12a, 8ab$

b. $x^2 - 9, x^2 - 2x - 3$

3 Refer to the application at the beginning of the lesson. Suppose Mrs. Reimer wants student groups of 4, 6, or 8. What is the minimum number of desks needed?

Find the LCM of 4, 6, and 8.

$4 = 2 \cdot 2$

$6 = 2 \cdot 3$

$8 = 2 \cdot 2 \cdot 2$

LCM $= 2 \cdot 2 \cdot 2 \cdot 3$ or 24 *Use each factor the greatest number of times it appears in any of the factorizations.*

Mrs. Reimer must have at least 24 desks in her classroom.

To add or subtract fractions with unlike denominators, first rename the fractions so the denominators are alike. Any common denominator could be used. However, the computation is usually easier if you use the **least common denominator (LCD)**. Recall that the least common denominator is the LCM of the denominators.

Examples

Write each pair of rational expressions with the same LCD.

4 $\dfrac{9}{2m^2}, \dfrac{5}{6m}$

First find the LCD. $2m^2 = 2 \cdot m \cdot m$

$6m = 2 \cdot 3 \cdot m$

LCD $= 2 \cdot 3 \cdot m \cdot m$ or $6m^2$

Then write each fraction with the same LCD.

$\dfrac{9}{2m^2} \cdot \dfrac{3}{3} = \dfrac{27}{6m^2}$ $\qquad$ $\dfrac{5}{6m} \cdot \dfrac{m}{m} = \dfrac{5m}{6m^2}$

5 $\dfrac{1}{x+2}, \dfrac{3x}{4x+8}$

First find the LCD. $x + 2 = x + 2$

$4x + 8 = 4(x + 2)$

LCD $= 4(x + 2)$

Then write each fraction with the same LCD.

$\dfrac{1}{x+2} \cdot \dfrac{4}{4} = \dfrac{4}{4(x+2)}$ $\qquad$ $\dfrac{3x}{4x+8} = \dfrac{3x}{4(x+2)}$

Your Turn

c. $\dfrac{6}{b^3}, \dfrac{7}{ab}$ $\qquad\qquad\qquad$ **d.** $\dfrac{x}{x-6}, \dfrac{x-3}{x-2}$

Use the following steps to add or subtract rational expressions with unlike denominators.

Step 1 Find the LCD.

Step 2   Change each rational expression into an equivalent expression with the LCD as the denominator.

Step 3   Add or subtract as with rational expressions with like denominators.

Step 4   Simplify if necessary.

Find each sum or difference. Write in simplest form.

6 $\dfrac{5}{6x} + \dfrac{7}{12x^2}$

Step 1   Find the LCD.
$$6x = 2 \cdot 3$$
$$12x^2 = 2 \cdot 2 \cdot 3 \cdot x \cdot x$$
$$\text{LCM} = 2 \cdot 2 \cdot 3 \cdot x \cdot x \text{ or } 12x^2$$

The LCD of $\dfrac{5}{6x}$ and $\dfrac{7}{12x^2}$ is $12x^2$.

Step 2   Rename each expression with the LCD as the denominator. The denominator of $\dfrac{7}{12x^2}$ is already $12x^2$, so only $\dfrac{5}{6x}$ needs to be renamed.

$$\dfrac{5}{6x} \cdot \dfrac{2x}{2x} = \dfrac{10x}{12x^2} \quad \textit{Why do you multiply } \dfrac{5}{6x} \textit{ by } \dfrac{2x}{2x}?$$

Step 3   Add or subtract.

$$\dfrac{5}{6x} + \dfrac{7}{12x^2} = \dfrac{10x}{12x^2} + \dfrac{7}{12x^2}$$

$$= \dfrac{10x + 7}{12x^2} \quad \textit{This expression is in simplest form.}$$

7 $\dfrac{m}{m^2 - 9} - \dfrac{3}{m - 3}$

$$m^2 - 9 = (m - 3)(m + 3)$$
$$m - 3 = m - 3$$
$$\text{LCM} = (m - 3)(m + 3)$$

The LCD of $\dfrac{m}{m^2 - 9}$ and $\dfrac{3}{m - 3}$ is $(m - 3)(m + 3)$.

$$\dfrac{m}{m^2 - 9} - \dfrac{3}{m - 3}$$

$$= \dfrac{m}{(m - 3)(m + 3)} - \dfrac{3}{m - 3} \cdot \dfrac{m + 3}{m + 3} \quad \textit{Multiply by } \dfrac{3}{m - 3} \textit{ by } \dfrac{m + 3}{m + 3}.$$

$$= \dfrac{m}{(m - 3)(m + 3)} - \dfrac{3m + 9}{(m - 3)(m + 3)} \quad \textit{The LCD is } (m - 3)(m + 3).$$

$$= \dfrac{m - (3m + 9)}{(m - 3)(m + 3)} \quad \textit{Subtract the numerators.}$$

$$= \dfrac{m - 3m - 9}{(m - 3)(m + 3)} \quad \textit{Distributive Property}$$

$$= \dfrac{-2m - 9}{(m - 3)(m + 3)} \quad \textit{Simplify.}$$

8 $\dfrac{5}{x-3} + \dfrac{7}{2x-6}$

$\dfrac{5}{x-3} + \dfrac{7}{2x-6} = \dfrac{5}{x-3} + \dfrac{7}{2(x-3)}$ *The LCD is $2(x-3)$.*

$= \dfrac{5}{x-3} \cdot \dfrac{2}{2} + \dfrac{7}{2(x-3)}$

$= \dfrac{10}{2(x-3)} + \dfrac{7}{2(x-3)}$

$= \dfrac{10+7}{2(x-3)}$ *Add the numerators.*

$= \dfrac{17}{2(x-3)}$

Your Turn

e. $\dfrac{3}{xy} + \dfrac{2}{y^2}$

f. $\dfrac{5a}{2a+6} - \dfrac{a}{a+3}$

Check for Understanding

Communicating Mathematics

Study the lesson. Then complete the following.

1. **Write** two rational expressions with unlike denominators in which one of the denominators is the LCD of the rational expressions.

2. **Explain** the steps you would take to find the LCD for the expression $\dfrac{3}{8x^2} - \dfrac{5}{12xy} + \dfrac{7}{6y^2}$. Then simplify the expression.

3. **YOU Decide** Ashley found the sum of $\dfrac{4}{x-1}$ and $\dfrac{5}{x}$ to be $\dfrac{9}{2x-1}$. Malik found the sum to be $\dfrac{9x-5}{x^2-x}$. Who is correct? Explain why.

Vocabulary

least common multiple
least common denominator

Guided Practice

Getting Ready   Find the LCM for each pair of numbers.

Sample: 10, 12   **Solution:** $10 = 2 \cdot 5$ $12 = 2 \cdot 2 \cdot 3$
                                                      $LCM = 2 \cdot 2 \cdot 3 \cdot 5$ or 60

4. 6, 10   **5.** 8, 18   **6.** 12, 15

Find the LCM for each pair of expressions.   *(Examples 1 & 2)*

7. $6xy, 15y^2$ **8.** $x+1, x^2+5x+4$

Write each pair of rational expressions with the same LCD.
(Examples 4 & 5)

9. $\dfrac{4}{a^2}, \dfrac{5}{a}$

10. $\dfrac{7}{t+3}, \dfrac{13}{2t+6}$

Find each sum or difference. Write in simplest form. *(Examples 6–8)*

11. $\dfrac{t}{6} + \dfrac{3t}{12}$

12. $\dfrac{2}{3a} - \dfrac{4}{9a}$

13. $\dfrac{2}{ab^2} + \dfrac{3}{ab}$

14. $\dfrac{6}{m} + \dfrac{5}{n}$

15. $\dfrac{x}{x^2 - 4} - \dfrac{4}{x + 2}$

16. $\dfrac{2}{3a - 1} - \dfrac{7}{15a - 5}$

17. Astronomy Earth, Jupiter, and Saturn revolve around the Sun about once every 1, 12, and 30 years, respectively. The last time Earth, Jupiter, and Saturn were lined up with the sun was in 1982, and Jupiter and Saturn could be seen close together, high in the sky at midnight. In what year will this happen again? *(Example 3)*

Exercises

• • • • • • • • • • • • • • • • • • •

Practice

Find the LCM for each pair of expressions.

18. $3t, 9t^2$

19. $12a^2, 2ab^2$

20. $16m, 6mn$

21. $y + 2, y^2 - 4$

22. $x^2 + 9x + 14, x^2 + 3x + 2$

23. $x^2 - 9, 3x^2 - 8x - 3$

Write each pair of rational expressions with the same LCD.

24. $\dfrac{4}{m^3}, \dfrac{1}{m}$

25. $\dfrac{7}{2t}, \dfrac{8}{10t}$

26. $\dfrac{3}{6ab}, \dfrac{14}{4a^2}$

27. $\dfrac{5}{xy}, \dfrac{6}{yz}$

28. $\dfrac{x}{x^2 - 1}, \dfrac{-2}{x - 1}$

29. $\dfrac{10k}{3k + 1}, \dfrac{k}{3 + 9k}$

Find each sum or difference. Write in simplest form.

30. $\dfrac{a}{5} - \dfrac{a}{15}$

31. $\dfrac{d}{2} - \dfrac{d}{5}$

32. $\dfrac{t}{3} + \dfrac{2t}{7}$

33. $\dfrac{9}{4x} + \dfrac{3}{2x}$

34. $\dfrac{5}{2a} - \dfrac{3}{6a}$

35. $\dfrac{m}{3t} + \dfrac{1}{t}$

36. $\dfrac{4}{a^3} - \dfrac{2}{a}$

37. $\dfrac{7}{3x} - \dfrac{1}{6x^2}$

38. $\dfrac{2t}{5n^2} - \dfrac{1}{3n}$

39. $\dfrac{2}{x} + \dfrac{x + 3}{y}$

40. $\dfrac{6z}{7w} + \dfrac{2z}{w^3}$

41. $\dfrac{1}{4xy} + \dfrac{y}{10x^2}$

42. $\dfrac{9}{y + 1} - \dfrac{3}{4y + 4}$

43. $\dfrac{4a}{2a + 6} + \dfrac{3}{a + 3}$

44. $\dfrac{7}{x - 2} + \dfrac{3}{x}$

45. $\dfrac{y}{y^2 - 2y + 1} - \dfrac{1}{y - 1}$

46. $\dfrac{3a}{a^2 + 4a + 4} - \dfrac{2}{a + 2}$

47. $\dfrac{5x + 2}{2x - 1} + \dfrac{2x - 3}{8x - 4}$

48. $\dfrac{7}{a^2 + 2a - 3} - \dfrac{5}{a + 5a + 6}$

49. $\dfrac{x^2 + 4x - 5}{x^2 - 2x - 3} + \dfrac{2}{x + 1}$

50. Add $\dfrac{2x + 3}{x^2 - 4}$ and $\dfrac{6}{x + 2}$.

51. Simplify $\dfrac{m}{m - n} - \dfrac{5}{m}$.

52. Entertainment The choreographer of a musical needs enough dancers so that they can be arranged in groups of 6, 9, and 12, with no one sitting out. What is the least number of dancers needed?

53. Fitness Cynthia jogs around an oval track in 150 seconds. Minya walks around the track in 210 seconds. Suppose they start at the same time. In how many minutes will they meet back at their starting point?

54. Critical Thinking Consider the numbers 18 and 20.
   a. Find the GCF and LCM of the numbers.
   b. What is the value of GCF · LCM?
   c. Describe the relationship between 18 · 20 and GCF · LCM.
   d. Describe how you could find the GCF of two numbers if you already know the LCM.

Mixed Review

Find each sum or difference. Write in simplest form. *(Lesson 15–4)*

55. $\dfrac{9}{a} + \dfrac{7}{a}$

56. $\dfrac{2}{5t} + \dfrac{3}{5t}$

57. $\dfrac{4n}{n+1} - \dfrac{2n-2}{n+1}$

58. Find $(m^2 + 6m - 7) \div (m + 7)$. *(Lesson 15–3)*

Simplify each expression. *(Lesson 14–4)*

59. $7\sqrt{19} + 2\sqrt{19}$

60. $6\sqrt{12} + 4\sqrt{3}$

61. $8\sqrt{5} - 3\sqrt{2}$

62. Geometry The length of the hypotenuse of a right triangle is $\sqrt{\dfrac{56}{25}}$ centimeters. Express the length as a radical in simplest form. *(Lesson 14–3)*

63. What is the solution of the system $y = 7 - x$ and $x - y = -3$? *(Lesson 13–3)*

64. Standardized Test Practice Determine which polynomial can be the measure of the area of a square. *(Lesson 10–5)*
   A $x^2 - 13x + 36$
   B $n^2 + 20n - 100$
   C $4a^2 + 12a + 9$
   D $4a^2 - 6a + 9$

Quiz 2   Lessons 15–3 through 15–5

Find each quotient. *(Lesson 15–3)*

1. $(x^2 + 6x - 16) \div (x - 2)$

2. $(2x^2 - 11x - 20) \div (2x + 3)$

Find each sum or difference. Write in simplest form. *(Lesson 15–4)*

3. $\dfrac{a}{9} + \dfrac{7a}{9}$

4. $\dfrac{5}{21t} + \dfrac{4}{21t}$

5. Parades At the Veteran's Day parade, members of the Veterans of Foreign Wars (VFW) found that they could arrange themselves in rows of 6, 7, or 8, with no one left over. What is the least number of VFW members in the parade? *(Lesson 15–5)*

What You'll Learn
You'll learn to solve rational equations.

Why It's Important
Aviation Pilots can use rational equations to find wind speed.
See Exercise 9.

Every week there is a 10-point quiz in math class. For the first five quizzes, Julia scored a total of 36 points, which gave her an average of 7.2. She is determined to get 10 points on each of the next quizzes until she brings her average up to an 8. On how many quizzes must she score 10 points in order to have an overall quiz average of 8 points?

Let x represent the number of quizzes on which she must score 10 points.

total number of quizzes $= 5 + x$

sum of scores $= 36 + 10x$

$$\text{average} = \frac{36 + 10x}{5 + x} \quad \begin{array}{l} \leftarrow \textit{sum of scores} \\ \leftarrow \textit{number of quizzes} \end{array}$$

$$8 = \frac{36 + 10x}{5 + x} \quad \textit{Julia wants to have an average of 8.}$$

The equation above is a **rational equation** because it contains at least one rational expression. *You will solve this problem in Exercise 39.*

There are three steps in solving rational equations.

Step 1 Find the LCD of each term.
Step 2 Multiply each side of the equation by the LCD.
Step 3 Use the Distributive Property to simplify.

Examples

Solve each equation. Check your solution.

1 $\dfrac{3a}{5} + \dfrac{3}{2} = \dfrac{7a}{10}$

$$\frac{3a}{5} + \frac{3}{2} = \frac{7a}{10} \qquad \textit{The LCD is 10.}$$

$$10\left(\frac{3a}{5} + \frac{3}{2}\right) = 10\left(\frac{7a}{10}\right) \qquad \textit{Multiply each side by the LCD.}$$

$$10\left(\frac{3a}{5}\right) + 10\left(\frac{3}{2}\right) = 10\left(\frac{7a}{10}\right) \qquad \textit{Distributive Property}$$

$$\overset{2}{\cancel{10}}\left(\frac{3a}{\cancel{5}}\right) + \overset{5}{\cancel{10}}\left(\frac{3}{\cancel{2}}\right) = \overset{1}{\cancel{10}}\left(\frac{7a}{\cancel{10}}\right)$$

$$6a + 15 = 7a$$

$$6a + 15 - 6a = 7a - 6a \qquad \textit{Subtract 6a from each side.}$$

$$15 = a$$

Check:

$$\frac{3a}{5} + \frac{3}{2} = \frac{7a}{10}$$

$$\frac{3(15)}{5} + \frac{3}{2} \stackrel{?}{=} \frac{7(15)}{10} \qquad \textit{Replace a with 15.}$$

$$9 + \frac{3}{2} \stackrel{?}{=} \frac{21}{2}$$

$$\frac{21}{2} = \frac{21}{2} \quad \checkmark$$

2 $\frac{2}{3x} - \frac{1}{2x} = \frac{1}{6}$

$$\frac{2}{3x} - \frac{1}{2x} = \frac{1}{6} \qquad \textit{The LCD is 6x.}$$

$$6x\left(\frac{2}{3x} - \frac{1}{2x}\right) = 6x\left(\frac{1}{6}\right) \qquad \textit{Multiply each side by the LCD.}$$

$$6x\left(\frac{2}{3x}\right) - 6x\left(\frac{1}{2x}\right) = 6x\left(\frac{1}{6}\right) \qquad \textit{Distributive Property}$$

$$\overset{2}{\cancel{6x}}\left(\frac{2}{\underset{1}{\cancel{3x}}}\right) - \overset{3}{\cancel{6x}}\left(\frac{1}{\underset{1}{\cancel{2x}}}\right) = \overset{1}{\cancel{6x}}\left(\frac{1}{\underset{1}{\cancel{6}}}\right)$$

$$4 - 3 = x \qquad \textit{Simplify.}$$

$$1 = x$$

Check:

$$\frac{2}{3x} - \frac{1}{2x} = \frac{1}{6}$$

$$\frac{2}{3(1)} - \frac{1}{2(1)} \stackrel{?}{=} \frac{1}{6} \qquad \textit{Replace x with 1.}$$

$$\frac{2}{3} - \frac{1}{2} \stackrel{?}{=} \frac{1}{6} \qquad \textit{The LCD is 6.}$$

$$\frac{4}{6} - \frac{3}{6} \stackrel{?}{=} \frac{1}{6}$$

$$\frac{1}{6} = \frac{1}{6} \quad \checkmark$$

3 $\frac{3x}{x+2} + \frac{1}{x+2} = 4$

$$\frac{3x}{x+2} + \frac{1}{x+2} = 4 \qquad \textit{The LCD is x + 2.}$$

$$(x+2)\left(\frac{3x}{x+2} + \frac{1}{x+2}\right) = (x+2)4 \qquad \textit{Multiply each side by the LCD.}$$

$$(x+2)\left(\frac{3x}{x+2}\right) + (x+2)\left(\frac{1}{x+2}\right) = 4x + 8 \qquad \textit{Distributive Property}$$

$$(\overset{1}{\cancel{x+2}})\left(\frac{3x}{\underset{1}{\cancel{x+2}}}\right) + (\overset{1}{\cancel{x+2}})\left(\frac{1}{\underset{1}{\cancel{x+2}}}\right) = 4x + 8$$

$$3x + 1 = 4x + 8 \qquad \textit{Simplify.}$$

(continued on the next page)

$$3x + 1 - 3x = 4x + 8 - 3x \qquad \textit{Subtract 3x from each side.}$$
$$1 = x + 8$$
$$1 - 8 = x + 8 - 8 \qquad \textit{Subtract 8 from each side.}$$
$$-7 = x \qquad \textit{Check the solution.}$$

④ $\dfrac{3}{r} - \dfrac{1}{r-1} = \dfrac{1}{r-1}$

$$\frac{3}{r} - \frac{1}{r-1} = \frac{1}{r-1} \qquad \textit{The LCD is } r(r-1).$$

$$r(r-1)\left(\frac{3}{r} - \frac{1}{r-1}\right) = r(r-1)\left(\frac{1}{r-1}\right) \qquad \begin{array}{l}\textit{Multiply each side}\\ \textit{by the LCD.}\end{array}$$

$$r(r-1)\left(\frac{3}{r}\right) - r(r-1)\left(\frac{1}{r-1}\right) = r(r-1)\left(\frac{1}{r-1}\right) \qquad \textit{Distributive Property}$$

$$\overset{1}{\cancel{r}}(r-1)\left(\frac{3}{\cancel{r}}\right) - r(\cancel{r-1})\left(\frac{1}{\cancel{r-1}}\right) = r(\cancel{r-1})\left(\frac{1}{\cancel{r-1}}\right)$$

$$(r-1)3 - r = r$$
$$3r - 3 - r = r \qquad \textit{Distributive Property}$$
$$2r - 3 = r$$
$$2r - 3 - r = r - r \qquad \textit{Subtract r from each side.}$$
$$r - 3 = 0$$
$$r - 3 + 3 = 0 + 3 \qquad \textit{Add 3 to each side.}$$
$$r = 3 \qquad \textit{Check the solution.}$$

Your Turn

a. $\dfrac{4x}{3} + \dfrac{7}{2} = \dfrac{9x}{12}$ **b.** $\dfrac{18}{b} = \dfrac{3}{b} + 3$ **c.** $\dfrac{4}{n+1} - \dfrac{2}{n} = \dfrac{5}{n}$

Recall that **uniform motion problems** can be solved by using the formula below.

$$\underbrace{distance}_{d} = \underbrace{rate}_{r} \cdot \underbrace{time}_{t}$$

Example

Travel Link

Real World

⑤ **The Milbys rented a houseboat on the Sacramento River. The maximum speed of the boat in still water is 8 miles per hour. At this rate, a 30-mile trip downstream took the same amount of time as an 18-mile trip against the current. What was the rate of the current?**

Explore Let c = the rate of the current.

Let $8 + c$ = the rate of the boat traveling downstream with the current. *When the boat is traveling with the current, you add the speed of the current to the speed of the boat.*

Let $8 - c$ = the rate of the boat traveling upstream against the current. *When the boat is traveling against the current, you subtract the speed of the current from the speed of the boat.*

Plan   Since $d = rt$, then $t = \frac{d}{r}$.

	d	r	$t = \frac{d}{r}$
Downstream	30	$8 + c$	$\frac{30}{8 + c}$
Upstream	18	$8 - c$	$\frac{18}{8 - c}$

Solve

$$\frac{30}{8 + c} = \frac{18}{8 - c}$$ *The time downstream equals the time upstream.*

$$(8 + c)(8 - c)\left(\frac{30}{8 + c}\right) = (8 + c)(8 - c)\left(\frac{18}{8 - c}\right)$$ *The LCD is $(8 + c)(8 - c)$.*

$$(8 \overset{1}{+} c)(8 - c)\left(\frac{30}{8 + c}\right) = (8 + c)(8 \overset{1}{-} c)\left(\frac{18}{8 - c}\right)$$

$$(8 - c)(30) = (8 + c)(18)$$
$$240 - 30c = 144 + 18c$$
$$240 - 30c + 30c = 144 + 18c + 30c$$ *Add 30c to each side.*
$$240 = 144 + 48c$$
$$240 - 144 = 144 + 48c - 144$$ *Subtract 144 from each side.*
$$96 = 48c$$
$$\frac{96}{48} = \frac{48c}{48}$$ *Divide each side by 48.*
$$2 = c$$

The rate of the current was 2 miles per hour.

Examine   Check the solution to see if it makes sense. The houseboat goes downstream at $8 + 2$ or 10 miles per hour. A 30-mile trip would take $30 \div 10$ or 3 hours. The houseboat goes upstream at $8 - 2$ or 6 miles per hour. An 18-mile trip would take $18 \div 6$ or 3 hours. Both trips take the same amount of time, so the solution is correct.

Check for Understanding

Communicating Mathematics

Study the lesson. Then complete the following.

1. **List** two differences between linear equations and rational equations.

Math Journal

2. **Describe** two different ways in which $\frac{4}{x + 1} = \frac{8}{x - 1}$ can be solved. Then determine whether there are other steps that you could use to solve the examples in this lesson.

> **Vocabulary**
> rational equation
> uniform motion problems

Solve each equation. Check your solution.   (*Examples 1–4*)

3. $\dfrac{2x}{7} + \dfrac{1}{2} = \dfrac{x}{14}$

4. $\dfrac{6}{c} = \dfrac{10}{c} + 4$

5. $\dfrac{5}{m+3} - \dfrac{2}{m+3} = -9$

6. $\dfrac{a+1}{a} + \dfrac{a+4}{a} = 6$

7. $\dfrac{2x}{x+3} + \dfrac{3}{x} = 2$

8. $\dfrac{r-1}{r+1} - \dfrac{2r}{r-1} = -1$

9. Transportation   An airplane can fly at a rate of 600 miles per hour in calm air. It can fly 2520 miles with the wind in the same time it can fly 2280 miles against the wind. Find the speed of the wind.   (*Example 5*)

Exercises • • • • • • • • • • • • • • • • •

Practice

Solve each equation. Check your solution.

10. $\dfrac{a}{6} + \dfrac{2a}{3} = -\dfrac{5}{2}$

11. $\dfrac{1}{4} + \dfrac{5r}{8} = \dfrac{r}{4}$

12. $\dfrac{b}{5} - \dfrac{2b}{15} = 1$

13. $\dfrac{x}{7} = \dfrac{x+3}{10}$

14. $\dfrac{t-1}{4} = \dfrac{t}{3}$

15. $\dfrac{2a-3}{6} = \dfrac{2a}{3} + \dfrac{1}{2}$

16. $\dfrac{4}{3y} + \dfrac{1}{y} = 7$

17. $\dfrac{6}{x} - \dfrac{3}{2x} = \dfrac{1}{2}$

18. $\dfrac{1}{2a} + \dfrac{3}{4a} = \dfrac{1}{4}$

19. $\dfrac{m+1}{m} + \dfrac{m+3}{m} = 5$

20. $\dfrac{t+1}{t} + \dfrac{t+4}{t} = 6$

21. $\dfrac{5}{5-p} - \dfrac{1}{5-p} = -2$

22. $\dfrac{5}{2x} - \dfrac{1}{6x} = 2$

23. $\dfrac{1}{4x} + \dfrac{1}{6x} = 5$

24. $\dfrac{3}{x} + \dfrac{4x}{x-3} = 4$

25. $\dfrac{n-3}{n} = \dfrac{n-3}{n-6}$

26. $\dfrac{3}{r+4} - \dfrac{1}{r} = \dfrac{1}{r}$

27. $\dfrac{5}{x+1} + \dfrac{1}{x} = \dfrac{2}{x^2+x}$

28. $\dfrac{5}{n-2} + \dfrac{2}{n} = \dfrac{1}{n}$

29. $\dfrac{6}{t+1} - \dfrac{3}{4t+4} = \dfrac{3}{4}$

30. $\dfrac{1}{2a+6} + \dfrac{3a}{a+3} = \dfrac{1}{2}$

31. $\dfrac{7}{a-1} = \dfrac{5}{a+3}$

32. $\dfrac{1}{m+1} + \dfrac{5}{m-1} = \dfrac{1}{m^2-1}$

33. $\dfrac{j}{j+1} + \dfrac{5}{j-1} = 1$

34. $\dfrac{6}{z+2} + \dfrac{3}{z^2-4} = \dfrac{7}{z+2}$

35. $\dfrac{1}{4m} - \dfrac{2}{m-3} = \dfrac{2}{m}$

36. $\dfrac{x+2}{2x-1} + \dfrac{3x-6}{8x-4} = \dfrac{9}{4}$

37. What is the value of w in the equation $\dfrac{2w-3}{w-3} - 2 = \dfrac{12}{w+3}$?

38. Solve $\dfrac{3n}{n^2-5n+4} = \dfrac{2}{n-4} + \dfrac{3}{n-1}$.

Applications and Problem Solving

39. Quizzes   Refer to the application at the beginning of the lesson. Solve $8 = \dfrac{36 + 10x}{5 + x}$ to find the number of quizzes on which Julia must score 10 points in order to have an overall quiz average of 8 points.

40. Cycling   A long-distance cyclist pedaling at a steady rate travels 30 miles with the wind. She can travel only 18 miles against the wind in the same amount of time. If the rate of the wind is 3 miles per hour, what is the cyclist's rate without the wind? (*Hint:* Use the formula $d = rt$.)

41. **Critical Thinking**   What number would you add to the numerator and denominator of $\frac{2}{11}$ to make a fraction equivalent to $\frac{1}{2}$?

Mixed Review

42. **Space**   Scientists can predict when an asteroid might hit Earth or come very close to Earth by studying their orbits. Toro is an asteroid that orbits the sun every 584 days. Earth orbits the sun every 365 days. Suppose Earth and Toro are lined up with the sun today. Find the LCM of 584 and 365 to determine how long it will be until Earth and Toro arrive together back at that same point in their orbits. *(Lesson 15–5)*

43. Find the sum of $\frac{11}{4n^2}$ and $\frac{7}{4n^2}$. Write your answer in simplest form. *(Lesson 15–4)*

Name the set or sets of numbers to which each real number belongs. Let N = natural numbers, W = whole numbers, Z = integers, Q = rational numbers, and I = irrational numbers.   *(Lesson 14–1)*

44. $\frac{16}{2}$ 45. -7 46. $\sqrt{15}$

47. **Farming**   In order to have enough time in the growing season, a farmer has at most 16 days left to plant his corn and soybean crops. He can plant corn at a rate of 10 acres per day and soybeans at a rate of 15 acres per day. Suppose he has at most 200 acres available and he wants to plant both crops.   *(Lesson 13–7)*

   a. Let c represent the number of days that corn will be planted and let s represent the number of days that soybeans will be planted. Write a system of inequalities to represent this situation.

   b. Graph the system of inequalities.

   c. List two ordered pairs that are possible solutions and explain what each ordered pair represents.

Solve each system of equations by graphing.   *(Lesson 13–1)*

48. $y = 2x - 1$
    $y = -x + 5$

49. $x = 4$
    $x + y = 3$

50. **Open-Ended Test Practice**   Write an irrational square root whose graph is between points A and B on the number line below. *(Lesson 8–6)*

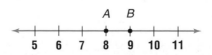

Lesson 15–6 Solving Rational Equations   **673**

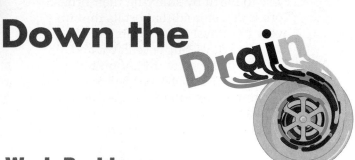

Down the Drain

Materials

calculator

You could also use a spreadsheet for your calculations.

Work Problems

Suppose you have a big job to do, like painting a house. How much faster could you get the job done if you had help? In this investigation, you will use algebraic techniques to solve **work problems**.

Investigate

1. A public swimming pool holds 100,000 gallons of water and has two drain outlets. If only outlet A is open, the pool will drain in 20 hours. If only outlet B is open, the pool will drain in 16 hours.

   a. The drainage rate of outlet A is 100,000 gallons ÷ 20 hours or 5000 gallons per hour. What is the drainage rate of outlet B?

   b. Make a table like the one below and continue filling in columns 2, 3, and 4 until the total water drained in column 4 is more than 100,000 gallons.

Number of Hours	Water Drained by Outlet A (gal)	Water Drained by Outlet B (gal)	Total Water Drained (gal)
1	5000	6250	11,250
2	10,000	12,500	22,500

   c. Approximately how long will it take to drain the pool with both outlets open?

2. The problem above also can be solved using an algebraic equation.

   a. Since outlet A can drain the pool in 20 hours, it can drain $\frac{1}{20}$ of the pool in 1 hour. How much of the pool can outlet B drain in 1 hour?

   b. Use the table at the right and the following formula to write rational expressions for the work done w by outlets 1 and 2 in t hours.

   $$\underbrace{rate\ of\ work}_{r} \cdot \underbrace{time}_{t} = \underbrace{work\ done}_{w}$$

	r	t	w
Outlet A	$\frac{1}{20}$	t	?
Outlet B	$\frac{1}{16}$	t	?

c. If both outlets are open, then the following equation is true.

$$\frac{\text{work of}}{\text{outlet A}} + \frac{\text{work of}}{\text{outlet B}} = \frac{\text{the whole}}{\text{job}}$$

Write an equation to represent the time it takes the pool to drain if both outlets are open. Assume that 1 represents the whole job. *Why?*

d. Solve the equation in part b to find the how long it takes to drain the pool if both outlets are open. Compare this answer to the estimate you made in Exercise 1d.

3. Micheal and Gus work for a ski resort. Their job is to prepare Bear Tooth Run for skiers. It takes Micheal 4 hours to prepare the run and it takes Gus 5 hours. How long would it take them to prepare the run if they work together?

a. In the swimming pool problem, you knew the total amount of gallons to be drained. In this problem since you do not know the size of the trail, write the work as a fraction done per hour. Micheal's rate is $\frac{1}{4}$ of the job per hour. What is Gus' rate?

b. Write an algebraic equation to represent the work done if Micheal and Gus work together.

c. How long would it take them both to prepare the run?

Extending the Investigation

Use a spreadsheet or write an algebraic equation to solve each problem.

- Ian and Mandy own a lawn care business. Ian can mow, trim, and fertilize one particular yard in 2 hours. Mandy can complete the same tasks for that yard in 4 hours. How long would it take them to complete this yard working together?

- In the summer, Sheila and her son, Cory, paint house exteriors. To paint a typical house, it takes Sheila about 45 hours. To paint the same size house, it takes Cory 55 hours. How long would it take them to paint one house working together?

Presenting Your Conclusions

Here are some ideas to help you present your conclusions to the class.

- Design a creative bulletin board displaying a problem from this investigation. Show two ways in which to solve the problem.

- Write a problem similar to those in this investigation. Solve the problem by using a spreadsheet and by using an equation. Write a one-page paper discussing the advantages and disadvantages of using each method to solve work problems.

 Investigation For more information on work problems visit: www.algconcepts.glencoe.com

Understanding and Using the Vocabulary

After completing this chapter, you should be able to define each term, property, or phrase and give an example or two of each.

*inter*NET
CONNECTION **Review Activities**
For more review activities, visit:
www.algconcepts.glencoe.com

excluded value *(p. 638)*
least common denominator *(p. 663)*
least common multiple *(p. 662)*
rational equation *(p. 668)*

rational expression *(p. 638)*
rational function *(p. 638)*
uniform motion problems *(p. 670)*
work problems *(p. 674)*

Choose the correct term or expression to complete each sentence.

1. An excluded value is any value assigned to a variable that results in a (denominator, numerator) of zero.

2. A rational expression is an algebraic fraction whose numerator and denominator are (rational numbers, polynomials).

3. To find $\dfrac{9}{x + y}$ divided by $\dfrac{3x + 6y}{x}$, multiply $\dfrac{9}{x + y}$ and $\left(\dfrac{3x + 6y}{x}, \dfrac{x}{3x + 6y}\right)$.

4. The excluded values of $\dfrac{4x}{x^2 - 36}$ are (0 and 36, 6 and -6).

5. To simplify a rational expression, divide the numerator and denominator by their (greatest common factor, least common multiple).

6. The rational expression $\dfrac{x + 2}{x + 4}$ (is, is not) in simplest form.

7. The reciprocal of $a + b$ is $\left(-a - b, \dfrac{1}{a + b}\right)$.

8. To add or subtract rational expressions, first rename the expressions using the (greatest common factor, least common multiple) as the denominator.

9. The least common multiple of x^2 and $2x$ is ($2x^2$, $2x^3$).

10. The divisor in $(x^2 + 8x + 15) \div (x + 5) = (x + 3)$ is ($x^2 + 8x + 15$, $x + 5$).

Skills and Concepts

Objectives and Examples	Review Exercises

• **Lesson 15–1** Simplify rational expressions.

$$\frac{z^2 - 3z + 2}{z^2 - 2z} = \frac{(z - 1)\overset{1}{\cancel{(z - 2)}}}{z\cancel{(z - 2)}_1}$$

$$= \frac{z - 1}{z}$$

Simplify each expression.

11. $\dfrac{4x^2yz}{16xy^3}$

12. $\dfrac{a^2 - 5a}{a - 5}$

13. $\dfrac{n^2 - 16}{n^2 + 2n - 8}$

14. $\dfrac{x^3 - 2x^2 - 3x}{x^2 + x - 12}$

Objectives and Examples

- **Lesson 15–2** Multiply and divide rational expressions.

$$\frac{6x}{x^2 + x - 2} \cdot \frac{x^2 - 4}{2x^2}$$

$$= \frac{\overset{3\,1}{\cancel{6x}}}{\cancel{(x+2)}(x-1)} \cdot \frac{(x-2)\overset{1}{\cancel{(x+2)}}}{\cancel{2x^2}}$$

$$= \frac{3(x-2)}{x(x-1)}$$

$$\frac{a^3}{2c} \div \frac{a^2}{c^3} = \frac{\overset{a}{\cancel{a^3}}}{\underset{1}{\cancel{2c}}} \cdot \frac{\overset{c^2}{\cancel{c^3}}}{\underset{1}{\cancel{a^2}}} \quad \textit{The reciprocal of } \frac{a^2}{c^3} \textit{ is } \frac{c^3}{a^2}.$$

$$= \frac{ac^2}{2}$$

Review Exercises

Find each product or quotient.

15. $\dfrac{4x^2y}{y^2z} \cdot \dfrac{z}{6y}$

16. $\dfrac{4a - 4b}{a} \cdot \dfrac{a^3}{8a - 8b}$

17. $\dfrac{x + 2}{x + 3} \cdot \dfrac{x}{x^2 - x - 6}$

18. $\dfrac{6m^2n}{10p} \div 3m$

19. $\dfrac{y^2 - 4y + 4}{3} \div \dfrac{y^2 - 4}{9}$

20. $\dfrac{x^2 - 9}{2x} \div \dfrac{6 - 2x}{8x^2}$

- **Lesson 15–3** Divide polynomials by binomials.

$$
\begin{array}{r}
2x - 3 \\
x + 2\overline{)2x^2 + x - 6} \\
\end{array}
$$

$\begin{array}{ll} & 2x^2 \div x = 2x \\ (\;\;)\,2x^2 + 4x & \textit{Multiply } 2x \textit{ and } x + 2. \\ \underline{} & \\ -3x - 6 & \textit{Subtract.} \\ (-)\,-3x - 6 & -3x \div x = -3 \\ \underline{} & \\ 0 & \textit{Subtract.} \end{array}$

Find each quotient.

21. $(a^2 + 6a) \div (a + 6)$

22. $(x^2 - 5x - 6) \div (x + 1)$

23. $(4y^2 + 8y + 5) \div (2y + 3)$

24. $(x^3 - 1) \div (x - 1)$

- **Lesson 15–4** Add and subtract rational expressions with like denominators.

$$\frac{x^2}{x + 3} - \frac{9}{x + 3} = \frac{x^2 - 9}{x + 3}$$

$$= \frac{\overset{1}{\cancel{(x+3)}}(x - 3)}{\underset{1}{\cancel{x+3}}}$$

$$= x - 3$$

Find each sum or difference. Write in simplest form.

25. $\dfrac{2x}{8} + \dfrac{4x}{8}$

26. $\dfrac{11}{15m} - \dfrac{2}{15m}$

27. $\dfrac{3x}{x + 4} + \dfrac{2x}{x + 4}$

28. $\dfrac{8x}{2x - 3} - \dfrac{2x + 9}{2x - 3}$

Objectives and Examples

- **Lesson 15–5** Add and subtract rational expressions with unlike denominators.

$$\frac{y}{y+2} - \frac{2}{y} \quad \text{The LCD is } y(y+2).$$

$$= \frac{y}{y+2} \cdot \frac{y}{y} - \frac{2}{y} \cdot \frac{y+2}{y+2}$$

$$= \frac{y^2}{y(y+2)} - \frac{2(y+2)}{y(y+2)}$$

$$= \frac{y^2 - 2(y+2)}{y(y+2)}$$

$$= \frac{y^2 - 2y - 4}{y(y+2)}$$

Review Exercises

Find each sum or difference. Write in simplest form.

29. $\dfrac{3}{4x} + \dfrac{5}{8x^2}$

30. $\dfrac{5}{x} - \dfrac{x+1}{3x}$

31. $\dfrac{4}{a} + \dfrac{3}{a-2}$

32. $\dfrac{y+2}{y^2-4} - \dfrac{2}{y+2}$

- **Lesson 15–6** Solve rational equations.

$$\frac{x}{4} - \frac{x}{6} = \frac{1}{4}$$

$$12\left(\frac{x}{4} - \frac{x}{6}\right) = 12\left(\frac{1}{4}\right)$$

$$\overset{3}{\cancel{12}}\left(\frac{x}{\cancel{4}}\right) - \overset{2}{\cancel{12}}\left(\frac{x}{\cancel{6}}\right) = \overset{3}{\cancel{12}}\left(\frac{1}{\cancel{4}}\right)$$

$$3x - 2x = 3$$

$$x = 3$$

Solve each equation. Check your solution.

33. $\dfrac{x}{3} - \dfrac{3x}{4} = \dfrac{1}{12}$

34. $\dfrac{6}{3x} - \dfrac{3}{x} = 1$

35. $\dfrac{x+2}{x} + \dfrac{2}{3x} = \dfrac{1}{3}$

36. $\dfrac{m}{m+1} + \dfrac{5}{m-1} = 1$

Applications and Problem Solving

37. **Carpentry** Determine the number of pieces of $\frac{3}{8}$-inch plywood that are in a stack 30 inches high. *(Lesson 15–2)*

38. **Boating** The top speed of a boat in still water is 5 miles per hour. At this speed, a 21-mile trip downstream takes the same amount of time as a 9-mile trip upstream. Find the rate of the current. *(Lesson 15–6)*

39. **Geometry** The volume of a rectangular prism is $4x^3 + 14x^2 + 10x$ cubic inches. The height of the prism is $2x + 5$ inches, and the length of the prism is $2x$ inches. *(Lesson 15–3)*

    a. Find the polynomial that represents the width.

    b. If $x = 2$, find the length, width, height, and volume of the prism.

1. **Explain** why $x = -3$ is an excluded value for $\dfrac{2x}{x+3}$.

2. **Identify** and correct the error that was made while finding the difference.

$$\dfrac{4}{x+1} - \dfrac{x-3}{x+1} = \dfrac{4-x-3}{x+1}$$
$$= \dfrac{1-x}{x+1}$$

3. **Find** the least common denominator of $\dfrac{5}{6a}$ and $\dfrac{2}{3a^2}$.

Find the excluded value(s) for each rational expression.

4. $\dfrac{8}{x}$

5. $\dfrac{6m}{m(m-2)}$

6. $\dfrac{2n}{n^2-9}$

Simplify each expression.

7. $\dfrac{9a^3b^2}{15ab^5}$

8. $\dfrac{x^3-x^2}{x-1}$

9. $\dfrac{x-2}{x^2-5x+6}$

Find each sum, difference, product, or quotient. Write in simplest form.

10. $\dfrac{z^3}{8} \cdot \dfrac{10x^2}{z^3}$

11. $\dfrac{x^2-1}{3x} \div \dfrac{1-x}{9x}$

12. $\dfrac{x^2-x-2}{x^2-4} \cdot \dfrac{x+2}{x^2+4x+3}$

13. $\dfrac{4a+4b}{a} \div \dfrac{10a+10b}{a^2}$

14. $\dfrac{5}{8x} + \dfrac{11}{8x}$

15. $\dfrac{t}{t+5} - \dfrac{t-6}{t+5}$

16. $\dfrac{6}{5y} + \dfrac{7}{10y^2}$

17. $\dfrac{2}{x+4} - \dfrac{x}{x^2-16}$

Find each quotient.

18. $(y^2 + 10y + 16) \div (y + 2)$

19. $(x^3 + x^2 - x - 1) \div (x - 1)$

20. $(a^2 - 10) \div (a + 3)$

21. $(x^2 - 5x + 8) \div (x - 2)$

Solve each equation. Check your solution.

22. $\dfrac{n+2}{3} + \dfrac{n}{2} = \dfrac{1}{2}$

23. $\dfrac{5}{x+2} - \dfrac{1}{4x+8} = \dfrac{1}{4}$

24. **Geometry** Find the measure of the area of the rectangle in simplest form.

$$\dfrac{x+7}{x^2-25}$$
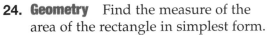
$$\dfrac{x^2+10x+25}{x^2-49}$$

25. **Transportation** A tugboat pushing a barge up the Mississippi River takes 1 hour longer to travel 36 miles up the river than to travel the same distance down the river. If the rate of the current is 3 miles per hour, find the speed of the tugboat and barge in still water.

Preparing for Standardized Tests

Right Triangle Problems

You'll need to know how to identify right triangles and how to apply the Pythagorean Theorem.

If a and b are the lengths of the legs of a triangle and c is the length of the hypotenuse, then $a^2 + b^2 = c^2$.

Proficiency Test Example

A 32-foot telephone pole is braced with a cable from the top of the pole to a point 7 feet from the base. What is the length of the cable, rounded to the nearest tenth?

A 31.2 ft   **B** 32.8 ft

C 34.3 ft   **D** 36.2 ft

Hint Since telephone poles are vertical and the ground is horizontal, the pole forms a right angle at the base.

Solution Draw a figure.

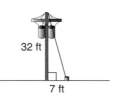

32 ft

7 ft

The cable forms the hypotenuse of a right triangle. Use the Pythagorean Theorem to find the length of the hypotenuse.

$a^2 + b^2 = c^2$ *Pythagorean Theorem*

$32^2 + 7^2 = c^2$ *Replace a with 32 and b with 7.*

$1024 + 49 = c^2$

$1073 = c^2$

$\sqrt{1073} = \sqrt{c^2}$ *Take the square root of each side.*

$32.8 \approx c$ *Use a calculator.*

To the nearest tenth, the length of the cable is 32.8 feet.

The answer is B.

ACT Example

In the figure, $\overline{MO}$ is perpendicular to $\overline{LN}$, LO is equal to 4, MO is equal to ON, and LM is equal to 6. What is MN?

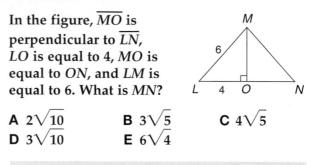

A $2\sqrt{10}$ **B** $3\sqrt{5}$ **C** $4\sqrt{5}$

D $3\sqrt{10}$ **E** $6\sqrt{4}$

Hint Identify the types of triangles shown in the figure.

Solution $\triangle LMO$ is a right triangle. Use the Pythagorean Theorem to find MO.

$a^2 + b^2 = c^2$ *Pythagorean Theorem*

$a^2 + 4^2 = 6^2$ *Replace b with 4*

$a^2 + 16 = 36$ *and c with 6.*

$a^2 + 16 - 16 = 36 - 16$ *Subtract 16.*

$a^2 = 20$

$\sqrt{a^2} = \sqrt{20}$ *Take the square root.*

$a = \sqrt{4(5)}$ or $2\sqrt{5}$

$\triangle MON$ is an isosceles triangle. Use the Pythagorean Theorem to find MN.

$a^2 + b^2 = c^2$ *Pythagorean Theorem*

$\left(2\sqrt{5}\right)^2 + \left(2\sqrt{5}\right)^2 = c^2$ $a = 2\sqrt{5}, b = 2\sqrt{5}$

$20 + 20 = c^2$

$\sqrt{40} = \sqrt{c^2}$

$2\sqrt{10} = c$

The answer is A.

After you work each problem, record your answer on the answer sheet provided or on a sheet of paper.

1. Quadrilateral $ABCD$ is a rectangle. $\overline{AD}$ is 5 centimeters long, and $\overline{CD}$ is 12 centimeters long. What is length of $\overline{AC}$?

   A 13 cm

   B 17 cm

   C 30 cm

   D 169 cm

2. What is the distance between points G and H in the graph?

   A 4

   B $4\sqrt{2}$

   C 8

   D $8\sqrt{2}$

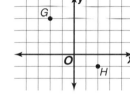

3. The hypotenuse of an isosceles right triangle has a length of 20 units. What is the length of one of the legs of the triangle?

   A 10   **B** $10\sqrt{2}$ **C** $10\sqrt{3}$

   D 20   **E** $20\sqrt{2}$

4. The streets on a jogging route form a right triangle. How long is the jogging route?

   A 10 km

   B 14 km

   C 24 km

   D 100 km

5. What is the equation of the graph?

   A $y = -x^2 + 2$

   B $y = x^2 - 2$

   C $y = (x + 2)^2$

   D $y = x^2 + 2$

6. What is the length of $\overline{BC}$?

   A 6

   B $4\sqrt{3}$

   C $2\sqrt{13}$

   D 8

   E $2\sqrt{38}$

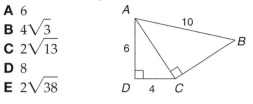

7. The base of a 25-foot ladder is placed 10 feet from the side of a house. How high does the ladder reach?

   A 12.5 ft   **B** 15 ft   **C** 22.9 ft   **D** 35 ft

Quantitative Comparison

8.   **Column A**   **Column B**

   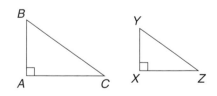

   $$\boxed{\sqrt{3} + \sqrt{4}} \qquad \boxed{\sqrt{3} \times \sqrt{4}}$$

   A if the quantity in Column A is greater;

   B if the quantity in Column B is greater;

   C if the two quantities are equal;

   D if the relationship cannot be determined from the information given.

Open-Ended Questions

9. **Grid-In** Triangles ABC and XYZ are similar right triangles. If $\angle B$ measures $55°$, what is the degree measure of $\angle Z$?

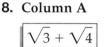

10. You want to fit a square cake inside a circular cake carrier that is 12 inches in diameter.

    Part A Draw a diagram showing side s of the *largest* square cake that will fit inside the carrier.

    Part B Use your diagram to write an equation and then find the value of s.

Student Handbook

Skills Handbook

Reference Handbook

Prerequisite Skills Review

Operations with Decimals

- To add or subtract decimals, line up the decimal points. You may want to *annex*, or place zeros at the end of the decimals, to help align the columns. Then add or subtract.

Examples   **①** Find 18.39 + 6.4.

$$\begin{array}{r} 18.39 \\ + 6.4 \\ \hline \end{array} \quad \rightarrow \quad \begin{array}{r} 18.39 \\ + 6.40 \\ \hline 24.79 \end{array} \quad \textit{Annex a zero to align the columns.}$$

② Find 5 − 2.17.

$$\begin{array}{r} 5 \\ - 2.17 \\ \hline \end{array} \quad \rightarrow \quad \begin{array}{r} 5.00 \\ - 2.17 \\ \hline 2.83 \end{array} \quad \textit{Annex two zeros to align the columns.}$$

- To multiply decimals, multiply as with whole numbers. Then add the total number of decimal places in the factors. Place the same number of decimal places in the product, counting from right to left.
- To divide decimals, move the decimal point in the divisor to the right, and then move the decimal point in the dividend the same number of places. Align the decimal point in the quotient with the decimal point in the dividend.

Examples   **③** Find 7.3(0.61).

$$\begin{array}{r} 7.3 \quad \leftarrow \textit{1 decimal place} \\ \times 0.61 \quad \leftarrow \textit{2 decimal places} \\ \hline 73 \\ 438 \\ \hline 4.453 \quad \leftarrow \textit{3 decimal places} \end{array}$$

④ Find 8.12 ÷ 5.8.

$$\begin{array}{r} 1.4 \\ 5.8\overline{)8.12} \\ \underline{5\ 8} \\ 2\ 32 \\ \underline{2\ 32} \\ 0 \end{array} \quad \textit{Move each decimal point right 1 place.}$$

Find each sum or difference.

1. 4.8 + 1.1	**2.** 10.25 + 0.16	**3.** 36.5 − 11.2
4. 13.6 + 20.41	**5.** 42.99 − 12.63	**6.** 5.37 − 2.47
7. 24 + 7.3	**8.** 7.28 − 1.1	**9.** 0.7 + 1.682
10. 6.2 − 4.01	**11.** 4.02 + 5.9 + 0.03	**12.** 18 − 16.39

Find each product or quotient.

13. 8(0.4)	**14.** 5.72 ÷ 2	**15.** 1.6(1.9)
16. 7.2(3.5)	**17.** 24(0.86)	**18.** 4.86 ÷ 0.3
19. 11.5 ÷ 4.6	**20.** 9.3(30.23)	**21.** 0.5(2.41)(6.7)
22. $\dfrac{50.4}{36}$	**23.** $\dfrac{6.46}{0.68}$	**24.** $\dfrac{9.264}{0.24}$

Simplifying Fractions

- A fraction is in simplest form when the greatest common factor (GCF) of the numerator and the denominator is 1.
- To write a fraction in simplest form, divide both the numerator and the denominator by the GCF.

Examples   Write each fraction in simplest form.

❶ $\frac{6}{9}$

The GCF of 6 and 9 is 3.

$$\frac{6}{9} \overset{\div 3}{\underset{\div 3}{=}} \frac{2}{3}$$ *Divide 6 and 9 by 3.*

❷ $\frac{4}{20}$

The GCF of 4 and 20 is 4.

$$\frac{4}{20} \overset{\div 4}{\underset{\div 4}{=}} \frac{1}{5}$$ *Divide 4 and 20 by 4.*

- To change an improper fraction to a mixed number, divide the numerator by the denominator. Write the remainder as a fraction in simplest form.
- To change a mixed number to an improper fraction, multiply the whole number by the denominator and add the numerator. Write this sum over the denominator.

Examples   **❸** Write $\frac{8}{5}$ as a mixed number.

$\frac{8}{5}$ means $8 \div 5$.

$$\begin{array}{r} 1 \\ 5\overline{)8} \\ -5 \\ \hline 3 \end{array} \rightarrow \quad 1 \text{ R3 or } 1\frac{3}{5}$$

❹ Write $2\frac{3}{4}$ as an improper fraction.

$$2\frac{3}{4} = \frac{(2 \times 4) + 3}{4}$$
$$= \frac{8 + 3}{4}$$
$$= \frac{11}{4}$$

Write each fraction in simplest form.

1. $\frac{4}{8}$ 2. $\frac{3}{9}$ 3. $\frac{2}{12}$ 4. $\frac{9}{15}$ 5. $\frac{8}{20}$

6. $\frac{6}{21}$ 7. $\frac{25}{30}$ 8. $\frac{30}{40}$ 9. $\frac{12}{16}$ 10. $\frac{16}{36}$

Write each improper fraction as a mixed number.

11. $\frac{7}{6}$ 12. $\frac{5}{3}$ 13. $\frac{9}{2}$ 14. $\frac{12}{5}$ 15. $\frac{10}{4}$

16. $\frac{11}{2}$ 17. $\frac{21}{10}$ 18. $\frac{16}{6}$ 19. $\frac{24}{14}$ 20. $\frac{30}{8}$

Write each mixed number as an improper fraction.

21. $2\frac{1}{3}$ 22. $6\frac{1}{2}$ 23. $2\frac{4}{5}$ 24. $4\frac{2}{5}$ 25. $6\frac{1}{7}$

26. $3\frac{5}{8}$ 27. $2\frac{7}{10}$ 28. $3\frac{5}{12}$ 29. $1\frac{10}{11}$ 30. $5\frac{8}{9}$

Multiplying and Dividing Fractions

- To multiply fractions, multiply the numerators and multiply the denominators.
- When a numerator and denominator have a common factor, you can simplify before multiplying.

Examples   Find each product.

❶ $\dfrac{1}{2} \cdot \dfrac{3}{5}$

$$\dfrac{1}{2} \cdot \dfrac{3}{5} = \dfrac{1 \cdot 3}{2 \cdot 5}$$
$$= \dfrac{3}{10}$$

❷ $\dfrac{4}{9} \cdot \dfrac{3}{5}$

$$\dfrac{4}{9} \cdot \dfrac{3}{5} = \dfrac{4}{\overset{3}{\cancel{9}}} \cdot \dfrac{\overset{1}{\cancel{3}}}{5} \qquad \textit{Divide by the GCF, 3.}$$
$$= \dfrac{4 \cdot 1}{3 \cdot 5}$$
$$= \dfrac{4}{15}$$

- To divide fractions, multiply by the reciprocal of the divisor.

Examples   Find each quotient.

❸ $\dfrac{5}{6} \div \dfrac{6}{7}$

$$\dfrac{5}{6} \div \dfrac{6}{7} = \dfrac{5}{6} \cdot \dfrac{7}{6} \qquad \textit{Multiply by the reciprocal.}$$
$$= \dfrac{5 \cdot 7}{6 \cdot 6}$$
$$= \dfrac{35}{36}$$

❹ $\dfrac{2}{5} \div 4$

$$\dfrac{2}{5} \div 4 = \dfrac{2}{5} \div \dfrac{4}{1}$$
$$= \dfrac{2}{5} \cdot \dfrac{1}{4} \qquad \textit{Multiply by the reciprocal.}$$
$$= \dfrac{\overset{1}{\cancel{2}}}{5} \cdot \dfrac{1}{\underset{2}{\cancel{4}}} \qquad \textit{Divide by the GCF, 2.}$$
$$= \dfrac{1 \cdot 1}{5 \cdot 2}$$
$$= \dfrac{1}{10}$$

Find each product or quotient.

1. $\dfrac{1}{3} \cdot \dfrac{1}{2}$ 2. $\dfrac{2}{5} \div \dfrac{1}{2}$ 3. $\dfrac{1}{4} \div \dfrac{4}{5}$ 4. $\dfrac{2}{7} \cdot \dfrac{1}{3}$

5. $\dfrac{1}{8} \div \dfrac{1}{5}$ 6. $\dfrac{4}{9} \cdot \dfrac{2}{3}$ 7. $\dfrac{2}{7} \div 7$ 8. $\dfrac{6}{7} \cdot \dfrac{1}{5}$

9. $\dfrac{1}{2} \cdot \dfrac{2}{7}$ 10. $\dfrac{2}{3} \cdot \dfrac{1}{6}$ 11. $\dfrac{5}{9} \div \dfrac{2}{3}$ 12. $\dfrac{1}{8} \div \dfrac{1}{2}$

13. $\dfrac{2}{9} \div \dfrac{4}{7}$ 14. $\dfrac{3}{4} \cdot \dfrac{5}{6}$ 15. $\dfrac{5}{12} \cdot \dfrac{3}{7}$ 16. $\dfrac{3}{4} \div \dfrac{5}{6}$

17. $\dfrac{2}{9} \div \dfrac{2}{3}$ 18. $\dfrac{2}{3} \div 4$ 19. $\dfrac{7}{9} \cdot \dfrac{6}{7}$ 20. $\dfrac{3}{10} \cdot \dfrac{5}{18}$

21. $\dfrac{5}{12} \cdot \dfrac{8}{25}$ 22. $\dfrac{3}{8} \div 12$ 23. $\dfrac{9}{20} \cdot \dfrac{4}{15}$ 24. $\dfrac{8}{21} \div \dfrac{12}{15}$

Adding and Subtracting Fractions

- To add or subtract fractions with like denominators, add or subtract the numerators. Simplify if necessary.

Examples Find each sum or difference.

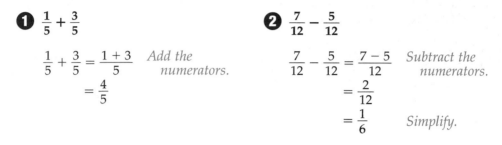

① $\frac{1}{5} + \frac{3}{5}$

$$\frac{1}{5} + \frac{3}{5} = \frac{1+3}{5} \quad \textit{Add the numerators.}$$
$$= \frac{4}{5}$$

② $\frac{7}{12} - \frac{5}{12}$

$$\frac{7}{12} - \frac{5}{12} = \frac{7-5}{12} \quad \textit{Subtract the numerators.}$$
$$= \frac{2}{12}$$
$$= \frac{1}{6} \quad \textit{Simplify.}$$

- To add or subtract fractions with unlike denominators, first find the least common denominator (LCD). Rewrite each fraction with the LCD, and then add or subtract the numerators. Simplify if necessary.

Examples Find each sum or difference.

③ $\frac{1}{3} - \frac{2}{9}$

The LCD of 3 and 9 is 9.

$$\frac{1}{3} - \frac{2}{9} = \frac{3}{9} - \frac{2}{9} \quad \textit{Rewrite } \frac{1}{3} \textit{ as } \frac{3}{9}.$$
$$= \frac{3-2}{9} \quad \textit{Subtract the numerators.}$$
$$= \frac{1}{9}$$

④ $\frac{3}{10} + \frac{1}{4}$

The LCD of 10 and 4 is 20.

$$\frac{3}{10} + \frac{1}{4} = \frac{6}{20} + \frac{5}{20} \quad \textit{Rewrite the fractions using the LCD, 20.}$$
$$= \frac{6+5}{20} \quad \textit{Add the numerators.}$$
$$= \frac{11}{20}$$

Find each sum or difference.

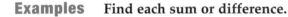

1. $\frac{1}{5} + \frac{2}{5}$ 2. $\frac{7}{8} - \frac{2}{8}$ 3. $\frac{1}{9} + \frac{4}{9}$ 4. $\frac{6}{7} - \frac{2}{7}$

5. $\frac{1}{15} + \frac{3}{15}$ 6. $\frac{4}{13} + \frac{1}{13}$ 7. $\frac{5}{6} - \frac{4}{6}$ 8. $\frac{9}{11} - \frac{3}{11}$

9. $\frac{3}{8} + \frac{1}{8}$ 10. $\frac{5}{6} - \frac{1}{6}$ 11. $\frac{7}{10} - \frac{3}{10}$ 12. $\frac{1}{12} + \frac{7}{12}$

13. $\frac{1}{3} + \frac{1}{5}$ 14. $\frac{5}{8} - \frac{1}{4}$ 15. $\frac{4}{9} + \frac{1}{3}$ 16. $\frac{9}{10} - \frac{3}{5}$

17. $\frac{3}{7} + \frac{1}{2}$ 18. $\frac{3}{4} - \frac{2}{5}$ 19. $\frac{2}{9} + \frac{2}{3}$ 20. $\frac{9}{10} - \frac{3}{4}$

21. $\frac{2}{15} + \frac{7}{15}$ 22. $\frac{1}{2} + \frac{1}{6}$ 23. $\frac{5}{6} - \frac{3}{4}$ 24. $\frac{3}{4} + \frac{1}{12}$

25. $\frac{4}{7} - \frac{1}{6}$ 26. $\frac{2}{15} + \frac{2}{3}$ 27. $\frac{5}{6} + \frac{1}{8}$ 28. $\frac{7}{8} - \frac{3}{12}$

Fractions and Decimals

- To change a fraction to a decimal, divide the numerator by the denominator.
- A **terminating decimal** is a decimal like 0.75, in which the division ends, or terminates, when the remainder is zero.
- A **repeating decimal** is a decimal like 0.545454 . . . whose digits do not end. Since it is impossible to write all the digits, you can use bar notation to show that 54 repeats. We can write 0.545454 . . . as $0.\overline{54}$.

Examples   Write each fraction as a decimal.

❶ $\frac{2}{5}$

$$
\begin{array}{r}
0.4 \\
5\overline{)2.0} \\
2\,0 \\
\hline
0
\end{array}
\qquad \frac{2}{5} = 0.4
$$

❷ $\frac{4}{9}$

$$
\begin{array}{r}
0.44 \\
9\overline{)4.00} \\
36 \\
\hline
40 \\
36 \\
\hline
4
\end{array}
\qquad \frac{4}{9} = 0.\overline{4}
$$
The pattern is repeating.

- Every terminating decimal can be expressed as a fraction with a denominator of 10, 100, and so on.
- Every repeating decimal can be expressed as a fraction.

Examples   Write each decimal as a fraction.

❸ 0.8

$$
0.8 = \frac{8}{10}
$$
$$
= \frac{4}{5}
$$

❹ $0.\overline{1}$

Let $N = 0.\overline{1}$ or 0.111 . . . .
Then $10N = 1.\overline{1}$ or 1.111 . . . .

$$
\begin{array}{r}
10N = 1.111 \ldots \quad \text{\textit{Subtract.}} \\
- 1N = 0.111 \ldots \quad \text{\textit{N = 1N}} \\
\hline
9N = 1
\end{array}
$$
$$
N = \frac{1}{9} \quad \text{So, } 0.\overline{1} = \frac{1}{9}.
$$

Write each fraction as a decimal.

1. $\frac{3}{4}$ 2. $\frac{1}{5}$ 3. $\frac{3}{8}$ 4. $\frac{2}{9}$

5. $\frac{4}{11}$ 6. $\frac{7}{10}$ 7. $\frac{2}{15}$ 8. $\frac{1}{6}$

9. $\frac{4}{15}$ 10. $\frac{3}{20}$ 11. $\frac{5}{6}$ 12. $\frac{5}{8}$

Write each decimal as a fraction in simplest form.

13. 0.9   14. 0.6   15. 0.25   16. $0.\overline{3}$

17. $0.\overline{5}$ 18. 0.4   19. 0.16   20. $0.\overline{6}$

21. 0.125   22. 0.35   23. $0.\overline{8}$ 24. $0.\overline{7}$

Decimals and Percents

- To express a decimal as a percent, first express the decimal as a fraction with a denominator of 100. Then express the fraction as a percent.

Examples   Write each decimal as a percent.

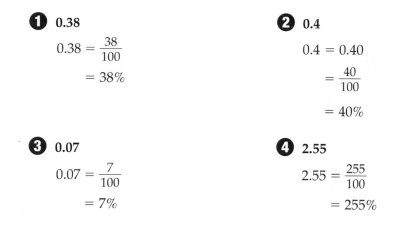

1 0.38

$$0.38 = \frac{38}{100}$$
$$= 38\%$$

2 0.4

$$0.4 = 0.40$$
$$= \frac{40}{100}$$
$$= 40\%$$

Shortcut

To write a decimal as a percent, multiply by 100 and add the % symbol.
0.62 = 0.62 = 62%

3 0.07

$$0.07 = \frac{7}{100}$$
$$= 7\%$$

4 2.55

$$2.55 = \frac{255}{100}$$
$$= 255\%$$

- To express a percent as a decimal, rewrite the percent as a fraction with a denominator of 100. Then express the fraction as a decimal.

Examples   Write each percent as a decimal.

5 26%

$$26\% = \frac{26}{100}$$
$$= 0.26$$

6 9%

$$9\% = \frac{9}{100}$$
$$= 0.09$$

Shortcut

To write a percent as a decimal, divide by 100 and remove the % symbol.
62% = 62% = 0.62

7 87.5%

$$87.5\% = \frac{87.5}{100}$$
$$= 0.875$$

8 125%

$$125\% = \frac{125}{100}$$
$$= 1.25$$

Write each decimal as a percent.

1. 0.62	**2.** 0.99	**3.** 0.14	**4.** 0.20
5. 0.15	**6.** 0.8	**7.** 0.5	**8.** 0.06
9. 0.42	**10.** 0.03	**11.** 0.1	**12.** 1.76
13. 0.08	**14.** 3.10	**15.** 1.05	**16.** 2.6

Write each percent as a decimal.

17. 12%	**18.** 37%	**19.** 86%	**20.** 51%
21. 30%	**22.** 90%	**23.** 2%	**24.** 55%
25. 5%	**26.** 12.5%	**27.** 73.6%	**28.** 134%
29. 208%	**30.** 120%	**31.** 60.5%	**32.** 200%

Fractions and Percents

- To express a percent as a fraction, write the percent as a fraction with a denominator of 100 and simplify.

Example ❶ **Express 24% as a fraction in simplest form.**

$$24\% = \frac{24}{100} \quad \text{\textit{Write as a fraction with a denominator of 100.}}$$

$$= \frac{\overset{6}{\cancel{24}}}{\underset{25}{\cancel{100}}} \quad \text{\textit{Divide by the GCF, 4.}}$$

$$= \frac{6}{25}$$

- To express a fraction as a percent, write a proportion and solve.

Example ❷ **Express $\frac{3}{5}$ as a percent.**

$$\frac{3}{5} = \frac{n}{100} \quad \text{\textit{Write a proportion.}}$$

$$3 \times 100 = 5 \times n \quad \text{\textit{Find the cross products.}}$$

$$300 = 5n$$

$$\frac{300}{5} = \frac{5n}{5} \quad \text{\textit{Divide each side by 5.}}$$

$$60 = n$$

So, $\frac{3}{5} = \frac{60}{100}$ or 60%.

Write each percent as a fraction in simplest form.

1. 25% **2.** 80% **3.** 70% **4.** 15%

5. 6% **6.** 56% **7.** 40% **8.** 98%

9. 19% **10.** 35% **11.** 22% **12.** 64%

Write each fraction as a percent.

13. $\frac{3}{10}$ **14.** $\frac{1}{5}$ **15.** $\frac{3}{4}$ **16.** $\frac{13}{25}$

17. $\frac{37}{50}$ **18.** $\frac{1}{20}$ **19.** $\frac{19}{20}$ **20.** $\frac{42}{50}$

21. $\frac{2}{25}$ **22.** $\frac{9}{10}$ **23.** $\frac{8}{32}$ **24.** $\frac{22}{40}$

Comparing and Ordering Rational Numbers

- To compare rational numbers, it is usually easier and faster if you write the numbers as decimals. In some cases, it may be easier if you write the numbers as fractions having the same denominator or as percents.

Examples   Replace each ● with <, >, or = to make a true sentence.

❶ $\frac{3}{5}$ ● 0.65

$\frac{3}{5} = 0.60$

Since $0.60 < 0.65$, $\frac{3}{5} < 0.65$.

❷ 0.18 ● 2%

$2\% = 0.02$

Since $0.18 > 0.02$, $0.18 > 2\%$.

- To order rational numbers, first write the numbers as decimals. Then write the decimals in order from least to greatest and write the corresponding rational numbers in the same order.

Example   **❸** Write $\frac{1}{2}$, 55%, and 0.20 in order from least to greatest.

Write each number as a decimal.

$\frac{1}{2} = 0.50$ $55\% = 0.55$ $0.20 = 0.20$

Write the decimals in order from least to greatest. $0.20 < 0.50 < 0.55$
                               ↓   ↓   ↓

Write the corresponding rational numbers in the same order. $0.20 < \frac{1}{2} < 55\%$

The numbers in order from least to greatest are 0.20, $\frac{1}{2}$, and 55%.

Replace each ● with <, >, or = to make a true sentence.

1. 0.35 ● $\frac{1}{4}$

2. 65% ● 0.7

3. 80% ● $\frac{4}{5}$

4. $\frac{1}{10}$ ● 1%

5. 12.5% ● $\frac{1}{8}$

6. 36% ● 3.6

7. $\frac{2}{3}$ ● 0.7

8. 0.2 ● 20%

9. $\frac{5}{6}$ ● $\frac{5}{7}$

10. $\frac{3}{8}$ ● 0.375

11. 0.9 ● 10%

12. 30% ● 3.9%

13. 0.9 ● $0.\overline{9}$

14. 0.15 ● $\frac{2}{15}$

15. 51% ● 51

16. 78% ● $\frac{7}{8}$

17. $\frac{6}{11}$ ● 50%

18. $0.\overline{4}$ ● $\frac{4}{9}$

Write the numbers in each set in order from least to greatest.

19. $0.52, 5\%, \frac{1}{2}$

20. $\frac{1}{3}, 40\%, \frac{1}{4}$

21. $20\%, 0.1, \frac{3}{10}$

22. $0.19, \frac{1}{5}, 15\%$

23. $63\%, \frac{2}{3}, 0.06$

24. $\frac{1}{9}, 0.\overline{2}, 19\%$

Lesson 1–1 *(Pages 4–7)* Write an algebraic expression for each verbal expression.

1. 4 less than x

2. the difference of 3 and m

3. 6 more than the quotient of b and 5

4. 2 less than 7 times k

Write a verbal expression for each algebraic expression.

5. $b + 4$

6. $12x$

7. $6 - k$

8. $5 - 2k$

9. $\dfrac{3}{x}$

10. $7 - \dfrac{m}{x}$

11. $\dfrac{6}{m} + 7$

12. $\dfrac{7}{2} + r$

Write an equation for each sentence.

13. Six times s plus seven equals fourteen.

14. Three minus k equals twelve plus two times j.

15. Five divided by m is equal to two times n minus three.

Lesson 1–2 *(Pages 8–13)* Find the value of each expression.

1. $32 + 4 \cdot 2$

2. $6 - 3 \cdot 4 + 7$

3. $16 \div 4 + 3$

4. $6 \cdot 2 \cdot 4 - 1$

5. $25 \div 5 - (3 + 2)$

6. $\dfrac{3(7 + 1)}{3 \cdot 4}$

7. $[2(6 - 1)] \div 2$

8. $2(5 + 3) - 6$

Name the property of equality shown by each statement.

9. If $3y + 4 = 5x$ and $5x = 10$, then $3y + 4 = 10$.

10. If $13 + y = 18$, then $18 = 13 + y$.

11. $9 - 7x = 9 - 7x$

12. $2(6 + 4) = 2(10)$

Find the value of each expression. Identify the property used in each step.

13. $6 + 5 \cdot 2 - 1$

14. $9(8 \div 2 - 3)$

15. $24 - 1(3 + 15)$

16. $5(6) + 6(14 - 14)$

17. $11 + (7 - 28 \div 4)$

18. $\dfrac{7 + 21 \div 3}{3(2 - 1) + 4}$

Evaluate each algebraic expression if $m = 6$ and $b = 3$.

19. $8m + 2b$

20. $m(b - 3)$

21. $m - 3b + 4$

22. $\dfrac{6 \cdot 2 + b \cdot m}{m - 4}$

23. $\dfrac{m(6 - 2)}{b + 1}$

24. $(2m + 1)(6 - b)$

Lesson 1–3 *(Pages 14–18)* Name the property shown by each statement.

1. $(7 + 7) + 2 = 7 + (7 + 2)$

2. $(3 \cdot m) \cdot 6 = 3 \cdot (m \cdot 6)$

3. $9 \cdot 13 = 13 \cdot 9$

4. $6 \cdot 3 \cdot k = k \cdot 6 \cdot 3$

5. $k + (p + q) = (k + p) + q$

6. $m + 5 = 5 + m$

Simplify each expression. Identify the properties used in each step.

7. $5(6r)$

8. $(k + 1) + 4$

9. $11 \cdot p \cdot 2$

10. $(s \cdot 4) \cdot 3$

11. $7 + y + 1$

12. $13 + w + 12$

Lesson 1–4 *(Pages 19–23)* Simplify each expression.

1. $6b + 3b$
2. $4(3k - 2k)$
3. $4a + 2b + 8a + 5b$
4. $9 + 12s - 2s$
5. $3 + 4p - 2p$
6. $8 + k - 7 + 6k$
7. $ab - b + 3ab$
8. $8(m + 3n) + 3m$
9. $6st - st + t$
10. $6(3 + 2m)$
11. $5(7 + 2r) + 2(r - 3)$
12. $4 + (7 + 5x)3$

13. Write $6m + (3k + m) + 2k + 4(m + 1)$ in simplest form. Indicate the property that justifies each step.

14. What is the value of $4x$ multiplied by the quantity $2x$ minus 3 if x equals 4?

Lesson 1–5 *(Pages 24–29)* Solve each problem. Use any strategy.

1. Simone has 16 players on her soccer team. There are 6 more boys than girls. How many girls and how many boys are there?

2. Dillon put $350 into a bank account that pays 6% simple interest. How much money will he have in four years?

3. How many ways can you make 50¢ using dimes, nickels, and pennies?

4. Kate lives 24 miles from school. Jake lives 6 miles closer to school than Kate does. How far does Jake live from school?

5. Nine paintings are for sale at an art gallery. Tara must choose two of the paintings.
   a. Make a chart or diagram to represent the problem.
   b. How many different combinations are there in all?
   c. How many different combinations would there be if there were eight paintings?

Lesson 1–6 *(Pages 32–37)* Determine whether each is a good sample. Describe what caused the bias in each poor sample. Explain.

1. Every other student in physical education class is surveyed to determine how many hours per week students exercise.

2. Every student at a school is asked to name their favorite subject in school.

3. Surveys are placed on every fifth car in a mall parking lot concerning the construction of a new lot.

4. Six people at a music concert are randomly chosen to find out their opinion of the concert.

Refer to the chart at the right.

5. Make a frequency table to organize the data.

6. What number of hours is most frequent?

7. How many students use the Internet for less than two hours per week?

8. How many more students use the Internet six or more hours than those who use it less than four hours?

Time on the Internet per Week (h)		
4	9	4
2	5	2
6	1	10
10	8	4
3	2	9
10	4	6
1	7	1

Lesson 1–7 *(Pages 38–43)* **People in three major cities were surveyed to see what transportation they used to get to work. The table shows the results.**

Type	New York, NY	Los Angeles, CA	Baltimore, MD
drive alone	24.0%	65.2%	75.5%
carpool	8.5%	15.4%	14.2%
public transit	53.4%	10.5%	2.7%
other	14.0%	8.9%	7.6%

Source: *Time Almanac 2000*

1. Make a histogram for each city that shows the results of the survey.
2. How do the histograms compare?
3. Make some assumptions to support your findings.

The record low temperatures for select states are shown.

4. How many states are represented?
5. What is the lowest record low temperature?
6. Which temperature occurs most frequently?
7. How many states have record low temperatures of less than $-50°C$?

Stem	Leaf
-6	2
-5	7 2 1 1
-4	6 4 4 3 0 0 0 0
-3	8 8 8 7 6 4 3
-2	8 7 7 7 6
-1	9 4 $-1|4 = -14°C$

Lesson 2–1 *(Pages 52–57)* **Name the coordinate of each point.**

```
        D   A       B       F       E       C
   ◄────┼───●───┼───●───┼───●───┼───●───┼───●───►
       -5  -4  -3  -2  -1   0   1   2   3   4   5
```

1. A
2. B
3. C
4. D
5. E
6. F

Graph each set of numbers on a number line.

7. $\{-1, 0, 4\}$
8. $\{-3, -2, -1\}$
9. $\{0, 3, 5\}$
10. $\{-4, -2, 1, 2\}$
11. $\{-1, 0, 1, 2\}$
12. $\{-4, 2, 3, 4\}$

Replace each ● with < or > to make a true sentence.

13. $-3 ● 0$
14. $-4 ● -3$
15. $-8 ● 2$
16. $0 ● -1$
17. $|6| ● -2$
18. $-2 ● |-3|$
19. $|-7| ● |-4|$
20. $|16| ● -12$

Evaluate each expression.

21. $|7|$
22. $|-5|$
23. $|-7| - 5$
24. $-11 + |4|$
25. $|4| - |-12|$
26. $|-8| + |-3|$
27. $|5| - 11$
28. $|18| + |-21|$

Lesson 2–2 *(Pages 58–63)* **Write the ordered pair that names each point.**

1. G
2. H
3. I
4. J
5. K
6. L

Graph each point on a coordinate plane.

7. $P(5, -1)$
8. $Q(0, 0)$
9. $R(1, 2)$
10. $S(-2, 4)$
11. $T(-3, -2)$
12. $U(0, -3)$

Name the quadrant in which each point is located.

13. $(6, 4)$
14. $(-2, 1)$
15. $(-3, -3)$
16. $(0, 0)$

Lesson 2–3 *(Pages 64–69)* Find each sum.

1. $2 + 14$
2. $8 + 4$
3. $-16 + (-7)$
4. $-13 + (-26)$
5. $(-21) + 14$
6. $-19 + 17$
7. $-5 + (-13) + (-24)$
8. $2 + 14 + (-12)$
9. $-16 + (-4) + (-11) + 5$
10. $12 + (-7) + 6 + (-18)$

Simplify each expression.

11. $6b + 2b$
12. $-8y + 23y$
13. $17k + (-18k)$
14. $7m - 7m$
15. $-8r + (-4r) + 12r$
16. $-6p + (-3p) + (-1p)$
17. $7c + 11c + (-12c)$
18. $-5x - 6x$

Lesson 2–4 *(Pages 70–74)* Find each difference.

1. $16 - 7$
2. $16 - 21$
3. $-10 - (-9)$
4. $-4 - 6$
5. $3 - 9$
6. $7 - (-12)$
7. $-12 - (-4)$
8. $0 - 11$

Evaluate each expression if $x = 6$, $y = -8$, $z = -3$, and $w = 4$.

9. $y - z$
10. $3 - z$
11. $6 + y$
12. $w - y$
13. $14 - y - x$
14. $6 + x - z$
15. $y + z + w$
16. $w - z + 11$

Simplify each expression.

17. $7k - 4k$
18. $6m + (-3m)$
19. $3y - 7y$
20. $-14n - 17n$
21. $13s - (-2s)$
22. $9r - (-2r) - 4r$
23. $8c - 12c + 2c$
24. $b + 2b - 3b$

25. Corey spends \$112 on movies each month. He spends \$80 less on books. How much does he spend on books each month?

Lesson 2–5 *(Pages 75–79)* Find each product.

1. $2(4)$
2. $3(-12)$
3. $-4(-13)$
4. $-5(8)$
5. $10(-3)(2)$
6. $-4(2)(-3)1$
7. $6(-1)(-1)(-2)$
8. $-8(-4)2$

Evaluate each expression if $x = -2$, $y = 4$, and $z = -6$.

9. $6x$
10. $2yz$
11. xyz
12. $-8xy$
13. $4x - 3z$
14. $6y - z$
15. $2x + z$
16. $10x + 5y$

Simplify each expression.

17. $-6(2b)$
18. $7(-4m)$
19. $-10(-3k)$
20. $4(5c)$
21. $(6g)(9h)$
22. $8m(-6n)$
23. $(-2r)(-3s)$
24. $(-5p)2q$

Lesson 2–6   *(Pages 82–85)*   **Find each quotient.**

1. $49 \div 7$
2. $-12 \div 6$
3. $-36 \div (-12)$
4. $21 \div (-3)$
5. $36 \div (-6)$
6. $-20 \div 4$
7. $-16 \div (-4)$
8. $-32 \div (-4)$
9. $\dfrac{8}{2}$
10. $\dfrac{-54}{-9}$
11. $\dfrac{63}{-7}$
12. $\dfrac{-64}{8}$

Evaluate each expression if $x = -5$, $y = -3$, $z = 2$, and $w = 7$.

13. $25 \div x$
14. $-42 \div w$
15. $3 \div y$
16. $2x \div z$
17. $-3x \div y$
18. $x \div (-1)$
19. $xyz \div 10$
20. $yzw \div 2w$
21. $\dfrac{3y}{-3}$
22. $\dfrac{6 - y}{y}$
23. $\dfrac{w}{-7}$
24. $\dfrac{w - x}{y}$

Lesson 3–1   *(Pages 94–99)*   **Replace each ● with <, >, or = to make a true sentence.**

1. -2 ● -5
2. -3.4 ● -4
3. -0.32 ● -0.3
4. $4(-2)(-1)$ ● $2(-3)$
5. 0.6 ● $\dfrac{3}{5}$
6. $\dfrac{2}{3}$ ● 0.7
7. $\dfrac{7}{9}$ ● $\dfrac{7}{8}$
8. $-\dfrac{3}{18}$ ● $-\dfrac{2}{9}$

Write the numbers in each set from least to greatest.

9. $\dfrac{5}{6}, 0.3, \dfrac{1}{3}$
10. $0.4, -\dfrac{2}{3}, \dfrac{3}{4}$
11. $-\dfrac{4}{5}, -\dfrac{6}{7}, -0.7$
12. $\dfrac{1}{2}, -\dfrac{7}{10}, -\dfrac{2}{3}$
13. $0.\overline{4}, \dfrac{3}{8}, \dfrac{4}{6}$
14. $-\dfrac{1}{8}, \dfrac{2}{9}, -\dfrac{3}{12}$

15. Compare the numbers $\dfrac{6}{7}$ and $\dfrac{7}{8}$ using an inequality.

16. Using a number line, explain why $0.7 > 0.6$.

Lesson 3–2   *(Pages 100–103)*   **Find each sum or difference.**

1. $-13.1 + (-21.4)$
2. $-8.14 - 0.13 + 1.11$
3. $6.2 - (-3.4)$
4. $-11.12 - 2.15 + 5.28 - 3.12$
5. $15.9 + 6.25 - 3.48 - 2.13$
6. $-17.6 + 0.3 - 3.7$
7. $-\dfrac{4}{5} + \dfrac{2}{3}$
8. $\dfrac{6}{7} - \dfrac{1}{8}$
9. $-\dfrac{3}{4} + \dfrac{2}{3}$
10. $3\dfrac{6}{7} - \left(-4\dfrac{5}{8}\right)$
11. $-6\dfrac{1}{6} + 6\dfrac{5}{6}$
12. $2\dfrac{8}{9} - 1\dfrac{1}{3}$

13. Find the value of $z - 0.25$ if $z = 0.5$.

14. Evaluate $m + 5\dfrac{3}{4}$ if $m = -3\dfrac{1}{2}$.

Lesson 3–3 (Pages 104–109) **Find the mean, median, mode, and range for each set of data.**

1. 8, 3, 2, 8, 1, 5, 8, 5
2. 11, 5, 18, 10, 14, 14, 18, 18, 14, 17, 15
3. 1.2, 5.4, 2.3, 3.2, 1.2, 5.3
4. 123, 153, 123, 114, 148, 114, 135
5. 47, 29, 77, 99, 50 47, 29
6. 6.5, 7.8, 5.7, 3.9, 9.9, 3.8, 5.5, 5.7

7.

8.
Stem	Leaf
11	0 1 7
12	3 4 4
13	5 5 6 7
14	2 2 $11\mid0 = 110$

Write a set of data with six numbers that satisfies each set of conditions.

9. The median is equal to the mean.
10. The median is less than the mean.

Lesson 3–4 (Pages 112–116) **Find the solution of each equation if the replacement sets are $x = \{-2, -1, 0, 1\}$, $m = \{-1, 0, 1, 2\}$, and $d = \{6, 7, 8, 9\}$.**

1. $x + 3 = 2$
2. $2d = 16$
3. $6m + 15 = 21$
4. $7m - 11 = \dfrac{3m}{2}$
5. $\dfrac{14 - 2}{x} = -6$
6. $\dfrac{2}{3(1 + 5)} = \dfrac{6 - 5}{d}$

Solve each equation.

7. $-17.8 - 12.2 = p$
8. $6.5 - 13.2 = y$
9. $x = 7 \cdot 2 + 1$
10. $j = 16 - 6 \cdot 2 \div 3$
11. $7 \cdot 4 - 4 \div 2 = r$
12. $w = 6.1 - 3.6 \cdot 4$
13. $n = 10.3 + 4.2 \div 3 - 1.2$
14. $[12 - (4 + 1)] = z$
15. $t = -4.2 - 3.3 \div 3$
16. $\dfrac{7 \cdot 3 + 3}{(3 - 1) \cdot 2} = k$
17. $\dfrac{-13 - 5}{3 \cdot 2} = s$
18. $\dfrac{6 - 3 \cdot 2}{12 \div 3 + 1} = m$
19. $x = \dfrac{5 \div (2 + 3)}{2 \cdot 2 - 3}$
20. $\dfrac{36 \div 9 + 1}{2 \cdot 5} = a$
21. $h = \dfrac{15 + 10 \div 2}{-3 \cdot 1 + 2}$

22. Find the solution of $6 - 3(8) = r$.
23. What is the value of k if $k = \dfrac{16 + 2}{-9}$?

Lesson 3–5 (Pages 117–121) **Solve each equation. Use algebra tiles if necessary.**

1. $g + 13 = 7$
2. $-8 = w + -6$
3. $m - 8 = -23$
4. $-3 + q = 9$
5. $-21 = a + 15$
6. $t - 12 = -3$
7. $d - (-6) = 14$
8. $s - 4 = 12$
9. $k - 9 = 18$
10. $c - (-3) = -8$
11. $13 + y = 4$
12. $h - 11 = 5$
13. $-24 = x + 7$
14. $16 = 9 + p$
15. $6 = -12 + b$
16. $-5 = -12 + b$
17. $m + 15 = 9$
18. $-8 = w + (-6)$

19. What is the value of y if $y - 4 = -15$?
20. When x is divided by 8, the result is 7. Find the value of x.

Lesson 3–6 *(Pages 122–127)*   **Solve each equation. Check your solution.**

1. $-18 = w + 3$
2. $m + (-5) = 6$
3. $-16.4 = r - 7.4$
4. $m - 16 = -13$
5. $5 = x - (-3)$
6. $x + (-8) = -7$
7. $45 + j = -27$
8. $-56 = g - 32$
9. $n - (-26) = 41$
10. $6.4 + k = -3.2$
11. $-2.8 + k = 3.1$
12. $s + 2.5 = -1.3$
13. $13.4 + p = -2.4$
14. $17.3 = -2.7 + d$
15. $y + 2.17 = 5.67$
16. $\frac{5}{7} = z - \frac{6}{21}$
17. $n - \frac{5}{8} = -\frac{3}{4}$
18. $-\frac{2}{3} = j + \frac{7}{12}$

19. Solve for m if $-20 + m = -7$.
20. What is the solution of $h - 4 = 14$?

Lesson 3–7 *(Pages 128–131)*   **Solve each equation. Check your solution.**

1. $|m| = -3$
2. $4 = |x|$
3. $|r| + 4 = 8$
4. $|y| - 8 = 11$
5. $|x + 5| = 10$
6. $|t + 8| = -2$
7. $|d - 4| = 10$
8. $6 = |-3 + p|$
9. $-4 = |5 + s|$
10. $|n - 11| = 15$
11. $-4 = |-2 + j|$
12. $|-2 + p| = 0$
13. $|k - (-2)| = 5$
14. $12 = |n + (-3)|$
15. $2 + |s + 3| = 8$
16. $5 = |-4 + g| + 17$
17. $16 = |z - 4| + 3$
18. $|s + 4| - 3 = 7$

19. How many solutions exist for $-3 = |4 + d|$?
20. How many solutions exist for $8 = |2 + k|$?

Lesson 4–1 *(Pages 140–145)*   **Find each product.**

1. $3(8.2)$
2. $-7.3(3)$
3. $-6.2(3.5)$
4. $(2.1)(-1)$
5. $-3.2(-0.5)$
6. $-2.1(3)(-2)$
7. $16.2(0)$
8. $8.5(-1)(-2.2)$
9. $\frac{2}{5}\left(\frac{1}{9}\right)$
10. $\left(-\frac{1}{2}\right)\left(\frac{3}{8}\right)$
11. $-\frac{1}{4}\left(-\frac{6}{5}\right)$
12. $-1 \cdot \frac{1}{8}$
13. $-\frac{6}{7}(0)$
14. $4\left(-\frac{3}{4}\right)$
15. $\left(-\frac{3}{4}\right)\left(\frac{5}{6}\right)$
16. $\frac{1}{2}\left(-3\frac{1}{8}\right)$

Simplify each expression.

17. $-7.2(2p)$
18. $(-9y)(-4.1)$
19. $(6.2k)(-3)$
20. $-5(0.5j)$
21. $2x(-3.4y)$
22. $5.4m(-2n)$
23. $\frac{7}{13}p(26)$
24. $\left(\frac{3}{5}r\right)\left(-\frac{1}{3}\right)$
25. $\left(\frac{1}{2}x\right)\left(\frac{2}{3}y\right)$
26. $\frac{5}{6}k(6)$
27. $\left(-\frac{2}{3}s\right)\left(\frac{4}{5}r\right)$
28. $-\frac{1}{4}w(-3t)$

Lesson 4–2 (Pages 146–151) Find the number of possible outcomes by drawing a tree diagram.

Men	Women
Troy	Ann
Malik	Lorena
Ben	Ellen
Aaron	

Exercise 2

1. tossing a coin twice
2. choosing different teams of one man and one woman
3. At a dinner party, you can choose either chicken or beef for your main dish, soda or juice for your drink, and pie, ice cream, or cake for your dessert. How many different meals are possible?

4. When choosing auto insurance, you have many choices, as shown in the table. How many different types of coverage are possible? Find the number of possible outcomes by using the Fundamental Counting Principle.

Liability	Underinsured Drivers	Collision	Comprehensive
$50,000	$50,000	$100	$100
$100,000	$100,000	$500	$500
$150,000	$150,000	$1000	$1000
	none		

5. choosing four cards from a standard 52-card deck
6. There are 5 multiple-choice problems on a quiz (with possible answers of a, b, c, and d). How many different combinations of answers are possible?

Lesson 4–3 (Pages 154–159) Find each quotient.

1. $12 \div 1.5$
2. $3.1 \div (-3.1)$
3. $-8.2 \div (-4.1)$
4. $-18.6 \div 6.2$
5. $0 \div 7.3$
6. $-9.3 \div (-3.1)$
7. $1.6 \div (-16)$
8. $-0.4 \div 0.2$
9. $-\frac{1}{6} \div \left(-\frac{1}{5}\right)$
10. $\frac{1}{3} \div (-5)$
11. $-4 \div \frac{2}{3}$
12. $8 \div \left(-\frac{1}{4}\right)$
13. $6 \div \left(-\frac{3}{4}\right)$
14. $-\frac{2}{5} \div \left(-\frac{7}{8}\right)$
15. $4\frac{5}{8} \div \frac{5}{8}$
16. $-\frac{4}{3} \div \frac{5}{6}$

Evaluate each expression if $j = -\frac{1}{8}$, $k = \frac{3}{4}$, and $m = \frac{1}{6}$.

17. $\frac{8}{j}$
18. $\frac{k}{2}$
19. $\frac{m}{5}$
20. $\frac{4}{m}$
21. $\frac{j}{k}$
22. $\frac{2m}{k}$
23. $\frac{j}{8}$
24. $\frac{km}{j}$

Lesson 4–4 (Pages 160–164) Solve each equation.

1. $6k = -54$
2. $0 = 8p$
3. $0.5y = 12$
4. $3.2m = -6.4$
5. $7.5 = -1.5w$
6. $2.1t = -6.3$
7. $-3n = 2$
8. $-7 = -3s$
9. $\frac{1}{3}x = -4$
10. $-3 = -\frac{5}{8}v$
11. $\frac{4}{5} = \frac{4}{9}r$
12. $-\frac{2}{3}z = \frac{1}{8}$
13. $-\frac{1}{2}a = 3$
14. $10 = \frac{c}{4}$
15. $\frac{21}{13} = -7j$
16. $-\frac{1}{18} = \frac{17}{18}b$

17. Solve $-1.5x = 18$. Then check your solution.
18. What is the solution of $\frac{s}{16} = \frac{1}{4}$?

Lesson 4–5 *(Pages 165–170)*   **Solve each equation. Check your solution.**

1. $6 + 2y = 8$
2. $3 - 3r = 4$
3. $4 + (-0.5y) = 8$
4. $-7s - 3 = 18$
5. $0 = 1.5a + 3$
6. $3.1 = 0.7w + 1$
7. $6.3 = 0.4t - 2.1$
8. $6c + 1.5 = 3$
9. $\frac{r}{5} - 1 = 5$
10. $6 = -\frac{y}{3} + 3$
11. $10 = -\frac{x}{2} + 6$
12. $8 - \frac{b}{4} = 2$
13. $8 = -\frac{x}{3} + 2$
14. $\frac{6 - p}{2} = -3$
15. $-\frac{7}{8}c + 1 = 4$
16. $\frac{8 + m}{3} = -8$

17. What is the solution of $3 = \frac{r + 10}{-3}$?

18. Find the value of w in the equation $6 + \frac{2}{3}w = 12$.

Write an equation and solve each problem.

19. Three less than half a number x is 12. Find the number.

20. Start with a number k. If you add 3, multiply by 7 and subtract 4, you get 18. What is the value of k?

Lesson 4–6 *(Pages 171–175)*   **Solve each equation. Check your solution.**

1. $6a = 3a + 3$
2. $16 + 5r = 3r$
3. $8x + 6 = 12 - 3x$
4. $7k + 3 = 2k$
5. $1.4y = -0.6y - 2$
6. $4 + 0.8p = -3.2p$
7. $6.4 + m = m - 2.3$
8. $\frac{1}{3}f = \frac{2}{3}f + 4$
9. $\frac{1}{8}t + 3 = -4 - \frac{1}{6}t$
10. $\frac{1}{2}w - 2 = \frac{2}{3}w$
11. $1 - \frac{7}{8}h = \frac{1}{4}h$
12. $\frac{2}{5}j + 3 = -\frac{3}{5}j$

13. Find the solution of $1.7x + 11 = -6.3x - 13$.

14. Six times a number k is 17 less than 2 times the number.
    a. Write an equation to represent the problem.
    b. Find the number.

Lesson 4–7 *(Pages 176–179)*   **Solve each equation. Check your solution.**

1. $16 = -2(p - 1)$
2. $6(x + 5) = 12$
3. $8x + 6 = 3(4 - x)$
4. $8(6 - q) - 5 = -21$
5. $6t = 3(2t + 5)$
6. $-7 = 3k - 2(2k)$
7. $7(z + 3) = 2z + 4$
8. $-1(r + 1) - 2 = 2 + (r - 1)$
9. $5x = 3(1.5x - 3)$
10. $-4(m - 2.2) = -2(m + 1.4)$
11. $6 + \frac{1}{3}(9x + 3) = 6x + 2$
12. $4\left(\frac{1}{2}x + \frac{1}{2}\right) = 2x + 2$

13. Find the solution of $4[8 + 2(m - 2)] = 8$.
14. What is the value of r in $7(8 - r) + 3(2 + 2r) = 9$?

Lesson 5–1 *(Pages 188–193)* Solve each proportion.

1. $\dfrac{9}{8} = \dfrac{3}{10}$

2. $\dfrac{6}{4} = \dfrac{y}{10}$

3. $\dfrac{b}{12} = \dfrac{32}{8}$

4. $\dfrac{1}{3} = \dfrac{5}{r}$

5. $\dfrac{21}{28} = \dfrac{h}{7}$

6. $\dfrac{4}{5} = \dfrac{a}{100}$

7. $\dfrac{25}{3} = \dfrac{500}{x}$

8. $\dfrac{p}{24} = \dfrac{8}{4}$

9. $\dfrac{z}{6} = \dfrac{9}{2}$

10. $\dfrac{6.2}{s} = \dfrac{3.1}{2}$

11. $\dfrac{0.3}{0.4} = \dfrac{x}{8}$

12. $\dfrac{6+m}{m-3} = \dfrac{1}{2}$

13. $\dfrac{7+b}{4} = \dfrac{3}{5}$

14. $\dfrac{7-d}{8+d} = \dfrac{4}{9}$

15. $\dfrac{6}{d-2} = \dfrac{7}{d}$

16. $\dfrac{k+1}{3} = \dfrac{k}{6}$

17. Are $\dfrac{18}{4}$ and $\dfrac{9}{2}$ equivalent ratios? Explain your reasoning.

18. Find the value of x that makes $\dfrac{17}{3} = \dfrac{x}{9}$ a proportion.

Convert each measurement as indicated.

19. 7200 pounds to tons

20. 4.5 feet to inches

21. 2.1 quarts to pints

22. 460 meters to kilometers

23. 13 grams to milligrams

24. 16 milliliters to liters

Lesson 5–2 *(Pages 194–197)* On a map, the scale is 1.5 inches = 100 miles. Find the actual distance for each map distance.

	From	To	Map Distance
1.	Paris, France	Brussels, Belgium	3.15 inches
2.	Amsterdam, Holland	Frankfurt, Germany	4.5 inches
3.	Luxembourg, Luxembourg	Frankfurt, Germany	$2\frac{1}{4}$ inches
4.	Luxembourg, Luxembourg	Paris, France	$3\frac{3}{5}$ inches

5. **Puzzles** The picture on a puzzle box shows the puzzle to be 4 inches tall. The actual height is 12 inches. What is the scale of the picture?

Lesson 5–3 *(Pages 198–203)* Express each fraction or ratio as a percent.

1. 11 to 44

2. 36 out of 18

3. 8 out of 32

4. 12 to 24

5. $\dfrac{12}{20}$

6. $\dfrac{27}{9}$

7. $\dfrac{10}{8}$

8. $\dfrac{7}{21}$

9. 4 to 5

10. 3 out of 10

11. 6 to 8

12. 2 out of 20

13. $\dfrac{18}{27}$

14. 5 out of 25

15. 3 to 8

16. $\dfrac{13}{100}$

17. 12 to 3

18. 6 out of 12

19. 10 to 5

20. 5 out of 6

21. One out of three people at the conference agreed with the speaker.

22. Two fifths of the students passed the test.

Use the percent proportion to find each number.

23. What number is 30% of 120?

24. 125% of what number is 15?

25. What percent of 21 is 7?

26. Find 10% of 12.

27. 50 is what percent of 200?

28. 16 is 25% of what number?

Lesson 5–4 *(Pages 204–209)* **Use the percent equation to find each number.**

1. What number is 110% of 36?

2. 30 is 300% of what number?

3. Find 25% of 120.

4. What number is 30% of 200?

5. What number is 10% of 25?

6. Find 200% of 18.

7. 15 is 75% of what number?

8. 16 is 40% of what number?

9. Find 5% of 120.

10. What number is 500% of 5?

11. 35 is 20% of what number?

12. Find 150% of 80.

13. 45 is 80% of what number?

14. What number is 18% of 324?

15. **Banking** How long will it take Kristin to earn $180 if she invests $4000 at a rate of 6%?

16. **Banking** How much interest will Michael earn if he invests $575 at a rate of 7% for 3 years?

Lesson 5–5 *(Pages 212–217)* **Find the percent of increase or decrease. Round to the nearest percent.**

1. original: 20
new: 22

2. original: 125
new: 100

3. original: 18
new: 9

4. original: 600
new: 750

5. original: 30
new: 10

6. original: 28
new: 21

7. original: 12
new: 15

8. original: 50
new: 70

The cost of an item and a sales tax rate are given. Find the total price of each item to the nearest cent.

9. painting: $600; 6.5%

10. shoes: $85; 4%

11. book: $24.95; 3.5%

12. shirt: $29.99; 6%

13. piano: $1600; 5%

14. sweater: $45.99; 7%

The original cost of an item and a discount rate are given. Find the sale price of each item to the nearest cent.

15. jacket: $120; 40%

16. hat: $23.99; 30%

17. basketball: $25; 10%

18. video game: $49.99; 5%

19. bicycle: $425; 15%

20. snow skis: $225; 25%

21. What number is 20% less than $75?

22. Find the percent of increase from $75 to $85.

Lesson 5–6 *(Pages 219–223)* **Refer to the application on page 219. Find the probability of each outcome if a pair of dice are rolled.**

1. 2 odd numbers

2. a sum of 7

3. a sum of 4

4. even number on the first die

5. a sum greater than 11

6. a sum of less than 10

Find the odds of each outcome if the spinner at the right is spun.

7. greater than 10

8. blue

9. blue or yellow

10. not red

11. an even number

12. not a two

13. an odd number

14. less than 4

15. What is the probability that you select a seven at random from a standard deck of cards?

Lesson 5–7 *(Pages 224–229)* A card is drawn from a deck of ten cards numbered 1 through 10. The card is replaced in the deck and another card is drawn. Find the probability of each outcome.

1. P(9 and then a 7)
2. P(6 and a 4)
3. P(an even and then a 2)
4. P(two numbers less than 6)
5. P(8 and then an odd number)
6. P(two odd numbers)

7. What is the probability of tossing a coin three times and getting two heads and a tail?

8. What is the probability of tossing a coin three times and getting a head, then a tail, then a head?

Determine whether each event is *mutually exclusive* or *inclusive*. Then find each probability.

9. There are 18 cars in a lot. There are 6 red cars, 8 blue cars, 3 white cars, and 1 purple car. What is the probability of randomly choosing a white or blue car?

10. In rolling a die, what is the probability that it is either an even or a four?

Lesson 6–1 *(Pages 238–243)* Express each relation as a table and as a graph. Then determine the domain and the range.

1. {(6, 3), (2, 4), (3, 2), (5, 5)}
2. {(5.9, −3), (−2, 3.1), (0, 0), (−1, 7)}
3. {(−1, 6), (2, −5), (−2, 9), (0, 4)}
4. {(1, −3), (2.4, 6), (0, −5.1), (−4, 4)}
5. $\left\{\left(\frac{1}{4}, \frac{1}{6}\right), \left(-\frac{2}{3}, \frac{4}{5}\right), \left(-1, \frac{1}{3}\right)\right\}$
6. $\left\{(3, 1), \left(\frac{3}{4}, 2\right), (2, 0), \left(-\frac{2}{3}, -3\right)\right\}$

Express each relation as a set of ordered pairs.

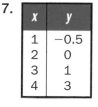

7.

x	y
1	−0.5
2	0
3	1
4	3

8.

x	y
−3	4
−2	3
−1	2
0	1

9.

x	y
1	−3
2	−2
−3	−1
−1	0

Lesson 6–2 *(Pages 244–249)* Which ordered pairs are solutions of each equation?

1. $6x + 2y = 12$ **a.** (1, 3)   **b.** (1, 1)   **c.** (2, 0)   **d.** (2, 2)
2. $3a + 6 = 7b$ **a.** (4, 3)   **b.** (1, 0)   **c.** (2, 3)   **d.** (5, 3)
3. $2m + n = 7$ **a.** (1, 6)   **b.** (3, 1)   **c.** (0, 2)   **d.** (0, 6)
4. $3r - 15 = -6s$ **a.** $\left(\frac{1}{3}, \frac{7}{3}\right)$ **b.** (1, 2)   **c.** (2, 1)   **d.** $\left(\frac{1}{3}, \frac{1}{2}\right)$

Solve each equation if the domain is {−2, −1, 0, 1, 2, 3}. Graph the solution set.

5. $y = 6x$
6. $y = -2x + 1$
7. $y = x - 8$
8. $x - y = 5$
9. $y = -3x - 4$
10. $3x - 2y = 6$
11. $4 - x = y$
12. $7 = 6x + 4y$

Find the domain of each equation if the range is {−1, 0, 1, 2}.

13. $y = 3x + 3$
14. $7 - 4x = y$
15. $6y = 2x$
16. $x = 5y - 2$
17. $2y = 3x + 1$
18. $5x - 15 = y$
19. $2x = y - 4$
20. $2y - 2x = 4$

Lesson 6-3 *(Pages 250–255)* Determine whether each equation is a linear equation. Explain. If an equation is linear, identify *A*, *B*, and *C*.

1. $4xy = 3$
2. $7x - 2y = 3$
3. $2xy + x = y$
4. $16y = 4x$
5. $x = 2y$
6. $3y = x + 4$
7. $y - 2 = x$
8. $x + 3y = 9$
9. $6x + 3 = y$
10. $2x^2 + y^2 = 8$
11. $4y + 8x = 0$
12. $y + xy = 1$

Graph each equation.

13. $x = y - 1$
14. $-3x + 2 = y$
15. $7y = 14x - 2$
16. $x + y = 4$
17. $2x + 3y = 0$
18. $x - y = 2$
19. $16 = 2x + 4y$
20. $5x - y = 15$

Lesson 6-4 *(Pages 256–261)* Determine whether each relation is a function.

1. $\{(3, 6), (4, -2), (-1, -3), (0, 6)\}$
2. $\{(-1, 8), (0, 3), (3, 3), (-2, 2)\}$
3. $\{(-1, 5), (3, 1), (5, 4), (-1, 3)\}$
4. $\{(-2, 4), (0, 0), (4, 2), (3, 0)\}$
5. $\{(0, 3), (1, 1), (1, -1), (2, 4)\}$
6. $\{(3, 3), (-3, 1), (4, 2), (-2, 3)\}$

7.

x	y
−4	2
−3	3
−2	2
−4	3
−5	4

8.

x	y
1	1
3	3
−1	1
2	2
0	0

9.

x	y
−1	−3
2	−2
−3	−3
0	−1
−1	0

Use the vertical line test to determine whether each relation is a function.

10. 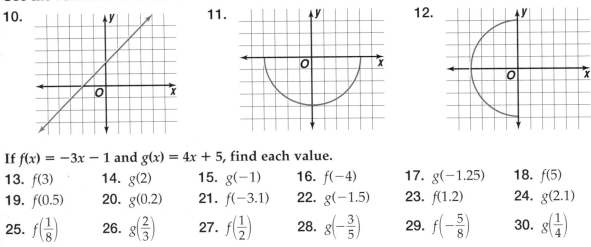 11. 12.

If $f(x) = -3x - 1$ and $g(x) = 4x + 5$, find each value.

13. $f(3)$
14. $g(2)$
15. $g(-1)$
16. $f(-4)$
17. $g(-1.25)$
18. $f(5)$
19. $f(0.5)$
20. $g(0.2)$
21. $f(-3.1)$
22. $g(-1.5)$
23. $f(1.2)$
24. $g(2.1)$
25. $f\left(\frac{1}{8}\right)$
26. $g\left(\frac{2}{3}\right)$
27. $f\left(\frac{1}{2}\right)$
28. $g\left(-\frac{3}{5}\right)$
29. $f\left(-\frac{5}{8}\right)$
30. $g\left(\frac{1}{4}\right)$

Lesson 6-5 *(Pages 264–269)* Determine whether each equation is a direct variation. Verify the answer with a graph.

1. $x = y + 2$
2. $y = 3x$
3. $y = 4x + 2$
4. $y = 2x + 2$
5. $y = 2x$
6. $x = -2$

Solve. Assume that *y* varies directly as *x*.

7. If $y = -3$ when $x = -6$, find *x* when $y = 16$.
8. If $y = 9$ when $x = 5$, find *y* when $x = 10$.
9. Find *x* when $y = 24$ if $y = 16$ when $x = 6$.
10. Find *x* when $y = 3$ if $y = 15$ when $x = 4$.
11. If $x = 13$ when $y = 7$, find *y* when $x = 26$.
12. Find *y* when $x = 2.5$ if $y = 10$ when $x = 6$.

Lesson 6–6 (Pages 270–275) Solve. Assume that *y* varies inversely as *x*.

1. Suppose $y = 14$ when $x = 7$. Find x when $y = 18$.
2. Find y when $x = 8$ if $y = 15$ when $x = 12$.
3. If $y = 2.5$ when $x = 6.5$, find y when $x = 13$.
4. Suppose $y = \frac{5}{8}$ when $x = \frac{1}{3}$. Find y when $x = -\frac{1}{2}$.

Find the constant of variation. Then write an equation for each statement.

5. *y* varies inversely as *x*, and $y = 7$ when $x = 4$.
6. *y* varies directly as *x*, and $y = 8$ when $x = -1$.
7. *y* varies inversely as *x*, and $y = -3.5$ when $x = 7$.
8. *y* varies directly as *x*, and $y = 4$ when $x = -12$.

Lesson 7–1 (Pages 284–289) Determine the slope of each line.

1.
2.
3.

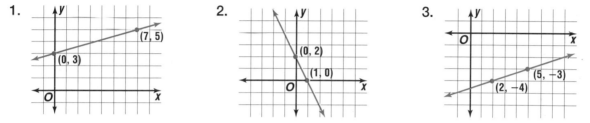

Determine the slope of the line passing through the points whose coordinates are listed in each table.

4.
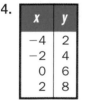

x	y
−4	2
−2	4
0	6
2	8

5.

x	y
−1	12
0	6
1	0
2	−6

6.
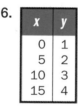

x	y
0	1
5	2
10	3
15	4

Lesson 7–2 (Pages 290–295) Write the point-slope form of an equation for each line passing through the given point and having the given slope.

1. $(6, 4),\ m = -3$
2. $(-7, 3),\ m = 2$
3. $(2, 2),\ m = -1$
4. $(3, -2),\ m = 6$
5. $(1, 5),\ m = -5$
6. $(1, 2),\ m = -4$
7. $(-7, 2),\ m = \text{none}$
8. $(-3, -3),\ m = \frac{1}{2}$
9. $(-4, -2),\ m = \frac{4}{5}$
10. $(4, 0),\ m = \frac{1}{3}$
11. $\left(-\frac{1}{2}, -1\right),\ m = 4$
12. $\left(\frac{3}{4}, 4\right),\ m = -2$

Write the point-slope form of an equation for each line.

13. the line through points at $(8, 4)$ and $(7, 6)$
14. the line through points at $(0, -2)$ and $(-4, 2)$
15. the line through points at $(4, 7)$ and $(-1, 0)$
16. the line through points at $(2, 1)$ and $(1, 2)$
17. the line through points at $(4, -8)$ and $(6, -2)$
18. the line through points at $(-3, 1)$ and $(2, -3)$

19. Write an equation in point-slope form of a line that has a slope of 1.5 and passes through the point $(16, -5)$.

Lesson 7–3 *(Pages 296–301)* **Write an equation in slope-intercept form of the line with each slope and *y*-intercept.**

1. $m = -3, b = 8$
2. $m = 2, b = -3$
3. $m = -1, b = -6$
4. $m = 3, b = 2$
5. $m = 0, b = 4$
6. $m = -2, b = 1$
7. $m = 6, b = 4$
8. $m = -1.4, b = 7$
9. $m = \frac{5}{7}, b = -7$
10. $m = -\frac{2}{3}, b = 9$
11. $m = -\frac{4}{3}, b = -3$
12. $m = -\frac{1}{4}, b = 6$

Write an equation in slope-intercept form of the line having the given slope and passing through the given point.

13. $m = 4, (1, -2)$
14. $m = 3, (-5, -3)$
15. $m = -2, (-3, 0)$
16. $m = -4, (-2, 6)$
17. $m = -1, (0, 4)$
18. $m = 0, (7, 8)$
19. $m = 6, (12, -6)$
20. $m = 5, (-8, 3)$
21. $m = -\frac{4}{5}, (3, 6)$
22. $m = \frac{3}{8}, (-1, 1)$
23. $m = \frac{1}{3}, (0, 5)$
24. $m = -\frac{2}{3}, (-6, 1)$

Write an equation in slope-intercept form of the line passing through each pair of points.

25. $(2, 4)$ and $(3, 1)$
26. $(0, 1)$ and $(3, 0)$
27. $(-4, -3)$ and $(2, 1)$
28. $(5, 4)$ and $(7, -8)$
29. $(-2, 3)$ and $(9, -4)$
30. $(8, 12)$ and $(-2, -8)$
31. $(5, 3)$ and $(6, -8)$
32. $(3, 0)$ and $(8, -2)$
33. $(7, 4)$ and $(4, 7)$
34. $(-2, -2)$ and $(3, -3)$
35. $(8, 2)$ and $(-4, -1)$
36. $(6, 2)$ and $(3, -4)$

Lesson 7–4 *(Pages 302–307)* **Determine whether each scatter plot has a *positive* relationship, *negative* relationship, or *no* relationship. If there is a relationship, describe it.**

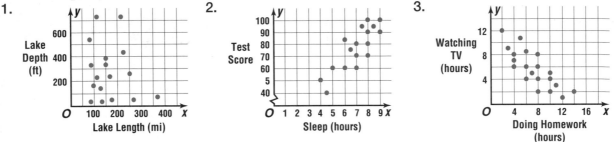

Determine whether a scatter plot of the data for the following would show *positive*, *negative*, or *no* relationship between the variables.

4. study time and score on a test

5. your shoe size and your age

Lesson 7–5 *(Pages 310–315)* **Determine the *x*-intercept and *y*-intercept of the graph of each equation. Then graph the equation.**

1. $2x + 8y = 16$
2. $x + 2y = 2$
3. $x + y = 3$
4. $2x - 9y = 18$
5. $6x - y = 4$
6. $7y - x = -3$
7. $6x + y = -2$
8. $4y + 3x = -4$
9. $x + \frac{2}{3}y = 6$
10. $\frac{1}{2}x + 4y = 1$
11. $\frac{2}{3}x + \frac{1}{3}y = \frac{4}{3}$
12. $3y - 2x = -4$

Determine the slope and *y*-intercept of the graph of each equation. Then graph the equation.

13. $y = 8 - x$
14. $y = 5x - 2$
15. $y = 3x + 6$
16. $y = 4x + 8$
17. $y = -\frac{1}{2}x - 3$
18. $y = \frac{4}{3}x - 2$
19. $y = \frac{1}{6}x + 5$
20. $y = -\frac{3}{4}x + 6$
21. $-2y + 4x = 8$
22. $7x + 8y = -56$
23. $9y + 3x = 6$
24. $-5x + y = 6$

Lesson 7-6 (*Pages 316–321*)   Graph each pair of equations. Describe any similarities or differences and explain why they are a family of graphs.

1. $y = 3x + 2$
   $y = 3x$

2. $y = -2x + 1$
   $y = 2x + 1$

3. $y = 3x$
   $y = 4x$

4. $y = x + 4$
   $y = 5x + 4$

5. $y = -4x + 1$
   $y = -4x - 2$

6. $-\frac{3}{4}x + 2 = y$
   $-x + 2 = y$

7. $y = \frac{1}{3}x$
   $y = 3x$

8. $y = \frac{1}{2}x + \frac{1}{2}$
   $y = 2x + 2$

Compare and contrast the graphs of each pair of equations. Verify by graphing the equations.

9. $y = 2x$
   $y = 2x + 1$

10. $y = 0.25x + 1$
    $y = 4x + 1$

11. $y = -3x - 2$
    $y = 3x - 2$

12. $y = x - 4$
    $y = x + 6$

13. $y = \frac{1}{2}x + 2$
    $y = x + 2$

14. $y = -\frac{2}{3}x + 2$
    $y = -\frac{2}{3}x$

15. $y = -5x + 4$
    $y = -\frac{1}{2}x + 4$

16. $y = \frac{1}{3}x$
    $5 + x = 3y$

Change $y = 0.5x - 5$ so that the graph of the new equation fits each description.

17. y-intercept is 2, same slope

18. negative slope, y-intercept is -3

19. shifted up 2 units, slope is 2

20. less steep positive slope, same y-intercept

Lesson 7-7 (*Pages 322–327*)   Determine whether the graphs of each pair of equations are *parallel*, *perpendicular*, or *neither*.

1. $y = 3x + 2$
   $y = -\frac{1}{3}x + 5$

2. $y = 3x + 4$
   $y = 3x + 2$

3. $y = 3x + 2$
   $y = \frac{1}{5}x - \frac{7}{5}$

4. $y = 2x + 4$
   $3y = 6x + 6$

5. $y = \frac{1}{2}x + 4$
   $4x - 3y = 12$

6. $2y = 10x - \frac{2}{5}$
   $\frac{1}{5}x + y = 3$

7. $y = \frac{2}{3}x - 6$
   $2x - 4 = 3y$

8. $2x + y = 6$
   $y = \frac{1}{2}x + 4$

Write an equation in slope-intercept form of the line that is parallel to the graph of each equation and passes through the given point.

9. $y = 8x + 5$; (0, 4)

10. $y = 3$; $(-1, 2)$

11. $x = 2$; (4, 3)

Write an equation in slope-intercept form of the line that is perpendicular to the graph of each equation and passes through the given point.

12. $y = x + 5$; $(-2, 1)$

13. $y = 4x$; (0, 0)

14. $6x - 2y = 3$; (0, 1)

Lesson 8-1 (*Pages 336–340*)   Write each expression using exponents.

1. $7 \cdot 7 \cdot 7$

2. $(-3)(-3)(-3)$

3. $5 \cdot 5 \cdot 5 \cdot 6 \cdot 4 \cdot 4$

4. 6 squared

5. $m \cdot m \cdot n \cdot n \cdot n$

6. 2 cubed

7. $4 \cdot r \cdot r \cdot r \cdot s \cdot s$

8. $(-2)(k)(k)(j)(j)$

Write each power as a multiplication expression.

9. 2^3

10. $(-4)^4$

11. $8^4 2^3$

12. k^5

13. $w^2 z^2$

14. $-6xy^2$

Evaluate each expression if $x = 4$, $y = -1$, $z = -2$, and $w = 1.5$.

15. y^3

16. $w(yz + 4)$

17. $z^3 + 2xy$

18. $2x^2 - z^5$

19. $-2(y^3 + w)$

20. wxy

Lesson 8–2 *(Pages 341–346)* **Simplify each expression.**

1. $2^2 \cdot 2^5$
2. $6^3 \cdot 6^4$
3. $x^4 \cdot x^4$
4. $y \cdot y^5$
5. $(a^3)(a^4)$
6. $(g^6h^3)(g^4h^3)$
7. $(r^2t^2)(rt^3)$
8. $(2k^2j)(-3kj)$
9. $(6m^3)(7m^2)$
10. $(2a^2b)(9ab^5)$
11. $(6a^3b)(4ac)$
12. $(8y^4)(3y^4)$
13. $(-3p^5q^2)(-pq^7)$
14. $(3xyz)(-4x^2y^3)$
15. $\dfrac{4^5}{4^2}$
16. $\dfrac{8^8}{8^7}$
17. $\dfrac{15m^{12}}{3m^5}$
18. $\dfrac{-4r^4s^5}{-rs^4}$
19. $\dfrac{-16m^5n^9p^{10}}{2m^2n^3p^6}$
20. $\left(\dfrac{2}{3}x^5y\right)(-12x^3y^2)$

21. Evaluate m^0.
22. Find the product of mn and $-3m^2n$.

Lesson 8–3 *(Pages 347–351)* **Write each expression using positive exponents. Then evaluate the expression.**

1. 8^{-3}
2. 2^{-4}
3. 10^{-3}
4. 5^{-1}
5. 9^{-3}
6. 7^{-2}
7. 3^{-5}
8. 2^{-6}

Simplify each expression.

9. $(m^6)(m^{-2})$
10. $r^0s^1t^4$
11. $6x^{-2}y^{-4}z$
12. $(a^{-3})(a^2)$
13. $\dfrac{1}{k^{-4}}$
14. $\dfrac{m}{m^{-5}}$
15. $\dfrac{6b^4}{3b^{-2}}$
16. $(r^{-3})(s^4)$
17. $-\dfrac{r^2s^5}{rs^6}$
18. $\dfrac{2a^2b^5}{ab^7}$
19. $\dfrac{-7cd^0}{14c}$
20. $-\dfrac{10a^7b^4}{2ab}$
21. $-\dfrac{4w^5z^4}{10z^7}$
22. $-\dfrac{2p^2q^7}{8p^4q}$
23. $\dfrac{56yz^4}{14y^5z^4}$
24. $\dfrac{x^{-3}y^4z^{-2}}{xy^2z^{-2}}$

25. Evaluate $7m^{-2}n^{-3}$ if $m = 4$ and $n = 2$.
26. Find the value of $5a^3b^{-1}$ if $a = -1$ and $b = 3$.

Lesson 8–4 *(Pages 352–356)* **Express each measure in standard form.**

1. 1.5 megaohms
2. 168 billion dollars
3. 2 megahertz
4. 76 milliamperes
5. 400 nanoseconds
6. 1.2 million dollars
7. 18 kilobytes
8. 93 micrograms

Express each number in scientific notation.

9. 178
10. 0.0098
11. 0.032
12. 106,000
13. 13.8
14. 269.3
15. 0.0000083
16. 100
17. 0.000016
18. 1.2
19. 0.3
20. 400,300
21. 17
22. 1852
23. 1900
24. 0.000000103

Evaluate each expression. Express each result in scientific notation and standard form.

25. $(6 \times 10^3)(4 \times 10^5)$
26. $(7 \times 10)(3.5 \times 10^2)$
27. $(2.1 \times 10^3)(1 \times 10^4)$
28. $(4 \times 10^{-3})(7 \times 10^4)$
29. $(1.5 \times 10^5)(6 \times 10^{-3})$
30. $(5 \times 10^{-1})(2.5 \times 10^{-4})$

Lesson 8–5 *(Pages 357–361)* Simplify.

1. $\sqrt{64}$
2. $-\sqrt{9}$
3. $\sqrt{121}$
4. $-\sqrt{225}$
5. $\sqrt{\dfrac{144}{81}}$
6. $-\sqrt{\dfrac{225}{100}}$
7. $\sqrt{\dfrac{256}{16}}$
8. $\sqrt{\dfrac{36}{64}}$
9. $-\sqrt{\dfrac{225}{441}}$
10. $-\sqrt{\dfrac{121}{289}}$
11. $-\sqrt{0.36}$
12. $\sqrt{0.81}$
13. $\sqrt{0.0025}$
14. $-\sqrt{0.0049}$
15. $\sqrt{0.0289}$
16. $-\sqrt{0.000196}$

17. Find the negative square root of 25.
18. If $x = \sqrt{1024}$, what is the value of x?

Lesson 8–6 *(Pages 362–365)* Estimate each square root to the nearest whole number.

1. $\sqrt{5}$
2. $\sqrt{10}$
3. $\sqrt{11}$
4. $\sqrt{15}$
5. $\sqrt{18}$
6. $\sqrt{24}$
7. $\sqrt{61}$
8. $\sqrt{126}$
9. $\sqrt{153}$
10. $\sqrt{412}$
11. $\sqrt{483}$
12. $\sqrt{504}$
13. $\sqrt{555}$
14. $\sqrt{621}$
15. $\sqrt{709}$
16. $\sqrt{981}$
17. $\sqrt{70.3}$
18. $\sqrt{81.4}$
19. $\sqrt{121.6}$
20. $\sqrt{153.2}$
21. $\sqrt{9.35}$
22. $\sqrt{13.6}$
23. $\sqrt{0.021}$
24. $\sqrt{0.29}$

25. Tell whether 11 is closer to $\sqrt{119}$ or $\sqrt{125}$.
26. Which is closer to $\sqrt{285}$, 16 or 17?

Lesson 8–7 *(Pages 366–371)* If c is the measure of the hypotenuse and a and b are the measures of the legs, find each missing measure. Round to the nearest tenth if necessary.

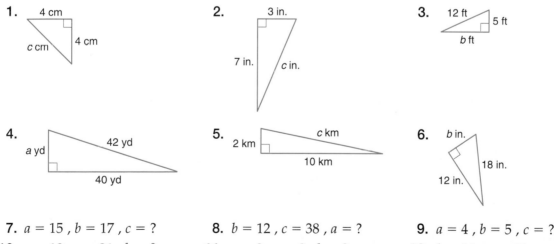

7. $a = 15$, $b = 17$, $c = ?$
8. $b = 12$, $c = 38$, $a = ?$
9. $a = 4$, $b = 5$, $c = ?$
10. $a = 10$, $c = 21$, $b = ?$
11. $a = 3$, $c = 8$, $b = ?$
12. $b = 20$, $c = 40$, $u = ?$

The lengths of three sides of a triangle are given. Determine whether each triangle is a right triangle.

13. 5 ft, 12 ft, 13 ft
14. 4 mi, 5 mi, 7 mi
15. 7 cm, 11 cm, 17 cm
16. 56 in., 70 in., 84 in.
17. 19 m, 24 m, 28 m
18. 31 mm, 37 mm, 49 mm
19. 3 ft, 6 ft, 12 ft
20. 2 in., 8 in., 10 in.
21. 6 mi, 8 mi, 10 mi
22. 10 cm, 24 cm, 26 cm
23. 20 m, 21 m, 29 m
24. 18 in., 26 in., 32 in.

Lesson 9–1 *(Pages 382–387)* **Determine whether each expression is a monomial. Explain why or why not.**

1. $6x$ **2.** $\frac{4}{k}$ **3.** $13b + 2a$ **4.** $7a^2b^{-1}$

5. m^2 **6.** $3x - y$ **7.** s^4 **8.** $-3x^2yz^3$

State whether each expression is a polynomial. If it is a polynomial, identify it as a *monomial*, *binomial*, or *trinomial*.

9. $11y$ **10.** $9xy^2 - y^3 + x^2$ **11.** $r^2 - r$ **12.** $\frac{1}{4}x - 2$

13. $3x - 11 + y - 2$ **14.** $6j^3k^2\ell - 7j^2$ **15.** $2.3mn^2$ **16.** $7r^2s^3t$

Find the degree of each polynomial.

17. y^2 **18.** $6x^3$ **19.** 4 **20.** $3j^2 - 2k + m$

21. $16m^2n + 14m^5n^3$ **22.** $11p^5q^2 - 6pq^7$ **23.** $x^2 + 2x + 3x^6$ **24.** $7a^2b^3 - 2ab$

Arrange the terms of each polynomial so that the powers of x are in descending order.

25. $4x^2 + x^3 - x$ **26.** $2x - 6x^5 - 3$ **27.** $5x^5 - 4x^2 + 3x - 2$

28. $6mx^5 - 4m^2x^6 + 4mx$ **29.** $3x^4y^5 + 2x^2y^3 + 6x^5y$ **30.** $2x - x^2 + 2$

Lesson 9–2 *(Pages 388–393)* **Find each sum.**

1. $\begin{array}{r} 7x - 2 \\ (+)\ x + 4 \\ \hline \end{array}$ **2.** $\begin{array}{r} 6x^2 - 2x - 1 \\ (+)\ 3x^2 - 4x - 7 \\ \hline \end{array}$ **3.** $\begin{array}{r} 2xy - 3x + 2 \\ (+)\ 4xy\quad\ - 7 \\ \hline \end{array}$

4. $(-3y + 7) + (4y - 2)$ **5.** $(7ab - 2a) + (2ab + 3b)$ **6.** $(4x^2 - 2xy + y^2) + (xy + 3y^2)$

Find each difference.

7. $\begin{array}{r} 3x - 4 \\ (-)\ 2x + 3 \\ \hline \end{array}$ **8.** $\begin{array}{r} 5m^2 - m + 2 \\ (-)\ 2m^2 + 2m + 5 \\ \hline \end{array}$ **9.** $\begin{array}{r} 11a^2 - 2a - 1 \\ (-)\ 5a^2 + 6a + 2 \\ \hline \end{array}$

10. $(13x - 2) - (9x + 4)$ **11.** $(7m + 4n) - (2m - n)$ **12.** $(6a + b) - (2b + 4)$

Find each sum or difference.

13. $(2x + 3) + (4x^2 + x - 7)$ **14.** $(mn + 2pn - mp) - (2pn - mp)$

15. $(y^2 + 2y - 3) + (2y^3 - y - 3)$ **16.** $(x^2y + 5xy - 9y^2) - (3x^2 - 2y^2)$

17. $(3x^2 + 2x - 4) + (5x^2 - 9)$ **18.** $(x^2y + 9xy - y^2) - (2x^2y - y^2)$

Lesson 9–3 *(Pages 394–398)* **Find each product.**

1. $4(3x + 2)$ **2.** $-3(7b - 4)$ **3.** $a(2a - 4)$ **4.** $2n(n - 5)$

5. $-y(10y - 5)$ **6.** $-3x(5x + 8)$ **7.** $7p(p^2 - 3)$ **8.** $-5m(3m^3 - 2m)$

9. $2r^3(7 - r + r^2)$ **10.** $-6x^3(x^2 - 4x)$ **11.** $0.75k(8k^3 - k^2)$ **12.** $2.4z^2(2z^2 - 3)$

Solve each equation.

13. $-3(4 - x) = 18$ **14.** $21 = 7(y - 11)$ **15.** $10x - 9 = 4(x - 1) + 1$

16. $7(r + 8) - 4 = -4(-7 - r)$ **17.** $6(s - 7) = 3(4s + 4)$ **18.** $-a + 6(a + 4) = 3(a + 5) + 1$

Lesson 9-4 *(Pages 399–404)* Find each product. Use the Distributive Property or the FOIL method.

1. $(m + 3)(m + 2)$
2. $(x - 4)(x + 6)$
3. $(y - 5)(y - 7)$
4. $(r + 9)(r - 3)$
5. $(2s - 3)(s + 5)$
6. $(a - 2)(4a + 8)$
7. $(7p - 3)(4p + 2)$
8. $(2k + 2)(2k + 1)$
9. $(d + 7)(3d - 5)$
10. $(8x + 2y)(4x + 3y)$
11. $(3m + n)(m - 3)$
12. $(7r + 8s)(5r - 3s)$
13. $(6x - 1)(x - 2)$
14. $(7p - 2)(3p + n)$
15. $(2p - 1)(p - 3)$
16. $(b + a)(a - b)$
17. $(5c + d)(3c - 2d)$
18. $(p + 1)(p + m)$
19. $(j^2 - 2)(j + 5)$
20. $(z^2 + r)(z^2 - r)$
21. $(n^2 + 1)(2n^2 - 3)$

Lesson 9-5 *(Pages 405–409)* Find each product.

1. $(m + 5)^2$
2. $(n - 3)^2$
3. $(2x + 3)^2$
4. $(3y - x)^2$
5. $(4a + b)(4a - b)$
6. $(2k + 2p)^2$
7. $(4x - 2)(4x + 2)$
8. $(1 + p)^2$
9. $(a - 2b)(a + 2b)$
10. $(6 + 3m)^2$
11. $(2 - 4t)^2$
12. $(r - 2s)(3r + 2s)$
13. $(2x - 7y)^2$
14. $(m + 3n)^2$
15. $2(p - q)^2$
16. $4(r + s)^2$
17. $k(2 + k)^2$
18. $4s(s - 1)^2$
19. $3r(r - 2)^2$
20. $y(y + 1)(y - 2)$
21. $p(p + 2)(2p - 3)$
22. $2(j + 3)(j - 3)$
23. $(x + 3)(x - 1)(x + 2)$
24. $6(m + 5)(m - 1)$

25. The area of a circle is given by the formula $A = \pi r^2$, where r is the radius of the circle. Suppose a circle has a radius of $k - 4$ inches.

   a. Write an equation to find the area of the circle.

   b. Find the area to the nearest hundredth if $k = 6$.

Lesson 10-1 *(Pages 420–425)* Find the factors of each number. Then classify each number as *prime* or *composite*.

1. 57
2. 22
3. 65
4. 17
5. 104
6. 18
7. 81
8. 73

Factor each monomial.

9. $12x^2$
10. $28m^2n$
11. $-33j^2k^2$
12. $54p^3$
13. $81ab^2$
14. $75xy$
15. $-13p^2q^2$
16. $105r^4$

Find the GCF of each set of numbers or monomials.

17. 13, 33
18. 50, 75
19. 32, 84, 144
20. 32, 64, 96
21. $-21, 15xy$
22. $4x^2, 2x, 8x$
23. $-3r^2s^2, -17rs$
24. $14rs, 12rst, 6t$
25. $7kr, 21k^2, 2kr$
26. $-16c^2d, -4cd, -8cd^2$
27. $24m^2n, 51m, 63m^2n^2$
28. $-1x^3yz, -7x^2y^2, -2x^3yz$

Lesson 10–2 *(Pages 428–433)* **Factor each polynomial. If the polynomial cannot be factored, write** *prime.*

1. $4m + 12$

2. $13n + n^2$

3. $2k^2 + 6k$

4. $7s^2t + 3$

5. $9pq^3 - 21pq^2$

6. $x^2y + 7y$

7. $3k - 4j$

8. $14m^2n - 18m$

9. $17cd - 14mn$

10. $16a^2b^2 - 3ab$

11. $16b^2c - 2ac + 4bc$

12. $12mn - 14m^2 + 16n$

13. $6ab - 7bc + 12ac$

14. $m^2n^3p + mn - mp$

15. $x^2y + 15xy + 5x$

Find each quotient.

16. $(16x + 4y^2) \div 4$

17. $(2rs + r) \div r$

18. $(7xy - 3x) \div x$

19. $(15a + 3b^2) \div 3$

20. $(6m^2n - 9m) \div 3m$

21. $(18cd^2 - 9cd) \div 9cd$

22. $(21a^2b - 14a^2) \div 7a^2$

23. $(16xy - 12xy) \div 4xy$

24. $(12a^2b + 4b) \div 4b$

25. $(36st^2 - 9st) \div 9st$

26. $(15rs^2 + 12r^2st) \div 3rs$

27. $(5xy^2z + 10x^2z) \div 5xz$

28. $(13r^2s - 26r^2) \div 13r^2$

29. $(32np + m^2np^2) \div np$

30. $(20r^2st + 15rs) \div 5rs$

Lesson 10–3 *(Pages 434–439)* **Factor each trinomial. If the trinomial cannot be factored, write** *prime.*

1. $x^2 + 4x + 4$

2. $n^2 + 6n + 9$

3. $t^2 - 8t + 16$

4. $w^2 + 5w - 2$

5. $r^2 + r - 12$

6. $p^2 - 8p + 4$

7. $s^2 - 7s - 8$

8. $d^2 + 21d + 110$

9. $y^2 - 10y + 3$

10. $q^2 - 3q - 28$

11. $4z^2 - 16z - 18$

12. $6c^2 - 3c - 3$

13. $2m^2 - 3m - 20$

14. $8y^2 + 6y - 2$

15. $3r^2 - 12r - 15$

Lesson 10–4 *(Pages 440–444)* **Factor each trinomial. If the trinomial cannot be factored, write** *prime.*

1. $5x^2 + 13x + 6$

2. $4w^2 + 7w + 3$

3. $2a^2 - 9a + 4$

4. $12t^2 + 18t + 2$

5. $4c^2 - 8c - 3$

6. $3r^2 + 15r + 12$

7. $8y^2 - 16y + 6$

8. $20n^2 - 40n + 20$

9. $10m + 3 + 3m^2$

10. $13 + 2m^2 + 14m$

11. $3q^2 - 4 - 4q$

12. $12d + 8 - 8d^2$

13. $2m^2 - 3mn - 2n^2$

14. $2a^2 + 4ab + 2b^2$

15. $12p^2 + 10mp + 2m^2$

16. $18x^2 + 15xy + 3y^2$

17. A rectangle has dimensions of $(y + 5)$ inches and $(y - 4)$ inches.

  a. Express the area as a trinomial.

  b. If y units are removed from the length, express the new area.

Lesson 10–5 *(Pages 445–449)* Determine whether each trinomial is a perfect square trinomial. If so, factor it.

1. $x^2 - 14x + 49$
2. $a^2 - 4a + 4$
3. $y^2 - 16y + 3$
4. $d^2 + 10d + 25$
5. $r^2 + 24r + 144$
6. $9c^2 + 12c + 4$
7. $4k^2 - 28k + 49$
8. $16m^2 - 8m + 1$

Determine whether each trinomial is the difference of squares. If so, factor it.

9. $x^2 - 9$
10. $4 - 49z^2$
11. $36x^2 - 4$
12. $17 - 3p^2$
13. $s^2 - 16r^2$
14. $9 - m^2n^2$
15. $32k^2 - 50$
16. $12b^2 - 48$

Factor each polynomial. If the polynomial cannot be factored, write *prime*.

17. $a^2 - a - 12$
18. $2s^2 - rs - r^2$
19. $2r^2 - 3$
20. $4m^2 + 48m + 144$
21. $6m^2 - 16m - 6$
22. $2xy - 8x$
23. $3a^2b^3 - 27b$
24. $16x^2 - 8xy + y^2$

Lesson 11–1 *(Pages 458–463)* Graph each quadratic function by making a table of values.

1. $y = 6x^2$
2. $y = 8x^2$
3. $y = -2x^2$
4. $y = -3x^2$
5. $y = 2x^2 + 1$
6. $y = x^2 - 3$
7. $y = -3x^2 + 5$
8. $y = -4x^2 - 3$
9. $y = x^2 + 3x - 6$
10. $y = -2x^2 - x + 5$
11. $y = 3x^2 - 2x$
12. $y = 0.25x^2 - 2x - 4$

Write the equation of the axis of symmetry and the coordinates of the vertex of the graph of each quadratic function. Then graph the function.

13. $y = 2x^2$
14. $y = 3x^2$
15. $y = 8x^2$
16. $y - x^2 + 2x$
17. $y = 3x^2 - 6x$
18. $y = -2x^2 + 1$
19. $y = 4x^2 - x + 2$
20. $y = -x^2 + 6x - 4$
21. $y = \frac{1}{2}x^2 - x + 1$
22. $y = -\frac{1}{2}x^2 - 2x + 2$
23. $y = 2x^2 + 3x - 1$
24. $y = -x^2 - 2x - 1$

Match each function with its graph.

25. $y = (x + 2)^2 - 2$
26. $y = -2x^2 - x - 1$
27. $y = -2x^2 + x + 2$

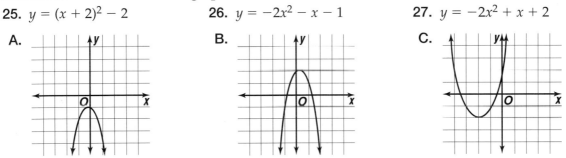

A. B. C.

Lesson 11–2 *(Pages 464–467)* Graph each group of equations on the same screen. Compare and contrast the graphs.

1. $y = x^2$
   $y = 2x^2$
   $y = 4x^2$
2. $y = x^2 + 1$
   $y = x^2 - 4$
   $y = x^2 - 8$
3. $y = (2x - 1)^2$
   $y = (2x - 2)^2$
   $y = (2x - 3)^2$

Describe how each graph changes from the parent graph of $y = x^2$. Then name the vertex of each graph.

4. $y = 8x^2$
5. $y = -2x^2$
6. $y = \frac{1}{2}x^2$
7. $y = (x - 1)^2$
8. $y = (x + 3)^2$
9. $y = (3x + 2)^2$
10. $y = -(x + 3)^2$
11. $y = -2x^2 - 3$
12. $y = 0.25x^2 + 0.5$
13. $y = -3x^2 + 8$
14. $y = (4x - 1)^2 + 3$
15. $y = (x + 2)^2 - 2$

Lesson 11–3 *(Pages 468–473)* Solve each equation by graphing the related function. If exact roots cannot be found, state the consecutive integers between which the roots are located.

1. $-x^2 - 2x + 24 = 0$ **2.** $x^2 + 4x - 5 = 0$ **3.** $x^2 - 4x - 2 = 0$ **4.** $x^2 + 7x + 3 = 0$

5. $x^2 + 7x + 5 = 0$ **6.** $x^2 - 6x + 6 = 0$ **7.** $-x^2 - 3x - 2 = 0$ **8.** $4x^2 + 2x + 1 = 0$

9. $-x^2 - 6x - 6 = 0$ **10.** $x^2 - 11x + 4 = 0$ **11.** $-2x^2 - 2x + 10 = 0$ **12.** $x^2 + 12x + 20 = 0$

13. $-x^2 - 12x - 3 = 0$ **14.** $-3x^2 - x + 8 = 0$ **15.** $x^2 + 3x + 4 = 0$ **16.** $x^2 + 5x - 9 = 0$

Use a quadratic equation to determine the two numbers that satisfy each situation.

17. Their sum is 21 and their product is 104.

18. Their difference is 8 and their product is 20.

19. Their sum is 13 and their product is 22.

20. Their sum is 32 and their product is 135.

Lesson 11–4 *(Pages 474–477)* Solve each equation. Check your solution.

1. $2r(r - 4) = 0$ **2.** $4k(k + 5) = 0$ **3.** $(s + 4)(s - 3) = 0$

4. $(m - 4)(m - 5) = 0$ **5.** $(3x - 4)(x - 2) = 0$ **6.** $(2y + 2)(2y - 4) = 0$

7. $(t + 2)(6t + 1) = 0$ **8.** $n^2 + n - 6 = 0$ **9.** $k^2 + 4k + 4 = 0$

10. $p^2 - 5p + 6 = 0$ **11.** $q^2 - 2q - 15 = 0$ **12.** $x^2 + 2x - 3 = 0$

13. $j^2 + 9j + 20 = 0$ **14.** $r^2 - 16r = 0$ **15.** $z^3 - 25z = 0$

For each problem, define a variable. Then use an equation to solve the problem.

16. Find two integers whose sum is 15 and whose product is 36.

17. The length of a swimming pool is 15 feet longer than it is wide. The area in square feet is 1350. Find the dimensions of the pool.

18. Find two integers whose difference is 12 and whose product is 13.

Lesson 11–5 *(Pages 478–482)* Find the value of c that makes each trinomial a perfect square.

1. $r^2 - 16r + c$ **2.** $k^2 + 12k + c$ **3.** $p^2 - 4p + c$ **4.** $n^2 + 2n + c$

5. $f^2 - 8f + c$ **6.** $s^2 + 18s + c$ **7.** $x^2 + 20x + c$ **8.** $r^2 + 14r + c$

9. $w^2 + 30w + c$ **10.** $h^2 + 10h + c$ **11.** $z^2 - 2z + c$ **12.** $m^2 - 6m + c$

13. $q^2 + 26q + c$ **14.** $t^2 + 28t + c$ **15.** $y^2 + 22y + c$ **16.** $z^2 + 24z + c$

Solve each equation by completing the square.

17. $z^2 + 10z + 12 = 0$ **18.** $h^2 - 8h - 15 = 0$

19. $y^2 + 3y + 1 = 0$ **20.** $w^2 + 15w = 5$

21. $m^2 + 2m = 0$ **22.** $t^2 + 2t = 18$

23. $r^2 - 20r + 24 = 0$ **24.** $p^2 - 2p = 32$

25. $q^2 - 7q + 12 = 0$ **26.** $n^2 - 4n - 16 = 0$

27. $x^2 + 10x = 12$ **28.** $r^2 + 12r = 0$

Lesson 11–6 *(Pages 483–488)* Use the Quadratic Formula to solve each equation.

1. $j^2 + 3j - 4 = 0$ **2.** $w^2 + 9w + 20 = 0$ **3.** $m^2 - 7m + 12 = 0$ **4.** $n^2 + 5n - 6 = 0$

5. $k^2 + 2k - 15 = 0$ **6.** $z^2 - 4z + 4 = 0$ **7.** $d^2 - 3d - 18 = 0$ **8.** $s^2 + 8s - 14 = 0$

9. $x^2 - 5x - 24 = 0$ **10.** $t^2 + 5t + 6 = 0$ **11.** $r^2 + 6r + 9 = 0$ **12.** $y^2 - 2y - 8 = 0$

13. $y^2 - y - 6 = 0$ **14.** $d^2 + 4d - 21 = 0$ **15.** $s^2 + 3s + 2 = 0$ **16.** $m^2 - 11m = -18$

17. $4k^2 - 4 = 8k$ **18.** $-3r^2 - 15r = -18$ **19.** $3p^2 + 5p + 2 = 0$ **20.** $6x^2 + 8x = 8$

21. $2p^2 + 5p = 3$ **22.** $4q^2 - 2 = -2q$ **23.** $-2c^2 - 4c - 2 = 0$ **24.** $-8m^2 - 10m = 2$

Lesson 11–7 *(Pages 489–493)* Graph each exponential function. Then state the y-intercept.

1. $y = 2^x$ **2.** $y = 5^x$ **3.** $y = 3^x - 1$ **4.** $y = 4^x + 4$

5. $y = 2^x - 3$ **6.** $y = 2^x + 3$ **7.** $y = 4^x + 3$ **8.** $y = 2^x - 6$

9. $y = 3^x - 3$ **10.** $y = 2^{4x} + 1$ **11.** $y = 4^{0.5x}$ **12.** $y = 5^{3x} + 1$

13. $y = 4^{3x} - 2$ **14.** $y = 3^{3x} - 2$ **15.** $y = 2^{0.5x} - 1$ **16.** $y = 2^{3x} - 4$

Find the amount of money in a bank account given the following conditions.

17. initial deposit = $6000, annual rate = 6.5%, time = 3 years

18. initial deposit = $1000, annual rate = 12%, time = 15 years

19. initial deposit = $2500, annual rate = 2%, time = 6 years

20. initial deposit = $5100, annual rate = 9%, time = 4 years

Lesson 12–1 *(Pages 504–508)* Write an inequality to describe each number.

1. a number less than 10

2. a number that is at least -4

3. a number greater than -2

4. a number less than or equal to 5

5. a number greater than 3

6. a number more than 12

7. a minimum number of 7

8. a number less than -1

9. a maximum number of 8

10. a number greater than -6

11. a minimum number of -8

12. a number more than -11

Graph each inequality on a number line.

13. $m < 8$ **14.** $n \le -4$ **15.** $x > 2$ **16.** $z < -7$

17. $r \ge 15$ **18.** $m \le -5$ **19.** $s > 8.4$ **20.** $y > 6.2$

21. $w \le 1.3$ **22.** $\ell \ge -2.4$ **23.** $p < -3.2$ **24.** $j > 4.3$

25. $t < \dfrac{1}{3}$ **26.** $r \le -2\dfrac{1}{4}$ **27.** $y \le \dfrac{1}{2}$ **28.** $q \ge 3\dfrac{5}{8}$

Write an inequality for each graph.

29. 5 6 7 8 9

30. 0 1 2 3 4

31. -5 0 5 10 15

32. 6 7 8 9 10

33. -6 -5 -4 -3 -2

34. -5 -4 -3 -2 -1

35. 4 5 6 7 8

36. -4 -3 -2 -1 0

37. 10 11 12 13 14

EXTRA PRACTICE

Lesson 12–2 (*Pages 509–513*) Solve each inequality. Check your solution.

1. $r + 3 < 8$
2. $m + 4 > -2$
3. $j - 3 > 5$
4. $k - 6 < -13$
5. $-12 + w < 15$
6. $p + 11 \geq 5$
7. $0.4 + p \geq 1.2$
8. $x - 6.2 < 4$
9. $4.3 < 2.1 + y$
10. $\frac{1}{4} + r \leq 3\frac{3}{4}$
11. $k + 10.6 \geq -3.4$
12. $\frac{1}{6} + s \leq \frac{1}{3}$

Solve each inequality. Graph the solution.

13. $2n > n + 3$
14. $8 + 5x > 6x$
15. $3y - 2 < 4y$
16. $5d < 8 + 4d$
17. $9 \geq 3t - 4 - 2t$
18. $-2a < -3(a - 2)$
19. $6u \leq 5(u + 1)$
20. $11p \leq 2(5p + 4)$
21. $7s < 3(2s - 3)$
22. $3(x - 3) \leq 4x$
23. $7b < 4(2b - 3)$
24. $2c + 3 \leq 3(c - 5)$

Lesson 12–3 (*Pages 514–518*) Solve each inequality. Check your solution.

1. $-3k < 15$
2. $2r > -10$
3. $-4d \geq -12$
4. $\frac{x}{5} \geq 6$
5. $-\frac{y}{4} < 8$
6. $\frac{p}{7} \leq -3$
7. $-9\ell < 27$
8. $2s > 20$
9. $-t > 11$
10. $-\frac{d}{2} > 12$
11. $\frac{m}{6} \leq -5$
12. $-\frac{b}{4} < -9$
13. $-3p < 2$
14. $-8y \geq -4$
15. $7v \leq 3$
16. $2 < \frac{2}{3}j$
17. $\frac{5}{8}t \geq -5$
18. $-\frac{1}{4}w < 6$
19. $0.01r \leq 8$
20. $2.1m > 6.3$
21. $-2.25j \geq 9$
22. $\frac{k}{7.2} < -3$
23. $-\frac{r}{2.5} > 4$
24. $\frac{y}{0.4} \geq 2$

Lesson 12–4 (*Pages 519–523*) Solve each inequality. Check your solution.

1. $2a + 7 \leq 11$
2. $5r - 3 > 27$
3. $6 - 4m > 10$
4. $1 - 3s \geq 13$
5. $3 + 5d \leq -12$
6. $-8x - 1 < 15$
7. $6 + 2b \geq -2.4$
8. $5.1x - 2.4 < -7.5$
9. $3.3 + 4k > -8.7$
10. $\frac{6-r}{8} < 7$
11. $\frac{j}{3} + 7 > 9$
12. $\frac{4 + 3m}{7} \leq -5$
13. $2x - 1 < 8x + 2$
14. $11 \leq -(t + 4)$
15. $4(3 - 6r) > 18$
16. $\frac{2}{3}(k - 6) > 4$
17. $\frac{1}{3}(y + 3) < \frac{1}{2}(y - 2)$
18. $\frac{1}{8}(m + 3) \geq \frac{1}{4}(m - 3)$

Write and solve an inequality for each situation.

19. Three fifths times the sum of a number and 5 is greater than 15.
20. Four times the difference of a number and 3 is less than 24.

716 Extra Practice

Lesson 12–5 (Pages 524–529) Write each compound inequality without using *and*.

1. $m > 2$ and $m < 6$
2. $j > -12$ and $j < 4$
3. $r < 6$ and $r \geq -1$
4. $x < 3$ and $x \geq 2$
5. $y \leq 5$ and $y \geq -3$
6. $s \geq -7$ and $s < -4$

Graph the solution of each compound inequality.

7. $w > 6$ or $w < 2$
8. $a < -4$ or $a \geq 4$
9. $z > 12$ and $z \leq 15$
10. $h \leq 20$ and $h \geq -3$
11. $s < -7$ or $s > -5$
12. $f \leq 8$ or $f > 9$

Solve each compound inequality. Graph the solution.

13. $4 < 2n < 10$
14. $-4 \leq x + 7 < 9$
15. $12 > v - 6 > -2$
16. $9 \leq 3q \leq 21$
17. $24 > b - 5 > -2$
18. $8 > c + 4 \geq 5$
19. $t + 3 > 15$ or $t - 5 < -12$
20. $-12 \leq -3d < 9$
21. $5.4 < 0.2k < 6$
22. $u - 12 > -3$ or $u + 11 < 2$
23. $-4 \leq g - \frac{1}{3} < 1$
24. $\frac{p}{3} > 3$ or $\frac{p}{3} \leq -2$

Lesson 12–6 (Pages 530–534) Solve each inequality. Graph the solution.

1. $|d + 7| > 15$
2. $|5c| > 30$
3. $|a - 2| < 17$
4. $|z + 3| > 12$
5. $|j - 12| < 10$
6. $|\ell + 8| \leq -14$
7. $|t - 6| < 9$
8. $|m - 4| \leq 3$
9. $|4x| < -20$
10. $|s + 5| < 8$
11. $|n + 1| \geq 7$
12. $|-2v| > 14$
13. $|y - 3| > 16$
14. $|r - 7| \leq -11$
15. $|6b| > 18$
16. $|w - 4| \leq 1.5$

For each graph, write an inequality involving absolute value.

17.
```
<-+--+--●==●==●==●--+--+--+--+->
 -7-6-5-4-3-2-1 0 1 2 3
```
18.
```
<-+--○==+==+==+==+==+==+==○--+->
   -3-2-1 0 1 2 3 4 5 6 7
```
19.
```
<-+--+--+--+--+--●==●==●--+--+->
 -1 0 1 2 3 4 5 6 7 8 9
```

Write an inequality involving absolute value for each statement. Do not solve.

20. Alli's quiz score was within 5 points of her average of 85.
21. The 5-inch-wide picture frame was made with an accuracy of 0.1 inch.

Lesson 12–7 (Pages 535–539) Graph each inequality.

1. $x < -3$
2. $y > -6$
3. $y \geq x - 3$
4. $y < x + 2$
5. $y \geq -4x$
6. $6 \geq 2x + 4y$
7. $y > -2x + 6$
8. $x + 5y \geq 15$
9. $x + y \leq -2$
10. $-3x + 2 > y$
11. $y \leq -2(x - 1)$
12. $8 \geq y - 4x$
13. $y > -x - 8$
14. $7 + 3y \leq x$
15. $2x - y \geq -6$
16. $x - y \leq 5$
17. $3(6x + y) < 4$
18. $y \leq 4(x - 2)$
19. $-(4x - 3) \geq 2y$
20. $-2(3x + y) < 1$

For Exercises 21–24, write an inequality and graph the solution.

21. The difference of a number and three is less than or equal to eight.
22. Three times a number is greater than negative six.
23. One half the sum of a number and eight is greater than or equal to twelve.
24. A number minus three is less than another number.

Lesson 13–1 *(Pages 550–553)* Solve each system of equations by graphing.

1. $x = -2$
  $y = 3$

2. $x = 5$
  $y = x$

3. $x = -1$
  $y = x - 2$

4. $x = -3$
  $y = x + 4$

5. $y = -x - 4$
  $y = x + 4$

6. $y = x + 2$
  $y = 2x - 1$

7. $y = -x - 1$
  $y = x - 1$

8. $x = 2$
  $x + 2y = 4$

9. $y = 4x + 1$
  $y = 3x$

10. $y = -3x - 2$
  $2x - y = 2$

11. $y = 2$
  $x - y = 3$

12. $x - y = 6$
  $2x + y = 3$

13. $y = -\frac{1}{2}x$
  $y = 3x + 7$

14. $y = \frac{1}{2}x - 3$
  $x - y = 6$

15. $\frac{1}{4}x - 2 = y$
  $x = -4y$

16. $2x + 3y = 12$
  $4x + y = 4$

Lesson 13–2 *(Pages 554–559)* State whether each system is *consistent and independent*, *consistent and dependent*, or *inconsistent*.

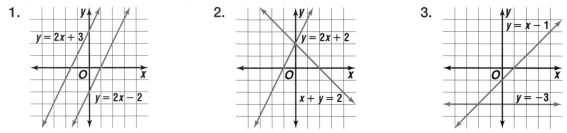

1. $y = 2x + 3$, $y = 2x - 2$

2. $y = 2x + 2$, $x + y = 2$

3. $y = x - 1$, $y = -3$

Determine whether each system of equations has *one* solution, *no* solution, or *infinitely many* solutions by graphing. If the system has one solution, name it.

4. $y = 2x + 1$
  $y = -x - 2$

5. $x = 5$
  $x - 4y = 1$

6. $y = 3$
  $x - y = 2$

7. $y = x + 7$
  $x = 7 - y$

8. $y = 2x - 6$
  $y = 2x + 4$

9. $y = 10x - 16$
  $y = 4x - 4$

10. $x + 2y = 5$
  $x + y = 4$

11. $x + 4y = 5$
  $2x + 6y = 6$

12. $y = \frac{1}{4}x - 1$
  $y = -x + 4$

13. $\frac{1}{3}y = -x - \frac{1}{3}$
  $y = -3x - 1$

14. $\frac{1}{3}x - 6y = 9$
  $x - 18y = 27$

15. $y = \frac{3}{2}x$
  $y = -\frac{2}{3}x + 13$

16. $x + y = 12$
  $2x - y = 3$

17. $x - 2y = -3$
  $-2x + 4y = 8$

18. $4x + 6y = 12$
  $2x - 6 = -3y$

19. $3x + 3y = 6$
  $4x - y = 3$

Lesson 13–3 *(Pages 560–565)* Use substitution to solve each system of equations.

1. $y = -x$
  $x - y = -12$

2. $x = 5 - y$
  $-3x + y = -3$

3. $y = x + 1$
  $-x + 3y = -15$

4. $y = x + 2$
  $2x - y = 1$

5. $y = 4x$
  $x + y = 3$

6. $x = 3y$
  $x + 3y = 4$

7. $y = 2x$
  $2x - y = -1$

8. $y = 3x - 1$
  $x + y = 4$

9. $y = 6x - 7$
  $-2x - y = 3$

10. $x = 6 - y$
  $x - 2y = 1$

11. $x + 4y = -8$
  $3x - 6y = 0$

12. $x = 3 + y$
  $3y + 2 = x$

13. $x - 7y = 0$
  $2x + y = 0$

14. $x - 4y = 8$
  $6y + 8 = 2x$

15. $6x - 3y = -2$
  $y + 1 = x$

16. $x - \frac{1}{2}y = 14$
  $x + \frac{1}{2}y = 2$

Lesson 13–4 (Pages 566–571) Use elimination to solve each system of equations.

1. $x + y = 4$
$x - y = 16$

2. $x + 2y = 4$
$x + 3y = 6$

3. $x = 13 - y$
$x - y = -3$

4. $x - 6y = -1$
$x + 3y = -10$

5. $x - 3y = 7$
$x + 2y = 8$

6. $x - 4y = 5$
$-5x + 10y = -5$

7. $x - y = -2$
$2x + 2y = 16$

8. $x - y = 14$
$3x - 6y = 15$

9. $x - 13y = -2$
$7y - x = 5$

10. $x - 6y = 9$
$4x - 8y = 12$

11. $x + y = 8$
$16 + 3x = y$

12. $x - y = 9$
$4x - 2y = -8$

13. $x = 5 - y$
$6x - 4y = 0$

14. $x = 6 - y$
$2x - 8y = 2$

15. $4x + 3y = 12$
$8x - 4y = -12$

16. $6x - 3y = 18$
$6x + 3y = 12$

Lesson 13–5 (Pages 572–577) Use elimination to solve each system of equations.

1. $x + 5y = 10$
$x + y = -6$

2. $x + 4y = 5$
$x - 2y = -7$

3. $x - 6y = 9$
$x + 3y = -9$

4. $-x - y = 4$
$x + y = -4$

5. $2x + y = -2$
$2x - y = 4$

6. $6x + y = -9$
$3x - y = 6$

7. $3x - y = 0$
$4x + 2y = 10$

8. $2x + 12y = 24$
$x + 7y = 18$

9. $4x - 8y = 0$
$x + 3y = -10$

10. $2x - 8y = 0$
$-x + 7y = -3$

11. $x - 8y = 5$
$-2x + 8y = -2$

12. $15x - y = 4$
$-6x + y = 5$

13. $4x - 16y = 24$
$2x + 2y = 12$

14. $x - 13y = 4$
$-2x + 10y = -4$

15. $-7x - y = 6$
$-3x + y = 4$

16. $6x - 3y = 18$
$10x + y = -12$

Lesson 13–6 (Pages 580–585) Solve each system of equations by graphing.

1. $x = 4$
$y = x - 1$

2. $x = -2$
$y = x^2$

3. $x = -3$
$y = x^2 + 3$

4. $y = x^2$
$y = 3x$

5. $y = x^2$
$y = 2x$

6. $y = x^2 + 2$
$y = x + 4$

7. $y = x^2 + 1$
$y = -x + 1$

8. $y = x^2 + 3$
$y = 5x - 3$

9. $y = 4x^2$
$y = 8x$

10. $y = x^2 + 12$
$y = -x - 8$

11. $y = x^2 + 4$
$y = 3x + 2$

12. $y = -x^2 + 3$
$y = 3x + 5$

Use substitution to solve each system of equations.

13. $y = 2x^2$
$x = -2$

14. $y = -x^2$
$y = 5x + 6$

15. $y = -2x^2$
$y = 4x - 6$

16. $y = -3x^2$
$x = -5$

17. $y = \frac{1}{2}x^2 - 1$
$x = 4$

18. $y = -\frac{1}{2}$
$y = \frac{1}{2}x^2 - \frac{5}{2}$

19. $y = -\frac{1}{2}x^2 + \frac{7}{2}$
$x = 2x + 1$

20. $y = \frac{1}{4}x^2 - 2$
$y = 2$

21. $y = -x + 4$
$y = x^2 + 2$

22. $y = x^2 + 3$
$y = -7x^2 + 5$

23. $y = x^2 + x + 4$
$y = 6$

24. $y = 3x^2 + 6x - 9$
$y = -12$

Lesson 13–7 *(Pages 586–591)* Solve each system of inequalities by graphing.

1. $y > -2$
$y < 3$

2. $x \le -4$
$y \ge -1$

3. $y < 0$
$x \ge -1$

4. $y \ge -2$
$x > 2$

5. $y > x$
$y > x + 1$

6. $y < x - 2$
$y > -1$

7. $y < -2$
$y > -2x$

8. $y \ge x$
$y \le -3x$

9. $x < 3$
$y < x + 1$

10. $y \ge -2$
$y \le -2x - 1$

11. $x + y > 0$
$2x + y \le 1$

12. $y \ge x + 1$
$y < 2x + 2$

13. $y - 2 \le x$
$y + 4 \ge x$

14. $y \le 3x - 1$
$2x + 4 < y$

15. $2x + 4 > y$
$y - 2 \le x$

16. $-3y - x > 2$
$2y + x < 0$

Lesson 14–1 *(Pages 600–605)* Name the set or sets of numbers to which each real number belongs. Let N = natural numbers, W = whole numbers, Z = integers, Q = rational numbers, and I = irrational numbers.

1. $\frac{21}{7}$

2. $\sqrt{13}$

3. $\sqrt{121}$

4. $47.13013001\ldots$

5. $-\sqrt{38}$

6. 0.631

7. $\frac{1}{4}$

8. $-\frac{8}{2}$

9. -3

10. $-\frac{1}{6}$

11. $0.949949994\ldots$

12. $-\sqrt{64}$

Find an approximation, to the nearest tenth, for each square root. Then graph the square root on a number line.

13. $-\sqrt{5}$
14. $\sqrt{14}$
15. $\sqrt{18}$
16. $\sqrt{29}$
17. $\sqrt{63}$
18. $-\sqrt{71}$
19. $\sqrt{82}$
20. $-\sqrt{93}$
21. $\sqrt{102}$
22. $\sqrt{145}$
23. $-\sqrt{201}$
24. $\sqrt{305}$

Determine whether each number is *rational* or *irrational*. If it is irrational, find two consecutive integers between which its graph lies on the number line.

25. $\sqrt{4}$
26. $-\sqrt{19}$
27. $-\sqrt{62}$
28. $\sqrt{33}$
29. $-\sqrt{54}$
30. $\sqrt{49}$
31. $-\sqrt{225}$
32. $-\sqrt{15}$
33. $\sqrt{196}$
34. $-\sqrt{152}$
35. $-\sqrt{181}$
36. $\sqrt{8}$

Lesson 14–2 *(Pages 606–611)* Find the distance between each pair of points. Round to the nearest tenth, if necessary.

1. $Q(2, 16), R(3, 15)$
2. $C(-4, -2), D(-6, -5)$
3. $P(-8, 1), Q(10, -7)$
4. $A(3, -9), B(1, -2)$
5. $G(-6, -14), H(-7, 2)$
6. $X(0, 0), Y(9, 9)$
7. $E(12, -2), F(-3, -4)$
8. $M(15, 3), N(8, -11)$
9. $V(-4, 4), W(6, -6)$

Find the value of a if the points are the indicated distance apart.

10. $A(a, 3), B(6, 5); d = 2$
11. $G(-1, 5), H(-8, a); d = \sqrt{85}$
12. $X(9, a), Y(5, -2); d = 4$
13. $P(6, 1), Q(a, -7); d = \sqrt{113}$
14. $C(-9, -2), D(0, a); d = \sqrt{90}$
15. $Q(a, -1), R(4, 5); d = 10$
16. $E(7, a), F(-2, 4); d = \sqrt{90}$
17. $M(a, 3), N(-1, 5); d = \sqrt{8}$
18. $V(-3, -3), W(a, 4); d = \sqrt{50}$

Lesson 14–3 (*Pages 614–619*) Simplify each expression. Leave in radical form.

1. $\sqrt{54}$
2. $\sqrt{80}$
3. $\sqrt{75}$
4. $\sqrt{300}$

5. $\sqrt{3} \cdot \sqrt{12}$
6. $\sqrt{4} \cdot \sqrt{8}$
7. $2\sqrt{6} \cdot \sqrt{18}$
8. $\sqrt{10} \cdot \sqrt{15}$

9. $\dfrac{\sqrt{18}}{\sqrt{9}}$
10. $\dfrac{\sqrt{12}}{\sqrt{4}}$
11. $\dfrac{\sqrt{20}}{\sqrt{8}}$
12. $\dfrac{\sqrt{75}}{\sqrt{5}}$

13. $\dfrac{4}{2 + \sqrt{12}}$
14. $\dfrac{6}{3 - \sqrt{8}}$
15. $\dfrac{3}{5 + \sqrt{3}}$
16. $\dfrac{5}{4 + \sqrt{10}}$

Simplify each expression. Use absolute value symbols if necessary.

17. $\sqrt{18a^2b}$
18. $\sqrt{21m^8}$
19. $\sqrt{120x^2y}$
20. $\sqrt{50jk}$

21. $\sqrt{16r^2s^3}$
22. $\sqrt{17n^3}$
23. $\sqrt{32c^4d^3}$
24. $\sqrt{12xy^2}$

25. $\sqrt{53m^2n^4}$
26. $\sqrt{51g^2h^2}$
27. $\sqrt{66mn^4}$
28. $\sqrt{48j^6k^2}$

29. $\sqrt{8a^2bc^2}$
30. $\sqrt{75x^6y^6}$
31. $\sqrt{30rs^2t^4}$
32. $\sqrt{196a^6}$

Lesson 14–4 (*Pages 620–623*) Simplify each expression.

1. $5\sqrt{3} + 7\sqrt{3}$
2. $4\sqrt{6} - 2\sqrt{6}$
3. $5\sqrt{7} + 2\sqrt{7}$

4. $13\sqrt{5} - 5\sqrt{5}$
5. $-11\sqrt{3} + 6\sqrt{2}$
6. $-6\sqrt{5} - 5\sqrt{5}$

7. $4\sqrt{2} + 3\sqrt{2} - 5\sqrt{2}$
8. $-3\sqrt{3} - 2\sqrt{3} - 4\sqrt{3}$
9. $2\sqrt{7} + 8\sqrt{7} - 14\sqrt{7}$

10. $4\sqrt{12} - 7\sqrt{3}$
11. $-2\sqrt{24} + 3\sqrt{6}$
12. $-6\sqrt{32} + 4\sqrt{8}$

13. $4\sqrt{2} - 3\sqrt{3} + 6\sqrt{2}$
14. $-7\sqrt{5} + 6\sqrt{3} - 2\sqrt{3}$
15. $\sqrt{105} - \sqrt{12} - \sqrt{18}$

16. $-8\sqrt{24} + 6\sqrt{12} - 3\sqrt{2}$
17. $4\sqrt{27} - 2\sqrt{48} + 3\sqrt{20}$
18. $\sqrt{52} - \sqrt{18} + \sqrt{120}$

19. If an equilateral triangle has a side of length $5\sqrt{6}$, what is the measure of the perimeter?

Lesson 14–5 (*Pages 624–629*) Solve each equation. Check your solution.

1. $\sqrt{m} = 4$
2. $\sqrt{y} = -2$
3. $\sqrt{2d} = 4$

4. $-\sqrt{5k} = 15$
5. $\sqrt{8t} = 4$
6. $\sqrt{r - 9} = 9$

7. $\sqrt{t + 5} = 7$
8. $\sqrt{n - 3} = 2$
9. $\sqrt{h + 5} = 0$

10. $\sqrt{2x + 6} = 6$
11. $\sqrt{4m + 5} - 6 = 2$
12. $\sqrt{3r - 2} + 5 = 4$

13. $r = \sqrt{r + 12}$
14. $k = \sqrt{3k + 10}$
15. $x - 10 = \sqrt{x + 2}$

16. $m = \sqrt{9m + 4} - 2$
17. $2 + \sqrt{4d + 4} = d$
18. $5 - \sqrt{5r - 6} = 9 - r$

Lesson 15–1 *(Pages 638–643)* **Find the excluded value(s) for each rational expression.**

1. $\dfrac{y}{y-6}$

2. $\dfrac{2m}{3m+6}$

3. $\dfrac{-n}{-8+4n}$

4. $\dfrac{j+1}{2j-6}$

5. $\dfrac{6}{r(r-2)}$

6. $\dfrac{8k}{k(k+1)}$

7. $\dfrac{2x}{(x-3)(x+1)}$

8. $\dfrac{7m}{(m+1)(m+1)}$

9. $\dfrac{2s}{s^2+2s}$

10. $\dfrac{q-2}{q^2-25}$

11. $\dfrac{3+p}{p^2-p-2}$

12. $\dfrac{x^2-2x+3}{x^2+x-6}$

Simplify each rational expression.

13. $\dfrac{8}{22}$

14. $\dfrac{6}{24}$

15. $\dfrac{30a}{36b}$

16. $\dfrac{2y^2}{4y}$

17. $\dfrac{2(x+1)}{x(x+1)}$

18. $\dfrac{(m-1)(m+2)}{(m+2)(m+3)}$

19. $\dfrac{2d^2-2d}{d^2-1}$

20. $\dfrac{-4j+12}{j^2-9}$

21. $\dfrac{n^2+2n-15}{n^2+3n-10}$

22. $\dfrac{2y^2+2y}{y^2-y-2}$

23. $\dfrac{x^2+2x-8}{x^2-7x+10}$

24. $\dfrac{r^2-r-6}{r^3-6r^2+9r}$

Lesson 15–2 *(Pages 644–649)* **Find each product.**

1. $\dfrac{xy}{x} \cdot \dfrac{z}{y}$

2. $\dfrac{2s}{3t} \cdot \dfrac{6}{4s^2}$

3. $\dfrac{7b^2}{c^2} \cdot \dfrac{3c}{b^2}$

4. $\dfrac{2p+1}{p^2} \cdot \dfrac{2p^2}{4p+2}$

5. $\dfrac{x(x+5)}{3} \cdot \dfrac{3}{x(2x+10)}$

6. $\dfrac{6r}{r+2} \cdot \dfrac{4r+8}{18}$

7. $\dfrac{3n+6}{n} \cdot \dfrac{n^2}{n^2+4n+4}$

8. $\dfrac{d^2+8d+16}{d^3} \cdot \dfrac{d^2}{d+4}$

9. $\dfrac{m}{m^2+4m+3} \cdot \dfrac{m+1}{m}$

Find each quotient.

10. $\dfrac{m}{n^2} \div \dfrac{2}{mn}$

11. $\dfrac{cd}{3a^2b} \div \dfrac{c^2}{3a^2}$

12. $6rs \div \dfrac{3r^2}{s}$

13. $\dfrac{2}{t-1} \div \dfrac{t}{2t-2}$

14. $\dfrac{y^2-2y+1}{2y} \div (y-1)$

15. $\dfrac{c^2-4}{3} \div \dfrac{c-2}{c+2}$

16. $\dfrac{x^2-9}{x^2} \div \dfrac{x+3}{x}$

17. $\dfrac{d}{d^2-16} \div \dfrac{6d}{d^2+2d-8}$

18. $\dfrac{b^2-4}{16b^2} \div \dfrac{2b+4}{4b}$

Lesson 15–3 *(Pages 650–655)* **Find each quotient.**

1. $(6x-3) \div (2x-1)$

2. $(m^2+4m) \div (m+4)$

3. $(16a^2+8a) \div (4a+2)$

4. $(k^2-5k-6) \div (k+1)$

5. $(s^2-5s+6) \div (s-3)$

6. $(2y^2-2y-24) \div (2y-8)$

7. $(m^2-6m+9) \div (m-3)$

8. $(r^2+3r-18) \div (r-3)$

9. $(s^2-3s-10) \div (s+2)$

10. $(4r^3-12r^2) \div (r-3)$

11. $(6p^3-10p^2) \div (3p-5)$

12. $(x^2+12) \div (x-3)$

13. $(t^2+8) \div (t-2)$

14. $(y^2+16) \div (y+4)$

15. $(4p^3-5p-3) \div (2p+3)$

Lesson 15-4 *(Pages 656–661)* Find each sum or difference. Write in simplest form.

1. $\dfrac{9}{r} + \dfrac{6}{r}$

2. $\dfrac{2a}{3} + \dfrac{5a}{3}$

3. $\dfrac{4}{j} - \dfrac{3}{j}$

4. $\dfrac{2b}{4} - \dfrac{4b}{4}$

5. $\dfrac{3}{2s} + \dfrac{1}{2s}$

6. $\dfrac{6}{3b} - \dfrac{9}{3b}$

7. $\dfrac{2a}{a} + \dfrac{7a}{a}$

8. $\dfrac{y}{16y} - \dfrac{y}{16y}$

9. $\dfrac{15}{8x} + \dfrac{1}{8x}$

10. $\dfrac{2}{9k} + \dfrac{5}{9k}$

11. $\dfrac{4}{6p} + \dfrac{10}{6p}$

12. $\dfrac{14}{25r} - \dfrac{19}{25r}$

13. $\dfrac{11}{10s} - \dfrac{12}{10s}$

14. $\dfrac{9}{2y} - \dfrac{3}{2y}$

15. $\dfrac{6}{6+m} + \dfrac{m}{6+m}$

16. $\dfrac{6s}{2-s} + \dfrac{3}{2-s}$

17. $\dfrac{2p}{3+j} - \dfrac{p}{3+j}$

18. $\dfrac{16}{m-5} - \dfrac{10}{m-5}$

19. $\dfrac{3r}{r+3} + \dfrac{9}{r+3}$

20. $\dfrac{-2x-4}{x+7} + \dfrac{2x+11}{x+7}$

21. $\dfrac{8}{6m-3} - \dfrac{4}{6m-3}$

22. $\dfrac{4x+5}{2x+6} + \dfrac{2-4x}{2x+6}$

23. $\dfrac{8-6r}{5r-2} + \dfrac{11r-10}{5r-2}$

24. $\dfrac{12t+1}{t+1} + \dfrac{3-8t}{t+1}$

Lesson 15-5 *(Pages 662–667)* Find the LCM for each pair of expressions.

1. $4mn, 6m^2$

2. $8x^2y, 20xy$

3. $4cd, 18d$

4. $d^2 - 9, d + 1$

5. $a^2 + a - 2, a^2 + 5a - 6$

6. $x^2 + x - 6, x^2 - x - 12$

Write each pair of rational expressions with the same LCD.

7. $\dfrac{2}{6mn}, \dfrac{1}{30}$

8. $\dfrac{5}{4p}, \dfrac{3}{2p}$

9. $-\dfrac{1}{6rs}, \dfrac{3}{4st}$

10. $\dfrac{5a}{3a^2b}, -\dfrac{4}{2a}$

11. $\dfrac{1}{y-2}, -\dfrac{3}{y^2-4}$

12. $\dfrac{m^2}{2m+4}, \dfrac{6m}{4m+8}$

Find each sum or difference in simplest form.

13. $\dfrac{r}{8} - \dfrac{r}{3}$

14. $\dfrac{2p}{6} - \dfrac{p}{5}$

15. $-\dfrac{r}{4} + \dfrac{3r}{7}$

16. $\dfrac{6t}{9} + \dfrac{t}{6}$

17. $\dfrac{2m}{5p^2} + \dfrac{2}{6p}$

18. $\dfrac{5x}{x-1} + \dfrac{7}{x}$

19. $\dfrac{6m}{j^2} + \dfrac{7m}{3j}$

20. $\dfrac{2t}{3+r} - \dfrac{9}{r^2}$

21. $\dfrac{4}{b^3} + \dfrac{3}{b}$

22. $\dfrac{6y+1}{2y+1} - \dfrac{1}{4y+2}$

23. $\dfrac{1}{x^2+x-2} + \dfrac{3}{x+2}$

24. $\dfrac{4}{b^2-16} + \dfrac{8}{b+4}$

Lesson 15-6 *(Pages 668–673)* Solve each equation. Check your solution.

1. $2 - \dfrac{2x}{5} + \dfrac{x}{10}$

2. $\dfrac{y}{5} + \dfrac{3y}{5} = \dfrac{4}{5}$

3. $\dfrac{m}{8} - \dfrac{3m}{4} = \dfrac{1}{2}$

4. $\dfrac{d}{5} = \dfrac{6d}{5} - \dfrac{3}{10}$

5. $\dfrac{8}{7v} + \dfrac{3}{4v} = -2$

6. $\dfrac{s+1}{2} + \dfrac{s}{4} = -2$

7. $\dfrac{2b-1}{2} = \dfrac{b}{3} - \dfrac{1}{4}$

8. $\dfrac{m-1}{3} + \dfrac{m+4}{3} = -5$

9. $\dfrac{4}{c+1} + \dfrac{2}{c+1} = -3$

10. $\dfrac{7}{3d} - \dfrac{5}{9d} = -3$

11. $\dfrac{1}{2m} + \dfrac{3}{m} = \dfrac{1}{4}$

12. $\dfrac{1}{r} + \dfrac{2}{r-5} = \dfrac{3}{r}$

13. $\dfrac{1}{3a+5} - \dfrac{4a}{6a+10} = \dfrac{1}{2}$

14. $\dfrac{1}{k-3} + \dfrac{k}{2} = \dfrac{k-2}{2}$

15. $\dfrac{r+2}{r^2-4} + \dfrac{1}{r-2} = \dfrac{1}{r}$

Graphing Calculator Tutorial

General Information

- Any yellow commands written above the calculator keys are accessed with the 2nd key, which is also yellow. Similarly, any green characters or commands above the keys are accessed with the ALPHA key, which is also green. In this text, commands that are accessed by the 2nd and ALPHA keys are shown in brackets. For example, 2nd [QUIT] means to press the 2nd key followed by the key below the yellow QUIT command.

- 2nd [ENTRY] copies the previous calculation so it can be edited or reused.

- 2nd [ANS] copies the previous answer so it can be used in another calculation.

- 2nd [QUIT] will return you to the home (or text) screen.

- 2nd [A-LOCK] allows you to use the green characters above the keys without pressing ALPHA before typing each letter. (This is handy for programming.)

- Negative numbers are entered using the (−) key, not the minus sign, − .

- The variable x can be entered using the X,T,θ,n key, rather than using ALPHA [X].

- 2nd [OFF] turns the calculator off.

Basic Keystrokes

Some commonly used mathematical functions are shown in the table below. As with any scientific calculator, graphing calculators observe the order of operations.

Mathematical Operation	Example	Keys	Display
evaluate expressions	Evaluate 2 + 5.	2 [+] 5 [ENTER]	$2 + 5$ 7
multiplication	Evaluate 3(9.1 + 0.8).	3 [(] 9.1 [+] .8 [)] [ENTER]	$3(9.1 + .8)$ 29.7
division	Evaluate $\frac{8 - 5}{4}$.	[(] 8 [−] 5 [)] [÷] 4 [ENTER]	$(8 - 5)/4$ .75
exponents	Find 3^5.	3 [∧] 5 [ENTER]	$3 \wedge 5$ 243
roots	Find $\sqrt{14}$.	[2nd] [√] 14 [ENTER]	$\sqrt{}$ (14 3.741657387
opposites	Enter −3.	[(−)] 3	-3
variable expressions	Enter $x^2 + 4x - 3$.	[X,T,θ,n] [x²] [+] 4 [X,T,θ,n] [−] 3	$x^2 + 4x - 3$

Key Skills

Each Graphing Calculator Exploration in the Student Edition requires the use of certain key skills. Use this section as a reference for further instruction.

A: Entering and Graphing Equations

Press [Y=]. Use the [X,T,θ,n] key to enter *any* variable for your equation. To see a graph of the equation, press [GRAPH].

B: Setting Your Viewing Window

Press [WINDOW]. Use the arrow or [ENTER] keys to move the cursor and edit the window settings. Xmin and Xmax represent the minimum and maximum values along the *x*-axis. Similarly, Ymin and Ymax represent the minimum and maximum values along the *y*-axis. Xscl and Yscl refer to the spacing between tick marks placed on the *x*- and *y*-axes. Suppose Xscl = 1. Then the numbers along the *x*-axis progress by 1 unit. Set Xres to 1.

C: The Standard Viewing Window

A good window to start with to graph an equation is the **standard viewing window.** It appears in the [WINDOW] screen as follows.

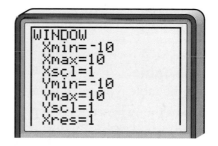

To easily set the values for the standard viewing window, press [ZOOM] 6.

D: Zoom Features

To easily access a viewing window that shows only integer coordinates, press [ZOOM] 8 [ENTER].

To easily access a viewing window for statistical graphs of data you have entered, press [ZOOM] 9.

E: Using the Trace Feature

To trace a graph, press [TRACE] . A flashing cursor appears on a point of your graph. At the bottom of the screen, x- and y-coordinates for the point are shown. At the top left of the screen, the equation of the graph is shown. Use the left and right arrow keys to move the cursor along the graph. Notice how the coordinates change as the cursor moves from one point to the next. If more than one equation is graphed, use the up and down arrow keys to move from one graph to another.

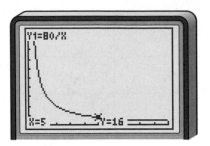

F: Setting or Making a Table

Press [2nd] [TBLSET]. Use the arrow or [ENTER] keys to move the cursor and edit the table settings. Indpnt represents the x-variable in your equation. Set Indpnt to *Ask* so that you may enter any value for x into your table. Depend represents the y-variable in your equation. Set Depend to *Auto* so that the calculator will find y for any value of x.

G: Using the Table

Before using the table, you must enter at least one equation in the [Y=] screen. Then press [2nd] [TABLE]. Enter any value for x as shown at the bottom of the screen. The function entered as Y_1 will be evaluated at this value for x. In the two columns labeled X and Y_1, you will see the values for x that you entered and the resulting y-values.

H: Entering Inequalities

Press [2nd] [TEST]. From this menu, you can enter the $=$, $\neq$, $>$, $\geq$, $<$, and $\leq$ symbols.

Chapter	Page(s)	Key Skills
2	61	B
3	106	I
5	214	K
6	272	A, B, E
7	317	A, D, E
8	307	A, B, E
10	422	K
11	471, 491	A, D, E, F, G, I, J
12	521	A, D, E, H
13	551	A, D
14	625	A, B
15	638–639	A, B, E

I: Entering and Deleting Lists

Press [STAT] [ENTER] . Under L_1, enter your list of numerical data. To delete the data in the list, use your arrow keys to highlight L_1. Press [CLEAR] [ENTER] . Remember to clear all lists before entering a new set of data.

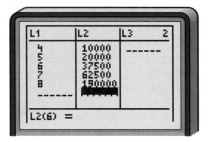

J: Plotting Statistical Data in Lists

Press [Y=] . If appropriate, clear equations. Use the arrow keys until Plot1 is highlighted. Plot1 represents a Stat Plot, which enables you to graph the numerical data in the lists. Press [ENTER] to turn the Stat Plot on and off. You may need to display different types of statistical graphs. To set the details of a Stat Plot, press [2nd] [STAT PLOT] [ENTER]. A screen like the one below appears.

At the top of the screen, you can choose from one of three plots to store settings. The second line allows you to turn a Stat Plot on and off. Then you may select the type of plot: scatter plot, line plot, histogram, two types of box-and-whisker plots, or a normal probability plot. For this text, you will mainly use the scatter plot, line plot, and histogram. Next, choose which lists of data you would like to display along the x- and y-axes. Finally, choose the symbol that will represent each data point. To see a graph of the statistical data, press [ZOOM] 9.

K: Programming on the TI–83 Plus

The TI–83 Plus has programming features that allow you to write and execute a series of commands to perform tasks that may be too complex to perform otherwise. Each program is given a name. Commands begin with a colon (:), followed by an expression or an instruction. Most calculator features are accessible from the program mode.

When you press [PRGM], you see three menus: EXEC, EDIT, and NEW. EXEC allows you to execute a stored program by selecting the name of the program from the menu. EDIT allows you to edit or change an existing program. NEW allows you to create a new program.

The following example illustrates how to create and execute a new program that stores an expression as Y and evaluates the expression for the designated value of X.

1. Press [PRGM] [▶] [▶] [ENTER] to create a new program.

2. Type EVAL [ENTER] to name the program. (Be sure that the A-LOCK is on.) You are now in the program editor, which allows you to enter commands. The colon (:) in the first column of the line indicates that it is the beginning of a command line.

3. The first command line will ask the user to choose a value for x. Press [PRGM] [▶] 3 [2nd] [A-LOCK] "ENTER X" [ENTER]. (To enter a space between words, press the 0 key when the A-LOCK is on.)

4. The second command line will allow the user to enter any value for x into the calculator. Press [PRGM] [▶] 1 [X,T,θ,n] [ENTER].

5. The expression to be evaluated for the value of x is $x - 7$. To store the expression as Y, press [X,T,θ,n] [−] 7 [STO▶] [ALPHA] [Y] [ENTER].

6. Finally, we want to display the value for the expression. Press [PRGM] [▶] 3 [ALPHA] [Y] [ENTER]. At this point, you have completed writing the program. It should appear on your calculator like the screen shown below.

```
PROGRAM:EVAL
:Disp "ENTER X"
:Input X
:X-7→Y
:Disp Y
:
```

7. Now press [2nd] [QUIT] to return to the home screen.

8. To execute the program, press [PRGM]. Then press the down arrow to locate the program name and press [ENTER] twice. The program asks for a value for x. Input any value for which the expression is defined and press [ENTER]. To immediately re-execute the program, simply press [ENTER] when the word *Done* appears on the screen. To break during program execution, press [ON].

9. To delete a program, press [2nd] [MEM] 2 7. Use the arrow and [ENTER] keys to select a program you wish to delete. Then press [DEL] 2.

While a graphing calculator cannot do everything, it can make some tasks easier. To prepare for whatever lies ahead, you should try to learn as much as you can. The future will definitely involve technology. Using a graphing calculator is a good start toward becoming familiar with technology.

Glossary

A

absolute value   The absolute value of a number is its distance from zero on a number line. *(p. 55)*

Addition Property of Equality   For any numbers a, b, and c, if $a = b$, then $a + c = b + c$. *(p. 122)*

additive inverses   Two numbers are additive inverses if their sum is 0. *(p. 65)*

algebraic expression   An expression consisting of one or more numbers and variables along with one or more arithmetic operations. *(p. 4)*

Associative Property   For any numbers a, b, and c, $(a + b) + c = a + (b + c)$ and $(ab)c = a(bc)$. *(p. 14)*

axis of symmetry   The vertical line containing the vertex of a parabola. *(p. 459)*

B

base   **1.** The number that is divided into the percentage in the percent proportion. *(p. 199)* **2.** In an expression of the form x^n, the base is x. *(p. 336)*

binomial   A polynomial with two terms. *(p. 383)*

boundary   A line that separates the coordinate plane into half-planes. *(p. 535)*

C

circle graph   A graph that shows the relationship between parts of the data and the whole. *(p. 200)*

Closure Property   A set of numbers is closed under an operation if the result of that operation on two numbers is included in the same number system. *(p. 16)*

coefficient   The numerical part of a term. *(p. 20)*

Commutative Property   For any numbers a and b, $a + b = b + a$ and $ab = ba$. *(pp. 14–15)*

complements   Two events are complements if the sum of their probabilities is 1. *(p. 223)*

completing the square   To add a constant term to a binomial of the form $x^2 + bx$ so that the resulting trinomial is a perfect square. *(p. 478)*

composite numbers   A whole number that has more than two factors. *(p. 420)*

compound event   Two or more simple events that are connected by the words *and* or *or*. *(p. 224)*

compound inequality   Two or more inequalities that are connected by the words *and* or *or*. *(p. 524)*

conjugates   Two binomials of the form $a\sqrt{b} + c\sqrt{d}$ and $a\sqrt{b} - c\sqrt{d}$. *(p. 616)*

consecutive integers   Integers in counting order. *(p. 167)*

consistent   A system of equations is said to be consistent when it has at least one ordered pair that satisfies both equations. *(p. 554)*

constant of variation   The number k in equations of the form $y = kx$ and $xy = k$. *(p. 264)*

coordinate   The number that corresponds to a point on a number line. *(p. 53)*

coordinate plane   The plane containing the x- and y-axes. *(p. 58)*

coordinate system   The grid formed by the intersection of two perpendicular number lines that meet at their zero points. *(p. 58)*

counterexample   An example showing that a statement is not true. *(p. 16)*

cross products   When two fractions are compared, the cross products are the products of the terms on the diagonals. *(p. 95)*

cumulative frequency histogram   A histogram organized using a cumulative frequency table. *(p. 39)*

cumulative frequency table   A table in which the frequencies are accumulated for each item. *(p. 33)*

D

data   Numerical information. *(p. 32)*

degree   **1.** The degree of a monomial is the sum of the exponents of its variables.   **2.** The degree of a polynomial is the greatest of the degrees of its terms. *(p. 384)*

dependent   A system of equations is said to be dependent when it has an infinite number of solutions. *(p. 554)*

dependent variable   The variable in a relation whose value depends on the value of the independent variable. *(p. 264)*

difference of squares   Two perfect squares separated by a subtraction sign. *(p. 447)*

$$a^2 - b^2 = (a + b)(a - b)$$

digit problems   Problems that explore the relationships between digits. *(p. 569)*

dimensional analysis   The process of carrying units throughout a computation. *(p. 190)*

direct variation   An equation of the form $y = kx$, where $k \neq 0$. *(p. 264)*

discount   The amount by which the regular price of an item is reduced. *(p. 213)*

Distance Formula   The distance d between any two points with coordinates (x_1, y_1) and (x_2, y_2) is given by the formula $d = \sqrt{(x_2 - x_1)^2 + (y_2 - y_1)^2}$. *(p. 607)*

Distributive Property   For any numbers a, b, and c, $a(b + c) = ab + ac$ and $a(b - c) = ab - ac$. *(p. 19)*

Division Property for Inequalities   For all numbers a, b, and c, the following are true. *(p. 514)*

1. If c is positive and $a < b$, then $\frac{a}{c} < \frac{b}{c}$, and if c is positive and $a > b$, then $\frac{a}{c} > \frac{b}{c}$.

2. If c is negative and $a < b$, then $\frac{a}{c} > \frac{b}{c}$, and if c is negative and $a > b$, then $\frac{a}{c} < \frac{b}{c}$.

Division Property of Equality   For any numbers a, b, and c, where $c \neq 0$, if $a = b$, then $\frac{a}{c} = \frac{b}{c}$. *(p. 160)*

domain   The set of all first coordinates from the ordered pairs in a relation. *(p. 238)*

elimination   The elimination method of solving a system of equations uses addition or subtraction to eliminate one of the variables and solve for the other variable. *(p. 566)*

empirical probability   The most accurate probability based upon repeated trials in an experiment. *(p. 220)*

empty set   A set with no members. *(p. 130)*

equation   A mathematical sentence that contains an equals sign, =. *(p. 5)*

equation in two variables   An equation that contains two unknown values. *(p. 244)*

equivalent equations   Equations that have the same solution. *(p. 122)*

equivalent expressions   Expressions whose values are the same. *(p. 20)*

evaluating   To find the value of an expression when replacing the variables with known values. *(p. 10)*

event   A subset of the possible outcomes in a counting problem. *(p. 147)*

excluded value   A value is excluded from the domain if it is substituted for a variable and the result has a denominator of 0 or the square root of a negative number. *(p. 638)*

experimental probability   What actually occurs when conducting a probability experiment. *(p. 220)*

exponent   In an expression of the form x^n, the exponent is n. It tells how many times x is used as a factor. *(p. 336)*

exponential function   A function that can be described by an equation of the form $y = a^x$, where $a > 0$ and $a \neq 1$. *(p. 489)*

factoring   To express a polynomial as the product of monomials and polynomials. *(p. 428)*

factors   In a multiplication expression, the quantities being multiplied are called factors. *(p. 4)*

family of graphs   Graphs and equations of graphs that have at least one characteristic in common. *(p. 316)*

FOIL method   To multiply two binomials, find the sum of the products of

F   the First terms,
O   the Outside terms,
I   the Inside terms, and
L   the Last terms. *(p. 401)*

formula   An equation that states a rule for the relationship between quantities. *(p. 24)*

frequency table   A table of tally marks used to record and display how often events occur. *(p. 33)*

function   A relation in which each element of the domain is paired with exactly one element of the range. *(p. 256)*

functional notation   In functional notation, the equation $y = x + 5$ is written as $f(x) = x + 5$. *(p. 258)*

functional value   The element in the range that corresponds to a specific element in the domain. *(p. 258)*

Fundamental Counting Principle   If event M can occur in m ways and is followed by event N that can occur in n ways, then the event M followed by event N can occur in $m \times n$ ways. *(p. 147)*

graph   To draw, or plot, the points named by certain numbers or ordered pairs on a number line or coordinate plane. *(p. 53)*

greatest common factor (GCF)   The greatest common factor of two or more integers is the product of the prime factors common to the integers. *(p. 422)*

H

half-plane   The region of the graph of an inequality on one side of a boundary. *(p. 535)*

histogram   A graph that displays data from a frequency table over equal intervals. *(p. 39)*

hypotenuse   The side of a right triangle opposite the right angle. *(p. 366)*

I

identity   An equation that is true for every value of the variable. *(p. 172)*

inclusive   Two events that can occur at the same time are inclusive. *(p. 227)*

inconsistent   A system of equations is said to be inconsistent when no ordered pair satisfies both equations. *(p. 555)*

independent   A system of equations is said to be independent if it has exactly one solution. *(p. 554)*

independent events   The outcome of one event does not affect the outcome of the other event. *(p. 224)*

independent variable   The variable in a function whose value is subject to choice. *(p. 264)*

inequality   A statement used to compare two nonequal measures. *(p. 95)*

integers   The set of numbers { . . . , $-3, -2, -1, 0,$ $1, 2, 3, . . .$ }. *(p. 52)*

intersection   The intersection of two inequalities is the set of elements common to both inequalities. *(p. 524)*

inverse variation   An equation of the form $xy = k$, where $k \neq 0$. *(p. 270)*

irrational number   A number whose decimal value does not terminate or repeat. *(p. 362)*

least common denominator (LCD)   The least common multiple of the denominators of two or more fractions. *(p. 663)*

least common multiple (LCM)   The least number that is a common multiple of two or more numbers. *(p. 662)*

legs   The sides of a right triangle that form the right angle. *(p. 366)*

like terms   Terms that contain the same variables raised to the same exponents. *(p. 20)*

linear equation   An equation whose graph is a straight line. *(p. 250)*

line graph   Numerical data displayed to show trends or changes over time. *(p. 38)*

M

maximum   The highest point on the graph of a curve. *(p. 459)*

mean   The mean of a set of data is the sum of the data divided by the number of items of data. *(p. 104)*

measures of central tendency   Numbers known as measures of central tendency are often used to describe sets of data because they represent a centralized, or middle, value. *(p. 104)*

measures of variation Measures of variation are used to describe the distribution of the data. *(p. 106)*

median The middle number when data are arranged in numerical order. *(p. 104)*

minimum The lowest point on the graph of a curve. *(p. 459)*

mixture problems Problems in which two or more parts are combined into a whole. *(p. 206)*

mode The item of data that occurs most often in the set. *(p. 104)*

monomial A number, a variable, or a product of numbers and variables that have only positive exponents and no variable exponents. *(p. 382)*

Multiplication Property for Inequalities For all numbers a, b, and c, the following are true. *(p. 515)*

1. If c is positive and $a < b$, then $ac < bc$, and if c is positive and $a > b$, then $ac > bc$.

2. If c is negative and $a < b$, then $ac > bc$, and if c is negative and $a > b$, then $ac < bc$.

Multiplicative Inverse Property For every nonzero number $\frac{a}{b}$, where a, $b \neq 0$, there is exactly one number $\frac{b}{a}$ such that $\frac{a}{b} \cdot \frac{b}{a} = 1$. *(p. 154)*

multiplicative inverses Two numbers are multiplicative inverses if their product is 1. *(p. 154)*

Multiplicative Property of −1 The product of −1 and any number is the number's additive inverse. *(p. 143)*

Multiplicative Property of Equality For any numbers a, b, and c, if $a = b$, then $a \cdot c = b \cdot c$. *(p. 161)*

Multiplicative Property of Zero For any number a, $a \cdot 0 = 0 \cdot a = 0$. *(p. 10)*

mutually exclusive events Two events that cannot occur at the same time. *(p. 226)*

negative exponent For any nonzero number a and any integer n, $a^{-n} = \frac{1}{a^n}$. *(p. 342)*

negative number Any number that is less than zero. *(p. 52)*

number line A line with equal distances marked off to represent numbers. *(p. 52)*

numerical expression An expression containing only numbers and mathematical operations. *(p. 4)*

odds The ratio of the number of ways an event can occur (successes) to the number of ways the event cannot occur (failures). *(p. 221)*

open sentences Mathematical statements with one or more variables. *(p. 112)*

opposites The opposite of a number is its additive inverse. *(p. 65)*

order of operations
1. Find the values of the expressions inside grouping symbols, such as parentheses, brackets, and braces, and as indicated by fraction bars. Start with the innermost grouping symbols.

2. Evaluate all powers in order from left to right.

3. Do all multiplications and/or divisions from left to right.

4. Do all additions and/or subtractions from left to right. *(pp. 8, 338)*

ordered pair A pair of numbers used to locate any point on a coordinate plane. *(p. 58)*

origin The point of intersection of the two axes in the coordinate plane. *(p. 58)*

outcomes All possible combinations of a counting problem or the results of an experiment. *(p. 146)*

parabola The general shape of the graph of a quadratic function. *(p. 458)*

parallel lines Lines in the plane that never intersect and have the same slope. *(p. 322)*

parent graph The simplest of the graphs in a family of graphs. *(p. 318)*

percent A ratio that compares a number to 100. *(p. 198)*

percentage The number that is divided by the base in a percent proportion. *(p. 199)*

percent equation Percentage = Base · Rate *(p. 204)*

percent of decrease The ratio of an amount of decrease to the previous amount, expressed as a percent. *(p. 212)*

percent of increase The ratio of an amount of increase to the previous amount, expressed as a percent. *(p. 212)*

percent proportion $\frac{\text{percentage}}{\text{base}} = \frac{r}{100}$ *(p. 199)*

perfect square The product of a number and itself. *(p. 336)*

perfect square trinomial A trinomial which, when factored, has the form $(a + b)^2 = (a + b)(a + b)$ or $(a - b)^2 = (a - b)(a - b)$. *(p. 445)*

perpendicular lines Lines that meet to form right angles. *(p. 324)*

point-slope form An equation of the form $y - y_1 = m(x - x_1)$, where m is the slope and (x_1, y_1) is any point on a nonvertical line. *(p. 290)*

polynomial A monomial or sum of monomials. *(p. 383)*

population A large group of data usually represented by a sample. *(p. 32)*

power An expression of the form x^n. *(p. 336)*

prime factorization A whole number expressed as a product of factors that are all prime numbers. *(pp. 358, 421)*

prime number A whole number whose only factors are 1 and itself. *(p. 420)*

prime polynomial A polynomial that cannot be written as a product of two polynomials with integral coefficients. *(p. 420)*

probability The ratio that compares the number of favorable outcomes to the number of possible outcomes. *(p. 219)*

product In a multiplication expression, the result is called the product. *(p. 4)*

proportion An equation of the form $\frac{a}{b} = \frac{c}{d}$ stating that two ratios are equivalent. *(p. 188)*

Pythagorean Theorem If a and b are the measures of the legs of a right triangle and c is the measure of the hypotenuse, then $c^2 = a^2 + b^2$. *(p. 366)*

Q

quadrant One of the four regions into which the x- and y-axes separate the coordinate plane. *(p. 60)*

quadratic equation An equation of the form $ax^2 + bx + c = 0$, where $a \neq 0$. A quadratic equation is one in which the value of the related quadratic function is 0. *(p. 468)*

Quadratic Formula The roots of a quadratic equation in the form $ax^2 + bx + c = 0$, where $a \neq 0$, are given by the formula $x = \dfrac{-b \pm \sqrt{b^2 - 4ac}}{2a}$. *(p. 484)*

quadratic function A function that can be described by an equation of the form $y = ax^2 + bx + c$, where $a \neq 0$. *(p. 458)*

quadratic-linear system of equations A system of equations involving a linear and a quadratic function. *(p. 580)*

R

radical equations Equations that contain radicals with variables in the radicand. *(p. 624)*

radical expression An expression that contains a square root. *(p. 358)*

radical sign The symbol $\sqrt{}$, used to indicate the positive square root. *(p. 357)*

random When all outcomes have an equally likely chance of happening. *(p. 220)*

range 1. The set of all second coordinates from the ordered pairs in a relation. *(p. 238)* 2. The difference between the greatest and the least values of a set of data. *(p. 106)*

rate 1. The ratio of two measurements having different units of measure. *(p. 190)* 2. In the percent proportion, the rate is the decimal form of the percent. *(p. 204)*

rate problems Problems involving distance, rate, and time. The formula $d = rt$ is used to solve rate problems. *(p. 266)*

ratio A comparison of two numbers by division. *(p. 188)*

rational equation An equation that contains at least one rational expression. *(p. 668)*

rational expression An algebraic fraction whose numerator and denominator are polynomials. *(p. 638)*

rational function A function that contains rational expressions. *(p. 638)*

rationalizing the denominator A method used to remove or eliminate radicals from the denominator of a fraction. *(p. 615)*

rational numbers A number that can be expressed in the form of a fraction $\frac{a}{b}$, where a and b are integers and $b \neq 0$. *(p. 94)*

real numbers The set of rational numbers and the set of irrational numbers together form the set of real numbers. *(p. 600)*

reciprocal The multiplicative inverse of a number. *(p. 154)*

relation A set of ordered pairs. *(p. 238)*

replacement set A set of numbers from which replacements for a variable may be chosen. *(p. 112)*

roots The solutions of a quadratic equation. *(p. 468)*

sales tax A tax added to the cost of an item. *(p. 213)*

sample A group that is used to represent a much larger population. *(p. 32)*

sample space The list of all possible outcomes of a counting problem. *(p. 146)*

sampling A method used to gather data in which a small group, or sample, of a population is polled so that predictions can be made about the population. *(p. 32)*

scale drawing A drawing that represents an object too large or too small to be drawn at actual size. *(p. 194)*

scale model A model that represents an object too large or too small to be built at actual size. *(p. 194)*

scatter plot Two sets of data plotted as ordered pairs in the coordinate plane. *(p. 302)*

scientific notation A number of the form $a \times 10^n$, where $1 \leq a < 10$ and n is an integer. *(p. 353)*

set-builder notation A notation used to describe the members of a set. For example, $\{y \mid y < 17\}$ represents the set of all numbers y such that y is less than 17. *(p. 510)*

simple interest The amount paid or earned for the use of money. The formula $I = prt$ is used to solve simple interest problems. *(p. 205)*

simplest form An expression is in simplest form when there are no longer any like terms or parentheses. *(p. 20)*

simplify In an expression, eliminate all parentheses and then add, subtract, multiply, or divide. *(p. 15)*

slope The ratio of the change in the y-coordinates to the corresponding change in the x-coordinates as you move from one point to another along a line. *(p. 284)*

slope-intercept form An equation of the form $y = mx + b$, where m is the slope and b is the y-intercept of a given line. *(p. 296)*

solution A replacement for the variable in an open sentence that results in a true sentence. *(p. 112)*

solving an open sentence Finding a replacement for the variable that results in a true sentence. *(p. 112)*

square root One of two equal factors of a number. *(p. 357)*

statement Any sentence that is either true or false, but not both. *(p. 112)*

stem-and-leaf plot In a stem-and-leaf plot, each piece of data is separated into two numbers that are used to form a stem and a leaf. The data are organized into two columns. The column on the left contains the stems and the column on the right contains the leaves. *(p. 40)*

substitution The substitution method of solving a system of equations uses substitution of one equation into the other equation to solve for the other variable. *(p. 560)*

Subtraction Property of Equality For any numbers a, b, and c, if $a = b$, then $a - c = b - c$. *(p. 124)*

system of equations A set of equations with the same variables. *(p. 550)*

system of inequalities A set of two or more inequalities with the same variables. *(p. 586)*

tally marks A type of mark used to display data in a frequency table. *(p. 33)*

term A number, a variable, a product, or a quotient of numbers and variables. *(p. 20)*

theoretical probability The probability that should occur in an experiment. *(p. 220)*

tree diagram A diagram used to show the total number of possible outcomes. *(p. 146)*

trinomial   A polynomial with three terms. *(p. 383)*

U

uniform motion problems   When an object moves at a constant speed, or rate, it is said to be in uniform motion. *(p. 670)*

union   The union of two inequalities is the set of elements in each inequality. *(p. 525)*

unit cost   The cost of one unit of something used to compare the costs of similar items. *(p. 97)*

unit rate   A simplified rate with a denominator of 1. *(p. 190)*

V

variable   Symbols used to represent unknown numbers. *(p. 4)*

Venn diagrams   Diagrams that use circles or ovals inside a rectangle to show relationships. *(p. 53)*

vertex   The maximum or minimum point of a parabola. *(p. 459)*

vertical line test   If any vertical line passes through no more than one point of the graph of a relation, then the relation is a function. *(p. 257)*

W

whole numbers   The set of numbers {0, 1, 2, 3, . . . }. *(p. 16)*

x-axis   The horizontal number line on a coordinate plane. *(p. 58)*

x-coordinate   The first number in an ordered pair. *(pp. 59, 238)*

x-intercept   The coordinate at which a graph intersects the x-axis. *(p. 296)*

y-axis   The vertical number line on a coordinate plane. *(p. 58)*

y-coordinate   The second number in an ordered pair. *(pp. 59, 238)*

y-intercept   The coordinate at which a graph intersects the y-axis. *(p. 296)*

Z

zero pair   The result of a positive algebra tile paired with a negative algebra tile. *(p. 65)*

Zero Product Property   For all numbers a and b, if $ab = 0$, then $a = 0$, $b = 0$, or both a and b equal zero. *(p. 474)*

zeros   The roots, or x-intercepts, of a function. *(p. 468)*

Selected Answers

Chapter 1 The Language of Algebra

Pages 6–7 Lesson 1–1
1. Sample answer: $3 + 7$, $10(6)$, $4 - 1$; $2x$, $5 + a$, gh
5. Sample answer: $7m$ **7.** Sample answer: 7 divided by q **9.** $6 + m = 17$ **11.** Sample answer: 5 increased by r is 15. **13.** $9w + 3$ **15.** rs **17.** $5a + 3$
19. $1 - n$ **21.** $10 + h \cdot 1$ or $10 + h$ **23.** Sample answer: the product of 9 and x **25.** Sample answer: the difference of 6 and y **27.** Sample answer: 8 less than the quotient of 3 and r **29.** $3 + w = 15$
31. $2 = \dfrac{7}{x}$ **33.** $3 - 5y = 2z$ **35.** Sample answer: 3 times r is 18. **37.** Sample answer: h equals 10 minus i. **39.** Sample answer: The quotient of t and 4 is 16. **41.** Sample answer: Let x — the variable; $4x - 7 = 15 + c + 2x$ **43a.** $20 + 10(15 - 1)$
43b. $20 + 10(m - 1)$

Pages 11–13 Lesson 1–2
1. parentheses, brackets, fraction bar **5.** Multiply 6 and 2. **7.** Subtract 4 from 10. **9.** 3 **11.** Transitive
13. $8(4 - 8 \div 2)$
$\quad = 8(4 - 4)$ *Substitution*
$\quad = 8(0)$ *Substitution*
$\quad = 0$ *Multiplicative Property of 0*
15. 8 **17.** 13 **19.** 14 **21.** 26 **23.** 40 **25.** 21 **27.** 8
29. Reflexive **31.** Substitution **33.** Transitive
35. $8(9 - 3 \cdot 2) = 8(9 - 6)$ *Substitution*
$\quad\quad\quad\quad\quad = 8(3)$ *Substitution*
$\quad\quad\quad\quad\quad = 24$ *Substitution*
37. $10(6 - 5) - (20 \div 2)$
$\quad = 10(1) - 10$ *Substitution*
$\quad = 10 - 10$ *Multiplicative Identity*
$\quad = 0$ *Substitution*
39. $6(12 - 48 \div 4) + 7 \cdot 1$
$\quad = 6(12 - 12) + 7$ *Substitution*
$\quad = 6(0) + 7$ *Substitution*
$\quad = 0 + 7$ *Multiplicative Property of 0*
$\quad = 7$ *Additive Identity*
41. 50 **43.** 23 **45.** 40 **47.** 5 **49.** 26 **51.** $311,392
53. 44 feet **55.** Sample answer: Eight more than x is 12. **57.** Sample answer: The quotient of 25 and n is 5. **59.** $3c = 27$ **61.** $b + 10 - 1 = 18$
63a. $4(20) + 7$ **63b.** 87

Page 13 Quiz 1
1. $2v - 5$ **3.** 13 **5.** 56 ft

Pages 17–18 Lesson 1–3
1. For any whole numbers that are multiplied, the product is a whole number. **3.** Jessie; the Associative Property can be applied to addition

or to multiplication, but not to both at the same time. **5.** Associative $(+)$ **7.** $(3 + 47 + p)(7 - 6)$; $50 + p$; Commutative $(+)$ **9a.** 188 **9b.** Use the Commutative Property of Addition to change the order of the numbers to $69 + 31 + 80 + 8$. Then use the Substitution Property, adding 69 and 31 first. **11.** Commutative $(+)$ **13.** Commutative $(\times)$ **15.** Associative $(\times)$ **17.** $r \cdot (30 \cdot 5)$; $150r$; Associative $(\times)$ **19.** $(6 + 3) + y$; $9 + y$; Associative $(+)$ **21.** $2 \cdot 7 \cdot j$; $14j$; Commutative $(\times)$
23. false; sample counterexample:
$\quad (4 \div 2) \div 2 \overset{?}{=} 4 \div (2 \div 2)$
$\quad\quad\quad 2 \div 2 \overset{?}{=} 4 \div 1$
$\quad\quad\quad\quad\quad 1 \neq 4$
25. 100 board feet
27. Sample answer:
$\quad (8 - 5) - 3 \overset{?}{=} 8 - (5 - 3)$
$\quad\quad\quad 3 - 3 \overset{?}{=} 8 - 2$
$\quad\quad\quad\quad 0 \neq 6$
29. 6 **31.** 7 **33.** A

Pages 21–23 Lesson 1–4
1. Sample answer: $1 + 2x + 3x + ab + 5ab$ **3.** b, e; Write all the expressions in simplest form and find two that are the same. **5.** $8p$, $9p$ **7.** $6bc$, bc **9.** $y + 2$ **11.** $12 - 18m$ **13.** $7a + 7t$ **15.** $21f$ **17.** $3r$
19. $2g + 6$ **21.** $6x + 5y$ **23.** $15am - 12$ **25.** $7a + 5b$ **27.** $4y$ **29.** $46xy$ **31.** $3r + 4s$
33. $5(2n + 3r) + 4n + 3(r + 2)$
$\quad - 10n + 15r + 4n + 3r + 6$ *Dist. Prop.*
$\quad = 10n + 4n + 15r + 3r + 6$ *Commutative $(+)$*
$\quad = (10 + 4)n + (15 + 3)r + 6$ *Dist. Prop.*
$\quad = 14n + 18r + 6$ *Substitution*
35. 11 **37.** $52 **39a.** 179.8 lb **39b.** 31 lb
41a. 96 ft² **41b.** $4(8 + 10 + 6)$ or 96
43. Commutative $(\times)$ **45.** Symmetric
47. 6 **49a.** $d + 30$ **49b.** 170

Pages 27–29 Lesson 1–5
1. Sample answer: You should look back to determine whether the answer fits the problem, whether there are other possible answers, or whether there might be a better way to solve the problem. **3.** no **5.** $n + 2$ **7a.** 25 $5 bills, 32 $1 bills **7b.** Sample answer: Use an equation. Let x represent the number of $1 bills. Then $x - 7$ represents the number of $5 bills. Write and solve the equation $11(10) + 1x + 5(x - 7) = 267$. **9.** Craig, 16; mother 40; sample explanation: Let m represent the age of the mother. Then $m - 24$ represents Craig's age. Solve the equation $m + (m - 24) = 56$.
11. 7 craft books, 13 cookbooks

13a. Sample answer:

Person	1	2	3	4	5	6
Number of Handshakes	5	4	3	2	1	0

13b. 15   **13c.** 66   **15.** 9:00 P.M.   **17a.** $d = rt$
17b. 703 mi   **19.** 20 people   **21.** $8b$ **23.** $24a + 45$
25. A

Pages 34–37   Lesson 1–6
1. A frequency table shows the number of times a single event occurs. A cumulative frequency table shows the number of times an event occurs plus the previous events.   **3.** No, the sample is not large enough.

5.

Number of Goals	Tally	Frequency
1	III	3
2	IIII	5
3	I	1
4	IIII	4
5	III	3
6	III	3
7	I	1
8	I	1

7. 1 time   **9a.** 100 s
9b.

Cycle (s)	Frequency	Cumulative Frequency
80	33	33
90	42	75
100	60	135
110	25	160

9c. 75 times   **11.** Yes; the sample is random and representative of drivers in the area.   **13.** No; the sample is not representative of the entire country.
15. No; people might prefer a golfer who is from the same area.
17a.

Quiz Score	Tally	Frequency
6	II	2
7	IIII	4
8	IIII IIII I	11
9	IIII IIII	9
10	IIII	4

17b. 8   **17c.** Rather than adding all of the numbers, this formula uses multiplication to find the sum. Then the sum is divided by the number of scores, 30.

17d. 8.3   **19.** That is a typical breakfast time and they want the business.   **21.** Sample answers: a stadium, restaurant, school campus   **25.** 275 miles
27. nine more than x

Page 37   Quiz 2
1. $11 + 6 + 2a$, Commutative (+); $17 + 2a$
5.

Eye Color		
Color	Tally	Frequency
brown	IIII IIII	10
blue	IIII	4
green	IIII I	6
hazel	IIII	5

Blue occurs the least.

Pages 41–43   Lesson 1–7
1. Line graphs usually show change over time, while histograms compare quantities of similar nature.   **2.** A correct graph has a title, horizontal and vertical axes labeled, and equally spaced units on both axes.   **3.** Manuel; line graphs are usually used to show trends over time. Histograms are usually used to how frequency of items in sets of data.   **5.** 1997–1998   **7.** 14–15 cm
9.

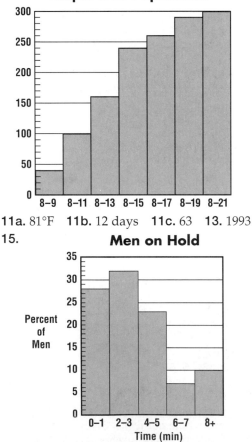

Leaf Lengths in a Maple Tree Population

11a. 81°F   **11b.** 12 days   **11c.** 63   **13.** 1993
15.

Men on Hold

17. Sample answer: The histograms show that a greater percent of men than women are willing stay on hold for 0–1 minutes   **19.** 123   **21.** 71, 73, 92, and 100 occur three times.   **23.** Sample answer: More students in second period scored 80 or more, so second period did better   **25.** Yes; it is a random sample.   **27.** $8x - 3y$

Pages 44–46   Chapter 1 Study Guide and Assessment

1. term   **3.** algebraic expression   **5.** counterexample   **7.** sample   **9.** frequency tables   **11.** $5n$ **13.** $2y - 6 = 14$ **15.** 7   **17.** 14   **19.** 9   **21.** Commutative ($\times$), $20c$ **23.** Commutative ($\times$), $9x$ **25.** Associative ($+$), $g + 3$ **27.** $7v - 7$ **29.** $2h + ah$ **31.** $22 - 3d$ **33a.** 1200   **33b.** 80   **35.** 0

37.

Number	Cumulative Frequency
0	6
1	9
2	13
3	15
4	19

39. 10   **41.** 10–11

43.

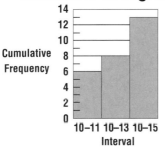

Cumulative Histogram

45. 8   **47a.** 100; 62   **47b.** 88   **47c.** 10

Page 49   Preparing for Standardized Tests
1. C   **3.** D   **5.** C   **7.** A   **9.** 234

Chapter 2   Integers

Pages 55–57   Lesson 2–1
1. Sample answer: temperature, elevation
3. Neither; 0 is neither negative nor positive.
5. -6 **7.** $+150$ **9.** 1
11.
![number line]
$-5-4-3-2-1\ 0\ 1\ 2\ 3\ 4\ 5$
13. $<$ **15.** 10   **17.** $-22, -17, -14, -7, -1, 9, 10, 13, 22, 26, 35$ **19.** 2   **21.** 0   **23.** -4
25.
![number line]
$-5-4-3-2-1\ 0\ 1\ 2\ 3\ 4\ 5$
27.
![number line]
$-5-4-3-2-1\ 0\ 1\ 2\ 3\ 4\ 5$

29.
![number line]
$-5-4-3-2-1\ 0\ 1\ 2\ 3\ 4\ 5$
31. $>$ **33.** $>$ **35.** $<$ **37.** $>$ **39.** 6   **41.** 2   **43.** 9   **45.** -3 **47.** 78, 14, -14, -25, -36 **49a.** $-36°$ **49b.** $-26°$ **49c.** $-36°$

51.

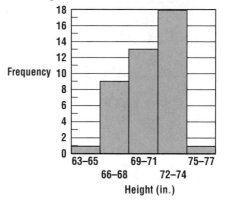

Heights of U.S. Presidents

53. 19 presidents   **55.** $6a$ **57.** $7m + 2n$ **59.** $7x + 11y$

Pages 61–63   Lesson 2–2
1. Start at the origin. Move 5 units to the left. Then move 1 unit up and draw a dot.
3.

Quadrant II ($-$, $+$)	Quadrant I ($+$, $+$)
Quadrant III ($-$, $-$)	Quadrant IV ($+$, $-$)

5. $(3, -2)$

7.

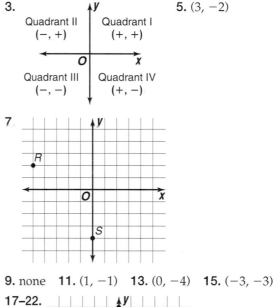

9. none   **11.** $(1, -1)$ **13.** $(0, -4)$ **15.** $(-3, -3)$

17–22.

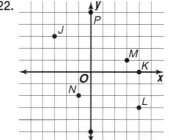

23. III   **25.** none   **27.** II   **29.** origin   **31.** I   **33.** IV
35a. $3, $9, $15
35b–c.

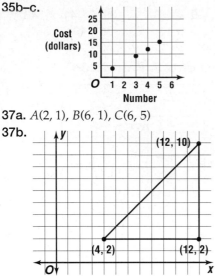

37a. $A(2, 1)$, $B(6, 1)$, $C(6, 5)$
37b.

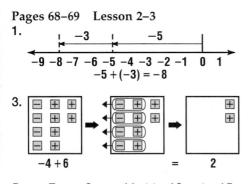

37c. The triangles have the same shape, but the second is larger.   **39.** -8685 **41.** Commutative $(+)$ **43.** Closure $(+)$ **45.** A

Page 63   Quiz 1
1. $>$ **3.** $<$ **5.** 11   **7.** III   **9.** II

Pages 68–69   Lesson 2–3
1.

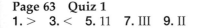
$-5 + (-3) = -8$

3.

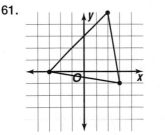
$-4 + 6$ $=$ 2

5. $+$ **7.** $-$ **9.** $-$ **11.** 16   **13.** -1 **15.** -5
17. $2x$ **19.** $2a$ **21.** 12   **23.** 21   **25.** -11
27. -8 **29.** 9   **31.** -13 **33.** -8 **35.** -21
37. -55 **39.** 13   **41.** -22 **43.** -4 **45.** -1
47. -3 **49.** $-15x$ **51.** $3m$ **53.** 0   **55.** $8y$
57. -9 **59.** -1
61.

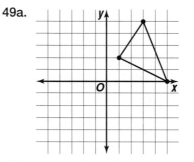

63. IV   **65.** II   **67.** -5 **69.** -10 **71.** 0   **73.** C

Pages 73–74   Lesson 2–4
1. To subtract an integer, add its additive inverse.
3. $10 + (-3)$ **5.** $-4 + (-8)$ **7.** -2 **9.** 11   **11.** 5
13. 8   **15.** 13   **17.** 7   **19.** -8 **21.** -13 **23.** -11
25. 16   **27.** -6 **29.** -9 **31.** 12   **33.** 7   **35.** -17
37. 3   **39.** -10 **41.** -12 **43.** $25n$ **45.** 50
47. $-34°F$ **49a.** false; $2 - 5 \neq 5 - 2$ **49b.** false;
$(6 - 2) - 3 \neq 6 - (2 - 3)$ **49c.** true   **51.** -20
53. -27 **55.** -8 **57.** $>$ **59.** $<$

Pages 77–79   Lesson 2–5
1. $3(-3) = -9$ **3.** -12 **5.** 80   **7.** 36   **9.** 28   **11.** $-18x$

13a.

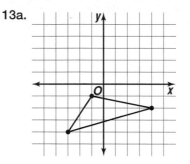

13b. It was reflected over the x-axis.   **15.** -48
17. -9 **19.** 45   **21.** 0   **23.** -24 **25.** -39 **27.** 27
29. -84 **31.** -1 **33.** 99   **35.** -8 **37.** 30   **39.** 13
41. $-8a$ **43.** $32mn$ **45.** -22 **47.** $-100(5) = -500$

49a.

49b. It was reflected over both the x- and y-axis.
51. 4   **53.** -5 **55.** -8 **57.** -2 **59.** 0   **61.** C

Page 79   Quiz 2
1. -7 **3.** 32   **5.** 13   **7.** -45 **9.** -9

Pages 84–85   Lesson 2–6
1. $-10 \div 2 = -5$, $-10 \div (-5) = 2$ **3.** -5 **5.** -5
7. 4   **9.** -8 **11.** 12   **13.** 1   **15.** 6   **17.** -6 **19.** 5
21. -9 **23.** 10   **25.** -8 **27.** -5 **29.** -12 **31.** -6
33. -5 **35.** -3 **37.** 6   **39.** -7 **41.** -2 **43a.** -119
farms   **43b.** 1000 acres or more   **45.** 74   **47.** -54
49. 16   **51.** 80   **53.** -2 **55.** 4   **57.** B

**Pages 86–88   Chapter 2 Study Guide
and Assessment**
1. negative numbers   **3.** opposites, additive inverses, or zero pairs   **5.** absolute value   **7.** integers
9. ordered pair   **11.** $>$ **13.** $<$ **15.** $-4, -3, -2, 0, 4, 7$ **17.** $(3, 2)$ **19.** $(-4, -3)$ **21.** I   **23.** none

25. -6 **27.** -10 **29.** 5 **31.** -5 **33.** $2x$ **35.** $4m$
37. -8 **39.** 9 **41.** -4 **43.** -1 **45.** -1 **47.** 35
49. 36 **51.** 27 **53.** $-42m$ **55.** -7 **57.** -6
59. -8 **61.** $213 **63.** $-40(2) = -80$

Page 91   Preparing for Standardized Tests
1. A **3.** B **5.** B **7.** E **9.** -1

Chapter 3   Addition and Subtraction Equations

Pages 97–99   Lesson 3–1
1. Sample answer: $-\frac{1}{2}, \frac{4}{3}$, and 0.9 **5.** 0.75 **7.** 0.625
9.

$$\underset{-5\ -4\ -3\ -2\ -1\ \ 0\ \ 1\ \ 2\ \ 3\ \ 4\ \ 5}{\longleftrightarrow}$$

Since -3 is to the left of $-2.\overline{6}$ on the number line,
-3 is less than $-2.\overline{6}$. **11.** $<$ **13.** $=$ **15.** $\frac{7}{10}, \frac{6}{8}, \frac{4}{5}$
17. $<$ **19.** $<$ **21.** $<$ **23.** $>$ **25.** $<$ **27.** $>$
29. $-2.002 > -2.02$
31.

$$\underset{-1\quad\ -\frac{8}{10}\qquad\qquad\qquad\qquad 0}{\overset{-\frac{5}{6}}{\longleftrightarrow}}$$

Since $-\frac{5}{6}$ is to the left of $\frac{8}{10}$ on the number line,
$-\frac{8}{10} > -\frac{5}{6}$. **33.** $\frac{1}{2}$, or 0.6, $\frac{5}{8}$ **35.** $-\frac{2}{6}, -\frac{1}{4}, -\frac{1}{8}$
37. $\frac{3}{8}, \frac{3}{5}, 0.\overline{6}$ **39.** a dozen eggs for $1.59;
$0.13 < $0.14 **41.** a 25-yd by 12-in. roll for $2.99;
$0.0004 and $0.0003 are both $>$ $0.0002
43a. Firefly, Japanese Beetle, Mealworm
43b. Elm Leaf Beetle **45.** -7 **47.** -4 **49.** 48
51. 3822 ft below sea level

Pages 102–103   Lesson 3–2
1. Add and subtract in the order in which they are
given; or group the positive numbers together and
the negative numbers together and then add.
3. 0.9 **5.** -15.05 **7.** 0 **9.** $\frac{1}{3}$ **11.** -13.3
13. -5.82 **15.** -1.26 **17.** -29.48 **19.** $-\frac{3}{8}$
21. $-5\frac{3}{10}$ **23.** $\frac{19}{35}$ **25.** $3\frac{1}{8}$ **27.** $-2\frac{3}{20}$ **29.** 6.9
31. $-6\frac{1}{6}$ **33.** $-$153.58 or a decrease of $153.58
35. $\frac{1}{4}$ foot above normal **37.** -20 **39.** $-12bc$
41. $11y + 14$

Pages 107–109   Lesson 3–3
1. When a set of data has one or more extremely
high or low values, the mean does not accurately
describe the values in the set of data. Consider
the set of data 3, 10, 5, 16, 148. The mean is 36.4.
3. Sonia; a few really high or really low values
could influence the mean so that it is not

representative of Eric's scores. **5.** 42 **7.** 0.151
9. 47.25 **11.** 15.5; 12.45; 12.5; 21.1 **13.** 62.3; 60;
72; 17 **15a.** Ms. Diaz: 201.625; 200.6; 210.0; 19.1;
Mr. Cruz: 201.625; 200.6; 210.0; 59.1 **15b.** Ms. Diaz;
her sales are more consistent **17.** 9; 8; none; 19
19. 0.25; 0.3; 0.3; 0.2 **21.** 60.8; 24; $\underline{20}$; 180
23. 10.375; 10.1; 10.2; 3.8 **25.** 10.83; 10.5; 10; 2.5
27. 19.5; 19.2; 18.3, 19.0, 21.0; 2.7 **29.** Sample
answer: 3.1, 1.2, 1.0, 1.5 **31.** Sample answer: 1, 7, 7,
7, 7, 7 **33.** 96 **35a.** $33,425; $29,050; none
35b. mean; highest value **35c.** median; lowest
value **37.** -9.9 **39.** -1
41.

$$\underset{-5\ -4\ -3\ -2\ -1\ \ 0\ \ 1\ \ 2\ \ 3\ \ 4\ \ 5}{\longleftrightarrow}$$

Pages 114–116   Lesson 3–4
1. An open sentence is neither true nor false; a
statement is either true or false. **3.** Replace n with
each element of the replacement set to see which
replacement results in a true statement. **5.** 6
7. 11.7 **9.** $\frac{4}{5}$ **11.** 4 **13.** -3 **15.** 3 **17.** 4 **19.** -5
21. 2 **23.** 42 **25.** 11 **27.** 6 **29.** -1 **31.** 18
33. $5\frac{1}{4}$ **35.** 2 **37.** 6 h **39.** 65 days **41.** 41.2; 40;
40; 8 **43.** $2\frac{1}{3}$ **45.** $-4\frac{1}{12}$ **47.** D

Page 116   Quiz 1
1. $<$ **3.** $=$ **5.** 1.69 **7.** 2.8; 2.85; 3.0; 3.6 **9.** -2

Pages 120–121   Lesson 3–5
1.

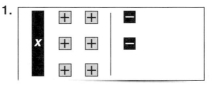

3. $x + (-2) = 12$ **5.** $7 = -6 + x$ **7.** -4 **9.** -4
11. 6 **13.** 3 **15.** -6 **17.** -7 **19.** 8 **21.** 0
23. -3 **25.** 1 **27.** 2 **29.** 16 **31.** -7 **33.** -21
35. 11 **37a.** $t - 8 = -2$ **37b.** 6°F **39.** 2 **41.** 1
43. 54 **45.** Commutative ($+$)

Pages 125–127   Lesson 3–6
1. Replace m with -4 to see if $12 + m$ is equal to 8
3. 15; In the first equation, add 5 to each side to
get $x = 17$. By replacing x with 17 in the second
equation, the result is $17 - 2$ or 15. **5.** Sample
answer: -21 **7.** Sample answer: $+9$ **9.** Sample
answer: -9.3 **11.** -15 **13.** 1.1 **15.** $-\frac{1}{6}$
17a. Sample answer: $x - 1792 = 67$ **17b.** 1859
19. -4 **21.** 3 **23.** 13 **25.** -13 **27.** -48 **29.** 34
31. -3.8 **33.** $\frac{2}{3}$ **35.** $\frac{10}{9}$ or $1\frac{1}{9}$ **37.** 13
39. $m - (-15) = 20$; $m = 5$ **41a.** $c + 27.16 = 32.50$
41b. $5.34 **43.** 0 **45.** 3.25 **47.** 6 **49.** $6.76
51. D

Page 127   Quiz 2
1. −4   **3.** 23   **5a.** Let c = cost of Chicago meal; $3.89 = c − 0.62$.   **5b.** $4.51

Pages 130–131   Lesson 3–7
1. Sample answer: $|a − 2| = −5$ **3.** Sample answer: The value inside the absolute value sign can either be positive or negative. You have to consider both cases, so there may be up to 2 solutions.   **5.** {−3, 3}   **7.** ∅   **9.** {−7, 13}
11. $|s − 145| = 6$; {139, 151}   **13.** {−2, 2}
15. {−6, 6}   **17.** ∅   **19.** {1}   **21.** ∅   **23.** {−21, 1}
25. {−19, 7}   **27.** {−9, 13}   **29.** ∅   **31.** 0
33. $|p − 165,000| = 10,000$; {155,000, 175,000}
35. A   **37.** 48   **39.** −14   **41.** B

Pages 132–134   Chapter 3 Study Guide and Assessment
1. median   **3.** range   **5.** rational number
7. inequality   **9.** rational number   **11.** >   **13.** >
15. <   **17.** $−1.1, 0, \frac{1}{8}, 0.25$ **19.** 10.5   **21.** −1
23. $5.\overline{72}$; 6; 5; 7   **25.** 12.92; 10; 10; 28   **27.** 0
29. −1   **31.** 45   **33.** −2   **35.** −4   **37.** 0
39. $x + 4 = 11$; $x = 7$ **41.** 12   **43.** 17   **45.** 9   **47.** $\frac{1}{2}$
49. $x + 12 = −108$; −120   **51.** {−8, 8}   **53.** {−4, 8}
55. ∅   **57.** $70.625; $72.5; $50, $75; $45

Page 137   Preparing for Standardized Tests
1. E   **3.** C   **5.** B   **7.** A   **9.** 0.6 mi

Chapter 4   Multiplication and Division Equations

Pages 143–145   Lesson 4–1
1. Always, because rational numbers are closed under multiplication.   **3.** $\frac{1}{4} \cdot \frac{5}{6} = \frac{5}{24}$ **5.** −9
7. −14.7   **9.** −2.3   **11.** $−\frac{9}{4}$ or $−2\frac{1}{4}$ **13.** $\frac{3}{5}y$ **15.** $6cd$
17. 24.4   **19.** −7.1   **21.** −51   **23.** −29   **25.** 57.6
27. $\frac{5}{8}$ **29.** 0   **31.** −1   **33.** $\frac{21}{8}$ or $2\frac{5}{8}$ **35.** $−3x$
37. $8.8rs$ **39.** $\frac{6}{25}ab$ **41.** $\frac{5}{4}s$ or $1\frac{1}{4}s$ **43.** −4   **45.** $135
47. $−\frac{2401}{40}$ m or $−60\frac{1}{4}$ m   **49.** ∅   **51.** {−12, −6}
53. 1.4   **55.** $−\frac{5}{8}$ **57a.** 73 in.   **57b.** 64 in.

Pages 148–151   Lesson 4–2
1. Sample answer: the number of ways two or more events can occur is the product of the number of ways each event can occur.   **3.** Ling; three outcomes have 2 heads: HHT, HTH, THH. Three outcomes have 2 tails: TTH, THT, HTT.   **5.** sample space   **7.** outcome   **9.** 36

11. 8 outcomes:

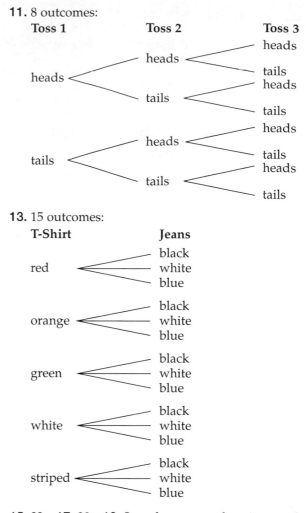

13. 15 outcomes:

15. 32   **17.** 30   **19.** Sample answer: choosing a red, green, or white car with or without a sunroof
21. 20   **23a.** 64   **23b.** 729   **25.** −16.2   **27.** $\frac{6}{35}$
29. $−\frac{5}{12}$ **31.** B

Pages 156–159   Lesson 4–3
1. No; the product of a negative number and a positive number can never equal 1.
5. $−\frac{3}{2}$ **7.** $−x$ **9.** 0.2   **11.** $−\frac{3}{14}$ **13.** $\frac{36}{5}$ or $7\frac{1}{5}$
15. $−\frac{15}{2}$ or $−7\frac{1}{2}$ **17.** 16 students   **19.** −2.5
21. −0.1   **23.** 12   **25.** 0   **27.** −1   **29.** $\frac{14}{15}$ **31.** 45
33. $−\frac{3}{32}$ **35.** −21   **37.** 12   **39.** $\frac{1}{10}$ **41.** $−\frac{2}{5}$
43. $−\frac{4}{5}$ **45.** 18   **47.** 2 min   **49.** $12.50
51. 48   **53.** 6; 6; 8; 5   **55.** 20.5; 20; 20; 18   **57.** =
59. $\frac{1}{2}$-pound bag of cashews for $3.15; $6.30 < $6.92
61. 22-ounce bottle for $1.09; $0.06 > $0.05   **63.** A

Page 159   Quiz 1
1. -18.8 **3.** $\frac{20}{8}$ or $2\frac{1}{2}$ **5.** 27   **7.** 2.7   **9.** $\frac{28}{9}$ or $3\frac{1}{9}$

Pages 163–164   Lesson 4–4
1. Divide each side by 4.   **3.** Sample answer:
$-8a = 56$ **5.** 8   **7.** 45   **9.** 4   **11.** 6   **13.** 15
15. -7 **17.** 28   **19.** 5   **21.** 64   **23.** -70 **25.** 27
27. $-\frac{5}{2}$ or $-2\frac{1}{2}$ **29.** 2   **31.** -8 **33.** $8x = 112$; 14
35. $\frac{3}{5}$ yr $= -9$; -15 **37a.** $6n = 288$ **37b.** 48
39. 110 mph   **41.** 20   **43.** $\frac{7}{12}$ **45.** 0   **47.** 57 ft
49. D

Pages 168–170   Lesson 4–5
1. Sample answer: If necessary, first undo
addition and/or subtraction of constants, then
undo multiplication or division of variables.
3. Jean; Soto did not multiply each side by 4 in the
second step.   **5.** Add 5 to each side.   **7.** 2   **9.** -5
11. -20 **13a.** $y = 6(x - 1) + 21$ **13b.** $\frac{1}{3}$ yr or
4 months   **15.** -1 **17.** -7 **19.** 0   **21.** 6   **23.** 9.6
25. 8.1   **27.** 30   **29.** -14 **31.** 68   **33.** -7.2
35. $\frac{n-2}{-3} = \frac{1}{3}$; 1   **37.** 3   **39a.** $8c + 15 = 143$
39b. $16   **41a.** $3x + 12 = 39$ **41b.** 9   **43.** 3
45. 6   **47.** 4   **49.** -81

Pages 173–175   Lesson 4–6
1. Add $2x$ to each side, subtract 5 from each side,
divide each side by 5.   **3.** An equation that is
an identity is true for every value of the variable;
an equation with no solution is never true.
5. Subtract $0.5p$ from each side.   **7.** 2   **9.** identity
11. -5 **13a.** $3n = 5n - 24$ **13b.** 12   **15.** 19
17. 2   **19.** $\frac{5}{2}$ or $2\frac{1}{2}$ **21.** no solution   **23.** 0.7
25. -1 **27.** 3.4   **29.** -20 **31.** 60   **33.** 4.2
35a. $3w - 18 = 2w$ **35b.** 18 cm by 36 cm
37. Sample answer: $2x = x + 6$ **39.** -9 **41.** 12
43. 264 numbers   **45.** 31   **47.** 3.3

Page 175   Quiz 2
1. -6 **3.** 63   **5.** 72   **7.** 12   **9.** no solution

Pages 178–179   Lesson 4–7
1. Use the Distributive Property to remove the
parentheses.   **5.** identity   **7.** -4 **9.** 8   **11.** 6
13. 5   **15.** -10 **17.** $\frac{1}{3}$ **19.** $\frac{7}{2}$ or $3\frac{1}{2}$ **21.** 2.5
23. no solution   **25.** 42   **27.** identity   **29.** -2
31. base 1 $= 20$ in.; base 2 $= 8$ in.   **33.** 17, 19   **35.** 1
37a. $\frac{1}{2}x - 4 = 7$ **37b.** 22   **39.** $-382, -369, -364,$
$-292, -229, 867$

Pages 180–182   Chapter 4 Study Guide and
                              Assessment
1. event   **3.** tree diagram   **5.** identity
7. outcomes   **9.** outcomes   **11.** 22.8
13. $-\frac{3}{4}$ **15.** $-10.8y$ **17.** $-\frac{6}{7}s$
19. 6 outcomes:

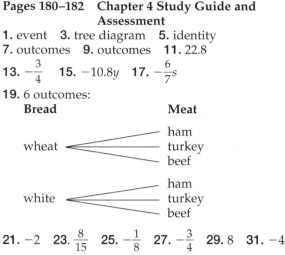

21. -2 **23.** $\frac{8}{15}$ **25.** $-\frac{1}{8}$ **27.** $-\frac{3}{4}$ **29.** 8   **31.** -4
33. 12.3   **35.** 9   **37.** -2 **39.** 0   **41.** 2   **43.** -25
45. 6   **47.** 7.3   **49.** 1   **51.** 2   **53.** identity
55. 5 times   **57.** 6 in.

Page 185   Preparing for Standardized Tests
1. A   **3.** C   **5.** E   **7.** D   **9.** 2

Chapter 5   Proportional Reasoning
            and Probability

Pages 191–193   Lesson 5–1
1. Multiply 12 and 5. Then divide by 6.   **3.** $75(4) =$
$100(3)$ **5.** $8(15) = 12(10)$ **7.** 60   **9.** 3.8   **11.** 5
13. 2 kg   **15.** 4.5 gal   **17.** 15   **19.** 4   **21.** 75
23. 9   **25.** 10.5   **27.** 3.5   **29.** 2.1   **31.** 5   **33.** 6
35. 37   **37.** Yes; the cross products are equal.
39. 60 in.   **41.** 3000 mL   **43.** 4 gal   **45.** 0.52 km
47. 9 pt   **49.** no   **51.** 750 lb, 1050 lb   **53.** It
decreases.   **55.** -6 **57.** -2 **59.** $-\frac{1}{2}$ **61.** 12
63. $-\frac{2}{5}$

Pages 196–197   Lesson 5–2
1. Sample answer:

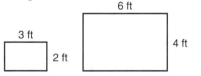

3. 150 mi   **5.** 1 m $= 35.2$ m   **7.** 120 mi
9. 100 mi   **11.** 30 mi   **13.** 1 in. $= 15$ in.
15. 0.5 c   **17.** 1 in. $= 30.2$ mi   **19.** $\frac{4}{3}$ **21.** -4

Pages 201–203   Lesson 5–3
1. P is the percentage, B is the base, and r is
the percent or the rate per hundred.
3. $\frac{12}{24} = \frac{r}{100}$ **5.** $\frac{P}{60} = \frac{45}{100}$ **7.** 60%   **9.** 70%
11. 25%   **13.** 16.4   **15a.** 45%, 55%

15b.

Senators in 106th Congress

Democrats, 45%   Republicans, 55%

17. 62.5%   **19.** 80%   **21.** 75%   **23.** 150%   **25.** 60%
27. 6%   **29.** 36%   **31.** 190   **33.** 12.5   **35.** 60
37. 0.5%   **39.** 9.3   **41.** 50%

43.

Mother's Day Flowers

Roses 23%

Mixed Bouquets, 47%

Single-stem flowers, 14%

Carnations, 16%

45. 1 in. = 70 mi   **47.** 2500 m   **49.** −5.2   **51.** $\frac{2}{5}$

Page 203   Quiz 1
1. 27   **3.** 100 ft   **5.** 950

Pages 207–209   Lesson 5–4
1. r represents the percent rate; R is the rate written in decimal form.   **3.** Nikki; in the percent equation, the R is the decimal form of the percent.   **5.** 41.8
7. 550   **9.** $75   **11.** 112   **13.** 117   **15.** 24   **17.** 625
19. 88   **21.** 64   **23.** 62.5   **25.** $50   **27a.** $P = 0.29(150)$
27b. 44   **29.** 7 adults   **31.** $x = y$ **33a.** 180 mi
33b. 210 mi   **33c.** 135 mi   **35.** −31   **37.** 5

Pages 215–217   Lesson 5–5
1. First method: find the amount of the tax, then add the tax to the cost of the item. Second method: add the tax rate to 100%, then find the total cost.
5. $\frac{4}{14} = \frac{r}{100}$ **7.** 20% decrease   **9.** $94.50   **11.** $1080
13. 83%   **15.** 12% increase   **17.** 16% decrease
19. 133% increase   **21.** $42.80   **23.** $39.52
25. $26.64   **27.** $266   **29.** $29.75   **31.** $14.21
33. 100%   **35.** 1996 and 1997; The amount of increase is larger and the base is smaller between these two years.   **37.** less than   **39.** 62.5%
41. 35%   **43.** 1.5 kg

Page 217   Quiz 2
1. 25   **3.** 20   **5.** $140.81

Pages 222–223   Lesson 5–6
1. Both theoretical and experimental probability are ratios that compare the number of favorable outcomes to the number of possible outcomes. Theoretical probability is based on known characteristics. It tells what should happen.

Experimental probability is the result of an experiment or simulation.   **3.** Yes; when the number of favorable outcomes is greater than the number of unfavorable outcomes, the odds of the outcome are greater than 1.   **5.** $\frac{1}{36}$

7. 1:1   **9.** $\frac{1}{2}$ **11.** $\frac{1}{9}$ **13.** 0   **15.** 1:1   **17.** 7:1

19. 5:3   **21.** $\frac{1}{4}$ **23a.** 9:00, $\frac{1}{20}$; 1:00, $\frac{1}{40}$; 4:00, $\frac{1}{8}$

23b. 4:00   **25a.** $\frac{1}{2}$ **25b.** 9:11   **27.** $36   **29.** $180

31. A

Pages 227–229   Lesson 5–7
1. $P(A$ and $B)$ is the probability that both A and B occur; $P(A$ or $B)$ is the probability that either A or B occur.   **3.** $\frac{1}{24}$ **5.** mutually exclusive, $\frac{2}{13}$

7. inclusive, $\frac{4}{13}$ **9.** 20.1%   **11.** $\frac{1}{4}$ **13.** $\frac{1}{20}$ **15.** $\frac{3}{20}$

17. mutually exclusive, $\frac{2}{3}$ **19.** 0.016   **21a.** 34.2%

21b. 28%   **21c.** 88.3%   **23.** $\frac{1}{3}$ **25.** $\frac{3}{5}$ **27.** D

Pages 230–232   Chapter 5 Study Guide and Assessment
1. proportion   **3.** impossible   **5.** odds   **7.** unit rate
9. Discount   **11.** 45   **13.** 4   **15.** 10 mi   **17.** 30 mi
19. 1 in. = 30.5 ft   **21.** 60   **23.** 150%   **25.** 50
27. 8.1   **29.** 150   **31.** 25% increase   **33.** $33.\overline{3}$%
decrease   **35.** 100%   **37.** $\frac{1}{11}$ **39.** $\frac{7}{11}$ **41.** $\frac{1}{2}$

43. mutually exclusive, $\frac{2}{3}$ **45.** $339.15

Page 235   Preparing for Standardized Tests
1. B   **3.** C   **5.** A   **7.** C   **9.** 2

Chapter 6   Functions and Graphs

Pages 241–243   Lesson 6–1
1. ordered pairs, table, graph   **3.** The domain is the set of all first coordinates from the ordered pairs. The range is the set of all second coordinates from the ordered pairs.

5.

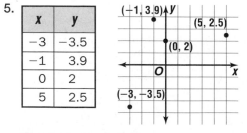

x	y
−3	−3.5
−1	3.9
0	2
5	2.5

(−1, 3.9)   (5, 2.5)   (0, 2)   (−3, −3.5)

domain: {−3, −1, 0, 5},
range: {−3.5, 3.9, 2, 2.5}

7. {(−3, −2), (1, 2), (−4, −2)}

x	y
−3	−2
1	2
−4	−2

domain: {−3, 1, −4},
range: {−2, 2}

9.

x	y
4	3
−2	3
−2	4
4	4

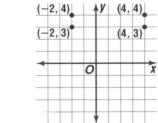

domain: {4, −2},
range: {3, 4}

11.

x	y
−2	0
3	−7
2	−5
−6	3
1	5

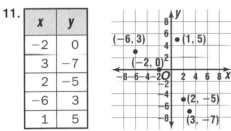

domain: {−2, 3, 2, −6, 1},
range: {0, −7, −5, 3, 5}

13.

x	y
$-\frac{1}{2}$	$\frac{1}{2}$
1	0
$-\frac{1}{2}$	−5

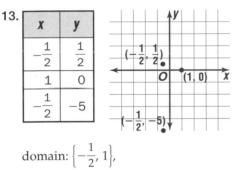

domain: $\left\{-\frac{1}{2}, 1\right\}$,

range: $\left\{\frac{1}{2}, 0, -5\right\}$

15. {(−3, 2), (−1, 1), (1, −2), (1, 3)}

x	y
−3	2
−1	1
1	−2
1	3

domain: {−3, −1, 1},
range: {2, 1, −2, 3}

17. {(−3, 2), (−2, 2), (−1, 4), (2, −2), (3, −2)}

x	y
−3	2
−2	2
−1	4
2	−2
3	−2

domain: {−3, −2, −1, 2, 3},
range: {2, 4, −2}

19. {(−3, −3), (−3, −1), (−1, 0), (2, −2), (3, 4)}

x	y
−3	−3
−3	−1
−1	0
2	−2
3	4

domain: {−3, −1, 2, 3},
range: {−3, −1, 0, −2, 4}

21. {(−1, 1), (0, 2), (1, 3), (2, 4)}
23. {(0, 0), (−1, −0.5), (1, 0.5), (−2, −1)}
25a. {(8, 4), (9, 3), (10, 2), (11, 1), (12, 0)}
25b.

x	y
8	4
9	3
10	2
11	1
12	0

27a. 3, 4, 5, 6, 7, 8, 9, 10, 11, 12, 13, 14, 15
27b. {(3, 1), (4, 1), (5, 2), (6, 2), (7, 3), (8, 3), (9, 4), (10, 3), (11, 3), (12, 2), (13, 2), (14, 1), (15, 1)}

29. 4:6 or 2:3 **31.** −24 **33.** B

Pages 247–249 Lesson 6–2
1. When the values are substituted in the equation, the equation is true. **3.** Dan; when using variables other than x and y, it can be assumed that the domain goes with the variable that comes first alphabetically. Thus, m is the domain. **5.** c, d
7. {(−2, 7), (−1, 2), (0, −3), (1, −8), (2, −13)}

9. $\left\{(-2, -5.5), (-1, -4), (0, -2.5), (1, -1), \left(2, \frac{1}{2}\right)\right\}$

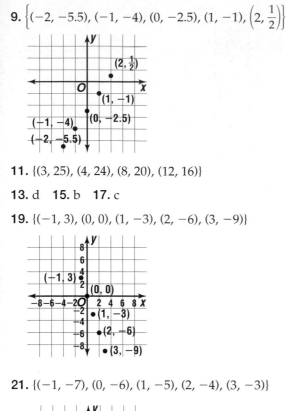

11. $\{(3, 25), (4, 24), (8, 20), (12, 16)\}$

13. d **15.** b **17.** c

19. $\{(-1, 3), (0, 0), (1, -3), (2, -6), (3, -9)\}$

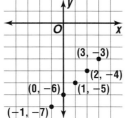

21. $\{(-1, -7), (0, -6), (1, -5), (2, -4), (3, -3)\}$

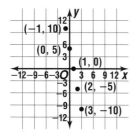

23. $\{(-1, 10), (0, 5), (1, 0), (2, -5), (3, -10)\}$

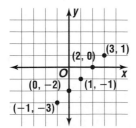

25. $\{(-1, -3), (0, -2), (1, -1), (2, 0), (3, 1)\}$

27. $\{(-1, 6), (0, 4), (1, 2), (2, 0), (3, -2)\}$

29. $\{(-1, -7), (0, -4), (1, -1), (2, 2), (3, 5)\}$

31. $\{0, 1, 2, 3\}$ **33.** $\{-1, 0, 1, 2\}$ **35.** 3

37a.

w	y
6.0	0.15
6.8	0.17
7.2	0.18
8.0	0.2
8.8	0.22

37b. $\{(6.0, 0.15), (6.8, 0.17), (7.2, 0.18), (8.0, 0.2), (8.8, 0.22)\}$

37c.

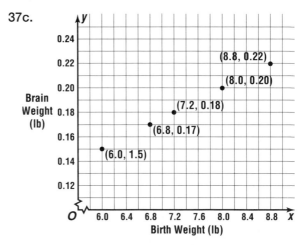

39a. $x + y = 90$; $x = 2y$ **39b.** $(60, 30)$
41. $10.99 **43.** 5 **45.** -1 **47.** C

Pages 254–255   Lesson 6–3
1. The graph of $x = -8$ is a straight line parallel to the y-axis. For every value of y, $x = -8$. **3.** no
5. yes; $A = 2$, $B = 0$, $C = 8$

7.

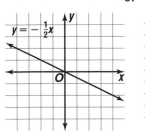
$y = -\frac{1}{2}x$

9.

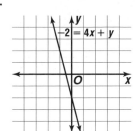
$-2 = 4x + y$

11.

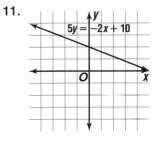
$5y = -2x + 10$

13. no **15.** yes; $A = -6, B = 2, C = 0$

17. yes; $A = -1, B = 1, C = 0$ **19.** no

21. yes; $A = 5, B = 1, C = 9$

23. **25.**

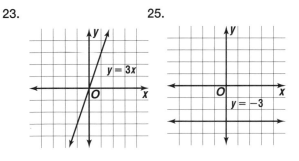
$y = 3x$

$y = -3$

27. **29.**

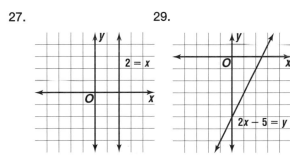
$2 = x$

$2x - 5 = y$

31. **33.**

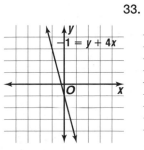
$-1 = y + 4x$

$3 = -3x + y$

35.

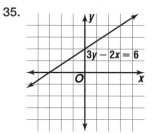
$3y - 2x = 6$

37.

x	y
0	−2
1	−1
2	0
3	1
4	2

39.

x	y
−2	−4
−1	−1
0	2
1	5
2	8

41a. $3x + 6y = 90$ **41b.** Sample answer: {(2, 14), (6, 12), (12, 9), (18, 6), (24, 3)}

41c.

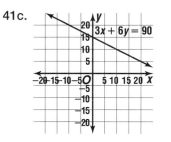
$3x + 6y = 90$

43. {−8, −1, 0, 2, 4} **45.** 6 **47.** 4 **49.** $67 + 2h = 79$

Page 255 Quiz 1

1.

x	y
1	−1
3	5
0	−2
3	3

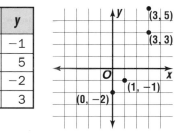
(3, 5)
(3, 3)
(1, −1)
(0, −2)

3. b, c

5a.

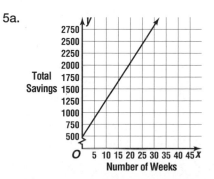
Total Savings
Number of Weeks

5b. 25 weeks **5c.** $y = 500 + 80x$
$$y = 500 + 80(25)$$
$$y = 500 + 2000$$
$$y = 2500$$

Pages 259–261 Lesson 6–4
1. Sample answer: A function is a relation. However, in a relation, each member of the domain can be paired with more than one member of the range. In a function, each member of the domain is paired with exactly one member of the range. **3.** Zina is correct. There are some graphs of straight lines that do not represent a function. The graph of $x = 2$ is a straight line. However, it is not a function since there are many y values for a single x value. **5.** yes **7.** no **9.** -15
11. $4b - 3$ **13.** 3.25 s **15.** yes **17.** no
19. yes **21.** no **23.** no **25.** yes **27.** yes
29. -1 **31.** 13 **33.** 3.4 **35.** -1.7 **37.** 3
39. $-3\frac{1}{2}$ **41.** $-3h + 4$ **43.** $6a + 4$ **45.** -12
47a. $p(a) = 0.5a + 110$ **47b.** 128

47c.
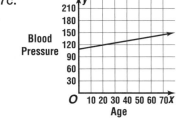

49a. -7

49b.
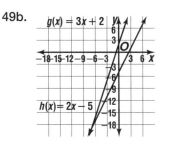

The lines intersect at $x = -7$.

51. {(2, 28), (4, 14), (7, 8)} **53.** 30% **55.** $\frac{3}{20}$ **57.** 1.2

Pages 267–269 Lesson 6–5
1. Divide each side of $y = kx$ by x, where $x \neq 0$
3. Sample answer: The length of the trip depends upon the amount of gasoline used. So, the length of the trip represents the dependent variable and the amount of gasoline used represents the independent variable. **5.** 5 **7.** $\frac{2}{5}$

9. yes
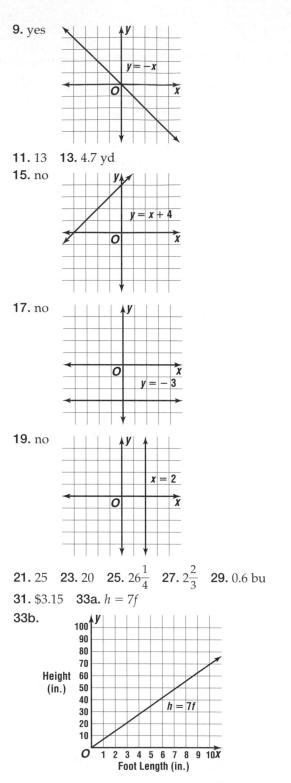

11. 13 **13.** 4.7 yd
15. no

17. no

19. no

21. 25 **23.** 20 **25.** $26\frac{1}{4}$ **27.** $2\frac{2}{3}$ **29.** 0.6 bu
31. $3.15 **33a.** $h = 7f$

33b.

33c. about 5.5 ft **35.** 11 **37a.** $x - 50 = 550$
37b. about 600 pounds **39.** 8 **41.** B

Page 269 Quiz 2
1. yes **3.** -7.4 **5.** 8 gal

Pages 273–275   Lesson 6–6

1. Multiply the value of x by the value of y.
3. Sample answer: In a direct variation, as x increases, y increases. In an inverse variation, as x increases, y decreases.   **5.** direct; 5   **7.** direct; $\frac{1}{3}$

9. direct; $\frac{1}{4}$ **11.** 6   **13.** 3.25 hours   **15.** 10

17. 0.09   **19.** $\frac{5}{8}$ **21.** $-\frac{1}{6}$; $y = -\frac{1}{6}x$ **23.** 0.3; $y = 0.3x$

25. a; inverse   **27.** c; direct   **29a.** 36   **29b.** 2.4 m
29c. $\ell w = 36$

29d.

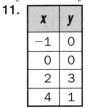

31. 27   **33.** -14.7 **35.** 457 m

Pages 276–278   Chapter 6 Study Guide and Assessment

1. linear equation   **3.** functional notation   **5.** inverse
7. $y = kx$ **9.** $y = 2x + 3$

11.

x	y
−1	0
0	0
2	3
4	1

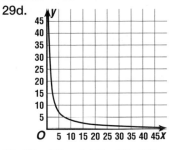

domain: {−1, 0, 2, 4},
range: {0, 3, 1}

13.

x	y
−1.5	2
1	3.5
4.5	−5.5

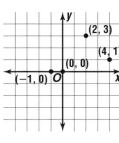

domain: {−1.5, 1, 4.5},
range: {2, 3.5, −5.5}

15. {(−4, 2), (−2, 1), (−2, −3), (0, −2), (1, 3)}

x	y
−4	2
−2	1
−2	−3
0	−2
1	3

domain: {−4, −2, 0, 1},
range: {2, 1, −3, −2, 3}

17. $\left\{\left(-1, \frac{1}{2}\right), (0, 0), (2, 1), (-4, 2)\right\}$

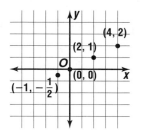

19. {(−1, 6.5), (0, 5), (2, 2), (4, −1)}

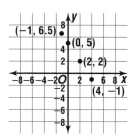

21. {−2, −1, 0, 1}   **23.** yes; $A = -1, B = -1, C = 0$
25. no

27.

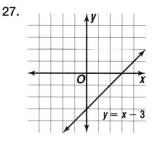

29.

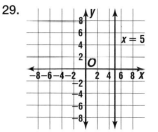

30.

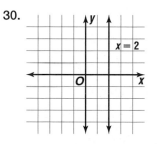

33. no   **35.** yes   **37.** 4.5   **39.** 21   **41.** 12.5
43. 5   **45.** 8; $xy = 8$ **47.** {(5, 1), (10, 2), (15, 3), (20, 4), (25, 5)} domain: {5, 10, 15, 20, 25}, range: {1, 2, 3, 4, 5}   **49.** 132 volts

Page 281 Preparing for Standardized Tests

1. A **3.** E **5.** E **7.** B **9.** $\frac{3}{4}$

Chapter 7 Linear Equations

Pages 287–289 Lesson 7–1

1a. Sample answer: **1b.** Sample answer:

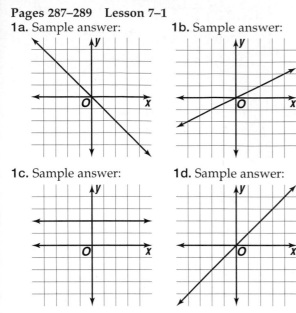

1c. Sample answer: **1d.** Sample answer:

3. Percy; any two points on the line can be used to find the slope. **5.** 2 **7.** $\frac{3}{4}$ **9.** 4 **11.** $-\frac{6}{5}$ **13.** $-\frac{5}{3}$

15. $\frac{1}{3}$ **17.** -4 **19.** -2 **21.** $\frac{1}{3}$ **23.** $-\frac{2}{5}$ **25.** $-\frac{3}{8}$

27. $\frac{7}{100}$ **29a.** As x increases by 1, $f(x)$ increases by 0.25. The slope is 0.25. **29b.** Sample answer: domain: positive integers; range: positive multiples of 0.25

31. 12 **33.** no **35.** -2

Pages 293–295 Lesson 7–2

1. the coordinates of a point on the line and the slope of the line, or the coordinates of two points on the line **3.** $m = 2$, $(1, 6)$ **5.** $m = -3$, $(-9, -5)$

7. $y - 4 = \frac{3}{4}(x + 2)$ **9.** $y - 1 = 0$ or $y = 1$

11. $y - 3 = -\frac{5}{6}(x)$ or $y + 2 = -\frac{5}{6}(x - 6)$

13. $y - 5 = -2(x + 2)$ or $y + 7 = -2(x - 4)$

15. $y - 4 = -3(x + 1)$ **17.** $y - 1 = 6(x - 9)$

19. $y + 2 = \frac{4}{5}(x + 2)$ **21.** $x = 4$ **23.** $y + 4 = -\frac{5}{2}\left(x - \frac{1}{2}\right)$ **25.** $y - 4 = \frac{1}{4}(x + 2)$ or $y - 5 = \frac{1}{4}(x - 2)$

27. $y = \frac{3}{4}(x + 2)$ or $y - 3 = \frac{3}{4}(x - 2)$ **29.** $y - 3 = -\frac{5}{2}(x + 3)$ or $y + 2 = -\frac{5}{2}(x + 1)$ **31.** $y - 5 = \frac{1}{2}(x - 4)$ or $y - 3 = \frac{1}{2}(x)$ **33.** $y + 4 = 3(x - 1)$ or $y - 5 = 3(x - 4)$ **35.** $y - 7 = -\frac{5}{8}(x + 3)$ or

$y - 2 = -\frac{5}{8}(x - 5)$ **37.** $y - 6 = -\frac{1}{7}(x - 3)$ or

$y - 7 = -\frac{1}{7}(x + 4)$ **39a.** Sample answer: $y - 5 = 5(x - 1)$ **39b.** Sample answer: domain: positive integers; range: positive multiples of 5 **41.** $(4, 8)$ **43.** $\{0, 1, 2\}$ **45.** $\{1, 2, 3\}$ **47.** 300% increase **49.** 8.4

Pages 299–301 Lesson 7–3

1. Sample answer:

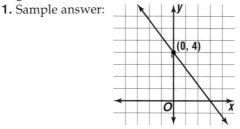

(0, 4)

3. Jacquie; the line $y = 2$ does not intersect the x-axis. **5.** $y = \frac{1}{4}x - 3$ **7.** $y = 4x - 7$ **9.** $y = -\frac{1}{2}x - 3$ **11.** $y = 2x - 8$ **13a.** $y = 25x + 100$

13b. the pounds of cans he plans to collect each week

13c. 400 lb **15.** $y = -2x + 5$ **17.** $y = \frac{3}{2}x + 7$

19. $y = -\frac{2}{5}x + 3$ **21.** $y = 4$ **23.** $y = 3x - 17$

25. $y = 5x$ **27.** $y = -2x + 13$ **29.** $y = -5x + 26$

31. $y = \frac{1}{2}x - \frac{9}{2}$ **33.** $y = -\frac{5}{4}x + 5$ **35.** $y = 4x - 12$

37. $y = -3$ **39.** $y = -\frac{4}{5}x + \frac{4}{5}$ **41.** $y = -\frac{3}{2}x + 6$

43. $y = -\frac{6}{5}x + \frac{8}{5}$ **45.** $y = -3x + 11$ **47a.** $y = 8x$

47b. cost per person **47c.** $320 **49.** $y - 5 = -2(x - 4)$

51.

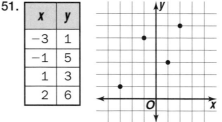

x	y
−3	1
−1	5
1	3
2	6

domain: $\{-3, -1, 1, 2\}$, range: $\{1, 5, 3, 6\}$
53. 9.144 km **55.** -5 **57.** A

Page 301 Quiz 1

1. $-\frac{4}{3}$ **3.** $y - 4 = -2(x - 5)$ **5a.** 90

5b. The slope represents how much each problem is worth and the y-intercept represents the extra points students get with the curve.

Pages 305–307   Lesson 7–4

1. Graph the two sets of data as ordered pairs, where the x-coordinate is a member of one data set and the y-coordinate is a member of the other data set. The set of x-coordinates is the domain and the independent variable. The set of y-coordinates is the range and the dependent variable.

5a.

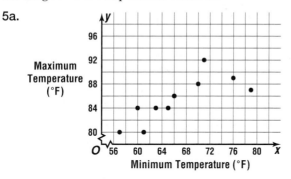

5b. Yes; as maximum temperature increases, so does the minimum temperature. There is a positive relationship.   **7.** Negative relationship; as speed increases, CO_2 decreases.

9.

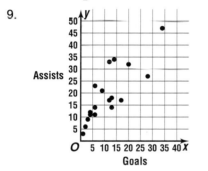

There is a positive relationship; as the number of goals increases, the number of assists increases
11. Sample answer: Have higher maximum speed limits because CO_2 emissions decrease as speed increases.   **13.** no   **15.** negative

17a.

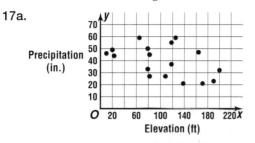

17b. No; there does not appear to be a relationship between elevation and precipitation.   **17c.** zero
19. $y = 5x + 2$ **21.** 1 cm = 5 m or 1:500
23. D

Pages 314–315   Lesson 7–5
1. two points, then draw a line through them

3. $-3, -3$

5. 2, 5

7. $m = 1, b = 1$

9. $m = \frac{1}{4}, b = -3$

11. $m = \frac{9}{5}, b = 32$

13. $-5, -5$

15. $3, -1$

17. $-3, -6$

19. $-4, 3$

21. $2, -7$

23. 4, 8

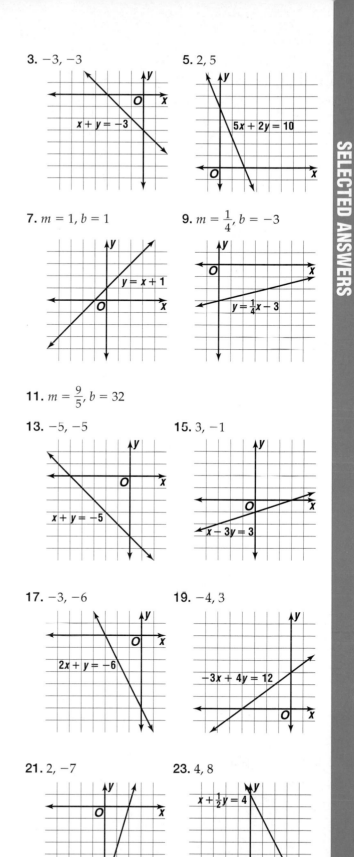

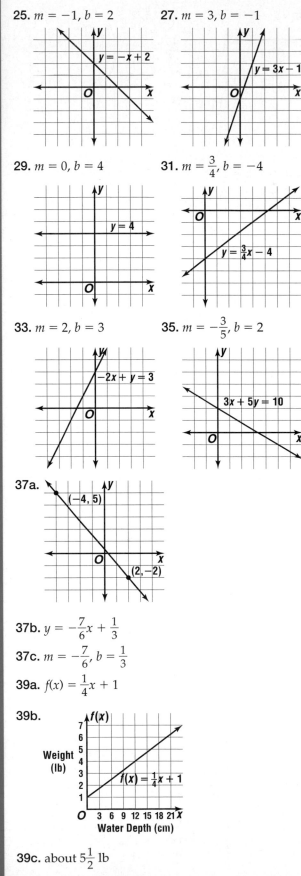

25. $m = -1$, $b = 2$

$y = -x + 2$

27. $m = 3$, $b = -1$

$y = 3x - 1$

29. $m = 0$, $b = 4$

$y = 4$

31. $m = \frac{3}{4}$, $b = -4$

$y = \frac{3}{4}x - 4$

33. $m = 2$, $b = 3$

$-2x + y = 3$

35. $m = -\frac{3}{5}$, $b = 2$

$3x + 5y = 10$

37a.

$(-4, 5)$

$(2, -2)$

37b. $y = -\frac{7}{6}x + \frac{1}{3}$

37c. $m = -\frac{7}{6}$, $b = \frac{1}{3}$

39a. $f(x) = \frac{1}{4}x + 1$

39b.

Weight (lb)

$f(x)$

$f(x) = \frac{1}{4}x + 1$

Water Depth (cm)

39c. about $5\frac{1}{2}$ lb

41. Sample answer:

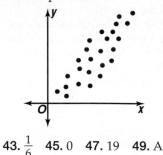

43. $\frac{1}{6}$ **45.** 0   **47.** 19   **49.** A

Pages 319–321   Lesson 7–6

1a. shifted down 4 units   **1b.** less steep
1c. steeper   **1d.** shifted up 3 units   **3a.** Slope represents rate.   **3b.** A greater rate results in a steeper positive slope.

5.

$y = \frac{1}{2}x - 4$

$y = \frac{3}{2}x - 4$

same y-intercept, different slopes

7. same y-intercept, different slopes   **9.** $y = x$

11.

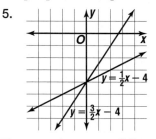

$y = 5x$ $y = 2x$

same y-intercept, different slopes

13.

$y = 4x$

$y = 4x - 2$

same slope, different y-intercepts

15.

$y = -3x$

$y = -\frac{1}{2}x$

same y-intercept, different slopes

17. same y-intercept, different slopes   **19.** same slope, different y-intercepts   **21.** same y-intercept, different slopes   **23.** $y = -\frac{4}{5}x$ **25.** $y = -\frac{4}{5}x + 2$

27. Sample answer: $y = -x + 6$ **29.** $y = 2x + 1$; Sample answer: $y = 2x$ **31a.** Yes; they have the same y-intercept, 0.   **31b.** miles per gallon   **31c.** A barge gets 500 miles per gallon of fuel; its graph has the steepest slope.   **33.** Yes; they have the same y-intercept, 7.   **35.** $m = -2, b = 0$ **37.** 48   **39.** C

Page 321   Quiz 2
1. Positive; as air temperature increases, the temperature of a glass of water would increase.
3. 4, 1

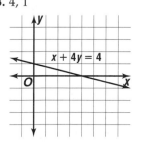

5a. Different slopes; they are a family because they have the same y-intercept, 0.   **5b.** cost per movie
5c. Store B

Pages 325–327   Lesson 7–7
1. Parallel lines have equal slopes. Perpendicular lines have slopes that are negative reciprocals.

3. $4, -\frac{1}{4}$ **5.** $\frac{2}{3}, -\frac{3}{2}$ **7.** neither   **9.** The lines are perpendicular.   **11.** $y = \frac{3}{4}x - \frac{3}{2}$ **13.** $y = \frac{8}{3}x + 4$
15. perpendicular   **17.** neither   **19.** perpendicular
21. parallel   **23.** perpendicular   **25.** $y = -\frac{1}{2}x$
27. $y = -4x - 7$ **29.** $x = 2$ **31.** $y = \frac{1}{4}x + \frac{3}{4}$
33. $y = -\frac{1}{2}x + 1$ **35.** $y = -\frac{5}{3}x - \frac{5}{3}$ **37.** neither
39. Yes; the product of their slopes is -1.
41. 4   **43.** same y-intercept, different slopes
45. mutually exclusive   **47.** mutually exclusive
49. 12.85

Pages 328–330   Chapter 7 Study Guide and Assessment
1. false; y-intercept   **3.** false; parallel   **5.** true
7. false; horizontal   **9.** true   **11.** $\frac{2}{3}$ **13.** -2
15. $y - 5 = 3(x - 2)$ **17.** $y - 3 = -5(x)$
19. $y - 4 = -\frac{1}{6}(x - 10)$ or $y - 7 = -\frac{1}{6}(x + 8)$
21. $y = 3x - 9$ **23.** $y = 2x$ **25.** $y = 11x - 6$
27. no relationship

29. 2, -4 **31.** $m = -2, b = 3$

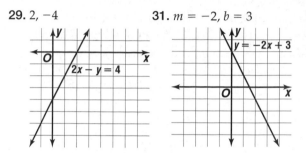

33. same y-intercept, different slope
35. same slope, different y-intercepts
37. neither   **39.** parallel
41a.

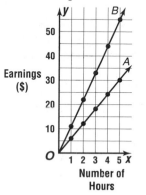

41b. It would be steeper than graph A and less steep than graph B.

Page 333   Preparing for Standardized Tests
1. B   **3.** E   **5.** C   **7.** A   **9.** 4

Chapter 8   Powers and Roots

Pages 339–340   Lesson 8–1
1. A perfect square is the product of a number and itself.   **3.** Becky; $(6n)(6n)(6n) = 6 \cdot 6 \cdot 6 \cdot n \cdot n \cdot n$, not $6n^3$.   **5.** a^5 **7.** $12 \cdot 12 \cdot 12 \cdot 12$ **9.** $m \cdot m \cdot m \cdot m \cdot n \cdot n \cdot n$ **11.** 162   **13.** $2^3 3^2 5$ **15.** $(-2)^4$
17. 7^3 **19.** $2^2 3^3 5$ **21.** $3x^2 y^2$ **23.** $3 \cdot 3 \cdot 3 \cdot 3 \cdot 3$
25. $2 \cdot 2 \cdot 2 \cdot 2 \cdot 3 \cdot 3$ **27.** $y \cdot y \cdot y$ **29.** $6 \cdot a \cdot b \cdot b \cdot b \cdot b$ **31.** -32 **33.** -128 **35.** 24   **37.** 0.5
39. 4   **41.** 486   **43a.** 200.96 ft^2 **43b.** 21 bags
45. $y = -2x + 6$
47.

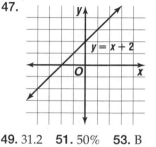

49. 31.2   **51.** 50%   **53.** B

Pages 344–345   Lesson 8–2
1. In the first case, the bases are the same, but in the second case, the bases are different.   **3.** y^7 **5.** t^5

7. $12a^5$ **9.** n^3 **11.** xy^3 **13.** b^5 **15.** $10^3 = 1000$
17. 5^4 **19.** d^6 **21.** a^3b^5 **23.** r^7t^8 **25.** $15a^2$
27. $8x^3y^5$ **29.** m^7b^2 **31.** $m^3n^3a^2$ **33.** 9 **35.** k^5
37. $-12ab^4$ **39.** 4 **41.** $15a^3c$ **43.** $2x^3$ **45.** $-16x^2$
47. 10^6 or 1,000,000 times **49a.** 10^{12} **49b.** 4^{15}
49c. x^8 **49d.** Multiply the exponents; $(x^a)^b = x^{ab}$.
51. 3 **53.** 15 **55.** 9.5 **57.** 3 **59.** D

Pages 349–351 Lesson 8–3

1. $\frac{1}{5^2} = \frac{1}{25}$ **3a.** Booker **3b.** Antonio **5.** $\frac{1}{3^3} = \frac{1}{27}$
7. $\frac{1}{n^3}$ **9.** $\frac{r^2}{pq^2}$ **11.** a^7 **13.** $\frac{c^3}{9a}$ **15.** $\frac{1}{2^5} = \frac{1}{32}$ **17.** $\frac{1}{4}$
19. $\frac{1}{r^{10}}$ **21.** a^2 **23.** $\frac{t^4}{s^2}$ **25.** $\frac{15r}{s^2}$ **27.** m^6 **29.** $\frac{1}{k^8}$
31. $\frac{a}{n^2}$ **33.** $\frac{1}{y^9}$ **35.** $3b^9$ **37.** $\frac{x^3}{4}$ **39.** $\frac{x^2}{7}$ **41.** $\frac{5}{8b^5c}$
43. 18 **45.** 2^{-4} **47.** radar microwaves
49. $-10a^7$ **51.** n^6 **53.** z^4 **55.** 9^1 **57.** B

Page 351 Quiz 1

1. 6^3 **3.** 10^1 **5.** 2^53 **7.** x^5 **9.** $-\frac{1}{3b^3}$

Pages 355–356 Lesson 8–4

1. no, 2.35×10^4 **5.** 68,000 **7.** 0.64 **9.** 0.0003
11. 0.0065 **13.** 5.69×10 **15.** 2.3×10^{-5}
17. $2.6 \times 10^3 = 2600$ **19.** 5,800,000,000
21. 0.0000000039 **23.** 0.009 **25.** 5.28×10^3
27. 2.683×10^2 **29.** 3.2×10^{-4} **31.** 4.296×10^{-3}
33. 1.2×10^{-2} **35.** 2.2×10^{-8} **37.** $6 \times 10^6 =$
6,000,000 **39.** $7.5 \times 10^2 = 750$ **41.** $2 \times 10^3 = 2000$
43. $5.5 \times 10^{-5} = 0.000055$ **45.** 5,000,000
47. Earth, Venus, Mars **49.** between 0.3 and
2 mm **51.** y^3 **53.** $-6c^6$ **55.** x^6 **57.** a^3b^3
59. $-6x^2y^6$ **61.** C

Pages 360–361 Lesson 8–5

1. 1, 4, 9, 16, 25, 36, 49, 64, 81, 100 **3.** $3^2 = 9$ and
$\sqrt{9} = 3$ **5.** 11^2 **7.** $2^2 \cdot 7^2$ **9.** 2 **11.** -11 **13.** $\frac{1}{2}$
15. 52.5 mi **17.** 10 **19.** -14 **21.** 23 **23.** -26
25. -22 **27.** 17 **29.** $-\frac{3}{10}$ **31.** $-\frac{1}{4}$ **33.** $\frac{6}{7}$
35. $\frac{3}{4}$ **37.** -0.05 **39.** 0.03 **41.** 6 **43.** 52 m
45. false; $-6(-6) \neq -36$ **47.** 6.3×10^4 **49.** $7.6 \times$
10^{-4} **51.** $\frac{1}{x^4}$ **53.** $\frac{1}{x^2y^3}$ **55.** Sample answer: $2x +$
$6 = x + 1$

Page 361 Quiz 2

1. 7×10^5 **3.** 21 **5.** square house, 30 ft $\times$ 30 ft

Pages 364–365 Lesson 8–6

1. Its decimal value does not terminate or repeat.
3. Sample answer: The area of a square garden is

200 square feet. Estimate the length of one side of
the garden. **5.** 81, 100 **7.** 484, 529 **9.** 8 **11.** 16
13. 2 **15.** 4 **17.** 6 **19.** 11 **21.** 20 **23.** 24
25. 8 **27.** 11 **29.** 12 **31.** 0 **33.** $\sqrt{34}$ **35.** 17 h
37. 31 mi^2 **39.** 10, 11, 12, 13, 14, 15 **41.** 16
43. $\frac{13}{11}$ **45.** 0.000000004 **47.** $y = 5x - 17$
49. $y = -5x + 29$ **51.** C

Pages 369–371 Lesson 8–7

1. Sample answer:

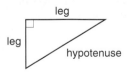

3. false **5.** false **7.** 14.1 **9.** 8.1 **11.** yes **13.** 13
15. 8.2 **17.** 4.2 **19.** 6.7 **21.** 21 **23.** 11.4 **25.** no
27. no **29.** yes **31.** yes **33.** 9.4 m **35.** 6.9 ft
37. 8 **39.** 14 **41.** 15 **43.** 10 **45a.** $100
45b. $5 **47.** C

Pages 374–376 Chapter 8 Study Guide and Assessment

1. prime factorization **3.** irrational numbers
5. exponent **7.** hypotenuse **9.** perfect squares
11. 9^5 **13.** 2^23^3 **15.** $12 \cdot 12 \cdot 12$ **17.** $y \cdot y$
19. $5 \cdot x \cdot x \cdot y \cdot y \cdot y \cdot z \cdot z$ **21.** b^7 **23.** a^4b^3
25. y^4 **27.** $3x^3$ **29.** $\frac{1}{2^3} = \frac{1}{8}$ **31.** $\frac{1}{x^4}$ **33.** y^7
35. $-5b^3$ **37.** 0.0000000015 **39.** 2.4×10^5
41. 4.88×10^9 **43.** 6×10^{11} **45.** 11 **47.** $\frac{2}{9}$
49. 4 **51.** 14 **53.** 7 **55.** 11.2 **57.** 10.4 **59.** 9 m

Page 379 Preparing for Standardized Tests

1. C **3.** B **5.** C **7.** D **9.** 25%

Chapter 9 Polynomials

Pages 385–387 Lesson 9–1

1. Sample answer: $\frac{2}{x}$; it includes division.
3. $2y^2 + 5xy + 3x^2$ **5.** yes; product of numbers and
variables **7.** yes; binomial **9.** no **11.** 1 **13.** 2
15a. 20 **15b.** 9 **17.** no; includes addition
19. yes; product of numbers and variables
21. yes; product of numbers and variables
23. yes; trinomial **25.** no **27.** yes; binomial
29. yes; binomial **31.** yes; monomial **33.** yes;
monomial **35.** 2 **37.** 2 **39.** 1 **41.** 3 **43.** 6
45. 5 **47.** $-x^5 + x^2 - x + 25$ **49.** $5x^7 - 10x^6 +$
$3wx^2 + 6w^3x$ **51.** $7 + 2x - x^2 + 5x^3$ **53.** $y^6 +$
$5xy^4 - x^2y^3 + 3x^3y$ **55.** false **57a.** $-0.006t^4 +$
$0.14t^3 - 0.53t^2 + 1.79t$ **57b.** 4 **59.** yes **61.** yes
63. 7 **65.** 11 **67.** $\frac{x}{y^3}$ **69.** B

Pages 392–393   Lesson 9–2

1. Arrange like terms in column form.
3. $-2a - 9b$ **5.** $-x^2 - 8x - 5$ **7.** $-4xy^2 - 6x^2y + y^3$
9. $8y + 2$ **11.** $3x^2 - x + 2$ **13.** $2x + 2$ **15.** $x^2 + 3x$
$- 3$ **17.** $x^2 - x + 5$ **19.** $4x^2 + 3x + 3$ **21.** $14x + y$
23. $3n^2 + 13n + 11$ **25.** $7n^2 - 8n + 5$ **27.** $5x + 8$
29. $-x^2 + 3x - 3$ **31.** $-3x + y$ **33.** $a^2 + 3a - 1$
35. $5x^2 - xy - 2y^2$ **37.** $2pq + pr$ **39.** $-2x^2y - 7x^2y^2$
41. $-2x^2 + 6x - 7$ **43a.** $2w + 4$ in. **43b.** 34 in.
45. 2 **47.** 3 **49.** x^5y^3 **51.** $10xy^2$

Pages 396–398   Lesson 9–3

1. Distributive Property **3.** Consuelo; Shawn
forgot to multiply $2x$ and 4. **5.** $x^2 + 2x$ **7.** $-4x + 8$
9. $4z^3 - 8z^2$ **11.** $24y^2 + 9y - 15$ **13.** 4 **15.** 2
17. $3x^2$ **19.** $-3y - 9$ **21.** $x^2 - 5x$ **23.** $3z^2 - 2z$
25. $8x^2 - 12x$ **27.** $-2a^2 + 4a$ **29.** $-30y + 10y^2$
31. $5d^3 + 15d$ **33.** $-10a^3 + 14a^2 - 4a$ **35.** $40n^4 +$
$35n^3 - 15n^2$ **37.** $-14a^2 + 35a - 77$ **39.** $1.2c^2 - 12$
41. $x^2 - 2x$ **43.** $4y^2 - 6y + 2$ **45.** 5 **47.** -2
49. -1 **51.** 2 **53.** -3 **55.** 17 **57.** $4x^3 - 8x^2 + 4x$
59. $-4a^3 - 10a^2 - 10a + 66$ **61.** $21n^3 - 6n^2 -$
$46n + 28$ **63.** $8t^2 + t$ **65a.** $7y^2 + 35y$ units^2
65b. 1050 in^2 **67.** $7x - 2$ **69.** $3x^2 - 2x - 3$

71. 5 **73.** -11 **75.** $\frac{3}{10}$ **77.** 1:1

Page 398   Quiz 1

1. 4 **3.** $-2x^2 + 8x - 2$ **5a.** x ft, $2x + 40$ ft
5b. $2x^2 + 40x$ ft^2

Pages 402–404   Lesson 9–4

1.

x	3
x^2	$3x$
$-2x$	-6

$(x - 2)(x + 3) = x^2 + x - 6$

3. The Distributive Property is used to multiply
both two binomials and a binomial and a
monomial. But when you multiply two binomials,
there are four multiplications to perform. With a
binomial and a monomial there are only two
multiplications to perform. **5.** $6x$ **7.** $13m$ **9.** $14y$
11. $w^2 - 2w - 35$ **13.** $2y^2 + y - 3$ **15.** $6y^2 +$
$11y - 35$ **17.** $2a^2 - ab - 10b^2$ **19.** $m^4 + 3m^3 -$
$2m^2 - 6m$ **21.** $x^2 + 12x + 32$ **23.** $a^2 + 4a - 21$
25. $n^2 - 16n + 55$ **27.** $z^2 + 2z - 24$ **29.** $3x^2 -$
$10x - 8$ **31.** $3z^2 - 17z + 20$ **33.** $5n^2 - 13n - 6$
35. $9a^2 + 6a + 1$ **37.** $12h^2 - h - 6$ **39.** $2x^2 + x - 21$
41. $6y^2 - 24yz + 24z^2$ **43.** $6x^2 + 13xy - 5y^2$
45. $x^3 - 2x^2 + x - 2$ **47.** $3x^4 + 8x^2 - 3$ **49.** $x^3 -$
$x^2 - 6x$ **51.** $x^2 - 9$ **53a.** $3x + 2$ **53b.** $3x^2 + 4x - 1$
55a. $4x^3 + 2x^2 - 20x$ in. **55b.** 450 in^3 **57a.** $10 - 2t$
57b. 6 in. **59.** 2^3x^3 **61.** $3^2(-2)^3$

63.

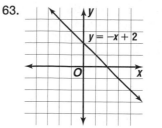

$y = -x + 2$

65. 84 students

Pages 408–409   Lesson 9–5

1a. ii **1b.** iii **1c.** i **3.** $y^2 + 4y + 4$ **5.** $m^2 - 18m +$
81 **7.** $x^2 - 49$ **9.** 90.25% **11.** $x^2 + 10x + 25$
13. $w^2 - 16w + 64$ **15.** $m^2 - 6mn + 9n^2$ **17.** $x^2 -$
$16y^2$ **19.** $25 + 10k + k^2$ **21.** $y^2 - 9z^2$ **23.** $9x^2 - 25$
25. $16 + 16x + 4x^2$ **27.** $a^2 - 6ab + 9b^2$ **29.** $x^3 +$
$2x^2 + x$ **31.** $x^2 - 4y^2$ **33.** $(40 - 1)(40 + 1) =$
$1600 - 1$ or 1599 **35a.** $\frac{1}{2}x^2 + \left(-\frac{9}{2}\right)$ **35b.** 8 square
units **35c.** about 9.5 units **37.** $3x^2 - 4x - 3$
39. $y = 3x + 4$ **41.** $y = \frac{2}{5}x - 3$ **43.** B

Page 409   Quiz 2

1. $c^2 + 10c + 16$ **3.** $c^2 - 2c + 1$ **5.** $12x$ cm^2

**Pages 412–414   Chapter 9 Study Guide and
Assessment**

1. i **3.** j **5.** c **7.** a **9.** e **11.** yes; monomial
13. yes; trinomial **15.** 0 **17.** 4 **19.** $7cd - d^2 + 3d^3$
21. $9x^2 - 3x + 6$ **23.** $7s^2 - 6s - 2$ **25.** $4g^2 + 8g$
27. $-10s^2t^2 + 2st^2$ **29.** $3x - 15$ **31.** $2m^3 + 8m^2$
33. $10x^2 + 15x - 10$ **35.** -3 **37.** $y^2 + 4y - 12$
39. $x^2 - 4x + 3$ **41.** $2m^2 - 9m + 4$ **43.** $12y^2 +$
$10y + 2$ **45.** $10x^2 - 7x - 12$ **47.** $4x^2 + 12x + 9$
49. $x^2 - 4x + 4$ **51.** $25m^2 - 30mn + 9n^2$
53. $4a^2 - 9$ **55.** $25a^2 + 10a + 1$ **57.** $10\frac{1}{2}x^2 +$
$22x - 16$ **59a.** $64x^2 - 32x + 4$ **59b.** $(48x - 12)$
$(48x - 12)$

Page 417   Preparing for Standardized Tests
1. C **3.** D **5.** B **7.** A **9.** $265

Chapter 10   Factoring

Pages 424–425   Lesson 10–1
1. 23, 29, 31, 37, 41, 43, 47 **3.** Write $8x^2$ and $16x$ as
the product of prime numbers and variables where
no variable has an exponent. Then multiply the
common prime factors and variables. The GCF is $8x$.
5. $3 \cdot 7$ **7.** $3 \cdot 17$ **9.** $2^2 \cdot 3^3$ **11.** 1, 2, 3, 6, 7, 14, 21,
42; composite **13.** $2 \cdot 2 \cdot 2 \cdot 3 \cdot x \cdot x \cdot y$ **15.** 5
17. 1 **19.** $7y^2$ **21.** 72 cm, 1 cm **23.** 1, 2, 4, 5, 10,
20; composite **25.** 1, 3, 5, 9, 15, 45; composite
27. 1, 7, 13, 91; composite **29.** $-1 \cdot 3 \cdot 5 \cdot a \cdot a \cdot b$
31. $2 \cdot 5 \cdot 5 \cdot m \cdot m \cdot n \cdot n$ **33.** $2 \cdot 3 \cdot 3 \cdot 5 \cdot y \cdot z \cdot z$
35. 4 **37.** 18 **39.** 9 **41.** $3y$ **43.** 9 **45.** 1 **47.** 1
49. 5 **51.** 5 **53.** 5 **55.** 12 in. by 12 in. **57.** Every
even number has a factor of 2. So, any even number

greater than 2 has at least 3 factors—the number itself, 1, and 2—and is therefore composite.
59. $4y^2 + 4y + 1$ **61.** $z^2 + 7z + 12$ **63.** $2n^2 + 9n + 4$ **65.** $6a - 2a^3$ **67.** 8.17×10^8

Pages 431–433 Lesson 10-2
1. Sample answer:

3. The Distributive Property is used to factor out a common factor from the terms of an expression. Examples: $5xy + 10x^2y^2 = 5xy(1 + 2xy)$, $2a^2 + 8a + 10 = 2(a^2 + 4a + 5)$ **5.** x **7.** $6y$ **9.** $5m^2n$
11. $2x(x + 2)$ **13.** prime **15.** $2ab(a^2b + 4 + 8ab^2)$
17. $c + 3$ **19.** $3(3x + 5)$ **21.** $2x(4 + xy)$ **23.** $3c^2d$ $(1 - 2d)$ **25.** $mn(36 - 11n)$ **27.** prime **29.** $y^3(12x + y)$ **31.** $3y(8x + 6xy - 1)$ **33.** $x(1 + xy^3 + x^2y^2)$
35. $2a(6xy - 7y + 10x)$ **37.** $9x^2 - 7y^2$ **39.** $a + 2$
41. $2xy^2 + 3z$ **43.** $2x^2 + 3$ **45.** $(4x + y)$ ft
47a. $8(a + b + 8)$ **47b.** $3(4a + b + 12)$ **49.** prime
51. composite **53.** composite **55.** $3x^2 - 2x - 12$
57. Sample answer: $x^2 + 3x - 4$

Page 433 Quiz 1
1. $2^3 \cdot 3$ **3.** x^3; $x^3(a + 7b + 11c)$ **5.** $8(x + 1)$

Pages 438–439 Lesson 10-3
1.

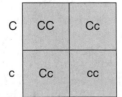

$(x + 1)(x + 6)$

3. $-10, -3$ **5.** $-4, -3$ **7.** $-6, 1$ **9.** $(x + 3)(x + 2)$
11. $(a - 1)(a - 4)$ **13.** $(x + 5)(x - 2)$ **15.** prime
17. $2(c - 7)(c + 1)$ **19.** $(b + 4)(b + 1)$ **21.** $(a + 3)(a + 4)$ **23.** $(y + 9)(y + 3)$ **25.** $(x - 5)(x - 3)$
27. $(c - 9)(c - 4)$ **29.** prime **31.** $(c + 3)(c - 1)$
33. $(r - 6)(r + 3)$ **35.** $(n + 15)(m - 2)$ **37.** $(x - 9)(x - 8)$ **39.** $(r + 24)(r - 2)$ **41.** $3(y - 4)(y - 3)$
43. $m(m + 1)(m + 2)$ **45.** $2a(a + 8)(a - 1)$
47. Sample answer: $x^2 + 7x - 9$ **49a.** $x^2 - 5x - 24$
49b. $(x - 8)(x + 3)$
51a.

	C	c
C	CC	Cc
c	Cc	cc

51b.

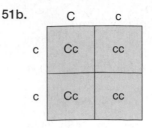

53. $2x^2 + 5y^2$ **55.** $3a^2 + 4ab - 3b^2$ **57.** 4 **59.** 1
61a. $y = 0.2x - 398$ **61b.** the average increase each year **61c.** $3 billion

Pages 443–444 Lesson 10-4
1a. $2x^2 - 5x - 12 = (x - 4)(2x + 3)$ **1b.** $6x^2 - 11x + 3 = (2x - 3)(3x - 1)$ **1c.** $5x^2 + 28x + 15 = (5x + 3)(x + 5)$ **3.** $(2a + 3)(a + 1)$ **5.** $(5x + 3)$ $(x + 2)$ **7.** $(2x + 7)(x - 3)$ **9.** $(5a - 2)(2a - 1)$
11. $2(3x + 5)(x + 1)$ **13.** $(2y + 1)(y + 3)$ **15.** $(2a + 1)(2a + 3)$ **17.** $(2q + 3)(q - 6)$ **19.** $(7a + 1)(a + 3)$
21. $(3x + 2)(x + 4)$ **23.** $(3x + 5)(x + 3)$ **25.** prime
27. $(7x - 1)(2x + 5)$ **29.** $(2m - 1)(4m - 3)$
31. $3(2x + 5)(x - 2)$ **33.** $x(2x - 3)(x + 4)$ **35.** $(6y - 1)(y + 2)$ **37.** $(2a - b)(a + 3b)$ **39.** $(3k + 5m)$ $(3k + 5m)$ **41.** $3x, 2x - 1, x + 3$ **43a.** $2(y^2 - 7y - 2)$
43b. $2(y - 9)(y - 1)$ **43c.** 56 in^2, 18 in^3 **45.** $(x + 16)$ $(x - 2)$ **47.** $3a(a^2 - 5a + 2)$ **49.** 25 ft **51.** 10

Page 444 Quiz 2
1. $(x + 5)(x - 2)$ **3.** $(2x + 7)(x + 1)$ **5.** $(x + 3)$ $(x - 2)$

Pages 448–449 Lesson 10-5
1. No; 7 is not a perfect square. **3.** $(y + 7)^2$
5. $(a - 5)^2$ **7.** $2(2x - 5y)(2x + 5y)$ **9.** $3(x^2 + 5)$
11. $3(y - 1)(y + 8)$ **13.** $(r + 4)^2$ **15.** $(a + 1)^2$
17. $(2z - 5)^2$ **19.** $(3a + 4)^2$ **21.** $(7 + z)^2$
23. $(a - 6)(a + 6)$ **25.** $(1 - 3m)(1 + 3m)$ **27.** no
29. $2(z - 7)(z + 7)$ **31.** Sample answer: $25x^2 - 4$; $(5x - 4)(5x + 4)$ **33.** $5(x^2 + 5)$ **35.** $(y - 3)(y - 2)$
37. $2(x - 6)(x + 6)$ **39.** $xy(8y - 13x)$ **41.** prime
43. $2(5n + 1)(2n + 3)$ **45.** $(4w - 3)(2w + 5)$
47. $5(x + 1)(x + 2)$ **49.** $2x(x - 4)(x + 4)$ **51.** 2, 4
53. 4 **55.** $(4x - 1)(x + 3)$ **57.** $(m - 7)(m + 2)$

59. a^2 **61.** $\frac{d}{3c^2}$ **63.** C

Pages 450–452 Chapter 10 Study Guide and Assessment
1. false, $2^2 \cdot 3$ **3.** true **5.** true **7.** false, $x^2 - 9$
9. true **11.** 5 **13.** 1 **15.** $9x$ **17.** $5(x + 6y)$
19. $6a(2b - 3a)$ **21.** $3xy(1 + 4xy)$ **23.** $5ab - 1$
25. $(x - 5)(x - 3)$ **27.** $(x - 4)(x + 2)$ **29.** $(x + 7)$ $(x - 5)$ **31.** $2(n - 6)(n + 2)$ **33.** $(3x + 5)(x + 1)$
35. $(3a - 2)(2a + 1)$ **37.** $(2y + 3)(y - 6)$
39. $5(3a - 1)(a - 1)$ **41.** $(a - 6)^2$ **43.** $(5x + 2)^2$
45. $(2x + 3)(2x - 3)$ **47.** $12(c + 1)(c - 1)$

49.

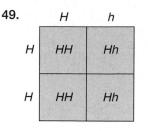

	H	h
H	HH	Hh
H	HH	Hh

Page 455   Preparing for Standardized Tests
1. C   **3.** D   **5.** D   **7.** B   **9.** 0.5

Chapter 11   Quadratic and Exponential Functions

Pages 461–463   Lesson 11–1
1. Sample answer: Quadratic functions have a degree of 2, but linear functions have a degree of 1. The graphs of quadratic functions are not straight lines as with linear functions.   **3.** the vertex
5. 1, 0, 4   **7.** 3, 4, 0
9.

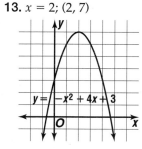

11. $x = 0$; $(0, 2)$

13. $x = 2$; $(2, 7)$

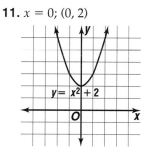

15a.

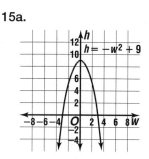

15b. 9 ft   **15c.** 6 ft

17.

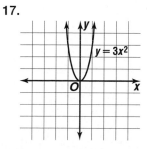

19.

21.   **23.**

25.   **27.**

29.   **31.**

33.   **35.**

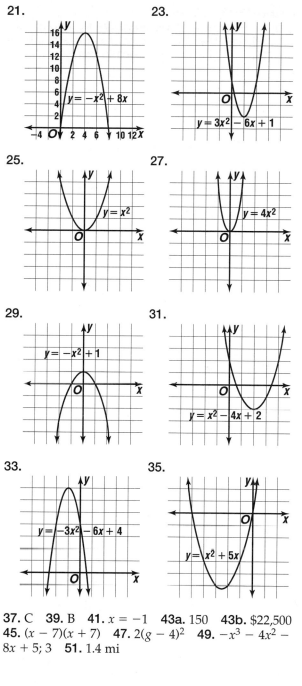

37. C   **39.** B   **41.** $x = -1$ **43a.** 150   **43b.** \$22,500
45. $(x - 7)(x + 7)$ **47.** $2(g - 4)^2$ **49.** $-x^3 - 4x^2 - 8x + 5$; 3   **51.** 1.4 mi

Pages 466–467   Lesson 11–2
1. extremely narrow with vertex at $(0, 0)$, opening up   **3.** B, A, C   **5.** widens; $(0, 0)$ **7.** left 4 units; $(-4, 0)$ **9.** $y = 0.5(x - 4)^2 + 3$ **11.** shifts left, opens up   **12.** same axis of symmetry, shifts down, opens down   **13.** narrows; $(0, 0)$ **15.** right 7 units; $(7, 0)$ **17.** opens down, narrows; $(0, 0)$ **19.** narrows, up 1 unit; $(0, 1)$ **21.** widens, down 8 units; $(0, -8)$
23. left 1 unit, down 5 units; $(-1, -5)$ **25.** $y = (x + 8)^2$ **27.** $h(t) = -4.9(t - 34)^2 + 80$ **29.** $y = (x - 2)^2 - 4$ **31.** $(-1.5, -1.5)$ **33.** $a^2 + 3u - 28$
35. B

Selected Answers   755

Pages 471–473   Lesson 11–3
1. Sample answer: The x-intercepts are the values for x where $f(x)$ equals 0. Before solving a quadratic function, you always set it equal to zero.
3. Sample answer:

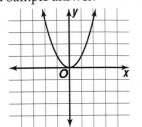

5. $-1, 4$ **7.** $2, 3$ **9.** between -2 and -1; between 2 and 3   **11.** -1 **13.** $-2, 7$ **15.** 5   **17.** between -5 and -4; between 0 and 1   **19.** between -1 and 0; between 2 and 3   **21.** $-4, -8$ or 4, 8
23.

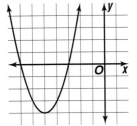

25. between 46 and 47 s   **27.** The value of the function changes from negative when $x = 1$ to positive when $x = 2$. To do this, the value of the function would have to be 0 for some value of x between 1 and 2. Thus, the value of x represents the x-intercept of the function. This is the root of the related equation.   **29.** up 4 units; $(0, 4)$ **31.** 4

Page 473   Quiz 1
1. $x = -3$; $(-3, -14)$ **3.** $x = -1$; $(-1, 7)$

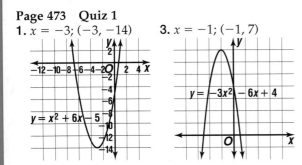

5. right 6 units; $(6, 0)$ **7.** 2, 3   **9.** between -1 and 0; between 2 and 3

Pages 476–477   Lesson 11–4
1. Any number times 0 is 0. Therefore, if two or more numbers are multiplied and the result is 0, at least one of those numbers has to equal 0.
3. They are both correct, but factoring can only be used when the equation is not prime.   **5.** 0, 5
7. 3   **9.** $-2, 4$ **11.** 0, -2 **13.** $-7, 6$ **15.** $-1, 4$
17. $-8, -2$ **19.** 3, 8   **21.** $-3, 4$ **23.** 0, 3   **25.** 2

27. $-1, 5$ **29.** 25 ft by 35 ft   **31.** 8, 11; $-11, -8$
33. 3 ft by 1 ft by 1 ft   **35a.** $-5, 2$ **35b.** Sample answer: $y = (x + 5)(x - 2)$ **37.** no real solution
39. The graph of $y = 0.25x^2$ is wider than the graph of $y = x^2$. The graph of $y = 4x^2$ is narrower than the graph of $y = x^2$.

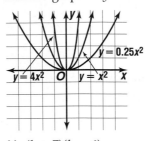

41. $(b + 7)(b - 6)$

Pages 481–482   Lesson 11–5
1. The equation is not factorable with integers and the coefficient of x^2 is 1.   **3.** yes   **5.** no   **7.** 1
9. $\frac{9}{4}$ **11.** $-3, -4$ **13.** 9 ft by 11 ft   **15.** 36   **17.** 1
19. 144   **21.** $\frac{25}{4}$ **23.** $-3, 1$ **25.** $-5, -1$ **27.** 0, 14
29. $1 \pm \sqrt{31}$ **31.** $3 \pm \sqrt{19}$ **33.** $-5, 2$ **35.** 3 s or 5 s   **37.** -18 or 18   **39.** 2, 5   **41.** -3
43a. $2b(a + 4b)$ **43b.** $2a + 12b$ **43c.** 9 in. by 2 in., 18 in^2, 22 in.

Pages 486–487   Lesson 11–6
1. $b^2 - 4ac < 0$ **3.** 48   **5.** 192   **7.** ± 5 **9.** $-3, -1$
11. $-6, -1$ **13.** 2, 3   **15.** no real solutions   **17.** 0, 8
19. $\frac{-1 \pm \sqrt{29}}{2}$ **21.** $\frac{-2 \pm \sqrt{7}}{2}$ **23.** $\frac{-7 \pm \sqrt{69}}{2}$
25. $\frac{5 \pm \sqrt{249}}{16}$ **27a.** 6.25 s   **27b.** about 1.02 s
29. 4   **31.** 81   **33.** B

Page 487   Quiz 2
1. $-6, 2$ **3.** 3, 7   **5.** $3 \pm \sqrt{2}$ **7.** no real solutions
9. $\frac{-5 \pm \sqrt{13}}{2}$

Pages 491–493   Lesson 11–7
1. Sample answer: The graph begins almost flat, but then for increasing x-values, it becomes more and more steep.   **3.** 256   **5.** 1.26
7. -5

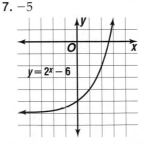

9a.

Year	Balance	Pattern
0	500	$500(1.02)^0$
1	$500(1.02)1.02 = 510$	$500(1.02)^1$
2	$(500 \cdot 1.02)1.02 = 520.20$	$500(1.02)^2$
3	$(500 \cdot 1.02 \cdot 102)1.02 = 530.604$	$500(1.02)^3$

9b. $T(x) = 500(1.02)^x$ **9c.** $714.12   **9d.** $1428.25
11. 1   **13.** 2

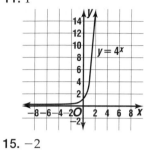

15. -2 **17.** 0

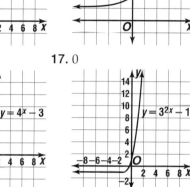

19. $5304.50   **21.** $3920.88   **23.** $7986.70

25a.

Day	Cents	Pattern
1	1	2^0
2	$2 \cdot 1$	2^1
3	$2 \cdot 2 \cdot 1$	2^2
4	$2 \cdot 2 \cdot 2 \cdot 1$	2^3
5	$2 \cdot 2 \cdot 2 \cdot 2 \cdot 1$	2^4

25b. $y = 2^{x-1}$ **25c.** January 17   **27.** $-2, 2$
29a. $(2x + 4)x = 30$ **29b.** 3 yd by 10 yd
31. $y - 6 = 0$

Pages 496–498   Chapter 11 Study Guide and Assessment

1. f   **3.** c   **5.** j   **7.** a   **9.** d   **11.** $x = 0; (0, -3)$
13. $x = -1; (-1, 20)$
15.

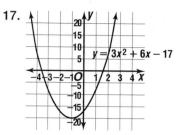

17.

19. shifts right; (8, 0)   **21.** widens; (0, 0)
23. shifts right and up; (1, 7)   **25.** $-3, 4$
27. no real solutions   **29.** $-1 < x < 0, 3$
31. $-4, 5$ **33.** -8 **35.** $-1, 1$ **37.** 4   **39.** $-3, 7$
41. $-1, -9$ **43.** $-\frac{1}{3}, 1$ **45.** $\dfrac{-5 \pm \sqrt{13}}{2}$
47. 1   **49.** 4

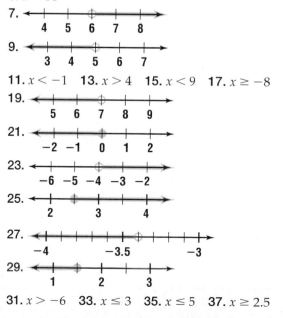

51. $5.50

Page 501   Preparing for Standardized Tests
1. D   **3.** A   **5.** C   **7.** C   **9.** 2

Chapter 12   Inequalities

Pages 506–508   Lesson 12–1
1. i. b; ii. a; iii. d; iv. c   **3.** The graph of $x \le 4$ has a bullet at 4. The graph of $x < 4$ has a circle at 4.
5. $x < 14$
7.
9.
11. $x < -1$ **13.** $x > 4$ **15.** $x < 9$ **17.** $x \ge -8$
19.
21.
23.
25.
27.
29.
31. $x > -6$ **33.** $x \le 3$ **35.** $x \le 5$ **37.** $x \ge 2.5$

39. $f < 3$;

41. $w \le 3600$;

43. 7 yr **45.** no real solutions **47.** $2^3 \cdot 3$; two
49. $-6x^2y$

Pages 511–513 Lesson 12–2
1. They are the same. **5.** $8 + n \le 12$ **7.** $\{v \mid v \ge 17\}$
9. $\{h \mid h \le 21\}$ **11.** $\{y \mid y < 1\frac{5}{6}\}$
13. $\{x \mid x < -1\}$;

15. $\{n \mid n > 3\}$ **17.** $\{b \mid b < -5\}$ **19.** $\{r \mid r < 19\}$
21. $\{w \mid w > 22\}$ **23.** $\{t \mid t \le 0\}$ **25.** $\{c \mid c > -4\}$
27. $\{d \mid d < 5.4\}$ **29.** $\{z \mid z \le -0.5\}$ **31.** $\{v \mid v \le 2\frac{1}{2}\}$
33. $\{g \mid g < 13\}$;

35. $\{x \mid x \ge -16\}$;

37. $\{t \mid t \ge 9\}$;

39. $x - 8 \le 12$; $x \le 20$ **41.** $t \ge 99$ **43.** $p \ge 99$
45. Yes; examples are $x > x$ and $2x + 1 \le 2x$.
47. $x < -8$ **49.** $x \ge 5$ **51.** $(3x + 2)(x - 5)$ **53.** 32

Pages 516–518 Lesson 12–3
1. For both, reverse the sign when dividing or
multiplying by a negative number.
3.

5. $\{x \mid x \ge -9\}$ **7.** $\{x \mid x \ge 20\}$ **9.** $\{a \mid a \le -30\}$
11. $\{h \mid h \ge -2\}$ **13.** $\{d \mid d \le -3\}$ **15.** $\{g \mid g > -36\}$
17. $\{x \mid x \ge 7\}$ **19.** $\{z \mid z > -3\}$ **21.** $\{c \mid c > -54\}$
23. $\{k \mid k > \frac{5}{3}\}$ **25.** $\{w \mid w > \frac{1}{2}\}$ **27.** $\{y \mid y \ge 6\}$
29. $\{v \mid v \ge 2\}$ **31.** $\{s \mid s \le 280\}$ **33.** $\{n \mid n < 105\}$
35. $x < 45$ **37.** at least 17 lawns **39.** $\{z \mid z \le 4\}$
41. $p > 600$ **43.** $4v^2 - 1$ **45.** $\frac{1}{400}$

Page 518 Quiz 1
1. $x < -1$
3. $\{x \mid x > 5\}$;

5. $\{z \mid z < -3\}$;

7. $\{x \mid x > -5\}$ **9.** $\{v \mid v \le \frac{15}{2}\}$

Pages 522–523 Lesson 12–4
1. subtraction, then division **3.** -7 **5.** -3.1
7. $\{y \mid y > 4\}$ **9.** $\{x \mid x < -1\}$ **11.** $\{w \mid w > -2\}$
13. $s \ge 92$ **15.** $\{b \mid b < 12\}$ **17.** $\{x \mid x \ge 4\}$
19. $\{g \mid g > 2\}$ **21.** $\{v \mid v \le 0.5\}$ **23.** $\{m \mid m \le -3\}$
25. $\{d \mid d < 1\}$ **27.** $\{x \mid x \le 2\}$ **29.** $\{y \mid y < 1\}$
31. $\{b \mid b < 11\}$ **33.** $\{x \mid x < 1\}$ **35.** at least \$260
37. $x > \$1484$ **39.** $\{p \mid p > 7\}$ **41.** $\{x \mid x \le -40\}$
43. $m \le \$19.50$ **45.** A

Pages 526–529 Lesson 12–5
1. an inequality made from two inequalities
3. intersection **5.** union **7.** $5 < \ell < 10$
9.

11. $\{c \mid 3 < c < 8\}$;

13. $\{v \mid 0 \le v \le 4\}$;

15. $\{j \mid j \le -1 \text{ or } j > 0\}$;

17. $9 \le t \le 11.5$;

19. $-8 < h < 8$ **21.** $2 \le g \le 5$ **23.** $6 < x \le 8$
25.

27.

29.

31. $\{x \mid 4 \le x \le 8\}$;

33. $\{w \mid 0 < w \le 3\}$;

35. $\{h \mid -3 \le h \le 3\}$;

37. $\{c \mid -5 < c < -4\}$;

39. $\{z \mid z \le -7 \text{ or } z > 4\}$;

41. $\{r \mid r \leq -1 \text{ or } r > 0\}$;

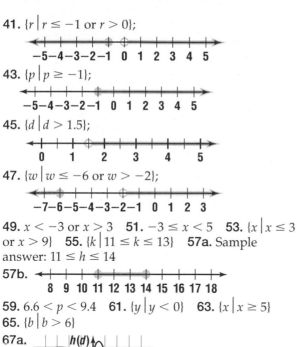

43. $\{p \mid p \geq -1\}$;

45. $\{d \mid d > 1.5\}$;

47. $\{w \mid w \leq -6 \text{ or } w > -2\}$;

49. $x < -3$ or $x > 3$ **51.** $-3 \leq x < 5$ **53.** $\{x \mid x \leq 3$ or $x > 9\}$ **55.** $\{k \mid 11 \leq k \leq 13\}$ **57a.** Sample answer: $11 \leq h \leq 14$

57b.

59. $6.6 < p < 9.4$ **61.** $\{y \mid y < 0\}$ **63.** $\{x \mid x \geq 5\}$
65. $\{b \mid b > 6\}$

67a.

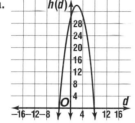

67b. 34 in. **69.** $3b - 18$ **71.** C

Pages 532–534 Lesson 12–6
1. $|x| < 7$ includes all numbers between -7 and 7. $|x| > 7$ includes all numbers to the left of -7 and to the right of 7. **3.** Mia; $|x| \leq 0$ includes only 0. $|x| \geq 0$ includes all real numbers. **5.** $x \leq 3$ and $x \geq -3$ **7.** $x > 8$ or $x < -8$
9. $\{x \mid -4 < x < 8\}$;

11. $\{t \mid t \geq 8 \text{ or } t \leq 2\}$;

13. $\{s \mid s \geq -1 \text{ or } s \leq -7\}$;

15. $\{m \mid -6 < m < 4\}$;

17. $\{z \mid -9 \leq z \leq -5\}$;

19. $\{p \mid -1 < p < 3\}$;

21. $\{t \mid -2 \leq t \leq 2\}$;

23. $\{k \mid -5.5 < k < 1.5\}$;

25. $\{y \mid y \leq -9 \text{ or } y \geq 3\}$;

27. $\{x \mid x < -2 \text{ or } x > 2\}$;

29. $\{r \mid r \leq 3 \text{ or } r \geq 5\}$;

31. $\{h \mid h < -6 \text{ or } h > 12\}$;

33. $|s - 90| < 4$ **35.** $|s - 65| \leq 3$
37. $|x - 1| \geq 1$
39. $\{x \mid 0 < x < 8\}$;

41. $\{x \mid x < -10 \text{ or } x > 8\}$;

43. $|m - 3.25| \leq 0.05$; $3.20 \leq m \leq 3.30$
45.

47.

49. $\{y \mid y \leq -1\}$ **51.** $-2, 6$

Page 534 Quiz 2
1. $\{x \mid x < 1\}$ **3.** $\{d \mid d \geq 1.2\}$
5. $\{n \mid -5 \leq n < 1\}$;

7. $\{y \mid 5 < y < 9\}$;

9. $\{x \mid -4 < x < 4\}$;

Pages 538–539 Lesson 12–7
1. Graph a solid line for $\leq$ and $\geq$. Graph a dashed line for $<$ and $>$.

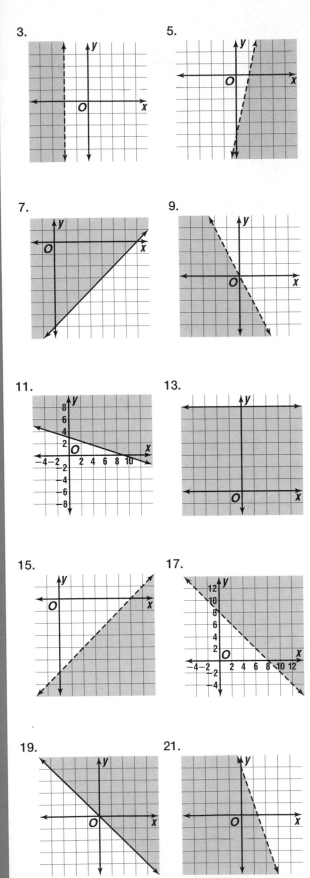

3.

5.

7.

9.

11.

13.

15.

17.

19.

21.

23.

25.

27.

29.

31.

$x + y > 4$

33a.

$a + t > 12$

33b. Sample answer: {(4, 9), (2, 11), (3, 16)}
33c. positive, whole

35.

37. {$h \mid 5 < h < 9$};

2 3 4 5 6 7 8 9 10 11 12

39. {$b \mid b < -2$ or $b > 1$};

−5 −4 −3 −2 −1 0 1 2 3 4 5

41. 8.483×10^2

Pages 542–544   Chapter 12 Study Guide and Assessment
1. union   **3.** half-plane   **5.** boundary
7. half-plane   **9.** set-builder notation
11.

−5 −4 −3 −2 −1

13.

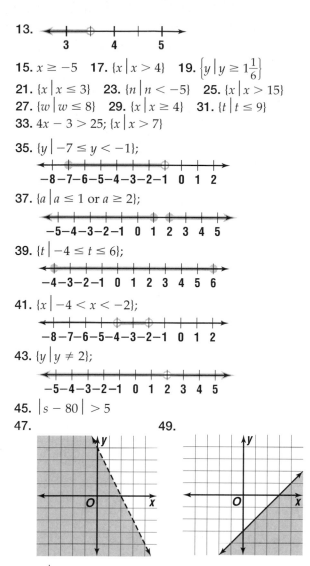

15. $x \geq -5$ **17.** $\{x \mid x > 4\}$ **19.** $\left\{y \mid y \geq 1\frac{1}{6}\right\}$

21. $\{x \mid x \leq 3\}$ **23.** $\{n \mid n < -5\}$ **25.** $\{x \mid x > 15\}$

27. $\{w \mid w \leq 8\}$ **29.** $\{x \mid x \geq 4\}$ **31.** $\{t \mid t \leq 9\}$

33. $4x - 3 > 25$; $\{x \mid x > 7\}$

35. $\{y \mid -7 \leq y < -1\}$;

37. $\{a \mid a \leq 1 \text{ or } a \geq 2\}$;

39. $\{t \mid -4 \leq t \leq 6\}$;

41. $\{x \mid -4 < x < -2\}$;

43. $\{y \mid y \neq 2\}$;

45. $|s - 80| > 5$

47.   **49.**

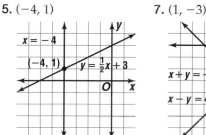

51. $\{x \mid 30 < x < 60\}$

Page 547   Preparing for Standardized Tests
1. A   **3.** C   **5.** A   **7.** B   **9.** 270

Chapter 13   Systems of Equations and Inequalities

Pages 552–553   Lesson 13–1
1. the ordered pair for the point at which the graphs of the equations intersect   **3a.** $(-4, 2)$
3b. $(2, -2)$ **3c.** $(2, 4)$

5. $(-4, 1)$ **7.** $(1, -3)$

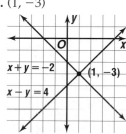

9. $(5, -4)$ **11.** $(0, 2)$

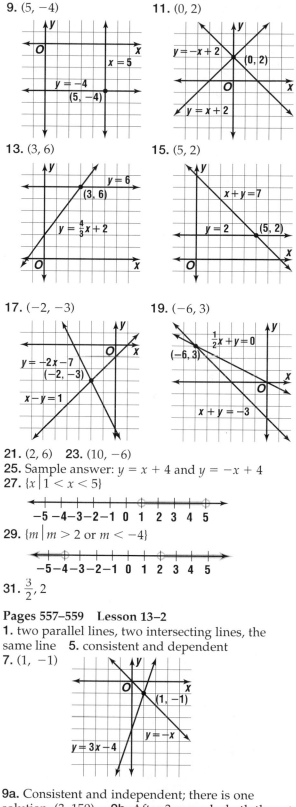

13. $(3, 6)$ **15.** $(5, 2)$

17. $(-2, -3)$ **19.** $(-6, 3)$

21. $(2, 6)$ **23.** $(10, -6)$
25. Sample answer: $y = x + 4$ and $y = -x + 4$
27. $\{x \mid 1 < x < 5\}$

29. $\{m \mid m > 2 \text{ or } m < -4\}$

31. $\frac{3}{2}, 2$

Pages 557–559   Lesson 13–2
1. two parallel lines, two intersecting lines, the same line   **5.** consistent and dependent
7. $(1, -1)$

9a. Consistent and independent; there is one solution, (3, 150).   **9b.** After 3 seconds, both the cat and dog will be 150 feet from the dog's original starting position.   **11.** inconsistent   **13.** consistent and dependent   **15.** consistent and independent

17. infinitely many   **19.** (5, −2)

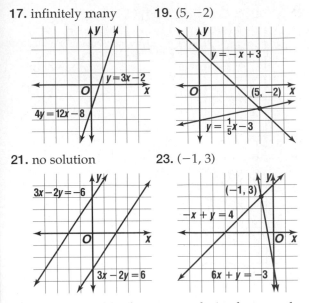

21. no solution   **23.** (−1, 3)

25. one; (5, 1)   **27.** There is no solution because the graphs of the equations are parallel lines. The second dog never catches up with the first.

29. Sample answer: $y = -\frac{1}{3}x$

31.

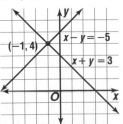

33. $4(x - 2)$ **35.** $3a(ab^2 + 2b - 3)$

Pages 564–565   Lesson 13–3
1. when the exact coordinates are not easily determined from the graph   **3.** Both are correct because you can substitute for either variable.
5. (4, 12)   **7.** (4, 0)   **9.** no solution   **11.** (1, 1)
13. (−3, −9)   **15.** no solution   **17.** (−4, 1)
19. (3, 4)   **21.** (2, 2)   **23.** (2, −10)   **25.** $\left(4, \frac{1}{3}\right)$
27. infinitely many   **29.** (13, 30)   **31a.** 5.5 hr
31b. 357.5 mi   **33.** No; the graphs of equations *A*, *B*, and *C* could form a triangle. Any two of the lines would be intersecting, but there is no point where all three lines intersect.
35. (−1, 4)

37. $-3 < x \le 2$ **39.** $460

Page 565   Quiz 1
1. (4, −1)   **3.** no solution

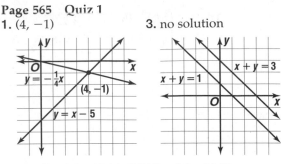

5. 12.5 lb of the $3.90, 37.5 lb of the $4.30

Pages 569–571   Lesson 13–4
1. when the coefficients of terms with the same variable are the same or additive inverses
3. addition   **5.** subtraction   **7.** (8, −1)   **9.** (2, 2)
11. $\left(\frac{16}{3}, 3\right)$ **13a.** $x + y = 42$ and $x = 2y + 3$
13b. 29, 13   **15.** (14, 5)   **17.** (−5, 15)   **19.** (0, 6)
21. (8, −1)   **23.** (−15, −22)   **25.** $\left(5, \frac{5}{2}\right)$
27. infinitely many   **29.** (−11, −4)   **31.** (2, 8)
33. $\left(2, \frac{1}{2}\right)$ **35a.** $13x + 2y = 137.50$ and
$9x + 2y = 103.50$ **35b.** $8.50 child, $13.50 adult
37a. $x + 2y = 29$ and $y = 3x - 10$ **37b.** 7 and 11
39. (11, −38)   **41.** consistent and dependent
43. $y < 4$ **45.** $x < -2$ **47.** $a^2 - 6a + 9$
49. $y - 1 = -5(x - 6)$ or $y + 4 = -5(x - 7)$

Pages 575–577   Lesson 13–5
1. when neither of the variables in a system of equations can be eliminated by simply adding or subtracting the equations   **3a.** substitution
3b. elimination (×)   **3c.** elimination (+)
3d. elimination (−)   **5.** Multiply second equation by 2. Then add.   **7.** (2, 2)   **9.** (−3, 2)   **11.** no
solution   **13a.** 2 mph   **13b.** 14 mph   **15.** (2, 0)
17. (−1, 3)   **19.** $\left(\frac{1}{2}, \frac{3}{4}\right)$ **21.** $\left(\frac{5}{2}, 1\right)$ **23.** (−6, −3)
25. infinitely many   **27.** (−2, −1)   **29.** $\left(\frac{7}{9}, 0\right)$
31. Sample answer: substitution or multiplication; (−2, 4)   **33.** Sample answer: graphing; no solution
35a. $6a + 10b = 108$ and $4a + 12b = 104$
35b. $8, $6   **37.** (7, −1)   **39.** $\left(\frac{3}{2}, 1\right)$ **41.** $m < -4$
43. $a \ge -4$ **45.** 5 s

Pages 583–585   Lesson 13–6
1. (−1, −7), (2, 0)   **3.** Sample answer: The following methods can be used to solve linear systems of equations: graphing, substitution, elimination using addition or subtraction, and elimination using multiplication. Graphing and substitution can be used to solve quadratic-

linear systems of equations. Graphing is useful when you want to estimate the solution. The other methods are useful when you want to find the exact solution.

5. (0, 0), (2, 4)

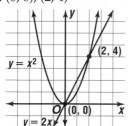

7. no solution

9. (2, 0), (0, −4)　　**11.** no solution

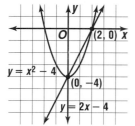

13. (−1, 4), (2, 1)

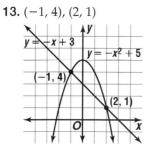

15. (−4, 7)　**17.** (8, 32), (−2, 2)　**19.** no solution
21. (3, 0), (0, −9)　**23a.** $A = 16$ and $A = 4x^2 − 28x + 40$　**23b.** $\ell = 8$ ft, $w = 2$ ft, $h = 1$ ft　**23c.** 16 ft^3
25. Side length 4 units; solve the quadratic-linear system of equations $p = s^2$ and $p = 4s$, where p represents perimeter and s represents side length. The solutions are (0, 0) and (4, 16).　**27.** (2, −1)
29. {−7, 7}　**31.** {4, −8}

Page 585　Quiz 2
1. (4, 2)　**3.** (6, 2)　**5a.** $x − y = 38$, $x = 3y − 2$
5b. 58, 20

Pages 589–590　Lesson 13–7
1. A system of linear equations may have at most one solution if the equations are distinct. A system of inequalities may have an infinite number of solutions.　**3.** Kyle; all of the points in region B satisfy both inequalities.　**5.** no

7.

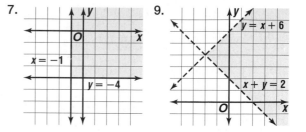

9.

11. Sample answer: 2 dozen sugar, 4 dozen choc. chip; 4 dozen sugar, 3 dozen choc. chip; 5 dozen sugar, 2 dozen choc. chip

13.　　　　　　　**15.**

17.　　　　　　　**19.** no solution

21.　　　　　　　**23.**

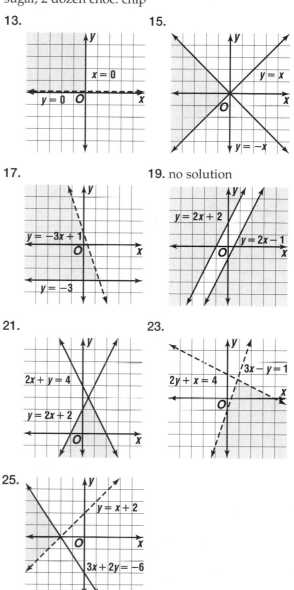

25.

27. People who use the phone less than 75 minutes a month should select the $0.15 per minute plan. People who use the phone more than 75 minutes a month should select the $0.05 per minute plan. The middle plan is never the cheapest.

29. $(2, 4), (-2, 4)$ **31.** no solution

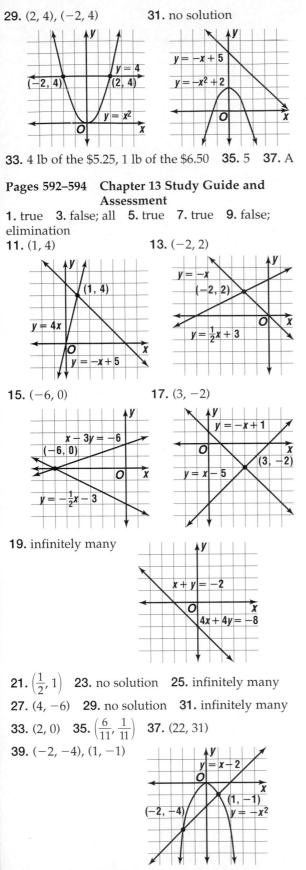

33. 4 lb of the \$5.25, 1 lb of the \$6.50 **35.** 5 **37.** A

Pages 592–594 Chapter 13 Study Guide and Assessment
1. true **3.** false; all **5.** true **7.** true **9.** false; elimination
11. $(1, 4)$ **13.** $(-2, 2)$

15. $(-6, 0)$ **17.** $(3, -2)$

19. infinitely many

21. $\left(\frac{1}{2}, 1\right)$ **23.** no solution **25.** infinitely many
27. $(4, -6)$ **29.** no solution **31.** infinitely many
33. $(2, 0)$ **35.** $\left(\frac{6}{11}, \frac{1}{11}\right)$ **37.** $(22, 31)$
39. $(-2, -4), (1, -1)$

41. no solution

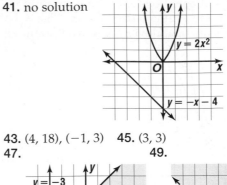

43. $(4, 18), (-1, 3)$ **45.** $(3, 3)$
47. **49.**

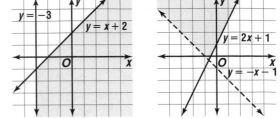

51. 8 lb of almonds, 12 lb of cashews **53.** 16 ft by 9 ft

Page 597 Preparing for Standardized Tests
1. D **3.** A **5.** B **7.** D **9.** 934

Chapter 14 Radical Expressions

Pages 603–605 Lesson 14–1
1. Sample answer: -4 **3.** Sample answer: $\sqrt{13}$
5. I **7.** Q
9. -4.5

11. 7.3

13. rational **15.** rational **17.** Q **19.** W, Z, Q
21. Z, Q **23.** N, W, Z, Q **25.** Z, Q **27.** Q
29. 1.7

31. -3.2

33. -6.3

35. 7.2

37. 10.4

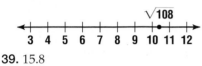

39. 15.8

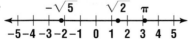

41. rational   **43.** irrational; 6 and 7   **45.** rational
47. irrational; -2 and -3 **49.** irrational; 11 and 12
51. rational
53.

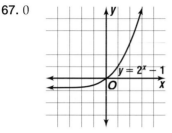

55. about 8.5 s   **57.** 5.2 ohms and 1200 watts or
4.6 ohms and 1500 watts
59a. $\sqrt{17}, \sqrt{18}, \sqrt{19}, \sqrt{20}, \sqrt{21}, \sqrt{22}, \sqrt{23},$
$\sqrt{24}$ **59b.** $\sqrt{19}$ and $\sqrt{20}$ **61.** (3, 11), (1, 3)
63. $\{x \mid x < -7\}$ **65.** $\{m \mid m \leq -1.8\}$
67. 0

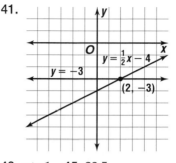

69. $4(a + 3)(a - 3)$ **71.** 4   **73.** A

Pages 608–611   Lesson 14–2
1. Pythagorean Theorem   **3.** Both are correct since
it does not matter which ordered pair is first when
using the Distance Formula as long as the
coordinates are used in the same order.   **5.** 5
7. 9 or -3 **9.** 9.5 mi   **11.** 12   **13.** $\sqrt{58}$ or 7.6
15. $\sqrt{185}$ or 13.6   **17.** $\sqrt{18}$ or 4.2   **19.** 17 or -13
21. -12 or 6   **23.** 2 or -14 **25.** $\sqrt{194}$ or 13.9
27. 8 or 32   **29.** No; no two sides have the same
measure.   **31.** 16.3 mi
33. 3.5

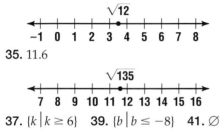

35. 11.6

37. $\{k \mid k \geq 6\}$ **39.** $\{b \mid b \leq -8\}$ **41.** $\varnothing$

Page 611   Quiz 1
1. W, Z, Q   **3.** I   **5.** 1.22, $1.\overline{2}$, $\sqrt{2}$ **7.** $\sqrt{98}$ or 9.9
9. 0 or 6

Pages 617–619   Lesson 14–3
1. to ensure nonnegative results   **3.** Greg is correct.
In its simplest form, the expression is written as
$\frac{3 - \sqrt{5}}{2}$ since the numbers 6, 2, and 4 in $\frac{6 - 2\sqrt{5}}{4}$
are all divisible by 2.   **5.** $2 - \sqrt{8}; -4$ **7.** $5\sqrt{3}$
9. $3\sqrt{10}$ **11.** $\frac{\sqrt{21}}{7}$ **13.** $6|x|\sqrt{y}$ **15.** $4\sqrt{2}$ units
17. $7\sqrt{2}$ **19.** $10\sqrt{5}$ **21.** 4   **23.** 15   **25.** 4
27. $2\sqrt{10}$ **29.** $\frac{2\sqrt{5}}{5}$ **31.** $\frac{8 + 2\sqrt{3}}{13}$ **33.** $\frac{24 + 4\sqrt{7}}{29}$
35. $2|m|\sqrt{10}$ **37.** $3|a|b\sqrt{6b}$ **39.** $6|x|y^2z^2\sqrt{xz}$
41a. $2\sqrt{3}$ mph   **41b.** 10 in.   **43.** 2 or -4 **45.** Z, Q
47. no solution   **49.** $\sqrt{79}$

Pages 621–623   Lesson 14–4
1. Sample answer: Add like terms.
3. The Distributive Property allows you to add
like terms. Radicals with like radicands can be
added or subtracted.   **5.** none   **7.** $8\sqrt{3}, -18\sqrt{3}$
9. $2\sqrt{5}$ **11.** $-5\sqrt{7}$ **13.** $-6\sqrt{2}$ **15.** $6\sqrt{3}$
17. $14\sqrt{5}$ **19.** $4\sqrt{7}$ **21.** $-2\sqrt{5}$ **23.** $-9\sqrt{2} +$
$8\sqrt{5}$ **25.** $7\sqrt{6}$ **27.** $14\sqrt{3}$ **29.** $-4\sqrt{2}$
31. $8\sqrt{2} + 6\sqrt{3}$ **33.** $-2\sqrt{5} - 6\sqrt{6}$ **35.** $5\sqrt{7}$ ft
37a. $4\sqrt{3}$ in.   **37b.** $6\sqrt{3} + 6$ in.   **39.** $3n|p|\sqrt{5mn}$
41. $\sqrt{101}$ or 10.0   **43.** $-1 \cdot 2 \cdot 13 \cdot a \cdot a \cdot a \cdot b \cdot b$
44. $2 \cdot 2 \cdot 3 \cdot 3 \cdot x \cdot y \cdot y \cdot y \cdot y \cdot z$ **45.** $a^2 - 4c^2$
46. A

Pages 627–629   Lesson 14–5
1. Isolate the radical.   **5.** $a + 5 = 4$ **7.** 121   **9.** 4
11. 35.5 ft   **13.** no solution   **15.** -12 **17.** 2
19. 75   **21.** 63   **23.** no solution   **25.** 2 and 3
27. 6   **29.** 10   **31.** 140   **33a.** 1.6 m/s^2
33b. 9.8 m/s^2 **35.** $-5\sqrt{5}$ **37.** $-\sqrt{6} - 3\sqrt{2}$
39. $\sqrt{5}$
41.

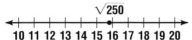

43. $x > 1$ **45.** 22.5

Page 629   Quiz 2
1. $3\sqrt{10}$ **3.** $\frac{7 - \sqrt{5}}{44}$ **5.** $12\sqrt{5}$ **7.** $7\sqrt{2}$ in.
9. no solution

Pages 630–632 Study Guide and Assessment
1. $\sqrt{(x_2 - x_1)^2 + (y_2 - y_1)^2}$ 3. radicals 5. $\sqrt{7ab}$
7. whole 9. no solution 11. Q 13. Z, Q
15. 2.2

$$\sqrt{5}$$
[number line from −5 to 5 with point marked between 2 and 3]

17. 10.5

$$\sqrt{111}$$
[number line from 5 to 15 with point marked between 10 and 11]

19. 5 21. $\sqrt{72}$ or 8.5 23. −1 or −5 25. $5\sqrt{2}$
27. $3\sqrt{3}$ 29. $\dfrac{4 + \sqrt{5}}{11}$ 31. $10\,|a|\,c^2\sqrt{3b}$
33. $7\sqrt{5}$ 35. 0 37. $22\sqrt{2}$ 39. $-14\sqrt{5}$ 41. 25
43. 9 45. no solution 47. 6 49. 150
51. integers, rational numbers 53. $10\sqrt{2}$ lb
55a. 3844 m 55b. 2601 m

Page 635 Preparing for Standardized Tests
1. A 3. A 5. A 7. C 9. 50

Chapter 15 Rational Expressions and Equations

Pages 641–643 Lesson 15–1
1. The denominator is 0 when $x = 2$. 3. Darnell; x is not a factor of $x + 3$ and $x + 4$. Therefore, it cannot be divided out. 5. $4x(4x - 5)$ 7. $(a - 5)(a + 4)$
9. $(x - 6)(x - 4)$ 11. 0, −6 13. $\dfrac{2}{5}$ 15. $\dfrac{2y}{3z^3}$ 17. $\dfrac{3}{4}$
19. $\dfrac{x - 2}{x + 1}$ 21. $\dfrac{-(x + 5)}{x + 6}$ 23. −5 25. 10 27. −2, 3
29. 0, −5 31. 4, −4 33. $\dfrac{1}{3}$ 35. $\dfrac{2c}{3d}$ 37. $-4bc$
39. $\dfrac{1}{5}$ 41. $\dfrac{x + 3}{x + 1}$ 43. $\dfrac{x}{2}$ 45. $\dfrac{r + 2}{3}$ 47. $\dfrac{y + 3}{y - 4}$
49. $\dfrac{y + 2}{y - 1}$ 51. $\dfrac{z - 5}{z + 3}$ 53. $\dfrac{8m}{m - 2}$ 55. $\dfrac{c - 5}{c(c + 6)}$
57. $\dfrac{-x}{4 + x}$ 59. $\dfrac{4}{3 + a}$ 61. 9 yr 63. 8 lumens/m²
65. 16 67. 19 69. $7\sqrt{3}$ 71. $6\sqrt{3}$ 73. 5 75. D

Pages 647–649 Lesson 15–2
1. The reciprocal of $\dfrac{a^2 - 25}{3a}$ was used instead of the reciprocal of $\dfrac{a + 5}{15a^2}$. 3. $\dfrac{4}{x^2}$ 5. $\dfrac{x + 5}{x - 4}$ 7. $\dfrac{a^2}{bd}$
9. $\dfrac{2}{x + 4}$ 11. $4a$ 13. $\dfrac{3x^2}{2}$ 15. 12 flags 17. $\dfrac{xy}{4}$
19. $3a$ 21. $\dfrac{2s}{3}$ 23. $\dfrac{x}{(x + 4)^2}$ 25. $\dfrac{2(x - 3)}{x + 3}$ 27. $\dfrac{n}{n + 2}$
29. $6a^2b$ 31. $2a$ 33. $\dfrac{y + 4}{y^2}$ 35. $\dfrac{(m + 1)(m - 1)}{2}$

37. $\dfrac{-(x + 2)}{x - 3}$ 39. $\dfrac{-3a^2(a + 3)}{2}$ 41. $\dfrac{1}{x - y}$ 43a. $\dfrac{x^4}{625}$
43b. $\dfrac{81}{625}$ or about 13% 45. $\dfrac{7b^3}{a}$ 47. $\dfrac{-4}{x - 2}$
49. (1, 3) 51. (−3, −4) 53a. (5, −75), (8, 0), (10, 50)
53b.

[graph with P vertical axis and n horizontal axis; points plotted at approximately (8, 50), (5, −75), (8, 0)]

Pages 653–655 Lesson 15–3
1. dividend: $3k^2 - 7k - 5$; divisor: $3k + 2$; quotient: $k - 3$; remainder: $\dfrac{1}{3k + 2}$ 3. The quotient is a factor of the dividend if the remainder is 0. 5. $3a$ 7. x
9. $x + 2 - \dfrac{1}{2x - 1}$ 11. $2x + 3$ in. 13. x 15. $3r^2$
17. $x + 4$ 19. $a + 5$ 21. $t - 2 - \dfrac{32}{3t - 4}$
23. $b + 2 - \dfrac{4}{2b - 1}$ 25. $x^2 - 3x + 9$
27. $4x^3 + 4x^2 + 2x + 3 + \dfrac{4}{x - 1}$ 29. $x - 3$
31a. 936 h 31b. 140 h 31c. 51 h 31d. 5 h
33. 27 35. 1 37. one solution 39. infinitely many 41. $y \geq -6$

Page 655 Quiz 1
1. 2; m 3. $\dfrac{1}{2}$ 5. $\dfrac{-2(a + 1)}{a - 1}$ 7. y 9. $x^2 + x + 1$

Pages 659–661 Lesson 15–4
1. They are additive inverses.
3.

a	b	$a + b$	$a - b$	$a \cdot b$	$a \div b$
$\dfrac{2}{t}$	$\dfrac{1}{t}$	$\dfrac{3}{t}$	$\dfrac{1}{t}$	$\dfrac{2}{t^2}$	2
$\dfrac{12}{y}$	$\dfrac{4}{y}$	$\dfrac{16}{y}$	$\dfrac{8}{y}$	$\dfrac{48}{y^2}$	3
$\dfrac{x}{x - 1}$	$\dfrac{1}{x - 1}$	$\dfrac{x + 1}{x - 1}$	1	$\dfrac{x}{x^2 - 2x + 1}$	x

5. $\dfrac{3}{4}$ 7. $\dfrac{7}{x}$ 9. −1 11. 2 13. $\dfrac{10y}{7x - 2y}$ in. 15. $\dfrac{7x}{8}$
17. $\dfrac{t}{2}$ 19. $\dfrac{7}{6r}$ 21. n 23. $\dfrac{x}{12}$ 25. $-\dfrac{1}{z}$ 27. 0
29. 2 31. 3 33. $\dfrac{2y}{y - 2}$ 35a. Sample answer:
$\dfrac{15{,}000}{5} - \dfrac{0.3(15{,}000)}{5}$ 35b. $2100 37. $a + 4$
39. $y - 5$ 41. $\dfrac{3axy}{10}$ 43. $9x^2 + 6xy + y^2$

Pages 665–667 Lesson 15–5

1. Sample answer: $\frac{1}{2x}, \frac{1}{4x}$ **3.** Malik; he found the LCD and then added the terms. Ashley incorrectly added the numerators and the denominators.

5. 72 **7.** $30xy^2$ **9.** $\frac{4}{a^2}, \frac{5a}{a^2}$ **11.** $\frac{5t}{12}$ **13.** $\frac{2+3b}{ab^2}$

15. $\frac{-3x+8}{x^2-4}$ **17.** 2042 **19.** $12a^2b^2$

21. $(y+2)(y-2)$ **23.** $(x+3)(x-3)(3x+1)$

25. $\frac{35}{10t}, \frac{8}{10t}$ **27.** $\frac{5z}{xyz}, \frac{6x}{xyz}$ **29.** $\frac{30k}{3(3k+1)}, \frac{k}{3(3k+1)}$

31. $\frac{3d}{10}$ **33.** $\frac{15}{4x}$ **35.** $\frac{m+3}{3t}$ **37.** $\frac{14x-1}{6x^2}$

39. $\frac{2y+x^2+3x}{xy}$ **41.** $\frac{5x+2y^2}{20x^2y}$ **43.** $\frac{2a+3}{a+3}$

45. $\frac{1}{(y-1)^2}$ **47.** $\frac{22x+5}{4(2x-1)}$ **49.** $\frac{x^2+6x-11}{(x+1)(x-3)}$

51. $\frac{m^2-5m+5n}{m(m-n)}$ **53.** 17.5 min **55.** $\frac{16}{a}$ **57.** 2

59. $9\sqrt{19}$ **61.** in simplest form **63.** (2, 5)

Page 667 Quiz 2

1. $x+8$ **3.** $\frac{8a}{9}$ **5.** 168

Pages 671–673 Lesson 15–6

1. Sample answer: In a linear equation, the variable cannot appear in the denominator and it cannot be raised to a power higher than 1. **3.** $-\frac{7}{3}$ **5.** $-\frac{10}{3}$

7. 3 **9.** 30 mph **11.** $-\frac{2}{3}$ **13.** 7 **15.** -3 **17.** 9

19. $\frac{4}{3}$ **21.** 7 **23.** $\frac{1}{12}$ **25.** 3 **27.** $\frac{1}{6}$ **29.** 6 **31.** -13

33. $-\frac{3}{2}$ **35.** $\frac{7}{5}$ **37.** 5 **39.** 2 **41.** 7 **43.** $\frac{9}{2n^2}$

45. Z, Q **47a.** $c+s \le 16$, $10c+15s \le 200$

47b.

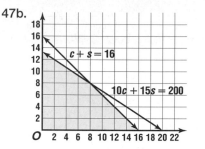

47c. Sample answer: (7, 8), he could plant corn for 7 days and soybeans for 8 days; (2, 12), he could plant corn for 2 days and soybeans for 12 days.
49. (4, −1);

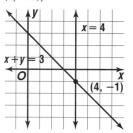

Pages 676–678 Study Guide and Assessment

1. denominator **3.** $\frac{x}{3x+6y}$ **5.** greatest common factor **7.** $\frac{1}{a+b}$ **9.** $2x^2$ **11.** $\frac{xz}{4y^2}$ **13.** $\frac{n-4}{n-2}$

15. $\frac{2x^2}{3y^2}$ **17.** $\frac{x}{(x+3)(x-3)}$ **19.** $\frac{3(y-2)}{y+2}$ **21.** a

23. $2y+1+\frac{2}{2y+3}$ **25.** $\frac{3x}{4}$ **27.** $\frac{5x}{x+4}$ **29.** $\frac{6x+5}{8x^2}$

31. $\frac{7a-8}{a(a-2)}$ **33.** $-\frac{1}{5}$ **35.** -4 **37.** 80 pieces

39a. $x+1$ **39b.** 4 in., 3 in., 9 in., 108 in^3

Page 681 Preparing for Standardized Tests
1. A **3.** B **5.** D **7.** C **9.** 35

Newman/PhotoEdit; **315** Ex Rouchon/ Explorer/Photo Researchers; **317** Renee Lynn/The Stock Market; **319** Aaron Haupt; **320** Jonathan Selig/Photo 20-20; **323** Mark Burnett; **326** Roberto Soncin Gerometta/ Photo 20-20; **334–335** Corbis Images; **337** (l)SSEC/UW-Madison, (r)file photo; **340** SuperStock; **345** Rob Atkins/Image Bank; **346** SuperStock; **347** Nick Nicholson/ Image Bank; **349** (l)SuperStock, (r)Alfred Pasieka/Peter Arnold, Inc.; **351** SuperStock; **352** Courtesy of Toshiba; **354** Albert Normandin/Masterfile; **356** Hal Horwitz/ CORBIS; **357** ©Tribune Media Services, Inc. All Rights Reserved. Reprinted with permission; **359** Bill Brooks/Masterfile; **360** Corbis Images; **361** SuperStock; **363** Michelle Garrett/CORBIS; **364** NASA/ The Stock Market; **365** Index Stock Photography; **368** SuperStock; **369** Icon Images; **371** Mark Tomalty/Masterfile; **373** Icon Images; **380–381** Aaron Haupt; **385** Syd Greenberg/Photo Researchers; **387** Doug Martin/Photo Researchers; **393** Courtesy Carol McAdoo; **398** Jon Eisberg/FPG; **400** Aaron Haupt; **405** Vladimir Pcholkin/FPG; **410** J.A. Kraulis/Masterfile; **411** SuperStock; **418–419** Toyohiro Yamada/ FPG; **425 427** Mark Burnett; **431** Aaron Haupt; **432** SuperStock; **434** Lynn M. Stone; **439** Mak-1; **445** L.D. Franga; **456–457** Gisela Damm/Leo de Wys; **458** Aidan O'Rourke; **461** Ron Chapple/FPG; **464** The Kobal Collection/TS2/©Disney/Pixar; **467** Garry Black/Masterfile; **468** G. Randall/FPG; **472 480** Aaron Haupt; **482** Mark Burnett; **485** Aaron Haupt; **486** Tony Freeman/ PhotoEdit; **488** (t)Aidan O'Rourke, (b)Mark Burnett; **491** R. Ashworth/Robert Harding Picture Library; **493** E. Alan McGee/FPG; **494–495** Mark Burnett; **502–503 504** Aaron Haupt; **506** Mark Newman/Photo 20-20; **507** Mark Burnett; **509** Telegraph Colour Library/FPG; **513** Courtesy Push America/ Pi Kappa Phi Fraternity/Journey of Hope; **514** Nickelsberg/Liaison Agency; **516** Aaron Haupt; **517** Jeri Gleiter/FPG; **519** AGE Fotostock/FPG; **523** Ron Holt/Aristock;

528 Mark Burnett; **529** (t)Aaron Haupt, (b)Peter Christopher/Masterfile; **530** Aaron Haupt; **532 534** Mark Burnett; **535** Richard Pasley/Stock Boston; **537 539** Mark Burnett; **541** Aaron Haupt; **548–549** Christie's Images/©2000 Artists Rights Society(ARS), New York/ADAGP, Paris; **555** David R. Frazier Photolibrary; **557** Josph DiChello; **558** Animals Animals/Robert Maier; **561** Rag Productions/FPG; **563** Photo 20-20; **566** K.L. Gay/©Arizona State Parks; **570** Aaron Haupt; **571** Mak-1; **574** Washington Metropolitan Area Transit Authority, Washington, DC; **575** David R. Frazier Photolibrary/Photo Researchers; **576** Janet Foster/Masterfile; **579** Sarah Jones (Debut Art)/FPG; **581** Ken Ross/FPG; **585** VCG/FPG; **587** Frank & Marie-Therese Wood Print Collections, Alexandria VA; **589** Aaron Haupt; **590** Carl Schneider/ FPG; **591** Mug Shots/The Stock Market; **598–599** Bachmann/PhotoEdit; **600** Lance Nelson/The Stock Market; **603** Aaron Haupt; **604** David Parker/Science Photo Library/ Photo Researchers; **608** Mark Reinstein/ FPG; **609** Mark Burnett; **610** Gordon R. Gainer/The Stock Market; **611** Miles Ertman/Masterfile; **613** Mark Burnett; **614** FOX TROT ©1998 Bill Amend. Reprinted with permission of UNIVERSAL PRESS SYNDICATE. All rights reserved; **623** CLOSE TO HOME ©1994 John McPherson. Reprinted with permission of UNIVERSAL PRESS SYNDICATE. All rights reserved; **628** Peter Brock/Liaison; **629** William J. Weber; **636–637** Tim Davis/Stone; **639** Ken Reid/FPG; **640** Aaron Haupt; **642** DUOMO/William Sallaz; **644** Aaron Haupt; **646** John Zimmerman/FPG; **654** Richard Price/FPG; **657** Lee Balterman/FPG; **661** Robert Brenner; **666** JPL; **668** Aaron Haupt; **671** Roberto Soncin Gerometta/Photo 20-20; **673** Science Photo Library/Photo Researchers; **675** David Lissey/FPG; **683** (t)(br)Aaron Haupt, (bl)Patrix Ravaux/ Masterfile, (bc)Gary Randall/FPG; **724** Aaron Haupt.

Index

INDEX

Q

Properties

Substitution (=)	If $a = b$, then a may be replaced by b.
Reflexive (=)	$a = a$
Symmetric (=)	If $a = b$, then $b = a$.
Transitive (=)	If $a = b$ and $b = c$, then $a = c$.
Additive Identity	For any number a, $a + 0 = 0 + a = a$.
Multiplicative Identity	For any number a, $a \cdot 1 = 1 \cdot a = a$.
Multiplicative (0)	For any number a, $a \cdot 0 = 0 \cdot a = 0$.
Multiplicative (−1)	For any number a, $-1 \cdot a = -a$.
Additive Inverse	For any number a, there is exactly one number $-a$ such that $a + (-a) = 0$.
Multiplicative Inverse	For any number $\frac{a}{b}$, where $a, b \neq 0$, there is exactly one number $\frac{b}{a}$ such that $\frac{a}{b} \cdot \frac{b}{a} = 1$.
Commutative (+)	For any numbers a and b, $a + b = b + a$.
Commutative (×)	For any numbers a and b, $a \cdot b = b \cdot a$.
Associative (+)	For any numbers a, b, and c, $(a + b) + c = a + (b + c)$.
Associative (×)	For any numbers a, b, and c, $(a \cdot b) \cdot c = a \cdot (b \cdot c)$.
Distributive	For any numbers a, b, and c, $a(b + c) = ab + ac$ and $a(b - c) = ab - ac$.
Comparison	For any numbers a and b, exactly one of the following sentences is true: $a < b, a > b$, or $a = b$.
Addition (=)	For any numbers a, b, and c, if $a = b$, then $a + c = b + c$.
Subtraction (=)	For any numbers a, b, and c, if $a = b$, then $a - c = b - c$.
Division and Multiplication (=)	For any numbers a, b, and c, with $c \neq 0$, if $a = b$, then $ac = bc$ and $\frac{a}{c} = \frac{b}{c}$.
Product Property of Square Roots	For any numbers a and b, with $a, b \geq 0$, $\sqrt{ab} = \sqrt{a} \cdot \sqrt{b}$.
Quotient Property of Square Roots	For any numbers a and b, with $a \geq 0$ and $b > 0$, $\sqrt{\frac{a}{b}} = \frac{\sqrt{a}}{\sqrt{b}}$.
Zero Product	For any numbers a and b, if $ab = 0$, then $a = 0$, $b = 0$, or both a and b equal 0.
Addition (>)*	For any numbers a, b, and c, if $a > b$, then $a + c > b + c$.
Subtraction (>)*	For any numbers a, b, and c, if $a > b$, then $a - c > b - c$.
Division and Multiplication (>)*	For any numbers a, b, and c, 1. if $a > b$ and $c > 0$, then $ac > bc$ and $\frac{a}{c} > \frac{b}{c}$. 2. if $a > b$ and $c < 0$, then $ac < bc$ and $\frac{a}{c} < \frac{b}{c}$.

* *These properties are also true for $<, \geq$, and $\leq$.*